# Behavioural Travel Modelling

**Edited by
David A. Hensher
and Peter R. Stopher**

**Croom Helm London**

1979 David A. Hensher and Peter R. Stopher
Croom Helm Ltd, 2-10 St John's Road, London SW11

British Library Cataloguing in Publication Data

Behavioural travel modelling.
1. Transportation - Mathematical models
I. Hensher, David A  II. Stopher, Peter Robert
380.5'01'84      HE199.9

ISBN 0-85664-819-1

Printed and Bound in Great Britain by
Redwood Burn Limited, Trowbridge & Esher

# CONTENTS

Foreword                                                        1
*Derek Scrafton, Director General of Transport,*
*South Australia*

Preface                                                         3

Acknowledgements                                                7

PART ONE

1. Behavioural Travel Modelling                                11
   *David A. Hensher and Peter R. Stopher*

PART TWO

2. New Approaches to Understanding Travel Behaviour:           55
   The Human Activity Approach
   *Peter M. Jones*

3. Urban-Travel Linkages: A Review                             81
   *Susan Hanson*

4. An Activity Model and its Validation                       101
   *Kenzo Kobayashi*

5. New Approaches to Understanding Traveller Behaviour        116
   *K. Patricia Burnett and Nigel J. Thrift*

PART THREE

6. Six Notes of Equilibrium and their Implications for        137
   Travel Modelling Examined in an Aggregate Direct
   Demand Framework
   *Marc Gaudry*

7. Equilibrium and Transport System Dynamics                  164
   *Alan G. Wilson*

8. An Equilibrium Model for Integrated Regional               187
   Development
   *A.E. Andersson and A. Karlquist*

9. Equilibrium Modelling                                      207
   *E.R. Ruiter and R.B. Dial*

PART FOUR

10. Market Segmentation: A Tool for Transport          219
    Decision-Making
    *Ricardo Dobson*

11. Market Segmentation in Behavioural Travel Modelling  252
    *Kenneth Heathington and David J. Barnaby*

12. Consumer Segmentation                             262
    *William B. Tye*

PART FIVE

13. Quantitive Methods for Analysing Travel Behaviour  279
    of Individuals: Some Recent Developments
    *Daniel McFadden*

14. Some Developments in Transport Demand Modelling    319
    *Andrew J. Daly*

15. Measuring the Value of Travel Time Savings from     334
    Demand Functions
    *Nils Bruzelius*

16. A Model Based on Non-Homogeneity in Allocation      355
    Problems
    *John F. Brotchie*

17. Theoretical and Conceptual Developments in          378
    Demand Modelling
    *Richard D. Westin and Charles F. Manski*

PART SIX

18. The Role of Disaggregate Modelling Tools in the     395
    Policy Arena
    *Stein Hansen and Ken G. Rogers*

19. Roles of Behavioural Traveller Models in Urban      410
    Policy Analysis
    *Fred A. Reid*

20. The Role of Disaggregate Travel Models in           446
    Transport Policy Analysis
    *Alistair Sherret*

PART SEVEN

21. Values of Time, Modal Split and Forecasting         459
    *Michael E. Beesley*

22. Behavioural Modelling: An Evaluator's Perspective    484
    *John K. Stanley*

23. Relationship between Behavioural Models,            512
    Evaluation, Forecasting and Policy
    *Robin Carruthers*

PART EIGHT

24. Urban Goods Movement: Process, Planning Approach     525
    and Policy
    *Peter J. Rimmer and Stuart K. Hicks*

25. Urban Goods Movement: Behavioural Demand     553
    Forecasting Procedures
    *Paul O. Roberts and Brian C. Kullman*

26. Indicators for Urban Commodity Movements     577
    *Marc R. Wigan*

27. Urban Goods Movement: Research Review     612
    *Keith J.G. Smith*

28. The Applicability of Behavioural Modelling     624
    to the Analysis of Goods Movement
    *Arnim H. Meyburg*

PART NINE

29. Behavioural Modelling, Accessibility, Mobility     639
    and Need: Concepts and Measurement
    *M. Quasim Dalvi*

30. Disaggregate Travel and Mobility Choice Models     654
    and Measures of Accessibility
    *Moshe Ben-Akiva and Steven R. Lerman*

31. Mobility, Accessibility and Travel Impacts of     680
    Transportation Programs for the Elderly and
    Handicapped
    *Ronald F. Kirby and Robert G. McGillivray*

32. Behavioural Modelling, Accessibility, Mobility     698
    and Travel Needs
    *Martin Wachs and Jan G. Koenig*

PART TEN

33. Application of Psychological Measurement and     713
    Modelling to Behavioural Travel-Demand Analysis
    *Jordan J. Louviere, Eugene M. Wilson and
    Michael Piccolo*

34. Attitude-Behaviour Relationships in Travel     739
    Demand Modelling
    *Thomas F. Golob, Abraham D. Horowitz and
    Martin Wachs*

35. The Development of Attitudinal Modelling     758
    Approaches in Transport Research
    *Irwin P. Levin*

36. Attitudes, Attitudinal Measurement, and the     782
    Relationship between Attitudes and Behaviour
    *Jordan J. Louviere*

PART ELEVEN

37. Evaluating the Social and Environmental Impacts       797
    of Transport Investments
    *Donald Appleyard*

38. Behavioural Modelling and the Evaluation of           815
    Social and Environmental Impacts of Transport
    Investment
    *Peter Hills and Ross King*


PART TWELVE

39. Summary of the Conference Findings and               831
    Recommendations
    *Peter R. Stopher and David A. Hensher*


APPENDIX I List of Workshop Members                      842


APPENDIX II The Third International Conference on         859
    Behavioural Travel Modelling
    *John Paterson*

The development of a broad set of behavioural travel models, to improve our understanding of the parameters affecting the demand for travel and to provide a better explanation of the process of choice, is a continuing requirement. The report of the Third International Conference on Behavioural Travel Modelling demonstrates the progress that has been made over the two-year period: model development is well established; behavioural models have advanced from being a research topic to a useful tool for practical problem solving; the validity of choice-based models is accepted; and choice-based sampling is not only as effective as traditional sampling techniques for considerably smaller sample sizes, but the necessary statistical measures of accuracy have been developed to enable more confident model formulations to be devised.

The challenge from the Second International Conference to researchers to apply their models to 'real world problems' was met in the Third Conference where researchers, practising transport planners and policy advisers sought to consolidate their work and responsibilities. Unfortunately an awesome list of inter-relationships, interdependencies and trade-offs resulted! In the model field, the significance of validity, cost and relevance reinforce the plea for simplicity. Accessibility, mobility and travel needs, themselves inter-related, need to be considered against such factors as service level, socio-economic characteristics, families' and individuals' tastes. However, this is how it is in the 'real world' that is being modelled, so the state of the art of behavioural modelling is close to reality, all of which tends to support the recommendation from Workshop B for priority to be given to modelling the interdependencies.

The Third International Conference highlighted problem areas which exist and enabled definition of future directions in which research should proceed. Behavioural researchers must expand their work to investigate land-use planning and economic planning processes and problems and attempt to model them to the same level of sophistication found in transport planning. Attempts must be made to integrate all aspects of planning into a single dynamic process since each component is interdependent and cannot be considered in isolation. Increased use of activity analysis of multi-trip and multi-purpose journeys should facilitate such research just as a clearer understanding of perceived impacts will permit research to more effectively undertake evaluation of social and environmental impacts.

For obvious reasons, most transport planning (and model development) is concerned with urban passenger movements. Yet some significant problems occur outside metro areas: intermodal freight issues, the role of airlines, regulatory

1

problems, etc.  Transport planners must widen their perspectives and view the transport system as an inter-related
subsystem and strive to solve problems in terms of the system
rather than particular market segments or modes.  Behavioural
modelling has the potential to enable such analysis to be
undertaken and this potential should be developed.  It is
heartening to see a workshop at the Third International Conference dealing with urban goods movement, although whether
behavioural models can be applied to goods movement was not
resolved.

The complexity of the decision-making process and the role
of the researcher underpinning the process are demonstrated
by the increasing emphasis on modelling personal decisions to
better formulate corporate or government decisions which in
turn can become constraints in the personal decisions!  While
this also supports the recommendation for a closer working
relationship and more communication between decision-maker,
policy adviser, planner and researcher, I believe the more
the barriers between these job classifications can be broken
down, the more the interface between the individuals and
organisations will be promoted and the more use the final
decisions will be.  One encouraging feature is the simplicity
of modern models which makes them easily understood by 'lay-
men' decision-makers.

There were local and, for me, personal benefits in the
Third International Conference on Behavioural Travel Modelling.  The gathering of researchers enabled people fully to
appreciate the governmental and cultural constraints which
have had an influence on the form of models developed in
different countries, e.g. the United States, Canada, France,
Israel, United Kingdom, Sweden, Germany and Japan.
Australians gained a great deal from the presence of so many
international researchers visiting the Barossa Valley.  The
Conference was an opportunity for Australian research staffs
to work alongside many experienced specialists in travel
modelling from many disciplines, particularly geographers,
economists and mathematicians.  The visitors were welcome to
South Australia; I thank the Organising Committee for holding
the Conference in Australia and commend its proceedings,
conclusions and report as an example of constructive international cooperation in transport research.

DEREK SCRAFTON
Director General of Transport,
South Australia.

This book comprises the proceedings of the Third International Conference on Behavioural Travel Modelling, which was held in Tanunda, South Australia, from 2 April to 7 April 1977. This conference was the immediate successor of the Second International Conference on Behavioural Travel-demand Models held in Asheville, North Carolina, in 1975 (1), the latter following on from a Conference on Travel Behaviour and Values held in South Berwick, Maine, in 1973 (2). The genesis of this conference series can be traced back to two important occurences: the Williamsburg (USA) Conference on Travel Forecasting (3), held in December 1972; and the work in the early 1970s of Neil Mansfield, in particular, at the Department of the Environment (UK) on values of travel time savings. One of the results of these two activities was a rapid increase of interest and research in an emerging area of travel forecasting and evaluation that was then referred to as 'disaggregate, behavioural, probabilistic choice models', now more generally described as 'individual choice models'.

With the increase of activity in research on individual choice models, a need was seen to hold a conference that would act as a communication mechanism between researchers and as a procedure for helping to direct research effort in the near future. Such a conference was sponsored by the Transportation Research Board and the Engineering Foundation in July 1973. The title of that conference was 'The International Conference on Behavioural Travel Demand Modelling and the Valuation of Travel Time', the proceedings of which were subsequently published by the Transportation Research Board.

The Second International Conference on Behavioural Travel Demand was organised under the sponsorship of the US Department of Transportation and the Engineering Foundation. These two conferences were both characterised by several common factors. Both had sought to bring together members of the worldwide research community engaged in research into various aspects of individual choice models. Their success in this respect can be judged, in part, from the fact that 40 per cent of the delegates to each of the conferences were from outside North America. Second, each conference adopted a working format, centred around concurrent workshops, from which certain products were requested and a relatively low emphasis given to the presentation of any formal papers, a format that encourages extensive interaction among the participants. Third, both conferences were designed to produce recommendations on future research directions, with a principal goal of preventing potentially wasteful duplication of effort by several researchers and the early identification of both successful advances and unproductive lines of research.

3

At the Asheville conference, it was decided that a third conference was needed and that such a conference should be held outside North America, in order that greater participation and the local effects of the conference in increasing interest and activity in individual choice research could be obtained for other countries. As a result of the offer of sponsorship, it was decided that the Third International Conference on Behavioural Travel Modelling should be held in Australia in April, 1977.

When the first conference was held in South Berwick in 1973, it was principally seen as a mechanism for advancing individual travel-choice research which was then being done by a very small minority of transport researchers. As interest and activity has grown in the meantime, the principal objective has changed to reflect more and more a concern with the appropriate directions of research and the implementation of the research products into current transport-planning practice. The conferences are all seen, however, as a central mechanism for bringing together the state of the art of individual choice modelling in transport and in co-ordinating and directing future research and development. They represent a major communication process that circumvents the usual delay between the completion of research work and its publication and that also permits reporting on unproductive research approaches that would probably not be published. This communication process generally has led to much more rapid advances in the development of the field than would have been likely otherwise.

In terms of the effects of the first two conferences, it is apparent that the first conference, together with the Williamsburg Conference on Travel Forecasting, resulted in a considerable increase, in the United States, in the allocation of research and development funds for further work on individual travel choice models. In turn, this has resulted in significant progress on a wide range of issues, such as the development of implementable models of individual choice behaviour, the inclusion of a logit-model calibration procedure in the transit planning package of the Urban Mass Transportation Administration of USDOT, research into new approaches for understanding traveller behaviour, etc. The second conference has continued to help direct the ongoing research and further research funding in the USA, particularly. It has also generated a specific product in the form of a user manual for applying individual choice models to current transport problems by describing the model structures in simpler terms than before and documenting illustrative case studies(4).

At South Berwick, the delegates were almost entirely drawn from the research community and were asked to identify key areas of research in individual travel-choice modelling, with an allocation of appropriate priorities to each research task. In 1975, at the Asheville Conference, a small number of practising transport planners were included among the delegates and the objectives were changed somewhat to give a stronger emphasis to research that might lead to early implementation of individual choice models. In the Third International Conference at Tanunda, both transport planners and senior

government advisers were included among the delegates and the
objectives were altered further to a major emphasis on applications and policy.

The general objectives of this conference were:

1  To define those research areas that are likely to impact
   policy as opposed to those that may be principally of
   theoretical interest and among those areas that are
   likely to impact policy, to define the extent of that
   impact.

2  To identify those aspects of individual choice models
   that are ready for implementation, to identify which of
   these will be most relevant to policy and to determine
   the likely basis for implementation, i.e. project or sub-
   area planning, or strategic or areawide planning.

3  To identify the highest priority areas for research in
   terms of those that will impact or influence policy,
   particularly in the short run.

A total of 114 delegates attended the conference, 45 of whom
were from outside Australia.  The delegates were each assigned
to one of ten concurrent workshops for the duration of the
conference.  Prior to the conference, resource papers were
commissioned for each workshop, with the idea of providing
both a summary of the state of the art and the identification
of issue areas, controversial points and potentials for
policy relevance.  The resource papers were, in most cases,
distributed to the members of a workshop prior to the conference.  The topics of the workshops were:

1  New approaches to understanding traveller behaviour
2  Equilibrium modelling
3  Consumer segmentation
4  Theoretical and conceptual developments in demand modelling
5  The role of modelling tools in the policy arena
6  Relationship between behavioural models, evaluation,
   forecasting and policy
7  Goods movement
8  Behavioural modelling, accessibility, mobility and travel
   need
9  Attitudes, attitudinal measurement and the relationship
   between behaviour and attitudes
10 Behavioural modelling and evaluation of social and
   environmental impacts of transport investment

The conference commenced with a plenary session, at which
each of the workshop chairmen outlined the specific objectives that they saw for their workshop.  This session was
designed, in part, to permit all delegates to obtain some
understanding of the likely directions of the other workshops
and, in part, to help identify and resolve potential areas of
extensive overlap between two or more workshops.

Following the plenary session, the workshops met for seven
sessions of about three hours each, with either the afternoon
or evening of each day set aside for informal interaction

outside the workshop groupings. In several instances, workshops convened additional meetings during the informal sessions, in order to be certain of meeting the conference objectives. The seven workshop sessions took place over three-and-one-half days of the conference.

The format of the workshop sessions was left open to each chairman to define. In certain instances, the chairman commenced with a fairly formal presentation of one or more resource papers by the author, while others used the resource papers as a framework around which to build an informal discussion. In either case, the first two workshop sessions were spent generally in defining terms and setting up a framework for achieving the conference objectives.

The conference concluded with a plenary session at which each of the workshop chairmen presented a summary of the findings and recommendations reached by their workshop.

This book includes the position papers, the summary of discussion of each workshop and a bibliography of the major references in the area of behavioural travel modelling. Although the general area of behavioural travel modelling is moving ahead at a quick pace, the contributions in this book represent a major research resource of value for a significant period of time. It represents the published state of the art.

REFERENCES

1  Stopher, P.R. and Meyburg, A.H. (eds) (1976) *Behavioural Travel Demand Models*, Lexington Books, Lexington.

2  Stopher, P.R. and Meyburg, A.H. (eds) (1974) *Behavioural Demand Modelling and Valuation of Travel Time*, Transportation Research Board Special Report No.149.

3  Brand, D. and Manheim, M. (eds) (1973) *Urban Travel Demand Forecasting*, Transportation Research Board Special Report No.143.

4  Spear, B.D. (1977) *Applications of New Travel Demand Forecasting Techniques to Transportation Planning*, Urban Planning Division, Federal Highway Administration, US Department of Transportation.

# ACKNOWLEDGEMENTS

Members of the Advisory Committee - Mr D. Gendell, Mr I. Kingham, Dr A.H. Meyburg, Dr S. Reichmann, Mr M.G. Richards, Dr D. Scrafton, Mr G. Searle and Dr J. Taplin - are ·thanked for their contribution to the planning of the conference.

Conducting a major international conference in Australia is an expensive exercise, and without the generous financial assistance of the Commonwealth Bureau of Roads, the South Australian Department of Transport, the Australian Department of Transport, the Australian Road Research Board, University of Melbourne, Monash University, Macquarie University and Shell Oil (Australia), such an event would not have been possible. The research community is truly grateful.

The support of the Transportation Research Board, through their Committee on Traveller Behaviour and Values, has been an essential ingredient to the recognition of the series of conferences to which this book belongs.

Finally, we would like to thank Margaret Starrs and Derek Scrafton of the Office of the Director-General of Transport (South Australia), who spent a considerable amount of time in co-ordinating the planning in South Australia, and in making available pre-conference copies of the resource papers to all participants.

A SODDIT TO TRANSPORT

or

AN(Y) ODE FOR A MODE

by Parkes, Stimson & Thrift

To *populate* we *activate*
In order to *facilitate*.
Transport planners *indicate*
That it's passé to *aggregate*,
So *concentrate* while we *postulate*
Our needs and wants to *aggravate*
Stopher and Hensher *fascinate*
With reasons why we *congregate*.

We came here to *initiate*
A 'Brand' new way to *disaggregate*.
If they Kant weight and *simulate*
We will simply *stimulate*.

Logit models *generate*,
And modal-splits make us *late*;
Probability *perpetuates* the likelihood
To *terminate* our ability to *mensurate*.
While transport planners *replicate*
We at least can *innovate*
Observe how people *motivate*,
(Pause), *disaggregate* and then, *rotate*.

But! If we cannot meet the *date*,
Remember!
It's never 2 late 2 *calibrate*.

PART ONE

Chapter 1

BEHAVIOURAL TRAVEL MODELLING

David A. Hensher and Peter R. Stopher

GENERAL

The purpose of this initial chapter in a state-of-the-art
book is to highlight the advances in theory, method and
application that have been made since the previous comprehensive overview which outlined developments up to mid-1975(124).
In the last two years, not only has there been a proliferation of research in the relatively established topics (e.g.
valuation of travel-time savings, goodness-of-fit measures,
model structure, estimation techniques), there has also been
a significant reappraisal of the conceptual framework in
which individual traveller behaviour and values are studied.
The diversity of ongoing research is a healthy indicator of
the importance of the new philosophy on travel modelling.
Together with these advances in theory and method that are
increasing the understanding of the behavioural process,
there is also a translation of the research tools into real-
world planning tools. The ultimate success of any models is
their usefulness in everyday planning as part of the process
leading to advice on policy formulation and implementation.
The usefulness has to be assessed in terms of existing modelling approaches, the issues confronting planning agencies and
the relative costs per predictive accuracy and contribution
to the evaluation process.

The strength of the behavioural approach to travel modelling is that it is formulated on a set of hypotheses related
to the decision-making unit (the individual) and tested with
models which adopt the decision-making unit as the unit of
analysis, even though at present most individual-choice
models seek a set of parameters to describe a group of individuals. Any major doubts about the current behavioural
models relate to the hypotheses of behaviour rather than the
models *per se*; and hence in recent years, there have been a
wide range of alternative hypotheses on behaviour. Some of
the new hypotheses can be tested within the same conceptual
framework as that within which the familiar multinomial logit
model operates. These hypotheses include those derived from
notions of habit, uncertainty and thresholds. Other hypotheses have led to the development of alternative conceptual
frameworks, as discussed later.

In one sense, this chapter provides a guide to the topic
areas of the following chapters. However, additionally the
chapter is designed to provide a single reference source on
the major emphases of developments in individual travel-choice

11

modelling during the period mid-1975 to mid-1977, and future
research directions.  Cross-referencing to other chapters is
for the benefit of those wishing to follow up particular
issues in more detail.  Readers unfamiliar with the area are
recommended to read the Proceedings of the Second International Conference on Behavioural Travel-Demand Models (124)
in order to have enough background to get maximum value out
of the present book.

Consistent with the principal objectives of the Third
International Conference,* particular consideration is given
to the policy relevance of individual-choice models, in addition to the synthesis of recent developments in theory, method
and application, noted above.  The chapter concludes with a
perception of the most likely achievements for the next two
years and the most needed areas for research and development.

THEORETICAL DEVELOPMENTS

*Introduction*

Theoretical research in the last two years can be classified
initially according to the conceptual framework in which
approaches to understanding traveller behaviour have been incorporated.  There is a well-defined division between the
economic-psychological contemporary framework, primarily
associated with McFadden (90) and Luce (88) and the more
speculative (yet potentially stronger) framework being
developed by Jones (66,67) and Fried and Havens (34) which
owe much of their influence to Hagerstrand (43,44) and Chapin
(15).  Essentially, the former framework is conditioned on
the utility-maximisation assumption with the travel task
being modelled as a choice process in isolation from the
wider set of human activities, involving the separability
notions of Leontief (74).  The latter framework, also conditioned on the utility-maximisation assumption but with a
broader behavioural base, assumes that travel is one of a
range of complementary and competitive activities operating
in a continuous pattern or sequence of events in space and
time (see Chapter 2).  This framework has the property that
travel represents the procedure by which individuals trade
time to move location in space in order to partake in

---

* The general objectives of the conference were:

 1. To define those research areas that are likely to impact
    policy as opposed to those that may be principally of
    theoretical interest; and among those areas that are
    likely to impact policy, to define the extent of that
    impact.
 2. To identify those aspects of individual-choice models
    that are ready for implementation, to identify which of
    those will be most relevant to policy, and to determine
    the likely basis for implementation, i.e. project or
    sub-area planning, or strategic or area-wide planning.
 3. To identify the highest priority areas for research in
    terms of those that will impact or influence policy,
    particularly in the short run.

successive activities. There is one school of thought within this latter framework which argues that activities are seen as the outcome of choices (15). However, the dominant belief or hypothesis is that time and space are seen as resources and that the constraints which operate on individuals are the main dictates of individual experience (136) which may in many instances remove the choice element.

Within the contemporary framework, the theoretical developments can be divided into two groups. First are those associated with the efforts to improve the multinomial logit (MNL) formulation of individual-choice models (in particular the general-share structure that results from the independence-of-irrelevant alternatives property (IIA) ). Second are those attempts to examine alternative model structures of travel behaviour, in particular Markov and semi-Markov processes (11,37,76) and threshold models (137,70,26, 27).

*Extensions to the Multinomial-logit Formulation*

1. The MNL model of choice assumes that the ratio of probabilities of choosing one alternative over another (where both alternatives have a non-zero probability of choice) is unaffected by the presence or absence of any additional alternatives in the set. The existence of the IIA property, implicit in any general-share structure, can result in two potential major problems: (a) Failure to ensure that all alternatives are distinct may lead to biased estimation of model parameters; (b) Proscribed market-share changes may occur when any existing alternative is altered or the set of alternatives is changed by addition or deletion.

Three major procedures have been proposed to deal with the IIA problem. These are given in historical sequence.

(a) Modifications of the MNL model that reduce or eliminate most of the problems resulting from violation of the IIA property. These modifications result in a 'structured-logit' model (91,148) where similar alternatives in a choice set are modelled together initially and then redefined as a single alternative in the next stage of the structure by either the maximum method, the cascade approach or the Mclynn adjustment (98).

(b) Tests to determine when the IIA property is violated (18). An extensive analysis has been undertaken by Charles River Associates in the context of the MNL model.

(c) The search for alternative formulations of individual-choice models that are capable of producing information to assess the degree of similarity among choice alternatives. The relative similarity of alternatives cannot be defined in the share-model formulation. In particular there is the multinomial probit formulation (MNP) (77,48,26,46). The model does not belong to the family of share models.

Although there is general support for the contention that problems of data measurement and variable specification are more important sources of error than the IIA property, especially in individual models (93), it is useful to test the IIA assumption during application: 'The independence assumption is *a priori* neither desirable nor undesirable, but should be accepted or rejected strictly on empirical grounds depending on the circumstances' ((18)p.D-104). Charles River Associates (18) suggest that violations of IIA fall into two basic classes:*

Case 1: Correlation of the unobserved components of utility among the alternatives ...

Case 2: Observed and unobserved attributes of utility are not independent of one another ((18)D-48).

The diagnostic tests for the IIA property have been developed by McFadden, Tye and Train for Charles River Associates. When violation of the IIA assumption is thought to exist then the following tests should be undertaken for each case:**

*Case 1(a) tests on residuals using an unbiased parameter estimate* by either estimating another data set or estimating a random subsample of the existing data set. The formula for transformed*** residuals ((18)p.D-109) is:

$$Y_{jn} = D_{jn} - d_{1n}\sqrt{P_{jn}}\,(1 - \sqrt{P_{2n}})\ /\ (1 - P_{1n})\ \text{and}$$

$$D_{jn} = (S_{jn} - R_n\,P_{jn})\ /\!\sqrt{R_n\,P_{jn}}$$

where n = number of individuals, j = number of alternatives, $R_n$ = repetitions for selection for case n, $S_{jn}$ = selections of alternative j, $P_{jn}$ = estimated probabilities. Hypothesising corrections of the estimated model ($Y_{jn}$ are IID), if the test statistic differed significantly from zero for the mean test, the hypothesis of the model can be rejected. Similarly for the variance test, if the unit variance of $Y_{jn}$ under the null hypothesis is true, we can accept the hypothesis if chi-square falls in the acceptance region for the chi-square test.

*Case 1(b) tests on estimated coefficients to assess bias:* by either rearranging data as in case 1(a) and comparing coefficients with those of the full data set, and/or comparing estimated models which are separated according to dominance and recessiveness of an attribute in the alternative known to be independent. The likelihood-ratio tests can be used for testing hypotheses.

---

*    This section draws heavily on the research of CRA ((18) App.D). It is referred to with permission.
**   We only outline tests that are likely to be of relevance to the most common data configurations.
***Because of statistical dependence among residuals.

Case 2(a) tests, based on restricted choice sets, for
aggregation of heterogeneous market segments as a result of
failure to separate market segments with significantly
different valuations of attributes. (Case 1 violation from
this source can also be tested using 1(a) and 1(b) tests.)
Additional dummy variables can be introduced into the full
data set to test an hypothesis that the valuation of a particular attribute is different between groups.   The same
statistic as used in 1(b) can be used to estimate the significance of the between-group difference in the value of the
attribute.

Case 2(c) tests based on the full data set where the
correlation is not due to aggregation of heterogeneous market
segments.   New data that permits unobserved attributes to be
tested as observed attributes is required to assess dependence.

If the test suggests violation, then the following
modifications (within the logit framework) to the model
should be carried out:

1. improve model specification to eliminate dependence.
Individual behaviour assumptions may be incorrect, relevant
variables may be missing (there are non-systematic variations
in unobserved attributes across alternatives), market segmentation may be required to overcome the lack of an appropriate
average behaviour of individuals in a group with similar
observed attributes and facing similar choices;

2. change the model estimation procedure to account for
dependence;

3. change the model itself, in particular use the maximum
model.

Remedies for case 1(a) and (b) are the addition of explanatory variables and the elimination of choice
alternatives.   In the latter instance, McFadden's 'maximum
method' (91) is proposed, which assigns the probability
between the two dependent alternatives according to a binary
logit model separately estimated; with the probability(ies)
of the independent alternative(s) being obtained directly
from the multinomial logit model.   The justification for the
maximum method is well summarised by CRA ((18)p.D-131):

(it) is the ability to factor the unobserved attributes
into two types of factors - those that are independent
between modes 2 and 3 and unimportant between 1 and 2 (or
3) and those that are independent between 1 and 2 (or 3)
and unimportant between 2 and 3, where 2 and 3 are the
dependent modes.   According to the assumptions of this
model, independence holds between 1 and 2 (or 3) and 2
and 3, but not for the joint model including 1, 2 and 3.

Remedies for case 2 include the use of new choice alternatives to break up the correlation (as was the situation in
a study by Hensher (57) of the effect of reopening the Tasman
Bridge in Hobart, Australia); converting an unobserved attribute to an observed attribute or segmenting the sample for

estimation purposes to allow for any non-quantifiable biases; and modifying the estimation procedure.  Westin and Gillen (149) have recently shown the existence of a major model mis-specification, and illustrated their contribution using instrumental variables in a simultaneous model of parking location and modal choice.  They criticise the assumption of choice from an homogeneous bundle of attributes in discrete choice decisions, and propose the existence of endogenous attributes determined by an auxiliary regression equation on a reduced-form model.  The normal quantal-choice model has a random-utility index entered conditional on the value of the reduced-form generalised-price function.  The empirical model studies simultaneously the joint effects of parking policies on mode choice and the redistribution of parking within the central area; where the latter is modelled for the individual as a location choice with a trade-off of parking cost and walk time to the destination.  The generalised price of a given parking location is defined as the sum of the parking-cost function (which is itself a function of distance to final destination), and the product of walking distance from the traveller's parking location to his final destination, the value of walking time, and the inverse of the individual's walking velocity.  Westin and Gillen have demonstrated convincingly the inadequate manner in which parking has been represented in mode-choice models and hence the general non-significance of an intuitively significant determinant of mode choice.

2. Most of the early work on logit models depends upon two statistical tests alone - the t-test on coefficients of the model and the likelihood-ratio test $(-2\log\lambda)$ for the model as a whole.  These measures were found to be somewhat inadequate as a general method for assessing models and provided no sound basis for comparison among different models particular-ly among models of different specifications.  The correlation ratio and its associated F statistic (119) provided an additional measure; however, this measure was not useful for an absolute assessment of goodness of fit, but only for com-paring models of different specifications.

During the past two years, several additional measures have been put forward to improve this. A pseudo-$R^2$ has been developed and used to assess and compare models (5,94). One of these, due to McFadden (see Chapter 13 of this book), is also termed a 'likelihood-ratio index', because it is formed from the likelihoods used in computing the $\chi^2$ value, as shown in Equation (1).

$$\rho^2 = 1 - L_0/L_1 \tag{1}$$

where $L_1$ = likelihood for fitted model
$\quad L_0$ = likelihood for the equal-shares or market-shares hypothesis
$\quad \rho^2$ = likelihood-ratio index.

When the fitted model adds no new information, the two like-lihoods, $L_1$ and $L_0$, will be approximately equal, their ratio

will be close to 1, and $\rho^2$ will tend to zero. As the fitted
model adds more information, the likelihood for the model
becomes larger than $L_0$ and the ratio becomes smaller. As $L_1$
becomes very much larger than $L_0$, the ratio tends to zero and
$\rho^2$ tends to 1. Values of $\rho^2$ of 0.2 to 0.4 represent excellent fit.

McFadden has also developed a classification test, based
upon assessing the number of individuals for whom the highest
estimated probability coincides with their actual, observed
choice. Suppose that the number of individuals who are both
predicted to and choose alternative k is denoted $N_{kk}$. This
number is given by Equation (2).

$$N_{kk} = \sum_n S_{kn} P_{kn} \tag{2}$$

where $N_{kk}$ = number of successful predictions for alternative
            k

       $S_{kn}$ = 1 if alternative k is chosen by individual n, 0
            otherwise

       $P_{kn}$ = probability from the model of individual n choos-
            ing alternative k.

A 'success index' can then be calculated as shown in Equation
(3).

$$\sigma_k = \frac{N_{kk}}{N_k} - \frac{N_k}{N} \tag{3}$$

where $\sigma_k$ = success index

      $N_k$ = observed number of individuals choosing alterna-
           tive k

      N = total population.

The ratio $N_k/N$ represents the success index for the market-
shares hypothesis, so that $\sigma_k$ is adjusted to represent the
marginal success over the market-shares hypothesis.

An overall success index can be formed by summing $\sigma_k$ over
all alternatives, k. Also, McFadden indicates a modification
that forces $\sigma_k$ and also $\sigma$ (overall success index) to lie between zero and one. This normalised overall success index is
given by Equation (4).

$$\sigma = \sum_k \left[ \frac{N_{kk}}{N} - \left[ \frac{N_k}{N} \right]^2 \right] \bigg/ \left[ 1 - \sum_k \left[ \frac{N_k}{N} \right]^2 \right] \tag{4}$$

Westin (149) and others have developed procedures, based upon
the likelihood-ratio test, for comparing models of differing
specifications. Suppose, for example, that two models are
built from one data set. The first model uses a set of independent variables, M, while the second uses a subset of
these, M', such that M' M. Suppose, also, that both models
contain the same set of (T - 1) alternative-specific constants. The significance of the model with M variables,

relative to that with M  variables is given by the test of Equation (5).

$$-2\log\lambda = -2 \; L'_M - L'_M \qquad\qquad (5)$$

where $L'_M$ = log likelihood for model with M' variables

$L'_M$ = log likelihood for model with M variables.

As for the standard likelihood ratio test, $-2\log\lambda$ is distributed like $\chi^2$, but with (M-M') degrees of freedom.  Similar tests may be conducted with different subsets of variables and alternative-specific constants.  Also, with adjustments for unequal segment populations, tests may be conducted between identically-specified models from different market segments.  These tests are generally much stronger than the comparable test for overall model performance and have been found to be useful in a number of applications (e.g. (126, 128)).

Finally, work has been pursued in using information theory as a basis for assessing the models, in which one can use the concept to determine both the information added by a model and that added by improving the specification of the model (47).  Measures can be developed that show the information added by the model, relative to that available in the data and from a simple hypothesis, such as market shares.

*Alternative Model Structures*

*Semi-Markov processes and learning.*  While improvements have been sought in the multinomial-logit model, efforts have also been made to examine alternative model structures that may offer distinct improvements over the MNL model.  One of the primary candidates that has been examined in this context is the Markov process model, and particularly the semi-Markov process.  In this general family of models, it is necessary to define a set of states that an individual might occupy at various points in time.  The process then defines either the probability of an individual being in a given state at time t, or, given the total number of individuals in various states at time (t - 1), the model will estimate the number in each state at time t, based on a transition-probability matrix.

Attempts have been made, most notably by Gilbert (36,37) and Burnett (11,12), to apply the Markov process to travel forecasting.  In particular, their efforts have examined a semi-Markov process, in which the transition probabilities are defined exogenously, instead of being determined by simulating actual transitions from observed behaviour.  The basic notion of the Markov process is an attractive one, since the model is dynamic in concept, operates in a chain process, and would not necessarily have the properties of the Independence of Irrelevant Alternatives (IIA) (89) that are troublesome when violated (as they frequently are in reality).  However, results so far have been disappointing, despite some operational models developed by Gilbert, Peterson and Linne (37).

The Markov model has two inherent drawbacks. First, the transition probabilities are assumed to be independent of previous states occupied by the individual and are assumed to be stable with time. Second, there are problems in defining the states that the individual occupies. In the work of Gilbert, the states were defined as a series of camping points in a major recreation area, where most·stays were measured in days. In this case, the state space is readily defined and the Markov process replicates movement through the area quite well, allocating individuals to new camps based only on information on the current camp occupied. Burnett attempted to apply the concept to destination choice in the urban area. Destinations are not appropriate as states in the Markov model, particularly since many journeys involve visiting only one non-home destination (perhaps only 30 per cent of trips involve multiple non-home destinations). Burnett (11,12) applied the procedure by defining states of learning about destination opportunities and then using the individual's state of knowledge to define which of the overall choice models is the appropriate one to use. The approach seemed promising but did not yield substantive improvements over the MNL model, while adding complexity to the process.

Interest still remains in the development of suitable applications of the Markov or semi-Markov process to travel-choice processes. Despite the problems of the process, it still has much intrinsic appeal as a potential procedure for travel modelling because of the appealing condition of maintaining the sequencing and direction of travel. However, further research is clearly needed to establish it as a viable alternative to the MNL model.

*Threshold models and elimination by aspects*. While work has continued on the development of multinomial logit and probit models, other model structures have also been considered that are based upon different assumptions of the choice process of individuals, although still basically utility-maximising procedures. A principle departure from the logit model is the consideration that the individual does not react in a continuous fashion to attribute changes, but rather that he/she reacts only when the changes are sufficiently large to cross a threshold. This threshold may be a threshold of awareness or a threshold of acceptability. In addition, thresholds may not be fully compensatory.

The existence of perceptible thresholds has been acknowledged in the literature for many years, and was formalised some time ago by Weber (147) in psychology and Georgescu-Roegen in economics. Georgescu-Roegen introduced the notion of the threshold of insensitivity into the theory of consumer behaviour nearly 40 years ago (35), suggesting that a choice will be considered only when the positive range or threshold of insensitivity is overcome. It has its origin in psychophysics and the general theory of signal detection (129), dealing with relations between experienced intensity and physical intensity based on Weber's Law which sought to answer the question of how much intensities of stimulation must differ before the difference can be noticed. Simon(112)

formalised the threshold notion in his definition of the individual as a satisficer not a maximiser; he hypothesised
that an individual seeks to achieve a minimum acceptable
utility and does not seek to achieve a higher utility once
this minimum is achieved.  Devletoglou (31,32) extended
Simon's hypothesis in a theory in which stimulation is normally capable of eliciting reportable human responses only if
a certain threshold or positive range of insensitivity/indif-
ference is first overcome.

One model structure that has begun to receive attention
from travel-behaviour researchers is the Elimination-by-
Aspects (EBA) model put forward by Tversky (137,138).  This
model postulates a hierarchy of attributes (aspects), each of
which must meet a threshold value in an alternative, in order
for that alternative to be considered or chosen.  Some preliminary investigation of this model for travel choices has
been undertaken by Golob and Nicolaidis (39) with results
that suggest that further investigation of the model would be
worthwhile.  So far, however, this appears to have been the
only transport application of the structure.

It is also worth noting that if the thresholds of individuals are distributed through the population on a Weibull or
normal distribution, the results, in aggregate, may not look
too different from the modelling point of view from the type
of process modelled by the logit model.  The principal difference might occur in the use of such models for forecasting,
where the logit model may over-predict changes in choice compared with reality and the threshold model.  This is particularly likely to be true where the changes in factors
affecting choice are small.

Besides the work on the EBA model, a number of isolated
transport studies have been completed in recent years, all
concerned with various facets of thresholds in choice.  De
Janosi (29), Goldberger and Lee (38), Uhler and Cragg (139),
Degenais (27,28) have developed threshold choice models for
the purchase of private cars; Wilson((150) and Chapter 7 of
the present book) has investigated the choice of mode using
catastrophe theory (where sudden jumps can occur from one
equilibrium value to another); Hensher (53) and Goodwin (41)
have explored the influence of habit on the trade-off between
the generalised costs of alternative modes; and Krishnan (70)
has recently tested a modified logit model incorporating consideration of 'just-noticeable differences' on traveller's
choice of mode for accessing commuter rail services.
Krishnan's contribution is most relevant to developments of
operational individual-choice models using the multinomial
logit formulation, in that he takes the theory of threshold
choice and hypothesises the role of thresholds in the utility-
maximisation framework underlying the logit model.  He
hypothesises that: 'the "satisficer" would ... become a
utility maximiser when a sufficiently more attractive alternative is made available to him' (p.1).  This is very similar
to the concept of 'transfer price' developed by Hensher (50,
52) and Dalvi (22,71) in the context of valuation of travel-
time savings, which represents the amount of change in a
particular journey attribute that would have to occur before
an individual would consider an alternative (i.e. before that

alternative would enter his perceptible reaction space).
Whereas the transfer-price approach has, to date, been based
on attitudinal (behavioural-intent) response, and has been
incorporated in the specification of the variables themselves,
Krishnan builds the threshold constraint directly into the
utility-maximisation theory underlying the derivation of the
estimable model. The only concern with this approach appears
to be the difficulty of defining the empirical level of 'just-
noticeable differences' across the sample.

The 'just-noticeable difference' (jnd) concept is built
into Krishnan's logit model to overcome the present restrictive assumption that *the probability of attaining a state of
indifference* (when two utilities are equal) *is zero*. To
quote Krishnan (p.4): 'experiments in psychophysics ... show
that the probability of indifference is not necessarily zero.
In fact, when two stimuli are similar, the experiments ...
have demonstrated that there is a finite probability of confusion between them.' Quandt (107) in a relatively neglected
paper had demonstrated this in economics with an indifference
surface comprising probability bands of indifference around
levels of utility. Leibenstein (72) has recently proposed a
theory of inert areas based on a common set of experiences in
which it is possible to improve a situation in some respects,
yet it is not worth the effort to do so. The action required
to move from a lower to a higher utility level involves a
utility cost that is not compensated for by the gain in
utility. This effect is analogous to the threshold concept,
suggesting a net utility band of indifference. Goodwin's
notion of an effort variable (40,59) has relevance here,
since it adds an additional dimension to the generalised cost
of alternatives. Not only is x minutes of walk time equal to
2x minutes of in-vehicle time (from a dis-utility viewpoint),
but the relationship between effort in walking (defined in
k/calories per minute) and effort in in-vehicle travel is not
proportionately the same, making the traditional generalised
cost trade-off approach deficient in a likely important
dimension.

Krishnan's new probability conditions (using the accepted
notation) are:

$$\text{Prob } (e_1 - e_2 > V_1 - V_2 + \delta) = \text{Prob } (A_1 > A_2) =$$

$$\text{Prob } (U_1 > U_2 + \delta)$$

$$\text{Prob } (e_1 - e_2 < V_1 - V_2 - \delta) = \text{Prob } (A_2 > A_1) =$$

$$\text{Prob } (U_2 > U_1 + \delta)$$

$$\text{Prob } (V_2 - V_1 - \delta \leq e_1 - e_2 \leq V_2 - V_1 + \delta) =$$

$$\text{Prob } (A_1 \sim A_2) = \text{Prob } (|U_1 - U_2| \leq \delta)$$

is the 'minimum perceivable difference'. The probability
distributions for the logit formulation become:

$$P_1 = 1/ [1 + \alpha \exp\{\beta(\delta + V_2 - V_1)\}]$$

$$P_2 = 1/\ [1+ \ exp\{\beta(\delta+V_1-V_2)\}]$$

$$P_{12}= [\alpha\ ^2 exp\ (2\beta\delta)-1]\ p_1 p_2, \text{where } p_{12} = 1-(p_1 + p_2)$$

If indifference exists, then $A_1$ and $A_2$ will be chosen randomly with probabilities $\theta$ and $(1-\theta)$ respectively; hence the overall probabilities of choosing $A_1$ and $A_2$ are:

$$P_1 + \theta P_{12} \text{ or } P_1[1 + \theta\{exp\ (2\delta) - 1\}P_2]$$

and $P_2 + (1-\theta)\ P_{12}$ or $P_2[1+(1-\theta)\{exp\ (2\delta)-1\}P_1]$

Minus log of the likelihood is then minimised subject to the minimum perceivable difference being greater than or equal to zero, and within the range zero to unity. The empirical results (see Krishnan) suggest biased results using the unconstrained logit model ( $\delta$ = 0) when the utility difference does not exceed the minimum perceivable difference.

This approach has important implications for choice-set determination, since it is conceivable (in the light of no evidence at present) that many physically existing 'alternatives' in space and time which individuals indicate are not presently valid alternatives (because assessment is based on levels of attributes of alternatives) may in fact be within the band of indifferences (when we also include the effort idea of Leibenstein), rather than be totally at odds on generalised cost with the chosen alternative.  Krishnan's model could be usefully extended to include a maximum perceivable difference, beyond which an alternative is not relevant, and the probability of it being chosen is zero.   In a sense, Tversky's elimination-by-aspects approach is consistent with this direction of thought.

*Other Theoretical Issues - Valuation of Journey Attributes*

Individual-choice models have been used extensively to derive values of journey attributes, in particular the value of travel-time savings.  The issue of traders v. non-traders has been clarified by Daly (23), concluding that all individuals in any data set are traders, regardless of how many explanatory variables are specified.  Additionally, the theoretical relationship between the derivation of values of journey attributes, from standard individual-choice models and alternative model specification which is concerned with a direct estimate of the value of time, has been determined; the conclusion being that the two approaches can be derived from the same theory, the main difference being that the latter (transfer-price approach) provides an additional piece of information to identify the trade-off margin.   The benefits of the additional information have to be assessed in the light of the problems of measurement by hypothetical contexts (30).

Guttman has clarified the relationship between the economic theory of the allocation of time and the way in which empirical values of journey attributes are derived.  He

argues that it is the constraints in the real world that
distort the theoretical uniformity of the value individuals
place on time, which at the margin should be the same at all
times.  Given this, the most appropriate manner of segmenting
the population in order to obtain empirical values of journey
attributes, in particular travel-time savings, is that which
accounts for the greatest between-segment variation in constraints and the least within-segment variation in
constraints.  The time of day seems the most appropriate
basis for segmentation, and not journey purpose, although the
latter could be a second-stage stratifier if required.  This
seems intuitively plausible given that the circumstances
under which travel time is spent are more variable throughout
a day than between journey purposes at a given time of the
day.  Guttman has evidence to support this.  The value of
time savings for off-peak commuter trips is similar to that
of off-peak shopping-recreational trips, and very different
from the peak commuter value of time savings.

The issue of valuation of journey attributes is treated in
greater depth by Hensher and Dalvi (62), to which the reader
is referred for further information on this topic.

*Alternative Approaches to Understanding Travel Behaviour*

During the last ten years there has been a major development
of thought into the true mechanism by which the actual and
optimal interaction of individuals in time and space occurs.
We have already distinguished between choice and constraint.
The essential emphasis, however, is on the relationship
between social, biological, technical, economic, and physical
processes and conditions, with particular interest in the
revelation of the structure of systems of constraints.  The
sequential courses of events at both an individual and an
aggregate level are emphasised.  This is in contradistinction
to the contemporary framework of individual-choice models
which, in a very real sense, utilise the separability condition (with limited empirical support, albeit operational
convenience) to simplify (or some could say distort) the
realities of the world of human interaction.  Sequencing and
direction of travel activities is lost in the present travel-
choice models.  Hagerstrand (45 p.10) sees this new approach
as the appropriate representation of 'the interwoven
distribution of states and events in coherent blocks of
space-time'.

We could not possibly do justice to the potential of this
alternative approach in the present chapter (see Chapters 2
to 5 for a comprehensive outline of the approach).  However,
it is worth pointing out that an immediate implication of the
research into the arrangement of human activities in time and
space for individual-choice modelling is that rearrangement
of activities (at origins and destinations) by modifying time
and space constraints, can work conjunctively with the linking of trips to increase accessibility to opportunities and
conserve energy (both human and non-human).

The application of the space-time approach defines the
unit of analysis as households of various compositions

carrying out programmes of typical everyday activities.
These programmes (or sets of activities) are defined by
performance time for a representative individual on an
origin-destination point basis.  The spatial locations in
which activities occur are mapped in time and space, as is
the multimodal transport system on a modal basis.  Lenntorp
(73) has developed a program for evaluating alternative
sample paths (PESASP) to show the number of possible ways in
which a given activity programme may be carried out.  The
level of aggregation of the basic unit of analysis (homogeneous person type) permits aggregation on any spatial,
temporal or other basis.

The procedure has the capability of predicting the optimal
arrangement of activities and appropriate form of transport
service in time and space for a variety of time-space constraint environments for members of a community.  At this
stage, however, it does not have the capability of accounting
for consumer behaviour beyond the economic assumption that
preferences are revealed through the manner in which
individuals currently undertake their activities in time and
space, although this use of revealed preference theory is
clearly the most realistic application to date.  There seems
no reason why other bases of determining the dynamics of
preference cannot be included in this framework, so that the
true relationship between choice ('constrained preference')
and preference can emerge (55).


METHODOLOGICAL DEVELOPMENTS

The greatest amount of development has occured in methodology,
which is broadly defined to include not only contributions to
improving particular features of models (e.g. specification
of functional form, measurement of variables, aggregation),
but also extensions of individual choice models beyond urban
mode choice, the major empirical context for the greater
proportion of research and development before mid-1975.
Significant steps have since been taken in extending the
individual choice approach to other decisions, in particular
residential location (75,4), destination choice (127,13), car
ownership decisions (14), inter-city mode choice (126) and
inter-city freight (109,24).


*A Method of Measurement*

Mathematical psychologists have recently formalised ((85) and
Chapter 32 below) a paradigm of human judgement that has been
implicit in the research of others (59,127,13).  Whereas this
paradigm has been mainly used in experimental design contexts
by psychologists (86,80), its value is now seen in real world
behavioural settings as a logical process in identification
of the relevant attributes influencing particular behaviour.
This area of research is concerned with the following
question (see Chapter 32), 'Given a set of attributes of
alternatives hypothesised to affect choice, how do individuals and groups combine attributes to value (assign a
worth or a utility) each alternative?'  The primary
concern, then, is with the functional form of the marginal

and joint utility functions.  Following Louviere (based on
Anderson (2,3) and Birnbaum (8) we have:

(1)  $x_{im} = \Psi(X_{im})$    (psychological )         (physical    )
                       (measures of  )         (measures of)
                       (physically   ) = f     (variable    )
                       (measurable   )         (im           )
                       (variables (im))

where m is the physically measurable subset of independent
variables which can have i different values.  (for example,
m = time, with value i = 5, i = 10, i = 15).

(2)  $I_i = I(x_{ij})$    Unobservable    = f   (subjective )
                       overall evalu-          (values of  )
                       ation of ith            (i different)
                       entity in               (values of j)
                       question

(3)  $R_i = R(I_i)$    Response in a    = f   (unobservable )
                       psychological           (overall       )
                       dimension (as           (evaluation of)
                       outcome of n            (ith entity   )
                       different
                       combinations
                       of the $x_{ij}$)

(4)  $B_i = B(R_i)$    Corresponding    = f   (observed)
                       behaviour of            (overall )
                       observed over-          (response)
                       all response

Louviere *et al.* demonstrate that a simple relationship does
exist between psychological measures and behaviour.  Given
this, and an existing estimated model, it is possible to
assess the effect of a change in a physical measure on
behaviour via the paradigm.  The emphasis is on how individuals evaluate alternatives (modes, destinations, etc.) as
viable choices, how they evaluate independent variables
hypothesised to affect choices, and the relationship between
evaluations of independent variables and corresponding
physical measures.

    A linear and a multiplicative version of a general
behavioural equation relating to the components of the
paradigm have been tested by Louviere *et al.* (87).  The
linear version is:

$$B_i = a_{m+2} + b_{m+2} \{ a_{m+1} + b_{m+1} [ \sum_m (a_m + b_m x_{im}^{Cm}) + \sum_l (X_{il}) ]^{Cm+1} \}^{Cm+2}$$

The multiplicative version is:

$$B_i = a_{m+2} + b_{m+2} \{ a_{m+1} + b_{m+1} [ \prod_m (a_m + b_m x_{im}^{Cm}) \prod_l (X_{il}) ]^{Cm+1} \}^{Cm+2}$$

The empirical evidence (summarised by Louviere, Wilson and
Piccolo in Chapter 32 of the present volume) suggests that
multiplication is the logical combination rule for choice
behaviour, and that information-integration theory has
empirical validity.  The main strengths of information

integration theory are: (i) it has an error theory, unlike
conjoint measurement; (ii) it is flexible in that even if the
assigned utilities represent only a ranking, they can be
transformed to a cardinal value; (iii) it provides a simple
way of obtaining both marginal and joint utility functions.

The empirical studies conducted by psychologists are a mix
of experimental design and real-world validation.  Without
the benefit of experimental design, an analogous paradigm was
developed in Australia for obtaining physical measures of
attributes influencing choice, based on an economic theory of
utility maximisation ((59) and Stanley, Chapter 22 below).
The implicit model is:

    choice    = f (utility)
    utility   = g (attitudes or behavioural intent)
    attitudes = h (attributes)

The main emphasis of this research, however, was on identification of the extended set of attributes influencing choice
in units that are sensitive to policy changes.  The
attitudinal step was seen as an appropriate intervening
mechanism for extending the scope of behavioural models,
because of the underlying behavioural theory and the ability
of the attitudinal dimension to provide an initial metric for
assessing the likely relevance of an extended set of
variables (comfort, convenience, etc.) in the individual's
utility function.  From an operations viewpoint, it does not
appear to be worthwhile to follow these procedures every time
a new problem is posed.  What is required is a basis for
inferring transferability of relationships.  The stage is set,
but the evidence is still being accumulated.

It should be noted that Louviere *et al.* also developed
separate models for each individual by using factorial
designs whereby a behavioural intent (or probability) is
obtained for each combination of attribute levels; then the
estimated true individual model is used to identify the
probability for the alternatives presently available.

The paradigm itself is separate from the complex issue of
the measurement of attributes.  The relationship between
reported-perceived and measured values of attributes, and the
appropriateness of each for analysis of individual choice
behaviour and evaluation of projects is still a major area of
research.  Little has been done to advance beyond the position that was popular two years ago; namely that the
reported-perceived (as the best estimate of perceived) is the
appropriate dimension for explanation and prediction of
behaviour, and that for evaluation a measure of the real
resource dimension of the attribute should be calculated
independently.  This poses particular problems when the
resource valuation of a journey attribute cannot be derived
directly from a resource calculation but has to be obtained
as a modification of the appropriate behavioural value of the
journey attribute.  The only known work that has derived a
resource value from a separate model to the behavioural model
for passenger travel is that of Hensher (56) for business
inter-city air travel (both local and international).  Recent
work using this new method by R. Travers Morgan for the UK

Department of the Environment on commercial vehicle time savings (which have traditionally been equated with the average wage rate plus allowance for marginal wage increments) concluded that the value is equal to 85 per cent of the wage.

Talvitie (134) has recently compared the various ways of measuring journey attributes, and used a wide variety of statistical tests. The results are generally inconclusive. The main concern was not to test whether the reported-perceived approach is the correct one for assessment of behaviour, but rather to see if there is any systematic relationship between the two measurement approaches. It is exceedingly difficult to test the appropriateness of one measurement approach (in a sense this is a value position); clearly if any systematic relationship exists between alternative measurement approaches, then it would be a simple exercise to compare the behavioural predictions of both as a useful sensitivity test.

*Aggregation and Aggregate Demand*

Researchers have recognised for some time the need to provide aggregate information, such as traffic-flow volumes, bus-ridership estimates, etc. from the individual-choice models. All of the conferences in this series, for example, have been concerned with aggregation issues (100,51,124). A number of advances have been made in this area in the last two years, advancing the field considerably in this time. Work has included both advancement in how to develop aggregate estimates and development of knowledge about the effect of using aggregate values in disaggregate models themselves.

In the matter of aggregation, Koppelman has raised the question of the extent to which one needs to consider aggregation error. While earlier work (133,148,145,68) had tended to reject naive aggregation on the grounds that errors from it were unacceptable, Koppelman has suggested subsequently that even the errors of naive aggregation are small compared with other sources in model building. This suggests that, perhaps, it is unnecessary to use complex correction procedures, such as Westin's $S_B$ distribution correction or Talvitie's Taylor-series correction to remove aggregation errors. Nevertheless, work has continued with a view to reducing aggregation errors as far as possible through simple procedures, or to finding procedures that allow the estimation of design volumes without recourse to expensive surveys or the need for large quantities of back-up data. One such procedure that simultaneously reduces aggregation error and obviates the need for extensive back-up data has been put forward by several researchers (117,102,58,57,75). This procedure, with minor variations, consists of delaying aggregation to the last possible moment, by aggregating the *results* of the model application rather than the *model* itself. Thus, the model is applied to a small sample of data points, selected at random for the prediction issue of concern, and the prediction results are then aggregated to the total population of concern. Stoner (117) has shown that quite small samples are adequate for this process and yield very reliable results.

Another concern has been with the use of aggregate measures of attributes as the basis of forecasting with disaggregate models (96). The use of such measures has been shown to result in the production of biased forecasts, although partial corrections can be made to these estimates.

A basic principle of aggregation is that within-group variance should be minimised and between-group variance maximised. Since market segmentation is a procedure aimed at finding homogeneous groupings of the population (Dobson, Chapter 10 below), it is clear that market segmentation holds out possibilities for reducing the errors of aggregation. This has been recognised in a number of processes for aggregation. Koppelman (69) has proposed group prediction as a means of reducing the naive aggregation error, while aggregation by market segmentation is also the backbone of the post-model aggregation processes referred to above, where the identification of appropriate segments can reduce substantially the required sample sizes for accurate prediction aggregation.

*The Broader Set of Mobility and Travel Decisions*

At the time of the Asheville conference (125) most uses of the logit model were still confined to mode choices in the urban area for work trips. A few extensions were noted to other trip purposes and to other choices, but such extensions were, at that time, minimal. In the time that has elapsed since then, an increasing number of applications to other trip purposes and other choices have been undertaken. In addition, most of the models developed prior to Asheville were binary choice models and multinomial choice models were only just beginning to be developed. In the last two-and-a-half years, there has been a major increase in the development of multinomial logit models.

A major area of development has been that of destination choice. Following the work of Charles River Associates (16) and Burnett (9,10), destination-choice models were developed by Adler and Ben-Akiva (1), CRA (17), Cambridge Systematics (14), Burnett (13), and Stopher et al. (120,127). Adler and Ben-Akiva (1), CRA (17), and Cambridge Systematics (14) all used simple measures of destination attractiveness, such as employment and used the destinations within a specified cost and time range as the choice set. Stopher et al. (127) have attempted to develop notions of attractiveness of destinations from measurements of perceptions and preferences, following some of the earlier work of Burnett (10). The third approach has been to examine alternative model structures and the development of choice sets from the notion of learning (12,13). Work has also progressed on applying the techniques to inter-city situations. Prior to Asheville, Watson (144) had developed a model of inter-city mode choice for a short inter-city trip (about 85 km). More recently, work had been undertaken to develop mode-choice models for the US inter-city market (126,128).

Other efforts have been concerned with urban choices outside the usual set of choices embodied in the conventional urban transport planning process. These include the work of Lerman (75) on residential-location choices, the development of auto-occupancy models (104), and auto-ownership choice models (6). A further area of development has been in the area of two-stage models of the form of those of Westin and Gillen (199), who developed models of parking and mode choice. These models follow on from some two-stage models developed earlier, such as those of Talvitie and Liou (82) on access- and main-mode choices.

These various developments are summarised in Table 1.1. While these do not represent an exhaustive listing of such extensions, they provide a good indication of the breadth of extensions that have been undertaken.

TABLE 1.1 Summary of recent model developments into other purposes and choices

| Authors | Functional form | Journey purpose | Choices |
|---|---|---|---|
| CRA | Multinomial logit | Work and shopping | Frequency, destination, mode, time of day |
| Adler and Ben-Akiva | MNL | Shopping | mode, destination |
| Cambridge Systematics | MNL | Work and shopping | mode, destination, auto-ownership |
| Burnett | Markov and MNL | Shopping | destination |
| Stopher et al. | MNL and regression | Shopping | destination |
| Clark | MNL | Shopping | mode |
| Stopher and Prashker | MNL | Business/ non-business | inter-city mode |
| Lerman | MNL | | residence |
| Pratt | MNL | All | mode, auto-occupancy |
| Ben-Akiva and Atherton | MNL | | auto-ownership |
| Gillen and Westin | MNL | Work | parking, mode |
| Ansah | regression | | destination |

*Data Management*

Another area of methodological development has been that of data collection (sampling) and data use. With the increase of interest in practical applications of individual-choice modelling, questions have been raised concerning both practical issues of data collection for choice models and the extent to which standard data, already collected in many urban areas, can be used in the models. In general, it has been suggested (17) that existing data, with some augmentation, can be used for the application of existing individual-choice models. For the development of models, however, new data must be collected. This follows out the experiences of various researchers elsewhere and embodies the general conclusions of the profession at this time. If, then, new data must be collected for individual-choice models, some attention appears to be warranted to the issue of data collection. Such attention has been given in the last two or three years.

Before proceeding further, it is necessary to draw a clear distinction between sampling method and survey procedure, since the two are often confused in transport work. The sampling method is concerned with the process of generating a representative or other sample of a population, while the survey procedure is concerned with the technique used to administer the survey itself to the selected sample. The concern here is principally with the sampling method.

Standard transport-planning practice has been to use a simple random sample or a systematic sample across an entire urban area as the means to generate the required data. For reasons of cost, the sample is generally small (of the order of 2 per cent to 7 per cent of the urban area population) and has, therefore, relatively large sampling errors associated with it, particularly at the level of the individual trip interchanges. In the development of individual-choice models, a new method of sampling was developed specifically as an efficient research tool. This method is termed 'choice-based sampling' and is a form of non-random sampling. This sampling technique involves sampling individuals, when they are carrying out the activity of concern for the choice model under development. In examining the data requirements of individual-choice models, choice-based sampling techniques have been formalised and adjustments to remove the biases introduced by these methods have been developed (78,79).

To apply choice-based sampling to the estimation of individual-choice models and remove the biases in the estimates, it is necessary, principally, to determine the actual market shares for each of the alternatives in the model. Thus, suppose that one were to carry out choice-based sampling for modal choice to Central Business District (CBD) workplaces. The sample could be obtained either by intercepting people on their work trips (on-board public transport surveys, roadside surveys), or at their workplaces. In either case, it is only necessary to know the actual mode shares for the market sampled in order to estimate an unbiased model.

Manski and Lerman (78) have shown that a simple weighting
process of each data point yields consistent and asymptotically efficient estimators in the logit model.  The weight for
each observation is given as $Q(i)/H(i)$, where $Q(i)$ is the
fraction of the total population choosing alternative i and
$H(i)$ is the fraction observed in the choice-based sample
choosing alternative i.  If weighting is not carried out,
then the maximum-likelihood estimators for the logit model
will generally be inconsistent from choice-based samples.
However, McFadden has shown (78) that if a full set of
alternative-specific constants (dummies) are included in a
choice model and that the conditional logit model characterises choice, then only the alternative-specific constants are
inconsistent.  In this case, maximum-likelihood estimation
may be applied to the unweighted data and the correct
estimates of the alternative-specific constants can be
recovered from Equation (5).

$$\gamma^*_i = \hat{\gamma}_i - \delta_i \tag{5}$$

where $\gamma^*_i$ = the correct estimate of the alternative-specific
constant for alternative i

$\hat{\gamma}_i$ = the estimated value of the alternative-specific
constant for alternative i

$\delta_i = \log_e (Q(i)/H(i))$

Whatever the survey procedure used, it is clear that choice-
based sampling offers major cost savings over simple random
sampling or systematic sampling, provided that the overall
population values of the choice proportions are known or
obtainable without great expense.  First, many of the choice
situations of interest to the planner will occur with only a
moderate frequency in the population.  This might necessitate
drawing a much larger population (household) sample in order
to generate a sufficient sample about the choice in question.
For example, suppose the issue is inter-city mode choice and
it has been decided to collect data on an inter-city trip
made in the last week.  The proportion of the population
making an inter-city trip in any one week is clearly rather
small, so that a very large sample would be required.
Sampling on the inter-city trip clearly produces the required
sample much more efficiently.  Second, the distribution of
questionnaires or the distribution of interviewers will be
much more efficient in choice-based sampling, because of the
geographic concentration of the survey locations.  Finally,
individual-choice modelling generally requires the collection
of much more detailed data than has been required customarily
from urban transport planning studies.  Recall and accurate
reporting could, potentially, be a much more serious problem
for individual-choice models than for more conventional
models.  However, the advantage of choice-based sampling is
clear in the matter of recall and reporting accuracy.  If
data are collected during or immediately after the activity
of concern, then recall and reporting accuracy should be at
their best.  In summary, choice-based sampling is likely to
be cheaper in sample generation and in survey distribution
and will be likely to be more accurate than conventional
methods of sampling.

In a similar manner, work has recently focused on issues of survey procedures. Much of the research-based data collection has used the self-administered questionnaire, in contrast to the conventional transport-planning process that has used home interview almost exclusively. Two principal developments should be noted here. Formal work has been done in comparing the results of home interviews and self-administered questionnaires (113). The conclusions of this work are, first that self-administered questionnaires are generally less biased and more accurate than home interviews, on a comparison between questionnaires. Second, it has been shown (113) that high response rates can be obtained from self-administered surveys, by means of good questionnaire design and appropriate follow-up procedures. With high response rates, the overall response accuracy of self-administered surveys can be equal to or better than results from interview surveys. These results tend to bear out the experiences of many researchers in individual-choice models, who have reported high response rates from self-administered surveys (e.g. (50,118,127)).

*Updating and Transferring Estimated Models*

A related issue to data collection is that of the use of multiple data sets, either for updating existing models or for transferring models from one location to another. In particular when a model is calibrated on very small samples, the reliability of the coefficient estimates may not be as high as desired. Two probable situations arise that suggest the usefulness of a simple procedure for updating models. In the first, prior data exist (for example, from a previous transport study), the use of which to supplement a small choice-based sample may benefit the accuracy of the resulting model considerably. Lerman, Manski and Atherton (79) have shown that such merging can be done, regardless of whether either data set was obtained from random (systematic), stratified, or choice-based samples. The two data sets may also have been collected using different sampling methods from each other. The second situation is one in which the initial data collection is of a small sample, but where subsequent planning efforts involve further small-sample data collection. In this case, the second small sample may add precision to the original model estimates. Again, the two data sets may be merged (79) to provide improved model estimates, regardless of the sampling methods used for each sample.

In each of these two cases, an important feature of the merging process is the avoidance of a need to merge data sets and recalibrate a model. The information required to update a model with a second sample is simply the parameter estimates for the models from each of the two data sets and the variance-covariance matrices for the parameter estimates. The procedure effectively combines the weighted parameter estimates, where the weights are the inverse of the appropriate variances and covariances. For a model with a single parameter, the updating procedure is given in Equations (6) and (7) (79).

$$\text{Updated Parameter} = \cfrac{\left(\cfrac{\text{Old Parameter}}{\text{Its Variance}} + \cfrac{\text{New Parameter}}{\text{Its Variance}}\right)}{\cfrac{1}{\text{Old Parameter Var}} + \cfrac{1}{\text{New Parameter Var}}} \qquad (6)$$

$$\text{Updated Variance for 1 Parameter} = \left(\cfrac{1}{\text{Old Parameter Var}} + \cfrac{1}{\text{New Parameter Var}}\right) - 1 \qquad (7)$$

The procedure can also be extended fairly easily to the case of models with multiple parameters. It can be seen, from the above expressions, that a model with large variances and covariances (i.e. relatively unreliable parameter estimates) will be dominated by a model with small variances and co-variances. The procedure outlined by Lerman, Manski and Atherton (79) depends on the assumption that the parameter estimates from both data sets are normally distributed. The extent to which this assumption may be violated and the effect of violations on the procedure do not appear to have been determined.

Attempts have also been made to develop procedures for transferring models between geographical locations. Several instances of transfers of models have now been documented, e.g. System Design Concepts Inc. and Cambridge Systematics Inc. (130), Wattanatada (146). Transferring models principally requires adjustments to the alternative-specific constants so that they reflect the different overall alternative shares found in the new location, compared with the one where the model was calibrated.

These methodological advances add considerably to the power and value of individual-choice models. The processes for updating models with subsequent data sets opens up tremendous possibilities for the use of small samples and simultaneous surveys of different types (e.g. a home-based survey supplemented by an on-board survey), while the transfer of models permits uses where either data collection is infeasible and prior data of the required type do not exist, or a choice of concern does not currently exist.

APPLICATIONS TO CONTEMPORARY PLANNING ISSUES

*Issues in Application*

With the increasing acceptance of individual-choice models for travel modelling, a number of instances can be documented in which models of this genre have been developed or applied to contemporary planning issues. There are several different levels at which planning is undertaken. These may be described in the following terms.

(i) National - planning for national networks, such as inter-
     state highways, rail systems, etc. General focus is

on inter-city travel and the size of urban travel
volumes in aggregate.

(ii) State or regional - planning for transport networks at
the state or regional level.  At the state level, the
emphasis will again be on inter-city travel.  At the
regional level, emphasis shifts to travel within
urbanised areas, together with some inter-city content.

(iii) Metropolitan - area-wide planning within an existing
urbanised area, together with anticipated urban or
suburban growth areas.

(iv) Sub-area - planning of portions of transport networks or
networks for local-area transport only, within a
metropolitan area.

(v) Corridor - planning of a transport facility or section of
a facility within a metropolitan corridor.

(vi) Project - planning for a specific scheme or project,
below the level of sub-area or corridor, e.g. location
or size of park-and-ride lots, station-closing
decisions, etc.

Individual-choice models have been applied at all of these
planning levels in at least one instance in the United States,
as well as at several levels in other countries.  Before
reviewing each of these various applications, a few general
comments may be made about the practical use of individual-
choice models.

Applications may take one of two forms.  In the first, an
original model, or models are developed for the application,
using estimation data from the problem area.  In the second,
a model estimated elsewhere is adjusted for use in the
particular problem area for the application.  In the former
case, as mentioned earlier, it is generally necessary to
collect special data for the individual-choice models.  This
has been done in several applications, as is noted below.
The data required from estimation should, ideally, contain
the following elements:

(i) reported (perceived) measures of level of service
and other attributes of the chosen options

(ii) an indication of other options (at least one)
considered

(iii) reported (perceived) measures of the levels of
service and other attributes of one or more of the
rejected options

(iv) selected characteristics of the chooser (traveller).

These data are clearly more extensive than those usually
collected in urban transport-planning studies, and will,
therefore, entail a higher unit cost for any given survey
method.  These additional costs, however, are more than
offset by the lower sample size requirements of individual-
choice models (83).  In addition, the development of the
models requires the availability of computer software to
calibrate the models.  The principal model form used is the
logit model (either binary or multinomial), for which there
are now several non-proprietary programs available, such as
XLOGIT (95), ULOGIT (140), PROLO (21) and QUAIL (97).  While
many of these programs are machine-specific, there are
programs available at this time for running on most of the

major makes of computer. All of these procedures have in
common the fact that they are relatively inexpensive to run
for even fairly large samples (1,000 to 2,000 observations).
In addition, the estimated models can be applied quite
frequently without the aid of a computer, by use of one of
the many pocket calculators or small desk computers that are
now widely available.

In the case of using an existing, calibrated model, the
only data needed are of market shares, to permit adjustment
of the model coefficients to the new situation. As before, a
computer may not be necessary to apply the models. The costs
and data demands of this situation are low, but there may be
less than the desired accuracy until more models have been
developed, tested and compared.


*A Summary of Applications*

The number of applications of individual-choice models to
contemporary planning issues has burgeoned in the last two or
three years. A substantial number of the US applications are
documented by Spear (114) and some are described by Spear in
detail as case studies to show different types of policy
applicability in the US situation. It should be noted that
most of these applications use zonal averages of level-of-
service variables. This procedure is likely to generate
errors, particularly where aggregate data are used to
calibrate a model (96,133,134), and is not recommended for
general practice. It has come about, generally, from force
of circumstances in the early applications. Nevertheless,
experience with the applications tends to suggest that, even
with such errors present, the models perform well and often
better than more conventional methods.

While it is not useful to repeat Spear's summaries here,
it seems useful to classify the applications according to the
planning levels discussed above. This classification is
shown in Table 1.2. The table also includes a number of
non-US applications, in an attempt to cover all current
applications. In this table, the categories of sub-area and
project have been combined, because of the difficulty of
distinguishing these in some of the documentation of specific
applications.

TABLE 1.2   A summary of applications

| Level | Location | Model(s) | Pur-pose(s) | Year | Reference |
|---|---|---|---|---|---|
| National | US | Mode | All | 1977 | Stopher and Prashker |
| State or regional | New York State | Mode | All | 1975 | Liou, Hartgen and Cohen |
| | New York State | Mode | All | 1975 | Howe and Cohen |

| Level | Location | Model(s) | Pur-pose(s) | Year | Reference |
|---|---|---|---|---|---|
| | Pretoria South Africa | Mode | All | 1977 | PWV |
| Metro-politan | San Diego | Mode | Work | 1972 | PMM |
| | Tallahassee | Mode | Work Other | 1975 | AMV |
| | Denver | Mode | Four trip types | 1977 | Stone and Thorstad |
| | Atlanta | Mode | Three trip types | 1974 | AMV |
| | Minneapolis-St Paul | Mode Auto-occup | Three trip types | 1976 | Pratt |
| | Washington DC | Mode Auto-occup | Three trip types | 1973 | Pratt |
| | San Francisco | All choices | All | 1974 | MTC |
| | | | | 1975 | Comsis |
| | Eindhoven | Mode | All | 1974 | Ben-Akiva and Richards |
| | San Francisco | Mode | Work Other | 1974 | McFadden |
| | Los Angeles | Frequency, mode, destination | Work, shop | 1975 | CRA |
| | Washington DC | Mode, frequency, destination, auto-ownership | Work Non-work | 1975 | CSI |
| | Various | Mode | Work | 1975 | Heaton et al. |
| Sub-area and Project | Chicago Loop | Mode Access | Work | 1974 | Lisco and Tahir |
| | Chicago Suburbs | Mode | Work | 1972 1973 | Hovind Tahir and Hovind |
| | | Access | | 1976 | Sajovec and Tahir |

| Level | Location | Model(s) | Pur-pose(s) | Year | Reference |
|-------|----------|----------|-------------|------|-----------|
| | Chicago Suburbs | Mode | Work | 1973 | Illinois |
| | Chicago Suburb | Access. mode | Work | 1974 | Tahir |
| | Chicago Suburb | Mode | Work | 1973 | Schindel |
| | New York State Communities | Mode | Work | 1974 | Liou |
| | Portland, Oregon | Mode | Work | 1974 | SDC & CSI |
| | Schaumburg/ Hoffman Estates, Illinois | Mode | Work | 1975 | Stopher, Pfefer and Stopher |
| | Minneapolis CBD | Mode and parking | Work | 1977 | Barton-Aschman |
| | Sydney, Australia (Green Valley) | Mode and car ownership | Work | 1976 | Hensher, Smith and Hooper |
| Corridor | Los Angeles | Car ownership | Work | 1977 | Dobson & Tischer |
| | Hobart, Tasmania | Mode and route | All | 1977 | Hensher |

From Table 1.2, it is notable that the majority of the
applications so far have been at the metropolitan planning
level, in most of which cases a mode-choice model has been
developed to fit into a more standard travel-forecasting
procedure.  This particular use of the individual-choice
modelling technique is, nevertheless, one that poses substantial problems of interface.  Typically, the data required
for the mode-choice model would not be forthcoming directly
from the earlier model steps and the output is not suitable
directly from subsequent modelling steps.  It seems likely
that the popularity of individual-choice models in this
application attests more to the dissatisfaction with conventional models of modal split than to the inherent
advantages of the individual-choice models.  Indeed, in one
of the regional studies (106), the individual-choice model
was the only one that could provide a statistically
significant result for mode choice.  Conventional aggregate
models simply would not calibrate significantly on any
available variables.

The next most frequent use of the models is for sub-area
or project applications, where the inherent advantages of the
models are exploited more readily.  Another area of great
potential is that of corridor analysis, although only one US

application of this type, by Dobson and Tischer (33) to the
car-pool lane on the Santa Monica Freeway, could be found.
Possibly, planning in the US has still been too little
concerned with corridor analyses for more applications to
have been realised as yet.

Spear (114) has noted that individual-choice models appear
to be applicable to most of the policy issues in planning in
the US at present.  His classification of applications is
based upon the type of planning issue.  Despite the breadth
of applications documented here and in Spear's report (114),
there are, nevertheless, some applications which seem less
useful than others.  First, given the level of detail
inherent in the approach, regional or state applications seem
rather inappropriate.  The variation of travel situations
over such areas is large and may result in serious aggregation problems for such large areas.  In addition, the type of
planning to be carried out in such situations is unlikely to
call for the level of detail and precision achievable with
individual-choice models.  Applications at the national level
may sometimes be appropriate and at other times be inappropriate.  For example, estimation of inter-city market shares
is similar to an urban-corridor application, which is well
suited to the individual-choice model.  On the other hand,
application to a national highway needs study would seem
quite inappropriate, both because of the lower level of
precision required from the planning and the wide variation
in travel characteristics.

Finally, it may be noted that Table 1.2 reflects the
general direction of model development in this field.  By far
the majority of applications are concerned with mode choice,
with a few developments of auto-occupancy, auto-ownership,
and some of the other travel decisions within the total set
(e.g. trip frequency, destination, etc.).  Clearly, more
research and development is needed in other choice areas for
applications to appear.  A number of problems also arise in
undertaking applications, some of which can only be solved by
pragmatic approximations at present.  A few of these are
mentioned here.

As noted earlier, a number of developments have been made
in the area of aggregation.  This still remains an area
needing further attention for successful and useful applications of individual-choice models.  Another issue of concern
is the determination of captives in any population, since the
models are all *choice* models.  Little attention has been
given, thus far, to determining and forecasting captives, yet
this is key to the successful application of individual-
choice models either in geographical transfer situations or
for any but very short-run forecasts.  Determination of
choice sets is, likewise, of considerable importance to
successful application, particularly since it is known that
incomplete or 'over-complete' choice sets can bias the models
to varying extents.

POLICY IMPLICATIONS AND IMPLEMENTATION

Taking the work of Warner (143) as the first attempt at
individual-choice modelling, the field has seen fifteen years
of research and development in the United States and at least
ten years in the United Kingdom (108). It is reasonable to
conclude, therefore, that more extensive attempts should be
made to develop procedures, based on individual choices, for
assisting in the formulation of transport policy and testing
policies. Up to now, applications of this type have been
scarce, most being concerned with testing projects and
policies which have been formulated without the aid of the
insights of the individual-choice models themselves.

Recognising this, a major thrust of the conference was to
explore the potential areas of immediate application of
individual-choice models and of policy impact of the models.
In past transport planning, travel-forecasting models have
been used as an aid to the development of policy as well as
for simple forecasting and testing of alternative strategies
and schemes. So far, individual-choice models have been
applied more to the latter tasks than the former. However,
it is clear that since the individual-choice models offer
much greater insights into travel behaviour than the traditional models, their potentials for informing policy
development must be that much greater than traditional
travel-forecasting techniques.

In the sense being used here, policy formulation would
arise from an examination of the choice models and an
identification of those attributes of the transport system
that affect certain choices to the greatest extent. For
example, if a general policy concern was to reduce car use
into the central city, and if a mode-choice model revealed
parking costs to be a significant factor, this might suggest
a policy of increasing parking charges significantly as a
means of inducing shifts to public transport, *ceteris paribus*.
A destination-choice model, however, might indicate that such
a policy of increasing parking costs would have the effect of
diverting workers from the city centre to other locations,
thereby damaging the economic life of the city centre. With
this information available, the decision-maker may then
decide that such a parking charge increase is not the best
policy to adopt for changing the modal split to the central
city. In this way, the individual-choice models may be able
to inform decision-makers of the full spectrum of consequences of any policy that may be considered.

Two levels of policy formulation can be identified. At
the higher level, policy is formulated directly from a
consideration of the goals and objectives for the planning
area. Such policy is not formulated from models, although it
may be informed by the results of models. At the lower level,
policies are formulated for achieving the higher-level
policies. These are the policies that are most likely to be
developed from a consideration of individual-choice models.

Conversely, models must be developed that are more
responsive to contemporary policy issues, to allow a wider
range of policies to be tested adequately. The need for such

development is being recognised, as is shown by some of the
applications, referenced in the previous section of this
chapter, where efforts have been made to develop models to
test pollution control strategies, car-pooling programmes,
and fuel restrictions on various aspects of trip-making.
Such developments have, however, occurred largely in reaction
to emerging policy issues and have not led the development of
policy.  As a result, they have tended to lag behind the
development of policy and have been unable to guide policy-
making.  For individual-choice models to be used in policy
formulation, it is necessary that researchers anticipate
future policy issues and construct models that will be able
to respond to a wide range of such issues.  It would be
desirable for more general models to be developed, so that
there is no longer a need to develop a specific model for
each application.  Furthermore, it is necessary for the
researcher to become better informed about the policy-makers'
needs and tailor his models to meet these needs.  In this way,
individual-choice models may become a significant force in
policy development.

In the short term, it is necessary for research and
development to become more attuned to the potentials of the
models for influencing policy and to use this factor as one
of the determining factors for immediate directions for
development.  Coupled with this is the need to identify those
directions of application that have little or no pay-off for
policy and accord those a relatively low priority.  In this
category, many theoretical and methodological improvements
may fall, although several such improvements are also likely
to have extensive policy impact potential.  It also seems
appropriate to note here, as was discussed in the previous
section, that there are a number of applications where
individual-choice models are unlikely to be appropriate.  In
general, individual-choice models should not be regarded as
being likely to provide a wholesale replacement for the more
traditional travel forecasting macro-models.  This fact
should be recognised and policy impacts in these areas left
to the more suitable macro-models (as argued by Andersson and
Karlquist in Chapter 8 of this book).

It is clear that individual-choice models are ripe for
application in many areas, as is shown by the number of
applications so far.  It is also clear that these techniques
can often provide insights into policy formulation that have
never been achieved before with any type of model.  With this
potential, the time is now ripe for more use to be made of
the techniques in the policy and applications arenas.  Those
professionals engaged in the continuing development and
research of these models must, therefore, become better
informed about the emergence of policy issues and the needs
of the policy-maker.  Without this, policy-making and
decision-making will be unable to capitalise on the potential
offered by the models.

FUTURE DIRECTIONS

'If analytical methods are to play a part in shaping
decisions they must address themselves to policy matters and
evolve concurrently with policy issues' (115).  Although
particular statistical techniques, especially multinomial
logit are often a central issue in contributions to the
behavioural travel-modelling literature, more appropriately
they form part of a subdiscipline concerned with the
development (and translation) of realistic approaches to use
in addressing relevant policy matters.  The major objective
is where increased understanding is expected to lead to
improved prediction capability, via the generation and
testing of more relevant hypotheses on human behaviour.  Of
course, identification of consumer preferences and choices
does not necessitate or guarantee the meeting of such
preferences by the responsible authorities, but it does
provide an invaluable source of information to advise on
'what is expected to happen if a particular policy change is
implemented'.  It is easy to sympathise with decisions taken
in the past on major investment in the public and private
sectors, when relatively crude demand-forecasting and
explanatory tools were used in the generation of advice,
although one can be critical of the way in which the
professional contribution (regardless of techniques
available) is communicated to the senior advisers to the
decision-makers.

   This introduction to future directions emphasises the
need for both advancement in method and dissemination of
output.  A detailed summary of the findings and recommendations arising out of the third international conference on
behavioural travel modelling is given in Chapter 39 (see
also (60)), although we offer some additional suggestions
for future directions to conclude this initial chapter.

   The main emphases of the next five years seem to be the
dissemination of the skills presently available to the
privileged few, in particular by empirical demonstration
of their relevance at various levels of spatial planning.
Since the conference (April 1977), major applications at the
urban area-wide level have been undertaken in Australia
(Hobart, Adelaide, Newcastle), Holland (Amsterdam), USA
(San Francisco) and South Africa (Pretoria).  The applications at the sub-area level are even more numerous (e.g.
access to the Sydney domestic and international airport
complex).

   Within the present conceptual framework in which
individual-choice models are being applied (and developed), a
major constraint is the nature of existing data. Applications
in the foreseeable future will hopefully have the advantage
of improved data sources collected with individual-choice
models in mind.  In addition, updating area-wide data sources
will provide a mechanism for adjustments of choice-based
samples.  Collecting entirely new data sets is necessary for

developmental research.*

Although the reduced sample-size requirement of the new approach has been promoted as a major benefit, this may mainly be valid for estimation of a model for relatively homogeneous contexts.  Sample sizes of 200 to 300, as generally recommended, limit considerably the ability to apply market-segmentation procedures which have been shown to be important for improved prediction.  If a reasonable number of market segments can be identified either *a priori* or *a posteriori*, sample-size requirements will generally still be much below conventional procedures.  For example, 300 observations in each of 10 market segments still only requires 3,000 observations, where 10,000 to 20,000 may be required in conventional processes.  Also, while it is feasible to introduce dummy variables into a single model to allow for heterogeneity of non-representativeness of utility for initial observed attributes; in the translation from the predictions of a segmented sample to the population, lack of compatible data at the same level of segmentation can diminish the benefits of sophisticated sample segmentation.  This has a modifying effect on ambitious segmentation, yet carries the risk of substantial prediction error.

It seems appropriate to summarise results in trip tables on an origin-destination basis.  Small samples reduce the ability to predict on each O-D sample pair and then adjust to the population on each O-D pair.  This is not a problem, however, if it can be shown that the estimates of the parameters used are invariant to the characteristics of spatial setting of the study (or variables to allow for the influence of spatial variation can be included in the model).  Then with a random sample (or adjusted choice-based sample) it is possible to apply an average area-wide sample population conversion factor.  More research into this important source of prediction error is required (57).

The emphasis of this chapter is on micro-behavioural approaches to identifying the demand for transport services.  Anderson and Karlquist (Chapter 8) remind us that the equilibrating mechanism of supply and demand is still relatively neglected in research.  However, in a disturbing way, they conclude that a model of traffic flows cannot be

---

* When a major supply change is known in advance of a study, then it is worth considering the costs of collecting data for each individual, related to the perceived level-of-service characteristics of the 'new' alternative. Particularly where changes are very much non-marginal with respect to generic variables, then it is suggested (on the basis of evidence from one recent study (57)) that the estimates of the parameters of the so-called 'generic variables' are significantly different between models in which the 'new' alternative is included in the estimation of the model, and those in which it is excluded in the estimation.  The Tasman Bridge (a previously dominant mode/ route for the majority of trips in Hobart) collapsed in 1975 and the reopening in 1977 provided a classical context for testing this condition.

*based* on a micro-economic approach.   Instead macro-dynamic growth and welfare-equilibrium models are proposed for choosing optimal policy parameters.   The interface between individual-choice models and the macro-equilibrium modelling research is unclear and needs urgent clarification.   To date, the only use made of individual-choice models in equilibrium-theory modelling is to obtain values of journey attributes (especially value of time savings) for determining the generalised costs of travel.   We believe that this is a misuse of the proven potential of individual choice models.

With energy conservation being an important concern (regardless of one's belief in the extent of resource scarcity), procedures that reflect policies capable of reducing vehicle miles travelled are being called for.   Until the activity approach has demonstrated its superiority with operational models (see Chapter 2 (Jones) and Chapter 4 (Kobayashi) ), it seems appropriate to develop logit-type models to allow for the possibilities of trip linking and multi-purpose journeys when activity rearrangement can be achieved.   This whole area is closely tied with the require-ments for accessibility to opportunities and the trip frequency and destination decisions (57,76).   Finally we should support the need for modelling of the freight task, although point out that greater understanding of the process and interaction of decision-making units is required before major sources of funds are put into the modelling phase (see Chapter 27 below).

REFERENCES

1. Adler, T.J. and Ben-Akiva, M.E., 'Joint choice model for frequency, destination and travel mode for shopping trips', *Transportation Research Record*, no. 569 (1976).
2. Anderson, N.H., 'Functional measurement and psycho-physical judgement', *Psychological Review*, vol. 77 (1970).
3. Anderson, N.H., 'Information integration theory: a brief survey' in *Contemporary Developments in Mathematical Psychology*, Krantz, D.H., Atkinson, R.C., Luce, R.D. and Suppes, P. (eds), vol. 2, W.H. Freeman, San Francisco (1974).
4. Ansah, J.A., 'Destination choice modelling', *Transportation Research*, vol. 11, no. 2 (1977).
5. Ben-Akiva, M.E. (1973), 'Structure of Passenger Travel Demand Models', Ph.D. Dissertation, Department of Civil Engineering, Massachusetts Institute of Technology.
6. Ben-Akiva, M.E. and Atherton, T.J. 'Choice model predictions of carpool demand: methods and results', *Transportation Research Record*, no.637 (1977).
7. Ben-Akiva, M.E. and Richards, M.G. (1974) *Disaggregate and Simultaneous-Travel Demand Models: A Dutch Case Study*, prepared for the Dutch Ministry of Transport.
8. Birnbaum, M.H., 'Using contextual effects to derive psycho-physical scales', *Perception and Psychophysics*, 15, 89-96 (1974).
9. Burnett, K.P., 'A Bernoulli model of destination choice', *Highway Research Record*, no. 527
10. Burnett, K.P., 'The dimensions of alternatives in spatial choice processes', *Geographical Analysis*, 5, pp. 181-204 (1973).

11. Burnett, K.P., 'A three-state markov model of choice behaviour within spatial structures', *Geographical Analysis*, 6, pp. 53-68 (1974).

12. Burnett, K.P., 'Tests of a linear learning model of destination choice: application to shopping travel by heterogeneous population groups', *Geografiska Annaler* (1976).

13. Burnett, K.P., 'Toward dynamic models of travel behaviour and point patterns of traveller origins', *Economic Geography*, 52, pp. 30-47 (1976).

14. Cambridge Systematics, Inc. (1975) *A Behavioural Model of Automobile Ownership and Mode of Travel*, vols 3 and 4, prepared for US Department of Transportation, Washington DC.

15. Chapin, F.S. (1974) *Human Activity Patterns in the City: Things People Do in Time and Space*, John Wiley and Sons, New York.

16. Charles River Associates, Inc. (1972) *A Disaggregated Behavioural Model of Urban Travel Demand*, final report to the Federal Highway Administration USDOT, Washington DC.

17. Charles River Associates, Inc. (1975) *The Effects of Automotive Fuel Conservation Measures on Automotive Air Pollution*, final report to the Environmental Protection Agency, Washington DC.

18. Charles River Associates, Inc. (1976) *Disaggregate Travel Demand Models*, vols I and II (Project 8-13 Phase 1 Report NHCRP), Charles River Associates, Boston.

19. Charles River Associates, Inc. (1977) *Guidelines for Using the Market Segmentation Technique to Apply Disaggregate Travel Demand Models with Census Data*, Charles River Associates, Working Paper, Boston.

20. Comsis Corporation (1975) *Travel Model Development Project Phase 2: Work Program*, prepared for the Metropolitan Transportation Commission, San Francisco.

21. Cragg, J.G. (1968), 'Some statistical models for limited dependent variables with application to the demand for durable goods', *Discussion Paper 8,* University of British Columbia, Vancouver, B.C.

22. Dalvi, M.Q. and Lee, N., 'Variations in the value of time: further analysis', *Manchester School*, 39, 3 (1971).

23. Daly, A., 'Issues in the estimation of journey attribute values', in *Determinants of Travel Choice*, Hensher, D.A. and Dalvi, M.Q. (eds), Teakfield, Farnborough, England (1978).

24. Daughety, A., 'Estimation of service-differentiated transportation demand functions', paper presented to the Transportation Research Board (January 1978).

25. Deganzo, C., Bouthelier, F. and Sheffi, Y., 'An efficient approach to estimate and predict with multinomial probit models', *Department of Civil Engineering*, MIT (1976).

26. Degenais, M.G., 'A threshold regression model', *Econometrica*, 37 (1969).

27. Degenais, M.G., 'Application of a threshold regression model to household purchases of automobiles', *Review of Economics and Statistics*, 57, 3 (1975).

28. Degenais, M.G., 'The determination of newsprint prices', *Canadian Journal of Economics*, IX, 3, 442-61 (1976).

29. De Janosi, P.E. (1956), 'Factors Influencing the Demand for New Automobiles: A Cross-section Analysis', unpublished Ph.D. thesis, University of Michigan, Department of Economics.

30. Department of the Environment (1976) *Valuation of Travel Time*, Economics and Statistical Note 22, London, Department of the Environment.
31. Devletoglou, N.E., 'Threshold and rationality', *Kyklos*, XXI, 4 (1963).
32. Devletoglou, N.E., 'Thresholds and transactions costs', *Quarterly Journal of Economics*, 85 (1971).
33. Dobson, R. de P. and Tischer, M., 'An empirical analysis of behavioural intentions to shift ways of traveling to work', paper presented to 56th Annual Transportation Research Board Meeting (1977).
34. Fried, M., Havens, J. and Thall, M. (1977) *New Approaches to Understanding Travel Behaviour*, Boston College, Boston.
35. Georgescu-Roegen, N., 'The pure theory of consumer's behaviour', *Quarterly Journal of Economics*, 1 (1936).
36. Gilbert, C.G., 'Semi-Markov processes and mobility', *Journal of Mathematical Sociology*, 3 (1973).
37. Gilbert, C.G., Peterson, G.L. and Linne, D.W., 'Toward a model of travel behaviour in the Boundary Waters Canoe Area', *Environment and Behaviour*, 4, 2, pp. 131-57 (1972).
38. Goldberger, A.S. and Lee, M.L., 'Toward a microanalytic model of the household sector', *American Economic Review*, 53 (1962).
39. Golob, T.F. and Nicolaidis, C.C., 'An elimination-by-aspects model of transportation behaviour', *General Motors Research Laboratories*, Warren, Michigan (1977).
40. Goodwin, P.B., 'Human effort and the value of time', *Journal of Transport Economics and Policy*, X, 1 (1976).
41. Goodwin, P.B., 'Habit and hysteresis in mode choice', *Urban Studies*, 14 (1977).
42. Goodwin, P.B. and Hensher, D.A., 'The transport determinants of travel choices; an overview', in *The Determinants of Travel Choice*, Hensher, D.A. and Dalvi, M.Q. (eds), Teakfield Publishing Limited, Farnborough, England (1978).
43. Hagerstrand, T., 'What about people in regional science?', *Papers of the Regional Science Association*, 24 (1970).
44. Hagerstrand, T., 'The impact of transport on the quality of life', *Fifth International Symposium on Theory and Practice in Transport Economics (Athens)*, OECD, Paris (1974).
45. Hagerstrand, T., 'Space, time and human conditions', in *Dynamic Allocation of Urban Space*, Lundquist, L. and Snickars, F. (eds), Saxon House Studios, Farnborough (1975).
46. Hartley, M.J., 'The tobit and probit models; maximum likelihood estimation by ordinary least squares', *Discussion Paper No. 374*, Department of Economics, State University of New York at Buffalo (1976).
47. Hauser, J.R., 'Testing the accuracy, usefulness and significance of probabilistic choice models: an information theoretic approach', *Operations Research* (forthcoming, 1978).
48. Hausman, J.A. and Wise, D.A., 'A conditional probit model for qualitative choice: discrete decisions recognising interdependence and heterogeneous preferences', *Econometrica*, 46(2), 403-426 (1978).
49. Heaton, C. *et al.* (1975) *Dual Mode Potential in Urban Areas*, Report No. DOT-OST-74-20, US DOT.
50. Hensher, D.A. (1972), 'The Consumer's Choice Function: A Study of Traveller Behaviour and Values', unpublished Ph.D. thesis, School of Economics, University of New South Wales (Australia).

51. Hensher, D.A., 'The problem of aggregation in disaggregate behavioural travel-choice models with emphasis on data requirements', *Transportation Research Board Special Report*, no. 149, pp. 85-102 (1974).

52. Hensher, D.A., 'Perception and commuter modal choice: an hypothesis', *Urban Studies*, 12 (1975).

53. Hensher, D.A., 'Valuation of commuter travel time savings: an alternative procedure' in *Modal Choice and the Value of Travel Time*, Heggie, I.G. (ed.), Oxford University Press, Oxford (1976).

54. Hensher, D.A., 'Value of commuter travel time savings: empirical estimation using an alternative valuation model', *Journal of Transport Economics and Policy*, X, 2 (1976).

55. Hensher, D.A., 'Letters to the editor: reply to N. Thrift on the structure of journeys and nature of travel', *Environment and Planning*, 8 (1976).

56. Hensher, D.A. (1977) *Valuation of Business Travel Time*, Pergamon Press, Oxford.

57. Hensher, D.A. (1977), 'The effect on mode and route for east-west travel of re-opening of Tasman Bridge', *School of Economic and Financial Studies Research Paper no.144*, Macquarie University (Sydney), Australia.

58. Hensher, D.A., Smith, R.A. and Hooper, P.G. (1976), *An Approach to Developing Transport Improvement Proposals*, Occasional Paper no. 2, Commonwealth Bureau of Roads, Melbourne, Australia.

59. Hensher, D.A. and Mcleod, P.B., 'Towards an integrated approach to the identification and evaluation of the transport determinants of travel choices', *Transportation Research*, 11, 2 (1977).

60. Hensher, D.A. and Stopher, P.R., 'Behavioural travel modelling', *Traffic Engineering and Control*, 19, 4 (1977).

61. Hensher, D.A. and Louviere, J.J., 'Behavioural intentions as predictors of very specific behaviour', *School of Economic and Financial Studies*, Research Paper No.146, Macquarie University (Sydney) (1977) (forthcoming in *Transportation*).

62. Hensher, D.A. and Dalvi, M.Q. (eds), (1978) *Determinants of Travel Choice*, Teakfield, Farnborough, England.

63. Hovind, M., 'Preliminary demand analysis for feeder bus services to the Lombard, Illinois commuter railroad station', *Technical Papers and Notes Series*, no.4, Illinois Department of Transportation (1972).

64. Howe, S.M. and Cohen, G.S., 'Statewide disaggregate attitudinal models for principal mode choice', *Preliminary Research Report 84*, New York State Department of Transportation, Albany, NY (1973).

65. Illinois Department of Transportation (IDOT) (1973), *State of Illinois Commuter Parking Program - Phase 1 Parking Demand Analysis*, IDOT, Chicago, Illinois.

66. Jones, P.M., 'A gaming approach to the study of travel behaviour, using a human activity approach', *Transport Studies Unit Working Paper no. 18*, University of Oxford, Oxford (1976).

67. Jones, P.M., 'Forecasting family response to changes in school hours: an explanatory study using 'HATS'', *paper presented to Urban Transport Studies Group Annual Conference*, Glasgow University (1977).

68. Koppelman, F.S., 'Prediction with disaggregate models: the aggregation issue', *Transportation Research Record*, no. 527 (1975).

69. Koppelman, F.S., 'Guidelines for aggregate travel prediction using disaggregate choice models', *Transportation Research Record*, no. 610 (1976).

70. Krishnan, K.A., 'Incorporating the concept of minimum perceivable difference in the logit model', *Management Science* (1977).

71. Lee, N. and Dalvi, M.Q., 'Variations in the value of travel time', *Manchester School*, 37, 3 (1969).

72. Leibenstein, H. (1976) *Beyond Economic Man: A New Foundation in Micro-economics*, Harvard University Press, Cambridge, Massachusetts.

73. Lenntorp, B. (1976) *Paths in Space-Time Environment - A Time Geographic Study of Movement Possibilities of Individuals*, Lund Studies in Geography Series B, no. 44, University of Lund (Sweden).

74. Leontief, W., 'Introduction to a theory of the internal structure of function relationships', *Econometrica*, 15, 4 (1947).

75. Lerman, S. (1975) 'A Behavioural Model of Urban Mobility Decisions', Ph.D. Dissertation, Massachusetts Institute of Technology.

76. Lerman, S., 'The use of disaggregate choice models in semi-Markov process models of trip chaining behaviour', *Working paper, Center for Transportation Studies*, MIT, Cambridge, Mass. (1977).

77. Lerman, S. and Manski, C., 'An estimator for the generalised multinomial probit choice model', *Transportation Research Record* (1977).

78. Lerman, S. and Manski, C., 'The estimation of choice probabilities from choice-based samples', *Econometrica*, 45 (1978).

79. Lerman, S., Manski, C. and Atherton, T.J. (1976) *Non-random Sampling in the Calibration of Disaggregate Choice Models*, no. PO-6-3-0021 for the US Department of Transportation, Federal Highway Administration, Washington, D C.

80. Levin, I., 'Information integration in transportation decisions', in *Human Judgements and Decision Processes: Applications in Problem Settings*, Kaplan, M.F. and Schwartz, S. (eds) Academic Press, New York (1977).

81. Liou, P.S., 'Comparative demand estimation for peripheral park-and-ride service', *Preliminary Research Report 71*, New York State Department of Transportation, Albany, New York (1974).

82. Liou, P.S. and Talvitie, A.P., 'Disaggregate access mode and station selection models for rail trips', *Transportation Research Record*, 526, pp. 42-52 (1974).

83. Liou, P.S., Cohen, G.S. and Hartgen, D.T., 'Application of disaggregate modal-choice models to travel demand forecasting for urban transit systems', *Transportation Research Record*, 534, pp. 52-62 (1975).

84. Lisco, T.E. and Tahir, N., 'Travel mode choice impact of potential parking taxes in downtown Chicago', *Technical Papers and Note Series*, no. 12, Illinois Department of Transportation, Chicago (1974).

85. Louviere, J.J., Ostresh, L.M., Henley, D. and Meyer, R., 'Travel demand segmentation: some theoretical considerations related to behavioral modelling', in *Behavioral Travel-demand Models*, Stopher, P.R. and Meyburg, A.H., Lexington, Mass: Lexington Books (D.C. Heath & Co) (1976).

86. Louviere, J.J. and Norman, K.L., 'Applications of information processing theory to the analysis of urban travel demand', *Environment and Behaviour* (1977).
87. Louviere, J.J., Piccolo, J.M., Meyer, R.J. and Duston, W.I., 'Theory and empirical evidence in real world studies of human judgement: three shopping behaviour examples', *Journal of Applied Psychology* (1978).
88. Luce, R.D. (1959), *Individual Choice Behaviour*, John Wiley and Sons, New York.
89. Luce, R.D. and Raiffa, H. (1957) *Games and Decisions*, John Wiley and Sons, New York.
90. McFadden, D., 'Conditional logit analysis of qualitative choice behaviour', in *Frontiers in Econometrics*, Zarembka, P. (ed.) Academic Press, New York (1974).
91. McFadden, D., 'The measurement of urban travel demand', *Journal of Public Economics*, 3 (1974).
92. McFadden, D., 'BART patronage and revenue forecasts for flat fares', *Working Paper no. 7407*, Travel Demand Forecasting Project, University of California, Berkeley (1974).
93. McFadden, D., 'Quantal choice analysis: a survey', *Annals of Economic and Social Measurement*, 5, National Bureau of Economic Research (1976).
94. McFadden, D., 'The theory and practice of disaggregate demand forecasting for various modes of urban transportation', *Working Paper no. 7623*, Travel Demand Forecasting Project, University of California, Berkeley (1976).
95. McFadden, D., Wills, H. and Brownstone, D., 'CDC XLogit 2. users' guide', *Working Paper no. 7412*, Travel Demand Forecasting Project, University of California, Berkeley (1975).
96. McFadden, D. and Reid, F., 'Aggregate travel demand forecasting from disaggregated behavioural models', *Transportation Research Record No. 534* (1975).
97. McFadden, J., Berkman, D., Brownstone, D. and Duncan, G.M. (1977) *QUAIL 3.0 Users' Manual*, Department of Economics, University of California, Berkeley.
98. McLynn, J., 'The simulator of travel choice behaviour without the independence of irrelevant alternatives hypothesis', paper presented at the Summer Simulation Conference, Washington D.C. (1976).
99. Metropolitan Transportation Commission (MTC) (1974) *Travel Forecasting Model Development Project*, Request for Proposal.
100. Miller, D.R., 'Aggregation problems', *Transportation Research Board Special Report No. 149*, pp. 25-30 (1974).
101. Peat, Marwick, Mitchell & Co. (PMM) (1972), *Implementation of the n-Dimensional Logit Model*, final report to the Comprehensive Planning Organization, San Diego County, California.
102. Pfefer, R.C. and Stopher, P.R., 'Transit planning in a small community: a case study', *Transportation Research Record No. 608*, pp. 32-41 (1976).
103. Pratt, R.H. and Associates, Inc. (1973) *Development and Calibration of the Washington Mode Choice Models, Technical Report No. 8*, prepared for the Washington Metropolitan Council of Government.
104. Pratt, R.H. and Associates, and DTM, Inc. (1975) *Interim Report 1, Calibration File Preparation and Development of Preliminary Tabulation of the Minneapolis - St Paul Mode Choice Model Development*, Metropolitan Council of Minneapolis-St Paul, Minnesota.

49 Notes to chapter 1

105. Pratt, R.H. and Associates, Inc. and DTM, Inc. (1976)
*Development and Calibration of Mode Choice Models for the
Twin Cities Area*, prepared for the Twin Cities Metropolitan
Council.
106. PWV Transportation Study (1977) *PWV White Modal Split
Investigation*, Report V 2.5/1 (VKE).
107. Quandt, R.E., 'A probabilistic theory of consumer
behaviour', *Quarterly Journal of Economics*, 30 (1956).
108. Quarmby, D.A., 'Choice of travel mode for the
journey to work: some findings', *Journal of Transport
Economics and Policy*, 1, 3, pp. 273-314 (1967).
109. Roberts, P.O., 'Forecasting freight flows using a
disaggregate freight demand model', *Report 76-1, MIT Center
for Transportation Studies*, MIT, Cambridge, Mass. (1976).
110. Sajovec, J.P. and Tahir, N. (1976) *Development of
Disaggregate Behavioral Mode Choice Models for Feeder Bus
Access to Transit Stations*, Illinois, Department of Transportation, Chicago, Illinois.
111. Schindel, S.E., 'Impact on station choice and access
mode choice due to the establishment of a commuter rail
station at Arlington Park, Illinois', *Technical Papers and
Notes Series no. 14*, Illinois Department of Transportation,
Chicago, Illinois (1973).
112. Simon, H.A. (1957) *Models of Man*, John Wiley and
Sons, New York.
113. Sozialforschung Brög, 'Considerations on the design
of behavioural orientated models from the point of view of
empirical social research', Paper presented to World Conference on Transport Research, Rotterdam (1977).
114. Spear, B.D. (1977) *Applications of New Travel Demand
Forecasting Techniques to Transportation Planning*, US
Department of Transportation.
115. Starkie, D.N.M. (1977) *Transportation Planning and
Policy Analysis*, Pergamon Press, Oxford.
116. Stone, T.J. and Thorstad, R.L., 'The Denver demand
modelling process', paper prepared for the 56th Annual
Transportation Research Board Meeting (1977).
117. Stoner, J. (1977), 'The Effects of Sampling Error on
the Use of Disaggregate Models in the Evaluation of
Alternative Transit Schemes', unpublished Ph.D. dissertation,
Department of Civil Engineering, Northwestern University.
118. Stopher, P.R., 'A probability model of travel mode
choice for the work journey', *Highway Research Record*, 283,
pp. 57-65 (1969).
119. Stopher, P.R., 'Goodness-of-fit measures for
probabilistic travel demand models', *Transportation*, 4, 1,
pp. 67-83 (1975).
120. Stopher, P.R., Watson, P.L. and Blin, J.M., 'A method
for assessing pricing and structural changes on transport
mode use', Transportation Center, Northwestern University,
Interim Report to the Office of University Research, US
Department of Transportation (1975).
121. Stopher, P.R., Spear, B.D. and Sucher, P.O., 'Toward
the development of measures of convenience for travel modes',
*Transportation Research Record*, 527, pp. 16-32 (1975).
122. Stopher, P.R. and Meyburg, A.H. (1975) *Urban
Transportation Modelling and Planning*, Lexington Books,
Lexington, Mass.
123. Stopher, P.R., 'Ridership estimates for alternative
system options', *Supplement to Technical Memorandum no. 3 for*

the Schaumberg/Hoffman Estates Transit Study, Jack E. Leisch and Associates, Evanston, Illinois (1975).
124. Stopher, P.R. and Meyburg, A.H., Behavioral travel-demand models', in *Behavioral Travel-Demand Models*, Stopher, P.R. and Meyburg, A.H. (eds) Lexington Books (D.C. Heath and Co.) Lexington, Mass. (1976).
125. Stopher, P.R. and Meyburg, A.H., (eds) (1976) *Behavioral Travel-Demand Models*, Lexington Books (D.C. Heath & Co.) Lexington, Mass.
126. Stopher, P.R. and Prashker, J.L., 'Inter-city passenger forecasting: the use of current travel forecasting procedures', *Transportation Research Forum Proceedings*, 17, 1, pp. 67-75 (1976).
127. Stopher, P.R., Koppelman, F.S., Peterson, G.L., Hauser, J.R., Blin, J.M. and Watson, P.L., 'Final report on a method for assessing pricing and structural changes on transport mode use', *Report on Project DOT-OS-4000 1*, US Department of Transportation, Washington, D.C. (1977).
128. Stopher, P.R., Prashker, J.L., Pas, E.I. and Smith, B.S., 'Final report to Ambrakal inter-city passenger forecasting from the National Travel Survey Data of 1972', The Transportation Center, Northwestern University, Evanston, Illinois (1976).
129. Swets, J.A., 'Detection theory and psychophysis' in *Decision Making*, Edwards, D.W. and Tversky, A. (eds), Penguin Modern Psychology, London (1967).
130. System Design Concepts and Cambridge Systematics, Inc. (1974) *Demand and Revenue Analysis for Proposed Light Rail and Express Bus Systems in Portland, Oregon*, Technical Memorandum prepared for the Governors Task Force on Transportation.
131. Tahir, N. and Hovind, M. (1973) *A Feasibility Study of Potential Feeder Bus Service for Homewood, Illinois*, Illinois Department of Transportation, Chicago, Illinois.
132. Tahir, N., 'Feeder buses as an alternative to commuter parking: an analysis of economic trade-offs', *Technical Papers and Notes Series No. 15*, Illinois Department of Transportation, Chicago, Illinois (1974).
133. Talvitie, A.P., 'Aggregate travel demand analysis with disaggregate or aggregate travel demand models', *Transportation Research Forum Proceedings*, 14, 1, pp. 583-603 (1973).
134. Talvitie, A.P., 'Disaggregate travel demand models with disaggregate data, not aggregate data, and how', *Working Paper No. 7615*, Urban Travel Demand Forecasting Project, University of California, Berkeley (1976).
135. Talvitie, A.P. and Leung, T., 'A parametric access network model', *Transportation Research Record No. 592*, pp. 45-9 (1976).
136. Thrift, N. (1977) *An Introduction to Time Geography*, Concepts and Techniques in Modern Geography no. 13, Geo Abstract Limited, University of East Anglia, Norwich.
137. Tversky, A., 'Elimination by aspects: a theory of choice', *Psychological Review*, 79, 4, pp. 281-99 (1972).
138. Tversky, A., 'Choice by elimination', *Journal of Mathematical Psychology*, 9, pp. 341-67 (1972).
139. Uhler, R.S. and Cragg, J.G. (1969) *The Demand for Automobiles*, Department of Economics, Discussion Paper no. 27, University of British Columbia.
140. Urban Mass Transportation Administration (1972),

*Transportation Planning System (UTPS) Course Notes*, US Department of Transportation, Washington, D.C.
141. Alan M. Voorhees (AMV) and Associates (1974) *Results of a Multimode Choice Model*, Prepared for the Metropolitan Atlanta Regional Commission.
142. Alan M. Voorhees (AMV) and Associates, 'TALUATS mode share model development', unpublished Technical Memorandum (1975).
143. Warner, S.L. (1962) *Stochastic Choice of Mode in Urban Travel: A Study in Binary Choice*, Northwestern University Press, Evanston, Illinois.
144. Watson, P.L. (1974) *The Value of Time, Behavioral Models of Modal Choice*, Lexington Books (D.C. Heath & Co), Lexington, Mass.
145. Watson, P.L. and Westin, R.B., 'Transferability of disaggregate mode choice models', *Regional Science and Urban Economics*, 5, pp. 227-49 (1975).
146. Wattanatada, T., 'An approach to aggregation of destination choices', unpublished memorandum, Transportation Systems Division, Department of Civil Engineering, Massachusetts Institute of Technology, Cambridge, Mass. (1975).
147. Weber, M. (1949) *The Methodology of the Social Sciences*, Glencoe Press, Illinois.
148. Westin, R.B., 'Predictions from binary choice models', *Journal of Econometrics*, 2 (1974).
149. Westin, R.B. and Gillen, D.W., 'Parking location and transit demand; a case study of endogenous attributes in disaggregate mode choice models', *Journal of Econometrics*, 8(1), pp.75-101 (1978).
150. Wilson, A.G., 'Catastrophe theory and urban modelling: an application to modal choice', *Environment and Planning A*, 8 (1976).

PART TWO

Chapter 2

NEW APPROACHES TO UNDERSTANDING TRAVEL BEHAVIOUR: THE HUMAN
ACTIVITY APPROACH

Peter M. Jones

INTRODUCTION

Transport investment and management decisions are usually
made in the context of a transportation planning process, in
which a set of travel forecasting and evaluation procedures
play a central role. The models used are normally aggregate
and descriptive, and are based on a very rudimentary understanding of the nature of the travel decision process (24,
46). Decision-making goals have changed substantially in
recent years and rather than continuing to plan for existing
trends, serious attempts are now being made in a number of
urban areas to alter these trends significantly, by restrict-
ing car use and encouraging cycling and public transport. At
the same time more attention is being paid to the wider
environmental and social implications of transport policy
(22,23) and a much broader range of strategies is being
considered.

The changing transportation planning environment has
invalidated many of the status quo assumptions inherent in
traditional models and has exposed the vulnerability of their
forecasts. Travellers' response to recent UK transport
policies has been complex and varied; policies to control
peak hour commuter car travel, for example, have resulted in
increased off-peak car travel, by other members of the family
(with a resulting deterioration in the environment and
falling public transport patronage), and more general city
centre car restraint may lead to substantial destination
switching or trip re-timing, consolidation of journeys, or
total abandonment (19). A recent proposal to stagger school
hours in an English town led to widespread protest because of
the interference it would cause to the daily living patterns
of affected families, and the scheme had to be dropped (28).

Many writers acknowledge the inadequacies of existing
procedures, which were a pragmatic solution to an urgent
problem, and there is now a growing interest in developing a
firmer conceptual modelling foundation, through a deeper
understanding of traveller behaviour. The best documented of
these new approaches is the development of disaggregate
behavioural travel-demand models, which are anchored to
theoretical bases in economics and psychology (38).

Research designed to improve our understanding of travel
behaviour is also being carried out under a UK Social Science

Research Council project at the Transport Studies Unit, Oxford University, where travel is being studied within the context of household daily activity patterns - 'things people do in time and space' (7). The research team believe that it is not simply the mathematical structure of the existing travel demand model which is at fault but, more fundamentally, the concept of travel which it embodies and hence have chosen this much broader framework for their research.

This chapter discusses the human activity approach to studying travel behaviour. It begins by briefly reviewing the history of human activity research and in the following sections describes how the approach may be used to study travel behaviour, both conceptually and practically. As an illustration, the application of the human activity approach to studies of journey structure (i.e. the incidence of multi-purpose journeys) is contrasted with traditional approaches.

## A REVIEW OF HUMAN ACTIVITY STUDIES*

Any study dealing with an aspect of human behaviour may in some sense be regarded as a study of human activity, but in this context the term is used more specifically to refer to studies which examine continuous sequences of human actions over a period of time, using some form of diary to record the information. In the urban context Chapin (6) has identified three component activity systems, centred on households, firms and institutions, and attention here is further restricted to studies of household activity systems. (For a review of studies of offices as activity systems, see Goddard (14).)

Human activity studies share a common concern with how people allocate their time among different activities, and 'time budget' diaries record basic information about what people are doing, when and where (for a discussion of diary techniques, see Hedges (18)). Depending on the nature of the study, supplementary questions may also be included about why, or with whom the activity was being carried out, under what constraints and with what benefit. Studies vary according to the level of detail with which they record activities and time periods, and by whether they simply describe activity patterns, or attempt to explain them. One useful distinction in the context of travel research is between studies which concentrate on time allocation, and those which examine the use of space and time.

### Time Allocation Studies

Studies of human time allocation began in the early decades of this century, when time budget information was obtained

---

*This brief review attempts to pick out some of the more salient material, but cannot do justice to the breadth and sophistication of recent conceptual and empirical developments and the interested reader is advised to refer to the references cited.

either for economic planning purposes (as in the Soviet
Union), or as part of a sociological or anthropological study
(47). This tradition has continued in economics and
sociology/anthropology up to the present day and recent
examples include a study of Stone Age economics (41), family
life in the London region (51) and comparisons of time budget
data between countries (48) and through time (39). Descriptive empirical analysis is normally presented in the form of
frequency counts (e.g. number of people participating in
leisure activities (43)), or daily time budget summaries of
average time spent on groups of activities (48). In general
the evidence suggests broad similarities in time allocations
between studies and across countries (see Table 2.1), with
certain exceptions which have a clear cultural basis (e.g.
eating habits).

TABLE 2.1  International time budget comparisons

| Activities | Population groups | | |
|---|---|---|---|
| | Married men | Married women | |
| | | Working | Not in paid work |
| Paid work and travel to work | 7.2(0.4) | 5.7(0.7) | 0.2(0.2) |
| Household tasks | 1.5(0.4) | 4.6(0.6) | 7.8(0.8) |
| Subsistence | 10.4(0.4) | 10.1(0.5) | 10.6(0.4) |
| Leisure | 4.4(0.5) | 3.1(0.6) | 4.7(1.0) |
| Non-work travel | 0.5(0.2) | 0.5(0.2) | 0.7(0.3) |

Mean time allocations in hours (standard deviation in
parentheses)

Source: Computed from M.Young and P. Willmott, The Symmetrical
Family: A Study of Work and Leisure in the London Region
(London: Routledge and Kegan Paul, 1973), Table A10.
Information derived from studies in 14 countries or cities:
Belgium, Bulgaria, Czechoslovakia, England (London region),
France, Hungary, Poland, USA (cities), USA (Jackson), USSR,
West Germany (National), West Germany (Osnabruck), Yugoslavia
(Kragujavac), Yugoslavia (Maribor).

Although much of the time budget literature is purely
descriptive, there have been several attempts to explain the
observed time allocations.  Cullen and Phelps (10), for
example, use multiple regression techniques to account for
time allocations to activities with socio-economic characteristics as dependent variables, while Bain (2) uses more
sophisticated Tobit probability techniques.  Sociological
investigations have continued the early work of Sorokin and
Berger (44) who studied the declared motives behind
activities, but studies now tend to incorporate more
sophisticated in-depth interviewing techniques (51).

Economists have characteristically viewed time as a resource to be allocated according to utility maximising principles and there have been a number of ambitious attempts to derive a mathematical theory of the allocation of time (3,32), with some limited attempts at empirical verification (8,13).

Studies of time allocation are providing valuable information about human activity patterns, but most workers in this area seem to have avoided studying continuous activity sequences over time; like their fellow workers in transport research they have tended to work with daily aggregations of behaviour, but mainly in the form of time budgets rather than activity occurrences or trip rates. It has been left to those studying human activity in space and time to explore more fully the temporal structure of activity patterns.

*Space-Time Allocation Studies*

Alongside the reports on time budget research, a body of literature has grown up in the last decade dealing with human activity patterns in time and space, mainly contributed by land-use planners, geographers and architects. This work is characterised by its study of *pattern*, in both time and space, and by a number of attempts to explain and model observed patterns. Modified time budget diary techniques are used for the collection of space-time activity data, but rather than condensing diary entries during analysis (as happens in time budget analysis) information is expanded to provide a continuous record of behaviour in space and time; Figure 2.1 illustrates this important distinction.

A number of research groups have contributed to the conceptual foundations of space-time activity research and distinct perspectives may be identified, based on the relative importance attached to choice and constraint as influences upon activity patterns. General reviews of space-time approaches have been published by Anderson (1), Gutenschwager (16), Ottensmann (36) and Thrift (49).

The approach adopted by Professor Chapin and co-workers at Chapel Hill, North Carolina, USA, may be characterised as an extension of the sociological/anthropological perspective within an urban planning framework. Human activity patterns represent the means by which people satisfy human needs and wants (7) and two broad groups are identified:

1  Subsistence needs (sleep, food, shelter, clothing and health care), plus activities which supply income to meet basic needs.
2  Culturally, socially and individually defined needs, which are met by engaging in a wide range of social and leisure activities.

The Chapel Hill team have chosen to concentrate on discretionary activity patterns and the choices which different groups exhibit. Although their data and theoretical framework incorporate both space and time, recent work has focused on activity preference and time allocation, and their work on life-styles of suburban communities (53) really belongs among the time allocation studies already discussed.

Figure 2.1   Contrast between time and space-time approaches

TIME STUDIES:

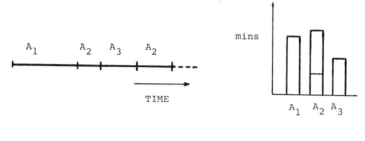

Observations                     Analysis

SPACE - TIME STUDIES:

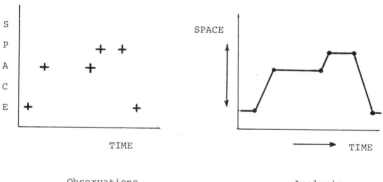

Observations                     Analysis

In strong contrast to the work of Chapin are a group of
Swedish studies carried out in the Department of Geography at
Lund, under Professor Hagerstrand; much of this work is only
available in Swedish and a good English summary has been
published by Thrift (49).  The Lund geographers adopt a
'physicalist' approach, emphasising the space-time constraints on choice imposed by physiological, economic and
cultural factors, and the nature of space itself.  Hagerstrand
(17) identifies three types of constraint:

1   Capability constraints: human capability is limited by
    biological factors and the capacity of the tools he
    commands (e.g. means available for travel or communica-
    tion).
2   Coupling constraints: arising from the need for certain
    people, tools and materials to come together at certain
    locations for given periods of time (e.g. work).
3   Authority constraints: which limit and control access to
    activity and travel facilities in space and time.

Two aspects of Hagerstrand's work which are of particular
interest here are the concept of an individual or other
entity having continuous existence in space-time, and the
recognition of the inter-dependence between space and time:
that space has temporal expression in the time it takes to
move from one location to another.  Transportation planners
have partially recognised this in using travel time instead
of distance as a measure of separation, but the full
implications have not been realised.

The constrasting approaches adopted by Chapin and
Hagerstrand reflect their different perspectives: Hagerstrand
is interested in understanding the operation of constraints
so that planning can relax them, whereas Chapin is concerned
with people's preferences so that as a planner he can provide
for them.  Taken together, these two perspectives provide a
powerful basis for understanding activity patterns: choice in
the context of constraints.  There are clearly semantic
problems in defining these terms (e.g. is habit a constraint
or a choice?) but this composite approach has been successfully adopted by several writers.

Concentrating on outdoor recreation activities, Maw (33)
has developed a complex conceptual model of recreation choice,
which incorporates a number of constraints that limit choice,
including time, money, energy, availability (i.e. access to
suitable facilities in space-time) and awareness.  The model
takes detailed account of space-time constraints by identifying people with blocks of free time of varying length at
different times of day and matching these against local
supply conditions (i.e. allowing for opening hours, minimum
time necessary for activity participation, and access times).

Another approach to understanding activity patterns in
space and time, which embodies concepts of constraint and
choice, has been developed by Ian Cullen at University
College, London.  Cullen's work differs from Maw's in that he
examines total daily activity patterns and the perceived
nature of commitment and choice through the day.  Instead of
looking at constraints surrounding specific decisions, he is

interested in their overall patterning ((9) p.8): 'Activities
to which the individual is strongly committed and which are
both space and time fixed, or just time fixed, tend to act as
pegs around which the ordering of other activities is arranged
and shuffled according to their flexibility ratings.' In addition to standard diary data, Cullen also collects information
about the space-time fixity of activities and the degree of
commitment involved (arranged, routine, planned, unexpected).
More recently he has extended his work to include the stress
experienced by people in scheduling their daily routines (10).

Space-time budget studies often use more sophisticated
forms of temporal analysis than is normal in studies only
concerned with the use of time, usually as a prerequisite to
model development.  In addition to summary statistics of
activity occurrences, durations and time budgets, information
is normally presented on activity occurrence by time of day.
Bullock et al. (5) present detailed frequency distributions
in the form of histograms,* while Cullen and Phelps (10) use
harmonic regression analysis to smooth out the envelope of
each distribution.  Maw (33) has examined the broad patterns
of time devoted to essential and optional activities through
the day and concludes that three superimposed trends are
apparent:

1  Optional activities progressively replace essential
   activities through the day, until bedtime.
2  Leisure activities increase to three plateaux at: 10:00,
   15:00 and 20:00.
3  Highly committed activities occur in three waves with
   peaks at: 12:00, 17:00 and 22:00.

By reducing human behaviour to two types of activity (committed and optional), Maw is able to account for interactions
between the two groups - since, within a daily 24-hour
budget, more of one implies less of the other.  A recent
paper by Shapcott and Wilson (42) extends the basic budget
trade-offs which Maw described to several activity groupings
within the 24-hour constraint.  They argue that because of
the budget limitation, more time on one activity implies less
time for others, so that one would expect negative correlations between activity durations.  Assuming maximum
independence between activities, one can compute a theoretical correlation matrix and compare it with the observed
matrix of correlations and examine patterns of residuals.
They found stronger than expected negative correlations
between sleep and domestic/paid work and positive
correlations between paid work and travel time, and sleep and
personal hygiene/private leisure.

Another way of looking at inter-relationships between
activities through time is to measure linkages using
transition probabilities, which specify the probability of
changing from one activity to another in a given period of
time (10).  This approach has also been used to examine
linkages between trip purposes on complex journeys (21), but

---

*The method of presentation is similar to that used to in-
dicate vehicle profiles of car park occupancy through the day.

it is a descriptive device not really suited to the study of
activity interactions.

In contrast to recent developments in the analysis of
temporal inter-relationships, work on spatial relationships
has been relatively neglected and variants of standard
transportation study techniques are normally used.  Simultaneous analysis of space-time patterns has proved
particularly difficult and this is commonly handled by
superimposed graphical plots of activity patterns or by the
use of simulation techniques.

Apart from these detailed studies of household activity
patterns, the recent geographic and planning literature has
also contained a debate about the nature of time and space
and the way in which they permeate the actions of society at
all levels (37).  One effect of this has been to make
researchers more aware of the context of the household
activity system within the overall urban activity system, and
to consider more seriously the extent to which individuals'
constraints represent society's choices.

THE HUMAN ACTIVITY APPROACH TO STUDYING TRAVEL BEHAVIOUR

The space-time activity framework described above provides a
general alternative basis for the study of travel behaviour.
Instead of regarding it as a phenomena which may be studied
in isolation, travel becomes one activity in a continuous
pattern or sequence of events in space-time; but with the
unique property that it represents the mechanism by which
people trade time to shift location in space, so that they
may take part in successive primary activities at different
places.  An example of the difference in emphasis is given in
Figure 2.2: the idea is simple, but the implications are
fundamental and far reaching.

*Components of the Activity-Travel Framework*

Within the literature, there are three space-time related
concepts which help to transform this general theoretical
approach into a more rigorous and precise framework with
many potential applications in transportation planning: (i)
travel choice options constrained in space-time; (ii) trip
generation as an outcome of activity choices; (iii) constant
time budgets.  These are summarised briefly below.

*Space-time prisms*.  Since it takes time to travel through
space, there is a limit to the range of destinations which
may be visited within a given period of time; clearly, travel
by a faster mode increases the distance range and hence the
number of practical opportunities.  This concept has been
formalised by Hagerstrand (17) using the idea of space-time
'prisms'.

Figure 2.3 illustrates the concept.  It assumes an
individual is committed to take part in activity $A_1$, at
location $l_0$ (say home), from $t_0 - t_1$, and to engage in
activity $A_2$, at $l_1$ from $t_2 - t_3$.  He thus has uncommitted

Figure 2.2  Travel and activity - travel approaches contrasted

TRAVEL PERSPECTIVE:

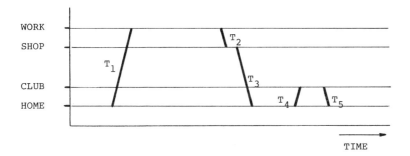

ACTIVITY - TRAVEL PERSPECTIVE:

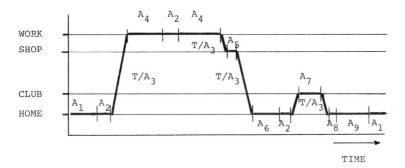

Figure 2.3   Concept of a space-time prism

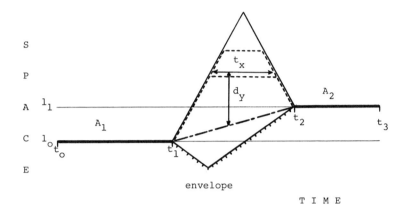

Travel modes:

| | |
|---|---|
| ———— | car |
| ⊤⊤⊤⊤⊤⊤⊤ | bus |
| —·—·—·· | walk |

time from $t_1$ to $t_2$, but the use he can make of it depends on the speed with which he can travel from $l_0$ to $l_1$. On foot, it may take $(t_2 - t_1)$, leaving him no time for an alternative primary activity; but faster modes enable the individual to reach an enlarged area of space and within this envelope he may trade off distance of intermediate destinations from the direct route $(d_y)$ against time spent there $(t_x)$. Public transport waiting times effectively shrink the prism and reduce the range of destination choice and the time which may be spent there. Space-time prisms thus define a potential range of choice.

*Activity trade-offs*. In behavioural terms, trip generation is interpreted as a trip/no-trip travel option, which is a function of socio-economic characteristics and (possibly) level of service offered by the transport system. Changes in trip rates (leading to trip suppression or the release of latent travel demand) appear to be a significant type of response to certain transport policies [19] but are very difficult to predict and understand conceptually. What do we really mean by latent demand?

Jones [26] has proposed that within an activity framework the trip/no-trip dichotomy can be replaced by an examination of activity choices; these lie between participating in activities at the decision-maker's current location (i.e. no trip), or in other activities which make use of facilities elsewhere (hence requiring travel). The decision has three main components: (i) the range of possible activities; (ii) the set of destinations offering suitable facilities for activity participation; (iii) the characteristics of the available transport systems. Choice is a function of the attributes of these three factors and is constrained in space-time by the appropriate prism. This formulation also leads to the explicit handling of travel as a derived demand.

*Constant time budgets*. Several writers have commented on apparent evidence for the existence of constant average travel time budgets, both at household and person levels for total travel and particular trip purposes. Rodgers [40], in a study of recreation in the UK observed coincidentally that the average journey to work times for samples in metropolitan areas, large and small urban areas and rural areas were each approximately 35 minutes per day. Kostyniuk [29] also observed a very close similarity between average travel time to grocery stores on the part of those who walked (9.9 min) or went by car (9.8 min).

Tomlinson *et al.* ([50] p.257) made a similar observation for mean daily travel time budgets of samples of students – regardless of accessibility to the university – Goodwin [15] has confirmed the stability of travel time budgets over a range of densities using UK National Travel Survey data. Zahavi [52] found a similar stability over space and through time, in the US context. Evidence presented in Table 2.1 indicates broad similarities in time budgets for other activities too, and the finding by Bain [2] that there is a fairly constant relationship between activity time at destination and travel time agrees nicely with this. A paper by Mogridge [35] has presented evidence for the existence of companion travel cost budgets.

*Synthesis of the Activity-Travel Framework*

The activity-travel framework view people as living on a daily basis by participating in a sequence of activities, in order to satisfy basic needs (such as sleeping and eating) and less essential preferences.   These activities use up a certain amount of the fixed daily time budget (24 hours), and may require facilities available at only a few locations in space (often only home), which may only operate at fixed times or between certain hours of the day.   By convention or necessity, certain activities are fixed in timing and/or location to a varying degree (e.g. sleep, work, meals) and provide fixed points in space-time around which more optional or flexible activities must be fitted in.   The degree of freedom of choice at such times is limited absolutely by the appropriate space-time prism (and often financial constraints) which is determined by the transport system, and choices lie between combinations of activity and location within the prism which are perceived as options by the individual.   At an aggregate level, factors operate which tend to control overall time allocations to activity and travel time placing a further limitation on activity and travel patterns.

The framework suggests that daily living is in some sense a giant jigsaw puzzle, with people juggling around with the pieces until the best fit is obtained.   Clearly life is not as complex as this, since many activities are of necessity fixed in space-time on a daily basis (e.g. school hours), or are treated as fixed by the individual, forming part of his 'daily routine'.   However, part of the stress of modern urban living may well reflect difficulties in easily scheduling daily activities, or coping with situations where schedules are disrupted by time-space 'failures' (e.g. shop closing early, business meeting lasting longer than expected, unanticipated road congestion).   Once an adjustment becomes necessary - by some individual miscalculation, or as a result of deliberate transport policy - adaptation may be very complex, involving the re-timing or reassignment of activities to other household members, or the substitution among activities or locations.   This approach at the same time both diminishes and enhances the role of travel: diminishes it because travel becomes just one of a series of daily activities performed by and essential to the well-being of the population, yet enhances it because it makes explicit the vital role travel plays as a lubricating mechanism, enabling the smooth operation of daily activity patterns in space-time.

Several writers have talked of activity patterns in the context of travel studies, but almost without exception this has been a semantic rather than a conceptual adjustment.   In their analyses, Hemmens (21) and Daws and McCulloch (11), for example, are effectively studying trip purposes under a different name, since in both cases they analyse trip-purpose linkages.   Writers such as Kutter (30) have gone one stage further by examining the total out-of-home space-time budgets, and so make much fuller use of available transport

study data.  By ignoring in home activities, however, most of
the travel-related activity research has failed to grasp the
full potential of the human activity approach: the ability to
study linkages between people and through the day, and the
facility for examining optional journeys as the outcome of
(typically) in-home v. out-of-home activity trade-offs.  Only
in its complete form will the approach be able substantially
to advance our understanding of the role of travel in
people's daily lives.

*A Reassessment of Existing Models*

Traditional travel demand models (comprising trip generation,
trip distribution, modal split and route assignment) embody a
number of well documented limitations, and are gradually
being replaced by disaggregate behavioural travel demand
models, particularly in the United States.  The proponents of
such models are very enthusiastic about their potential
capabilities (38,46) and cite a number of important
advantages: they are disaggregate and require smaller data
sets for calibration, they attempt to directly model travel
choice processes, and they incorporate a wider range of
explanatory variables; the more recent simultaneous forms of
model also allow for fuller interaction between sub-choices
and a common set of variables.

Although undoubtedly a considerable advance, viewed from
the perspective of the activity-travel framework the existing
disaggregate travel demand models contain a number of fundamental limitations and appear more as a refinement of
existing approaches than a radical departure from them.
Limitations include:

1  Assumption that travel is a choice process: the models
   take little account of the complex space-time constraints
   which may in certain instances entirely remove the choice
   element.
2  Assumption that behavioural response is continuous and
   not subject to discontinuities and threshold effects (20).
3  Inter-relationships between travel choices are handled
   inadequately at three levels: (i) trip sequences are
   subdivided into their component trips for analysis and
   the trip chain linkages in space-time are lost; (ii)
   trips made by an individual on one journey from home are
   not explicitly related to previous or subsequent excur-
   sions from home; (iii) trip-making behaviour of one
   person in a household is assumed to be unaffected by the
   actions of other household members (except that car
   availability is considered).
4  Travel is modelled as though demanded in its own right,
   and not explicitly as a derived demand; stratification by
   trip purpose does not really avoid the problem.
5  The models do not provide a means by which the inter-
   relationships between travel and non-travel aspects of
   life may be better understood.  Technological advances
   relating to certain activities (e.g. availability of
   frozen foods) may affect activity participation and so
   indirectly travel patterns; conversely, transport
   policies may have important non-travel impacts.

Some of these criticisms refer to the way the techniques are currently applied, rather than being a criticism of disaggregate techniques in themselves, although it is doubtful whether some of these problems could be overcome using such techniques. The purpose of these observations is not to argue that such models have no value, but to suggest that they should be applied with caution. At the present time they represent the best available approach, and the activity-travel framework can be used to indicate whether the simplifying assumptions embodied in such models are seriously violated in reality or not. The following sections discuss how the activity approach can be more directly used to assist in the solution of transport-related problems.

USING THE HUMAN ACTIVITY APPROACH

*Modelling Travel in an Activity Framework*

There are several examples in the literature of attempts to develop models of activity and travel behaviour, which predict how people will be affected by changes in the land-use pattern or transport system in space-time. They may be divided according to whether they model choice or define constraint, and by their use of entropy or simulation techniques.

*Forecasting constraints on choice.* Lenntorp (31) describes a detailed simulation mode, 'PESASP' which explores the implications of changes in the transport network or land-use pattern on individual activity-travel choices. In line with the approach adopted by Hagerstrand *et al.*, the purpose of the model is not to predict choice as such, but to illustrate the way in which altering factors in the environment (e.g. bus schedules) can constrain or expand the range of choice, by altering the operation of space-time prisms. A summary of the model is given in Thrift (49).

The model requires three inputs:

1   the relevant activity program, specifying which activity has to be fitted in between fixed activities in space-time (e.g. visiting a shop for t minutes en route from work to home).
2   details of the land-use pattern, recorded in space-time: locations and opening hours.
3   details of the transport network in space-time (actual road network and bus schedules).

The output indicates the number of options available to people resident in each area with access to alternative modes of transport; in general it appears that in a medium-sized town, decreasing freedom of choice is provided by: car, cycle, bus, walk. Although not a travel demand model in the traditional sense, it can provide information of great relevance to many transport decisions.

*Forecasting travel and activity choices.* The more common approach has been to attempt forecasts of actual behaviour, using either some form of simulation or entropy model formulation. The simulation approach has been more generally proposed, but less successfully applied. Brail (4), for example, discusses at some length the structure of a micro-level simulation model, but does not attempt practical construction. Jones (25) describes a simplified simulation model using aggregate data, which was applied using a trial data set but not validated in an empirical study. An attempt by Stephens (45) to develop a detailed micro-level simulation model proved unable to replicate the behaviour patterns of the sample student population on which it was based.

The most successful attempt to model activity-travel patterns to date involved the use of entropy maximisation procedures using a sample of students from two English towns (50). Inputs to the model included:

1   mean daily time allocation for each of 12 activity groups, including travel (summing to 24 hours)
2   locations at which each activity could be undertaken
3   timing restrictions on activities at different locations
4   information on travel times between locations.

The model predicts the most probable allocation of the total population to locations and activities for each successive 15-minute time period and a comparison of actual and pre-dicted distributions shows reasonable agreement; the model assumes independence between successive activities and was therefore unable to replicate certain associations, as between sleep and personal care activities. Due to problems of limited data, the model predicted use of locational type rather than actual location and so did not replicate actual travel patterns - although in theory this could have been attempted.

Results to date thus indicate mixed success from attempts to replicate observed activity-travel patterns and to fore-cast the effects of spatial or temporal change. The Swedish work appears most promising, but avoids the problem of forecasting actual behaviour; in order to do this success-fully, it appears that more information is first needed about the ways in which household activity-travel patterns are structured and how people respond to changes in personal circumstances or environmental conditions. Research at the Transport Studies Unit (TSU) has recognised the dangers of premature quantitative model development, and has concen-trated on understanding household behaviour using exploratory interview and modelling techniques.

*Exploratory Studies at the Transport Studies Unit, Using the Activity Approach*

The Transport Studies Unit (TSU), Oxford University, have been carrying out studies of travel behaviour using the human activity approach, at both conceptual and empirical levels since late 1974.

*Data collection.* Early studies confirmed the practicability of using an activity framework in field surveys (12). These demonstrated that respondents reacted favourably to discussing travel in an activity context; recall of trips was substantially more complete when described as part of a daily sequence of events, rather than as a separate trip listing. Respondents also seemed better able to explain their reported travel behaviour if reference had been made to the context in which they occurred.

Subsequent survey work has been couched in activity terms, and a dual approach has been adopted in which both qualitative and quantitative data have been collected.  Quantitative date is in the form of activity-travel diaries which record the time, location and nature of each activity or journey in sequence through the day.  The information is requested for each household member for periods up to one week, depending on the nature of the study, and parents are asked to complete diaries for babies and younger children.  A representative sample of families completing diaries are asked to take part in subsequent unstructured, in-depth interviews, which discuss matters relating to the temporal structure of activity-travel patterns, activity preferences, effect of family and external constraints, and knowledge of activity and travel opportunities in the surrounding area.  Interviews take place in the respondents' homes and are tape recorded for later analysis; they last up to one hour.

Activity-travel diaries contain a wealth of data, with almost endless possibilities for analysis; the in-depth interviews are designed to assist in identifying hypotheses for empirical examination.  Qualitative findings suggest that groups of households can be identified along life-cycle and life-style dimensions who share common sets of activity needs and preferences, although actual behaviour may show considerable variation because of local differences in the availability of activity and travel facilities and the voluntary adoption of different forms of activity structure.

A survey of 300 households was carried out in North Oxfordshire during Autumn 1976, using seven-day activity-travel diaries.  This information is now being coded and, together with a space-time land-use survey, will provide a substantial data base for testing the detailed hypotheses which have been generated by the in-depth surveys.

*Exploratory modelling and the use of 'HATS'.*  As a prerequisite to formal model development, it was felt necessary to obtain a better understanding of the way in which activity-travel patterns are structured and modified in response to external policy stimuli.  The TSU research team have been examining this problem in an exploratory way using an interactive gaming device called 'HATS' (Household Activity-Travel Simulator).  The model was designed with a number of objectives in mind: (i) simplicity of operation; (ii) flexibility and adaptability; (iii) incorporation of a wide range of dependent variables and based on a space-time metric; (iv) ability to allow for secondary impacts through the day; (v) incorporation of dynamic interaction between household members; (vi) possibility of discontinuities in response; (vii) incorporation of subjective factors in the decision process.  HATS is used in a household interviewing context and enables household members to create a physical representation of their daily activity-travel patterns, which can then be modified under guidance from the interviewer to explore response to different hypothetical external changes (e.g. revised bus schedules, staggered working hours).

Each household member is provided with a board, on which to set out their activity travel patterns (Figure 2.4). The lower part comprises three horizontal, parallel grooved sections marked out in 15-minute time periods through the day, which (from top to bottom respectively) are used to indicate periods of time spent on activities away from home, in travelling, or on in-home activities. The upper part of each board is in the form of a map display, on which the locations at which activities take place and the travel routes may be indicated. Coloured blocks of different sizes are provided, which fit into the grooved sections and represent the time periods spent on different types of activity; ten activity groups are distinguished by colour and up to six different modes of travel may be represented. The boards are faced with metal and the activity blocks and location indicators are magnetic.

Once each person's board has been correctly set up, changes are made in response to a given external space-time modification. The household member most directly affected first modifies his board and examines any likely secondary effects through his day; other household members then consider any indirect impacts they might experience and an iterative procedure is used to arrive at a solution acceptable to all. Discussion is encouraged at all stages of the procedure and is tape recorded for subsequent analysis. The apparatus used provides a clear visual display of daily behaviour and draws attention to the structure of the activity pattern and any anomalies which may arise from a tentative adjustment - such as being in two places at once, or having unaccounted for periods of time. The displayed information is readily convertible back to a diary format and may be quantitatively analysed using normal techniques.

HATS has been used to study the impact of proposed school hour changes in West Oxfordshire (28) and a second study is underway to look at the effects of rural bus service frequency reductions. In the school survey, the model identified three forms of repercussion:

1  changes in school journey
2  changes in other travel patterns and associated out-of-home activities
3  changes in in-home activity patterns

Non-marginal adjustments were frequently made, and the exercise generated much useful information about fixity and flexibility in daily activity schedules and the forms of decision role which household members adopt. Household response was generally very favourable and it was clear that HATS was able to improve considerably on intuitive forecasts of the impact of the change.

The human activity approach represents a way of viewing travel, not a rigid set of procedures. HATS is providing

Figure 2.4   Household Activity-Travel Simulator

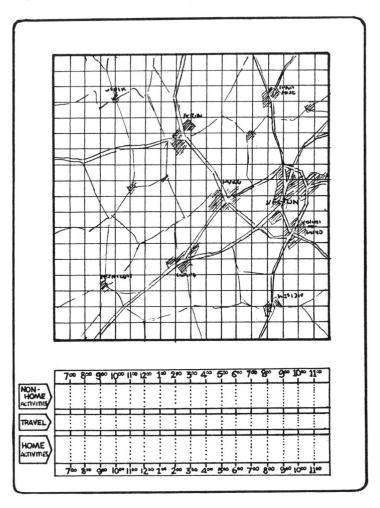

Source: P.M. Jones, 'A gaming approach to the study of travel
behaviour, using a human activity framework', *Oxford
University TSU Working Paper*, 18 (1976).

valuable information about behavioural adaptation, but is
only one way in which the approach may be applied and may be
viewed as a step along the way to more quantitative model
development.

## APPLICATION TO THE STUDY OF COMPLEX JOURNEY STRUCTURE

The analysis and forecasting of journey structure is receiving growing attention in the literature and is the subject of
two companion resource papers, by Hanson and Kobayashi. This
section illustrates how the alternative perspective of the
activity-travel approach can offer new solutions to a problem
which has not been very successfully handled within the
conventional framework.

### The Travel-Based Approach to Modelling Journey Structure

*Traditional model viewpoint.* In both the traditional four-
stage model and the recent disaggregate behavioural travel
demand models, journeys are separated into their component
legs or trips and each is analysed independently. Thus the
sequence: home - work - shop - home is treated as three
separate trips:

1 home-based work trip
1 home-based shopping trip
1 non-home-based trip

The models further ignore time of day and so remove travel
patterns from their space-time context, with consequent
dangers of wrongly inferring (and calibrating) travel choice
relationships. In the simple three-trip journey quoted
above, for example, the 'chosen' shopping location may have
been the only destination option lying within the traveller's
space-time prism and one which they would normally rate as
very unattractive; yet the depiction of this as the destination of a home-based shopping trip implies that it is
preferred to others in the vicinity of the traveller's home!
An additional weakness lies in the inflexibility of the
models; home-based trips are usually assumed to remain a
fixed proportion of total trips and so the effects of
alternative land-use arrangements on travel patterns cannot
be properly examined.

*More advanced analysis within a travel framework.* Several
writers have attempted to model journey structure as a series
of linked trips, using distribution models (for inter-zonal
interactions), factor analytic techniques (cells of interaction) or Markov chain processes (probability linkages in
space or time). The problem with these measures is that they
are essentially descriptive; they may be able to replicate
existing travel patterns (to varying degrees), but they do
not increase our understanding of *why* these patterns arise,
nor do they provide forecasts of how such patterns might
change in response to modifications to the land use or
transport systems.

   These methods suffer from the inherent and fundamental

limitation that, as they are conceived within a *travel* framework, they cannot easily accommodate major changes in journey structure since the basic unit of analysis itself (i.e. the trip) undergoes substantial modification. Significant changes in journey structure can only be handled within a broader framework, such as that provided by the human activity approach. As indicated previously, several so-called activity studies are in effect travel studies, and so also suffer from this conceptual limitation.

*The Problem Reformulated Under the Activity Approach*

Figure 2.5(a) illustrates a four-trip journey comprising the sequence:

Home - bank - shop - social visit - home

The diagram is set out in a way which indicates the traditional emphasis in travel modelling - i.e. *linkages* between specified *destinations*. Figure 2.5(b) attempts to show in simplified form how this would be represented in activity terms: as movement in space-time between a number of *locations* offering *facilities* at which certain *activities* may be pursued, for specified durations of *time*.

Since the purpose of travelling is to reach places where activities may be performed, the actual travel pattern is of secondary importance and represents only one of several ways in which the main objective of successful activity participation may have been met. Some examples of different travel patterns which could meet the *same* activity needs are given in Figure 2.5(c); depending on local circumstances, the three non-home activities could necessitate between two and six trips in all (and one to three journeys). It is not possible to handle such variability if the trip itself is used as the basic unit of analysis.

Further combinations are possible, by allowing for two other forms of adaptation by the individual: (i) *Activity consolidation*, whereby people engage in the same amount of an activity overall, but in different units of duration (e.g. substituting one major shopping trip for two or three small ones); (ii) *Activity Substitution*, whereby people satisfy the same need or want by participating in a different form of activity; this may often involve a home/non-home trade-off (e.g. restaurant meal vs. self-prepared meal, watching television vs. visiting the cinema), although the substitutes may all be out-of-home activities.

There is growing evidence that people respond to travel-related policy measures in complex ways which lie outside the choice domain of conventional approaches (20), but which follow logically when the problem is reformulated within the activity-travel framework.

*Modelling complex journey structure*. The most sophisticated operational model of journey structure in the literature is that by Lenntorp (31) and described in Hanson's companion paper. The model is able to take account of many

Figure 2.5  Analysis of Complex Journey Structure

5(a)  TRADITIONAL REPRESENTATION:

5(b)  ACTIVITY - TRAVEL REPRESENTATION:

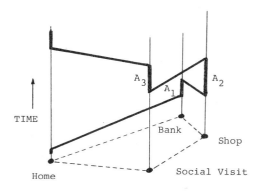

5(c)  SOME POSSIBLE REARRANGEMENTS:

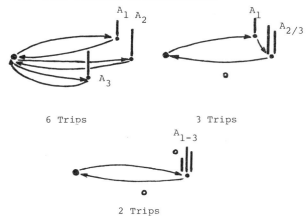

space-time constraints on choice and goes far beyond a simple
description of existing behaviour, although it does not
attempt to forecast actual behaviour.   In its present form it
takes home and work activities as fixed in space and time and
examines options for participation in a given activity en
route between the two; it thus assumes a fixed activity
structure.

The Transport Studies Unit have recently obtained a two-
year contract to develop a formal activity-based model to
forecast changes in journey structure, using mathematical
programming techniques.   The rationale behind the model is a
development of that described in Figure 2.5, and it comprises
the following components

1   A set of non-home destinations defined in space-time
    (i.e. location and opening hours), with facilities at
    which given sets of activities may be undertaken.
2   Network travel times between the home and destination
    nodes, by time of day; in the case of public transport,
    with precise schedules specified.
3   A set of non-home activities of given duration, to be
    undertaken by a particular type of individual, flexible
    in space-time.
4   Information on the space-time co-ordinates of fixed,
    compulsory activities inside and away from home, for the
    same person type.

These constraints define 'windows' of opportunity in space-
time, and the model determines the activity-travel pattern
that will minimise total travel time, on one or more
successive journeys from home, using network route optimisation procedures within a master control program.   Where
several options exist preference rankings are established, in
accordance with observed behaviour (e.g. uncommitted time
saved for later in the day).

The basic model operates with single-person households.
Subsequently, more sophisticated formulations will be
attempted, which allow for the varying attraction of destinations and which incorporate utility maximisation as the
objective function.   HATS will be used in the project to
assist in the formulation of realistic decision rules in a
later extension of the model dealing with multi-person
households.

IMPLICATIONS OF THE ACTIVITY APPROACH

The activity approach offers what amounts to a new paradigm
for the study of travel behaviour - a way of viewing the
travel phenomena rather than a specific set of procedures to
be followed.   As such it has basic implications for problem
recognition, and policy evaluation.

Examples of how problems are reformulated under an
activity approach have already been given; complex journey
structure is a reflection of activity scheduling and can be
better understood in this context.   Similarly, trip suppression and latent travel demand are given clearer definition

within an activity-travel framework.  The approach allows a
reassessment of other situations too; for example, the
concern to provide funds to maintain minimum bus service
levels in rural areas may in some situations be misguided.
Since the purpose is to enable people to reach destinations
in order to take part in certain activities, it may well
prove cheaper or more desirable to instead bring the
activities to the people - by subsidising village shops, or
mobile delivery services.

Probably the clearest advantages of the activity approach
are in the way it assists policy evaluation, because it is
likely to produce both more reliable and more complete
information about primary and secondary impacts.  This has
been clearly illustrated by the use of HATS to examine repercussions of proposed school hour changes described earlier.
It also offers the opportunity to take account of the impacts
of non-transport-related changes or policies on the demand
for travel.

In comparison with established views of travel, the
activity approach offers the following advantages:

1  Travel is specifically treated as a derived demand.
2  Travel linkages between people and over time can be
   incorporated.
3  Non-travel implications of transport policies may be
   assessed.
4  Travel implications of non-transport policies or
   technological developments can be examined.
5  Qualitative as well as quantitative understanding of the
   role of travel in people's daily lives may be obtained.

Taking a broader viewpoint, the activity approach provides an
obvious basis for integrating land use and transportation
planning; although they are often talked of as being closely
inter-related, this is rarely achieved in practice.  The use
of a common activity framework would also enable simplification of data collection* (since various surveys could be
replaced by one activity-travel survey) and ensure
consistency of forecasting.  In an evaluative context, the
approach could form the basis of a standard procedure against
which all policies affecting human behaviour could be
assessed.  Writers have already used time budget and other
activity data to give a measure of the overall quality of
life (34).

The activity approach thus appears to offer an attractive
alternative framework for studying travel behaviour, although
as yet many of its potential benefits have still to be
confirmed in practice.  Work in this field is in its infancy
and the workshop has provided a useful opportunity to make an
assessment of the approach and give guidance on its future
development.

---

*Time budget data has been collected in Hungary as a part of
the national population census.

REFERENCES

1. Anderson, J., 'Space-time budgets and activity studies in urban geography and planning', *Environment and Planning*, 3, pp. 353-68 (1971).
2. Bain, J.H. (1976), 'Activity Choice Analysis, Time Allocation and Disaggregate Travel Demand Modelling', Unpublished M.Sc. Dissertation, Department of Civil Engineering, Massachusetts Institute of Technology, Massachusetts.
3. Becker, G.S., 'A theory of the allocation of time', *Economic Journal*, 75, pp. 493-517 (1965).
4. Brail, R.K. (1969), 'Activity Systems Investigations: Strategy for Model Design', unpublished Ph.D. Thesis, University of North Carolina, Chapel Hill, North Carolina.
5. Bullock, N., Dickens, P., Shapcott, M. and Steadman, P., 'Time budgets and models of urban activity patterns', *Social Trends*, 5, pp. 45-63 (1974).
6. Chapin, F.S. (1965) *Urban Land Use Planning*, University of Illinois Press,Illinois, Chapter 6.
7. Chapin, F.S. (1974) *Human Activity Patterns in the City: Things People do in Time and Space*, John Wiley & Sons, New York.
8. Cohen, M.S. and Stafford, F.P., 'A life cycle model of the household's time allocation', *Annals of Economic and Social Measurement*, 3, pp. 447-62 (1974).
9. Cullen, I. and Godson, V. (1972) *The Structure of Activity Patterns*, Joint Unit for Planning Research, Research Paper No. 1, University College, London.
10. Cullen, I. and Phelps, E. (1975) *Diary Techniques and the Problems of Urban Life*, Final Report to the Social Science Research Council, London.
11. Daws, L.F. and McCulloch, M., 'Shopping activity patterns: a travel diary study of Watford', *Building Research Establishment, Current Paper CP 31/74* (1974).
12. Dix, M.C., 'Application of in-depth interviewing techniques to the study of travel behaviour: some preliminary results', *Oxford University, TSU Working Paper 9* (1975).
13. Evans, A.W. (1969) *A Revised Theory of the Allocation of Time*, Department of Social and Economic Research, University of Glasgow, Glasgow.
14. Goddard, J.H. (1975) *Office Location in Urban and Regional Development, Theory and Practice in Geography*, Oxford University Press, Oxford.
15. Goodwin, P.B., 'A hypothesis of constant outlay on travel', *PTRC Summer Conference*, Paper F29 (1973).
16. Gutenschwager, G.A., 'The time budget-activity systems perspective in urban research and planning', *Journal of American Institute of Planners*, 39, pp. 378-87 (1973).
17. Hagerstrand, T., 'What about people in regional science?', *Papers of the Regional Science Association*, 24, pp. 7-21 (1970).
18. Hedges, B.M., 'Time budgets', *Social Trends*, 5, pp. 35-44 (1974).
19. Heggie, I.G., 'Consumer response to public transport improvements and car restraint: some practical findings', *Oxford University, TSU Working Paper 2* (1976).
20. Heggie, I.G., 'Behavioural dimensions of travel choice', in *Determinants of Travel Choice*, D.A. Hensher and M.Q. Dalvi (eds), Teakfield, Farnborough (UK) (1978).

21. Hemmens, G.C., 'Analysis and simulation of urban activity patterns', *Socio-Economic Planning Sciences*, 4, pp. 53-66 (1970).

22. Hillman, M., Henderson, I. and Whalley, A., 'Personal mobility and transport policy', *Political and Economic Planning, Broadsheet*, 542 (1973).

23. Hillman, M., Henderson, I. and Whalley; A., 'Transport realities and planning policy', *Political and Economic Planning Broadsheet*, 567 (1977).

24. Hutchinson, B.G. (1974) *Principles of Urban Transport Systems Planning*, McGraw-Hill, New York.

25. Jones, P.M., 'An alternative approach to person trip modelling', *PTRC Summer Annual Meeting*, Paper N23 (1974).

26. Jones, P.M., 'Travel as a manifestation of activity choice: trip generation reinterpreted', *Oxford University TSU Working Paper 14* (1976).

27. Jones, P.M., 'A gaming approach to the study of travel behaviour, using a human activity framework', *Oxford University TSU Working Paper 18* (1976).

28. Jones, P.M., 'Forecasting family response to change in school hours: an exploratory study using 'HATS'', *paper presented at the University Transport Studies Group*, Glasgow University (1977).

29. Kostyniuk, L.P. (1975), 'A Behavioural Choice Model for the Urban Shopping Activity', unpublished Ph.D. Dissertation, Department of Civil Engineering, State University of New York at Buffalo, Buffalo.

30. Kutter, E.,' A model for individual travel behaviour', *Urban Studies*, 10, pp. 233-56 (1973).

31. Lenntorp, B. 'Paths in space-time environments - a time-geographic study of movement possibilities of individuals', *Lund Studies in Geography, Series B, Human Geography*, no. 44 (1976).

32. Linder, S.B. (1970) *The Harried Leisure Class*, Columbia University Press, New York.

33. Maw, R., 'Analysing demand for leisure facilities', *Built Environment*, 1, pp. 519-22 (1972).

34. Meier, R.L., 'Human time allocation: a basis for social accounts', *Journal of American Institute of Planners*, 25, pp. 27-33 (1959).

35. Mogridge, M.J.H., 'An analysis of household transport expenditures, 1971-5', *PTRC, Summer Annual Meeting*, Session G (1977).

36. Ottensmann, J.R., 'Systems of urban activities and time: an interpretative review of the literature', *Urban Studies Research Paper*, Chapel Hill, Centre for Urban and Regional Studies (1972).

37. Parkes, D.N. and Thrift, N., 'Timing space and spacing time', *Environment and Planning A*, 7, pp. 651-70 (1975).

38. Richards, M.G. and Ben-Akiva, M.E. (1975) *A Disaggregate Travel Demand Model*, Saxon House, Farnborough (UK).

39. Robinson, J.P., 'Social changes as measured by time budgets', *Journal of Leisure Research*, 1, pp. 75-7 (1969).

40. Rodgers, H.B. (1967) *Pilot National Recreation Survey: First Report*, British Tourist Authority and University of Keele, Keele.

41. Sahlins, M. (1972) *Stone Age Economics*, Tavistock, London.

42. Shapcott, M. and Wilson, C., 'Correlations among time uses', *Transactions of the Martin Centre, Cambridge University*, 1, pp. 73-94 (1976).

43. Sillitoe, K.K. (1969) *Planning for Leisure,* Government Survey, HMSO, London.

44. Sorokin, P.A. and Berger, C.Q. (1939) *Time Budgets of Human Behaviour,* Harvard University Press, Massachusetts.

45. Stephens, J.D., 'Daily activity sequences and space-time constraints', in *Time Space Budgets and Urban Research: A Symposium,* B.P. Holly (ed.), Department of Geography, Kent State University (Discussion Paper 1) (1976).

46. Stopher, P.R. and Meyburg, A.H. (1975) *Urban Transportation Modelling and Planning,* Lexington Books, Lexington.

47. Szalai, A., 'Trends in comparative time-budget research', *The American Behavioural Scientist,* 9, pp. 1-12, 21-31 (1966).

48. Szalai, A. (ed,) (1972) *The Use of Time,* Mouton, The Hague.

49. Thrift, N., 'An introduction to time geography', *Concepts and Techniques in Modern Geography,* 13, Geography Abstracts Ltd, University of East Anglia, Norwich (1977).

50. Tomlinson, J., Bullock, N., Dickens, P., Steadman, P. and Taylor, E., 'A model of students' daily activity patterns', *Environment and Planning A,* 5, pp. 231-66 (1973).

51. Young, M. and Willmott, P. (1973) *The Symmetrical Family, a Study of Work and Leisure in the London Region,* Routledge and Kegan Paul, London.

52. Zahavi, Y. (1974) *Travel Time Budgets and Mobility in Urban Areas,* Final Report US Department of Transportation, Federal Highway Administration, Washington, D.C.

53. Zehner, R.B. and Chapin, F.S. (1975) *Across the City Line: a White Community in Transition,* Saxon House, Farnborough (UK).

Chapter 3

URBAN-TRAVEL LINKAGES: A REVIEW

Susan Hanson

INTRODUCTION

Urban travel and communication linkages owe their critical
importance to the fact that cities comprise many relatively
specialised land uses that are spatially separated yet
functionally inter-dependent.  The intimate relationship
between transport and land use has escaped few students of
urban spatial structure, and many have commented upon the
dependency ties that underlie the existence of certain types
of urban establishments.  A special kind of dependency tie is
the travel linkage created by an individual travelling to a
sequence of different locations in the course of a single
journey.  Although such multi-purpose travel is common and
clearly involves combining a number of non-work purposes,
urban transport models have concentrated overwhelmingly upon
the work trip and have, moreover, conceptualised the journey
to work as a single purpose trip.  A large proportion of
urban travel, including the travel to work, entails the
traveller making more than one stop in the course of a
journey, yet the transport system, by and large, is designed
to meet the travel needs created by a single-purpose journey
to work.

The aim of this chapter is to draw together, from a
variety of research areas, the literature that has been
accumulating on urban-travel linkages and to examine the
potential utility to transport planners of abandoning their
single-minded dedication to the single-purpose trip in favour
of explicit recognition and consideration of multi-purpose
travel.  In order to carry out this aim, the chapter is
divided into four parts. The first examines the rationale for
obtaining a more thorough understanding of the multiple-
purpose trip. The second section reviews some of the
difficulties involved, particularly in the past, with studying multiple-purpose travel. In the third part various
methods of analysing travel linkages are reviewed, and in the
fourth part future avenues of investigation are outlined.

RATIONALE FOR INVESTIGATING MULTI-PURPOSE TRAVEL

Interest in multi-purpose travel springs from recognition of
the fact that the sequence of travel linkages etched between
destinations in the course of a multiple-purpose trip both
create and reflect the inter-dependence of selected urban
land uses.  Mitchell and Rapkin's (38) remarkable book was

81

the first attempt to delve deeply into the mutual dependence existing between traveller behaviour and the location of urban establishments. Subsequently, most who have contemplated the underpinnings of urban spatial structure have stressed the pivotal role of travel in linking together spatially separated establishments and have pointed to the absolute necessity of understanding the impact of spatial arrangements on the travel patterns of urban residents (2,8, 10,13,14,16,17,19,22,23,24,26,29,30,34,37,39,48,49,51,52,53).

Detailed investigation of multi-purpose travel would appear to be one tool suitable for prying open this circularly causal relationship between land use and travel. The circle can be broken by recognising that while movement patterns affect land-use patterns in the long run, in the short run the spatial distribution of land use is a major factor influencing observed movement patterns. It is necessary first, then, to place travel linkages in the position of the criterion variable, to understand the way in which travel linkages develop within different spatial environments so that we can establish the sensitivity of linkages to varying spatial arrangements and to other factors such as the characteristics of the traveller. Planners would then be in a position to make informed decisions that would be effective in the long run.

To summarise, the basic rationale for acquiring a thorough understanding of the travel linkages forged on multiple-purpose trips is to enable land-use and transport planners to assess the impact of alternative spatial arrangements or of alternative policies upon these linkages. The capacity to anticipate impacts on linkages is critical because it is equivalent to the capacity to anticipate the impacts on *individuals' activity patterns* and consequently on the *viability of certain establishments' or activities' locations*. Before exploring in greater detail the role of travel linkages on individuals' activity patterns and in the location of certain establishments, a brief aside concerning the nature of travel linkages is in order.

A multiple-purpose trip is conceptualised here (and elsewhere, e.g. (21,22,34)) as a home-to-home circuit during which the traveller makes more than one stop. Home-to-bakery-to-bank-to-department-store-to-home is one example of a multiple-purpose trip; home-to-work-to-restaurant-to-work-to-food-store-to-home is another example. A travel linkage is the spatial connection created when the individual moves from one of these locations to another. There are several characteristics of travel linkages that should be spelled out before we try to see more specifically why transport planners should pay attention to them. First, each link represents a *functional* connection between two urban activities; the nature of the functional connection is given by the two urban activities or functions that are joined by the travel linkage. Second, each link has a certain *intensity* which is reflected in the frequency with which it occurs. Also, a given travel linkage occurs at a particular *time*, and the activity pursued at the destination is undertaken for a given *duration*. Fifth, each link has a *spatial* dimension, measured, simply enough, by the distance travelled; a travel linkage may also

be distinguished by the *mode* of travel used.   Finally, as
Cullen (10) has pointed out, each link has a certain degree
of space-time *fixity* which refers to the degree of flexibility involved in the timing and location of the link.
These dimensions of travel linkages are enumerated not only
to clarify the terms used in the following discussion, but
also to lay the groundwork for the later section on
approaches to modelling multi-purpose travel since most
modelling efforts have focused upon subsets of these
dimensions.

We turn now to a closer inspection of the role of multi-
purpose travel in the activity pattern of the individual and
in the location of certain establishments.   The purpose of
the following discussion is to illustrate the need for
planners to be able to predict the impact of various proposed
alternatives upon travel linkages and to point to some of the
questions that require answers if impacts are to be correctly
foreseen.

*Impacts on Individuals' Activity Patterns*

Individuals frequently combine several purposes in the course
of a given journey in order to reduce the overall travel
costs associated with acquiring a given set of goods or
services (1,4,11,12,22,38,39,51) or in order to comparison
shop before making a purchase (22,39).   It is widely
recognised that individuals make travel decisions under a
variety of constraints, particularly those of time, money and
mode availability (8,10,19,32,34), and multi-purpose travel
can be viewed as an attempt by the traveller to pursue,
within the set of operating constraints, the sequence of
linkages that has the maximum utility for him or her at a
particular point in time.   The multi-purpose trip, therefore,
plays an important role in the individual's organisation of
time, in the individual's use of space, and consequently in
the individual's observed activity pattern.

Just how important multi-purpose travel is to urban
residents is difficult to pin down because very few studies
have addressed this question and those that have have used
widely differing data bases (described in the next section).
Estimates of the proportion of trips that are multi-purpose
range from about one-third (in Lansing, Michigan where only
vehicular trips made between origin-destination zones were
recorded) to nearly one-half (in Uppsala, Sweden where *all*
movements made by all modes were recorded).   Estimates of
what proportion of all intra-urban movements are associated
with multi-purpose travel also vary according to the nature
of the data used.   In the case of the Lansing, Michigan study
(52), 46 per cent of all trip segments were found to occur on
multi-purpose travel, while in the Uppsala study (36), 61 per
cent of all movements were associated with multi-purpose
travel.   In short, at this point, our knowledge of the role
of multi-purpose travel in urban areas is exceedingly
limited.

If, as many writers have urged (8,13,19,26,51), planners
are able to assess the impact of alternative policies and

plans upon the daily activity patterns of urban residents,
there are many questions concerning multi-purpose travel that
require answers.  For example, as Westelius (51) has observed,
planners should be aware of how different spatial arrangements affect people's overall time budgets as well as their
travel-time budgets.  We also need to know how important
different linkages are to individuals and how easily
different travel linkages are substituted by communication.
Similarly, which linkages (if any) tend to be the dominant
ones on a multi-purpose trip; that is, which trip segment or
purpose is the primary reason for travel (10,26,30,51)?  How
stereotyped, routine or repetitive are different sequences of
linkages?  What is the implication for the sequence of movements of snipping one link, especially a dominant link?  An
important set of questions revolves around the relationship
between the functional and the spatial aspects of different
linkages.  For example, how does the distance between
establishments or activity sites affect the intensity of the
linkage?  How far can a linkage be stretched over space
before it is damaged or destroyed (2,39)?  To what extent
would spatial reorganisation of activity sites induce
increases or decreases in travel linkage intensities?  A
final set of questions focuses on the relationship between
multi-purpose travel and the socio-economic and locational
characteristics of individuals.  How do travel linkage configurations vary with the age, location and income of a
household?  How is multi-purpose travel related to the
availability of travel modes?  Without answers to questions
such as these, it will be difficult for planners to evaluate
the probable impact of different proposals upon the daily
lives of urban residents.  It will be difficult, for example,
to assess the impact of energy-related policies or, indeed,
of energy shortages on individuals' travel patterns.  It
would also be hard to predict how different socio-economic or
locational groups would be affected by proposed changes in
the urban environment.

*Impact on Activity Sites*

An understanding of multi-purpose travel would enable
planners not only to forecast impacts on urban travel
behaviour, but also to assess the impact of proposed changes
upon certain activity sites or establishments.  Multi-purpose
travel clearly has an impact on the viability of certain
establishments' locations because the potential market of an
establishment includes those who are brought into the
vicinity of the outlet in the course of a trip to a nearby
place (11,38,51).  For many urban activity sites, then, stops
are generated by people travelling to proximate locations,
and the nature of the surrounding land use is critical to the
establishment's survival.  It should be noted, however, that
proximity does not always indicate a functional connection
(2,38); two establishments could be close spatially and be
completely independent of each other.  Therefore, it is
necessary to scrutinise travel linkages to establish which
types of activities (which types of urban functions) are most
heavily linked.  In order to assess how the viability of one
establishment's location would be affected by the removal of
a nearby establishment, we need to know to what extent travel

to one place is dependent upon travel to another.  On the
other hand, it would be useful to know what activities could
be moved with the least disruption of travel linkages and
therefore with the least damaging side effects on their own
well-being and that of other establishments (2).  An understanding of multi-purpose travel linkages would also provide
needed insight into the implications for certain establishments of selected travel linkages being replaced by
communication.

In sum, this portion of the chapter has argued for giving
the multi-purpose trip the recognition it deserves as an
essential ingredient in the daily activity patterns of
individuals and in the locational strategies of certain urban
functions.  Without a thorough understanding of multi-purpose
travel, planners will not be effective in predicting the
impacts of alternative plans on people's travel behaviour or
on the viability of certain activity sites.

## PROBLEMS ASSOCIATED WITH INVESTIGATING MULTI-PURPOSE TRAVEL

That the questions posed above have not been addressed
vigorously stems perhaps from the simple and obvious fact
that investigating multi-purpose travel is considerably more
complex than studying single-purpose travel.  The difficulties confronting anyone launching an attack on multi-purpose
travel can be traced to two sources: data requirements and
modelling challenges.  The aim of this section is to sketch
out the nature of these problems and to illustrate the
necessity of their solution if the kinds of issues raised in
the previous section are to be resolved.  It could be argued
that multiple-purpose travel is a familiar concept and that
well-established urban transport planning models take multi-
purpose trips into account, albeit rather surreptitiously.
The gravity model, for example, predicts flows on the basis
of the attractiveness of the destination, where attractiveness is frequently measured by the number of opportunities
per zone.  It is clear, however, that while such an approach
may inadvertently recognise multi-purpose travel, it does so
at such a gross scale that it leaves most (if not all) of the
basic questions concerning multi-purpose travel unanswered.

### Data-related Problems

Disaggregate data are absolutely essential to the serious
study of the travel linkages created by individuals travelling to a sequence of locations in the course of a single
journey, for the linkages observed are entirely a function of
the level of aggregation of the data.  Concerns here include
the concept of the trip, the level of geocoding, the
categorisation of travel purposes, the recognised travel
modes, and the length of time for which travel data are
recorded.

When urban travel is envisioned, as it was for many years,
in terms of aggregate flows between various areal units (or
zones) within the city, the travel links forged in the course
of multi-purpose journeys are often obscured.  One data

requirement, if travel linkages are not to be obscured, is
that the trip be conceptualised and recorded as a series of
movements that begin and terminate at the residence.  When a
trip is viewed as a single link between an origin and a
destination, the sequence of movements that may occur on a
single outing may be lost (9).  A second requirement is that
the origins and destinations of movements be locationally
coded as points (corresponding, for example, to street
addresses) rather than as origin-destination zones or census
tracts.  The use of zones as the spatial units of analysis
invites the vast majority of travel linkages to fall through
the spatial seive since most movements on a multi-purpose
trip are likely to be within a given zone.  Without spatially
disaggregated data pinning down the origins and destinations
of trip segments, it would be impossible to examine, for
example, the sensitivity of different linkages to being
stretched over space.

The categories of trip purposes should also be extremely
detailed if we are to dissect urban travel linkages successfully.  The nature of the linkages revealed and the
conclusions reached will obviously be influenced by whether
travel purposes are recorded in detail or whether such
general categories as 'work, shop, recreation and social' are
used.  For example, Wheeler (52) using O-D data from Lansing,
Michigan with ten trip-purpose categories, found a strong
linkage between shop and shop (that is, there was a high
probability that a movement to shop would be followed by
another movement to shop).  Hanson and Marble (22), however,
using disaggregate data from Cedar Rapids, Iowa with 99 trip-
purpose (land-use) categories, found strong linkages only
between certain types of retail establishments.  Many
shopping trips for certain kinds of goods were more likely to
be linked only to the home than to other retail establishments.  It would be particularly useful, in terms of
understanding urban activity patterns, to separate the land
uses visited from the activities undertaken.  'Trip purpose'
would then be recorded as a particular *activity* (e.g.
participate in outdoor sport) engaged in at a particular
*land-use* type (e.g. park).  This would enable explicit
recognition of the fact that a variety of activities can take
place at any given land use and would clearly provide a more
accurate picture of the nature of certain travel linkages.

If travel data are to be suitable for studying multiple-
purpose trips, a fourth requirement must be met: movements by
all modes must be recorded.  Excluding non-motorised modes of
transport from consideration means not only overlooking a
sizeable portion of linkages made on multiple-purpose trips,
but also excluding perhaps the most important class of multi-
purpose travel linkages: pedestrian movements.  The Uppsala
data (36), for example, reveal that 45 per cent of all movements made on multi-purpose trips were made on foot!  The
nature of the travel linkages observed by Hemmens (23,24) and
Wheeler (52) is a direct reflection of the omission of
pedestrian and bicycle movements (the only exception is that
Hemmens' data included walking trips to work).  It is the
exclusive reliance on vehicular travel that leads Wheeler to
find more linked trips at the city's periphery than near its
centre and to conclude 'It was here, at greater distances

from the city center, that the advantages of linking trips
may be particularly great, and a higher proportion of all
trips had more than a single stop' ((52) p.649). Similarly,
the combination of gross trip-purpose categories and only
vehicular movements led Hemmens (23) to conclude that multi-
purpose travel linkages constituted a rather small proportion
of urban travel.  Examples were cited earlier·of the way in
which the nature of the data base impacts the simple volume
of linkages that can be observed.

A final data characteristic that is desirable for the
study of multi-purpose travel linkages concerns the length of
the time period for which travel data are recorded.  In order
to evaluate the relative importance of different travel
linkages to different classes of individuals, daily travel
records should be kept for some extended·period of time; a
month is suggested as the bare minimum.  Longitudinal data
covering travel over several weeks, instead of the conventional 24-hour period, enables the individual's or household's
repetitive patterns of travel to emerge.  The relative
frequency with which different linkages occur can be observed
to provide an indication of the importance of different
linkages to people's daily activity patterns.

In sum, the nature of travel data available is critical to
any meaningful undertaking concerning multiple-purpose travel.
The problem, of course, lies in the fact that data sets
satisfying the requirements outlined here are extremely
costly and cumbersome to collect and therefore are exceedingly scarce.  Efforts such as Wheeler's (52) - purporting to
analyse urban activity linkages using ten trip-purpose
categories, only vehicular travel and origins and destinations geocoded to the census tract level - are clearly a
desparate attempt to get by with traditional transport data
sources.  Such studies by no means begin to address the kinds
of questions we must have answers to if we are to make
informed transport planning decisions.  Studies based on
inappropriate data do not even provide answers to such simple
questions as (1) what proportion of intra-urban travel is
multiple, as opposed to single, purpose; (2) how important is
the multi-purpose trip to different socio-economic groups; or
(3) what proportion of all intra-urban movements occur on
multi-purpose trips?  Detailed disaggregate data of the sort
described above are required if these and more difficult
questions are to be answered.  The dearth of suitable data
sources is not, however, the only problem facing the
investigation of multiple-purpose travel; even with the ideal
data set firmly tattooed on magnetic tape, the problems
associated with modelling spatial choice in the context of
multi-purpose travel remain to be faced.

*Problems of Modelling Multi-purpose Travel*

Most disaggregate models of urban travel behaviour have
assumed, either implicitly or explicitly, that the traveller
is embarking on a single-purpose journey.  Exceptions can be
found in the work of Bacon (1), who deductively examines the
impact of multi-purpose shopping trips on trip generation for
shopping, Stopher (45), who recognised that mode choice on

the journey to work is likely to be influenced by whether the
work trip is single- or multiple-purpose, and Hanson (20),
who acknowledged that destination choice is affected by the
number of purposes to be combined on a given trip.  Although
incorporation of multi-purpose travel into disaggregate
models of urban travel behaviour would necessarily impact the
modelling of all stages of the travel decision process - trip
generation, modal choice, destination choice and route
selection - it would add perhaps the greatest degree of
complexity to the modelling of destination choice.  For this
reason the problems of modelling multi-purpose travel are
outlined below primarily within the context of the implications for models of destination choice.

When a single-purpose trip is assumed, the selection of a
destination amounts to choosing one location from a set of
possible alternatives; in such a case, destination choice is
most frequently seen as the outcome of a trade-off between
attractiveness (size) and effort (distance) (41).  However,
in the case of a multiple-purpose trip, the modeller must
cope with the selection and sequencing of n purposes to be
carried out at m destinations, each of which represents an
element in a set of many alternatives.

If we ignore all factors other than distance that impinge
upon the destination choice process (i.e. ignore the personal
characteristics of the traveller and the non-locational
attributes of destinations), and assume that the traveller
exhibits rational behaviour, the complexity involved in
modelling destination choice on the multiple-purpose trip can
be demonstrated.  Under the assumptions of (1) rational
behaviour and (2) distance as the sole choice criterion,
destination choice on the single-purpose trip is a straightforward problem of distance minimisation; that is, the
nearest appropriate activity site is visited.  The only
question is which distance metric the traveller is minimising:
time-distance, cost-distance or raw distance.

On the multi-purpose trip, however, there are several
different strategies that the rational traveller might adopt.
Again, to simplify the problem, the role of the non-
locational attributes of a site in destination choice is not
taken into consideration.  Given that the individual must
combine a certain set of purposes on a trip, he or she might
opt to minimise the distance travelled on each leg of the
journey.  This myopic type of chain would be an appropriate
strategy if the selection of the $i + 1^{th}$ destination were
made at the $i^{th}$ destination on the journey.  Alternatively,
the traveller could choose destinations in order to minimise
the number of non-walk legs on the journey; such a strategy
might be entirely rational in terms of minimising the
traveller's effort.  A third scheme open to the rational
multi-purpose traveller is to select destinations such that
the total distance travelled on the tour is minimised.  The
latter two approaches are appropriate only for travellers
who plan the course of a multi-purpose journey before
leaving home.  Of course, the question raised above of which
distance metric is being minimised remains, and further
complications can arise with the third strategy when, for
example, order or locational constraints are placed on the

tour.  The problem then becomes one of minimising the total
distance travelled on the chain, subject to an order
constraint (purposes are to be pursued in a given order) or
subject to a locational constraint (certain activities can be
pursued only at specified locations, as in the case of
visiting the dentist).

In short, when multiple-purpose trips are openly
acknowledged, the modeller must cope with a considerable
increase in complexity.  As illustrated above, even the task
of specifying an optimal tour within a simple normative
framework is bedeviled by the existence of several perfectly
rational strategies for constructing an optimal path on a
multi-purpose trip.  When destinations within a given class
are permitted to vary in attractiveness, the challenge to the
modeller becomes gargantuan.  An important contribution in
this area has been made recently by Pipkin (40) who has
proposed a random utility model for predicting destination
choice on multi-purpose trips as a function of the utilities
assigned by the traveller to the different potential
destinations to be combined as well as to the distances
linking these sites.  Although a great deal of additional
work stands between our present position and a thorough
understanding of multi-purpose travel, considerable headway
has been made by researchers in the last two decades.  The
following section outlines the approaches researchers have
taken in attempting to analyse and to model multi-purpose
travel linkages.


ANALYSING MULTI-PURPOSE TRAVEL LINKAGES

Existing studies of multi-purpose travel have concentrated,
by and large, on a relatively small subset of the questions
raised in the first portion of this chapter.  Most work has
been aimed at shedding light upon the relative strength of
various linkages or upon the number of links in a trip.
Approaches to analysing multi-purpose travel linkages will be
examined within the context of these two basic questions to
which most studies have been addressed.


*Trip Length*

Empirical work on multi-purpose travel has shown that the
number of trips with n links decreases as n increases (14,23,
31,35,49,50,51).  Attempts to model this feature of multi-
purpose travel include Nystuen's (39) use of an exponential
function and Vidakovic's (50) model based upon a harmonic
series.  Building models of the number of expected stops (or
links) per trip, however, would seem to be an exercise of
relatively little value unless such work is extended to
relate this aspect of multi-purpose travel to other variables.
Yet disappointingly little has been done beyond the initial
step of modelling the number of stops per trip.  One
exception is the work of Vidakovic (49), who has related trip
length (the number of stops per trip) to the mixture of
travel modes on a given trip and to the number of different
purposes pursued on a given trip.  But this is only a start;
there are a host of questions remaining to be investigated.

For example, how do the number of stops per trip vary as a
function of the purposes to be combined?  How is the number
of links per trip affected by the spatial configuration of
land uses?  How is trip length related to varying travel
costs?  How are the socio-economic characteristics of the
traveller related to the propensity to make trips of varying
numbers of stops?  There is some evidence that stage in the
life-cycle (24,35) and income (24) do have an impact on trip
length, but we are far from having definite answers to any of
these questions.  We have barely begun to examine how the
number of stops per trip is related to other factors.  While
relatively little consideration has been given to trip
length, considerably more attention has been paid to
investigating the relative strength of different travel
linkages.

*Establishing the Strength of Linkages*

Research that has focused upon the strength of linkages has
tackled questions relating to the frequency with which
different establishments, land-use types, activities or zones
are tied together by travel or communication linkages.  In
most cases, the frequency of contact has been used as the
measure of linkage strength or intensity.  Straightforward
questions such as which land-use types rely most heavily on
multi-purpose travel (5,21), or which activities are most
likely to be pursued on multiple- rather than single-purpose
journeys (22)have been investigated simply by comparing stop
frequencies on single- and multiple-purpose trips.  In order
to examine an entire system of linkages or to identify sets
of urban activities rendered inter-dependent by virtue of
strong linkages among them, approaches based on flow matrices
have frequently been taken.

   Travel linkages, like other flows including trade between
countries or communications between firms, can be summarised
in the form of a square from/to flow matrix where the rows
represent origins and the columns destinations.  The origins
and destinations in such a matrix can be trip-purpose
categories (23,24,33,52), land-use types (22,39,51),types of
firms (2) or areal units (2,48); the cell entries indicate
linkage strength in that they give the number of times an
interaction occurred between origin i and destination j.
From such a matrix we can derive at a glance the relative
importance of dyadic flows; Mitchell and Rapkin's (38)
'paired linkage factor', for example, gives the proportion of
a firm's (or land use's or zone's) total interactions that
involve some other specific firm (or land use or zone).  The
proportion of a destination's arrivals that are not tied
directly to home is also easily generated from such a matrix.
But in order to comprehend the overall pattern of linkages
contained in a flow matrix, manipulation of the raw data
matrix is required.

   One method of simplifying flow matrices that has proved
effective in identifying 'important' interactions is transaction flow analysis (43).  The transaction flow model
involves comparing observed contact frequencies between
origin i and destination j with the contact frequency that

would be expected on the basis of destination j's share of
all movements.  This technique has been used in a number of
studies (2,23,52) to identify salient linkages, where
salience is defined in terms of some arbitrarily large
difference between observed and expected values.  Transaction-
flow analysis is useful, then, in pointing out particularly
strong linkages, but since it results in a dichotomous
classification of linkages (salient/non-salient), no shades
of strength are recognised.

Conceptualising multi-purpose travel in terms of a Markov
process has provided the basis for another way of simplifying
the flow matrix (15,23,27,28,31,33,39,42,52) in a manner that
does retain the relative strength of all linkages.  The
essence of the Markov model is the matrix of transition
probabilities in which the strength of a linkage is embodied
in the probability of moving from origin i to destination j.
These probabilities are empirically generated from observed
linkage data by calculating the row percentages for each
cell.  While the transition matrix itself shows the pattern
of direct linkages, manipulation of the matrix has the
advantage of yielding the pattern of indirect linkages.

Recently a number of writers, reviewing the use of Markov
chains in the modelling of multi-purpose travel have pointed
to the shortcomings of this approach (2,10,25,29,49).  One
problem is that the process is memoryless in the sense that
the probability of moving to state j is dependent only upon
the preceding state, i; places visited before i cannot
influence the process.  A second problem is that within any
one transition matrix the probabilities are assumed to be the
same for all travellers; variations in linkage patterns
between different population subgroups can be examined only
by constructing a separate transition matrix for each group
(28).  A third problem is that in the simple Markov model,
the duration of the stay in each state is the same for all
states and does not affect the transition probabilities;
Gilbert et al. (15) have suggested a semi-Markov process to
correct this deficiency.  However, Hemmens (25) found
activity duration to be dependent upon not only activity
type but also time of day and journey complexity (number of
stops on the trip) - complications with which the semi-Markov
model cannot cope.  A fourth problem is that the simple
Markov chain is an aspatial treatment of an inherently
spatial process (33); the model does not recognise that trip
purposes occur in space and that the relative location of
activities is likely to have an impact on observed linkage
patterns.  A possible solution to this problem could be in a
semi-Markov approach in which the transition probabilities
reflect, in part, the spatial separation of activities.  A
final criticism, leveled by Cullen et al. (10), is perhaps
the most devastating: they argue that Markov models are
inappropriate because a large proportion of an individual's
daily activities are fixed in time and space.  Hemmens (25)
echoes this criticism in his finding that the probability of
the occurrence of an activity is dependent on time of day.
In fact, Hemmens concluded after an extensive review of the
applications of the semi-Markov process to modelling urban
activity patterns, 'We find that the...lack of fit between
the behaviour of the phenomena to be modelled and the

assumptions of the semi-Markov process plagues these efforts
well.  All attempted applications of the semi-Markov process
to social systems that we found were abortions or distortions
of the semi-Markov process' ((25) p.48).  Despite the shortcomings of modelling multi-purpose travel within a Markovian
framework, the approach has provided some insight into the
relative strengths of direct and indirect travel linkages.

Another approach to simplifying flow matrices is factor
analysis (2,3,17,21,22,44,52), a method that has been helpful
in delineating sets of destinations with similar origin
linkage configurations (Q-mode) or sets of origins with
similar destination linkage configurations (R-mode).  Factor
analysis has been used to identify land use types or firms
with similar linkage patterns (2,22) as well as zones that
exhibit similar linkages (2,52).  While the technique is useful in identifying groups of origins or destinations that are
functionally inter-dependent or that have similar patterns of
linkages, the factor model's emphasis on producing statistically independent groups (i.e. orthogonal factors) renders
the technique inappropriate for identifying indirect linkages
or linkages between groups of inter-dependent urban functions
(2).  To obviate this difficulty Baker and Goddard (2)
employed Latent Profile Analysis, a method that is similar to
factor analysis but takes into account higher order correlations.  The factor analytic model, when applied to land-use
types or travel purposes, also suffers from being aspatial;
the method has been applied, by and large, to identify groups
of urban functions (aspatially defined) with similar linkage
patterns.  In this respect, the technique has proven itself
useful in pointing out those urban functions that are linked
primarily to the work place, those that are linked primarily
to the residence, and those that are linked principally to
other urban functions (22).

The methods outlined above are all aimed at making more
convenient the task of *describing* linkage patterns by identifying important linkages or by specifying sets of destinations
that are strongly interconnected.  Certainly without these
techniques and the descriptive studies that have relied upon
them, we would know even less about travel linkages than we
presently do.  There remain, however, additional questions
whose resolution requires examining the relationship between
travel linkage strength and other variables.  Prominent among
these questions are (1) the impact of spatial structure (the
spatial arrangement of establishments) on the strength of
travel linkages; (2) the relationship of the personal and
locational characteristics of the traveller to travel linkage
configurations; (3) the impact of travel mode on travel
linkages; and (4) the role of certain 'dominant' purposes on
multi-purpose trips.  Clearly, the methods briefly described
above are poorly suited to respond to these questions, all of
which call for specifying the sensitivity of travel linkages
to other factors.

Three recent empirical studies have used approaches that
enable travel linkages to be viewed as a function of other
variables.  The work of Tomlinson *et al*. (47), Westelius (51)
and Lenntorp (32) is cast in such a way that the variation in
travel linkages can be seen, at least to some degree, as a

function of the spatial arrangement of activities, the travel
mode used, the characteristics of the traveller and the
dominant role of certain activities.  With respect to this
last issue, a number of researchers (U,1U,26,30,51) have
stressed the need to distinguish between anchor, or non-
discretionary, or fundamental or fixed activities, on the one
hand, and discretionary, incidental, complementary or
substitutable activities on the other hand.  The idea here is
that activities in the former category can generate stops at
activities in the latter category, while the reverse is not
possible.

   Although Tomlinson *et al.* (47) set out to model activity
patterns rather than travel linkages *per se*, their entropy
maximising model could easily be interpreted in terms of
travel linkages.   In their scheme the amount of time to be
spent in each of a number of activities is given, and the
time budgets may vary according to the socio-economic
characteristics of the group.  Also given are certain key or
dominant activities that are fixed in time and space.  The
sequencing and locations of activities are permitted to vary
within the above constraints, and an entropy model is used to
predict the number of people engaged in different activities
at different locations in different time periods.  Although
only three locations (home, school or town) are recognised,
the structure of the model is such that the volume of flows
among activities at different locations between time periods
essentially measures the strength of the linkages involved.
Within the framework of the model one could examine the
sensitivity of travel linkages to varying spatial arrangements, differing levels of fixity in activities, and
differing time budgets.

   Westelius' (51) formulation also offers the opportunity to
examine variations in linkage strength as a function of
variations in other factors.  His rich conceptualisation is
based on the recognition that the combination of purposes
sought on any particular trip is a function of (1) which
purchase needs coincide in time, and (2) the relative location of activities.  Westelius draws a distinction between
activities that are *fixed* in time and space (e.g. work,
visits to the doctor or to friends) and activities that are
*substitutable* in the sense that the individual is free to
select, from among a number of alternative times and
locations, when and where the activity will be undertaken.
In addition, Westelius specifies *need* variables which vary
with the individual's distance from opportunities and which
could be made to vary with the socio-economic characteristics
of the individual as well.  At any given distance and for any
given purpose, there is a minimum need threshold required to
produce a trip.  The threshold of need for a substitutable
activity may be lowered when the individual is making a fixed
call in the vicinity of the substitutable activity; Westelius
thereby recognises explicitly the externalities implicit in
relative location.  Using these basic concepts, Westelius
simulates chains of movements and observes how travel-linkage
strength between pairs of land-use categories varies as a
function of the need variable.  The model also predicts the
proportion of chains that involve stops at fixed, as opposed
to only substitutable, activities.

The most recent of these modelling efforts is that of
Lenntorp (32), whose conceptual framework is rooted in
Hägerstrand's notions of individuals' space-time prisms.
Lenntorp constructs a highly interesting and extremely
detailed model that places multi-purpose travel firmly in the
context of the time and space constraints impinging on the
individual's activity pattern.  Given two activity episodes
that are fixed in time and space (he uses home and work) and
given the mode of travel to be used, as well as the functional class and the duration of the other stops to be made on
the trip, the model supplies the specific sites selected in
the course of the journey.  In testing the model, Lenntorp
restricted his investigation to chains involving stops at the
work place and at a food store before the day's final return
to home.  Since the time constraints and the mode of travel
are given, and since each mode is assigned a certain velocity,
the individual's choice of site is clearly limited by space-
time constraints.  Furthermore, since the road network and
the public transport lines as well as all food stores in the
study area (Vällingby, Sweden) are geocoded, Lenntorp is able
to compare actual choice behaviour with choices that would
have minimised total distance travelled or total time spent
away from home.  In sum, recent developments such as those
outlined here by Lenntorp, Tomlinson *et al*. and Westelius are
encouraging, and potentially productive in terms of beginning
to tie travel linkages to spatial arrangements, to travel
mode, to characteristics of travellers, and to constraints on
time-space budgets.

A final comment on the methods that have attempted to
establish the relative strength of travel linkages concerns
the virtually universal use of simple contact frequency as
the measure of strength.  This practice disregards the fact
that linkages which occur with the same frequency may vary in
importance; an alternative approach might be to weight
contacts according to expenditures made or length of time
spent at the destination.  Others (2,46) have suggested
distinguishing between contacts made on a regular basis and
contacts that are exploratory in nature.  Future studies
might profit from acknowledging that simple contact frequency
need not be the only measure of linkage strength employed.

In reviewing the various approaches that have been taken
in analysing multi-purpose travel linkages, this portion of
the chapter has revealed the past proclivity for descriptive
rather than predictive studies of urban-travel linkages.  In
analysing both the number of links per journey and the
relative strength of linkages, studies have tended to
describe rather than to view travel linkages as a dependent
variable.  Future studies will need to correct this
deficiency by relating specific aspects of travel linkages to
such items as the spatial arrangement of activities, the
locational and socio-economic characteristics of travellers,
the time budgets of individuals, and the nature of the
activities linked by travel.  The final section explores a
framework within which to conduct further studies of urban-
travel linkages.

SOME THOUGHTS ON FUTURE AVENUES OF INVESTIGATION

It is quite evident that the foregoing discussion highlights
the need for and urges the execution of predictive studies of
urban-travel linkages.  It is not the purpose here, however,
to present one grand predictive model as a panacea or as the
definitive statement on how to handle urban-travel linkages.
Indeed, it is my opinion that attempting to build a single,
all encompassing super-model of travel linkages would be
inappropriate at this time.  A more reasonable approach, it
would seem, is to construct any number of more modestly conceived models aimed at addressing selected aspects of urban-
travel linkages.  This is challenge enough.

The purpose of this final portion of the chapter is to
present a conceptual framework within which such modelling
efforts could go forward.  The proposed framework calls for
placing travel linkages within the context of spatial
behaviour, and for viewing multi-purpose travel as an
integral part of the overall out-of-home activity pattern of
the individual.  The argument is that by not building multi-
purpose travel into our models of spatial behaviour we are
most likely overlooking the fundamental process generating
observed behaviour patterns.  The conceptualisation offered
here places multi-purpose travel decisions at the heart of
spatial behaviour, and hinges on the union of two familiar
concepts: the cyclical nature of travel demands and
stereotyped behaviour.

Several theorists have noted the cyclical nature of travel
demand for a given purpose (11,14,51).  Certain activities,
such as the journey to work, are undertaken on a daily basis
while others occur at weekly or bi-weekly intervals.  The
temporal cycles with which various activities are pursued can
be viewed as cyclical frequency functions (one for each
purpose) '... whose amplitude, periodicity, and variance
define both the mean rate of daily trip-making and also the
stability and sequencing of such travel behaviour over time'
((14) p.36). Although Garrison and Worrall (14) examine the
cyclical frequency functions for a number of purposes, they
fail to consider the implications of *combining* the functions
for several purposes.  Curry (11) and Westelius (51) point
out that the cyclical frequency functions for different
activities are likely to covary or to coincide at certain
times.  The result would be the combining of activities on a
single journey (1,11,22,51).

It is important to note that since each activity may have
a unique cyclical frequency function, the possible number of
different sets of activities to be pursued on a given journey
is very large.  That is, there are many possible configurations.  The number and the selection of activities to be
combined on any given journey is related to the temporal
coincidence of the appropriate frequency functions.  It is
also important to note that the parameters of these cyclical
frequency functions are most likely related to the socio-
economic characteristics of the traveller (14), the spatial
arrangement of activity sites, the location of the traveller
*vis-à-vis* the activity sites (51), the availability of
different modes of transport, and such cross-cultural

variables as home inventory levels, which are in turn dependent, for example, on prevailing refrigerator size.

The second familiar concept drawn upon here is that of stereotyped behaviour.  Golledge (18) has suggested that when an individual is new to an environment, he/she searches out a variety of opportunities and visits many different activity sites on an experimental basis.  As the learning process progresses, fewer locations are visited, and in the final stages of the learning process the individual settles into a stereotyped pattern of behaviour, in which a relatively small set of locations is visited repetitively on a regular basis.  This equilibrium stage is characterised by a 'multiple response pattern' in which *several* activity sites or establishments of a given functional class are visited regularly (18).  In other words, stereotyped behaviour does not mean that the individual settles into a routine of visiting only one grocery store, or only one park; several of each of these are likely to be included in the stereotyped travel pattern.  Burnett (6) has documented the development and the existence of this multiple response pattern of stereotyped behaviour.  The notion of stereotyped behaviour is both intriguing and alluring because routine behaviour is readily susceptible to prediction, and because this type of behaviour implies the suitability of deterministic rather than probabilistic modelling techniques.

The proposal here is that multi-purpose travel, when viewed as the outcome of temporally coinciding needs, provides an explanation for the fact that *several* activity sites of a specific class are visited regularly once the individual settles into a routine pattern of behaviour.  It is proposed that the particular destination chosen (within a functional class) depends upon what other purposes (if any) have to be accomplished at a given time.  In this scheme, then, the utility the individual attaches to visiting a given destination varies according to which other destinations need to be visited on the same trip.

Burnett (7) has suggested that spatial-choice models should predict destination choice as the outcome of two processes: (1) the conditional spatial choice process; and (2) the activity sequencing process.  The suggestion here is that the former is very likely to be dependent upon the latter.  While most existing models of destination choice assume temporally stable subjective utility functions and stable preference structures, a more appropriate conceptualisation might be to allow the utility of a destination to vary over time in a cyclical manner; thus the probability of individual i visiting destination j is a function of the probability that destination k will also be visited on the same trip.  Since the latter probability depends on the parameters of the cyclical frequency functions of needs associated with destinations i and j, the probability of visiting destination j is related to the socio-economic characteristics of the traveller, the spatial arrangement of activity sites, the location of the traveller *vis-à-vis* the activity sites, and modal availability.  Of these, the availability of modes is particularly important as this variable, too, is likely to vary cyclically in the (very) short run.

There is one additional comment pertaining to the
conceptual framework within which the modelling of multi-
purpose travel should proceed.   This concerns the question
raised earlier of whether the decision-maker chooses
destinations *serially* as the tour proceeds or simultaneously
selects the *set* of destinations to be contacted.   Marble (34),
for example, had theorised a sequential decision process
whereby the traveller selects the $i+1^{th}$ destination from the
$i^{th}$ destination visited on the tour.   The results of the
Uppsala study (36), however, demonstrate that the vast
majority (95 per cent) of all stops are planned before the
individual leaves home.   This empirical evidence suggests
that travellers select destinations conjointly, and lends
support to the conceptualisation offered above.

Within the general framework proposed, there are numerous
specific questions to be addressed and models to be constructed.   Although the questions have been detailed
previously, it would seem worthwhile to summarise the earlier
discussions by specifically indicating appropriate variables
that future modelling efforts should focus upon.   Key
dependent variables are (1) trip length (e.g. the probability
of undertaking a trip with n links) and (2) linkage strength
(e.g. the probability of combining purpose i and purpose j or
site i and site j on the same trip).   Key independent
variables are (1) the nature of the trip purposes, in
particular, whether they are discretionary or not; (2) the
spatial arrangement of activity sites (e.g. the distance
between destination i and j); (3) the availability of
different travel modes; (4) travel costs; and (5) socio-
economic and locational characteristics of the traveller.
A battery of models employing these variables would extend
our knowledge of multi-purpose travel considerably.


SUMMARY AND CONCLUSION

If planners are to be able to plan successfully, they need to
have the capability of assessing the impact of alternative
proposals.   The relative benefits of different proposals
cannot be evaluated without information concerning how each
proposal would affect the daily activity patterns of
individuals and the viability of selected urban activity
sites.   A thorough understanding of multi-purpose travel is
necessary if such evaluations are to be made.

This chapter has outlined the problems that face the
researcher and planner in attempting to gain this needed
understanding of urban travel linkages.   The contributions
and shortcomings of existing studies have been presented, and
the difficulties of obtaining appropriate data and of
modelling multi-purpose travel have been examined.   A conceptual framework within which future modelling efforts could
be undertaken was proposed, and some suggestions for specific
models made.   The present need is for numerous predictive
studies of urban travel linkages to be undertaken in a
variety of urban environments.   Such studies would begin to
provide planners with the insight required to plan
effectively.

REFERENCES

1. Bacon, R.W., 'An approach to the theory of consumer shopping behaviour', *Urban Studies*, 8, pp. 55-64 (1971).

2. Baker, L.L.H. and Goddard, J.B., 'Inter-sectoral contact flows and office location in Central London', in *Patterns and Processes in Urban and Regional Systems*, A.G. Wilson (ed.), Pion Ltd, London (1972).

3. Berry, B.J.L., 'Commodity flows and the spatial structure of the Indian economy', *Research Paper III*, Department of Geography, University of Chicago (1966).

4. Bishop, W.R. and Brown, E.H., 'An analysis of spatial shopping behaviour', *Journal of Retailing*, 45, pp. 23-30 (1969).

5. Boal, F.W. and Johnson, D.B., 'Non-descript streets', *Traffic Quarterly*, 22, pp. 329-44 (1968).

6. Burnett, P., 'The dimensions of alternatives in spatial choice process', *Geographical Analysis*, 5, pp. 181-204 (1973).

7. Burnett, P., 'Disaggregate behavioral models of travel decisions other than mode choice: a review and contribution to spatial choice theory', in *Behavioral Demand Modelling and Valuation of Travel Time*, Transportation Research Board Special Report 149, National Research Council, Washington, D.C. (1974).

8. Chapin, F.S. Jr. (1974) *Human Activity Patterns in the City: Things People Do in Time and in Space*, John Wiley & Sons, New York.

9. Chicago Area Transportation Study, *Final Report*, 1, Chicago: Western Engraving and Embossing Co. (1959).

10. Cullen, I., Godson, V. and Major, S., 'The structure of activity patterns', in *Patterns and Processes in Urban and Regional Systems*, A.G. Wilson (ed.), Pion Ltd, London (1972).

11. Curry, L., 'The geography of service centres within towns', *Lund Studies in Geography*, Series C 24, 31-53 (1962).

12. Day, R.A., 'Consumer shopping behaviour in a planned urban environment', *Tjidschrift voor Econ. en Soc. Geografie*, 64, pp. 77-85 (1973).

13. Foley, D.L., 'An approach to metropolitan spatial structure', *Explorations into Urban Structure*, University of Pennsylvania Press, Philadelphia (1964).

14. Garrison, W.L. and Worrall, R. (1966) *Monitoring Urban Travel*, Transportation Center, Northwestern University, Evanston, Illinois.

15. Gilbert, G. *et al.*, 'Markov renewal model of linked trip travel behavior', *Transportation Engineering Journal*, 98, pp. 691-704 (1972).

16. Gilbert, G. and Dajani, J.S., 'Energy, urban form and transportation policy', *Transportation Research*, 8, pp. 267-76 (1974).

17. Goddard, J.B., 'Functional regions within the city centre: a study by factor analysis of taxi flows in Central London', *Transactions of the Institute of British Geographers*, 49, pp. 161-82 (1970).

18. Golledge, R.G., 'The geographical relevance of some learning theories', in *Behavioral Problems in Geography*, K.R. Cox and R.G. Golledge (eds), Northwestern University Press, Evanston, Ill. (1969).

19. Hägerstrand, T., 'What about people in regional science?', *Papers of the Regional Science Association*, 24, pp. 7-24 (1970).

20. Hanson, S., 'On assessing individuals' attitudes towards potential travel destinations: a research strategy', *Transportation Research Forum*, 15, pp. 363-70 (1974).

21. Hanson, S., 'The structure of the work trip in intra-urban travel behavior', paper presented at the *NATO Conference on Urban Life*, Munich, Germany (1975).

22. Hanson, S. and Marble, D.F., 'A preliminary topology of urban travel linkages', *East Lakes Geographer*, 7, pp. 49-59 (1971).

23. Hemmens, G.C. (1966) *The Structure of Urban Activity Linkages*, Urban Studies Monograph, Center for Urban and Regional Studies, University of North Carolina, Chapel Hill.

24. Hemmens, G.C., 'Analysis and simulation of urban activity patterns', *Socio-Economic Planning Science*, 4, pp. 53-66 (1970).

25. Hemmens, G.C., 'Simulation of urban activity patterns', mimeo, Department of Urban and Regional Planning, University of North Carolina at Chapel Hill (undated).

26. Hensher, D., 'The structure of journeys and nature of travel patterns', *Environment and Planning*, 8, pp. 655-72 (1976).

27. Horton, F.E. and Shuldiner, P.W., 'The analysis of land use linkages', *Highway Research Record*, 165, pp. 96-107 (1967).

28. Horton, F.E. and Wagner, W.E., 'A Markovian analysis of urban travel behavior: pattern response by socio-economic-occupational groups', *Highway Research Record*, 283, pp. 19-29 (1968).

29. Jones, P.M., 'The analysis and modelling of multi-trip and multi-purpose journeys', Oxford University TSU, *Working Paper No. 6* (1976).

30. Kofoed, J., 'Person movement research: a discussion of concepts', *Papers, Regional Science Association*, 24, pp. 141-56 (1970).

31. Kondo, K., 'Estimation of person trip patterns and modal split', in *Proceedings of the 6th International Symposium on Transportation and Traffic Theory*, D. Buckley (ed.), Sydney, Australia, pp. 715-42 (1974).

32. Lenntorp, B., 'Paths in time-space environments: a time geographic study of movement possibilities of individuals', *Lund Studies in Geography, Series B, Human Geography*, no. 44 (1976).

33. Marble, D.F., 'A simple Markovian model of trip structures in a metropolitan region', *Papers, Regional Science Association*, Western Section, pp. 150-56 (1964).

34. Marble, D.F., 'A theoretical exploration of individual travel behavior', in *Quantitative Geography Part I: Economic and Cultural Topics*, W.L. Garrison and D.F. Marble (eds), Northwestern University Studies in Geography No. 13, Northwestern University Press, Evanston, Ill. (1967).

35. Marble, D.F., Hanson, P. and Hanson, S., 'Intra-urban mobility patterns of elderly households: a Swedish example', *Proceedings of the International Conference on Transportation Research*, Bruges, Belgium, pp. 655-64 (1973).

36. Marble, D.F., Hanson, P. and Hanson, S., 'The Uppsala travel behavior study: design and methodology', *Geografiska Annaler*, Series B (1978).

37. Meier, R. (1962) *A Communications Theory of Urban Growth*, MIT Press, Massachusetts.

38. Mitchell, R.B. and Rapkin, C. (1954) *Urban Traffic:*

*A Function of Land Use*, Columbia University Press, New York.
39. Nystuen, J., 'A theory and simulation of intra-urban travel', in *Quantitative Geography Part I*, W.L. Garrison and D.F. Marble (eds), Studies in Geography No. 13, Northwestern University, Evanston, Illinois (1967).
40. Pipkin, J.S., 'Extension of psychologically-based models of destination choice to multi-purpose travel', paper presented at the *Annual Meetings of the Association of American Geographers*, New York City, April (1976).
41. Rushton, G., 'The scaling of location preferences', in *Behavioral Problems in Geography*, K.R. Cox and R.G. Golledge (eds), Northwestern University Press, Evanston, Illinois (1969).
42. Sasaki, T., 'Estimation of person trip patterns through Markov chains', *Proceedings of the Fifth International Symposium on the Theory of Traffic Flow and Transportation* (1971).
43. Savage, R. and Deutch, K.W., 'A statistical model of the gross analysis of transaction flows', *Econometrica*, 28, pp. 551-72 (1960).
44. Simmons, J., 'Interaction patterns', *Urban Affairs Quarterly*, Dec., pp. 213-32 (1970).
45. Stopher, P.R., 'Predicting travel mode choice for the journey to work', *Traffic Engineering and Control*, 9, pp.436-9 (1968).
46. Thorngren, B., 'How do contact systems affect regional development?', *Environment and Planning*, 2, pp. 409-27 (1970).
47. Tomlinson, J., Bullock, N., Dickens, P., Steadman, P. and Taylor, E., 'A model of students' daily activity patterns', *Environment and Planning*, 5, pp. 231-66 (1973).
48. Törnqvist, G., 'Flows of information and the location of economic activities', *Geografiska Annaler*, 50B, pp. 99-107 (1968).
49. Vidakovic, V., 'Modal mixture of the serial transportation demand: an explorative study', *Proceedings of the International Conference on Transportation Research*, Bruges, Belgium, pp. 679-83 (1973).
50. Vidakovic, V., 'A harmonic series model of the trip chains', *Proceedings of the 6th International Symposium on Transportation and Traffic Theory*, D. Buckley (ed.) Sydney, Australia, pp. 375-86 (1974).
51. Westelius, O. (1973) *The Individual's Way of Choosing Between Alternative Outlets*, Svensk Byggtjanst, Stockholm.
52. Wheeler, J.O., 'Trip purposes and urban activities linkages', *Annals of the American Association of Geographers*, 62, pp. 641-54 (1972).
53. Worrall, R.D., 'A longitudinal analysis of household travel', *Research Report, Transportation Center, Northwestern University*, Evanston, Illinois (undated).

Chapter 4

AN ACTIVITY MODEL AND ITS VALIDATION

Kenzo Kobayashi

INTRODUCTION

The aim of this chapter is to outline an empirical study
which draws its theoretical base from the activity approach
presented in Chapter 2 by Jones.  The background of the
particular model presented below is given in an earlier paper
(15).  The model, originating from the concept of time
budgets as suggested by Chapin (2,3), explains how urban
residents allocate their available time in the fulfilment of
various activities.  Although the choice mechanism does not
at this stage have the analytical rigour of the individual
choice models discussed in much of the present book, the
conceptual framework is more appealing in that the locational
activities are given prominence, and the relationship between
activities in time and space is modelled so as to account for
the relevance of direction and sequencing of journeys in the
travel choice process.  Hensher (11) has recently outlined an
alternative approach consistent with the spirit of the
empirical study presented below, which proposes a redefini-
tion of the operational representation of a journey in order
for the true mechanism of the inter-relationship between
trips, journeys and purposes to be modelled.  The following
definitions are used in the empirical study outlined in this
chapter:

1  A *trip* is a link which connects only two places, origin
   and destination.
2  A *home-based journey* is a composite of trips and visits;
   both origin and destination are home.
3  An *office/school-based journey* is the same as (2) except
   both O and D are office or school.
4  An *office/school-home journey* is an incomplete journey
   originating from office or school and ending at home.  In
   the empirical study described in the text, there were
   very few sample points in this category, but some of the
   home-based journeys were converted into this category as
   explained in the 'trip conversion' section of the
   chapter.

Given the recent evidence that important issues such as
energy restrictions have the effect of increasing trip
linking, car pooling and multi-purpose journeys, then policy
responsive models must have the capability of assessing the
likely rearrangement of travel patterns beyond the traditional travel choices (i.e. mode, route, destination, frequency
and timing).

AN OVERVIEW OF AN ACTIVITY MODEL

An activity in an urban area is defined as an interaction
between human behaviour and the environment and is an
evolutionary process of motivation-choice-activity (2).   In
this choice process, three-dimensional components of time,
space and activities are considered to be important factors.

A serial queueing model is developed as follows.   A route
from an origin area A to a destination area B is depicted in
Figure 4.1.   Starting from the origin (I), an individual
travels using a commuting transport system (1) to the first
destination (II) in the B area.   After some time at II, the
individual leaves for the next destination (III) using one of
the area transport systems (2) in the B area.   The approach
applies for the rest of the journeys, with area transport
used in undertaking trips (3) and (4) and commuter transport
for the trip to the origin (I).   An objective of the model is
to predict the number of visits to the destination places in
the B area and the number of journeys to the B area from the
origin area A.   Since the B area is a destination cluster of
similar functions such as shops, leisure facilities and
offices whose probability characteristics are postulated to
be identical, then Figure 4.1 can be simplified to Figure 4.2
with five queueing stages to represent activity and transport
systems.   The doubly circled queues 1 and 3 show the
activities at the origin and destination area respectively,
and singly circled numbers (2, 4 and 5) represent the transport activities within and between the two areas.

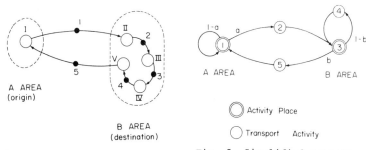

Fig. 1   Original activity model   Fig. 2   Simplified activity
                                              model

A transition probability b from queue 3 to queue 5 is
introduced to explain the phenomena that some users go to
only one destination at the B area but others go to many
destinations.   Then, the transition probability from queue 3
to queue 4 becomes (1 - b) and the expected number of visits
per journey to the destinations at the B area is 1/b, denoted
by b' hereafter.   Then Equation (1) defines the times spent
at the various locations.

$$T_3 = b't_3, \quad T_4 = (b' - 1)t_4, \quad T_1 = a't_1, \quad T_2 = t_2, \quad T_5 = t_5 \qquad (1)$$

where

$t_1$ is the staying time at activity place i,

$T_i$ is the total staying time at i.

In an earlier paper [15] a specially conceptualised cost-effectiveness function (SCEF) was introduced as the central notion in the development of a procedure for calculating the optimal visiting rate to activity locations.  The SCEF is defined in Equation (2).

$$SCEF = n\frac{ET_3}{C_c(T_2 + T_5) + C_aT_4} \tag{2}$$

where

n is the number of journeys attainable for a user within his available time $T_0$,

$T_0$ (= nT) is the total available time for a user each day or week,

T is the total time required for each journey (= $\sum T_i$),

$C_c$, $C_a$ are the trip costs of using the commuting or area transport system respectively per unit time,

E is the trip effect per unit time at the destination area.

An optimal visiting rate ($b'_{opt}$) is obtained by maximising the SCEF, as shown in Equation (3).

$$b'^2_{opt} = \frac{\{C_c(t_2 + t_5) - C_at_4\}(a't_1 + t_2 - t_4 + t_5)}{(t_3 + t_4)C_at_4}, \quad b'_{opt} \geq 1 \tag{3}$$

When data are unavailable to enable a distinction to be made between $C_c$ and $C_a$ and the staying time at the initial activity place is zero ($t_1 = 0$), then Equation (3) can be converted into the more simple form shown by Equation (4).

$$b'_{opt} = \frac{(t_2 + t_5 - t_4)}{\sqrt{(t_3 + t_4)t_4}} \quad \text{and} \quad b'_{opt} \geq 1. \tag{4}$$

Using this equation, an optimal trip pattern of a user under various transport circumstances can be calculated.  Six cases were examined numerically, based on tentative assumptions about the allocation of time to commuter travel, involvement in destination activities, and travel on the area transport system (see [15]).

THE EMPIRICAL VALIDATION STUDY

A data set was collected to identify the utilisation of time for various mixes of journey purposes during certain days in October 1976.  The main categories of journey purpose at the point of contact with a respondent were chosen to be daily shopping, non-daily shopping and business.  The interviews were conducted between 12 noon and 5 p.m. in order to

increase the possibility of selecting individuals who were on
the return trip of their journey and hence would be in a
position to provide reliable details on trips completed,
rather than have to predict details of trips not yet completed (or even decided upon).  Interviews were terminated at
5 p.m. to ensure that the number of sampled commuters was
minimised.

Since most non-daily shopping journeys in Japan take place
on Sunday, interviews were conducted for this journey purpose
on Sunday, 10 October.  The Umeda Terminal in central Osaka
was selected as the survey location because of the concentration of transport system interchanges (Figure 4.3), and the
high likelihood of a sample of such shoppers undertaking
multi-purpose journeys.

A weekday (Thursday, 7 October) was selected for a survey
of daily shopping journeys.  The study was conducted in
central Takatsuki city, 30 minutes travel from central Osaka
(Figure 4.4).  The business journey used the same sampling
frame as the non-daily shopping study, except a weekday
(Friday, 8 October) was chosen.

The interviews were conducted at the time of trip-making,
and hence the amount of information obtained was limited.
Emphasis was placed on the allocation of time, although a
number of other socio-economic characteristics were noted
(sex, age and occupation).  All participants were asked to
trace through the trips undertaken, the activities undertaken
(and their location) at each destination (both intermediate
and final) and the time associated with each trip and
activity (activity place codes were 1 = office or school
places; 2 = shopping places; 3 = eating and pleasure places;
4 = private duties and business places; 5 = home).  The
details of time allocation on the survey day were complemented by information on the number of journeys (by type) and
activities (by type) during the last month.  Random samples
were drawn from selected survey sites during the survey
period.  On average at each survey point, a person was
interviewed every two minutes.  The sample size and usable
response rate for each day are given in Table 4.1.  Chi-
square tests using the socio-economic variables indicate that
the sample is random (16).

TABLE 4.1   Sample details

|  | Survey day | | | |
|---|---|---|---|---|
|  | Thursday | Friday | Sunday | Total |
| Number of interviews | 473 | 521 | 491 | 1,485 |
| Number of usable interviews | 386 | 356 | 396 | 1,138 |

Each respondent was classified according to main journey
purpose (the three categories given above), sex and

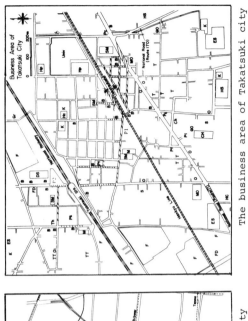

The business area of Takatsuki city

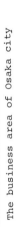

The business area of Osaka city

Symbols:

| | | | |
|---|---|---|---|
| B | Bank | HS | High School |
| Ch | Church | K | Kindergarten |
| CH | City Hall | L | Library |
| DS | Department-store | M | Market |
| ES | Elementary school | MO | Municipal Office |
| F | Factory | NP | News Paper Office |
| FD | Fire Department | O | Office |
| GO | Government Office | Pc | Police |
| Gr | Ground for Sports | Pk | Park |
| H | Hotel | Pt | Post Office |
| HC | Health Centre | S | Shrine |
| Hp | Hospital | SM | Super Market |
| | | ST | Stock Market |

| | |
|---|---|
| T | Temple |
| Th | Theatre or Hall |
| TT | Telephone and Tele-gramme office |
| Univ | University |
| ⋕ | Survey was imple-mented at this point. |
| P1 | Small observation is made (place 1) |
| P2 | place 2. |

employment status.   The particular segmentation adopted in
each empirical calculation is given in the tables in the
following section.   The activity model is primarily concerned
with the prediction of journeys associated with the chosen
journey purpose(s).   Hence although regular journeys such as
the commuter journey are not of concern, if a stopover on the
way home occurs (say for shopping), then the observation is
included in the model.

The central analytical unit is the journey, and hence the
data must be coded to accommodate the journey in the
empirical assessment of time allocation.   Activity place
codes are used in coding the sequencing of activities and
journeys.   An example of a place trace code is '5 1 3 4 5',
which represents a journey from home (5) to office (1), to
leisure (3), to private business (4) and home (5).   This
information on place trace is associated with a time trace.
For example, '30 (150) 20 (50) 15 (40) 30' represents 30
minutes travel time from the home (5) to the office (1), 150
minutes at the office place, then 20 minutes travel time from
the office to the leisure activity location (3) where the
individual stayed for 50 minutes; then a 15-minute trip to
complete private business, the latter taking 40 minutes and
finally the individual's trip to his final destination (5)
lasting 30 minutes.

Journeys were simplified in order to concentrate on the
three main purpose categories.   In the example above, the
commuter component (5 at the beginning of the place trace
code, and 30 (150) for the time trace code) was eliminated.
The model has been developed (at this stage) to explain
optimal trip rates for homogeneous trip purposes, and hence
it is assumed that the destination (or B) area contains trip
purposes that are relatively homogeneous.   Activity place
codes 2, 3 and 4 are assumed to represent an homogeneous trip
purpose.   The distribution of trip times, times at the
activity locations, and the frequency of visits for each
journey purpose are summarised in Figure 4.5.

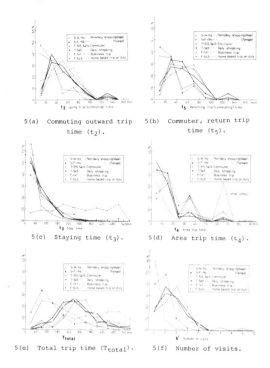

5(a)   Commuting outward trip
       time ($t_2$).

5(b)   Commuter, return trip
       time ($t_5$).

5(c)   Staying time ($t_3$).

5(d)   Area trip time ($t_4$).

5(e)   Total trip time ($T_{total}$).

5(f)   Number of visits.

Fig. 5   Distribution of trip times, times at activity locations,
         and number of visits for various journey purposes

THE VALIDATION TEST

The validation test involves a comparison of the actual
visiting rate ($b'_{act}$) and the calculated visiting rate ($b'_{cal}$),
using the activity data, $t_2$, $t_3$, $t_4$ and $t_5$.  $t_2$ and $t_5$ are
the times associated with use of the commuting transport
system, $t_3$ is the time associated with activity places, and
$t_4$ is the time spent using the area transport system.  The
basic data for each consumer segment are summarised in
Tables 4.2 and 4.3.  The actual visiting rate for the $i$th
segment is calculated as Equation (5).

$$b'_{act_i} = \sum_{j=1}^{n_i} v_j/n_i \qquad (5)$$

where $n_i$ is the number of individuals in the $i$th segment,
      $v_j$ is the number of places visited by $j$th individual
          in the $i$th segment.

TABLE 2   SOME BASIC DETAILS OF THE DATA SET

| | Pattern* | Explanation | No. of observations | $b'_{act}$ | $b'_{cal}$ | $|\Delta b'|$ |
|---|---|---|---|---|---|---|
| Sunday (S) | "S-M-Ns" | Male, non-student | 131 | 2.60 | 2.52 | 0.08 |
| | "S-F-Ns" | Female, non-student | 123 | 2.54 | 2.58 | -0.04 |
| | "S-M-Sd" | Male, student | 37 | 3.00 | 2.15 | 0.85 |
| | "S-F-Sd" | Female, student | 38 | 2.76 | 2.30 | 0.46 |
| | "S-la5" | Office to home journey | 29 | 1.45 | 2.08 | -0.63 |
| | "S-la1" | Office based journey | 14 | 2.00 | 2.05 | -0.05 |
| | "S-ala" | Office worker | 12 | 1.33 | 1.18 | 0.15 |
| | (Sub-group) | | (384) | (2.49) | (2.39) | (0.10) |
| Thursday (T) | "T-la5" | Office to home journey | 87 | 1.28 | 1.00 | 0.28 |
| | "T-515" or "T-5al5" | Commuters | 76 | 1.12 | 1.00 | 0.12 |
| | "T-141" | Office based journey on business | 39 | 1.77 | (0.65)+ | 0.36 |
| | "T-545" | Home based journey on business | 48 | 1.75 | 1.00 | 0.75 |
| | "T-5a5" | Home based daily shopping journey | 118 | 1.35 | (0.91)+ 1.32 | 0.03 |
| | (Sub-group) | | (368) | (1.38) | (1.08) | (0.30) |
| Friday (F) | "F-la5" | Office to home journey | 100 | 1.66 | 1.35 | 0.31 |
| | "F-515" or "F-5al5" | Commuters | 38 | 1.34 | 1.04 | 0.30 |
| | "F-141" | Office based journey on business | 85 | 2.59 | 1.79 | 0.80 |
| | "F-545" | Home based journey on business | 50 | 2.68 | 1.68 | 1.00 |
| | "F-5a5" | Home based non-daily shopping journey | 55 | 1.98 | 1.67 | 0.31 |
| | (Sub-group) | | (328) | (2.07) | (1.55) | (0.52) |
| Special | "T-F-1a1" | Office based non-compulsory journey | 32 | 1.03 | 1.00 (0.23)+ | 0.03 |
| | "others" | More than 3 hours journey (one way) or incomplete trip data | 26 | 1.54 | 1.25 | 0.29 |

+ b' cannot be less than unity.   See equation 4.

TABLE 3   MEASURES OF CENTRAL TENDENCY ASSOCIATED WITH TRIP AND ACTIVITY PLACE TIMES

| Pattern | $\bar{t}_2$ | $\sigma t_2$ | $\bar{t}_5$ | $\sigma t_5$ | $\bar{t}_3$ | $\sigma t_3$ | $\bar{t}_4$ | $\sigma t_4$ | $T_{total}$ |
|---|---|---|---|---|---|---|---|---|---|
| S-M-Ns | 44.8 | 25.2 | 45.1 | 24.8 | 66.1 | 63.5 | 12.2 | 12.4 | 280.9 |
| S-F-Ns | 45.4 | 26.0 | 46.2 | 26.4 | 66.7 | 53.8 | 12.0 | 13.6 | 279.9 |
| S-M-Sd | 37.7 | 19.6 | 42.8 | 22.3 | 70.1 | 60.8 | 12.2 | 16.1 | 315.4 |
| S-F-Sd | 38.0 | 19.6 | 39.9 | 22.4 | 62.0 | 54.8 | 11.4 | 15.4 | 269.3 |
| S-la5 | 28.1 | 29.9 | 47.8 | 31.4 | 60.2 | 80.4 | 12.7 | 10.3 | 168.8 |
| S-la1 | 18.9 | 15.8 | 19.6 | 11.5 | 23.0 | 13.5 | 7.5 | 3.8 | 92.1 |
| S-ala | 38.8 | 22.9 | 39.2 | 22.3 | 131.6 | 156.8 | 17.5 | 11.9 | 259.2 |
| (Sub-group) | (41.2) | | (43.8) | | (65.9) | | (12.0) | | (266.7) |
| T-la5 | 26.6 | 24.5 | 34.6 | 21.8 | 52.8 | 57.9 | 21.5 | 19.1 | 134.1 |
| T-515 | 44.1 | 23.2 | 46.9 | 22.2 | 272.7 | 155.5 | 29.4 | 24.9 | 399.5 |
| T-141 | 21.5 | 17.0 | 24.8 | 16.3 | 41.5 | 37.3 | 11.5 | 7.9 | 128.6 |
| T-545 | 27.7 | 19.9 | 28.4 | 16.6 | 91.1 | 101.7 | 17.1 | 12.5 | 228.3 |
| T-5a5 | 23.9 | 16.9 | 25.3 | 18.3 | 67.9 | 65.5 | 10.8 | 10.0 | 144.5 |
| (Sub-group) | (29.0) | | (32.3) | | (99.1) | | (15.6) | | (204.0) |
| F-la5 | 27.3 | 18.7 | 48.8 | 23.7 | 78.7 | 69.7 | 18.6 | 18.7 | 219.0 |
| F-515 | 50.4 | 25.9 | 50.4 | 27.5 | 254.1 | 193.7 | 21.2 | 25.3 | 449.1 |
| F-141 | 29.5 | 25.8 | 30.5 | 23.9 | 42.5 | 51.2 | 12.7 | 10.8 | 190.1 |
| F-545 | 48.4 | 25.3 | 47.4 | 21.9 | 73.3 | 75.2 | 21.0 | 16.8 | 327.5 |
| F-5a5 | 38.2 | 20.5 | 41.2 | 22.7 | 80.2 | 64.5 | 15.4 | 15.4 | 253.5 |
| (Sub-group) | (35.6) | | (42.8) | | (79.3) | | (16.5) | | (260.5) |
| T-F-1a1 | 18.6 | 21.1 | 15.6 | 19.2 | 38.2 | 11.6 | 25.0 | -- | 75.6 |
| others | 164.6 | | 157.2 | | 120.5 | 83.4 | 115.4 | 104.5 | 537.1 |

* Notation: S = Sunday, T = Thursday, F = Friday, M = Male, F = Female, Ns = non-student, Sd = student, a = anywhere.

The calculated number of visits is given by Equation (6).

$$b'_{cal_i} = \frac{(\overline{t}_{2_i} + \overline{t}_{5_i} - \overline{t}_{4_i})}{\sqrt{(\overline{t}_{3_i} + \overline{t}_{4_i}) \cdot \overline{t}_{4_i}}} \quad \text{and} \quad b'_{cal_i} \geq 1. \tag{6}$$

where $\overline{t}_{k_i}$ is the average required time at the $k^{th}$ activity place for $i^{th}$ segment.

The variance associated with the mean number of visits is obtained by rewriting ($b'_{cal}$) as Equation (7).

$$b'_{cal} = \frac{1}{\sqrt{1 + \overline{t}_3/\overline{t}_4}} \left( \frac{\overline{t}_2 + \overline{t}_5}{\overline{t}_4} - 1 \right) \tag{7}$$

and

$$\text{var}(b') = \frac{1}{1 + \overline{t}_3/\overline{t}_4} \cdot \text{var}\left( \frac{\overline{t}_2 + \overline{t}_5}{\overline{t}_4} \right) \tag{8}$$

This calculation can be simplified by assuming that the variance of $(\overline{t}_3/\overline{t}_4)$ is small relative to its mean value, and var(b') is dependent only on the variance of $(\overline{t}_2 + \overline{t}_5)/\overline{t}_4$, and $(\overline{t}_2 + \overline{t}_5)$ and $\overline{t}_4$ are independent. It can be shown, thus, that ([14] p.232)

$$\text{var}\left( \frac{\overline{t}_2 + \overline{t}_5}{\overline{t}_4} \right) = \left\{ \frac{E(\overline{t}_2 + \overline{t}_5)}{E(\overline{t}_4)} \right\}^2 \left\{ \frac{\text{var}(\overline{t}_2 + \overline{t}_5)}{E^2(\overline{t}_2 + \overline{t}_5)} + \frac{\text{var}(\overline{t}_4)}{E^2(\overline{t}_4)} \right\} \tag{9}$$

The final form of Equation (9) is given by Equation (10).

$$\text{var}(b') = \frac{1}{1 + \overline{t}_3/\overline{t}_4} \left\{ \frac{\text{var}(\overline{t}_2 + \overline{t}_5)}{E^2(\overline{t}_4)} + \frac{\text{var}(\overline{t}_4)}{\{E(\overline{t}_2) + E(\overline{t}_5)\}^2} \right\} \tag{10}$$

The relationship between the average number of visits (b') predicted from the model, and that present in the population is given in Table 4.4 for various values of ($b'_{act} - b'_{cal}$). Approximately 75 per cent of the individual observations are predicted by the model to within 20 per cent of the actual value of b'.

The best agreement (i.e. minimum $|\Delta b'|$) occurs with non-student shopping journeys on a Sunday and home-based daily shopping journeys. $|\Delta b'|$ is less than 0.1. Commuter and office-to-home journeys produce reasonable agreements, with

TABLE 4.4   Predictive Capability of Model

| $|\Delta b'|$ | Number of individuals | Percentage of total population (%) (1,138) |
|---|---|---|
| $\leq 0.1$ | 418 | 37 % |
| $\leq 0.2$ | 506 | 44 % |
| $\leq 0.3$ | 657 | 58 % |
| $\leq 0.4$ | 851 | 75 % |
| $\leq 0.5$ | 889 | 78 % |

$|\Delta b'|$ less than 0.3. Home-based non-daily shopping journeys and office-based non-compulsory journeys come close to 0.3. The least agreement occurs with journey patterns F-141, F-545, T-545 and T-141. This can be partially explained by the compulsory nature of such journeys, according to the dictates of an employer; and hence sub-optimality from the travellers' viewpoint might be expected. Most of the levels of good and bad agreement can be rationalised.

The variances associated with b' are given in Table 4.4 for each segment, as calculated from the actual data. The variance calculated from the model (Equation 10) is also shown in Table 4.5 and in Figure 4.6. A comparison of the two variances $\sigma_{act}$ and $\sigma_{cal}$ indicated their closeness for most segments except journey patterns T-5a5, S-F-Ns, S-1a1, S-1a5, and T-F-1a1. These five journey patterns comprise 28 per cent of the sample. Overall, the assessment of the model using the two validation tests seems to provide encouraging support for its appropriateness in predicting $b'_{act}$.

The correlations between the actual number of visits and the calculated number of visits for each segment and for the total sample are shown in Figure 4.7.* The overall correlation coefficient is 0.84; which suggests that the calculated optimal visiting rate ($b'_{cal}$) is related in a strong positive way to the actual visiting rate ($b'_{act}$). At least for non-compulsory journeys, we can conclude that the model has been validated by real-world data.

---

*The correlation coefficient (r) is given by the formula

$$r = \frac{\sum n_i (b'_{act_i} - b'_{act})(b'_{cal_i} - b'_{cal})}{\sqrt{\{\sum n_i (b'_{act_i} - b'_{act})^2\}\{\sum n_i (b'_{cal_i} - b'_{cal})^2\}}}$$

where

$b'_{act}$ is the mean of $b'_{act_i}$ and $b'_{cal}$ is the mean of $b'_{cal_i}$.

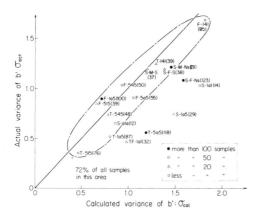

Fig. 6   Configuration of model agreement: $\sigma_{cal}$ vs. $\sigma_{act}$

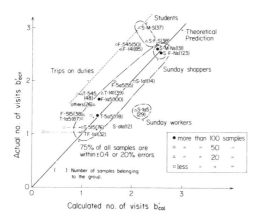

Fig. 7   Configuration of model agreement: $b'_{cal}$ vs. $b'_{act}$

TABLE 4.5   The variance and coefficient of variation
associated with b' (actual and calculated)

| Pattern | act (min) | $C_{act}$ ($= \sigma_{act}/b'_{act}$) | $\sigma_{cal}$ (min) | $t_2+t_5$ (min) | $\sigma_{t_2+t_5}$ (min) |
|---|---|---|---|---|---|
| 'S-M-Ns' | 1.22 | 0.47 | 1.47 | 89.89 | 45.17 |
| 'S-F-Ns' | 1.09 | 0.43 | 1.59 | 91.54 | 49.04 |
| 'S-M-Sd' | 1.20 | 0.40 | 1.21 | 80.54 | 38.35 |
| 'S-F-Sd' | 1.20 | 0.43 | 1.40 | 77.89 | 40.36 |
| 'S-1a5' | 0.74 | 0.51 | 1.49 | 75.86 | 45.36 |
| 'S-1a1' | 1.04 | 0.52 | 1.77 | 38.57 | 26.71 |
| 'S-a1a' | 0.65 | 0.49 | 0.87 | 77.92 | 44.39 |
| (Subgroup) | | | | | |
| 'T-1a5' | 0.52 | 0.41 | 0.82 | 60.84 | 32.05 |
| 'T-515' or 'T-5a15' | 0.36 | 0.32 | 0.47 | 91.05 | 43.74 |
| 'T-141' | 1.25 | 0.71 | 1.31 | 46.31 | 32.17 |
| 'T-545' | 0.73 | 0.42 | 0.78 | 56.15 | 33.35 |
| 'T-5a5' | 0.56 | 0.41 | 1.18 | 49.26 | 34.48 |
| (Subgroup) | | | | | |
| 'F-1a5' | 0.90 | 0.54 | 0.72 | 76.00 | 30.39 |
| 'F-515' or 'F-5a15' | 0.85 | 0.63 | 0.67 | 100.79 | 50.80 |
| 'F-141' | 1.69 | 0.65 | 1.80 | 60.00 | 47.62 |
| 'F-545' | 1.04 | 0.39 | 0.94 | 95.88 | 41.49 |
| 'F-5a5' | 0.91 | 0.46 | 1.06 | 79.36 | 40.56 |
| (Subgroup) | | | | | |
| 'T-F-1a1' | 0.47 | 0.46 | 0.99 | 34.22 | 39.23 |
| 'others' | 1.07 | 0.69 | | | |

DISCUSSION AND CONCLUSIONS

A theoretical model based on assumptions about human
behaviour has been developed in terms of the time required
for each activity, which includes both transport and other
spatial activities.   Within the limits  of data availability,
the model has been validated in terms of the average and
variance of the number of visits using real data.   As this is

a preliminary study of validation, there remain some untested points some of which were proposed in the original paper (15).

1 *The inter-dependence between the number of visits (b')
and number of journeys (n)*. In the original paper the number of visits (b') in a journey and the number of journeys n per day (or week) were related as $n = T_0/T$, where T is the total time for a journey and can be calculated by given $t_i$'s and b', and $T_0$ the total available time for the specified purpose per day (or week). Let us assume that $T_0$ includes the free time available per day, and n the number of journeys requested to complete, say, shopping commitments. In this case, the number of visits (b') and the number of journeys (n) can be related; however validation of this complication needs to be undertaken in future studies because of the inadequacies of available data.

2 *Applicability of the model for various values of $t_i$*. Although the theoretical model does not place limits on the values of the $t_i$'s, the real world data collected identified a relatively limited range of values. In Table 4.3, the average values of $t_2$ and $t_5$ range from 15 to 50 minutes, $t_4$ ranges from 7 to 25 minutes, and $t_3$ (staying time) ranges from 23 to 270 minutes. Average total time varies between 75 and 450 minutes. A knowledge of these average variations is useful in policy formulation, because it provides some idea of the room for adjustment in the rearrangement of the allocation of time. There appears to be plenty of scope for time re-allocation.

3 *Method of data acquisition*. The preferred method of collecting data on time and space utilisation is the 'diary method' (see Chapter 2 by Jones). Travel diaries involve a home location survey and are thus relatively more expensive, even if more detailed in information. With efficient and effective sampling, a substantial amount of data can be obtained by other procedures, such as that used in the present study. An analysis of the present data suggests that using the relatively less expensive method may produce a bias effect for the number of visits, but not for the visit duration time.

4 *Location and mode selection*. A major limitation of the present form of the model is its inability to predict the selection of activity location; for example the selection between two shopping locations. A possible selection rule is a 'Q' index defined as the product of $b'_{opt}$ and population size (16).

5 *The effect of technological and social change*. Changes in life-styles through technological progress and social position adjustments can be allowed for in the model. They are usually reflected in changes in residential location, employment location, and in the amount of time allocated to various types of visits. For example, the introduction of a new regional shopping centre ((11) p. 669) near the work place will show up as a decrease in travel time on the area transport system, which will tend to increase the number of visits per journey, and decrease the number of journeys per given time interval.

6 *Multi-purpose journeys*. The present version of the model assumes homogeneous visit purposes. It also assumes that the distribution of times spent at each visit location is an exponential distribution. It is however not really

appropriate in future studies to aggregate trip purposes as was done in the present series of validation tests. Further research is required before the model can handle a mix of trip purposes in a journey.


ACKNOWLEDGEMENTS

The author wishes to express his indebtedness to Dr David Hensher who spent a considerable amount of time assisting in the rewrite of this final version. He improved both the English presentation and the substantive content. Cordial thanks are given to Professor Yasushi Ishii of the University of Tokyo, Dr Fumio Baba, Dr Sadao Shimoji, Dr Keishi Kawamo of Mitsubishi Electric Corporation, and Dr Velibor Vidakovic who provided many useful comments on earlier versions of the paper.


REFERENCES

1. Bauer, H.J., 'User attitude surveys and transportation system development', *Automotive Engineering Congress*, SAE-720176 (1972).

2. Chapin, J.S., 'Activity systems and urban structure', *Journal of American Institute of Planners*, 34, pp. 11-18 (1968).

3. Chapin, J.S., 'Research report: free time activities and quality of life', *Journal of American Institute of Planners*, 37, pp. 411-17 (1971).

4. Clapp, R.G., Holligan, P.E., Jain, S.C. and Lippke, B.R., 'Quantifying the passenger selection process for accurate market projections', *Transportation Research*, 6, pp. 275-85 (1972).

5. Constantino, D.P., 'Attributes of transit demand', *Traffic Quarterly*, 29, pp. 243-57 (1975).

6. Demetsky, M.J. and Hoel, L.A., 'Modal demand: a user perception model', *Transportation Research*, 6, pp. 293-308 (1972).

7. de Neufville, R. and Stafford, J.H., 'The identification of transport demand', *ASCE National Meeting on Transportation Engineering*, Preprint 1261 (1970).

8. Golob, F., Canty, E. and Gustafson, R.L., 'An analysis of consumer preferences for a public transportation system', *Transportation Research*, 6, pp. 81-102 (1972).

9. Hartgen, D.T., 'Attitudinal and situational variables influencing urban mode choice: some empirical findings', *Transportation*, 3, pp. 377-392 (1974).

10. Hemmens, G.C., 'Analysis and simulation of urban activity patterns', *Socio-Economic Planning Sciences*, 4, pp. 53-66 (1970).

11. Hensher, D.A., 'The structure of journeys and nature of travel patterns', *Environment and Planning*, A 8, pp. 655-72 (1976a).

12. Hensher, D.A., 'Market segmentation as a mechanism in allowing for variability of traveller behaviour', *Transportation*, 5, pp. 257-84 (1976b).

13. Keller, W.F., 'A preliminary evaluation model for personal rapid transit', *Intersociety Conference on Transportation*, ASME, 73-ICT-66, 9 (1973).

14. Kendall, M.G. and Stuart, A. (1969) *The Advanced Theory of Statistics*, vol. 1, Charles Griffin Co. Ltd, London.

15. Kobayashi, K., 'An activity model: a demand model for transportation', *Transportation Research*, 10, pp. 105-10 (1976).

16. Kobayashi, K., 'An activity model and its validation. Utilization survey of urban facilities, *Research Laboratory Report* DRR77-025, Mitsubishi Electric Corporation (1977).

17. Kobayashi, K., Aoki, Y. and Tani, A., 'A method for evaluating urban transportation planning in terms of user benefit', *Transportation Research*, 9, pp. 67-79 (1975).

18. Lave, C.A., 'A behavioral approach to modal split forecasting', *Transportation Research*, 3, pp. 463-80 (1969).

19. Meyburg, A.H. and Stopher, P.R., 'Aggregate and disaggregate travel demand models', *Transportation Engineering Journal, Proceedings*, ASCE, 101, pp. 237-45 (1975).

20. Nicolaidis, G.C. and Dobson, R., 'Disaggregated perceptions and preferences in transportation planning', *Transportation Research*, 9, pp. 279-95 (1975).

21. Quandt, R.E. (ed.) (1970) *The Demand for Travel: Theory Theory and Measurement*, Heath Lexington Books, Lexington, Massachusetts.

22. Uematsu, T., 'Statistical analysis of the commuter's preference among available routes', *Annals of Institute of Statistics and Mathematics*, 13, pp. 93-134 (in Japanese) (1965).

23. Vickerman, R., 'The demand for non-work travel', *Journal of Transport Economics and Policy*, 6, pp. 176-210 (1972).

24. Wohl, M., 'A methodology for forecasting peak and off-peak travel volumes', *Highway Research Record*, no. 322, pp. 183-219 (1970).

Chapter 5

NEW APPROACHES TO UNDERSTANDING TRAVELLER BEHAVIOUR

K. Patricia Burnett and Nigel J. Thrift

CHARGE TO WORKSHOP

Approaches to understanding traveller behaviour have been
proposed which pay particular attention to the derived nature
of travel. With an emphasis on the activities undertaken in
time and space and action constraints, rather than on travel
per se, more realistic models of the individual behavioural
process may be formulated. This workshop should appraise the
present state of knowledge on new approaches and propose a
course of action for future research and application of those
approaches with promise.

SCOPE OF WORKSHOP

Given the objective of exploration of alternative approaches
to urban travel behaviour, the workshop did not pursue
modifications, extension and applications of current choice
theories from economics and psychology to associated mathematical models of travel demand. Accordingly, there was
little debate over the relative merits of, for instance, the
'strict' utility framework and the multinomial logit model or
information integration theory and its varying algebraic
formulations. Similarly, interesting questions concerning
mathematical specification, calibration, testing and application of the currently proliferating choice-based models were
laid aside. Instead, the continued experimentation with
choice-based models, and the difficult questions this has
raised, was interpreted as indicating the need for caution
with regard to the efficacy of such models and not just as a
tribute to the vitality of the subject area. The workshop
therefore addressed three principal issues: first, an
assessment of the general defects of choice-based models in
use at present; second, the identification of a complementary
theoretical framework within which to study urban travel
behaviour; and third, the formulation of future directions
for policy application, theoretical development and data
collection under the new rubric.

   The principal defect of current choice-based models is
that they only pay lip service to the notion of travel as a
derived demand. While it is widely recognised that travel is
rarely desired for its own sake (travel per se is of course
generally accorded a negative utility), few models relate the
demand for travel to the biological and social needs it fulfils in a simple, explicit way. Activities generated by

116

biological needs (in Hägerstrand's scheme (43,15) predominantly activities circumscribed by capability constraints)
are eating, sleeping and the like.  Other activities arise
from an individual's social role and location in the physical
environment (in Hägerstrand's scheme activities circumscribed
by authority and coupling constraints) and can therefore be
regarded as socially produced.  Thus, the basic forces which
generate travel are not identified, and the linkages of these
fundamentals with trip timing, length, frequency, mode,
destination and route choices are not understood.  This not
only matters in terms of the level of intellectual satisfaction we feel with our comprehension of important forms of
urban behaviour, but also in terms of the number of policy
questions which current models cannot satisfactorily address.
These questions are spelt out in detail below.

A second defect of the majority of current models is their
assumption of two-stop single-purpose journeys.  There is
increasing evidence that this type of journey represents only
about 60 to 70 per cent of current metropolitan travel (17),
a fact that is emphasised by the inclusion of movement on
foot made on multiple-purpose trips.  Multiple-stop journey
sequences constitute the remainder.  A framework is needed
which will help describe, explain and forecast the allocation
of trips by individuals over a daily sequence of stops.  This
will provide additional insight into spatial and temporal
variations in, and levels of, demand for urban transport
facilities.

A third and related limitation of many present models is
their assumption of static, equilibrium conditions.  For
example, the widely used multinomial logit model of mode or
destination choice assumes there is a single allocation of
trips between choice alternatives by a population group; this
allocation presumably arises from stable individual utility
functions.  The allocation of trips reaches a new equilibrium
following a shift in the socio-economic characteristics of
travellers or a change in mode or destination attributes.  It
is true that several recent models incorporate the notion of
constantly changing utility functions, for example, through
the reassessment of different modes or destinations in the
light of experience or learning (7).  However other factors
which can create temporal variations in utility functions
have usually been left aside, especially daily, weekly,
monthly and other cycles in the combinations of activities
which individuals wish to pursue (17).  These cycles affect
the choice of activity sets over the short run, and thus
choices of the frequency and timing of single- and multiple-
purpose trips of different kinds.  It is very difficult to
incorporate these travel choices within current models.
First, the choice alternatives (activity set, purpose
combination, frequency, timing) cannot easily be conceptualised as having attributes in the way that destinations and
modes do in these models.  Secondly, the list of purposes and
purpose combinations whose frequency and timing must be
predicted is very large, particularly when different
varieties of multiple-purpose shopping trips are included
(16).  The normal strategy of developing a model for each
population segment and each purpose combination becomes
extremely unwieldy.

In addition to short-run fluctuations in utility functions, there are also changes in the longer run which current models cannot easily allow for. Long-run alterations in travel needs or constraints can generate alterations in the individual's daily activity set, and hence in purpose, frequency, timing, mode, destination and route choice. For example, as leisure time or needs increase, or as work places and shopping areas increasingly decentralise, there could be complex changes in trip-purpose combinations and resultant mode and destination choices. Current models cannot handle these possibilities. In sum, what is required is a theoretical framework which will help relate short- and long-run variations in travel behaviour to variables other than population characteristics and transport system attributes.

Finally, current choice models from economics and psychology do not really explain individual travel behaviour. Although often mathematically specified for the individual, in practice such models are commonly used to describe, explain and forecast the behaviour of 'homogeneous' population groups. If behavioural models *are* applied to individual travel choices, then the individual is considered atomistically, that is, out of his/her context as a household member (for example, the application of information integration theory (29)). These two approaches are obviously the result of either aggregation needs, on the one hand, or, on the other, the need to conform to a particular brand of theory developed to describe laboratory responses of persons in very simple choice situations. The rationale for looking at *individual* travel decisions in a *'real-world'* context is of course well understood and need not be repeated here. However, at this point there is an additional reason for probing behaviour at the individual level. This scale of analysis could give some insight into reasons for the 10 per cent errors in forecasting of some current choice models; errors which have been attributed to individual differences within supposedly homogeneous population segments.

One theoretical framework which does not suffer from the limitations described above is the 'human activity' approach. This combines a number of concepts from time-allocation and space-time allocation studies. A review of this burgeoning literature and its application to travel is contained in the chapter in this volume by Jones. However, a simple outline may be given here.

Each individual in a household at the start of each time period (usually a day) selects an activity set, e.g. working, shopping, eating, sleeping. Activities are the outcome of biological needs, social role requirements and the physical realities of the environment. Thus some activities are obligatory, some are discretionary. Realisation of the activity set during a time period involves the selection of a time schedule for each, together with trip purpose(s), frequency, destinations, modes and routes. The travel choices which are normally modelled are therefore viewed as the outcome of the need of individuals to undertake a sequence of activities within the constraints of time and space. Other constraints, such as income and travel costs operate through their effects on the time available for activities and/or the distances over which travel can occur.

Such a simplistic introduction is still able to highlight the necessary conceptual reorientation which the framework provides. Urban travel behaviour is linked to the fulfilment of needs through the formation and accomplishment of activity sets. Travel is explicitly handled as a derived demand. The diurnal allocation of trips may be described and explained; single- and multiple-purpose journey generation may be understood. In addition short-run variations in the demand for urban transport facilities may be related to factors such as diurnal, weekly or monthly changes in activity sets, as well as to changes in learning about destinations, modes or routes. Then again long-run variations in demand may be associated with changes in socially-determined needs, and with changes in time and space constraints through alterations in such things as socio-economic status, technology or the characteristics of the physical environment (such as the locations and other attributes of destinations). Finally, the activity framework focuses on the individual in his/her social context.

Obviously, the activity framework is still a rudimentary state. Acceptance of the theory will depend, at least in part, on the kinds of policy questions it can address which current choice models cannot. In part, too, its acceptance will depend on the extent to which it can spawn more precise methods of defining the linkages between its key variables: needs, activities, constraints and travel choices. The remainder of this chapter is devoted to an exploration of key policy applications, and to suggestions of the possibilities of more rigorous theoretical and empirical research.

## SUMMARY OF RECOMMENDATIONS

*Impact on Policy*

The activity approach should become a focus of research because of its potential to address a broad range of contemporary policy issues which most current behavioural models cannot satisfactorily handle. These issues (some of which are raised in the companion papers in this volume by Hanson and Jones) include:

1   The effects on all travel choices (activities, purposes, timing, frequency, modes, destinations, routes), and thus on the demand for transport facilities, of current national policies which change travel constraints, for example energy conservation policies;
2   The ramifications for individual life-styles and quality of life of specific local or metropolitan transport policies, such as a policy of reduced fares for commuting trips;
3   Interactions between travel behaviour and land uses (the physical environment) in an urban setting, for transport and land-use planning;
4   Suggestions as to future policy to meet desirable, not necessarily legislated, social goals; for example, the kinds of transport facilities necessary to supply travel demands in a society where the goal of complete equality between races

or the sexes is reached, that is where socially-conditioned needs of individuals for activities and movement are radically altered.

Since each of the policy issues above is stated in very general terms, some specific applications of the activity framework are considered next, and the advantages of such a framework over current choice models are illustrated.

*Applications*

*Effects of national policies changing travel constraints.* Important policies which currently affect travel constraints in the industrialised world include energy pricing policies, schemes to redistribute income and alterations to working or school hours. The activity approach can capture the effects of changes in travel costs, disposable income and the timing of obligatory activities on the selection of daily activity sets, the allocation of daily activity sets between household members, the scheduling of the individual's activities, the individual's purpose, mode, destination and route possibilities, and his/her resultant travel selections and, in particular, response to changes in travel opportunity which is resolved by an activity other than travel. For instance truncated travel opportunities could lead to a trip being replaced by a telephone call. In the aggregate this could have ramifications not only for traffic flows but also telephone system loads. The possibility of all these responses is not incorporated in current choice models. The result is that these models cannot help to describe, explain or evaluate such things as the change in the activities of household members and the increase in number of changes in particular kinds of multiple-purpose trips which certain petrol price increases seem to cause (41).

*The ramifications of transport policies for individual life-styles and quality of life.* Transport policies are often designed to obtain a single desired alteration in travel behaviour. For example, alterations in bus headways and routing, low-cost transit schemes, or the development of suburban demand-responsive transit are often advocated as a means of altering urban modal split.

Current behavioural models can describe and possibly predict changes in modal split and other travel choices through alterations in the cost and/or level of service variables incorporated in them (40). However, other possible effects of transport policies cannot be discussed. And yet these may well include alterations in personal daily, weekly or monthly activity sets, activity scheduling, and the destinations, modes and/or routes open to the individual. For example, low-cost commuting trips by transit may cause the principal wage-earner of a one-car household to leave the car at home; this may substantially alter the activity and travel options open to any household members not in the paid labour force. In particular, it may increase the range of possible activities and the number of trip-timing, destina-tion, route and mode choices available to suburban housewives (37,15). Thus a transport policy may have effects on the

life-styles and quality of life of individuals which cannot
be explored using normal modelling frameworks.  It seems
important to take these effects into account when evaluating
transport policies, and the activity approach provides one
means of analysing them which can improve over intuitive
approaches.  In particular it highlights the need for more
multiple-impact transport policies in which it is realised
that not one policy to meet a problem but more often two or
three working in concert are needed for desired goals to be
reached.

But all the ramifications of a particular metropolitan
transport policy need exploring for other reasons.  Many such
policies are framed to meet broader social goals, for
example, mode switch policies can be envisaged as part of an
energy conservation programme.  If all the effects of say, a
transit fare experiment, are not taken into account, the
particular transport policy may well be self-defeating.  In
the case cited above, if the mode switch of principal wage-
earners causes a considerable increase in car travel by other
household members through a corresponding increase in the
number of activity, purpose and destination options, clearly
the energy conservation effects of the new fare policy are
questionable.

This case also illustrates the utility of the framework
for exploring latent travel demand.  The effects of a
particular transport policy on the range of options open to
the individual can be explored.  This includes options for
activities, trip-purpose combinations, trip scheduling,
modes, routes and destinations.  Gaming simulations of
responses to different transport policies (described in
detail in the chapter by Jones (23)) can help especially in
the identification of adjustments in the range of travel
options, together with possible interactions between, and
shifts in all kinds of travel choices.  Certainly there
seems to be no simple way in which current choice models can
identify these impacts.  Gaming simulation may even be helpful in the case of policies with supposedly simple effects,
such as a decision to offer bus rapid transit along a CBD-
suburban artery for the journey to work.  The use of an
activity framework might help identify a population group who
would become commuters only if their daily activity set could
be changed, for example if shopping, often combined with work
(16), can be allocated to some other household member.  Such
a segment could not easily be allowed for using conventional
choice models.

*Interactions between transport and land use.*  Aggregate and
disaggregate modelling of metropolitan land use-transport
interactions is in its infancy.  Yet it is a truism that
understanding these interactions is important for informed
transport and land-use planning.  The activity analysis
framework is useful in accounting for and helping to forecast
first, the viability of locations (the impact of travel on
land use) and secondly, the effects on travel of different
land-use configurations (17).  To illustrate the first case,
the shopping traffic attracted within a given time period to,
say, a declining downtown can be explained partly by activity
scheduling (trip-timing) decisions, partly by trip-purpose

combination decisions, and partly by mode, route·and
destination decisions.  An activity analysis of samples of
households would reveal the relative importance of each, and
thus target decisions for policies to increase trade.  The
application of conventional destination choice models for
shopping trips does not readily permit this, because the full
range of choices affecting viability cannot be readily
included in them.

In addition, because the locations of activity sites may
be specifically included as spatial constraints within the
activity framework, the impacts of different land-use configurations on travel choices can be explored directly ((39)
pp. 211-12,(32)).  In particular, the activity framework
appears to be the only approach which allows the ready
examination of the effects of 'stretching' the current
activity-site linkages generated by single- and multiple-
purpose travel.  Such 'stretching' can be accomplished by
proposing alterations in destination locations, such as the
relative locations of home, work and different shopping and
recreational facilities.  The activity framework permits
separation of pivotal activity sites from those whose
displacement has little effect on travel patterns, a distinction which can be invaluable for planning purposes.

*Future scenarios.*  Finally, application of the activity
framework can help with the development of future scenarios
of urban transport and behaviour.  These scenarios may be
designed to help set, reinforce and reach long-range social
goals.  Because travel demands are viewed within the activity
framework as arising from socially as well as biologically
conditioned needs which operate within time and space
restrictions, some attempt can be made to describe the
effects on travel of long-run changes in constraints and in
socially defined needs.  For example, workers at the
University of Lund have used 'a series of scenarios on future
economic growth and population trends', together with
projected alterations in factors affecting time and space
constraints, 'to work out what daily life will be like in the
year 2000' and 'to calculate life paths (movements in time
and space) of typical individuals at some future date and
even the number of trips that will be made in the selected
region for some representative day at that future date' ((43)
p. 30).  The objective of the travel predictions was to
suggest possible future environmental changes through the
application of public policies in intervening years.  The
proposed social goal which the research attempted to
reinforce was 'to increase freedom of choice', and 'to move
towards an equitable urban environment' ((43)p. 25).

A similar approach may be taken to specifying what the
city would be like and what its travel demands would be if
certain more specific social goals were met - for example,
long-term changes in energy consumption or equality between
the sexes.  The future conditions of institutions influencing
travel needs could first be determined (for example, class
structure, the family, the work force), together with factors
influencing constraints (such as the spatial distribution of
the population, transport technology or car ownership).
Alternative 'travel behaviour' scenarios could then be

delineated according to a series of sets of assumptions
concerning the nature of institutions and constraints.  This
could assist with the evaluation and assessment of priorities
for different social goals in terms of both short- and long-
range policy formation.  Certainly current disaggregate
models have yet to be developed for these tantalising kinds
of application.

*Future Directions for Research*

Given that the human activity framework has potential for the
preceding kinds of policy applications, directions are
recommended for future research in theory and data collection.
The aim of this research is to make the theoretical framework
and its applications more rigorous by isolating those parts
of it which can be expressed in quantitative rather than
qualitative terms.  The recommendations for future research
directions are summarised briefly; the remainder of the
report is then devoted to suggestions as to how these
recommendations might be fulfilled.

*Theoretical questions.*  The first requirement is the
development of *models of the formation of activity sets*, that
is, of the relations between the biologically and socially
conditioned needs of individuals, on the one hand, and the
selection of those activity sets within the mesh of time and
space constraints on the other.  This is necessary for all
the designated areas of policy application, since in each
case activity set selection may be tied to those travel
behaviours - such as trip purposes, timing, lengths, modes,
destinations and routes - which it is desired to forecast.
Only by focusing on the formation of activity sets in the
context of needs and constraints will the derived nature of
travel be clarified for policy purposes.

A second requirement is the development of *models of the
formation of space-time constraints.*  This is especially
necessary for first, the examination of effects on travel of
current national policies influencing income, energy costs
and working hours; secondly, study of latent demand; thirdly,
examination of the effects of metropolitan or local transport
policies which alter constraints (for example, mode pricing
policies); and lastly, the investigation of land-use and
travel-behaviour interactions.  The focus of attention is on
the identification of all factors influencing time and space
constraints and on the specification of how changes in
constraints affect travel choices.  This implies a much
greater emphasis on the specificity of the physical environment in space and time (that is, opening hours, etc.) than
has hitherto been found and of the conditions that the
environment places on individuals' activity patterns.  Of
special interest is the isolation of situations in which
constraints reduce the choice sets of individuals for travel
to one or perhaps to no possible alternatives.  This has
consequences not only in terms of obvious policy applications
but also for the way in which models of travel behaviour may
be formulated.  For some population groups in some situations,
'choice models' may be of limited value simply because
'choice' does not exist.  This can be particularly apposite

in terms of the information available to individuals.   A
large proportion of travellers are so badly informed about
objectively available alternatives that for them a subjective
situation of choice is non-existent.   Thus perceptual aspects
of behaviour can be as important as the constraints that the
physical environment, biological limits to action and
societal rules impose.   Other interesting conceptual
questions raised by examining space-time constraints are
taken up later.

Finally, the development of *models of how activities are
carried out within specified space-time constraints* is
needed.   Such models are especially necessary for the study
of the short-run impacts of specific metropolitan or local
transport policies on all kinds of travel decisions, and for
the study of the effects of existing land-use configurations
on travel behaviour.   Particular questions bearing on model
formulation are as follows: the separability or simultaneity
of activity and trip-purpose choices and other choice
decisions; how to describe the allocation of time to travel
and to the accomplishment of activities at different
destinations; how to model trade-offs in the selection of
destinations over a specified time period for the accomplishment of different trip purposes and purpose combinations;
how to identify the effects of 'stretching' trip linkages;
how to describe the development of stability or patterning of
individual travel behaviour within a given set of constraints.
Suggestions concerning how to answer these questions are
discussed below.

*Data collection questions*.   The application of the human
activity framework requires two kinds of data.   First,
information is required about the individual's travel choices
through time (activities, purposes, timing, modes, destinations, routes, generally over some short-run period).
Secondly, data are necessary concerning the factors governing
the individual's activity needs and his/her space-time
constraints.   Precisely what information is required in each
of these two categories will depend upon which of the
preceding kinds of models are being developed and tested.
This implies an important recommendation: that the activity
framework be fleshed out by letting empirical research be
guided by quantitative model development for designated
policy applications.

Despite the fact that different models will have different
data requirements, several techniques are versatile enough to
gather information on time and space use as manifested in
travel choices under most circumstances.   These include the
well-established travel diary and questionnaire (34) and the
newer gaming simulation device (10).   The advantages and
disadvantages of each technique require further investigation.   Until such time as these are known, it is recommended
that such techniques be applied in conjunction with each
other to cross-check results.

Procedures for collecting information on factors
influencing the individual's activity needs and space-time
constraints are in a more rudimentary state.   Although well-
established techniques exist for eliciting information on

some biological and social determinants of needs (for
example, age or income), the measurement of other possible
determinants (for example, psychological states) is far more
problematical.  Also, while some Swedish work has attempted
to measure factors influencing space-time constraints and
space-time constraints themselves ((43) pp. 25-32), more
research is necessary to provide 'hard data' on the spatial
and temporal situations in which individuals currently
operate and in particular on the physical environment in time
and space.  In addition, more data is needed on long-term
trends in factors governing spatial and temporal constraints
in order to be able to generate long-run predictions of
behaviour.  The absence of data and of well-tested data-
generating procedures in these fundamental areas leads to a
recommendation that considerably more research should be
carried out on data collection for the study of travel
behaviour in an activity framework.


DISCUSSION OF RECOMMENDATIONS

The remainder of this report is devoted to some recent
developments which promise a more rigorous means of fulfilling the recommendations for theoretical and empirical
research.  This helps to justify the call for research on
these specific topics; the general justification of the
development of the activity framework for a wide range of
policy applications has already been dealt with.


*Recommendations for Future Theoretical Research*

*The formation of activity sets.*  The relations between needs
and activity set formation within space-time constraints
have been explored in detail by Chapin (10,11), Fried *et al.*
(13), Hägerstrand *et al.*(15), and Kutter (26).  All four
emphasise the importance of socially-determined roles,
although to different degrees, as a major determinant in the
selection of the individual's activity sets.  Fried *et al.*
spell out some complex relations verbally ((17) pp. 85-95).
However, Kutter following Chapin, suggests that taxonomic
models may be used to explore the associations between
activity sets and selected demographic characteristics
related to roles (sex, stage in life-cycle, educational
attainment, family status, etc.).  The repeated application
of such models could give a solid foundation to the widespread assumption in human activity studies of archetypal or
typical individuals (e.g. employed women).  If social roles
and related activity sets are accepted as a source of travel
demand then the discovery of associations between demographic
characteristics and activity patterns also contributes
information on appropriate market segmentation for current
choice based models.  In addition, workers following
Hägerstrand's time-geographic model of society as a web of
interacting life-lines (for example, Martensson (31)) have
been able to portray graphically the relations between role,
space and time constraints, and the selection of activities
and activity sites.  This work leads into mathematical
simulation of the effects of long-term changes in roles and
constraints on activity selection and travel.  Thus Kutter

has envisaged the possibility of simulating changes in mathematically defined roles, spatial constraints and activity sets over time (27). This work, however, ultimately requires an analysis of the changing social conditions which produce roles and their allocation to individuals, that is, of the changing conditions which produce the changing division of labour by age, sex, race and class. At this point research on travel behaviour as a derived demand merges with inherently ideological speculations on the nature of state capitalist societies and their development over the next twenty years. None the less obvious connections exist between this research, the policy area of scenario generation and forecasting the long-term effects on travel behaviour of policies influencing roles and travel constraints (for example, income redistribution). Thus a rationale for pursuing this kind of future research is easily found.

Thus far, activity sets have been conceptualised as including activities allocated to members of a household, by role, to perform over some specified time period (working, studying, recreation, shopping, etc.). The precise scheduling of these activities over time has not been discussed, for example, the selection on a day-to-day or week-to-week basis of combinations of these activities or subtypes of these activities (shopping for groceries; visiting a theatre). Yet the choice of daily or weekly activity sets has obvious consequences for understanding and forecasting short-run spatial and temporal changes in travel behaviour. Two lines of research are important here (17). One is the fitting of probability distributions to model the frequency of each selected activity through time; the probability of activity combinations at any moment are then given by the joint probability of each activity being undertaken at that time. This approach has two deficiencies. First, the variables influencing the parameters of the probability distribution for an activity still need to be specified so that changes in activity sets can be determined through an assessment of policy impacts on controlling variables. Second, and more damaging, the probability distribution for a single activity over time may not be derived independently of the probability distribution for another activity; the selection of one activity for a day (work) may rule out another (a long drive for recreational purposes). More time devoted to one activity must imply less time devoted to another and since most statistical analyses in this area involve a constant time constraint (e.g. 24 hours), negative relationships between time uses can be expected. This is the 'closure' problem. If it is not taken into account, spurious correlations will result. However, for some activities at some scales of analysis (shopping for clothes and bread), the independence assumption may be true. Theoretical and empirical investigation should therefore be able to demonstrate which sets of activity choices can be modelled using the probability distribution approach, and which types of travel behaviour can consequently be predicted.

The second line of approach concerns the use of semi-Markov processes to model the choice of daily activity sets. Again, its application has been restricted to limited activity sets (different shopping purpose combinations) and it is

difficult to relate the Markov process to variables which
control it.  It seems clear that much ground-breaking
theoretical work needs to be done on the choice of activity
sets over a short-run time period.  This is especially
necessary in order to assess the short-run impacts of
metropolitan transport and land-use policies.

*Models of the formation of space and time constraints*.  The
identification and measurement of all the factors which place
spatial and temporal constraints on movement is fundamental
for the understanding of individual travel and for forecasting the impacts of many policy decisions.  Hand in hand with
the identification and measurement of such factors goes the
development of mathematical models to relate changes in
constraining factors to changes in travel choices.  Little
quantitative work has as yet been undertaken on these
problems, though several qualitative approaches have
suggested hypotheses for mathematical specification and
testing.  Fried *et al*. describe how spatial and temporal
constraints on movement may be related to socio-economic
states ((13) pp. 82,83,97), urban/suburban residential
location ((13) p. 87), 'the objective, environmental opportunity structure' ((13) p. 92), and 'perceptions of space and
of available activity and travel alternatives' ((13) pp. 97-
8).  Little attention is given to the precise mathematical
description, measurement and testing of such relationships
however.  Brög, Heuwinkel and Neumann's (4) formulations are
more impressive.  Factors affecting the individual's travel
options are divided into three categories - spatial and
temporal constraints placed by the physical environment (the
location of activity sites; the location, scheduling and cost
of different transport systems), constraints placed by the
socio-economic status of the individual (car availability,
time and money budgets for travel), and constraints placed by
the psychological disposition of the individual (attitudes
towards and motives for travel).  The operation of these
constraints determine the 'objective' range of options open
for each travel choice at any given time period; the 'real'
range of options, however, depends on information, that is,
on how many options are known and how well they are known.
Groups of individuals face 'typical' sets of real options for
travel choices ('situations'), which generate typical
patterns of travel behaviour.  This model has been successfully applied to the simple case of mathematically describing
modal split in West Germany.  It is clear that the application is also useful for the design of various mode split
policies: the effects on modal choice of altering any
constraint by an appropriate policy can be evaluated for each
population group in each 'situation'.

   Brög *et al*.'s model, however, ignores the complex
interactions that could occur between all travel choices in
response to transport or land-use policies which alter
temporal or spatial constraints.  Lenntorp has used a computer simulation to model some of these interactions: he
discusses the effects of change in route, speed and frequency
constraints of public transit on mode choice, activity
selection and destination choice ((43) pp. 25-31).  Others of
the Lund School have used simple mathematical combinatorial
techniques to model the effects of alternative land-use

configurations on activity site selection for selected
individuals whose paths are simulated one at a time (15,28).
Here the firm belief is that the physical environment is
paramount in dictating activity location once the activity is
chosen, and hence in choice of destination.   Certainly
results from the simulation model would seem to bear this
out.   Thus, it is important to stress that this model uses as
a data input not diaries but 'hard' environmental information
and gives as output the expected space-time budget of each
individual's simulated path.

    But clearly these are limited beginnings.   Many important
questions remain to be answered about space and time constraints themselves before changes in them can be effectively
related to changes in travel behaviour.   What are the
mathematical relations of the objective spatial distribution
of activity sites and the distribution of activity sites
which an individual knows?   Do characteristics, mathematically specifiable configurations of objective or perceived
activity sites exist for individuals in different population
groups (26,27)?   Do individuals optimise or satisfice or
adopt some other strategy when making travel decisions within
time or space constraints?   What criterion do individuals use
if they are optimisers - subjective travel time or objective
time ((4) p. 43)?   The total time of a round trip, where this
may be single or multiple purpose, or the time of individual
trip legs (17,21)?   What are the relations between subjective
and objective time (9)?   Are there violable and inviolable
constraints for different groups?   Are there thresholds of
change in some constraints (for example, gasoline costs)
which must be reached before alterations in travel choices
occur?   Do these thresholds vary for different population
groups?   The answers to most of these questions will have
implications not only for the modelling of travel within the
activity framework, but also for other travel models.   For
example, the mathematical specification of a characteristic
spatial configuration of activity sites for different population groups will help solve the problem of specifying
destination choice sets for some mathematical logit models
((6) pp. 213-14).   In addition, the question of whether
travel time is optimised or not and of what kind of time is
optimised has an obvious bearing on the formulation of
optimisation models of urban travel flows and facility
location (23).   Quite clearly the modelling of the formation
of space and time constraints and their relation to travel
could be a rewarding direction for future research.

*Models of behaviour within space-time constraints*.   The
previous section focused on the definition of the space and
time constraints within which travel behaviour occurs.   It
also discussed the identification of alterations in travel as
constraints change.   However, it will not be possible to
assess all the impacts of those policies which change constraints before an understanding is gained of how all kinds
of travel choices are made, given some set of space and time
restrictions.   A number of models have been developed which
illustrate how different facets of this question might be
pursued.

One of the first problems which must be addressed is the separability or simultaneity of activity and trip purpose choices and other travel decisions ((17) pp. 35-6). Thus far, it has been assumed for modelling purposes that activity sets and trip-purpose selections are always made prior to other travel decisions. It is obvious that this is not necessarily the case: this situation calls for the development and testing of models which predict all kinds of travel choices under different assumptions as regards their separability and simultaneity. Ben-Akiva (2) has pioneered models which handle traditional travel choices in this way (for mode and destination especially). It is doubtful, however, whether his utility-based model structures can be extended to activity and purpose selection. It is difficult to conceptualise activity and purpose as having attributes to which utilities can be attached for modelling purposes. Other model structures may need to be developed. Ben-Akiva's work has shown the importance of developing such models, since 'correct' assumptions about separability and simultaneity improve 'goodness of fit' and forecasting accuracy. However, as Williams (46) has shown, there is good reason to doubt that the structure of 'simultaneous' demand models as used at present is consistent with the utility function assumed to generate these models.

Another question which needs to be investigated is that of timing of activities at a given set of activity sites. Very little is known as yet about trip scheduling. Kobayashi (24, 25) illustrates how trip scheduling may be explored within the activity framework by using queueing theory. Although the assumptions of the model are unrealistic, it is clear that serial queue models can be applied not only to predict the duration of activities at destinations, but also 'the likelihood of multipurpose journeys and the number of activities per journey' ((24) p. 106). A disadvantage of the model is that it seems to be applicable only in very simple choice situations, for example travel choices linking home and CBD.

By far the greatest effort so far has gone into the modelling of trip linkages on multi-purpose trips. Hanson reviews most of the relevant literature in the paper in this volume. However, several additional comments about future possibilities in this area may be mentioned. First, little is known about the trade-offs which are made when destination chains are selected on multi-purpose travel. What are the relevant travel costs and benefits for each leg of the journey and how are they balanced against each other when a multiple-purpose trip is planned? An answer to this question is necessary in order to be able to identify the proper dependent variables for destination choice models for multi-purpose trips. The application of a conjoint measurement model might be helpful here (8). Secondly, some means must be found of isolating 'pivotal' from 'non-pivotal' activities on multi-purpose travel, that is, those activities whose displacement would greatly affect formation of trip chains, and those whose displacement would not (38). Markov and semi-Markov chains have been used for this purpose, but their disadvantages are now believed to outweigh their advantages (see the extensive critique by Hanson (17)). Lastly, a new type of utility-theory model should be developed for destination choice, allowing for variable utilities of destinations over time (17). These variable utilities would reflect the fact that destinations seem to be linked to each other in differing time-dependent patterns. Two possible sources for this have already been identified, namely, cyclic choices of activity sets by household members, and spatial learning. A time-dependent adaptive utility model incorporating these two elements would not only describe and help forecast trip chaining; it would also show more clearly how routinised patterns of destination choice are developed by individuals over time (12). The successful testing of such a model might suggest some additional population segments for whom separate travel models are required (for example, new residents, old residents). It would also assist our understanding and predictions of spatial and temporal variability in travel behaviour.

To sum up, suggestions have been made about mathematical models which could now be developed within the human activity framework; it is clear that the framework has considerable potential for the understanding of travel behaviour and for policy applications because of its integrative nature. However, this potential will not be fulfilled unless some specific problems with current data gathering techniques are resolved. It is to these problems that we finally turn.

*Recommendations for Data Collection*

The use of the travel diary and time-budget methodology raises many questions for further research (1,33). What, for example, is the time period (week, month) over which data should be collected? What is the response error which accrues because individuals are asked to report on their behaviour? What biases are built in under different administration procedures? What are the advantages and disadvantages of home interviews versus mail-outs? How should the very poor, the elderly and children be investigated? Is the use of statistical sampling and analysis methods appropriate? Should the locations of origins and destinations be recorded and coded as points corresponding to street addresses, rather than as zones or census tracts? Since there are still relatively few data sets for activity analysis in Europe and the USA, these questions remain largely unanswered. They are, of course, of considerable importance for future attempts to assess the costs versus the benefits of activity analysis for different policy applications.

The use of newer gaming simulation techniques raised other questions. The technique only seems to be operable with small samples. This is because an interviewer requires every sampled household member to simulate and record *all* response to a policy change (for example, activities, trip purposes, trip scheduling, modes, trip frequencies and destination choices). It thus appears to be an exploratory rather than a forecasting device but with undeniable advantages, not only as a means of obtaining citizen input into the planning process and as a way of obtaining much needed perceptual data but also as a means of allowing people to make more

satisfactory prognoses about their own situation regarding
degree of choice and constraint.  However more research is
required on the kinds of policy application for which it is
most appropriate.


CONCLUSION

This report justifies the use of the human activity framework
for modelling urban travel demand.  First, certain conceptual
deficiencies of current choice-based models were pointed out,
together with the ways these deficiencies are overcome by the
activity approach.  Secondly, certain important policy areas
were identified; it was argued that these areas could perhaps
be better handled within an activity framework than within
the frameworks provided by current disaggregate models, not
least because travel is then regarded as dependent upon other
activities for its existence.  Finally, future directions for
research were discussed: these related to the development of
more rigorous mathematical models and data collection
procedures.

   One important question has been left aside until this
point.  It concerns the aggregation procedures which are to
be used in activity analysis.  Several have been implied,
such as the use of household member archetypes, or the use of
the population grouped by similarity of activity patterns, by
similarity of choice situation, or by similarity of socio-
economic and other factors governing roles.  However, to date
there has been little investigation of how the different
aggregation procedures affect travel choice predictions, or
of how activity analysis can be used to explore individual
variations around group travel norms (the need to study
individual variation within 'homogeneous' groups using
activity analysis was mooted under 'Scope of Workshop').  At
the moment, indeed, there appears to be no clear relation
between the results of analysis of individual activity and
travel patterns and answers to such traditional questions as
aggregate mode split for the work trip for different metropolitan corridors.  It will be an important task for future
research to unearth and assess relevant aggregation techniques.

   In conclusion, it must be remarked that the activity
framework at this stage of development has the advantage of
simplicity over other modelling approaches.  This obviously
facilitates its adoption by planners.  The framework also
allows the exploration of many 'academic' questions about
human behaviour, besides those related to transport and urban
planning.  For example, Pred and others have discussed the
use of activity analysis to explore 'alienation' in an
industrialised, capitalist economy (39,32) and the relations
of power expressed in differing degrees of forced mobility.
This flexibility stems from a focus on the universals
governing all kinds of human behaviour (needs, space, time)
and, above all, from an emphasis on individuals as human
beings in the context of society.  It is this flexibility
which finally gives the activity approach so many possible
applications in the study of urban travel demand.

REFERENCES

1. Anderson, J., 'Space-time budgets and activity studies
in urban geography and planning', *Environment and Planning*,
3, pp. 353-68 (1971).
2. Ben-Akiva, M.E., 'Structure of passenger travel demand
models', *Transportation Research Record No. 526*, pp.26-42
(1974).
3. Brail, R.K. and Chapin, F.S., Jr., 'Activity patterns
of urban residents', *Environment and Behaviour*, 5, pp. 163-90
(1973).
4. Brög, W., Heuwinkel, D. and Neumann, K. (1977)
*Psychological Determinants of User Behaviour*, Round Table 34,
OECD, Paris.
5. Bullock, N., Dickens, P., Shapcott, M. and Steadman,
P., 'Time budgets and models of activity patterns', *Social
Trends*, 5, pp. 45-63 (1973).
6. Burnett, K.P., 'Disaggregate behavioural model of travel
decisions other than mode choice', in *Behavioural Demand
Modelling and Valuation of Travel Time*, Special Report 149,
Transportation Research Board, National Research Council,
Washington, D.C., pp. 207-22 (1974).
7. Burnett, K.P., 'A three-state markov model of spatial
choice', *Geographical Analysis*, 6, pp. 53-68 (1974).
8. Burnett, K.P., 'The application of conjoint measurement
to recurrent urban travel', in *Data Analysis in Multidimensional Scaling*, R.G. Golledge and J.Rayner (eds) Ohio State
University Press, Columbus, Ohio (1977).
9. Burnett, K.P., 'Time cognition and urban travel
behaviour', *Geografiska Annaler*, forthcoming.
10. Chapin, F.S., Jr. (1974) *Human Activity Patterns in
the City: Things People Do In Time and Space*, John Wiley and
Sons, New York.
11. Chapin, F.S., Jr., 'Human time allocation in the
city', in *Timing Space and Spacing Time: Vol. 2, Human
Activity and Time Geography*, T. Carlstein, D.N. Parkes and
N.J. Thrift (eds), Arnold, London (1977).
12. Cullen, I., 'The treatment of time in the explanation
of spatial behaviour', in *Timing Space and Spacing Time:
Vol. 2, Human Activity and Time Geography*, T. Carlstein, D.N.
Parkes and N.J.Thrift (eds), Arnold, London (1977).
13. Fried, M. *et al.* (1977) *Travel Behaviour - A Synthesised Theory*, Final Report, National Co-operative Highway
Research Program, Laboratory of Psychosocial Studies, Boston
College, Chestnut Hill, Massachusetts.
14. Hägerstrand, T., 'What about people in regional
science?', *Papers of Regional Science Association*, 24, pp. 7-
21 (1970).
15. Hägerstrand, T.,'The impact of transport on the
quality of life', Paper read at European Conference of
Ministers of Transport, 5th Symposium, October 1973, Athens,
*Department of Geography, University of Lund, Rapporter och
Notiser 10* (1974).
16. Hanson, S. and Marble, D.F., 'A preliminary typology
of urban travel linkages', *East Lakes Geographer*, 7, pp. 49-
59 (1971).
17. Hanson, S., 'Urban travel linkages', chapter in this
volume.
18. Hemmens, G.C., 'Analysis and simulation of urban
activity patterns', *Socio-Economic Planning Sciences*, 4, pp.
53-66 (1970).

19. Hitchcock, J.R., 'Daily activity patterns: an exploratory study', *Ekistics*, 33, pp. 323-7 (1972).

20. Holly, B.P. (cd.), 'Time-space budgets and urban research: a symposium', *Kent State University, Department of Geography Discussion Paper 1* (1976).

21. Jones, P.M., 'The analysis and modelling of multi-trip and multi-purpose journeys', *Oxford University, Transport Studies Unit Working Paper 6* (1976).

22. Jones, P.M., 'Travel as a manifestation of activity choice; trip generation reinterpreted', *Oxford University, Transport Studies Unit Working Paper 14* (1976).

23. Jones, P.M., 'New approaches to understanding travel behaviour: the human activity approach', chapter in this volume.

24. Kobayashi, K., 'An activity model: a demand model for transportation', *Transportation Research*, 10, pp. 67-79 (1976).

25. Kobayashi, K., 'An activity model and its validation', chapter in this volume.

26. Kutter, E., 'A model for individual travel behaviour', *Urban Studies*, 10, pp. 233-56 (1973).

27. Kutter, E., 'Individual needs and urban structure - determinants of behaviour oriented model construction in traffic planning', mimeo, Technischen Universitat Berlin, Berlin (1976).

28. Lenntorp, B., 'Paths in space-time environments - a time-geographic study of movement possibilities of individuals', *Lund Studies in Geography*, Series B, 44 (1976).

29. Louviere, J.J., Piccolo, J.M., Meyer, R.J. and Duston, W., 'Theory and empirical evidence in real-world studies of human judgement: three shopping behaviour examples', mimeo, Department of Geography, University of Wyoming, Wyoming, USA.

30. Marble, D.F., 'A theoretical explanation of individual travel behaviour', in *Quantitative Geography Part 1: Economic and Cultural Topics*, W.L. Garrison and D.F. Marble (eds), Northwestern University Studies in Geography 13, Northwestern University Press, Evanston, Illinois, pp. 33-53 (1967).

31. Martensson, S., 'Childhood interaction and temporal organisation', *Economic Geography*, 53, pp. 99-125 (1977).

32. Matzner, E. and Rusch, G. (eds), 'Workshop: transport as an instrument for allocating space and time - a social science approach', *Technical University Vienna, Institute of Public Finance Publication*, 11 (1976).

33. Michelson, W. (ed.), 'Time budgets and social activity: 1', *University of Toronto, Centre for Urban and Community Studies, Major Report*, 4 (1975).

34. Michelson, W. and Reed, P., 'The time budget', in *Behavioural Research Methods in Environmental Design*, W. Michelson (ed.), Dowden, Hutchinson and Ross, Stroudsbury, Pennsylvania, pp. 180-234 (1975).

35. Nystuen, J., 'A theory and simulation of intra-urban travel', in *Quantitative Geography Part 1, Economic and Cultural Topics*, W.L. Garrison and D.F. Marble (eds), Northwestern University Studies in Geography 13, Northwestern University Press, Evanston, Illinois, pp. 33-53 (1967).

36. Ottensman, J.R., 'Systems of urban activities and time: an interpretive review of the literature', *Chapel Hill, North Carolina, Centre for Urban and Regional Studies Research Paper* (1972).

37. Palm, R. and Pred, A., 'A time-geographic perspective

on problems of inequality for women', *Institute of Urban and Regional Development Working Paper*, no. 236, University of California, Berkeley, California (1974).

38. Parkes, D.N. and Wallis, W., 'Graph theory and the study of activity structures', in *Timing Space and Spacing Time, Vol. 2, Human Activity and Time Geography*, T. Carlstein, D.N. Parkes and N.J.Thrift (eds), Arnold, London (1977).

39. Pred, A., 'The choreography of existence: comments on Hägerstrand's time-geography and its usefulness', *Economic Geography*, 53, pp. 207-21 (1977).

40. Richards, M.G. and Ben-Akiva, M.E. (1975) *A Disaggregate Travel Demand Model*, Saxon House, Farnborough (UK).

41. Stearns, M.D., 'The behavioural impacts of the energy shortage: shifts in trip-making characteristics', *US Department of Transportation, Transportation Systems Center*, December (1975).

42. Tomlinson, J., Bullock, N., Dickens, P., Steadman, P. and Taylor, E., 'A model of students' daily activity patterns', *Environment and Planning A*, 5, pp. 231-66 (1973).

43. Thrift, N.J. (1977) *An Introduction to Time-Geography*, Concepts and Techniques in Modern Geography 13, Geographical Abstracts Ltd, Norwich.

44. Thrift, N.J., Time and theory in human geography 2, *Progress in Human Geography*, 1(3) (1977).

45. Vidakovic, V., 'Modal mixture of the serial transportation demand: an explorative study', *Proceedings of the International Conference on Transportation Research*, Bruges, Belgium, pp. 679-83 (1973).

46. Williams, H.C.W.L., 'The formation of travel demand models and user benefit measures', *Environment and Planning A*, 9, pp. 285-344 (1977).

PART THREE

Chapter 6

SIX NOTIONS OF EQUILIBRIUM AND THEIR IMPLICATIONS FOR TRAVEL
MODELLING EXAMINED IN AN AGGREGATE DIRECT DEMAND FRAMEWORK

Marc Gaudry

INTRODUCTION

This chapter purports to analyse the implications of various
notions of equilibrium used in the formulation and estimation
of transport models which explain observed amounts of travel
in terms of supply and demand considerations. We will
perform this analysis by developing spatially, in increasing
detail, an aggregate direct demand function for trips in an
area during period t:

$$T_t^d = f^d(S_T, \ PD_T, \ P_v, \ P_f, \ A_f)_t + e_t^d, \tag{1}$$

where the quantity of trips demanded depends on the service
level of transport modes $S_T$, on the price of trip distance
units $PD_T$, on the prices of commodities (including labour)
bought or sold in variable and fixed amounts, $P_v$ and $P_f$, and
on the amount of fixed commodity trades $A_f$. For the sake of
convenience, time subscripts on all variables and on the
stochastic term will be dropped in the first five sections
of the chapter in which static notions of equilibrium are
presented and equilibria are assumed to occur within this
implicit time period; the time subscripts will be reintroduced in the sixth section where time paths of adjustment of
variables are explicitly considered. We assume throughout
that the location (but not the level) of activities is
invariant. These time-period, level of aggregation and land-
use assumptions are meant to isolate a number of situations
in which short-run equilibrium notions have relevance to
transport modelling. We shall not address directly equilibrium situations in which the time span required for
adjustments to occur is very large, such as those of
interaction between transport networks and land use. Our
analysis is meant as a methodological document focused on six
equilibrium issues and not as a review of the literature of
equilibrium concepts and practices.

The first section will contain an examination of the
nature and dimensions of the 'commodity' involved in presumed
equilibria and explore the incorporation of all of these
dimensions in the formulation of the dependent variable of
travel demand functions. The second section will show that
the use of activity level variables to explain derived trip
levels introduces specification problems which might lead one
to expect negative correlations between observed trip levels

137

and activity levels in certain circumstances.  The third
section argues for greater understanding of money and time
income effects in trip analysis.  The fourth and fifth
sections explore the distinct and inter-related notions of
market and network equilibrium in the joint determination of
travel flows and service levels.  The final section introduces dynamic considerations.  We shall therefore examine
successively the role of each type of variable used in
Equation (1).  The notions of disequilibrium and the
estimation problems discussed at various occasions will not
be regrouped in specific sections.


THE NATURE OF TRAVEL OUTPUT

The first issue which naturally comes to mind is that of the
nature of the commodity which is the object of equilibrium,
or of the dimension of the dependent variable in Equation (1).
We wish to argue that better use might be made of these
dimensions in the formulation of models.  To that end, let us
specify the dimensions of transit trips in our prototype
demand equation.


*The Demand Side: Dimensions of Transit Trips*

Consider an individual living in an urban area and facing a
network of bus lines of identical frequency $S_{1b}$ and speed
$S_{2b}$.  Assume that the $i^{th}$ trip using one or more lines
consists of waiting time $t_{1bi}$, in-vehicle time $t_{2bi}$ and seat-
mileage $t_{3bi}$.  Under uniform price per mile and service level
conditions, the consumer's total demand for trip characteristics corresponding to $T_b$ trips can be expressed as

$$(T_{1b} \ T_{2b} \ T_{3b})^d \equiv T_b^d = f_b^d (S_{1b}, \ S_{2b}, \ PD_b, \ ETC) + e_b^d \tag{2}$$

where total seat-waiting time $T_{1b} = \sum_i t_{1bi}$, total seat-travel
time $T_{2b} = \sum_i t_{2bi}$ and total seat-mileage $T_{3b} = \sum_i t_{3bi}$.  In
this perspective, changes in the level of independent vari-
ables may induce the consumer to change the *number* of trips
$T_b^d$, their *length* $d_i$ or both: a higher bus speed $S_{2b}$ could
leave trip demand $T_b^d$ constant but affect the consumption of
trip characteristics in one of various ways (e.g. longer
trips ($T_{3b}$ increases) and more or less in-vehicle time ($T_{2b}$)).
In Equation (2), the *frequency* and *speed* properties of the
mode behave as time prices and *say nothing about actual trip
characteristic purchases per trip*.  In real contexts, price
and service levels are not identical on all lines.  Moreover,
Equation (2) does not tell us the location of the demands.
A number of approaches can be used to circumvent these
problems.

*The standard format.* The first one consists of disaggregating trips by origin-destination pair. This makes two simplifications possible. First, the vector $(S_{1b}\ S_{2b}\ PD_b)$ can be transformed into a vector of trip rates and Equation (2) can be written as

$$(T_{1b}\ T_{2b}\ T_{3b})^d_{ij} \equiv ((K_{1b}\ K_{2b}\ K_{3b})_{ij} \cdot T_{b_{ij}})^d$$

$$= f^d_b ((K_{1b}, K_{2b}, P_b)_{ij}, ETC_{ij}) + e^d_{b_{ij}}, \qquad (3)$$

where $P_b$ denotes the price for $K_{3b}$ units of distance on buses. It is important to realise that, even if the transformation results in waiting time $K_{1b}$ and travel time $K_{2b}$ per trip being used on both sides of the equation, the significance of these variables differs: on the left-hand side, they enter as commodities; on the right-hand side, they enter as resources or prices.

Secondly, the commodity-vector of demanded characteristics can be neglected and only $T_{b_{ij}}^d$ kept as dependent variable, as in

$$T_{b_{ij}}^d = f^d_b(K_{1b}, K_{2b}, P_b, ETC)_{ij} + e^d_{b_{ij}}. \qquad (4)$$

This second simplification can blur even further the distinction between commodity and resource. Utility is defined over $(K_{1b}\ K_{2b}\ K_{3b}) \cdot T_{b_{ij}}$, a vector of commodities, and not over $(K_{1b}, K_{2b}, P_b)$, which consists of prices. Appendix A of a Charles River Associates report (5) makes this point clearly. By contrast, it is not clear in the current rationale of the logit model (8), which does not use an indirect utility function, what is a price and belongs to a budget constraint and what is a commodity and belongs to the utility function.

*The joint consumption format.* Let us develop an alternative approach based on the simple idea that all of the constitutive elements of trips should be used as a dependent variable. Consider that the trip elements cannot be bought separately but have to be bought jointly; rewrite Equation (3), dropping i j subscripts, as

$$JZ_b^d = (\gamma_1\ T_{1b} + \gamma_2\ T_{2b} + \gamma_3\ T_{3b})^d$$

$$= f^d_b (K_{1b}, K_{2b}, P_b, ETC) + e^d_b, \qquad (5)$$

where the trip elements are weighted and the $\gamma$ weights sum to one. In this perspective, the individual's demand curve for JZ results from the *vertical* addition of demand curves for each of the trip component elements (16) in the spirit of the welfare conditions for jointly supplied goods (27). The aggregate demand curve is obtained by conventional *horizontal* addition over agents.

We recently compared transit demand elasticities obtained from a standard specification of the dependent variable, as in Equation (4), and from a joint consumption specification, as in Equation (5). We did not have disaggregate individual trip data and consequently made considerable errors of observation on the components of the dependent variable; the results for selected transit trip demand elasticities listed in Table 6.1 supported the standard specification because the in-transit time demand elasticity had an unreasonably high negative value under the joint consumption specification.

TABLE 6.1   Selected transit demand elasticities derived from different specifications of the dependent variable for Montreal adults[a]

|  | Standard | Joint consumption |
|---|---|---|
| Bus-metro fare | 0.18 (-  6.41) | 0.25 (-  0.60) |
| Bus-metro waiting time | 0.16 (-  2.15) | 0.79 (- 13.02) |
| Bus-metro in-transit time | 0.49 (- 12.69) | -1.18 (- 16.99) |
| Car in-transit time | -0.07 (   0.41) | 0.10 (-  0.18) |
| Income | -0.52 (   3.67) | -0.77 (   0.63) |

a. Numbers in parentheses are t-ratios and give the sign of the appropriate independent variable.   More details on model formulation and estimation are given in Appendix 6.A.

It was interesting to note that the weights (and t-ratios) of the components of the dependent variable were: -0.14 (-13.90) for waiting time, -0.02 (-18.43) for in-transit time and 1.16 for mileage; the calculated value for $JZ_b$ was negative which had the *prima facie* implication that derived demands yield disutility and would lead one to expect that, for a final good, JZ would be positive.   These results are interesting enough to suggest that the joint consumption specification should be tested with a better data set.   If trips really consist of a vector of commodities, there should be a way of entering these elements as dependent variables and of using that information directly; at present, the standard format of the dependent variable can be seen as a special case of the joint consumption formulation in which $(\gamma_1 K_1 + \gamma_2 K_2 + \gamma_3 K_3)$ is constrained to equal one.

*Extensions to Other Modes and to the Supply Side*

The vector of characteristics demanded from i to j can also be written for automobile trips:

$$(T_{1a} \ T_{2a} \ T_{3a})_{ij}^{d}, \tag{6}$$

where elements respectively denote seat-waiting time, seat-travel time and seat-mileage.

This definition of the dimension of trips is also applicable to the supply side; the quantity supplied is

$$(T_{1m} \ T_{2m} \ T_{3m})^S_{ij} \equiv ((K_{1m} \ K_{2m} \ K_{3m}) \cdot T_m)^S_{ij}, \tag{7}$$

where m denotes any mode and the elements denote the number of seats provided, $T_m$, multiplied by their waiting time $K_{1m}$, by their travel time $K_{2m}$ and by the distance $K_{3m}$ they cover in going from i to j. Other dimensions (comfort, safety, walk time, etc.) could in principle be added in similar fashion both on the demand and on the supply side. Whether they are used as dependent variables in supply equations or as explanatory variables in demand equations, the elements $K_{1m}$ and $K_{2m}$ are of course transformations of frequency and speed variables into time per trip variables. For the purposes of the following section, we shall use the standard format, $T_m$, of the dependent variable in demand functions.

## EQUILIBRIUM FOR DERIVED DEMANDS

A second issue of importance in the explanation of the amounts of trip making is that of the presence of activity level variables in trip generation and direct demand equations. These variables introduce the overall level of economic activity as an explanatory variable of the frequency of trips. A paradigm equation could be

$$T^d = f^d \ (K_{1T}, \ K_{2T}, \ P_T, \ P_A, \ A) \tag{8}$$

in which the money and time prices of travel, the price of other consumption activities and the level of these activities appear as arguments. Functions of that general form are normally estimated on the assumption that activity levels $A = (A_1, \ldots, A_n)$ are exogeneous and predetermined. Yet the underlying rationale for this procedure is far from clear. We shall argue that the presence of activity variables in numerous trip demand equations often introduces a simultaneous equations bias and can sometimes be inappropriate. Let us examine situations which may increase our understanding of the problem.

### With Variable Activity Levels

Certain classes of cases arise if the levels of various trades for labour and other goods are allowed to be determined by the prices of these activities and by travel prices. Work and shopping trips are examples.

*Work-related trips.* Consider the following simplistic version of the structure of an aggregate market for labour. The demand function for units of labour services L (say man-hours) is simply

$$L^d = \alpha_1 \, P_L + \alpha_2 \, OTHER_d \; , \qquad (9)$$

where $P_L$ denotes the price of labour, $OTHER_d$ is self-explanatory and the $\alpha$'s are parameters. Specify the supply function as

$$L^s + \beta_1 \, T^s = \beta_2 \, P_L + \beta_3 \, P_T + \beta_4 \, OTHER_s \; , \qquad (10)$$

where $T^s$ denotes the amount of travel workers are willing to do as they supply $L^s$ work services and $P_T$ is the price of travel. In this formulation $L^s$ and $T^s$ are jointly supplied and for that reason are jointly dependent variables. This formulation incorporates the fact that working involves supplying both man-hours on the job and work-related trips, elements which have certain relative weights (here the parameter of the first element has been 'normalised' or constrained to be equal to one) in the decision to supply one unit of 'working'. These jointly supplied elements are made a function both of $P_L$ and of $P_T$ because there is no reason to suppose that one earned dollar has the same effect as one dollar spent on travel (if one believed otherwise, one would simply write $\beta_2 = \beta_3$) on the supply of these jointly produced elements.

The market clearing equation is

$$L^s = L^d \; . \qquad (11)$$

If the amount of travel per unit of labour is not technologically fixed and if the labour market is in equilibrium, one can write the reduced form equation* for the jointly dependent pair of variables as

$$L + \beta_1 \, T^s = \frac{\beta_2}{\alpha_1} \, L + \beta_3 \, P_T + \beta_4 \, OTHER_s - \frac{\alpha_2 \beta_2}{\alpha_1} \, OTHER_d \; , \qquad (12)$$

or as

$$\left(1 - \frac{2}{1}\right) L + {}_1 \, T^s = {}_3 \, P_T + {}_4 \, OTHER_s - \frac{2 \; 2}{1} \, OTHER_d. \qquad (13)$$

To obtain a derived demand function for work trips, one simply relabels $T^s$ as $T^d$ and collects terms:

$$T^d = \frac{2^- \; 1}{1 \; 1} \, L + \frac{3}{1} \, P_T + \frac{4}{1} \, OTHER_s - \frac{2 \; 2}{1 \; 1} \, OTHER_d. \qquad (14)$$

_____

*In practice one would redefine $L^s + \beta_1 \, T^s$ as W and add that identity to the system consisting of the demand, supply and market clearing equations.

We note the following:

1  because L, an endogeneous variable of the system, appears on the right-hand side of Equation (14), an ordinary least-squares regression of T on L, $P_T$ and OTHER would produce *biased estimates* of the resulting parameters;

2  if these parameters are called $\gamma$'s and Equation (14) is rewritten as

$$T^d = \gamma_1 L + \gamma_2 P_T + \gamma_3 \text{ OTHER} , \tag{15}$$

the sign of $\gamma_1$ should be *negative* (given $\alpha_1 < 0$ ; $\beta_1$ , $\beta_2 > 0$): the fact that one expects the parameter of the employment level to be positive in such equations indicates the presence of an unresolved problem. By contrast, the sign of a variable included in both $\text{OTHER}_s$ and $\text{OTHER}_d$ will (given $\alpha_2 < 0$ ; $\beta_2 > 0$) depend on the size of each component of

$$\gamma_3 = \frac{\beta_4}{\beta_1} - \frac{\alpha_2 \beta_2}{\alpha_1 \beta_1} ;$$

3  although not all work trip generation equations have the appearance of Equation (15), many include its basic elements. This is as it should be: the variable $P_T$, for one, is important in the decision to supply work. Indeed the changes in the labour participation rate which accompany the business cycle are certainly partly due to the fact that a decrease (or an increase) in the number of jobs typically implies longer (or shorter) work trips to similar jobs. In fact such changes in the participation rate, notably of women, may give an important clue concerning the numerical importance of increases in the supply of labour which result from transport network improvements.

By contrast, if the amount of work trips per unit of labour is technologically fixed as in

$$T^s = kL^s , \tag{16}$$

where k is a technological constant, then the trip demand equation is

$$T^d = kL . \tag{17}$$

Under this rigid view, it would be inappropriate to use as a trip generation equation expression (14) or expression (15) because the former can be rewritten from (17) as

$$L + \beta_1 kL = \frac{\beta_2}{\alpha_1} L + \beta_3 P_T + \beta_4 \text{ OTHER}_s - \frac{\alpha_2 \beta_2}{\alpha_1} \text{OTHER}_d , \tag{18}$$

which implies that a regression of T on L, $P_T$ and OTHER would in effect be, up to a normalisation factor k, *a regression of L on itself*:

$$T^d = kL = \frac{\beta_2 - \alpha_1}{\alpha_1 \beta_1} L + \frac{\beta_3}{\beta_1} P_T + \frac{\beta_4}{\beta_1} OTHER_s - \frac{\alpha_2 \beta_2}{\alpha_1 \beta_1} OTHER_d . \tag{19}$$

The proper trip generation equation is either (17) or the relationship obtained after substituting in (17) for the value of L derived from (19):

$$T^d = kL = k \ [\frac{\beta_3}{\beta_1} \ \delta \ P_T + \frac{\beta_4}{\beta_1} \ \delta \ OTHER_s - \frac{\alpha_2 \beta_2}{\alpha_1 \beta_1} \ \delta \ OTHER_d] , \tag{20}$$

where $\delta = \dfrac{\alpha_1 \beta_1}{\alpha_1 \beta_1 k - \beta_2 + \alpha_1}$ .

It is clear that a regression of T on $P_T$ and OTHER would be a meaningful trip generation equation; *it does not include the activity level variable* L explicitly.

*Trips for other purposes.* Similar reasoning can be applied to trips for purposes other than work. If $R^s$ and $R^d$ denote the amount of goods supplied and demanded in retail establishments, the supply, demand and market clearing equations are

$$R^s = \alpha_1 P_R + \alpha_2 OTHER_s , \tag{21}$$

$$R^d + \beta_1 T^d = \beta_2 P_R + \beta_3 P_T + \beta_4 OTHER_d , \tag{22}$$

$$R^s = R^d . \tag{23}$$

Again, Equation (22) allows the price of retail goods and the cost of travel to have distinct effects on $R^d$ and $T^d$ which are bought jointly in varying proportions. As for Equation (14), regressing shopping trips on retail sales, the cost of travel and other variables would yield biased estimates of mixtures of structural parameters. Analogs of Equations (17) or (20) would be appropriate.

The presence of changing activity level variables in trip demand equations is therefore by no means unambiguous because these equations are in fact summary descriptors of more complex underlying equilibria. At the best, their presence introduces a simultaneous equations bias in equations of form (15): the strength of the bias depends on the constancy of coefficient k; at the worst, that form is tantamount to a regression of the activity variable on itself and on other variables: the proper format is then either Equation (17) or, going a step behind, Equation (20). The latter form includes as regressors the prices of travel and non-travel goods (under $OTHER_s$).

*With Fixed Activity Levels*

Suppose that the level of activity A is fixed. In the presence of a technological relationship between the level of A and trip making, the derived demand for A-linked trips is given by an analog of Equation (17):

$$T^d = k A + e^d \tag{24}$$

and there is no analog of Equation (20) which would explain the level of A. The aggregate demand for trips is given by the function

$$T^d = k A + f^d (P_T, P_A, \text{Other prices}) + e^d, \tag{25}$$

where $P_A$ is in effect a measure of income if A denotes work activity and a measure of disposable or discretionary income after fixed commitments are met if it denotes some other fixed activity. If there is no constant relationship between activity levels and trip making, Equation (25) reduces to

$$T^d = f^d (P_T, P_A, \text{Other prices}) + e^d \tag{26}$$

in which *activity level variables are absent*. Equations (25) and (26) are perfectly consistent with the general demand function derived in the Appendix of the Charles River Associates report (5).

*Implications*

Exceeding care should then be taken with the inclusion of activity level variables in trip generation and direct demand equations. Further work is clearly needed to circumscribe the situations in which these variables may be included with reasonable expectations as to their sign and expected role. It may be found that travel prices influence the number of trips per unit of activity for work activity, but not the level of activity itself and that, for other purposes, they influence both trip inputs per unit of activity and the level of activity. If it is found that some significant inter-dependence exists between activity variables and some of the travel prices, it may be appropriate to use estimation methods which re-establish this independence, at least asymptotically. In the meantime, if one uses equations like (24) or (26) to explain trip making for various purposes or by various groups, it may be useful to estimate these sets of generation equations in grouped fashion in order to use, à la Zellner (30), at least the information contained in the covariances of error terms across equations: the stochastic errors of work, shopping and social trip generation equations are certainly often correlated and the most efficient estimates should be obtained with Zellner's method.

The explicit treatment of most travel demands as derived demands in a general equilibrium context should progress as macro-economic demand models such as those of the Rotterdam

model (1) gradually allowing for explicit transport sectors
(e.g. (21)) and as systems of micro-economic equations.
Until these approaches have developed further, adjustments
must be made in an *ad hoc* and partial fashion: the correct
relationships among activity levels and travel must be
specified by relying more on direct intuitive formulations in
a partial equilibrium framework than on formulations strictly
derived from rigorous general equilibrium approaches. The
market balance is the underlying principle of the mechanism
which the incorporation of activity level variables in trip
frequency equations describes, and one must be conscious of
that in using simple formulations such as Equation (8).


## EQUILIBRIUM WITH RESPECT TO INCOME

A third issue in the explanation of travel behaviour is that
of the role of income, the exact impact of which remains
somewhat elusive due to the difficulty of the problem. We
shall argue that fundamental work is needed in this area and
that the current emphasis on mode split in the literature
gives a distorted picture of time and money income effects.


### General Considerations

Assume that consumer tastes are given. In standard demand
analysis, it is traditional to distinguish between gross and
net substitution (or complementarity) and income effects as
follows: if a demand function is written in general form as

$$T^d = f^d \ (P_T, \ P_{NT}, \ Y), \tag{27}$$

where the prices of travel and non-travel goods and income
are included, the gross substitution (or complementarity)
effects are defined as $\partial T^d/\partial P_T$ and $\partial T^d/\partial P_{NT}$ and the gross
income effect is defined as $\partial T^d/\partial Y$. The net effects arise
from a distinction between the part of a price change which
leaves consumers as well of as before, $SUB_{P_T}$ or $SUB_{P_{NT}}$ and
the part which amounts to a change of income, $\Delta Y_{P_T}$ or $\Delta Y_{P_{NT}}$.
When the price-specific income effect is thought to be small
(when the consumers' expenditure on the good constitutes a
small part of their total expenditures), the gross substitu-
tion $(\partial T^d/\partial P = SUB_p + \Delta \ Y_p)$ and income $(\partial T^d/\partial Y)$ effects,
estimated as coefficients of price and income of a market
demand function, should not differ much from the net effects
$SUB_p$ and $(\Delta Y_p + \partial T^d/\partial Y)$.

Travel demand functions have to be more complex than
Equation (27) because of the importance of time prices; an
analog of that equation could be

$$T^d = f^d \ (K_T, \ P_T, \ P_{NT}, \ Y, \ TB) \ , \tag{28}$$

where $K_T$ denotes time prices of travel and TB is the consumers' time budget. De Serpa (6) has examined some of the fundamental problems involved in obtaining appropriate formulations of Equation (28) and in deriving time prices from parameter estimates. Because travel time and money expenditures constitute a very significant proportion of consumers' total allocations, gross and net money or time effects might really differ. Fundamental work is needed to measure these income effects.

*Specific Considerations*

In practice it will be tempting to make strong assumptions on the utility function in order to derive testable hypotheses which contain a reasonable number of parameters or which constrain the functional form of demand models. This is a task which requires great care: one could not easily accept functional forms based on the assumption of additive separability of the utility function which implies (e.g.(19)) that all goods are *substitutes* (none are complements) and that all are *superior*. Indeed, these properties would be strong if one were interested in aggregate trip generation by mode among pairs of zones and if one wanted to know whether two modes are complements and whether public transit is an inferior good: they would mean that an improvement of one mode could never lead to an increase in demand for another mode and that inferiority $(\partial T_m^d/\partial y < 0)$ of any mode m would be ruled out by the rationale of the model.

On a more applied level, strong emphasis on modal choice at the expense of trip generation can give a misleading impression of the (gross) impact of a price or income change on the demand by mode. This can be seen most clearly in two-stage models which incorporate a generation and a modal share part as follows:

$$T_m^d = f_g \ (\overline{P}_T, \ \overline{K}_T, \ A, \ Y) \cdot f_s \ (P_T, \ K_T, \ Y) \ , \qquad (29)$$

where $\overline{P}_T$ and $\overline{K}_T$ denote average money and time costs of travel and $P_T$ and $K_T$ are vectors of money and time costs of the modes considered individually. Gross price and income effects are the sum of two elements

$$\frac{\partial T^d}{\partial X} = f_s \cdot \frac{\partial f_g}{\partial X} + f_g \cdot \frac{\partial f_s}{\partial X} \qquad (30)$$

which must both be considered in demand analysis. Even without using the distinction between gross and net effects, these elements might have opposite signs and the computation of the total impact of income (or price) changes on the demand for a particular mode $T_m^d$ should take into account the sign and size of the effect on modal split (if any) and of the effect on the total number of trips. It is interesting to note that some urban-direct demand models for transit work trips (7) or for transit trips for all purposes (20,14,10) suggest, in contrast with mode split models, that public transit is *not an inferior* good, results which may precisely be due to the fact that the explanation of the trip frequency decision is not isolated from the mode choice decision in these models and which imply that public transit is not necessarily doomed in a growing economy.

*Implications*

A first implication of these considerations is the importance of distinguishing in models between the role of income variables as market disaggregation categories which represent tastes and their use to estimate the effect of income proper. In the latter capacity, their absence from some trip genera- tion equations is a specification error which biases the estimates of the parameters of the retained variables.

A second implication is the need for precise measures of substitution and income effects both for money and for time prices. This will require fundamental work on the develop- ment and estimation of functional forms which permit these measurements without excessively strong assumptions on the utility function, a very difficult task indeed.

MARKET EQUILIBRIUM

Observed trip levels do not depend only on consumers' demands: they also depend on the behaviour of suppliers of trips. To bring into focus the importance of supply, we shall use the notion of market equilibrium. Although nobody thinks that it is a perfect notion, most might agree that it is a useful notion. We shall apply it to two modes, starting with the most difficult one, call for a greater recognition of the importance of distinguishing producers and consumers in explanations of observed trip levels and point to a number of implications of the notion for model estimation.

In order to specify the notion of market equilibrium, we shall assume that there exist single non-overlapping itineraries by automobile a and by bus b among all i, j pairs of zones under consideration. We shall assume further that there is no congestion on these itineraries. These assump- tions define unique money and time prices for each mode among all zone pairs. Illustrative market structures are then as follows.

*The Car-trip Market*

In the car-trip market, one specification of the demand, supply and market clearing conditions could be, if all $V_a$ cars have the same capacity $c_a$:

$$((K_{1a} \ K_{2a} \ K_{3a}) \ c_a \ V_a)^d =$$
$$f_a^d \ (K_{1a}, \ K_{2a}, \ P_a, \ K_{1b}, \ K_{2b}, \ P_b, \ \text{OTHER}_{da}), \qquad (31)$$

$$((K_{1a} \ K_{2a} \ K_{3a}) \ c_a \ V_a)^S = f_a^S \ (K_{2a}, \ P_a, \ OTHER_{sa}), \tag{32}$$

$$((K_{1a} \ K_{2a} \ K_{3a}) \ c_a \ V_a)^d = ((K_{1a} \ K_{2a} \ K_{3a}) \ c_a \ V_a)^S. \tag{33}$$

With cars, the waiting time $K_{1a}$ is difficult to define but we shall leave it in these functions. If the simpler standard format is used and the amount of seats demanded and supplied, $T_a^S$ and $T_a^d$, are assumed equal to $c_a \ V_a^S$ and $c_a \ V_a^d$, the market clearing condition (33) can also be written as

$$(c_a \ V_a)^S = (c_a \ V_a)^d, \quad \text{or} \quad T_a^S = T_a^d. \tag{34}$$

There are a number of ways of understanding these equations. If one is thinking about an interzone taxi market, $P_a$ is paid out to the taxi firm and $OTHER_{sa}$ includes vehicle and wage costs and other variables such as the type of regulation of the industry. In that case, the supply curves *proper* are given by the partial derivatives of the supply *vector* with respect to that market price $\partial [(K_{1a} \ K_{2a} \ K_{3a}) \ c_a \ V_a]^S / \partial P_a$ : one would expect the seat-waiting time element to be constant $[\partial (K_{1a} \ c_a \ V_a)^S / \partial P_a = 0]$ if, by definition, $K_{1a}$ is inversely related to $V_a^S$, and the other two elements to be positive. Slower travel speeds which increase production costs would have no impact on the seat-waiting time and would shift the other two supply curves to the left. Under the very strict definition of market clearing given above, most taxi systems suffer from excess supply because there is usually little pooling of taxi trips under prevailing money and time prices.

If one is thinking about a private car-trip market, the structure can be applied either within households or for groups of households. Given a cost sharing formula for out-of-pocket costs, all members of households demand car trips but only members with a driver's licence supply them and an increase in the latter's supply following an increase in their remuneration defines a number of standard supply curves (one for each element of the vector, as above). Because these drivers are not financially compensated for their driving time, an alternative definition of the supply function can be obtained by defining a generalised cost $GIC_{sa}$ as

$$GIC_{sa} = \lambda_1 \ R_a + \lambda_2 \ K_{2a}, \quad \lambda_1 < 0, \ \lambda_2 > 0, \tag{35}$$

where $R_a$ is the net financial compensation of drivers per trip. If one defines a similar notion for the demand side as

$$GIC_{da} = \theta_1 \ P_a + \theta_2 \ K_{1a} + \theta_3 \ K_{2a}, \quad \theta_1, \ \theta_2, \ \theta_3 > 0, \tag{36}$$

seat demand and supply curve graphs can be constructed as in

Figure 6.1. At GIC level 1, slightly more than half of the seats are occupied. A rise in out-of-pocket costs which would raise the GIC uniformly to level 2 (by raising $P_a$ and lowering $R_a$ in such a way that $\Delta\theta_1 P_a = \Delta\lambda_1 R_a$) would increase the car occupancy rate from AB/AC to A'B'/A'C'.

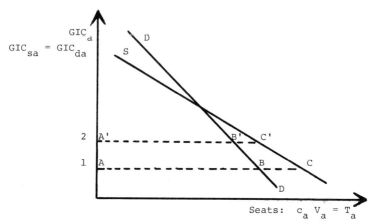

Figure 6.1  Generalised Itinerary Costs and the Car-trip Market

    Although basic work is necessary to obtain convincing specifications of the car trip structural equations for various levels of aggregation of models, the distinction between the supply and demand sides of that market is the basis of the explanation of car pooling, car occupancy rates and of the analysis of policy options which affect the supply side in a fashion distinct from the way in which they affect the demand side.


*The Bus-trip Market*

The structural equations for the bus market are:

$$[(K_{1b}\ K_{2b}\ K_{3b})\ c_b\ V_b]^d =$$

$$f_b^d\ (K_{1a},\ K_{2a},\ K_{1b},\ K_{2b},\ P_a,\ \tilde{P}_b,\ OTHER_{db})\ , \qquad (37)$$

$$[(K_{1b}\ K_{2b}\ K_{3b})\ c_b\ V_b]^s =$$

$$f_b^s\ (\tilde{P}_b,\ C(K_{2b},\ K_{3b}),\ OBJ_{bs},\ OTHER_{bs})\ , \qquad (38)$$

$$(c_b\ V_b)^s = (c_b\ V_b)^d\ ,\ or\ T_b^s = T_b^d\ . \qquad (39)$$

These equations are the same as those presented in Gaudry
(13) except for the dependent variable which is here presented in all of its dimensions instead of simply as capacity
demanded or supplied.

We have assumed that a bus monopoly with certain
objectives $OBJ_{bs}$ and identical vehicles of capacity $c_b$
supplies on each itinerary services, the unit costs of which
are a function of the speed and distance covered, $C(K_{2b} \; K_{3b})$,
in that itinerary. The supply curves are obtained as usual
by taking the partial derivative of the supply vector with
respect to the regulated price $P_b$. With identical buses the
waiting time will be inversely related to the number of
vehicles and it is reasonable to presume that, if the firm's
objective is to make zero profit, we should have

$$\partial (K_{1b} \; c_b \; V_b)^s / \partial \tilde{P}_b = 0 \; , \; \partial (K_{2b} \; c_b \; V_b)^s / \partial \tilde{P}_b < 0$$

and

$$\partial (K_{3b} \; c_b \; V_b)^s / \partial \tilde{P}_b > 0 \; .$$

An increase in wage costs would have no effect on the first
element and shift the other two supply curves to the left.
Strict market clearing occurs when the capacity supplied is
strictly equal to the capacity demanded over a given
itinerary.

In both the car- and the bus-trip markets, the classical
supply curve has simply become a supply vector. The apparent
paradox

$$\partial (K_{1a} \; c_a \; V_a)^s / \partial P_a \; , \; \partial (K_{1b} \; c_b \; V_b)^s / \partial \tilde{P}_b = 0$$

results from the fact that we have expressed the frequency
dimension of service in terms of a transformation of waiting-
time; had we kept measures of frequency of service $S_{1a}$ and
$S_{1b}$, the paradox would not have appeared. The proper
estimation of supply and demand functions for these two modes
will require meeting a number of conditions which we shall
presently examine.

*Implications*

*Market clearing: the existence of equilibrium.* In practice,
estimation of the full model for each mode does not require
that the strict market clearing conditions defined in
Equations (34) and (39) hold. What must exist is market
clearing in the weaker sense of

$$T_a^s \geqslant T_a^d \; , \; \text{and} \; T_b^s \geqslant T_b^d \; , \tag{40}$$

otherwise the demand relationships might not be observed: if

excess demand prevails, the amount supplied is observed but
the amount demanded is not.

Because of the difficulty of formulating convincing
structural equations for the car market, even this weak
notion of market equilibrium is very approximate: in the car
trip market, there is of course a difference between excess
supply of seats and excess supply of vehicle trips; there is
also a difference between aggregate and individual observations because excess supply in the aggregate is compatible
with excess demand for individual cars.  Yet the notion is
useful because it implies that an aggregate number of
observed trips by car does not represent the quantity truly
demanded or supplied.  This error of observation on the
*dependent* vehicle becomes a part of the stochastic term
associated with each car market relationship.

In the bus market, excess demand means that passengers
have to wait for more than one vehicle to find space.  If
waiting time $K_{1b}$ is measured from data on vehicle operations,
errors of observation on both that variable and the dependent
variable can occur.  The use of sophisticated disequilibrium
estimation techniques (9,25,26) might be required.  Such
situations might arise frequently if it does not pay the bus
firm to supply adequate capacity on certain itineraries.

*Identification: the structure of equilibrium.*  The existence
of an equilibrium is not sufficient for modelling.  If, in
each market, the list of variables included in the supply and
demand functions were identical (and forgetting the presence
of stochastic error terms), a statistical relationship
between an observed number of trips on a mode and the
variables of that list would be neither the demand nor the
supply relationship but a mixture of the two from which
neither could be 'unscrambled', to use a fitting expression
(2).  The resulting 'salad' relationship would not even be a
reduced-form equation.

It might be so difficult in practice to identify distinct
supply and demand equations in the car-trip market that one
may be obliged to use an aggregate 'salad' relationship which
one might call a 'derived demand' equation and formulate as

$$(c_a \ V_a)^{dd} = T_a^{dd} =$$

$$f^{dd}(K_{1a}, \ K_{2a}, \ P_a, \ K_{1b}, \ K_{2b}, \ P_b, \ OTHER_{da}, \ OTHER_{sa}) , \qquad (41)$$

and the parameters of which have no unambiguous interpretation.  It is fortunate that demand and supply relationships
for modes other than the car are probably over-identified
because quantities supplied and demanded typically depend on
different lists of variables in the time periods over which
the observations are defined.

*Simultaneity and pure structural relationships.*  But, even
with over-identified equations, consistent estimation of the
parameters of an equation might be difficult.  As an example,
note that, under regulated pricing, the service level

variables $K_1$ and $K_2$ are determined simultaneously by the behaviour of consumers and by that of producers. If, in a shuttle service, buses are added immediately when passenger queues are formed at the gate, a regression of the observed number of trips on service frequency and other determinants of demand would not yield a pure demand relationship because, with aggregate data, one would expect high frequencies to be correlated with high demands *even if passengers were indifferent to waiting time*. There are many situations in which line, area or time of day service levels are increased in anticipation of higher demands. Similarly, if higher ridership slows down vehicles, high travel times might be associated with high ridership *in spite of passengers' dislike of in-vehicle time*. Estimation procedures have to differentiate in demand equation parameter estimates between what reflects the suppliers' behaviour and what reflects the consumers'. Removing the simultaneous equations bias arising from supply-demand interactions may change the habitual expectations concerning money and time price demand elasticities estimated by limited-information techniques. Similar problems arise in estimating supply functions as soon as a measure of expected demand is used among the regressors. And even if the suppliers take no formal account of expected demand, the error term $e_s$ of the supply equation is likely to be correlated with the error term $e_d$ of the demand equation, a situation which reintroduces the simultaneous equations bias for different reasons.

*Other stochastic specification problems*. In most models, the size of the dependent variable varies considerably. This means that the error terms, particularly of the demand equations, are not likely to have a constant variance. In addition to being heteroskedastic, the error terms for different zones are likely to be correlated both spatially and over time. These forms of non-sphericity of the variance-covariance matrices of the error terms of each equation are not invariant with respect to the functional form used to estimate the equation (29) and should ideally be estimated jointly with the optimal form of the demand or supply function.

Within the appropriate time periods for which data are gathered, the existence of market clearing (in at least a weak sense), the identifiability of presumed relationship of interest, the explicitation of simultaneity in the structure and the correct specification of the stochastic error terms are essential requirements to the formulation and consistent estimation of travel supply and demand models. These conditions are important whether the presumed demand (or supply) relationship at hand results from a direct intuitive specification, from a formal link with utility theory or from analogy with a physical process (classical or relativistic, deterministic or stochastic) of any sort.

## NETWORK EQUILIBRIUM

Reconsider the 'derived demand' equation for car trips and the demand equation for bus trips of the previous section:

$$v_{ij_a}^{dd} = f_a^{dd}(K_{ij_a}, K_{ij_b}, OTHER_{ij_{da}}, OTHER_{ij_{sa}}) \qquad (42)$$

$$c_b\, v_{ij_b}^{d} = f_b^{d}(K_{ij_a}, K_{ij_b}, OTHER_{ij_{db}}) \qquad (43)$$

in which we have reintroduced i, j subscripts and written

$$K_{ij_a} = (P_{ij_a}\ K_{ij_{1a}}\ K_{ij_{2a}})$$

and

$$K_{ij_b} = (\tilde{P}_{ij_b}\ K_{ij_{1b}}\ K_{ij_{2b}}).$$

In that section, $K_{ij_a}$ and $K_{ij_b}$ were assumed to be unique
vectors of money and time prices.  The assumption was
apparently required to obtain, from appropriately *linear*
parts of demand equations, a vector of parameters $\pi_m =$
$(\overline{\pi_{Km}\ \pi_{other\,m}})$ the elements of which are respectively
associated with $K_{ij_m} = (K_{ij_a}\ K_{ij_b})$ and with $OTHER_m$ in the
$m^{th}$ equation.  In this section, we shall reverse the situation and examine the determination of the $K_{ij_m}$ vectors given
the existence of demand parameter vectors $\pi_m$; we shall then
conclude with a few words on the problem of the joint
determination of $\pi_m$ and $K_{ij_m}$ (this difficulty was pointed out
by Michael Florian).

*The Assumption of Generalised Itinerary Cost Minimisation*

The determination of the vectors $K_{ij_m}$ is not straightforward
because individual car drivers and transit passengers are in
general faced with many possible itineraries between any two
points; in fact, we will note that such vectors need not
exist as long as a unique weighted sum of their components
exists: that sum can be compatible with a great number of
different itineraries between the same two points.  Indeed a
natural approach to the explanation of how itineraries are
chosen is to consider travellers as producers of their
itineraries and to make hypotheses concerning their behaviour.
And one such hypothesis is that the $\pi_{Km}$ parameters are
'prices' used to weight money and time inputs required to go
from i to j and that, as any cost-minimising entrepreneur of
the standard theory of the firm, the traveller chooses the
itinerary r which minimises his total generalised cost
incurred on network links $\ell$, or $GIC_{ij_a} = (\pi_{Ka}\ K_{ij_a}^{'r})$ and
$GIC_{ij_b} = (\pi_{Kb}\ K_{ij_b}^{'r})$, where ' denotes transpose,

$K_{ij}^{r} = \sum_{\ell(ij)r} \delta_{\ell(ij)r} \; K_{\ell}$ and $\delta_{\ell(ij)r}$ equals one if the link is chosen and zero otherwise. The way in which the sets of feasible itineraries or input combinations $\left\{ K_{ij_a}^{r} \right\}$ and $\left\{ K_{ij_b}^{r} \right\}$ arise is very complex.

Consider car trips. The set of feasible input combinations $\left\{ K_{ij_a}^{r} \right\}$ depends on the set of link input requirements $\left\{ K_{\ell_a} \right\}$ which are functions of physical (length and number of lanes, design, etc.), regulatory (tolls, signals, police, etc.) and other link factors $N_{\ell}$ and of vehicle usage on the link $V_{\ell_a}$ :

$$\left\{ K_{ij_a}^{r} \right\} = F_{ij_a} \left\{ K_{\ell_a} \right\} , \text{ where } K_{\ell_a} = f_{\ell_a} (N_{\ell}, V_{\ell_a}) . \qquad (44)$$

For given values of $N_{\ell}$ and $V_{\ell_a}$, $\left\{ K_{ij_a}^{r} \right\}$ is the set of all possible itineraries from i to j. If cars are the sole users of a fixed supply of roads, Equation (44) is an appropriate description of input combination possibilities. A public transit authority, however, determines its link character- istics not only in relation to $N_{\ell}$ and $V_{\ell_b}$ but also, as we saw in the previous section, in relation to its objectives, subsidies, etc., on each line and we may write

$$\left\{ K_{ij_b}^{r} \right\} = F_{ij_b} \left\{ K_{\ell_b} \right\} , \text{ where } K_{\ell_b} =$$

$$f_{\ell_b} [f_{\ell_b}^{s}, f_{\ell_b}, (N_{\ell}, V_{\ell_b})] \qquad (45)$$

which is simply a link (or line) version of Equation (38) with endogeneous link speeds; in Equation (45) itineraries also consist of sets of links. One could envisage grouping Equations (44) and (45) to provide a full list of itineraries by mode when buses and cars interact on the road and even explaining the government's decisions on $N_{\ell}$ in a general explanation of the supply of feasible input combinations. (The idea of a general presentation of vectors of itinerary input possibilities embodying Equations (44) and (45) as special cases is Michael Florian's who also suggested calling these vectors performance vectors.)

Given input 'prices' $\pi_{Km}$ and trip demands by mode among all pairs of points $(V_{ij_a}^{dd})$ and $(c_b \; T_{ij_b}^{d})$, assignment algorithms derive specific values for $GIC_{ij_a}$ and $GIC_{ij_b}$ from the feasible sets $\left\{K_{ij_a}^{\ r}\right\}$ and $\left\{K_{ij_b}^{\ r}\right\}$ : for m = a and/or b, they bring about

$$\begin{bmatrix} GIC_{ij_a} \\ GIC_{ij_b} \end{bmatrix}_m \leftarrow \left( \pi_{km} , \left\{K_{ij_a}^{\ r}\right\} , \left\{K_{ij_b}^{\ r}\right\} , [v_{ij_a}^{dd}] , [c_b \ v_{ij_b}^{d}] \right). \quad (46)$$

Obtaining the set of itinerary costs which minimise user itinerary costs does not mean that a unique vector is chosen for each specific pair of points: it only means that the generalised itinerary cost of all chosen itineraries between these points is unique. This is possible because it is not only the inputs which matter (distance, time, money tolls, or other physical performance or economic descriptors, such as risk of accident) but also the way in which they are perceived, weighted or transformed into a generalised cost. Although road assignment procedures (22,23) have a general validity, they are mostly used only with the travel time component of $K_{ij_a}$ ; neither are transit assignment procedures often used with more than one element of $K_{ij_b}$ .   Elsewhere (15) we have discussed the automobile part of Equation (46): 'the computation of the cheapest input requirements for each output level is what we may call the *production* equilibrium problem'.  We now believe that this terminology was slightly incorrect because standard production problems, as Michael Florian points out, minimise the *total* cost of producing a certain output for a system as a whole, whereas, in this case, *individual* (user) itinerary costs are minimised: this means that, in the presence of congestion, where system and user solutions differ, the user-optimal solution is more akin to a *decentralised production* problem in which each unit (the origin-destination pair) is subject to decreasing returns to scale ($\partial GIC_{ij_m} / \partial V_{ij_m} > 0$) and produces externalities ($\partial GIC_{i'j'_m} / \partial V_{ij_m} \neq 0$) because each unit's costs depend on other units' output levels.  In Hicksian terminology, as Ake Andersson reminded us, the street network is a set of techniques: it consists of a group of linked small plants or machines used simultaneously to produce different outputs. The transit network is also a set of techniques among which travellers choose with the help of their input 'prices'; modifications of the network change the set of available techniques or, loosely speaking, the production 'function'.

*The Network Equilibrium Problem*

In reality it is not sufficient to consider trip demands as given and to find the GICs for those demands: changes in GICs feedback on trip demands.  If we write demand functions (42) and (43) for instance as

$$V^{dd}_{ij_a} = GIC_{ij_a} + GIC_{ij_b} + \pi_{other}\,OTHER_{ij_a} + e^{dd}_{ij_a} ,$$

$$(47)$$

$$c_d\,V^d_{ij_b} = GIC_{ij_a} + GIC_{ij_b} + \pi_{other}\,OTHER_{ij_b} + e^d_{ij_b} ,$$

the network equilibrium problem is to find values ($GIC^*_{ij_a}$, $GIC^*_{ij_b}$) and $V^{dd}_{ij_a}{}^*$ and $c_b\,V^d_{ij_b}{}^*$ which solve Equations (46) and (47) simultaneously when all other variables and parameters are held constant. Florian (11,12) has recently shown that, given demand parameters $\pi_m = (\pi_{km}\,\pi_{other\ m})$, it is possible to compute, for certain classes of demand functions, such network equilibria even if buses and cars interact on the common infrastructure.

If stochastic assignment procedures were ever used, one could envisage substituting the mean or expected value of GICs for deterministic values in the demand functions.

*Implications*

*Linearity in the form of the generalised cost function.*
Instead of using a weighted sum of variables which enter linearly in the generalised cost function, $GIC_{ij} = \sum_k \gamma_k\,K_{ij_k}$, one might wish to use a measure of generalised cost derived from a more general form of the demand function $GIC_{ij} = \sum_k \gamma_k\,K^{(\lambda)}_{ij_k}$, where $\lambda$ denotes the standard Box-Cox (4) transformation function. The growing use of these transformations in model estimation will imply new problems in the computation of user optimal paths in congested networks: further research is clearly needed to work out reasonable network equilibrium procedures in the context of generalised functional forms of demand equations (e.g. (18)).

*Market equilibrium and network equilibrium.* The distinction between supply-demand equilibrium and network equilibrium can be expressed as follows: empty buses (excess supply) and having to wait for more than one bus to obtain space (frequent excess demand) are perfectly compatible with the existence of a simultaneous solution for Equations (46) and (47). It is difficult to visualise in the case of excess demand because the solution is obtained through queues which cannot be measured by a *simple* transformation of data from vehicle operations. When variations in $V^{dd}_{ij_a}$ and in $c_b\,V^d_{ij_b}$ cause changes in link characteristics (because cars cause congestion or bus passengers slow down service) which imply changes in generalised itinerary costs, a new network equilibrium has to be computed. As such, that equilibrium is in general perfectly consistent with many degrees of aggregate excess supply in the car market (low occupancy rates) and in the bus market (empty buses). Shifts of network

supply or demand variables (legal car driving age, transit
management objectives, etc.), will imply a new network
equilibrium and an entirely different market disequilibrium.
This distinction between market and network equilibrium could
be made more spatially precise (link specific excess demand
on buses is consistent with system-wide excess supply, and so
on) and interesting situations could be derived from the fact
that parameter vectors $\pi_{Km}$ in general differ among mode
equations, but we shall instead belabour another implication
of network equilibrium procedures, namely the *natural
separability of demand estimation and network equilibrium
procedures*. We shall drop i, j subscripts, writing
$K_m = (K_a \; K_b)$.

*Demand estimation and the true itinerary input vector.*   In
the section on market equilibrium, we examined the problem of
estimation of $\pi_m = (\pi_{Km} \; \pi_{other \; m})$, given $K_m$. In this section
we have looked at the determination of $K_m$, given $\pi_m$.  Can
they be jointly determined?  Faced with this difficulty, as
he was estimating the demand function from observed link
flows, Wills (28) found that, using uncongested assignment
procedures of estimated origin-destination (O-D) demands, he
could determine $K_m$ and $\pi_{Km}$ separately from $\pi_{other \; m}$ by
iterating successively on $\pi_m$ and $K_m$, for constant demands by
origin-destination pair, and on $\pi_m$, for constant input
vectors $K_m$.  This result may have been due to the fact that,
with O-D demands of strongly varying sizes (including a few
very large ones), $K_m$ was in fact nearly constant or insensi-
tive to small changes in $\pi_{Km}$ in a certain domain.  He did not,
however, reach convergence when he tried to determine $K_m$ and
$\pi_m$ jointly.  This result may have been attributable to the
fact that this assignment procedure in the road network
caused a diversion on many paths, thus implying a different
value for some $K_a$s at each iteration; it may also have been
caused by changes in all-or-nothing itineraries on buses when
fare, departure frequency and travel time elements of $K_b$ were
considered in the assignment.  Had only one element been used
in all-or-nothing assignment procedures, there would have
been no clear reason to cycle.  It is interesting to conjecture on what would have happened if he had allowed
congestion to occur: convergence might not have been reached,
even for constant-pair demands.  The absence of a unique
solution might have been even more apparent than with
uncongested networks.  Further research in this area is
needed: note for instance that, if the criterion function
which is minimised in the estimation of $\pi_m$ is formulated in
terms of errors on pairs rather than on links, the problem
appears to be more difficult.  Such increased difficulty
might arise because, for given values of $\pi_{other \; m}$, a greater
number of combinations of $\pi_{Km}$ and $K_m$ would be compatible with
a certain value of the criterion function; in consequence,
the indeterminacy noted by Wills would be more likely to

appear, even under congested assignment procedures.

From our point of view, this problem of joint determination of $\pi_m$ and $K_m$ does not radically change the problem of estimation of the demand function because consistent estimation of $\pi_m$ *does not require actual observations* on the true value of $K_m$. All that is needed for consistency is an initial vector of instruments $\hat{K}_m$ correlated with the true and unknown $K_m$ and asymptotically independent of the error terms in Equation (47). As a first approximation it appears that nothing could be gained by subsequently iterating on $\hat{K}_m$ and $\pi_m$ but further research might show that a more efficient $\hat{K}_m$ might thus be obtained. Such a procedure amounts to taking into account an error of observation on $K_m$ and does not basically differ from those we referred to in our earlier discussion of simultaneous equations biases and errors of observation on service level variables. The procedure implies a separation of demand estimation from network equilibrium procedures in which the consistent estimates of $\pi_m = (\pi_{Km} \; \pi_{other \; m})$ obtained from the demand estimation are subsequently used in the network equilibrium problem.

## DYNAMIC EQUILIBRIUM

Changes in infrastructure in one period lead to changes in generalised itinerary costs over time; similarly, changes in other determinants of demand or of the behaviour of suppliers take time to make themselves felt. Demand function (1) can in fact be written as

$$T_t^d = f^d \; (\sum_n PD_{T_{t-n}} \, , \; \sum_n S_{T_{t-n}} \, , \; P_{v_t} \, , \; P_{f_t} \, , \; A_{f_t}) + e_t^d \, ,$$

$$n = 0, 1, \ldots, \qquad (48)$$

and it is part and parcel of the explanation of observed travel behaviour to determine these time paths which depend on the costs of acquiring information and other market imperfections. In fact all of the equilibria we have studied in the previous section may or may not occur within the time period over which observations are defined.

Although little is known in general about dynamic equilibrium, it should not be thought that these time paths are obvious. For instance, recent calculations [17] comparing fare and waiting time transit-demand elasticities in Montreal brought out a *variable-specific* response and a *market asymmetry*: as shown in Table 6.2, increases in adult fares resulted in lower passenger losses over two periods than over one period (e.g. some riders came back) but increases in waiting time resulted in higher losses over two periods than over one period (e.g. losses are permanent); the opposite appeared to be true for schoolchildren but the results for that market were statistically less sturdy than the results for the adult market.

TABLE 6.2   One-month and two-month transit demand elasticities and t-statistics for Montreal adults and schoolchildren[a]

|  | Adults | | Schoolchildren | |
|---|---|---|---|---|
|  | 1-period | 2-period | 1-period | 2-period |
| Net fare elasticity | 0.16 (-4.94) | 0.14 | 0.44 (-3.64) | 0.45 |
| Net waiting time elasticity | 0.36 (-3.77) | 0.48 | 0.21 (-0.75) | 0.05 |

a. Further details supplied in Appendix 6B below.  Results for
adults differ from those of Table 6.1 due to updating of
sample base.

Further analysis of the establishment of travel equilibrium
over time could reveal the existence of analogous differences
across income groups, trip purposes and other market segmentation criteria.  Models should clearly incorporate the
dynamic equilibrium features which transport system managers
have often felt relevant to policy decisions.  We have noted
(14) that the transit suppliers can react to past values of
demand and adjust service levels with a lag which implies
that market disequilibria can take time to disappear and
should make a difference to model formulation and estimation.


CONCLUSION

We have chosen paradigm equations to centre attention on six
notions of equilibrium and point out some of their implications for the explanation of travel behaviour.  Other
paradigms could have been chosen to discuss the same or other
issues.  We have not developed important topics such as that
of the relationship between car-seat trips and car-vehicle
trips or the problem of network aggregation which introduces
new errors of observation on the money and time itinerary
inputs; we have also abstained from discussing the
specification error arising from the standard consumer
homogeneity assumption of demand analysis and from the use of
variables to account for the actual heterogeneity of samples.
These topics and others, such as levels of aggregation of
models, are important to travel modelling but are perhaps
less closely connected to notions of equilibrium than those
we have selected in the hope of eliciting further discussions
of these issues and of their methodological implications.


APPENDIX 6A

The joint supply formulation of the dependent variable
arises out of the simple idea that if trip characteristics
constitute the travel good, then purchases of these
characteristics can be used as the dependent variable of
travel demand functions.  The models compared were based on
the following structure:

$$y_t = X_t \beta + u_t , \tag{49}$$

$$u_t = \rho_1 u_{t-1} + \rho_{12} u_{t-12} + e_t , \tag{50}$$

where all observations are monthly observations and $e_t$ is a
random vector of independent and identically distributed
error terms with zero constant variance $\sigma^2$. The matrix of
observation contains service levels of public transit and
competing modes, comfort levels, activity levels and other
variables. Under the standard format, the vector $y_t$ is
simply the number of adult passengers and, under the joint
supply format, it is $\alpha_1 K_{1_t} y_t + \alpha_2 K_{2_t} y_t + \alpha_3 K_{3_t} y_t$, where
$\alpha_1 + \alpha_2 + \alpha_3 = 1$. In the standard case the system consisting
of Equations (49) and (50) is estimated by a procedure (14)
which yields efficient estimates of the parameters $\beta$, $\rho_1$, $\rho_{12}$
and $\sigma^2$. In the joint consumption format all but one of the
elements of the dependent variable are put on the right-hand
side of Equation (49) and the presence of these endogeneous
variables as regressors biases the parameters estimates
obtained from the same limited-information technique. The
results were invariant with respect to the normalisation
procedure used. Table 6.1 presented some elasticities.
Further details can be found in Bilodeau (3).


APPENDIX 6B

The following equations for the schoolchildren market can be
added to the adult market Equations (49) and (50)

$$y_t = X_t \beta + u_t , \tag{51}$$

$$u_t = \rho_1 u_{t-1} + \rho_3 u_{t-3} + \rho_4 u_{t-4} + \rho_{12} u_{t-12} + e_t \tag{52}$$

and both systems can be estimated together using a generalised and iterated version of Park's method (24). To test for
lags in speed of adjustment of adults and schoolchildren to
fare and waiting time changes, lagged values of these
variables were added to the regressor lists of Equations (49)
and (51). In Table 6.2 t-statistics are given for the fare
and waiting time elasticities in the absence of adjustment
lags. No such statistics are given for the 2-period results
because these statistics exist only for each of the variables
the parameters of which were combined to yield the 2-period
adjustment results. Further details and results are
available in Gaudry (17) which also contains a summary of
market forecasts performed with estimates obtained under
various estimation techniques.


ACKNOWLEDGEMENTS

This chapter was made possible by a grant from the Research
and Development Centre of Transport Canada and benefitted from

comments on earlier drafts by Earl Ruiter and by Michael
Florian with whom fruitful discussions on the nature of
network equilibrium were also frequently held.  Michael Wills
gracefully permitted the reference to his forthcoming thesis.

REFERENCES

1. Barten, A.P., 'Evidence on the Slutsky conditions for
demand equations', *Review of Economics and Statistics*, 49,
pp. 77-84 (1967).
2. Baumol, W.C. (1965) *Economic Theory and Operations
Analysis*, Prentice-Hall, New Jersey, ch. 10.
3. Bilodeau, D., 'La formulation des équations de demande
de transport: un problème de consommation conjointe',
publication no. 74, *Centre de Recherche sur Les Transports*,
Université de Montréal (1977).
4. Box, G.E.P. and Cox, D.R., 'An analysis of transformations', *Journal of the Royal Statistical Society* B211,
pp. 211-43 (1964).
5. Charles River Associates, *A Disaggregated Behavioral
Model of Urban Travel Demand*, Report No. CRA-156-2, Cambridge,
Mass. (1972).
6. De Serpa, A.C., 'A theory of the economics of time',
*The Economic Journal*, 81, pp. 828-46 (1971).
7. Domencich, T.A., Kraft, G. and Valette, J.P.,
'Estimation of urban passenger travel behavior: an economic
demand model', *Highway Research Record*, 238 (1968).
8. Domencich, T.A. and McFadden, D. (1975) *Urban Travel
Demand*, North Holland, Amsterdam.
9. Fair, R.C. and Jaffee, D.M., 'Methods of estimation for
markets in disequilibrium', *Econometrica*, 40, pp. 497-514
(1972).
10. Fairhurst, M.H. and Morris, P.J., 'Variations in the
demand for bus and rail travel up to 1974', *Economic Research
Report R 210*, London Transport, London (1975).
11. Florian, M., 'A traffic equilibrium model of travel by
car and public transit modes', publication no. 32, *Centre de
Recherche sur Les Transports*, Université de Montréal (1975).
12. Florian, M., 'Urban travel demand models and multi-
modal traffic equilibrium', *Proceedings, Seventeenth Annual
Meeting, Transportation Research Forum*, pp. 184-90 (1976).
13. Gaudry, M. (1973)'The Demand for Public Transit in
Montreal and its Implications for Transportation Planning and
Cost-Benefit analysis', unpublished Ph.D. dissertation,
Princeton University, New Jersey.
14. Gaudry, M., 'An aggregate time-series analysis of
urban transit demand: The Montreal Case', *Transportation
Research*, 9, pp. 249-58 (1975).
15. Gaudry, M., 'A note on the economic interpretation of
delay functions in assignment problems', *Lecture Notes in
Economics and Mathematical Systems*, 118, Springer-Verlag,
New York, pp. 368-82 (1976a).
16. Gaudry, M. (1976b) *Passenger Demand Choice of Mode
Travel: Some Dimensions of Models*, Report EP 210 1-G-50,
Ministry of State for Urban Affairs, Ottawa.
17. Gaudry, M., 'Seemingly unrelated urban travel demands',
publication no. 61, *Centre de Recherche sur Les Transports*,
Université de Montréal (1977).
18. Gaudry, M. and Wills, M., 'Estimating the functional

form of travel demand models', publication no. 63, *Centre de Recherche sur Les Transports*, Universite de Montreal (1977).

19. Goldberger, A.S., 'Functional form and utility: a review of consumer demand theory', *Social Systems Research Institute Workshop Paper 6703*, University of Wisconsin (1967).

20. Huang, T.J. (1973) *An Econometric Analysis of the Determinants of Demand for Local Transit in Edmonton, 1961-1970*, Report No. 1, Finance Department, City of Edmonton.

21. Kresge, D.T. and Roberts, P.O. *et al.* (1971) 'Systems analysis and simulation models', in Meyer, J.R. (ed.) *Techniques of Transport Planning* II, The Brookings Institution, Washington, D.C.

22. Nguyen, S., 'An algorithm for the traffic assignment problem', *Transportation Science*, 8, no. 3, pp. 203-16 (1974a).

23. Nguyen, S. (1974b) 'Une approche unifiée des méthodes d'équilibre pour l'affectation du trafic', unpublished Ph.D. dissertation, publication no. 171, département d'informatique, Université de Montréal.

24. Parks, R.W., 'Efficient estimation of a system of regression equations when disturbances are both serially and contemporaneously correlated', *Journal of American Statistical Association*, 62, pp. 500-9 (1967).

25. Quandt, R.E., 'Testing hypotheses in disequilibrium models', *Econometric Research Program Memo no. 197*, Princeton University (1976a).

26. Quandt, R.E., 'Maximum likelihood estimation of disequilibrium models', *Econometric Research Program Memo no. 198*, Princeton University (1976b).

27. Samuelson, P.A., 'Contrast between welfare conditions for joint supply and for public goods', *Review of Economics and Statistics*, II, pp. 26-30 (1969).

28. Wills, M.J. (1978) 'Linear and Non-Linear Estimators of the O.D. Matrix', unpublished Ph.D. dissertation, Department of Geography, University of British Columbia, Vancouver, Chapter 6.

29. Zarembka, P., 'Transformation of variables in econometrics', in *Frontiers in Econometrics*, P. Zarembka (ed.), Academic Press, New York (1974).

30. Zellner, A., 'An efficient method of estimating seemingly unrelated regressions and tests for aggregation bias', *Journal of American Statistical Association*, 57, pp. 348-68 (1962).

Chapter 7

EQUILIBRIUM AND TRANSPORT SYSTEM DYNAMICS

Alan G. Wilson

INTRODUCTION

*The Nature of Demand-Supply Equilibrium in Transport Modelling*

It seems best to begin by spelling out explicitly, for a simple example, the nature of demand-supply equilibrium in transport modelling.  Whether micro (disaggregate) or macro (aggregate) demand models are used, the supply side of the transport model is essentially aggregative and is concerned with flows on links of networks.  Thus, initially we can consider aggregate demand variables, which can be generated directly or by aggregating micro-model variables, and we can relate these to supply side variables.

The outline to be sketched below is broadly speaking well known.  The notation used here is an extension of that used by Le Blanc, Morlok and Pierskalla (12), which is useful for exposition at a general level and for certain special problems, though the notation used by Evans (8) is more useful for proof of some of the relevant theorems.  Consider a spatial system divided into zones labelled i, j and so on in the usual way, and a network with links (r,s), where r and s label nodes.  Demand for travel between i and j, say $X_{ij}$, is usually considered to be a function of interaction cost, say $c_{ij}^I$ and other variables:

$$X_{ij} = X_{ij} \ (c_{ij}^I, \text{ other variables}).$$  (1)

The supply side is concerned with the provision of facilities for travel and, for convenience, the supply curves are best expressed in terms of the inverse function - costs of travel as functions of possible flows.  The complication of transport modelling arises from the relationship between interaction costs and link costs.  Let $\gamma_{rs}^L$ be link costs, and then

$$c_{ij}^I = \sum_{(r,s) \in R_{ij}^{min}} \gamma_{rs}^L$$  (2)

where $R_{ij}^{min}$ is the set of links forming the minimum cost route between i and j.  (For convenience, at present, we will assume that Wardrop's (26) first principle holds.)  Then, we could write the supply curves, formally, as

$$\gamma_{rs}^L = \gamma_{rs}^L \ (Q_{rs}^L, \text{ other variables}), \tag{3}$$

where $Q_{rs}^L$ is the flow on link $(r,s)$ and the 'other variables' would include representations of the nature of the facilities provided (as the parameter of the usual speed-flow curves for the highway network, or of such variables as vehicle frequency for a public transport network) and 'management' variables.

Transport models usually concern themselves with the equilibrium point - the intersection of supply and demand curves.  This is a reasonable assumption in relation to the transport system variables, since it implies simply that travellers respond quickly to any changes in those variables. It is less reasonable in relation to land-use variables, and at a later point in the chapter we return to the problem of modelling such disequilibria.  It can be shown that, in general, a unique equilibrium point exists.  At this point, let the flows be given by $x_{ij}^{rs}$, the flow from i to j carried on link $(r,s)$.  Then we will work mainly with the equilibrium demand variable $T_{ij}$ ($=x_{ij}$ at this point) given by

$$T_{ij} = \sum_{(r,j)\varepsilon R_{ij}} x_{ij}^{rj} \tag{4}$$

(or as $\sum_{(i,r)\varepsilon R_{ij}} x_{ij}^{ir}$) where $R_{ij}$ is the set of routes, each a collection of links used for $(i,j)$ trips, and the link loadings given by

$$Q_{rs} = \sum_{ij} x_{ij}^{rs}. \tag{5}$$

Let $\gamma_{rs}$ be the link costs at equilibrium and

$$c_{ij} = \sum_{(r,s)\varepsilon R_{ij}} \gamma_{rs} \tag{6}$$

again assuming that Wardrop's first principle holds.  At equilibrium, the basic transport model equations can be written

$$T_{ij} = T_{ij} \ (c_{ij}, \ O_i, \ D_j, \text{ other variables}) \tag{7}$$

$$\gamma_{rs} = \gamma_{rs} \ (Q_{rs}, \text{ other variables}), \tag{8}$$

together with Equations (5) and (6) for $Q_{rs}$ and $c_{ij}$.  In Equation (7), we have distinguished the usual trip end totals within the 'other variables' list.

Evans (8) has shown that a unique equilibrium exists for one of the standard entropy maximising distribution models together with a capacity restraint assignment algorithm by minimising

$$\sum_{rs} \Gamma_{rs} (Q_{rs}) + \frac{1}{\beta} \sum_{ij} T_{ij}(\log T_{ij}-1) \qquad (9)$$

subject to a suitable set of constraints, where

$$\Gamma_{rs} (Q_{rs}) = \int_{0}^{Q_{rs}} \gamma(q)\,dq \qquad (10)$$

and $\beta$ is the usual distribution model parameter. The first term in the summation can be written

$$\sum_{rs} Q_{rs} \, \bar{\gamma}_{rs} \qquad (11)$$

where

$$\bar{\gamma}_{rs} = \frac{1}{Q_{rs}} \int_{0}^{Q_{rs}} \gamma(q) \, dq \qquad (12)$$

so that $\bar{\gamma}_{rs}$ is a kind of average cost. Thus, in this formulation, users minimise their costs and satisfy Wardorp's first principle, while in the overall optimisation problem, a set of system-wide costs are minimised - provided $\bar{\gamma}_{rs}$ are taken as these costs - thus satisfying Wardrop's second principle in relation to these costs. This is an extension by Evans of a result cited by Potts and Oliver (17). Algorithms based on this theory have been developed by Murchland (16) and Florian, Nguyen and Ferland (9).

The essence of this discussion for present purposes is that demand-supply equilibrium can only be achieved in a transport model by matching demand, in Equation (7), against supply, in Equation (8), via a complicated set of relationships on a network implied by Equations (5) and (6). The complications have to be dealt with using iterative procedures or by setting up a large mathematical programme. This turns out to produce some problems for disaggregate models which will be confronted below.

Henceforth we will assume that this *kind* of framework exists for any model discussed below, even though the equilibrium existence proof cited above was for a particular model, and even though assignment procedures may be appropriate in some cases which do not satisfy Wardrop's first principle (cf. Van Vliet (25)).

*The Issues*

The complexities discussed above become very much more difficult to handle with disaggregation and the introduction of modes, person types and so on. So the first set of issues

are concerned with aspects of *modelling* and five such topics
are discussed below.  They deal with:

1   The structure of costs (or utilities or disutilities).
It is shown that particular care has to be taken to identify
the tree-like or other structures which carry the components
of costs and this is especially important since such costs
lie at the heart of the achievement of equilibrium, as we
have seen.
2   The nature of equilibrium in different kinds of models,
especially in relation to forecasting.  Because equilibrium
has not been considered sufficiently carefully in disaggregate models, forecasting with such models is not straightforward.  Further, it can be argued that land use-transport
system interactions remain an important consideration for
long-run dynamics and in this respect systems may be
permanently in disequilibrium.  Ultimately, parameter changes
should be related to this.  There are two issues here: first,
capacity restraint routines should be included in land-use
models; secondly, if land-use variables are not fixed, as in
transport models, there may be problems of uniqueness.
3   Evaluation measures.  Models are used in planning to
calculate the impact of change from one equilibrium to
another.  Some recent results in this field are reviewed and
it is shown that existing aggregate and disaggregate models
are often inconsistent with associated measures of consumers'
surplus.
4   Parameter changes and long-run equilibrium.  Is it
reasonable to assume that parameters in disaggregate models
are constant over time?
5   General considerations of dynamics.  Possible applications of catastrophe theory and related mathematics are
reviewed.  This helps us to deal with not only discrete
change but hysteresis effects and bifurcation which are
important for systems which are not in equilibrium.
Morphogenesis is discussed briefly.

Two aspects of the use of models are also considered:

1   How can disaggregate models provide more guidance for
policy-makers?  A recently developed technique for investigating the policy implications of a great variety of changes
is described.  It has been applied with an aggregate model
but could be applied to disaggregate models.
2   Optimal design.  Some recent results obtained by
embedding transport and related models within mathematical
programming models are described.


SOME RELEVANT MODELLING DEVELOPMENTS

*The Structure of Costs in Transport Models*

Within demand models, individuals, implicitly or explicitly,
make choices in relation to costs; and these costs measure
the supply of alternatives.  The supply-demand relationships
for transport models were explored in broad terms above.  The
choices involved are almost invariably of a complicated
nature, involving say, destination, j, mode k and route r at
the very least.  We will use these categories to illustrate

the problems of clearly defining and measuring cost structures, and we re-emphasise how essential this is, as they are typically the representatives of the supply side within the demand model (which is usually presented and estimated at the equilibrium point).

Consider first the choice (j,k), neglecting r.  In general, for an individual this may be characterised by a utility function u(j,k) made up of three components as follows:

$$u(j,k) = u_j^{(1)} + u_k^{(2)} + u_{jk}^{(3)}. \tag{13}$$

$u_j^{(1)}$ would refer to the properties of zone j - wages offered, attractiveness of shops, or whatever; $u_k^{(2)}$ to general properties of modes, such as comfort; and $u_{jk}^{(3)}$ to properties varying with j and k simultaneously, such as modal costs to particular destinations.  Costs are negative components of utility in this situation.  The utilities would be functions of attributes Z of j and k, perhaps combined linearly to give

$$u_j^{(1)} = \sum_s \alpha_s^{(1)} z_{js}^{(1)} \tag{14}$$

$$u_k^{(2)} = \sum_s \alpha_s^{(2)} z_{ks}^{(2)} \tag{15}$$

$$u_{jk}^{(3)} = \sum_s \alpha_s^{(3)} z_{jks}^{(3)}. \tag{16}$$

Much insight on cost and choice structures can be gained from random utility theory, drawing heavily on the recent work of Williams [28].  Each component of utility can be represented in the form

$$u = u^* + \varepsilon \tag{17}$$

where u* is some (observed) mean and ε is a random variable with mean zero.  Equation (13) then becomes

$$u(j,k) = u_j^{(1)*} + u_k^{(2)*} + u_{jk}^{(3)*} + \varepsilon_j^{(1)} + \varepsilon_k^{(2)} + \varepsilon_{jk}^{(3)}. \tag{18}$$

The resulting probabilistic choice model depends on the assumptions made about the distributions of $\varepsilon^{(1)}$, $\varepsilon^{(2)}$ and $\varepsilon^{(3)}$.  A common assumption is that $\varepsilon_j^{(1)}$ and $\varepsilon_k^{(2)}$ are zero and that the $\varepsilon_{jk}^{(3)}$ distributions are independent or that $\varepsilon_{jk}^{(3)}$ represents some overall random component.  This leads to the multinomial logit model

$$P_{jk} = \frac{e^{\Delta(u_j^{(1)*} + u_k^{(2)*} + u_{jk}^{(3)*})}}{\sum_{jk} e^{\Delta(u_j^{(1)*} + u_k^{(2)*} + u_{jk}^{(3)*})}} \tag{19}$$

where $\Delta$ is some suitable parameter.  This is the so-called simultaneous model much favoured by authors such as Ben-Akiva [1].  Costs (or disutilities) in such a model simply appear as attributes such as in Equations (14-16).  It can easily be shown however, that the assumptions generating this model are too simple and above all neglect the phenomenon of attribute correlation which is frequently present in transport modelling.  The model structure is inconsistent with the form of utility function assumed to generate it, in particular, if either $u_j^*$ or $u_k^*$ are zero, a recursive structure would be expected, but this does not follow from Equation (19).  Such a recursive model is explored further below.  We can now seek more guidance on composite costs.  Let us take one example. Define

$$P_{j*} = \sum_k P_{jk} \qquad (20)$$

It should be possible to view the choice of j in terms of some composite transport cost, say $\tilde{c}_{j*}$ (as well as other attributes) and that this $\tilde{c}_{j*}$ should be a function of the modal costs $c_{jk}$ (which for present purposes we can take as the measured cost).  Indeed one might expect the transport cost part of j-utility to be 'passed up' to the j-decision in terms of modal costs.  This does turn out to be the case. However, a modelling decision has to be made on the choice structure, simultaneous, sequential or something more complicated, and ultimately, this decision has to be tested in relation to actual traveller behaviour.

Suppose, for example, choice behaviour is governed by a utility function of the form

$$u(j,k) = u_j^{(1)*} + u_{jk}^{(3)*} + \varepsilon_j^{(1)} + \varepsilon_{jk}^{(3)} . \qquad (21)$$

Then

$$P_{jk} = P_j P_{k|j} \qquad (22)$$

It can be shown that this is equivalent to choice behaviour in two stages: choose j such that

$$u_j^{(1)*+} = u_j^{(1)*} + \tilde{u}_j^{(1)*} + \varepsilon_j^{(1)} + \tilde{\varepsilon}_j^{(1)} \qquad (23)$$

is a maximum, having chosen k such that

$$u_{jk} = u_{jk}^{(3)*} + \varepsilon_{jk}^{(3)} \qquad (24)$$

is a maximum for each j.  Equation (24) and an assumed Weibull distribution for $\varepsilon_{jk}$ produces

$$P_{k|j} = \frac{e^{\Delta u_{jk}^{(3)*}}}{\sum_k e^{\Delta u_{jk}^{(3)*}}} . \qquad (25)$$

Equation (23) is more tricky in general.  Williams suggests that, as an approximation,

$$\varepsilon_j^{(1)+} = \varepsilon_j^{(1)} + \tilde{\varepsilon}_j^{(1)} \tag{26}$$

should itself be assumed to have a Weibull distribution. Then,

$$P_j = \frac{e^{\beta(u_j^{(1)*} + \tilde{u}_j^{(1)*})}}{\sum_j e^{\beta(u_j^{(1)*} + \tilde{u}_j^{(1)*})}} . \tag{27}$$

But Williams shows that the joint, sequential decision can only be conceptualised in this form if

$$\tilde{u}_j^{(1)*} = \frac{1}{\Delta} \log \sum_k e^{\Delta u_{jk}^{(3)*}} \tag{28}$$

which is

$$e^{\Delta \tilde{u}_j^{(1)*}} = \sum_k e^{\Delta u_{jk}^{(3)*}} \tag{29}$$

the formula for the composite cost function.  The sequential model can now be formed by substituting from (25) and (27) into (22).  In effect, $u_j^{(1)*}$, represents the properties of zone j excluding transport attributes (except possibly ones which 'correlate' modal attributes) and $\tilde{u}_j^{(1)*}$ is the estimate of transport disutility transmitted to the j-decision level in the utility tree via Equation (28).  The parameters $\beta$ and $\Delta$ can be related to the standard deviation of $\varepsilon_j^{(1)+}$ and $\varepsilon_{jk}^{(3)}$.  When the standard deviation of $\varepsilon_j^{(1)}$ is zero, $\beta=\Delta$ and the model reduces to an indirect representation of the multinomial logit model - as for example in Equation (19), but with $u_k^*=0$ also.  It should be emphasised that other choice structures are possible and, indeed, Williams shows how to develop a general choice model which can be operation-alised at least in approximate form.  In each case, an appropriate concept of composite cost arises naturally.

   If tree-like utility and choice structures are assumed, the argument presented above can easily be extended up and down the hierarchy.  For more complicated structures, the argument becomes correspondingly more difficult, as hinted in the previous paragraph.  In the context of the (j,k,r) choice example, with origin i, say, this means that route costs are related to mode costs by a relationship of the form

$$e^{-\Delta c_{ij}^k} = \sum_r e^{-\Delta c_{ij}^{kr}} . \tag{30}$$

At higher levels in the tree, costs associated with such levels are incorporated directly and referred to here as $\delta_{ij}^k$, $\delta_{ij}$ and $\delta_i$. Then mode costs are related to distribution costs by

$$e^{-\lambda c}{}_{ij} = \sum_k e^{-\lambda (c_{ij}^k + \delta_{ij}^k)}. \tag{31}$$

It is also possible to get average origin costs, as

$$e^{-\beta c}{}_{i*} = \sum_j e^{-\beta (c_{ij} + \delta_{ij})}, \tag{32}$$

and total system costs as

$$e^{-\alpha c}{}_{**} = \sum_i e^{-\alpha (c_{i*} + \delta_i)}, \tag{33}$$

which turns out to be useful for evaluation purposes. Through the $\delta$'s, standard deviation is being added at each level so that $\alpha < \beta < \lambda < \Delta$ (since the parameters are inversely related to the standard deviation). The essential demand-supply equilibrium is that it is costs such as $c_{ij}^{kr}$ (and the $\delta$'s) which are measured, via related network link flows, say $\gamma_{r's'}^{kr}$, for link r s . They must represent an equilibrium point at the link level, and effects of congestion or whatever must then be transmitted to other levels of the model via relationships (29-32). This remark applies to aggregate and disaggregate models. It should also be noted that these concepts of composite costs are particularly useful in the context of multi-route assignment models where the routes can have common links. This arises because the common links give rise directly to attribute correlation between the routes containing them, and the general methods outlined above and given in detail by Williams, can handle such problems.

*The Achievement of Demand-Supply Equilibrium in Models*

It is already clear from the argument at the beginning of this chapter, that equilibrium can only be achieved either by continual iteration between distribution-modal choice and assignment submodels, or by setting up a large-scale joint distribution-modal choice-assignment mathematical programme. This applies to disaggregate as well as aggregate models because of the inherent aggregate nature of the supply-side model. Thus, a disaggregate model may be calibrated using observed data, but should then be embodied in a capacity-restraint assignment procedure of some kind. This is particularly important if route choice is in any case being modelled explicitly. Further, the costs at various levels of such models should be related according to the methods outlined above. Such procedures should be the basis of the use of disaggregate models in policy-making where they must

be run for some new situation.  It would be wrong either to
keep costs fixed, or to assume they can be given exogenously
without the demand function interacting with a supply-side
model.  There should be no problem, except possibly a
computational one, in relation to the highway network speed-
flow relationships which are well known and much studied (as
outlined in the paper by Le Blanc *et al.* (12)).  On the
public transport side, more work is now being done which
estimate supply functions (e.g. Morlok and Vandersypen (14),
Morlok (15), Gaudry (10)).  This style of work, in effect,
provides short-run supply functions; for the longer run,
account has also to be taken of the supply of new infrastructure.

*Some Notes on Evaluation Criteria*

The essence of economic evaluation of alternative plans lies
in the estimation of costs and benefits at the demand-supply
equilibrium of each plan and, usually, the comparison of each
with some base year; or possibly directly with each other.
This short section draws attention to some new and important
results in the context of this discussion of equilibrium and
the reader is referred elsewhere for details (Williams (27);
Champernowne, Williams and Coelho (3); and Williams (28)).
The most difficult problem is always the estimation of
consumers' benefits measured as the area under a demand
curve.  This measure can be defined as a path integral, and
is usually calculated in an approximate form using the well-
known rule-of-a-half.  The authors referred to above have
shown how, in many cases, the path integral can be calculated
explicitly and exactly, and that the answer is independent of
the path taken when Hotelling's integrability conditions are
satisfied.  These conditions do hold for many disaggregate
models, though not always exactly for some of the approximate
forms of general choice models and for some entropy maximising models.  It also turns out that the surplus is in effect
the highest level of composite utility or cost in hierarchic-
al·schemes - involving quantities such as $c_{**}$ in Equation
(33) above.  In many cases where land-use configurations
change, model balancing factors play an important role in the
surplus formula, being closely related to the dual variables
in a mathematical programming formulation of the corresponding model.  This simplifies computation enormously (and
achieves exact answers) but emphasises the point that
information must be passed up the cost hierarchy as properly
identified composites, and that the base-level costs *must* be
calculated at equilibrium.  Williams (28) writes: 'What is
assumed is that the dual variables $\underline{\alpha}$ and $\underline{\gamma}$ are computed with
level-of-service variables appropriate to a state in which
demand and supply are in equilibrium'.

*Parameter Change, Land Use and Long-run Equilibrium*

Random utility methods give new insights into the parameters
of disaggregate models: such parameters can be directly
related to the standard deviations of the random 'dispersion'
variables associated with corresponding terms of the utility
function.  Further, in hierarchical cases, as outlined

earlier where r, k and j are chosen in sequence, if the
parameters associated with the r,k,j choices are $\Delta$, $\lambda$ and $\beta$,
then it can be shown that $\beta < \lambda < \Delta$ since, because of transmitted
composite cost or utility components, the standard deviation
is being added to in progress up the utility tree.   Ideally,
all these parameters should be estimated for as fine as
possible categories of socio-economic groups within the
population.   It then follows that any change of parameters
over time should be related mainly to changes of taste and
that an assumption that they remain constant over time, the
typical one, may not be a bad one at least for the short run.
However, there are many possible reasons for the existence of
the dispersion of individual utilities (see Williams (28)
p. 19 for a list and discussion of these) and, of course,
changes in any one of these may change the model parameters.
Further, parameters are not usually in practice estimated for
very finely defined socio-economic groups.   And finally,
other changes may arise in particular at times of high
inflation and the rapid *relative* changes of price among many
goods; in other words, when utility is related to the money
price of goods, if relativities are not maintained,
preferences may change substantially.

Parameters in entropy maximising models are usually seen
from a different perspective.   They can be taken as measures
of another kind of dispersion of the 'distance' away from
maximum utility as represented in one of the constraint
equations and which can be calculated from a mathematical
programme with such a constraint replacing entropy as the
objective function (cf. Wilson and Senior (34)).   A simpler,
but related view, is that the parameters can be directly
related to the term on the 'right-hand side' of the
constraint equations, such as $\beta$ in a distribution model

$$T_{ij} = A_i B_j O_i D_j e^{-\beta c_{ij}}, \tag{34}$$

to C in

$$\sum_{ij} T_{ij} c_{ij} = C. \tag{35}$$

This has led to the suggestion (Wilson (30)), which has been
investigated by Southworth (22,23), that C could be disaggregated to $C_i^n$, where n is person type, and the parameter to $\beta_i^n$.
$C_i^n$ can then be estimated in terms of a small number of
exogenous variables and all the $\beta_i^n$s become endogenous to the
model.

Thus, from a random utility or entropy maximising
perspective, disaggregate or aggregate, some further insights
into the nature of model parameters can be obtained.   Much
more research is needed to take this knowledge a step further
and for us then to be more confident about parameter values
for forecasting.   The point to be emphasised in the context
of this discussion is that from any perspective, main model
parameters are likely to be functions of supply-side
variables computed at equilibrium, and a good deal of further
research should be devoted to this issue.

This is also an appropriate place to emphasise one further point: long-run equilibrium involves land use and associated activities as well as transport demand and supply. The search for effective links between land use and transport models should therefore be continued. A number of aggregate models now exist which incorporate land use-transport model feedbacks, some in considerable detail (for example, Putman (18), Mackett (13)). The time may now be ripe for corresponding developments with disaggregate models. There should be no difficulty in principle, for example, in developing a random-utility residential location model which simultaneously modelled the location variables and the journey to work, and have explicitly related transport model parameters to those associated with other choices. Some progress has been made with corresponding evaluation issues for land use as well as transport variables (3,4).

*Towards a Current Best-practice Model*

We noted in the previous section that random-utility theory suggests that choice parameters at higher levels of a hierarchy should be less than or equal to those at lower levels. For example, if $\beta$ is the distribution parameter and $\lambda$ the modal choice one, then $\beta < \lambda$. In practice, in aggregate models, $\beta > \lambda$, and often by a factor of 2 or 3. This has led to discussion of possible changes in modelling practice which are summarised here and reported in more detail elsewhere (21).

Consider a model of the form

$$T_{ij}^{kn} = A_i B_j O_i^n D_j e^{-\beta^n (u_{ij}^n + C_{ij}^n)} M_{ij}^{kn}, \tag{36}$$

where

$$M_{ij}^{kn} = \frac{e^{-\lambda^n c_{ij}^k}}{\sum_{k \epsilon \gamma(n)} e^{-\lambda^n c_{ij}^k}}. \tag{37}$$

$T_{ij}^{kn}$ is the number of trips from i to j by mode k by person of type n; $M_{ij}^{kn}$ is modal share; $O_i^n$ are trip productions in i by person of type n; $D_j$ are trip attractions in j; $c_{ij}^k$ are modal generalised costs. $C_{ij}^n$ are composite costs calculated, using Equation (31) as

$$e^{-\lambda^n C_{ij}^n} = \sum_{k \epsilon \gamma(n)} e^{-\lambda^n c_{ij}^k} . \tag{38}$$

$u_{ij}^n$ are utilities associated with locational characteristics of the (i,j) choice, such as residential quality at i and wage at j. $A_i^n$ and $B_j$ are the usual balancing factors and

$\gamma(n)$ is the set of modes available to type n people. This model can be derived as an entropy maximising model (cf. (29)) or (give or take the balancing factors) as an aggregated random-utility model for 'market segment' n. Usually, $u_{ij}^n$ would be taken as zero. It can now be argued that it is this latter practice which 'distorts' the best estimates of $\beta^n$ and $\lambda^n$ to produce $\beta^n > \lambda^n$ when it should be the other way round. In the long run, this difficulty could be overcome by research on the $u_{ij}^n$s. In the short run, the following procedure is suggested.

Take $u_{ij}^n$ as zero and calibrate the model in the usual way. Then write

$$e^{-\beta^n C_{ij}^n} = e^{-(\beta^n - \beta^{nl})\, C_{ij}^n}\, e^{-\beta^{nl} C_{ij}^n} \qquad (39)$$

where $\beta^{nl}$ is to be chosen so that $\beta^{nl} \leqslant \lambda^n$. It can then be argued that $e^{-(\beta^n - \beta^{nl})C_{ij}^n}$ is an estimate of $e^{-\beta^{nl} u_{ij}^n}$. That is

$$e^{-\beta^{nl} u_{ij}^n} = e^{-(\beta^n - \beta^{nl})C_{ij}^n} \qquad (40)$$

So far, this is a matter of definition, together with a conjecture about $u_{ij}^n$ at the base year. For forecasting, however, we can now say that in the absence of any new information, it is best to keep $u_{ij}^n$ at its base year value, but to change everything else in the usual way:

$$T_{ij}^{kn} = A_i^n B_j^n O_i^n D_j^n e^{-\beta^{nl} \bar{u}_{ij}^n}\, e^{-\beta^{nl} C_{ij}^n}\, M_{ij}^{kn} \qquad (41)$$

where the variables all take future values except $u_{ij}^n$, written as $\bar{u}_{ij}^n$ to emphasise this, and the parameters also retain their base year values. Note that if $\beta^{nl}$ is chosen to be equal to $\lambda^n$, then Equation (41) reduces to

$$T_{ij}^{kn} = A_i^n B_j^n O_i^n D_j^n e^{-\lambda^n \bar{u}_{ij}^n}\, e^{-\lambda^n c_{ij}^k} \qquad (42)$$

by using Equation (38) to simplify the last two terms.

*Aspects of Equilibrium and Dynamical Systems Theory*

In any discussion of equilibrium, it is appropriate to examine the contributions of dynamical systems theory. We look at three aspects - the application of catastrophe theory to the individual's change of state, the application of the theory of differential equations to modal competition, and an examination of the concept of morphogenesis in relation to higher order transport facilities (in effect, a fundamental change of state on the supply side).

Catastrophe theory has been applied to the study of an individual's equilibrium position in relation to choice of mode, and shows how this can change over time (31,7). The individual's state is characterised by a section of the folded surface which generates the cusp catastrophe, as shown in Figure 7.1. The dashed lines show possible trajectories in time for the individual's state as a function of changing $\Delta c$, the difference in modal costs. The important new, and probably realistic, phenomenon built into this model which does not appear in others is the *hysteresis* effect. If a passenger is 'lost' to a mode, it takes a reverse discrete shift $2\lambda$ in $\Delta c$ to win him back again. Further research is needed to integrate this model with the more traditional, say logistic, models.

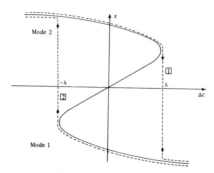

Fig. 1.   Modal choice and catastrophe theory

The second example builds on an approach to ecological modelling developed by Rescigno and Richardson (19) and described by Hirsch and Smale (11). Their application relates to two species of animals competing for a fixed food supply. It can equally well be applied to two modes competing for a fixed supply of passengers. At this very aggregative level, all we can hope for is new insight, but we do begin to achieve this in relation to systems in states of disequilibrium. The spirit of the approach is captured in a phase diagram relating modal shares $M_1$, $M_2$ in Figure 7.2. This represents solutions to differential equations of the form

$$\dot{M}_1 = K_1(M_1, M_2) M_1 \qquad (43)$$

$$\dot{M}_2 = K_2(M_1, M_2) M_2 \qquad (44)$$

where $K_1$ and $K_2$ are rates of change of modal patronage. They would be complicated functions of 'profits' to suppliers and surpluses to consumers, computed at a series of equilibrium

points (in transport costs - which may be states of
disequilibrium in modal shares). The analysis is given in
detail elsewhere for competing shopping centres (33). In
essence, however, such phase diagrams typically have two
stable points A and P and a phase space divided into two
regions by the dashed line shown. (This assumption can be
made without loss of generality: if A is unstable, then B is
stable, and vice versa.) Thus, if a system is in disequilibrium following some disturbance, perhaps a policy change,
it will then tend to A if it starts above the dashed line and
to P if it starts below. This illustrates *bifurcation* of
system behaviour: a small difference in initial state, as for
example $V_0$ and $V_1$, can lead to a major difference in final
state. Note also that one stable point (in this case P)
typically represents coexistence while the other the annihila-
tion of one mode. Can the downward public transport spiral
be interpreted in terms of trajectories in a phase diagram of
this kind?

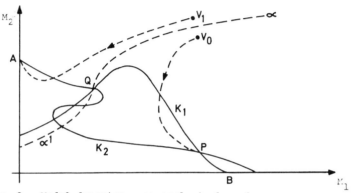

Fig. 2. Modal dynamics: an ecological analogy

The third example relates to morphogenesis and change of
state on the supply side. Consider the two situations shown
in Figure 7.3. We have a one dimensional channel on which
transport facilities have to be supplied at a certain minimum
density across the channel. Suppose there are roads, each of
width w. Then, typically, costs of building a road of width
w, say f(w) will have the form shown in Figure 7.4(a) and
capacity, C(w), that in Figure 7.4(b). This shows that for
low demand, D, a large number of the smallest roads would be
built, while at some point of greater demand, one 'trunk'
facility will be required - as in Figure 7.3(b) as compared
to Figure 7.3(a). This process can be represented by the
supply side mathematical programme

$$\text{Min} \sum_i n_i f(w_i) \tag{45}$$

such that

$$\sum_i n_i C(w_i) > D \tag{46}$$

$$\sum_i n_i > N \qquad\qquad\qquad\qquad (47)$$

where $n_i$ is the number of facilities of width $w_i$. As $D$ increases smoothly, at a certain point, a trunk facility will 'suddenly' be required. Thus, if the nature of the technology can be specified in functions such as $f$ and $C$, supply-side morphogenesis can in principle be predicted as a function of demand. Once again, realistic problems would be very much more difficult to handle, but the insight is useful. This problem is sketched out in more detail for different orders of shopping centre (7).

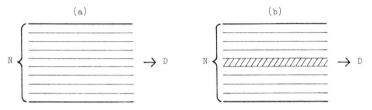

Fig. 3.   The evolution of 'trunk' structure

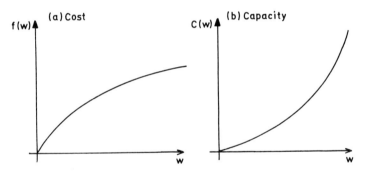

Fig. 4.   Cost and capacity curves for facility supply

## THE USE OF MODELS: MORE GUIDANCE FOR THE SUPPLY SIDE

*The Problem: Large Numbers of Alternatives*

Typically, models are used to assess the impacts of policy options represented as alternative plans. The policy options are the inputs to the models and the impact measurements are the outputs - each at some new equilibrium point. The problem is that it is rarely possible to consider an adequate range of alternatives and even more difficult to organise feedback to the designers for new alternative plans to be generated in the light of predicted impacts of earlier tests. This has occasionally been usefully attempted (for example,

see SELNEC Transportation Study (20)), but still the guidance
offered from the demand models to the supply side seems
inadequate.  The problems arise partly from the difficulty
and expense of running models for a sufficiently large number
of alternatives (and of interpreting the vast amount of
computer output!), and partly from difficulties in the design
process.  The two pieces of research reported below suggest
ways of tackling each of these problems in turn.

*A Method for Encapsulating a Large Amount of Policy-oriented
Information*

The method turns on a simple idea: to present information
graphically relating inputs and outputs (policies and
impacts) relative to an origin point - or base run - which is
that adjudged to be the *most likely future* (MLF) for the
planning horizon being considered.  The graphs are then
presented in two ways: first, choosing an output and plotting
percentage changes in that output against percentage changes
in various input variables; or conversely, focusing on an
input variable and plotting percentage changes in that
against percentage changes in a range of output variables.
It turns out that the curves can be sketched from a
relatively small number of model runs.  Two typical examples
are shown in Figures 7.5 and 7.6.  Given the first type, in
this case taking private mode trips to a central area, the
policy-maker can see which policy instrument or combination
of instruments may be most effective in achieving goals
related to that indicator.  Conversely, with Figure 7.6, he
can see the range of impacts of any particular policy
instrument.  Some caution has to be exercised in interpreting
the results as they have been obtained mostly using model
runs which vary one policy instrument at a time from the MLF,
and the consequences of changing two or more policy instruments simultaneously are not likely to be strictly additive.
Note finally, that slopes of the curves at the origin give a
measure of elasticities.

The method and results are reported in more detail
elsewhere for the application of a disaggregated macro-model
to West Yorkshire (2).  There is no problem in principle in
applying this method to disaggregate models, however.

One final point should be noted from this work in the
context of this particular discussion on equilibrium.  The
effects of using different assignment procedures to achieve
equilibrium were tested, and it was found that in some
instances, the resulting variance in model outputs was of a
comparable order of magnitude to some of the consequences of
policy changes which were being investigated.  So this is
another reason for caution in certain kinds of model-based
analysis and emphasises the need to carry out these kinds of
sensitivity tests in particular situations.

*Mathematical Programming and Aids to Design*

We have seen that one of the major problems of using models
in policy-making lies in the long-recognised combinatorial

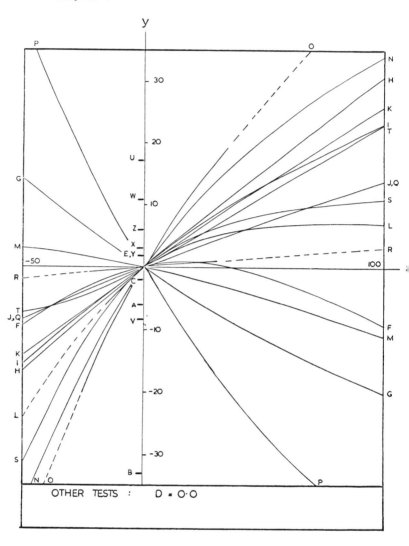

Figure 7.5   Private Transport Trips from Suburban Leeds to
Central Leeds: Impact of Possible Policies

*Key to Figure 7.5*

| | |
|---|---|
| x | % change in test parameter from MLF value |
| y | % change in $T^{1n}_{ij}$ from MLF value |
| | where i is all zones in category 2 (suburban Leeds) |
| | j is all zones in category 1 (central Leeds) |
| A,B | land-use plans |
| C,D,E | network alternatives |
| F | private running costs |
| G | parking charges |
| H | public transport fares |
| I | time spent walking to public transport |
| J | time spent waiting for public transport |
| K | time spent travelling on public transport |
| L | mean car occupancy |
| M | value of time |
| N | income level (with corresponding value of time change) |
| O | income level (without corresponding value of time change) |
| P | car prices |
| Q | trip rates of highest income category |
| R | mean income in a particular zone (17) |
| S | distance deterrence parameter for car-available people |
| T | distance deterrence parameter for non-car-available people |
| U | WYTCONSULT deterrence parameter |
| V | damped trip ends |
| W | use of post-distribution modal split model |
| X | Burrell-type assignment |
| Y | Dial-type assignment |
| Z | 'all or nothing' assignment |

*Key to Figure 7.6*

| | |
|---|---|
| x | % change in Leeds city centre park charge from MLF value |
| y | % change in indicators from MLF value |
| a | mean private transport speed in the Leeds area |
| b | private transport destinations in Leeds city centre |
| c | public transport destinations in Leeds city centre by people with cars available |
| d | destinations in Leeds city centre by people without cars available |
| e | private transport trips within central Leeds (zone group 1) |
| f | private transport trips from suburban Leeds (group 2) to central Leeds (group 1) |
| g | private transport trips from medium sized towns of the conurbation (group 3) to central Leeds (group 1) |
| h | private transport trips from rural areas and free standing towns (group 4) |
| i | private transport trips from large towns of the conurbation (group 5) |

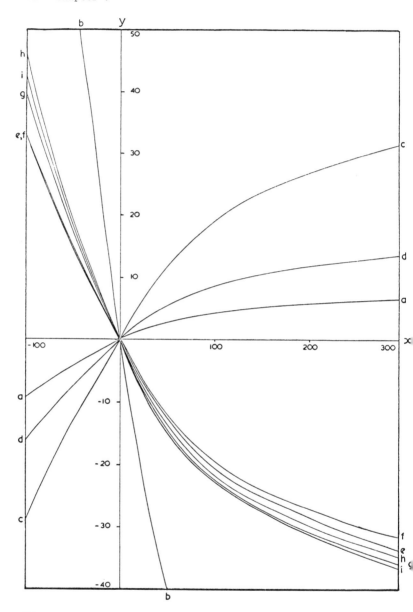

Figure 7.6  Parking Charges in Central Leeds: Impact on
Various Indicators (Test G)

problem of design: how to set up, search and evaluate the
impacts of the vast number of options in policy space.  In
the previous section, one heuristic approach to this problem
was outlined, in which demand variables could be directly
related to those supply variables which are policy options.
A more formal approach is to use mathematical programming.
Let the supply-side policy variables be $x$, the demand side
variables $y$ and exogenous variables and parameters $\alpha$.
Assume it is possible to construct an objective function
$F(x,y,\alpha)$; then the optimum policy is obtained from

$$\underset{x}{\text{Max}} \; F(x,y,\alpha) \tag{48}$$

such that

$$f_i(x,y,\alpha) = 0 \tag{49}$$

where the equations in $f_i$, Equation (49), represent the
behavioural demand model.  Such problems are usually highly
non-linear in both objective functions and constraints and
are computationally intractable.  Equation (49), for example,
may be the sort of transport demand model, computed at
equilibrium, much discussed in this chapter.  One recent
discovery has made at least a subset of these problems
tractable however.  If the model represented by Equation (49)
can itself be represented as a mathematical programme
involving a non-linear objective function but linear constraints (as with an entropy maximising model, but also with
some utility maximising models), say as

$$\underset{y}{\text{Max}} \; G(x,y,\alpha) \tag{50}$$

such that

$$g_i(x,y,\alpha) = 0 \tag{51}$$

then there is a related problem which can be written as

$$\underset{x,y}{\text{Max}} \; F(x,y,\alpha) + \lambda G(x,y,\alpha), \tag{52}$$

such that

$$g_i(x,y,\alpha) = 0 \tag{53}$$

for a suitable constant $\lambda$.  This then has a non-linear
objective function but wholly linear constraints and is often
computationally feasible.  This is *not* equivalent to (48) and
(49) (as erroneously stated in Coelho and Wilson (5), but it
can be shown that the variables $x,y$ satisfy equations of the
form (49) and that in many cases, this second and computationally tractable problem is the most interesting from a
planning viewpoint.  This method has been applied, for
example, to the calculation of optimum shopping centre sizes
$\{W_j\}$ in the standard Lakshmanan-Hansen model, assuming that
consumers' surplus is to be maximised (5).  There seems to be

no reason in principle why such a method should not be applied in the transport model case. The model Equation (49) would be a joint distribution-modal split-assignment model which Evans (8) has shown can itself be represented as a mathematical programme. The difficulty is that it is much harder to handle network-link optimisation problems in this way than shopping centre sizes (cf. (24)). Finally, we note that the objective function in such a programme need not be confined to consumers' surplus terms; other terms can be added with appropriate weights representing the objectives of government, producers and so on. It is then possible that smooth changes in such weights can be related through bifurcation theory to discrete changes in policy (32,33).


CONCLUSIONS

A number of conclusions follow directly from the argument presented above. Choice structures in disaggregate models have to be investigated in detail, theoretically and empirically; 'simultaneous' models, while easier to handle than others, are not usually adequate. This leads to some results on composite costs which underpin any discussion of demand-supply equilibrium. Then, the use of some kind of capacity-restraint assignment procedure is necessary even in disaggregate models, for equilibrium to be presented explicitly. There are related results on evaluation and some new insights on parameter values. A plea is made not to forget the importance of ongoing research on land use-transport long-run equilibrium. Further insights, though largely of a more speculative nature, are obtained from dynamical systems theory. The two pieces of research on the use of models show that progress can be made in connecting demand-model information with supply-side policy variables.

Perhaps two broad conclusions can be drawn. First, attempts to describe a variety of more or less best-practice transport models would in general lead to different models than those used by most transport planning authorities. Research has moved on very quickly in the last few years but has not been adequately synthesised into a new and suitably eclectic best practice. Secondly, the research frontier has perhaps already moved on towards more speculative areas. This should be encouraged, but such research should be kept in balance with the urgent task of consolidation so that the best available tools become available for current problems.


ACKNOWLEDGEMENTS

The author is grateful to Dr Huw Williams for a number of helpful discussions in relation to this chapter.


REFERENCES

    1. Ben-Akiva, M.E. (1973), 'Structure of Passenger Transport Demand Models', unpublished Ph.D. Dissertation, Department of Civil Engineering, MIT, Cambridge, Massachusetts.
    2. Bonsall, P.W., Champernowne, A.F., Mason, C.A. and

Wilson, A.G. (1977) *Transport Modelling: Sensitivity Analysis and Policy Testing*, Pergamon Press, Oxford.

3. Champernowne, A.F., Williams, H.C.W.L. and Coelho, J.D., 'Some comments on urban travel demand analysis, model calibration and the economic evaluation of transport plans', *Journal of Transport Economics and Policy*, 10, pp. 267-85 (1976).

4. Coelho, J.D. and Williams, H.C.W.L., 'On the analysis, evaluation and design of land use-transport plans', *School of Geography Working Paper*, University of Leeds (1977).

5. Coelho, J.D. and Wilson, A.G., 'The optimum location and size of shopping centres', *Regional Studies*, 10, pp. 413-21 (1976).

6. Coelho, J.D. and Wilson, A.G., 'Some equivalence theorems to integrate entropy maximising submodels within overall mathematical programming frameworks', *Geographical Analysis*, 9, pp. 72-83 (1977).

7. Dendrinos, D.S., 'Two applications of catastrophe theory in transportation planning and urban economics', *Paper presented at the National Meeting, Regional Science Association*, Toronto (1976).

8. Evans, S.P., 'Derivation and analysis of some models for combining trip distribution and assignment', *Transportation Research*, 10, pp. 37-57 (1976).

9. Florian, M., Nguyen, S. and Ferland, J., 'On the combined distribution assignment of traffic', *Transportation Science*, 9, pp. 43-53 (1975).

10. Gaudry, M., 'An aggregate time-series analysis of urban transit demand: the Montreal case', *Transportation Research*, 9, pp. 249-58 (1975).

11. Hirsch, M.W. and Smale, S. (1974) *Differential Equations, Dynamical Systems and Linear Algebra*, Academic Press, New York.

12. LeBlanc, L.J., Morlok, E.K. and Pierskalla, W.P., 'An efficient approach to solving the road network equilibrium traffic assignment problem', *Transportation Research*, 9, pp. 309-18 (1975).

13. Mackett, R.L., 'A dynamic integrated activity allocation-transport model for West Yorkshire', in *Urban Transportation Planning - Current Themes and Future Prospects*, Hills, P.J., Dalvi, M.Q. and Bonsall, P.W. (eds), Abacus Press, Tunbridge Wells (1977).

14. Morlok, E.K. and Vandersypen, H.L., 'Schedule planning and timetable construction for commuter railroad operations', *Transportation Engineering Journal*, 99, pp. 627-36 (1973).

15. Morlok, E.K., 'Supply functions for public transport: initial concepts and models', unpublished paper, Department of Civil and Urban Engineering, University of Pennsylvania (1974).

16. Murchland, J.D., 'Road network traffic distribution in equilibrium', paper presented at the conference *Mathematical Methods in the Economic Sciences*, Mathematisches Forschungs-institut, Obersolfach (1969).

17. Potts, R.B. and Oliver, R.M. (1972) *Flows in Transportation Networks*, Academic Press, New York.

18. Putman, S.H. (1975) *An Empirical Model of Regional Growth*, Regional Science Research Institute, Philadelphia.

19. Rescigno, A. and Richardson, I., 'Struggle for life 1: two species', *Bulletin of Mathematical Biophysics*, 29, pp. 377-88 (1967).

20. SELNEC Transportation Study (1972) *A Broad Plan for 1984*, Town Hall, Manchester.

21. Senior, M.L., Williams, H.C.W.L. and Wilson, A.G., 'Towards "best practice" transport planning models', *School of Geography Working Paper*, University of Leeds (1977).

22. Southworth, F., 'A disaggregated trip distribution model with elastic frequencies and expenditures', *School of Geography Working Paper 97*, University of Leeds (1974).

23. Southworth, F., 'Highly disaggregated modal split models', *School of Geography Working Paper 108*, University of Leeds (1975).

24. Steenbrink, P.A. (1974) *Optimisation of Transport Networks*, John Wiley, London.

25. Van Vliet, D., 'Road assignment', *Transportation Research*, 10, pp. 137-57 (1976).

26. Wardrop, J.G., 'Some theoretical aspects of road traffic research', *Proceedings of the Institution of Civil Engineerings, Part II*, 1, pp. 325-78 (1952).

27. Williams, H.C.W.L., 'Travel demand models, duality relations and user benefit analysis', *Journal of Regional Science*, 16, pp. 147-66 (1976).

28. Williams, H.C.W.L., 'On the formation of travel demand models and economic evaluation measures of user benefit', *Environment and Planning, A*, 9, pp. 285-344 (1977).

29. Wilson, A.G. (1970) *Entropy in Urban and Regional Modelling*, Pion, London.

30. Wilson, A.G., 'Further developments of entropy maximising transport models', *Transportation Planning and Technology*, 1, pp. 183-93 (1973).

31. Wilson, A.G., 'Catastrophe theory and urban modelling: an application to modal choice', *Environment and Planning, A*, 8, pp. 351-6 (1976a).

32. Wilson, A.G., 'Non-linear and dynamic models in geography: towards a research agenda', *School of Geography Working Paper 160*, University of Leeds (1976b).

33. Wilson, A.G., 'Towards models of the evolution and genesis of urban structure', in *Dynamic Analysis of Spatial Systems in Geographic and Regional Science*, Bennett, R.J., Martin, R.L. and Thrift, N.J. (eds), Teakfield Ltd, Farnborough (1977).

34. Wilson, A.G. and Senior, M.L., 'Some relationships between entropy maximising models, mathematical programming models and their duals', *Journal of Regional Science*, 14, pp. 207-15 (1974).

Chapter 8

AN EQUILIBRIUM MODEL FOR INTEGRATED REGIONAL DEVELOPMENT

Åke E. Andersson and Anders Karlqvist

INTRODUCTION

This chapter describes a model for balanced regional growth.
The model is based on a dynamic Leontief growth model and
extended to a regional context by the means of a transport
model which links different regions together. It is argued
that a model of regional transport flows cannot be based on a
micro-economic approach. Instead statistical methods are
used to derive likelihood estimates of flows. Three different
aspects of equilibrium are involved. The dynamic growth
equilibrium and the welfare equilibrium for choosing optimal
policy parameters. Some general points concerning different
notions of equilibrium are made in the first sections. The
chapter concludes with a few remarks on computational
experience.

THE NOTION OF EQUILIBRIUM

Equilibrium is one of the most deeply founded ideas of
economic theory. Few if any topics in economics have been
studied in greater depth than general equilibrium theory.
Although the idea of equilibrium has been much criticised for
its lack of realism it remains an important frame of
reference for analysing economic matters and it has also
become a rich source of procedures and mathematical methods
used in other parts of economics.

The concept of equilibrium has its roots in science. The
physical interpretation of equilibrium as a state of 'balancing forces' has not, however, served as a very fruitful
analogy for economics (6). Instead of dealing with balancing
forces of supply and demand the definition of an economic
equilibrium is usually directly related to the behaviour of
actors in the economic system. A market equilibrium refers
to an aggregate situation and is defined as a state, where
excess demand is less than or equal to zero. A competitive
equilibrium, which is a more general notion of equilibrium
not depending on aggregation, is a state in which no
participant of the decision process has an incentive to
change any of his decision variables. Such a definition
clearly needs further specifications in order to be meaningful. The incentives to change must be related to a
behavioural (choice) structure, external constraints must be
specified and assumed known and moreover the actual clearing
process of decisions must be made simultaneously and as if a

187

state of equilibrium prevails.

The last point has caused a great deal of argument about the relevance of general equilibrium theory. The Walrasian tatônnement process is an adjustment process without time. All adjustments of prices and quantities are made prior to the actual transactions which do not take place until an equilibrium is reached. In real life a process means a passage of time. Therefore it is difficult to give a realistic interpretation of the tatônnement process. The market participant must be endowed with extra-sensory perception (if acting simultaneously) or with supernatural premonition (if acting successively) (9). A dynamic approach would be more satisfactory. Not surprisingly the introduction of dynamics gives raise to a host of very difficult problems. As noted by Hicks (6) there is a qualitative difference between the past and the future. Events in the past enter as facts, events in the future as expectations. An equilibrium can be a local dynamic equilibrium, if it is an equilibrium with respect to facts and expectations at that particular point in time. At a later time some of these expectations may have turned into facts, others may not. Retrospectively the former equilibrium is not any longer regarded as an equilibrium. In order to get a global dynamic equilibrium the ex ante and the ex post evaluation of equilibria at every point in time must coincide. It can be concluded that a static equilibrium is a dynamic equilibrium, but obviously there are dynamic equilibrium processes which are not equilibria in a static sense. Optimal growth models provide an example.

The concept of equilibrium also plays an important role in welfare economics. While in a competitive equilibrium analysis the rational behaviour of the individuals is assumed to take place *before* the aggregation, the welfare equilibrium is defined as the rational decision by a collective decision-maker *after* the aggregation. In its simplest form the economy is represented by one actor (society), which seeks to optimise some social welfare function. The optimal choice constitutes the state of equilibrium. A social equilibrium need not have any interpretation at the micro level. An individual actor may very well have an incentive to change his situation but from a macro point of view an equilibrium persists if every individual decision is exactly counteracted by the actions of others. Analogous interpretations of equilibrium can be found in thermodynamics. An equilibrium state of constant temperature corresponds to many possible arrangements on the molecular level, but with temperature being an undefined concept at the micro level.

The distinction between social (welfare) and economic (competitive) equilibrium and their applicability in various contexts has been studied extensively. It seems to be a general conclusion that the Walrasian type of equilibrium analysis has its greatest merits when applied to perishable, continually endowed classes of goods (18) or to capital utilisation decisions (9). Durable goods and investments in large capital units are more appropriately handled within the framework of welfare equilibrium analysis. In the case of transport planning this distinction is of vital importance. The relevant context of transport planning covers a wide

range of decision-making processes from extremely long-term,
centralised decisions about infrastructure investment to
daily individual decisions about choices of routes and
travelling modes, etc.

The structure of a transport system is usually represented
by networks (although there are examples of continuous
representations of transport systems such as Angel's and
Hyman's research on velocity fields). The network property
implies certain kinds of inter-dependency of flows on the
links of the network. Equilibrium in network terms leads to
balancing flow conditions which are not recognised in the
classical framework of economic theory. An analogy can be
made with the idea of homeostasis in biology rather than with
mechanics. A principle of cost minimisation with respect to
route choice in a network (Wadrop's principle) does not only
depend on the supply-demand relations on each link separately
but a change of demand on one link indirectly affects the
demand on other links by the flow conservation laws. The
equilibrium problem of trip distribution according to gravity
principles and route assignment according to cost minimisation, where transport costs on the links are non-decreasing
functions of the flows, has attracted a lot of interest in
transport research. Unlike the case of Walrasian equilibrium
problems Kakutani's fixed point theorem cannot be directly
applied. However, it has recently been shown by Evans (3)
that the combined distribution-assignment equilibrium problem
can be elegantly solved by a direct application of
Rockefeller's conjugate duality methods. Under general conditions the existence and the uniqueness of an equilibrium
can be established. Such a flow pattern is sometimes
referred to as a selfish equilibrium. With a game-theoretical
terminology it represents a Nash point equilibrium.

Much of the results in general equilibrium theory concerns
existence (and uniqueness) of equilibria under various
assumptions. The methods employed, such as fixed-point
arguments, are as such non-constructive. Considerable
progress has been made in the effort of finding algorithmic
solutions. A most important contribution is the 'pivotal
methods' introduced by Scarf (17). These methods have been
attempted in various applications, primarily for comparative
static analysis (13). However the iterations of the
algorithms do not generally have any meaningful interpretation in the real world. They do not serve as models of
adjustment processes as they take place in the transport
system. The question of dynamics and hence also the question
of stability of equilibria and related issues remain
unanswered. Some aspects of this problem are treated later
in this chapter in relation to an inter-regional growth
model, with investment rate as a control variable and
adjustment parameter.

Stability is a phenomenon which has been thoroughly
analysed and explored in science. The singular perturbation
theory has been an important tool. A deep insight has been
gained into instability properties of physical and chemical
processes, where the dynamics of the system is given by
differential equations involving time scales of different
magnitudes. The impact of these ideas on the social sciences

has so far been rather small (for an interesting application, see Mees' article (14) on the growth of medieval cities). However, a qualitative picture of an inter-regional system exhibits features which might be appropriately described in these terms.  The time scale of the physical infrastructure and capital is very long and the process of change can be represented by a 'slow manifold'.  The pattern of trade has a different time scale.  It can change the process of adjustment as a 'fast foliation' in respect to the slow investment-production system (this terminology is presented and applied in the context of chemical reactions by Hahn (4)).  The 'inner solution' over a very short time span corresponds to an equilibrium where the slow variables related to the production and transport system can be regarded as given (time derivative equal to zero), i.e. an adiabatic process. The 'outer' solution for large time-periods gives the 'slow manifold' for which the trade pattern is always in equilibrium.

The asymptotic behaviour of the system depicted in the two extreme time perspectives is abruptly changed when the time scale is allowed to shift continuously from one time scale to the other.  At this preliminary stage, such a qualitative argument is of course only a way of formulating certain hypotheses, such as the switching of trade partners when production is changed or moved to another location.  A precise derivation of such behavioural changes must be built upon a quantitative description of the inter-regional production-trade system.  A step in this direction is taken in the following sections of this chapter.

## A BASIC FRAMEWORK OF INTEGRATED TRADE-TRANSPORT-LOCATION ANALYSIS

*Introduction*

A complete analysis of international or regional development must necessarily take into consideration the relationships between trade, transport and location.  Usually the problem of transport flows is not seen in its relation to the structure of production.  In trade theory, the link to location analysis is, with a few exceptions, very weak. Moreover if dynamics is properly introduced, determination of technology, growth and relocation, changes of trade patterns and stability properties of these processes should be incorporated into a spatial theory.

It should be obvious from these observations that any realistic analysis of development in space is an immensely complicated task.  To put these issues in a reasonable perspective we will use some simple illustrations and try to connect them into an integrated structural framework before a more formal theory is presented in the following section.

*Communication Networks, Technology and Choice of Technique*

At the outset we will assume that new technological inventions are additions to the known set of production techniques.

For the union of regions (or nations), there exists a
superset of techniques at each time-period.  To each region
(nation) only a subset is known (has been innovated) in each
time-period and the size of this subset depends among other
things on the location of this region in the networks of
communication.  The simple assumption is that if technology
has been developed as rectangular matrices (T**, A**)
distributed over the spatial network, then the size of the
technological subset known to the region (T*$^r$, A*$^r$) depends
upon the centrality of the region on the network.  This does
not imply that a central region necessarily uses the most
'advanced' techniques of production.  The actual choice of
techniques (T$^r$, A$^r$) is determined by the price of factors of
production and by the structure and size of demand for the
products occuring in the nodes.

We can summarise these arguments in the flow chart shown
in Figure 8.1.  A full fledged development theory should
contain models of all the illustrated stages of the flow
chart.  At the very least one should know which of these
factors are *not* included in an analysis of development.

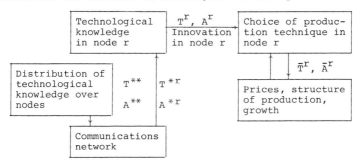

Figure 8.1  Relationship Between Some Elements of a Framework
for Integrated Trade-Transport-Location Analysis

*List of Symbols Used*

$_k t_{ij}$ = T = rectangular matrix of turnover times of
product i when used in production of product
j according to technique k

$_k a_{ij}$ = A = rectangular matrix of input-output coefficients denoting the use of product i as
current input in the production of product j
according to technique k

$_m t^r_{ij}$ = T$^r$ = rectangular matrix of turnover times of
product i when used in production of product
j according to technique m (=k) known in
region r

$_m a_{ij}$ = A$^r$ = rectangular matrix of input-output
coefficients denoting the use of product i as

current input in the production of product j according to technique m(=k) known in region r

$$\left[ {}_\ell t^r_{ij} \right] = T^r =$$ rectangular matrix of turnover times of product i when used in production of product j according to technique ( $\ell = m = k$ ) that has been technologically *adapted* to the environment of region r

$$\left[ {}_\ell a^r_{ij} \right] = A^r =$$ rectangular matrix of input-output coefficients denoting the use of product i as current input in the production of product j according to technique $\ell$ that has been technologically adapted to the environment of region r

$$\left[ \bar{t}^r_{ij} \right] = \bar{T}^r =$$ quadratic matrix of turnover times of product i when used in production of product j according to an optimal technical choice among the techniques available in $T^r$ for region r

$$\left[ \bar{a}^r_{ij} \right] = \bar{A}^r =$$ quadratic matrix of current input-output coefficients denoting the use of product i as current input in the production of product j according to an optimal technical choice among the techniques available in $A^r$ for region r

All coefficients can be regarded as functions of the structure of production (see the following sections).


TRANSPORT AND TRADE

From now on we assume that techniques of production have been chosen in each node (region or nation) and that the amounts to be sold to all nodes and purchased from all nodes are to be determined.

The amounts to be purchased and sold in other nodes must be related to the costs of communication and transport over the links to these nodes.  Furthermore, the trade and transport outcome is conditioned upon the structure of demand and supply of commodities in different nodes in the transport network.  In an equilibrium growth model the growth paths of demand and supply of commodities are, however, endogenously determined.  The solution of this growth model is dependent upon the pattern of trade in its turn.  This implies that trade is a function of location, while location is, at the same time, a function of trade.

Conventional trade theory has certain disadvantages as a starting point for a general analysis of transport and location.  According to the propositions by Ohlin and Hecksher, trade is a function of the relative abundance of resources in different regions.  The spatial structure is not explicitly recognised.  A notable exception to this rule is

given by Lefeber in his study on allocation in space (11).
Under economic equilibrium with the assumptions of deterministic, individual profit-maximising (or cost-minimising)
behaviour the conclusion is that no crosshauling can occur in
the system.  This conclusion is obviously at odds with any
empirically observed structure of trade and it casts serious
doubts on the usefulness of such trade equilibrium conditions.
Such an approach also loses sight of the fact that micro
costs of trade are determined by the macro solution for trade
patterns in any situation with given capacities of transport.

There are several reasons for the deviation from a strict
cost-minimising transport behaviour.  An economic theory
based on the assumption of perfect information is almost a
self-contradiction in a spatial context.  We know from many
empirical studies (with the pioneering work of Hägerstrand in
the 1950s) that diffusion of information is very much
dependent on spatial relations.  Thus, trade must rely on
incomplete information.  This introduces a random element
both in the coupling of buyers to sellers and in the choice
of transport services.  In the presence of uncertainties it
can also be assumed that the contracts between buyers and
sellers are not based on a strict optimisation principle, but
that in fact calculated risks are taken into account.  Hence
portfolio solutions can influence the outcome of the spatial
relations.

These arguments would indicate the necessity of using a
stochastic, rather than a deterministic model, to describe
and predict patterns of trade.  Another, more technical
problem adds to this picture, namely the problem of aggregation.  Aggregation over space obviously makes the trade
problem less dominant (that is if intra-regional spatial
variations can be neglected).  Other types of aggregation
over activities and goods and over time make the role of
crosshauling more apparent.  Even on a very fine level of
aggregation, small differences of quality, say cars of the
same make but of different colours, are enough to induce
crosshauling over great distances.  Temporal variations of
stocks may have the same effect.

The mainstream of non-optimisation approaches falls in the
category of gravitation or information-based models, and that
will be the focus of interest in this chapter.  However,
first some other lines of thought should be mentioned, which
are of potential interest in transport-location modelling.

Recently much research in this field has focused on the
handling of information.  It has been argued that the changing role of transport is a key characteristic of post-
industrial society.  The relative importance of transport of
goods is diminishing (at least as long as no drastic rise in
energy prices takes place) and the relative importance of
information processing and transfer are becoming greater.
Information which is transferred by face-to-face contacts
thus becomes a crucial factor in determining the transport
pattern.  This is particularly so since transfer of material
and information often appear in combination.  The effect of
the personal contact system cannot be described simply as a
cost-minimisation process.  Empirical evidence shows that

social norms, attitudes and habits, as well as legislation
have an impact on how contacts are made. This is particularly well demonstrated in studies of firms which have moved
from one region to another and which have kept their former
pattern of sub-contractors more or less intact in spite of
the possibility of establishing new, closer contacts.
Obviously ownership and other legal arrangements put restrictions on transport and trade which cannot be derived from
spatial considerations. Conversely transport flows which
take place within a firm but between units at different
locations are not always accessible for analysis since they
are not registered in trading accounts.

It is possible to avoid the great difficulties in working
with causal relations between location and transport on the
micro level by using a statistical or information theoretical
approach. The problem can typically be posed in the following way: certain flow conditions are given exogenously, e.g.
by supply-demand constraints in different regions. Sometimes
also some empirical observations of inter-regional flows are
at hand. The problem becomes one of estimating the complete
transport pattern and of predicting this pattern into future
situations. The tools are provided by information theory. A
standard example of this approach is the derivation of
gravity models for travel flows. The distribution of trips
derived as the maximum likelihood solution of a probability
distribution constrained by a set of statistics on origins,
destinations and on the total cost (or distance) of trips. A
weakness of this simple approach is that no *a priori* information about existing flows is taken into account. However, as
shown by Snickars and Weibull (20), it is possible to define
the micro states of the statistical distribution in such a
way that *a priori* information of that kind can be incorporated. This so called minimum information principle has been
shown by Hobson (7) to provide a generalisation of the
Shannon-Weaver entropy measure. Empirical tests on data from
work trips in the Stockholm region shows a significantly
better result using estimates based on prior traffic data
rather than applying the gravity model without *a priori* flow
information.

To obtain estimates of transport flows of goods between
regions a similar statistical approach can be applied. Let
$a_{ij}$ denote the usual Leontief input coefficient, i.e. to
produce an amount $x_j$ of goods j an input $a_{ij}x_j$ is needed of
goods i. $a_{ij}$ is a technological coefficient which is
assumed to be independent of volumes and prices (no scale
effect, no substitution) and which remains constant over
time. It seems to be a natural step to generalise this
input-output notion to a spatial context and to introduce a
regional input-output relation $a_{ij}^{rs}$, where rs denotes
deliveries from region r to region s. However as will be
clear from the sequel it is not possible to express the
regional input-output relations as a linear function of
production volumes x, and hence the definition of $a_{ij}^{rs}$ cannot
be made unambiguous.

In the case where spatial separation between regions r and s can be totally ignored an unbiased assumption of deliveries to production $x_j^s$ region s from sector i in region r would be

$$x_{ij}^{rs} = \frac{x_i^r a_{ij} x_j^s}{\sum_r x_i^r} , \qquad (1)$$

that is each unit of production i contributes with the same amount or with the same probability. If the number of delivering regions is r(=1), the expression above reduces to the usual input-output relations

$$x_{ij}^{rs} = a_{ij} x_j^s . \qquad (2)$$

In no other cases can the quadratic expression for flows $x_{ij}^{rs}$ be reduced to linear relationships. There is no reason to assume that the introduction of a distance factor would upset this observation.

The flows $x_{ij}^{rs}$ are subject to the following general conditions

$$\sum_r x_{ij}^{rs} = a_{ij} x_j^s \qquad \text{input constraint i, j=1,...s=1,.} \qquad (3)$$

$$\sum_{js} x_{ij}^{rs} = x_i^r \qquad \text{output constraint i=1,...r=1,.} \qquad (4)$$

(making the simplifying assumption that no deliveries go to capital and labour, and that final demand is treated as endogenous). It is easily seen that the maximum likelihood estimate of $x_{ij}^{rs}$, taking the two constraints above into account, leads to the formula for unbiased transport flows given above.

Equation (3) can be generalised to include investment flows. According to the acceleration principle of capital formation, the investment terms can be expressed as a linear function of the change of production $\Delta x_j^s$ (Equation (5)). Thus

$$\sum x_{ij}^{rs} = a_{ij} x_j^s + b_{ij} \Delta x_j^s \qquad (5)$$

This formulation corresponds to the dynamic version of Leontieff's input-output model. The distinction between flows and stocks is a relative one. It can be expressed in terms of durability, i.e. the time span during which a certain commodity is utilised. This provides an alternative formulation, whereby durabilities $t_i$ for capital i can be used to link input-output coefficients $a_{ij}$ to capital-output coefficients $b_{ij}$:

$$t^i a_{ij} = b_{ij}. \tag{6}$$

In its simpliest form the cost (c) of transport of goods $x_{ij}^{rs}$ can be expressed as a linear function of unit costs

$$c = x_{ij}^{rs} \cdot t^i \cdot d^{rs} \tag{7}$$

where $t^i$ denotes the cost per value of goods i per km and $d^{rs}$ denotes the distance in km.

The information about transport costs can be used in various ways to constrain the set of feasible transport flows and hence also to effect the most probable distribution of flows.  For example, if the capacity in terms of total transport costs between each pair of regions is known then the cost constraints read

$$\sum_{ij}\sum x_{ij}^{rs} t^i d^{rs} = c^{rs}. \tag{8}$$

Together with the two previous constraints (4) and (5) the maximum likelihood solution becomes

$$x_{ij}^{rs} = A_i^r x_i^r B_{ij}^s (a_{ij} x_j^s + b_{ij} x_j^s) \exp (-\gamma_{rs} t^i d^{rs}). \tag{9}$$

$A_i^r$ and $B_{ij}^s$ are balancing factors which are implicitly defined by the first two constraints.  They depend on the whole trade pattern

$$B_{ij}^s = \frac{1}{\Sigma A_i^r x_i^r \exp(-\gamma_{rs} t^i d^{rs})} \tag{10}$$

$$A_i^r = \frac{1}{\sum_{js}\Sigma B_{ij}^s (a_{ij} x_j^s + b_{ij} x_j^s) \exp(-\gamma_{rs} t^i d^{rs})}. \tag{11}$$

The cost constraint can be defined in other ways, e.g. in terms of total cost or total cost per type of goods.  The corresponding changes of parameters in the formula for $x_{ij}^{rs}$ are obvious, and will not be derived here.

As noted above there are methods to improve the estimate of the flow matrix $(x_{ij}^{rs})$ by using a *priori* information according to the minimum information principle [20]. Historical data of flows $x_{ij}^{rs}$ together with actual data (observed or exogenously determined) related to the demand and supply constraints (C) can be used to ensure an 'effective' statistical estimate (i.e. with the lowest information content),

$$\min \ -\Sigma x_{ij}^{rs} \ \log \frac{x_{ij}^{rs}}{\hat{x}_{ij}^{rs}} \qquad\qquad (12)$$

Subject to (C).

A Taylor-expansion of $\log \dfrac{x_{ij}^{rs}}{\hat{x}_{ij}^{rs}}$ in the neighbourhood of $x_{ij}^{rs}=\hat{x}_{ij}^{rs}$ shows that the expression above can be approximated with a measure of quadratic deviation (8)

$$\min \ - \ \Sigma \frac{1}{\hat{x}_{ij}^{rs}}(\hat{x}_{ij}^{rs}-x_{ij}^{rs})^2. \qquad\qquad (13)$$

In many cases the *a priori* information cannot be given in the form of a complete historical inter-regional flow table. Regional trade information is often available only for special branches and regions. However a flow matrix can be completed by using the same information-theoretic argument as above. Hence the matrix $(x_{ij}^{rs})$ is derived from the formula

$$\min \ -\underset{x\notin K}{\Sigma}x_{ij}^{rs}\log \ x_{ij}^{rs} \ - \ \underset{x\in K}{\Sigma}x_{ij}^{rs}\log\frac{x_{ij}^{rs}}{\bar{x}_{ij}^{rs}} \qquad\qquad (14)$$

where $K=(i,j,r,s \mid$ a priori value $\bar{x}_{ij}^{rs}$ exists$)$.

This corresponds to the assumption that for $x\notin K$ all micro events are assumed to be equally probable. It should be noted that *a priori* information $\hat{x}_{ij}^{rs}$ derived from a maximum likelihood estimation of a subset of the constraints does not add anything new. In that case we can put $\hat{x}_{ij}^{rs}$ equal to a constant.

Another possibility is to subtract the given $\bar{x}_{ij}^{rs}$ from the matrix and the constraints, and then treat the residual problem as a standard entropy-maximising problem without *a priori* information. However this method fails to take into account the information about the total distribution con- tained in the given $\bar{x}_{ij}^{rs}$. The discrepancy can be demonstrated to depend on the relative size of K and $\bar{K}$. The influence of the entropy distribution on $x\in K$ can be expressed as an exponential factor $\alpha x=\alpha\bar{x}$, which tends to one, $\alpha\to 1$ when

$$\text{Prob} \ (\underset{x\in\bar{K}}{\Sigma\bar{x}})\to 1. \qquad\qquad (15)$$

THE GENERAL EQUILIBRIUM THEORY OF GROWTH

*The Quantitative Equilibrium Problem*

An economic equilibrium has been defined as 'a state in which no participant has an incentive to change any of his decision variables'. Regarding an input-output economy from the *quantitative* aspect - in contrast to the *price* aspect of the same economy - participants' incentives to change their decision variables can be related to the problem of avoiding queues for commodities to be purchased or accumulation of stocks of commodities to be sold. The decisions to be taken are thus of two kinds, to sell commodities and to purchase commodities. Correspondingly there can be two types of *plans*, selling plans and purchasing plans. We assume that this economy has plans formulated for one basic planning period ahead, a planning period that could be assumed to be infinitely short. A seller lacks incentives to change his selling plan, if he does not observe any unsold quantity of the commodity or unfulfilled demands at the end of the basic planning period. The purchasers of the same commodity lack incentives to change their purchase plans when there are no queues for the commodity.

Regarding the problem from the standpoint of an individual seller, there is a planned volume of sales, $x_i$, that should be equal to the sum of purchase plans of all his buyers, $z_i$, if there should be no queuing for commodities and no unwanted growth of stocks of commodities.

$z_i$ is a sum of planned flows from the $i^{th}$ seller to the, $j=1,\ldots,n$, purchasers. This means that we can define $z_i = x_{ij}$ where $z_i$=the sum of purchase plans for commodity i, and $x_{ij}$= delivery of a commodity from seller i to purchaser j. We will now assume that each seller sells a separate commodity. If this is a *closed* system there will be as many sellers (n) as there are purchasers. Let us further assume that the static demand (planned purchase) is a continuous function of the supply (planned sales) of the individual seller (Equation (16)).

$$x_{ij} = f(x_j). \tag{16}$$

This expression can be rewritten in the form of input-output coefficients, which are functions of the scale of production (Equation (17)).

$$x_{ij} = a_{ij}(x_j)x_j. \tag{17}$$

In an economy *linearised through origo of the product space*, this static demand equation can be written:

$$x_{ij}a_{ij}(\overline{x}_j)x_j = a_{ij}x_j. \tag{18}$$

The intercept term is thus generally assumed to be zero for all demand equations. The introduction of dynamic demand can be shown in a similar way

$I_{ij} = b_{ij}(x_j)\dot{x}_j$ with a linearised counterpart

$$I_{ij} = b_{ij}(\overline{x}_j)\dot{x}_j = b_{ij}\dot{x}_j \qquad (19)$$

where $I_{ij}$ = delivery of a commodity from seller i to
purchaser j to be used by j for capital information,
$b_{ij}(x_j)$ = a scale-dependent capital-output ratio.

The first condition of a general equilibrium of the whole
economy is thus:

$$x_i = \sum_{j=1}^{n} a_{ij}(x_j)x_j + \sum_{j=1}^{n} b_{ij}(x_j)\dot{x}_j; \quad (i=1,..,n). \qquad (20)$$

With matrix notation

$$\left[a_{ij}(x_j)\right] \equiv A(x) \text{ and } \left[b_{ij}(x_j)\right] \equiv B(x) \qquad (21)$$

a general equilibrium is such that

$x=A(x)x + B(x)\dot{x}$: with a linearised dynamic input-output
system as a counterpart:

$$x = A(\overline{x})x + B(\overline{x})\dot{x} \qquad (22)$$

(supply)   (current     (investment
            demand)       demand)

It is well known that the linearised system has meaningful
equilibrium solutions [2,12]. It is also well known, that
these solutions are highly unstable [12], if computed from
the system

$$\dot{x} = B^{-1} (I - A)x. \qquad (23)$$

An important question is how such an equilibrium solution is
achieved by an actual economy, if it is working under
Equation (26) given below.

The theory given so far is rather rigid in its basic
structure. Demands are related to activity levels in a
fairly strict technological way. This is a reasonable
assumption in the case of immediate demands for inputs in the
production process. It is a much less defensible assumption
in the theory of investment demand. Since the time of Keynes
it has become an accepted fact that decisions on investments
are influenced by expectations of future demands. The
demands of the future cannot be known but must be guessed.
One factor influencing the expectations about future growth
is the growth that has been recorded in the past. Another
source of information is the observed rate of capacity use in
the economy. It has been argued by Keynesians that the formation of expectations is absurd in decentralised capitalist
economies. An observation of a poor growth record and
unemployment of resources easily gives rise to a revision of

expectations such that investment plans are scaled down to a
lower level than initially.  This reaction reduces factual
demand, and a combination of multiplier, accelerator and
expectation effects reduces total demand.  Such a 'principle
of revisions of expectations' can imply that an economy that
goes slightly below its maximum investment potential slides
to its minimum investment demand level.

The argument given can be introduced as $\lambda_0 x_0$ giving the
expected growth of demand at a uniform expectation of the
rate of growth of the economy at time period $(0, \lambda_0)$.  It is
then possible to formulate an expectation-revision function
in which the expected rates of growth are determined
successively by the observed rates of growth and observed
rates of capacity utilisation.  Nothing of a formal nature
would prohibit a sectorisation of these expectations, but
this matter will not be pursued in this context.  We will
instead assume expectations to be exogenously determined at
$\lambda_0$, which is assumed to be a constant over time.  We also
assume that beside the decentralised investment and production planners of the economy there is also a centralised
authority responsible for stabilisation of the economy as a
whole at a full employment level.  The basic principle of the
stabilisation policy is then of a fix price nature.  The
central authority observes the actual rate of capacity use in
each basic period (say a month, a quarter or even a year).
It then changes the investment demand by the use of subsidies
or direct purchases to increase growth of capacity, supply
and demand.  A stabilised full employment equilibrium can
eventually be achieved if a suitable stabilisation function
is used by the central authority.  The basic scheme of
analysis is given in Figure 8.2.

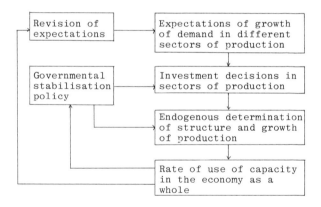

Figure 8.2  Stabilisation Schema

A formal expansion of the model (Equation (22)) is now
possible.  Let us assume that investment demand is determined
by expected growth of demand so that the investment function
can be written as  $Bx = \lambda_0 Bx$, where $\lambda_0$ is the planned increase

of capacity of the economy at a certain point of time $(t_0)$.
We can now introduce an excess demand indicator $\gamma$, which
indicates excess capacity, if lower than 1 and excess demand,
if higher than 1.

$$\gamma x = Ax + \lambda_0 Bx \equiv Qx. \tag{24}$$

In this form it is known that there is one and only one
positive $\gamma$ with all $x_i \geq 0$ if $q_{ij}$ of Q are all semi-positive.

Semi-positiveness turns out to be a natural condition of any
economic system. Reintroducing *periods* of decision-making,
the economy can be assumed to be stabilised through investment planning of the enlarged system.

$$\gamma_t x_t = Ax_t + \lambda_t Bx_t \tag{25}$$

$$\lambda_t = F(\gamma_{t-1}).$$

The algorithm used in computations has been based on the idea
of a policy institution responsible for stabilisation of the
economy, which has a stimulus-response pattern of the kind
often observed in situations of learning. If there is a
small deviation from the optimal rate of capacity utilisation
($\gamma=1$) there is a small response. When the stimulus $\gamma-1$
becomes larger, absolutely, the response grows more than
proportionally and goes towards infinity when capacity
utilisation goes to zero. $\gamma_t$ is now an indicator of use of

capacity in the economy. Let us assume that $\lambda$ is set low, $\gamma$
will then also be low which means that there is unemployment
of resources. If $\lambda$ is then adjusted upwards (investments are
stimulated) then employment will increase, etc.


*The Cost-revenue Problem*

There are always two inter-connected aspects of a market
economy. The first aspect, considered in the former section,
concerns the quantitative equilibrium of supply and demand of
commodities. They have to be balanced, if there should be no
inducements for the firms to expand or decrease production,
in such a way that a disequilibrium is generated in the
quantitative part of the economy. The problem can be
formulated in the following way: a certain sector i is
specified. A necessary condition of dynamic economic
equilibrium is the requirement that the total revenues of the
sector are equal to the total costs, at least when calculated
over a long time-period. Variable costs are defined to
include remuneration to managers for their services, while
capital costs are determined to be the 'loan rate of
interest' multiplied with the capital value. It is furthermore assumed that all costs are measured in deflated terms so
that the rate of interest is a real rate of interest. Pigou
effects and similar problems of a monetary economy are also
disregarded.

An equilibrium of a sector in a linearised economy can
thus be defined to be a state of no excess profits, i.e.

$$P_i x_i = \Sigma_j p_j a_{ij} x_i + r \Sigma_j p_j b_{ij} x_i \qquad (26)$$

(revenue)   (current          (capital cost)
            input cost)

where $p_i$ = price of commodity i

$\quad\quad x_i$ = production (sales) of commodity i

$\quad\quad r$ = real rate of interest.

A general equilibrium is thus a state such that:

$$p = pA + rpB. \qquad (27)$$

In any real economy r must be assumed to be *exogenously* determined by policy authorities. Hence (26) can only hold *accidently*, if there are no meaningful rules for the determination of r. The system given in Equation (26) can however be extended with an interest-rate determination rule that can be formulated as a function G ($\alpha$), where $\alpha$ is an excess-cost criterion. The argument for such a function can be formulated in the following way. Redefine (26) as a disequilibrium relation (Equation (28)).

$$(1 +\alpha) p_i x_i = \Sigma_j p_j a_{ji} x_i + r \Sigma_j p_j b_{ji} x_i \qquad (28)$$

or in the general system (Equation (29))

$$(1+\alpha) p = pA + rpB \qquad (29)$$

$$r = G(\alpha).$$

If we assume r to be larger than the equilibrating value $\bar{r}$, $\alpha$ must be larger than 0, if it is left to be determined as an endogenous *equilibrium value*. This implies that revenues are less than costs. If this short-term equilibrium state should be preserved, with fixed prices, the inducement to invest must necessarily drop and the economy would be caught in a state of unemployed capacity. If on the other hand r $\bar{r}$ there would be an $\alpha < 0$, implying excess profits. This would call forth a disproportionate expansion of the number of firms and thus an unbalanced quantitative growth. The interest-rate determination rule should then be of such a form that the rate of interest should be decreased when $\alpha > 0$, with r = 0 when $\alpha = 0$. A further constraint on the form of the G-function is the requirement that average excess profits over an infinite time horizon should be equal to zero.

$$(1+\alpha_t) p = pA + r_{t-1} pB \qquad \textit{Cost-revenue relation} \qquad (30)$$

$$r_t = G(\alpha_t), \quad r = g(\alpha) \qquad \textit{Interest rate determination}$$
$$\frac{dr}{dx} < 0 \qquad\qquad\qquad\quad \textit{rule}$$

$$\lim_{T \to \infty} \frac{1}{T} \int_0^T \left| (1+\alpha(t)) p(t) - p(t)A - r(t) p(t)B \right| x(t) dt = 0 \qquad (31)$$

$$\textit{Long-run excess profits condition}$$

$$p(t) \overset{\geq}{=} p(t)A \qquad \textit{'No-shutdown' condition} \qquad (32)$$

The 'no-shutdown' condition states that revenues must be larger than or equal to variable costs in each period in order to avoid destruction of real capital. It must be noted that this price equilibrium formulation only gives the necessary preconditions of a long-run equilibrium path of the economy. We have not presented any interest policy that would guarantee the achievement of a time path of investments required for quantitative equilibrium path. That question is discussed in the earlier section on quantitative equilibria.

## SPATIAL GROWTH THEORY

The theory of stabilised equilibrium growth should now be expanded to include spatial interactions. The transport model, presented above, gives the basic framework for such an extension. The transport model predicts the flows $(x_{ij}^{rs})$, if the transport distances, the capacities of the transport network and the activity levels $(x_i^r)$ are predetermined. The non-spatial growth model determines the time path of production in sectors $(x_i)$ if functions determining flows between sectors, $(x_{ij})$, are given. A discrete spatial growth theory by analogy predicts the time path of production in sectors, located to regions $(x_i^r)$, if functions determining flows between sectors in regions and sectors in regions, $(x_{ij}^{rs})$, are given.

The transport analysis provides us with such flow functions for a spatial system for the variables $(x_{ij}^{rs})$. The basic form of these flow functions is quadratic, with the parameters of the quadratic form determined by the production technique, the distances between regions, the capacities of the links, nodes and of the transport system as a whole. The scale of production in each node is, furthermore, a determinant. This implies that the transport flows and the growth pattern are *inter-dependently* determined, as illustrated in Figure 8.3.

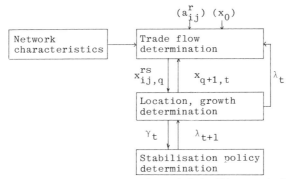

Figure 8.3   Flow Chart of Iterations (q) in Time-period, t

The regional growth theory is thus built on two different equilibrium principles. One is the economic equilibrium principle. It regulates the production (or sales) plans of the sectors to be in balance with the purchasing plans. The other, physical-statistical equilibrium principle regulates the structural use of the transport network. A total equilibrium is then such that both these equilibria hold true, simultaneously.

A spatial growth equilibrium also requires commodities used as capital to grow in a balanced fashion. Hawkins (5), Lange (10), Hicks (6) and Brody (2) have advocated the use of economic durability or turnover-time parameters to solve *this* capital accumulation problem. The transport section shows that

$$x_{ij}^{rs} = x_j^r q_{ij}^{rs}(1)x_j^s + x_i^r q_{ij}^{rs}(2)x_j^s \tag{33}$$

with a pre-specified location-allocation pattern, a given network and a given production technique.

The spatial growth model, to be used numerically is then of the following form:

$$\begin{cases} \gamma x = x^T Q_1 x + \lambda x^T Q_2 x \\ \lambda = F(\gamma) \end{cases} \tag{34}$$

$x = (x_i^r)$ = production in sector i, region r

$Q_1$ = a matrix of production interdependency coefficients connecting $x_i^r$ with $x_j^s$

$Q_2$ = a matrix of production - investment interdependency coefficients connecting $x_i^r$ with $x_j^s$

Superscript T indicates transposition

The elements of Q are *iteratively* determined by the transport model.

The model can be rewritten as:

$$\gamma x = x^T (Q_1 + \lambda_0 Q_2) x \equiv x^T Kx \equiv H(x) \tag{35}$$

with $\gamma$ as a variable if the rate of capacity growth has been exogenously determined; x is a non-negative vector and K is also non-negative in all its elements.

A theorem by Nikaido (16) can then be applied to ensure the existence of an equilibrium solution (1)

$$\gamma x = H(x) \text{ with } \gamma \geq 0 \text{ and } x \geq 0. \tag{36}$$

A stabilised general equilibrium in space is then a state such that

$$x = x^T Kx \tag{37}$$

with K determined as a set of parameters in the stochastic

trade-transport model with a constraining x-vector corresponding to the economic equilibrium x-vector in Equation (25).

## SOME QUALITATIVE ASPECTS

Equation (34) describes a control model for spatial economic growth.  The dynamics of the system are defined by the change of the state variable

$$\dot{x}=G(\lambda,x) \tag{38}$$

and the change of the control parameter (the reactive behaviour) as a function of the level of control

$$\dot{\lambda}=F(\lambda,x) \tag{39}$$

This represents a type of differential equation which was discussed in the introductory section on equilibrium (4).  The transition from a linear non-spatial theory of economic growth to a non-linear spatial theory implies a qualitative transition from a continuous system to a system with singularities.  Although these singularities can be characterised in general mathematical terms it remains for future research to establish how these properties can be interpreted in a concrete planning modelling context.  The numerical experimentation with this type of model to date has only concerned the regular behaviour.  These computational experiences are described briefly below.

## COMPUTATIONAL EXPERIENCE

Linear variants of this theory and a non-linear variant with a simplified transport model have been used on actual data for a regionalised national economy.  The largest computed model has been specified for an economy with 32 sectors and i regions (256 variables) (1).  The model did not include any stabilisation rule, nor did it include stochastic determination of trade on a transport network.  The stabilisation policy algorithm has recently been tried on a linear variant with more than 200 variables and behaves very efficiently.  A compact version of the model has been computed with an integrated stochastic trade model and an economy-wide network capacity constraint.

## REFERENCES

1. Andersson, A., 'A closed non-linear growth model for international and inter-regional trade and location', *Regional Science and Urban Economics*, 5, pp. 219-28 (1975).
2. Brody, A. (1970) *Proportions, Prices and Planning*, Amsterdam, Elsevier.
3. Evans, S., 'Derivation and analysis of some models for combining trip distribution and assignment', *Transportation Research*, 10, pp. 35-7 (1976).
4. Hahn, H. (1974) *Geometrical Aspects of the Pseudo Steady State Hypothesis in Enzyme Reactions*, Tübingen.
5. Hawkins, D., 'Some conditions of macro-economic stability', *Econometrica*, 16 (1948).

6. Hicks, J. (1965) *Capital and Growth*, Oxford University Press, London.

7. Hobson, A. (1971) *Concepts in Statistical Mechanics*, Gordon & Breach Inc., New York.

8. Kádas, S. and Klafszky, E., 'Estimation of the parameters in the gravity model for trip distribution', *Regional Science and Urban Economics*, 6, no. 4 (1976).

9. Koopmans, T., 'Is the theory of competitive equilibrium with it?', *The American Economic Review*, pp. 325-9 (1974).

10. Lange, O. (1959) *Introduction to Econometrics*, Warsaw.

11. Lefeber, L. (1968) *Allocation in Space - Production, Transport and Industrial Location*, North-Holland, Amsterdam.

12. Leontief, W. (1970) *The Structure of American Economy*, John Wiley, New York.

13. McKenzie, L., 'Why compute economic equilibria? *Paper presented at the Conference on Computing Equilibria: How and Why?*, Computation Center of the Polish Academy of Sciences, Torin, Poland (1970).

14. Mees, A.L., 'The revival of cities in medieval Europe', *Regional Science & Urban Economics*, 5, no.4 (1975).

15. Moses, L., 'The stability of interregional trading patterns and input-output analysis', *American Economic Review*, 14 (1955).

16. Nikaido, H. (1968) *Convex Structures and Economic Theory*, John Wiley, New York.

17. Scarf, H. (1973) *The Computation of Economic Equilibria*, New Haven.

18. Smale, S., 'Dynamics in general equilibrium theory', *American Economic Association*, pp. 288-94 (1976).

19. Snickars, F., 'Convexity and duality properties of a quadratic intraregional location model', *Regional Science and Urban Economics*, 7 (1977).

20. Snickars, F. and Weibull, J., 'A minimum information principle - theory and practice', *Regional Science and Urban Economics*, 7, 1 (1977).

Chapter 9

EQUILIBRIUM MODELLING

Earl R. Ruiter and Robert B. Dial

CHARGE TO WORKSHOP

The aim of this workshop is to discuss the requirements for
behaviourally appropriate modelling systems that contain
arguments on both the demand and the supply sides. Disequilibrium is the general state of affairs in the real
world. We need to take this point and assess the extent to
which modelling of supply and demand in an equilibrium type
of framework will give more respectable guidance on the real
role of demand modelling in improving on our understanding
and prediction of consequences of policy options.

SCOPE OF WORKSHOP

The workshop's discussions were directed by a constant
consciousness of and reference to a number of interdependencies in the modelling of equilibrium systems: (1)
inter-dependency of travel demand and facility levels of
service in transport systems; (2) inter-dependency of
demands and availabilities (at various costs) of urban land
in land-use systems; and (3) inter-dependency of location-
specific production factors and demands for final goods in
economic systems. The persistent attention of the workshop
to these inter-dependencies makes it clear that they became
the workshop's primary concern. The workshop's interest was
not just equilibrium within the various systems.

This concern extended to inter-dependencies among the
three systems themselves: transport, land use and the
economy. It is these inter-dependencies which cause
transport planners to be concerned about economic and land-
use systems in the first place; and this fact showed itself
clearly in discussions of each system: the emphasis throughout was upon those analyses of economic and land-use systems
that explicitly treat transport as a vital subsystem. The
notion of nested systems takes on considerable conceptual
importance: the transport system as a subsystem of the land-
use system, which is in turn a subsystem of the economic
system.

A number of important items were *not* discussed; in
particular the institutional, political and natural (or
ecological) systems. The importance of those systems is
recognised.

RECOMMENDATIONS

The proceedings of the equilibrium modelling workshop are
more easily, accurately and suggestively discussed in terms
of general systems and their inter-dependencies.  The terms
of equilibrium analysis and equilibrium modelling are themselves inadequate to an efficient description of the
workshop's discussions - given their synthetic nature and
their breadth.  Therefore, it is useful to organise the
recommendations by system, within two general classifications
based on the type of action being called for.  The first of
these classifications, 'immediately needed implementation and
development', includes recommendations which could also be
classified as either *applications* or *impacts on policy*.  The
second classification, 'conceptual and empirical needs'
straddles the areas of *impacts on policy* and *future
directions*.

*Immediately-needed Implementation and Development*

*Transport systems:*

1  Traffic due to urban freight should be incorporated in
current analysis procedures.
2  Using *ad hoc* extensions of available theory, network
equilibrium procedures should be developed that will include
the most advanced demand and supply models extant.  A richer
set of multi-modal capabilities is a special concern of the
workshop.
3  Work can begin on disaggregate assignment methods that
will match disaggregate demand models.  It is understood that
such methods cannot become practical without greater computational efficiencies that will require significant
algorithmic breakthroughs.
4  The present analysis of static network equilibrium should
be extended to analysis of dynamic equilibrium.  Time-of-day
variations in travel demand should be analysed as responses
to changes in work hours, such as flexitime.
5  As windowing and focusing methods become available to
reduce costs in network analyses, their adaptability to
studies of equilibrium should be investigated.

*Land-use systems:*

1  Standardised land-use model-application capacities,
analogous to those of the Urban Mass Transit Authority (UMTA)
and Federal Highway Administration (FHWA) packages for
network analysis are needed.  The first step is to choose a
class of land-use models for such a system (such as models of
the Lowry or Wilson type).  The second step is the coding of
programs for use in various ways within the general structure.
2  The work now underway at the US National Bureau of
Economic Research (11) and elsewhere on prototype econometric
land-use models should be followed up with additional model
development work to provide land-use models useful for a wide
range of land-use and transport studies.

*Conceptual and Empirical Needs*

*Transport systems:*

1  Gaudry's call in his resource paper (Chapter 6) for the simultaneous treatment of level-of-service weights in path-building and model development should be considered carefully.
2  A system is needed to collect behavioural data on travellers' route choices.  In studies of route choice, alternatives are rarely enumerated specifically and many are too non-independent.  A conceptual framework is needed for the testing and analysis of collected data.
3  Further mathematical analyses should be made, both of the classical network equilibrium problem and of its more useful extensions.  The goal is to increase the efficiency and the understanding of computational requirements and of the solution characteristics of more complex problems.
4  The structures of both demand and supply models developed using data for different years should be studied to determine their stability and, if necessary, to develop relations for the prediction of secular changes in the parameters of the model.
5  An important need, as network equilibrium procedures are expanded to deal with public transit systems, is for improved link supply functions (time versus volume and service characteristics) for public transit, analogous to functions for automobiles, as given by the *Highway Capacity Manual* (10). These functions should hold as fixed those changes in operating strategy made by the transit operator (on the assumption, of course, that such changes would be repeatedly tested in application).

*Land-use systems:*

1  Conceptual work should continue towards a better understanding of land use.  One framework worthy of consideration is the behavioural choice work which has recently been extended in new applications to urban residential behaviour (15).
2  The more specific representation of the dynamics of land use, using dynamic systems methods, should be investigated (3).

*Economic systems:*

Many of the new analysis methods should be investigated to determine their contributions, if any, to understanding economic systems.  These methods include, for example, information theory, game theory, catastrophe theory and dynamic system analysis.

BACKGROUND FOR THE WORKSHOP'S RECOMMENDATIONS

*The General Nature of Equilibrium*

Presentation of the three papers prepared as resource material for the workshop provided the background for a general discussion of the nature of equilibrium as the term has traditionally been applied by transport planners (see Chapters 6, 7 and 8).

Equilibrium is a phenomenon both of physical processes and of their mathematical representation.  In a physical process, considerations of equilibrium are important in the transient or stable balancing of mutually interactive and adaptive forces.  The mathematical representation of the physical process involves typically an iterative algorithm in which inter-dependent equations are successively evaluated.  If the representation is successful, the successive approximations will converge upon a unique solution.

An example of an equilibrium system exists in the flows which occur in transport networks, where travellers, on a daily basis, can change their behaviour, making different travel decisions.  An example of a non-equilibrium system is in urban land use: decisions are less frequent and their high cost (as in the case of fixed facilities) keeps changes from being quick responses to new conditions.  In land use, lagged changes are always occurring, but an equilibrium state is never reached before exogenous effects are felt.

Equilibrium analyses serve two very different ends.  For systems that adapt rapidly, the equilibrium *solution* is useful immediately in policy analysis.  When adaptation is slower, knowledge of the *process* toward equilibrium is an invaluable guide to understanding the dynamics of a real system.  An example of the process to a final state was offered.  A new, unique residential area at first attracts people who might be called pace-setters.  As the area ages, its demographic character is likely to stabilise near the norm of the metropolitan area in question.  The understanding of such dynamics is necessary in the planning of such facilities as shopping centres for the area.

Equilibrium solutions often differ from optimum solutions. This is well known in the problem of network equilibrium (30). These different types of solutions might be distinguished as *descriptive* and *normative*.  Broadly speaking, there are two kinds of equilibrium analyses.  The micro-economists see equilibrium as that state in which no person can better satisfy himself by a change in his actions.  (Network equilibrium is treated in these terms.)  Exponents of the 'physical forces' position see equilibrium as the balance of opposing forces, such as supply and demand.  Because these opposing forces are usually assumed to be static forces, this latter view is usually applied to problems of static rather than dynamic equilibria.

*Network Equilibrium*

The discussion of network equilibrium began with the identi-fication of a number of existing packages which can be used to compute approximations (based on varying simplifying assumptions) of network equilibrium.  This summary of the state of the art led to discussions of further implementation and research needs.

*Computer programs, packages and systems:*

1  *Capacity restraint assignment.*  Available in many

packages. Solutions can provide good approximations to equilibrium (despite the use of pre-specified trip-tables) when network loads are far enough below capacity [7].

2 *Dynamic assignment to subnetworks*. Deals with variations of flow in short time-periods (15 minutes to 1 hour) and queueing at bottlenecks. Fixed trip-tables are assumed [20, 35].                                              .

3 *Network equilibrium with fixed trip-tables*. TRAFFIC [26], UROAD [29] and ATIM [24].

4 *Traffic assignments to multi-modal networks with varying trip-tables*. UMITE: aggregate versions of disaggregate demand models incorporated in UROAD [36]; DODOTRANS: direct demand models [21]; EMME: aggregate mode choice models [8].

5 *Special analysis methods for sub-areas*. UTPS windowing and focusing methods now being implemented.

6 *Special assignment techniques*. STOCH [6]; STOCH plus later iterations using network equilibrium in UROAD: Burrell random link lengths [2].

7 *Network improvement models*. LeBlanc [13]; Control Analysis Corporation [5].

*Discussion*. In the discussion of needed development and conceptual activity, it was remarked that too little work incorporates variable demand. Perhaps this is because a mathematically complete solution can be developed only for much simpler demand relationships than those now typically estimated, either aggregatively or disaggregatively. However, certain *ad hoc* approaches, whose convergence and uniqueness remain unproven have been used in relationships of such complexity (DODOTRANS, UMITE). The consensus is that work should continue on the implementation of *ad hoc* varying demand equilibrium processes using the best demand relationships available - whatever their complexity. Participants gave several reasons for this. They doubt that further advances in mathematics and theory are in the offing. The general structure needed for a valid equilibrium algorithm is well understood and easily extended to more complex models. The possible occurrence of significantly different multiple solutions is of no practical importance. The convergence properties of algorithms can be observed after they are implemented, even if unknowable beforehand; and the existence of an equilibrium can be safely assumed.

Another task worth undertaking is the simplification of complex demand equations to make them compatible with the mathematical theory in a 'pivot point' approach. Detailed models would be used to compute the base point, before starting the network equilibrium process. Then, within the process, demand variations around the base point would be approximated with simple model forms derived from the elasticities of complex models. The Scarf [28] algorithm shows promise in solving a number of equilibrium problems. Talvitie used it to model equilibrium demand in a transport corridor. Kuhn and MacKinnon have developed improved algorithms [16].

The group took up points raised about network equilibrium in the resource papers delivered by Alan Wilson and Marc Gaudry: to ignore equilibrium in analysis of many network situations is to do no less than to render the solutions incorrect. Equilibrium models prove their importance by providing valid numbers for the evaluation of alternatives. Also, it has been shown that the highest-level composite costs relate to a valid measure of consumer surplus. Gaudry and others maintain that impedances used in network tracing should be consistent with those used in the demand model. Ruiter and others say that this is unimportant if the weights differ for path choice and other travel decisions. It was agreed that the sources of impedance figures should be kept in mind (since they can be so diverse, ranging from 'pure' times or monetary costs to composites weighted by linear or non-linear methods) to avoid misapplications, such as would occur if a composite and weighted variable were treated (in any part of an analysis) as though it were a simple monetary cost.

There is an important practical deterrent to the use of network equilibrium procedures with disaggregate demand models. Since congestion is an aggregate phenomenon, the determination of demands consistent with generalised origin-destination itinerary costs presents tremendous computational difficulties; and it poses conceptual questions which are by no means completely answered. Equilibrium methods are already avoided because of their expense; and if their expense increases they will become even less acceptable.

Throughout this discussion there was evidenced a new congruence and harmony of research efforts underway world-wide, a good omen of the continued rapid advancement of transport systems analysis.

*Land-use Equilibrium*

Here work designed to represent the relations of land use and transport was discussed. Among integrated land-use transport model systems, usually with lagged transport variables affecting land-use development, the following Lowry-type models were identified: IPLUM; models developed at the University of Leeds; DRAM and TOPAZ (1,17,27). On a larger scale, models concerned with the location, sizing and transport links to new towns were identified:

1 The Mills model (used simultaneously for transport and land-use design) minimising costs of city development and transport; based on continuous variables (9,22,23).
2 The TRANSLOC model, which deals with indivisibilities - not all continuous variables - and contains three levels: allocation of road links and towns; detailed location of towns with fixed networks; and prediction of transport flows (16).

There was some disagreement about the utility of catastrophe theory on the analysis of transport and land-use systems. The possible usefulness of the theory was said to show itself in such situations as auto-versus-transit mode shares and urban decentralisation, where there is the possibility of bifurcation points and hysteresis effects (31).

In our discussion of long-range equilibrium models, the

possibilities of model parameters changing significantly over time were discussed.  If these changes are observed, it raises the question of our ability to develop models which can be used to predict parameter changes in other models. While some in the group saw the need for such an approach, others preferred to recommend the combination of level-of-service and socio-economic variables in a single model which would directly deal with what otherwise would be time-varying parameters.

*General Economic Equilibrium*

The discussion touched unevenly upon a number of items of work of which the following is a selection:

1  Leontief's (14) input-output work: the conclusion was that transport is inadequately represented.
2  Moses' (24) work: the limitation to linear models is restrictive.
3  Wilson's and others' (4,18,19,32,33,34) work on entropy: based on linear models.
4  Kresge and Roberts (12): the Brookings model, which combines a macro-level regional input-output economic model with a more disaggregate transport model.

The workshop's general conclusion was that much is left to be done in the area of integrating transport into general economic equilibrium models.  We need models which incorporate diverse economic elements such as production, location, transport and growth.  It was remarked that deficiencies in the representation of transport and location in economic models are due to a lack of data to estimate regional input-output coefficients for capital.  In addition, integration of information theory and system modelling is needed, following on the work reported by Andersson and Karlquist in their resource paper.

## REFERENCES

1. Brotchie, J.F. and Sharpe, R., 'TOPAZ and its use in Australia', *Report of the First Binational (US/Australian) Urban Systems Symposium*, NTIS (PB241120), Washington, D.C. (1975).
2. Burrell, J.E., 'Multiple route assignment and its application to capacity restraint', *Fourth International Symposium on the Theory of Traffic Flow*, Heft 86, Strassenbau und Strassenverkehrstechnik, Karlsruhe (1968).
3. Coelho, J.D. and Williams, H.C.W.L., 'On the analysis, evaluation and design of land-use transport plans', *School of Geography Working Paper*, University of Leeds (1977).
4. Cripps, E.L. *et al.*, 'Energy and materials flows in the urban space economy', *Transportation Research*, 8, 293 (1974).
5. Dantzig, G.B. *et al.*, 'The application of decomposition to transportation network analysis', *Interim report prepared for Office of the Secretary, US Department of Transportation*, Report No. DOT-TSC-OST-76-26, Washington, D.C. (1976).
6. Dial, R.B., 'A probabilistic multipath traffic assignment model which obviates path enumeration', *Transportation Research*, 5, 83 (1971).

7. Federal Highway Administration (1977) *Planpac/Backpac: General Information*, US Department of Transportation, Washington, D.C.

8. Florian, M. *et al.*, 'EMME: A planning method for multi-modal urban transportation systems', Publication No. 62, *Centre de Recherche sur les Transports*, Université de Montréal (1977).

9. Hartwick, P. and Hartwick, J., 'Simulating a multi-nucleated city by market mechanisms: Mills's model extended', *Economics Department*, Queens University (1974).

10. Highway Research Board (1965) *Highway Capacity Manual*, Special Report 87, Washington, D.C.

11. Ingram, G.K. *et al.* (1972) *The Detroit Prototype of the NBER Urban Simulation Model*, National Bureau of Economic Research, Columbia University Press, New York.

12. Kresge, D.T., Roberts, P.O. *et al.*, 'Systems analysis and simulation models', in *Techniques of Transport Planning*, J.R. Meyer (ed.), vol. II, The Brookings Institution, Washington (1971).

13. LeBlanc, L.J., 'An algorithm for the discrete network design problem', *Transportation Science*, 9, 183 (1975).

14. Leontief, W. (1970) *The Structure of American Economy*, John Wiley, New York.

15. Lerman, S.R., 'Location, housing, automobile ownership and mode to work: a joint choice model', *Transportation Research Record 610* (1976).

16. Los, J. and Los, M. (1976) *Computing Equilibria: How and Why*, North-Holland, Amsterdam.

17. Mackett, R.L., 'A dynamic activity allocation-transportation model', in *Urban Transportation Planning: Current Themes and Future Prospects*, P.W. Bonsall, *et al.*(eds) Abacus Press, Tunbridge Wells (1977).

18. Magill, S.M., 'Alternative rectangular input-output models', *School of Geography Working Paper 182*, University of Leeds (1977).

19. Magill, S.M., 'The Lowry model as an input-output model and its extension to incorporate full inter-activity relations', *Regional Studies* (1977).

20. Makigami, Y. and Woodie, W.L., 'Freeway travel time evaluation technique', *Highway Research Record 321* (1970).

21. Manheim, M.L. and Ruiter, E.R., 'DODOTRANS I: A decision-oriented computer language for analysis of multimode transportation systems', *Highway Research Record 314* (1970).

22. Mills, E.S., 'Markets and efficient resource alloca-tion in urban areas', *Swedish Journal of Economics*, 74(2) (1972).

23. Mills, E.S., 'Mathematical models for urban planning', Department of Economics, Princeton University (1972).

24. Moses, L., 'The stability of interregional trading patterns and input-output analysis', *American Economic Review*, 8 (1955).

25. Murchland, J.D., 'Road network traffic distribution in equilibrium', paper presented at the conference. *Mathematical Methods in Economic Sciences*, Mathematisches Forschung-institut, Oberwolfach (1969).

26. Nguyen, S. and James, L., 'TRAFFIC - An equilibrium traffic assignment program', Publication no. 17, *Centre de Recherche sur les Transports*, Universite de Montreal (1975).

27. Putnam, S.H., 'Preliminary results from an integrated transportation and land use models package', *Transportation*, 3, 193 (1974).

28. Scarf, H. (1973) 'The computation of economic equilibria', conference (unpublished), New Haven.

29. Urban Mass Transportation Administration (1977) *Urban Transportation Planning System: Reference Manual*, US Department of Transportation, Washington, D.C.

30. Wardrop, J.G., 'Some theoretical aspects of road traffic research', *Proceedings of the Institution of Civil Engineers*, Part II, 1, 325 (1952).

31. Wilson, A.G., 'Catastrophe theory and urban modelling: an application to modal choice', *Environment and Planning, A*, 8, 351 (1976).

32. Wilson, A.G. (1970) *Entropy in Urban and Regional Modelling*, Pion, London.

33. Wilson, A.G., 'Further developments of entropy maximising transport models', *Transportation Planning and Technology*, 1, 183 (1973).

34. Wilson, A.G. and Senior, M.L., 'Some relationships between entropy maximising models, mathematical programming models and their duals', *Journal of Regional Science*, 14, 207 (1974).

35. Yagar, S., 'CORQ - A model for predicting flows and queues in a road corridor', *Transportation Research Record 533* (1975).

36. Young, S. (1976) The Application of Disaggregate Demand Models in Urban Travel Forecasting, unpublished S.M. Thesis, Department of Civil Engineering, Massachusetts Institute of Technology, Cambridge, Massachusetts.

PART FOUR

Chapter 10

MARKET SEGMENTATION: A TOOL FOR TRANSPORT DECISION-MAKING

Ricardo Dobson

INTRODUCTION

There is a need for a market-segmentation orientation in
transport analysis. This need stems from fundamental
similarities and differences among the entities with which
transport analysis is concerned and from the fact that these
similarities and differences often have critical implications
for user and non-user reactions to transport services and
facilities. Market-segmentation procedures represent a way
to process these similarities and differences for input to
comprehensive systems analyses which support transport
policy-making.

Market-segmentation procedures are relevant at several
unique levels of analysis in addition to having the potential
for direct application to transport policies. Segmentation
can be used to reveal actual and/or anticipated traveller
responses to alternative classes of transport services and
facilities. It is also possible to assess for non-travellers
potential and/or experienced benefits and disbenefits within
a market segmentation framework. At a more aggregate level,
geographic regions within an urbanised area, such as travel-
analysis zones, can be grouped to demonstrate the existence
of alternative transport requirements for different parts of
an area. At even further levels of aggregation, it is
possible to cluster entire urbanised areas into market
segments which are likely to react in common ways to the
demonstration and testing of innovative transport services,
facilities and managements actions. Finally, various transport policies can be partitioned into segments which are
likely to have common impacts with respect to travellers,
non-travellers, implementation and operation costs, and other
transport decision-making criteria.

Market segmentation has evolved from marketing research
and theory. Central concepts from this body of literature
are reviewed briefly with the aim of showing the relevance of
market segmentation to transport analysis and policy-making.
Prior applications and discussions of market-segmentation
approaches in transport planning and research literature are
also reviewed. In addition, various technical issues relating to segmentation-model specification, clustering
algorithms and tests of significance for different market
segmentations are mentioned. These reviews point out policy-
relevant information which can be produced through a
segmentation approach and procedures for deriving that

219

information.  The reviews are not meant to be exhaustive with
respect to any body of literature but rather to suggest
concepts which are discussed in varied bodies of literature.
A series of specific transport applications that are suitable
for a market-segmentation orientation are defined and
elaborated in a section which builds upon the information
base developed in prior sections.  It is hoped that the great
promise of segmentation procedures for transport analysis and
policy-making are highlighted through the contrasting of its
policy-relevant and technical aspects.

## CONCEPTS FROM MARKETING RESEARCH AND THEORY

### Defining Market Segmentation

Market segmentation is predicated on two key concepts.  The
first one is that meaningful groups of users of a product or
service can be identified.  The second one revolves around
the creative assessment of product and/or service requirements for different market segments.  The objective of this
activity is usually to maximise profits through rendering
services and/or products which appeal strongly to lucrative
market segments.  Market segmentation, as traditionally used
in business contexts, always refers to a market mechanism.

Within a transport sphere, it is likely that the manner of
use and objective of market segmentation procedures will be
somewhat different from traditional business applications
(62,38).  Hensher (52) discusses a process-oriented, as
opposed to a model-oriented approach for the incorporation of
market segmentation in transport analysis.  The concept of a
process as a problem-solving tool is familiar to transportation analysts.  Hensher also points to another key difference
when he remarks that the utility of a segmentation approach
extends beyond behavioural demand models.  In other words,
its scope of application is more inclusive than profit maximisation and related issues.  Some examples of more inclusive
topics are better utilisation of existing system resources,
decreased needs for costly new system additions, and/or
increased transit ridership even if that may imply reduced
profits.  Therefore, market segmentation in transport is used
in an explicitly different way from business that is more
generic than the traditional marketing definition of the
term.  In particular, market segmentation need not always
refer to a market mechanism.  Market segmentation, as
discussed herein, can facilitate the formulation, implementation and evaluation of transport policy.

A variety of criteria have been proposed for assessing
whether any particular group of market segments is
appropriate.  Kotler (62) notes that it is important for
segments to be measurable, accessible and substantial.  Wilkie
and Cohen (97) remark that 'true' market segments should be
homogeneous within groups and heterogeneous between groups,
useful as a behavioural correlate, and efficient to market
from production, distribution and/or pricing perspectives.
The key problem with the approach lies in choosing segmentation variables that relate to the purpose for which they are
being used.  Discussions and/or business applications which

describe successful market segmentations include the
following: Bass, Tigert and Lonsdale (3); Blattberg and Sen
(8); Bruno, Hustad and Pessimier (10); Haley (49); Johnson
(59); and Lessig and Tollefson (63)

*Segmentation Bases*

A pivotal aspect of market segmentation is the selection of
appropriate variables on which to divide entities, such as
individuals, households, analysis zones of an urbanised area,
or cities.  A set of variables selected to divide entities is
called a segmentation basis.  Since market-research literature
had generally ignored segmentation applications to other than
buyers and their households, most discussions of segmentation
bases are restricted to variables which characterise individuals.  These variables can be loosely classed under three
headings: (1) socio-demographic data; (2) consumption
patterns; and (3) individuals' subjective judgements.

*Socio-demographic data.*  These data comprise social, economic
and demographic indicators, such as age, income, family size
and auto-ownership.  This class of variables is interesting
as a segmentation basis because it is readily measured and it
is often provided through secondary sources.  In addition, it
is often possible to isolate substantial market segments
which are readily identifiable for access through a promotional campaign.  The key problem lies in determining which,
if any, of the socio-demographic variables are closely linked
with consumption for the product or service of interest.
Another critical issue surrounds the use of socio-demographic
data as a market-segmentation basis.  This problem is that
observed relationships between socio-demographic data and
purchase patterns are *generally not behaviourally based*.  The
implication of this problem is that socio-demographic-buyer
relationships can change even if socio-demographic data
remain invariant.  Nevertheless, socio-demographic data
remain a popular source for the composition of segmentation
bases.

     There are two common ways of using socio-demographic data
as a segmentation basis.  The more direct procedure is to
divide a market on the basis of a single variable.  For
example, a market might be conveniently and usefully divided
into low, medium and high income groups.  Other examples
include young, middle-aged, and old segments or male and
female markets.  A more sophisticated approach to the utilisation of socio-demographic data involves the concurrent use
of two or more variables.  This may be achieved through the
Cartesian product for categories of at least two socio-
demographic variables.  Cross-classification tables for trip
rates as a function of income and auto ownership represent an
example of the latter technique.  Another method for
concurrently employing more than one socio-demographic
variable involves the use of social composite indicators,
such as life-cycle or social class.  However, there is often
more than one way of defining these composite indicators, and
comparisons across studies have a potential for lacking
precision through the incorrect contrasting of incommensurate
indices.  Finally, multivariate-clustering algorithms exist

for the grouping of entities based on two or more variables. These procedures rely on natural covariance in a data set to form category boundaries for ground membership.

The success of socio-demographic data as determinants of buyer behaviour is mixed. It is reported that socio-demographic data are of little use as bases for segmentation with respect to frequently-purchased household products (38). On the other hand, Carman (13) found life-cycle and disposable income, among other socio-demographic variables, to influence significantly the percentage of family income spent for durable home furniture.

*Consumption patterns.* Early studies by Brown (9) and Cunningham (18) first aroused interest in the idea of using consumption patterns with respect to products, brands and/or stores as bases for segmentation. By understanding the social, economic and psychological determinants of loyal patrons, it would be possible to formulate promotional campaigns to reinforce their patterns and to convert customers who are loyal to competitors. Empirical investigations summarised by Frank, Massy and Wind (38) reveal correlations with buyer behaviour that correspond to socio-economic bases. That is, the average $R^2$ is 10 per cent or less. There are brands for which the $R^2$ rises above 30 per cent, but these seem to be situation-specific.

Another distinction in consumption patterns is for heavy versus light users of a product or service. It is occasionally the case that a relatively small percentage of customers may consume the majority of some commodity. In the latter situation, it might be profitable to orient marketing and product planning towards the requirements of this particular market segment. On the other hand, the desires and needs of light- and/or non-users of a product or service may signal important and profitable marketing opportunities. Which strategy is most appropriate depends on background factors, such as the proportion of the heavy users in the total market and the degree of choice that heavy users have with respect to the product or service. If the proportion of heavy users is small, it may be well to attend closely to the needs and desires of light- and/or non-users. Conversely, if heavy users of a brand could easily be converted to another one and heavy users constitute a large percentage of the total market, then the preferences of the heavy-user segment may be paramount.

*Individuals' subjective judgements.* Data based on the subjective judgements of individuals are used in market research for multiple purposes, including as a segmentation basis. At least two papers in the transport-research literature review market-research applications of individuals' subjective judgements (41,45). One of these papers, Golob and Dobson (45), specifically discusses the notion of segmentation based on perceptions and preferences within the context of a general schema for predicting traveller decision-making. More recent papers review alternative ways in which individuals' subjective judgements can be used to facilitate transport planning and analysis requirements, but the theme of market segmentation is not strongly developed (21,22,23).

Personality and/or life-style indices provide a general product measure of likely customer reaction. The measure is specific to given individuals but generic across product classes or brands. If given personality types differentially prefer and purchase some products or brands, then marketing opportunities exist with respect to the design of products or services and the development of promotional campaigns. Personality indices which correlate with buyer patterns are behaviourally based, and they thus admit the possibility of causal relationships. Nevertheless, a personality or life-style variable could correlate with some observed buyer pattern because of a more critical, underlying variable. There are research methodologies capable of testing such hypotheses (24).

In a manner which parallels socio-demographic data, personality and life-style variables have mixed relationships with purchases. For example, Evans (34) finds relatively few differences between Chevrolet and Ford owners with respect to the Edwards Personality Preference Schedule. On the other hand, Birdwell (6) reports a consistent pattern of relation-ships between self-perceptions and the perception of one's car. Grubb and Hupp (47) report compatible findings with those of Birdwell. Perceptions and preferences represent an alternative class of subjective judgements from individuals. Unlike personality measures, these indices are not likely to remain invariant across product classes or brands. The distinction between general and specific customer character-istics is developed by Frank, Massy and Wind (38) and used as a major basis for describing key aspects of two schools of segmentation analysis by Wilkie and Cohen (97). Yankelovich (99) was among the first to specify subjective customer characteristics, such as preferences and perceptions, as a basis for segmentation studies.

A special case of the use of perceptions and preferences for segmentation has been designated benefit segmentation by Haley (49). Haley asserts:

The belief underlying this segmentation strategy is that the benefits which people are seeking in consuming a given product are the basic reasons for the existence of true market segments. Experience with this approach has shown the benefits sought by consumers determine their behavior more accurately than do demographic characteristics or volume of consumption.

Wilkie (96) examined the benefit-segmentation concept to find that benefit segments were moderately correlated with brand purchases but more strongly related to underlying measures, such as user satisfactions and purchase intentions. The General Motors Research Laboratories produced a series of reports examining the benefit-segmentation concept without using that terminology. It was found that benefit segments utilised different decision-making criteria to determine anticipated user satisfactions or purchase intentions, that anticipated benefits corresponded meaningfully to socio-demographic characteristics of respondents, and that insight-ful benefit segments could be derived by a wide variety of alternative psychometric and statistical procedures (17,30, 69,28).

There is not total agreement on the nature of the
correlation between perceptions and preferences (commonly
referred to as attitudes) and behaviour.  A widely referenced
paper by Wicker (95) reviews 33 studies to conclude that
attitudes are only slightly, at best, related to overt
behaviours, such as the purchase of a product or service.
However, Fishbein and Ajzen (36) argue that attitudes do
predict behaviour provided that the attitudinal measure is
appropriate for the type of criterion being predicted.  They
are not surprised that attitudes about an object, such as a
product, fail to correlate with behaviour towards that object.
They assert that purchasing a product can be better accounted
for by beliefs about purchasing the product rather than just
beliefs about the product.  Frank, Massy and Wind (38)
support Fishbein's position when they conclude that 'intention data provide strong expressions of intended behaviour
and can be used as a basis for market segmentation'.

*Research Strategies for Market Segmentation*

The wide diversity of potential segmentation bases, along
with similar heterogeneity with respect to segmentation-
research designs and objectives, has resulted in the
development of alternative schools of thought on research
strategies for market segmentation.  For example, Frank,
Massy and Wind (38) distinguish between behavioural and
decision-oriented market-segmentation strategies.  Wilkie and
Cohen (97) also define two alternative streams of market
segmentation analysis.  These are labelled the correlation
and product-instrumentality streams.

The behavioural versus decision-oriented dichotomy for
segmentation analyses is based principally on distinctions
between basic and applied research.  For instance, the
behavioural market-research approach is utilised primarily by
academicians.  This approach attempts to document group
differences in consumer reactions.  On the other hand, the
decision-oriented approach assumes that group differences
exist, and it focuses on the clearest ways to define those
groups.  The decision-oriented approach, which is used most
often by practising marketing researchers, is concerned with
uncovering profitable market segments, whether or not those
segments contribute to theory or consumer reactions.

The correlational versus product-instrumentality dichotomy
is predicated on distinctions in theoretical orientations and
related issues.  The correlational stream is based on micro-
economic theory.  Emphasis is placed on outputs compatible
with the development of a promotional campaign, and the
principal mode of analysis is multiple regression.  The
product-instrumentality stream relies on a mix of broadly
based behavioural science theories, and it implements these
theories through advanced scaling and clustering algorithms
The product-instrumentality stream attempts to develop
outputs relevant to product positioning as well as promotional campaigns.

These alternative segmentation strategies complement each
other.  For example, the decision-oriented school serves as

a market for the theory and results of the behavioural research approach. The availability of correlational and product-instrumentality streams of market-segmentation analysis permits the selection of a research approach which is compatible with the objectives of particular market-segmentation studies.

## Alternative Marketing Approaches

When a segmented-market structure can be presumed to exist, several marketing approaches for selling the product or service are possible. Kotler (62) describes several marketing strategies. For example, information on the differential desires and needs of alternative market segments can be ignored. Another approach involves attending to the special needs of all segments. Finally, a third approach gives concentrated attention to the requirements of a few selected segments. If either of the latter two strategies is adopted, then a promotional campaign must be developed that transmits different information about alternative products or services to known market segments.

The undifferentiated marketing plan, which ignores information about market segments, is attractive for cost and simplicity considerations. Undifferentiated marketing costs less because only a single product or service is provided. Economies of scale in the manufacture and delivery of goods and services are more readily realised with a single product or service. An undifferentiated marketing plan must of necessity be directed at the broadest segment of the market. However, the profitability of this segment may be relatively low because of intense competition.

Differentiated marketing involves a decision to satisfy the needs and desires of alternative market segments through separate products and/or marketing programmes for each segment. One purpose for the adoption of such a marketing strategy is to enhance customer loyalty and to generate repeat purchasing because goods and/or services have been tailored to the customer's wishes. While a differentiated marketing plan is likely to result in high sales volume, it will also incur product customising costs for market segments and extra market-research costs to support the segmentation analyses. As a consequence, such a plan is not suitable for profit maximisation unless its special costs are less than the profits from enhanced sales volume.

The final marketing plan, called concentrated marketing, does not attempt to consider the entire market structure as does differentiated and undifferentiated marketing. The concentrated marketing plan attempts to capture a large share of one or a few market segments. This marketing approach achieves operating economies because of specialisation in the manufacture and/or delivery of goods and promotional campaigns.

When either a differentiated or concentrated marketing plan is adopted, there will be a need to transmit selectively promotional information to relatively homogeneous groups of

consumers.  Frank, Massy and Wind (38) compare and contrast
promotion plans based on segments and media versus those
based on segments, descriptors of segments and media.  Their
conclusion is that considering descriptors significantly
reduces the complexity and potentially reduces the costs of a
media-promotion element in a market-segmentation plan.  These
advantages are fully realised when the descriptors are
maximally correlated with the media and the segments.

Another element of a promotional campaign can be direct
mailing.  Reed (78) notes the high level of precision
contacts which can be achieved through direct-mail companies
with these four sample listings:

A cross-section of people who fall in the grandparents age
category.  Each one has been identified as being in a
financial position to express grandparental love and
affection.  80 per cent men.  State selection and
sequencing available.

Compilation of professional politicians who have com-
peted for federal, state and municipal offices ... winners
and losers for offices from senator to dog catcher.  95
per cent men.

Selection made from birth lists eliminating all high
income ZIP codes.  New parents in age groups 19-39.

Waiters and waitresses at home addresses.

Direct-mail campaigns are compatible with market segmentation
because they permit the transmission of information at
relatively low costs to well-defined groups of consumers
without the contamination of alternative market segments with
this same information.


MARKET SEGMENTATION IN TRANSPORT ANALYSIS

*Overview*

The concept of market segmentation within a consumer-research
framework was introduced into transport analysis in the early
and middle part of the 1970s.  Three major review papers
specifically address market segmentation and how it can be
used to facilitate transport analysis (52,65,78).  Transport
analysts have long grouped households on the basis of geographic proximity (i.e., households are assigned to travel-
analysis zones).  However, the introduction of the
market-segmentation concept has pointed to the fact that
spatial arrangement is not the sole basis for aggregating
households or individuals to a desirable level.  Golob and
Dobson (45) have suggested that perceptions and preferences
may serve as a useful transport basis for grouping households
or individuals.  Furthermore, market segmentation seems
appropriate in a variety of transport analysis contexts, some
of  which do not involve the aggregation of households but
rather of travel-analysis zones, urbanised areas, special-
interest groups or transport management actions.

Transport-analysis empirical studies utilising a market-segmentation strategy have employed user characteristics, socio-demographic data, and perceptions and/or preferences as segmentation bases. User characteristics typically designate a category of transport system use, such as travellers who normally commute to work by car-pools or individuals who journey from home to work three or more times per week. Socio-demographic data are regularly used in transport analysis. Income and auto availability, for example, are known to be related to both trip generation and mode choice (98,51). Perceptions and preferences have been included in transport analysis primarily in a research context. Golob (41) discussed the development of attitudinal models for travel behaviour. Dobson was subsequently involved in a series of papers which illustrate actual or potential applications of perceptions and preferences in transport analysis (45,21,22,23). All of those papers illustrate means of collecting, analysing, and interpreting subjective data from respondents that are suitable for market-segmentation studies.

The vast majority of the market-segmentation studies in the transport-analysis literature emphasise the empirical demonstration of a research concept instead of the practical illustration of a research application in an operational setting. It is likely, however, that the incidence of segmentation analyses in operational settings will increase markedly with the more widespread dissemination of information on how to implement a successful market-segmentation study. Another factor likely to influence the diffusion of market-segmentation procedures is information on the results which can be obtained through segmentation studies. Brief highlights of findings with perceptions and preferences, user characteristics and socio-demographic data as segmentation bases are noted below. More detailed discussions of segmentation results for travel-demand analysis, special-interest group studies and groupings based on varied geographic units are subsequently presented.

A broad array of empirical analyses have found interpretable relationships between travellers' perceptions and/or preferences for transport-system attributes (30,25,44). Other investigations (e.g. Nicolaidis and Dobson (69)) have demonstrated that these findings are generally robust in the face of fairly divergent research methodologies. Finally, it has been noted that segmentations based on perceptions and preferences develop insights for mode choice and destination choice (28,85).

In comparison to segmentations predicated on perceptions and preferences, those based on user or socio-demographic characteristics are often oriented more towards substantive than methodological issues. This practical thrust is well illustrated by Byrd's comparison (12) of Atlanta transit users and non-users. He notes that perceptions of transit convenience sharply differentiate these two groups. Gustafson and Navin (48) and Lovelock (66) find this pattern of positive perceptions of transit alternatives extends beyond the standard bus service provided by the Metropolitan Atlanta Rapid Transit Authority. Behnam and Thomson (4) in another practical study support some common conclusions, such as the need for low cost and comfortable transit service for the elderly and the concern of female travellers with personal safety. Nicolaidis, Wachs and Golob (70) in a paper with a stronger methodological component find socio-demographic characteristics based on auto and transit availability to be excellent correlates of varied behavioural aspects of travel.

*Travel-demand Analysis*

The review of travel-demand studies with a market-segmentation element will consider three classes of reports. In the first class to be mentioned, the emphasis will be on predicting the subjective reaction of travellers. The second class will include studies concerned with the analysis of objective choices for transport-related alternatives. The third class will describe some practical implementations of segmentation procedures in transit-marketing environments.

*Predicting subjective reactions.* Studies directed at the prediction of subjective reactions are most often concerned with innovative or evolutionary transport systems. Respondents are generally asked to say how satisfied they would be or how frequently they would use the new system. Within the transport-analysis literature, these types of responses are often predicted through a mix of perception and preference scores for transport-system attributes.

Costantino, Dobson and Canty (17) tested the hypothesis that a more thorough understanding of mode choice could be obtained by stratifying a sample into homogeneous groups rather than by considering respondents as an undifferentiated set. Through the use of AID (Sonquist, Baker and Morgan (84)) and a linear-regression procedure, they supported this hypothesis. Models which stratified the sample on either socio-economic characteristics or importance judgements for system attributes produced larger multiple coefficients of determination than those which considered respondents as an undifferentiated set. Furthermore, the separate strata of respondents or market segments emphasised different system attributes as determinants of their mode choice for innovative urban-transport modes. This information enhanced the understanding of the choice process for different market segments.

Alpert and Davies (1) divided respondents based on their intention to use the Austin, Texas transit system. They were not able to differentiate on the basis of a mix of perception and preference responses between those who replied definitely yes they would use transit if some improvements were made from those who gave another response. However, Tischer and Dobson (88) scored behavioural intention to switch on the basis of fourteen responses whereas Alpert and Davies scored this concept on the basis of a single response. It is possible that this methodological distinction explains the variance in substantive findings.

Dobson and Kehoe (28) investigated the convergent-discriminant validity of market segments based on perceptions

about system attributes.  This study was implemented by
comparing and contrasting two individual-differences scaling
models.  It was found that distinct market segments could be
defined which related in interpretable ways to the socio-
economic characteristics and activity patterns of respondents.
It was possible to account for anticipated satisfactions with
innovative transport modes on the basis of their perceptual
dimensions, and one market segment's dimensions were better
at estimating its satisfactions than other segments'
dimensions were.  The latter finding, which was statistically
significant for one model, supported discriminant validity
across market segments.

*Predicting objective choices*.  Studies concerned with
predicting objective travel choices are most often concerned
with analysing the operation of an existing transport system.
When these studies are implemented within a market-research
context, their goal is frequently the diagnosis of existing
functions in a search for system enhancements.  When these
studies are performed in a planning-research environment,
their goal is often to devise a better forecasting model.

Hensher (53) has analysed the trip frequency of weekly
household shopping trips within a market-segmentation framework.  He used AID and linear regression to uncover basic
socio-economic determinants of shopping-trip frequency.
Among the principal variables were number of household
members, occupation of housewife and income of main household
earner.  By interpreting both the AID and linear-regression
findings, he concluded that there were three major socio-
economic dimensions for shopping-trip frequency: (1) employment commitment, defined by such indicators as number of
hours working outside the home and total household income;
(2) size and composition of household; and (3) age distribution of the travellers in a household.

Recker and Golob (76,77) restrict variables for clustering
responses to perceived auto accessibility, perceived bus
accessibility, and reported number of bus transfers to make a
work trip.  These variables are labelled choice constraints
and a market-segmentation structure is built from them.
Typical segment designations are carless, busless and mobile.
It is found that segments with different choice constraints
emphasise different classes of system attributes as determinants of their commute-to-work mode.  In addition, the
average percentage of correct mode-choice predictions across
segments is 85, but it is only 81 per cent when the choice-
constraint segmentation structure is ignored.

Dobson and Tischer (31) segmented Los Angeles, California,
central business district commuters on the basis of the mode
they usually took to work (i.e., single-occupant auto, bus or
car-pool).  It was found that individuals who use a mode view
that mode as more favourable than those who do not.  In
addition, Dobson and Tischer reported mode-choice models
which were based solely on system perceptions that were
statistically significant beyond the 0.01 level and which
predicted 63 per cent correctly.  It was noted that convenience was the prime determinant of mode choice.

*Segmentation in transit marketing.*  In *Pricing-Transit Marketing Management Handbook*, a marketing publication issued by the Urban Mass Transportation Administration (1976), the theory and practice of transit pricing is discussed with respect to elasticity and fare structure and collection procedures.  Some case study histories also illustrate various market-segmentation concepts.

For example, the San Diego, California Transit Commission instituted a successful programme to increase ridership through a reduced fare programme initiated in August 1972. However, it was subsequently necessary to increase revenues. A segmented fare structure was adopted.  The general fare was raised 10 cents to a total of 35 cents.  Two groups were sheltered from the fare hike.  Senior citizens and handicapped patrons had their fare set at 15 cents, and the student fare remained at 25 cents.  In response to this segmentation policy, the transit revenues increased.  While general patronage declined slightly, there were substantial increases in degree of bus use among groups which were spared the fare increase.

The transit systems of New York City and Chicago, Illinois introduced a segmented fares policy based on the time of travel.  New York City permitted a round-trip for the price of a one-way trip on Sundays only.  Sunday ridership increased 30 per cent during the first year, but Sunday revenues declined overall by 12 per cent.  The transit price reduction in Chicago was also restricted to Sundays with similar consequences in terms of revenues and ridership.

*Special Interest Group Studies*

Within the context of this paper, special interest group studies are used to refer to two classes of transport analyses.  The first class of analyses refers to efforts designed to incorporate the viewpoints of various citizens and/or citizen groups on either specific transport projects or the general goals and objectives of a transport system in a region.  The second class of analyses designates a special transport study of a particular group of citizens, such as the elderly or the handicapped.

Lichfield (64) was among the first to propose a form of an interest-group impact theory, under the title 'balance sheet of development'.  Schermer (80) summarises two practical implementations to the interest group theory.  The fundamental tenet of the theory as described by Schermer is that a decision-making situation can be reduced to a set of identifiable groups with clearly defined objectives and concerns.  Such notions are central to market segmentation as well.  Therefore, through market-segmentation concepts it should be possible to aid in the evaluation of alternative courses of action with respect to public works projects such as major transport facilities.

Using a less elaborate framework than interest-group impact theory, Simpson and Ismart (83) describe an effort to update the transport goals, objectives and policies for the

Puget Sound Governmental Conference. This Conference
includes a four-county region which contains the cities of
Seattle and Tacoma, Washington. Their methodology relied
heavily on a mail survey which was balanced to census
statistics on several key variables. Among the principal
findings of the study, which was segmented along geographic
lines, was that Seattle residents were more likely to use and
support transit than residents of the somewhat more rural
city of Tacoma.

The elderly are an important segment in transport analyses
for legislative as well as other reasons. For example,
section 16(a) of the US Urban Mass Transportation Act of
1964, as amended, states, 'that special efforts shall be made
in the planning and design of mass-transport facilities and
services so that the availability to elderly and handicapped
persons of mass transport which they can effectively utilize
will be assured'. To support this general objective
Edelstein (33) performed an in depth analysis of the travel
patterns and attitudes of elderly in Buffalo, New York.
Wachs (e.g., Wachs and Blanchard (93)) has also examined
transport planning practices for the elderly. It is his conclusion that 'planning and forecasting needs for the future
needs of the elderly should not be based upon the transport
patterns and needs of those who are currently elderly, but
rather should focus more upon those who are now in their
thirties and forties and who will become the elderly of the
future'. This philosophy of planning for elderly transport
needs is based on a segmentation principle with an age and
cohort basis.

*Segmentation for Geographic Units*

Segmentation for geographic units, either within an urbanised
area or between different urbanised areas, has emerged as a
transport-analysis topic which builds upon prior experiences
in urban ecology (73,58,40) and social-area analysis (89,5,
82). Such segmentations are typically specified and interpreted within a three-stage process. In the first stage, the
geographic units are analysed using some variation of
singular decomposition (32,57,26,27). This analysis defines
the geographic units with respect to a set of common factors
which are often derived to be mutually orthogonal. The
geographic units are subsequently grouped into homogeneous
clusters based on the common factors. Finally, the clusters
of homogeneous geographic units are interpreted for their
relevance to travel behaviour, socio-demographic variables,
and/or transport policy.

At least two reports describe efforts to develop market
segments for different urbanised areas (42,43). These
studies are concerned with the revealing of structural
relationships between the characteristics of urbanised areas
and transport systems for requirements analysis and/or market
research. A total of 80 urbanised areas were clustered in
terms of 15 factors based on a larger set of 53 variables.
The variables measured social, physical and transport aspects
of urbanised areas, and they included such items as land
area, population and total proposed miles of freeways. The

Wilks-lambda criterion of the Friedman and Rubin (39) algorithm used to form clusters did not indicate an optimal cluster size, but 8- and 9-cluster solutions were discussed most fully. While geographic proximity was important for cluster formation, it was by no means the sole basis for areas being grouped together. For example, group 6 in the 8-cluster solution includes the following four urbanised areas among others: Sacremento, California; Milwaukee, Wisconsin; Syracuse, New York; and Richmond, Virginia. Various procedures were sketched for using the cluster results in the analysis and planning of transport-systems demonstration projects.

While it is reasonable that different urbanised areas are heterogeneous with respect to their requirements for transport systems, it is also likely that large urbanised areas are not internally spatially homogeneous with respect to transport demand, supply and impact phenomena. Golob, Hepper and Pershing (46) and Hepper and Gustafson (54) studied the Detroit urbanised area to uncover geographic units, such as minor civil divisions and central-city sub-communities, which exhibited homogeneous travel patterns and interpretable links to socio-demographic measures on those geographic units. Despite the fact that the level of analysis is substantially different, the general methodological approach is generally similar to the market segmentation of different urbanised areas (42). The search for homogeneous geographic travel units within an urbanised area is based upon the statistical decomposition of origin-destination data between units. Through this decomposition, it is possible to determine groups of origin units which distribute trips to destination units in a common way. Golob, Hepper and Pershing (46) relate homogeneous groups of geographic units to economic theories of central cities, suburban industrial centres and satellite cities. Hepper and Gustafson (54) show how a market segmentation built on geographic units can be used to reveal differences in homogeneous population segments, such as different racial groups.

Hepper and Pershing (55) reanalyse the data previously discussed by Golob, Hepper and Pershing (46) and Hepper and Gustafson (54). However, while the latter two analyses were only indirectly concerned with methodological issues, the Hepper and Pershing report is primarily directed at the methodological contrasting of three different singular-decomposition procedures. It is concluded that the three different procedures, which are differentiated with respect to how the raw data is pre-processed before decomposition, are not substantially different in the results which are produced. Nevertheless, the procedure which involves no preprocessing of the origin-destination flow matrices prior to decomposition is recommended because it may simplify forecasting studies and lead to easier interpretation of residuals.

ANALYTICAL-DESIGN CONCEPTS

*Overview*

Analytical-design concepts appropriate for market-
segmentation studies depend on a broad range of issues.
These issues include: (1) the context of the study; (2) the
nature of the dependent variable(s); (3) the independent
variable(s); (4) the unit of analysis; and (5) the class of
research procedures which are compatible with study objectives.  Study guidelines for specific market-segmentation
studies can be deduced from a consideration of these general
issues and their interactions.  Brief descriptions are
provided below to facilitate an understanding of the issues
and their interactions.  In addition, a more detailed review
of central concepts relevant to market-segmentation research
is presented.

*The study context.*  The purpose of a study is determined by
its context.  As noted above, three common market-segmentation study contexts in transport include travel-demand
analysis, geographic analysis and special-interest and/or
citizen group analysis.  These contexts relate to different
aspects of transport planning and they, therefore, require
different dependent variables, independent variables,
analysis units, research procedures, and display formats for
research findings.

   Travel-demand analysis and geographic analysis generally
support technical elements underlying the urban transport-
planning process.  Travel demand typically refers to
traveller choices with respect to trip frequency, destination,
mode and route of travel.  These elements describe several
important decisions which individuals must address in travelling between points in space to satisfy their daily needs.
Geographic analysis has been performed at inter- and intra-
urban levels.  At the inter-urban level, market segmentation
can be used to designate sets of urbanised areas which have
common transport-system requirements or which are likely to
respond in a common way to various transport-system requirements or which are likely to respond in a common way to
various transport innovations.  The empirical definition of
super districts can be achieved through the application of
market segmentation at the intra-urban level.

   Special-interest and/or citizen group analysis relates to
more qualitative aspects of urban transport planning.  These
aspects include means of incorporating the needs and wants of
diverse publics into systems and project-level elements of
transport planning.  These publics may include, but are not
restricted to, the elderly and handicapped, minority racial
or ethnic groups, or the business interests in a community.
In addition, the expressed needs of individuals from different political jurisdictions within an urbanised area can
be compared and contrasted conveniently.

*The dependent and independent variables.*  Independent and
dependent variables both contribute to the design of a
market-segmentation study.  The dependent variable can
delimit the range of analytical procedures which are

appropriate, or it may suggest the relevance of one class of independent variables in comparison to others.  The independent variables can show a greater or lesser alignment with an econometric orientation in contrast to a more diverse range of behavioural-science perspectives.

Dependent variables which have been used for market-segmentation analyses include both reports of objective behaviour and descriptions of subjective states.  An example of reports of objective behaviour include travel or purchase diaries.  These diaries constitute panel data; that is, they represent repeated measurements on sampling units over time. Such data are readily susceptible to time-series analyses and before-after study designs.  Other dependent variables are linked with subjective states of an individual, such as behaviour intentions to use a proposed type of transit service.  As noted above, behavioural intentions have been shown to be correlated with purchases of houses, autos and home improvements (68).

Independent variables in a market-segmentation analysis are the basis where the term basis is used to denote one or more variables for grouping entities according to a criterion of interest.  The entities are often, but not always, consumers.  The consumers can be actual or potential users of a transport service or facility.  The criterion of interest is designated by the dependent variable.  In the sphere of transport analysis, relevant criteria of interest include trip frequency, mode split and political-social acceptability.

Three possible segmentation bases were identified above. These were socio-demographic data, consumption patterns and individuals' subjective judgements.  These bases are compatible with the assumption that consumers are to be segmented.  Some alternative bases which do not necessarily suggest that assumption, are zonal productions and/or attractions and value structure of different citizen groups. These bases imply, respectively, zones and citizen groups as the entities to be segmented.

*The unit of analysis*.  As indicated, market segmentation can be applied at several different levels of analysis.  The most common level is the individual or household consuming unit.  In this case, consumers are grouped into homogeneous clusters with respect to one or more criteria of interest. On the other hand, market segmentation has also been applied with respect to geographic units or interest groups within a community.  For example, segmentation procedures with a subjective dependent variable designating social desirability and objective independent variables characterising politically relevant population segments can be used to clarify, understand and compromise conflicting values of different community interest groups.  This basic notion underlies Lichfield's 'balance sheet of development' approach which was mentioned above.

*The research procedures*.  Research procedures constitute a critical element of any market-segmentation study design.  As suggested above, it is likely to be partly determined by the study context, the dependent and independent variables, and

the unit of analysis.  Research procedures, in turn, place
restrictions on other elements of the study design.

One principal aspect of any market-segmentation study
context involves whether it is to develop a classification
system or to assign entities to an established classification
system.  Market-segmentation frameworks are 'calibrated' in a
sense by the development of a classification system.  Such a
system can be predicated on a simple income split of households (e.g. Stopher and Lavender (86)) or involve the
sequential use of complex multivariate statistical techniques
with many different basis variables (e.g. Nicolaidis and
Dobson (69)).  For forecasting and analysis purposes, it is
often useful to assign entities to an existing classification.
The actual grouping of entities can be implemented by
hierarchical, stepwise methods or by non-hierarchical,
simultaneous techniques.  Hierarchical procedures emerged
from biometric analysis in which it was reasonable to assume
a genealogy of species within genera and genera within
families of plants or animals.  In quantitative psychology,
hierarchical procedures have been widely used because of the
feasibility of nested concepts and the ready availability of
computer programs to implement Johnson's hierarchical-
clustering schemes (1967).  Simultaneous segmentation
procedures do not require a hierarchy of relationships among
the basis variables.

After a classification system is established, it is
usually advisable to assess its effectiveness with respect to
one or more criteria of interest.  Dobson (24) has described
a statistical technique, which he calls GLANOVA, for evaluating the fit of a market-segmentation structure.  GLANOVA can
be used to compare and contrast alternative segmentation
structures for the same data set.  Horowitz and Sheth (56)
have applied this kind of statistical approach in a psychosocial analysis of ride-sharing to work.  Split-sample
procedures provide a means of assessing the stability of
models built on a market-segmentation structure.  Through the
appropriate joint application of GLANOVA and split-sample
methods, it is possible to evaluate both the homogeneity of a
set of market segments and their relationship to some
external criterion of interest.

*Classification and Assignment*

Since market-segmentation structures are 'calibrated' through
the development of a classification system, procedures for
developing these classifications are central to market-
segmentation research.  The term calibration is used because
the selection of a classification system is comparable to
estimating the parameters of a model.  After a decision is
made about basis variables, the selection of critical values
for these variables which classify entities represents a
calibration task.  Assignment refers to the process by which
entities are sorted into categories according to an
established market-segmentation structure.  Classification-
system development and assignment complement each other with
respect to the calibration and application phases of
transport analysis.

*Classification.* One of the simplest ways of clustering consumers is through a set of arbitrarily defined clusters. By an arbitrary definition, it is meant that category boundaries are formed with respect to a consideration other than the primary variable of interest. For example, we might divide households into high-, middle- and low-income households when the primary variable of interest is the desirability of better widget service. It may be that for widget service only two income categories are necessary or that three categories should be used but with different cut-off values for low and high income. Arbitrary procedures for selected basis variables and category boundaries are generally not helpful for other than trivial market-segmentation problems because of such considerations.

A wide range of procedures is available for grouping entities according to one or more basis variables. These procedures are described, compared and contrasted in a variety of sources including Blashfield (7), Overall and Klett ((71), Chapter 8), and Tyron and Bailey (90). Hierarchical-clustering procedures assume a set of nested relationships among the entities which are being formed into groups. Agglomerative-hierarchical procedures, such as those discussed by Blashfield, define a proximity index on the set of entities being clustered into market segments. All stepwise agglomeration is performed with respect to the proximity index. AID (Sonquist, Baker and Morgan (84)) assumes a set of nested relationships among entities, but it splits the sample of entities on the basis of variables which sequentially maximise between-group variance, instead of a proximity index.

Other procedures for developing a classification framework also rely on variance-related indices. For example, in Q-type factor analysis, persons replace variables. Since Q-type factor analysis is mechanically like any other factor analysis in all other respects, the standard factor-analysis literature is a sufficient resource material for implementation and interpretation guidelines (50,57,26,27). The central notion behind Q-type analysis is that person factors constitute 'ideal types' or market segments and that most individuals have a high loading on only a single factor. The Friedman and Rubin method (39) uses an iterative reclassification technique (79) to maximise the between-group variance relative to the within-group variance. A comparative study of Friedman and Rubin versus Q-type procedures (60) finds them both of comparable effectiveness in cluster formation, but Q-type analysis was between 100 to 33 per cent less expensive to implement. Also, Q-type analysis is readily available at many different computer installations because it is based on factor-analysis computational algorithms.

*Assignment.* After a classification system is developed, useful insights can frequently be derived by comparing the composition of different market segments with respect to other than basis and dependent variables. In addition, it may be desirable to apply a calibrated market-segmentation structure to another sample of entities to help assess policy implications of alternative transport plans for some future design year or in another urbanised area. While the

assignment needs of the former requirement are often
satisfied by the output of cluster algorithms, the latter
requirement is not addressed by procedures for developing
classification frameworks.

The problem of assigning new entities to an existing
segmentation structure reduces to a classic choice problem,
such as that considered in mode-choice analysis.  Clusters
are analogous to modal alternatives and basis variables
correspond to independent variables describing individuals
who select among modes.  The market-segmentation calibration
process will frequently designate each entity as belonging to
a specific cluster.  Logit and/or discriminant analysis
(McFadden (67), Overall and Klett (71), Chapters 9 and 10)
can then be applied to the basis variables to check their
ability to map entities into clusters.  This operation will
provide one or more criteria for assessing the goodness of
fit of the cluster analysis, and it will estimate logit and/
or discriminant coefficients for the basis variables.  If the
goodness of fit is satisfactory, these coefficients can subsequently be used to assign within the previously calibrated
market-segmentation structure different entities subject only
to these entities having measurements in the original basis
variables.

Overall and Klett ((71) Chapters 12-16) describe and
implement a wide variety of assignment procedures which
emanate principally from computational algorithm research to
enhance the diagnosis of psychiatric and neurotic disorders.
These assignment procedures rely on maximum likelihood or
Bayesian models and empirical or multivariate normal
probability density functions.  The central notion of all
these methods is that market segments, or disorder types, can
be summarised by average profiles with respect to a set of
variables.  In addition, any new entity can also be described
by a profile on the same variables.  Assignment is achieved
by placing entities in the market segment to which their
profile of measurements is most similar.

*Hierarchical and Non-hierarchical Clustering*

As mentioned above, hierarchical methods assume nested
relationships among entities grouped and/or the basis
variables for a segmentation analysis.  Hierarchical clustering algorithms are most appropriate when a data set has a
natural structure of nested relationships, such as genealogy
lines for plant and animal subspecies.  Even in situations
where a nested structure of relationships is not strictly
necessary, hierarchical methods will uncover a cluster
structure.  However, the general procedure has a fault in
that a solution of N clusters is always dependent upon a
previous one of N-1 clusters.  It is entirely possible with
non-hierarchical data that the most appropriate set of N
clusters does not follow directly from the best set of N-1
clusters.  Results reported by Johnson (60) show non-
hierarchical algorithms to derive better clusters than a
hierarchical procedure for data sets without an obvious
nested-relationship structure.  Nevertheless, hierarchical
procedures have been meaningfully used in diverse social
science contexts (11,19,75).

A variety of tools are available for implementing hierarchical clustering. For example, Veldman (92) presents a computer program for an analysis method proposed by Ward (94). A widely distributed computer package (20) provides several variations within a hierarchical clustering framework. When it can be assumed that a data set describes relationships at no more than the ordinal level, then a non-metric procedure, such as that proposed by Johnson (61), is most appropriate. All of the above-mentioned tools agglomerate entities solely with respect to basis variables. AID (84) clusters entities with respect to the effect of basis variables on a dependent variable of interest. Its output can be represented in the tree-like structure of other hierarchical clustering tools.

Non-hierarchical-agglomerative tools do not assume nested relationships among entities and/or basis variables being clustered. All of the assignment procedures (i.e. logit and discriminant analysis and the profile maximum-likelihood and Bayesian models) are examples of tools implementing the approach. Overall and Klett (71) provide computer programs for all of the above techniques except logit analysis. However, logit analysis is widely discussed in literature commonly used to support transport analysis (87,14,67). Among non-assignment methods, Friedman and Rubin's classification building algorithm (39) is a simultaneous or non-hierarchical clustering tool. Q-type factor analysis is another example of this latter type of tool.

*Statistical Considerations*

Most statistical research for market segmentation is directed at algorithm design for classification development and assignment. Relatively little effort is devoted to the specification and application of statistical tests for goodness of fit or stability of market-segmentation structure. Two such procedures, namely GLANOVA and split-sample methods, are mentioned here briefly.

GLANOVA, which is an acronym for general linear analysis of variance, is well known to theoretical and applied behavioural statisticians (e.g. Park and Federer (73), Rao (74), Appelbaum and Cramer (2), Cohen (16), Overall and Klett (71)). Dobson (24) described two potential GLANOVA applications which generally relate to market segmentation in transport analysis. It is shown that GLANOVA represents a statistical test of cross classification and that it can also be used to compare the effectiveness of a market-segmentation structure based on one set of independent variables with other continuously coded independent variables in accounting for a dependent variable of interest, such as mode choices. Dobson and McGarvey (29) illustrate the empirical application of GLANOVA to trip-generation analysis in which market segments specified in terms of the basis variables of income and auto ownership are found to be statistically different in terms of trip frequency.

GLANOVA-procedure results are identical to the simpler category or cross-classification models which have been used

and recommended for trip-generation analysis (98,15,91).
However, GLANOVA permits the computation of statistical
indices which are not at all, or at least not readily available with category or cross-classification models.  For
example, GLANOVA facilitates the computation of an overall
goodness-of-fit index for a market-segmentation structure
through the familiar $R^2$ statistic.  In addition, it can be
used to test for the effect of adding another variable to a
market-segmentation basis.  Finally, it can test the relative
efficacy of alternative critical values of basis variables
for designating membership in market segments.

Split-sample methods provide a means of assessing the
stability of market-segmentation structures and their
correspondence with dependent variables.  An agglomerative
algorithm will always find groups of entities, even if the
data do not have a true cluster or market-segmentation
structure.  When sample sizes are sufficiently large in comparison to the estimated number of market segments, it is
possible to compare cluster structures across two different
halves of the same sample.  If the data do have a natural
market-segmentation structure, then independent clustering of
the two halves should result in similar relationships among
the basis variables in the two halves.  However, even if the
structure varies across halves, the two different structures
may have invariant relationships to the dependent variable in
both halves.  That is, a travel-choice model built on the
structure of the first half with respect to the dependent-
variable observations in the first half may yield a
goodness-of-fit index of approximately equal magnitude for
the dependent-variable observations of the second half.  By
using these two variations of split-sample methods, an understanding of market-segmentation structure stability can be
established.

PROTOTYPICAL TRANSPORT APPLICATIONS

*Overview*

It should be clear that there are numerous ways in which
market segmentation can be used to support transport policy-
making through enhancing transport analysis.  Two broad areas
of potential application include the segmentation of
traveller units and the agglomeration of geographic units.
Other specific topics relate, but are not restricted, to
citizen participation, the analysis of the determinants of
mode shifting, and the segmentation of transport-system
management actions.  This section synthesises and highlights
various aspects of some prototypical market-segmentation
applications which are mentioned above.

*Segmentation for Traveller Units*

*Mode choice*.  Mode choice represents an important policy-
analysis subject which is susceptible to study via market-
segmentation procedures.  Among the least complicated ways to
organise behavioural factors influencing mode choice is to
segment by user group.  For example, in the fashion of Dobson

and Tischer (31), it is possible to categorise travellers on
the basis of whether they usually take a bus, car-pool or
drive by themselves to work.  Such a segmentation is based on
a pattern of revealed preferences.

After a user-group segmentation is implemented, several
sorts of analyses are possible to help guide the design and/
or evaluation of policies to influence modal usage.  By
studying how perceptions of modes with respect to system
attributes vary across segments, it is possible to identify
salient transport-service features.  Differences in perceptions among user groups can serve two purposes.  In the first
place, these differences help to diagnose a system by
highlighting strengths and weaknesses associated with
alternative transport modes.  Secondly, system modifications
to facilities and/or operating policies can be suggested by
the perceptual differences among transport modes for various
user groups.  Behavioural principles for guiding the selection of policies on the basis of perceptual factors have been
described by Dobson (21).

When the primary focus of a study is towards innovative,
evolutionary and/or non-existent transport options, it is
often useful to segment directly on the basis of perceptions
or preferences for service attributes (28,30).  Such a segmentation analysis helps to point out key attributes which
might influence the use of planned, innovative transport
services or facilities.  This information can then be
considered in the design, specification and delivery of the
services and facilities.  Furthermore, by studying the
socio-demographic composition of groups with specified sets
of preferences for service attributes, it is possible to
formulate a promotional campaign which selectively directs
specific information to well-defined segments of the
travelling public.

*Trip-generation analysis.*  Category or cross-classification
models, which rely on market-segmentation principles, have
been recommended as a trip-generation analysis technique (91,
98,37).  These models segment households on the basis of two
or more socio-demographic variables, such as income and auto
ownership.  The average trip frequency is subsequently computed for each segment which is correctly defined by two or
more socio-demographic variables.  When these models are used
for forecasting purposes, it is necessary to predict the
relative frequency of segments, such as households with
common income and auto-ownership characteristics, in the
geographic units used for forecasting.  Some procedures for
accomplishing this goal are discussed in *Trip Generation
Analysis* (91).  It is also possible that some variations of
the maximum-likelihood and Bayesian-assignment models
discussed by Overall and Klett ((71) Chapters 12-16) could be
adapted to satisfy this objective.

Category models are simple in concept and execution since
they have been espoused partly as a reaction against more
obfuscatory and statistical approaches to trip-generation
analysis.  Until recently, this advantage has been diminished
by the lack of a readily available goodness-of-fit index.
Dobson (24) however, has described a new technique of

calibrating category models.  This new technique, labelled
GLANOVA, produces results which are identical to the simpler
method of computing cell means, but it also estimates a
goodness-of-fit index that can be tested for statistical
significance via an F statistic.  An attractive aspect of the
general approach from an operational planning perspective is
that category models can now be calibrated by GLANOVA but
explained and conceived of in the simpler terms of average
trip rates for income-auto ownership or other combined socio-
demographic, segments of the travelling public.  Dobson and
McGarvey (29) have empirically and theoretically demonstrated
the inter-changeability of the two methods.


*Agglomeration of Geographic Units*

*Grouping cities*.  It is widely recognised that there is an
interaction between transport systems and urban form.  Some
cities heavily concentrate social and economic opportunities
at one or two locations within them.  However, other urbanised areas have a more diverse arrangement of economic,
recreational and cultural activity sites.  These alternative
spatial distributions of activity generators have strong
implications for transferability across urbanised areas of
transport facility and service concepts.  Market segmentation
of urbanised areas with respect to a wide range of urban-form
and transport-system basis variables provides a means of
summarising quantitatively the degree to which concepts are
transferable.  At least two reports in the transport research
literature describe efforts to develop market segments for
different urbanised areas (42,43).

Any of the basic approaches to classification development
are appropriate for clustering cities.  It is probably
reasonable to assume that cities do not have a nested set of
relationships among themselves.  Therefore, a non-hierarchical agglomeration procedure, such as Q-type factor analysis or
the Friedman-Rubin iterative reclassification technique, is
most appropriate.  In order to assess the transferability of
facility and service concepts across cities, two sets of
proximity indices are desirable.  The first of these is the
set of intersegment distances, computed according to a
Euclidean or other suitable metric.  If the segmentation
structure is relevant to the transferability of a concept,
the distance between two segments should be inversely related
to the probability that a concept will yield identical consequences in two cities from the different segments.  In
addition, through test marketing a concept in the most and
least representative cities of a segment, it should be
possible to estimate a likely reaction for other cities in a
segment.  In any event, bounds on the range of a reaction can
be clearly specified through a knowledge of extreme reactions
for the least and most representative entities in a cluster.

*Grouping analysis zones within an urbanised area*.  Clustering
intra-urban geographic units, such as traffic-analysis zones
or districts, establishes a new set of objectives for
marketing-segmentation research.  The emphasis shifts away
from transferability and towards matters which influence the
degree of travel between zones.  For example, it is possible

to identify key attractor zones for classes of trip-
production zones.  The spatial arrangement of these
production and attraction zones defines natural traffic
corridors.  It may be possible to use these spatial arrangements to suggest optimal locations for park-and-ride lots.
Market segmentation applied to intra-urban geographic units
can also be used to identify groups of units that mutually
feed each other with trips.  Groups of units which exhibit
this pattern are likely to be better served by some form of
ubiquitous point-to-point transport system as opposed to a
limited, high-level system which serves a few points
extremely well but ignores many other points.  Once groups of
geographic units with common traffic flow tendencies are
specified, it is of interest to note the socio-demographic
characteristics of travellers and the land-use features of
the zones which are correlated with the traffic-flow
tendencies.  This information can have design implications
for the amenities of transport systems and their consequent
social acceptability.

The basic data matrix for clustering geographic units
within an urbanised area is a production-attraction matrix,
such as that which results from the trip-distribution phase
of the travel-demand forecasting process.  It will sometimes
be helpful to form separate matrices for different trip
purposes and/or socio-demographic groups so that the segmentation structures corresponding to these different matrices
can be compared and contrasted.  The segmentation structure
can be derived through any standard classification-building
procedure, but Q-type factor-analysis procedures relate
production and attraction zones to each other while they also
group geographic units.  Golob, Hepper and Pershing (46) and
Hepper and Gustafson (54) applied variations of Q-type factor
analysis to traffic-flow matrices in order to accomplish
several of the objectives described above.

*Miscellaneous Transport Segmentation Applications*

Market segmentation is a potentially important transport-
analysis tool because of the diverse set of transport-
decision-making topics to which it is relevant.  In addition
to the consumer demand and geographic aggregation issues
mentioned above, market segmentation can contribute meaningful input to citizen participation, conflict resolution with
respect to classes of transport-system management actions,
and the analysis of determinants influencing mode shifting
among other policy-making concerns.

In the area of citizen participation, market segmentation
can be used in at least two ways.  Schermer (80) describes
how interest-group impact theory, a form of market segmentation reduces public works decision-making situations to
identifiable affected groups with clearly defined objectives
and concerns.  She asserts that the method aids the evaluation of alternative courses of action, and she reviews two
practical applications of it.  In conjunction with a public
opinion survey, market segmentation can also be used to
update the goals and objectives of transport policy-making in
an area.  By segmenting the respondent replies on the basis

of political jurisdictions, or other relevant basis variables, contrasting opinions and values can be observed and considered in the formulation of regional transport goals and objectives.

It is possible to develop through multi-dimensional scaling and cluster analysis (Shepard (81)) a classification scheme for transport-system management actions which is sensitive to any particular set of transport options, attributes underlying the effect of those options and actor groups evaluating the options with respect to the attributes. Dobson (21) briefly mentions one procedure for achieving this goal. The differing perspectives with respect to transport options of alternative actor groups in the planning process can be objectively compared by the procedure, and it is entirely possible that this objective comparison of subjective viewpoints can form a basis for conflict resolution.

Factors underlying mode shifting can be induced indirectly by calibrating a model based on existing mode-usage patterns and then adjusting independent variables to predict mode shifting. Alternatively, determinants of mode shifting can be deduced directly by segmenting travellers on behavioural intentions and then trying to explain segment membership on the basis of socio-demographic and/or perceptual variables. Alpert and Davies (1) and Tischer and Dobson (88) have implemented the latter approach. While it appears that further research is necessary to refine the methodology of the technique, their procedure is interesting because of its directness and simplicity. The direct approach is particularly attractive from an application perspective when studying new transport modes which do not currently exist.


SUMMARY AND CONCLUSION

The primary objective of this paper has been to show the relevance of market segmentation to transport analysis and policy-making. Towards this end, it has been indicated that market segmentation provides useful insights about consumer behaviour, such as mode choices for the journey to work. In addition, it was noted that market segmentation facilitates citizen-participation analyses and that segmentation applied to geographic units can have implications at inter- or intra-urban levels. The appeal of market segmentation is enhanced by the wide range of procedures which can be used to implement it. Market segmentation is a flexible tool which adapts easily to any particular set of study goals.

Figure 10.1 reveals a conceptualisation of the integration between market-segmentation application topics and procedures used to implement segmentation analyses. Various policy and/ or technical issues create the need for a market-segmentation study. Typical issues emanate from travel-demand analysis requirements, special interest group studies, and the need to group geographic units. One or more of these issues can lead to the formulation of a set of study objectives. These objectives, in turn, combined with other considerations result in the execution of segmentation procedures, such as AID, Q-factor analysis or discriminant analysis.

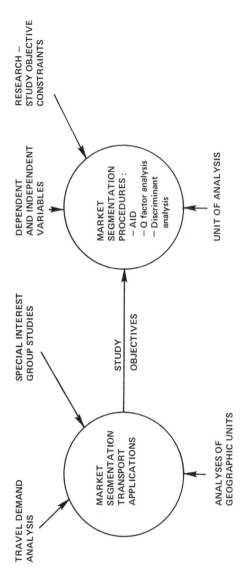

Figure 10.1  A Conceptualisation of the Integration Between Market
Segmentation Application Topics and Analysis Procedures

Additionally, considerations relating to the independent and dependent variables, the unit of analysis, and the interface between study objectives and research resource constraints have an effect on which segmentation procedures are executed. Figure 10.1, therefore, represents several salient factors which highlight the flexibility of market segmentation with respect to transport analysis and policy-making.

Another related objective herein has been to stimulate the more widespread use of market segmentation in transport analysis. One step taken to achieve this end was to summarise previous market-segmentation studies in the transport-planning and research literature. Research was reviewed to highlight practical findings, methodologies and legislative factors justifying analyses. It was also noted in at least two studies (17,76) that a market-segmentation strategy accounted for traveller choices better than one which failed to differentiate among groups of travellers. A second step taken to encourage market-segmentation studies in transport analysis has been to identify a variety of prototypical studies. These studies were selected to indicate that market segmentation is relevant to an extremely wide range of issues in the area of transport analysis and policy-making. Descriptions of the prototypical applications were directed towards showing insights which could be developed through their implementation.

An overall goal of the paper has been to provide some perspectives on future research for and applications of market segmentation in transport analysis and policy-making. With respect to research goals, it can be noted that there is a need for market-segmentation paradigms which are compatible with the requirements of transport planners in operational environments. These paradigms should illustrate simplified procedures for achieving market-segmentation in transport-planning contexts. Efforts can also be directed fruitfully at the validation of market-segmentation procedures by demonstrating their ability to predict traveller choices with high degrees of accuracy and by revealing the robustness of the segmentation approach with respect to different planning requirements and methodologies. It is, of course, important that continued research be conducted in the comparison and contrasting of different basis variables and agglomeration techniques for their suitability in transport analysis. Particular attention should be devoted to validity, cost effectiveness and simplicity of use.

In the area of practical applications, there is a need for the institutionalisation of a market-research plan. This plan should identify information to be transmitted to given population segments through specified communication media. Since effective market-segmentation practice depends on more than just advertising, it is important that effective liaison be established between the market-research and service-and-facility planning units of transport agencies. This liaison will facilitate the design of services and facilities which have maximum consumer appeal because of obvious concerns for the preferences of various segments of the travelling public. Finally, transport-market research should rely more on models which use traveller preferences and perceptions. It is known

from a vast body of market-segmentation literature that
attitudes are strongly correlated with consumer choice.

ACKNOWLEDGEMENT

The first draft of this paper was prepared while the author
was employed by the Federal Highway Administration, US
Department of Transportation.  Subsequent modifications,
which were completed at Charles River Associates Inc., were
facilitated through review comments by David S. Gendell,
David Hensher, Christopher Lovelock and Mary Stearns.
However, the author is solely responsible for matters of fact
and interpretation.

REFERENCES

1. Alpert, M.I. and Davies, C. Shane, *Segmentation of a
Market by Determinant Attributes*, presented at 1974 American
Psychological Association meeting, Austin, Texas: Council for
Advanced Transportation Studies, University of Texas (1974).
    2. Appelbaum, M.I. and Cramer, E.M., 'Some problems in
non-orthogonal analysis of variance', *Psychological Bulletin*,
81, pp. 335-43 (1974).
    3. Bass, F.M., Tigert, D.G. and Lonsdale, R.F., 'Marketing
segmentation: group versus individual behavior', *Journal of
Marketing Research*, 5, pp. 264-70 (1968).
    4. Behnam, J. and Thomson, T.L., 'Transit service improve-
ments as seen by non-users', *Traffic Engineering*, 46, pp. 18-
21 (1976).
    5. Bell, W., 'Social areas: typology of urban neighbor-
hoods', in *Community Structure and Analysis*, M.B. Sussman
(ed.), Crowell, New York City (1959).
    6. Birdwell, A.E., 'A study of the influence of image
congruence on consumer choice', *Journal of Business*, 41,
pp. 76-88 (1968).
    7. Blashfield, R.K., 'Mixture model tests of cluster
analysis: accuracy of four agglomerative hierarchical
methods', *Psychological Bulletin*, 83, pp. 377-88 (1976).
    8. Blattberg, R.C. and Sen, S.K., 'Market segmentation
using models of multidimensional purchasing behavior',
*Journal of Marketing*, 38, pp. 17-28 (1974).
    9. Brown, G., 'Brand loyalty-fact or fiction?', *Advertising
Age*, 23, pp. 53-5 (1952).
    10. Bruno, A.V., Hustad, T.P. and Pessimier, E.A., 'Media
approaches to segmentation', *Journal of Advertising Research*,
pp. 35-42 (1973).
    11. Burton, M., 'Semantic dimensions of occupation names',
in A.K. Romney, R.H. Shepard and S.B. Nerlove (eds), *Multi-
dimensional Scaling*, 2, pp. 55-72, Seminar Press, New York (1972).
    12. Byrd, J.P., 'Characteristics, attitudes, and percep-
tions of transit non-users in the Atlanta region',
*Transportation Research Record*, 563, pp. 29-37 (1976).
    13. Carman, J.M., *Studies in the Demand for Household
Equipment*, Berkeley, California: University of California,
Institute of Business and Economic Research, Research Program
in Marketing (1965).
    14. Charles River Associates, *A Disaggregated Behavioral
Model of Urban Travel Demand*, Federal Highway Administration,

US Department of Transportation, Washington, D.C. (1972).

15. Chatterjee, A. and Khasnabis, S., 'Category models: a case for factorial analysis', *Traffic Engineering*, 44, pp. 29-33 (1973).

16. Cohen, J., 'Multiple regression as a general data-analytic system', *Psychological Bulletin*, 70, pp. 426-43 (1968).

17. Costantino, D.P., Dobson, R. and Canty, E.T., 'An investigation of modal choice for dual mode transit, people mover, and personal rapid transit systems', *Proceeding, International Conference on Dual Mode Transportation*, Transportation Research Board, Washington, D.C. (1974).

18. Cunningham, R., 'Measurement of brand loyalty', *Proceedings of the American Marketing Association Conference*, 39-45 (1955).

19. D'Andrade, R.G., Quinn, N.R., Nerlove, S.B., Romney, A.K., 'Categories of disease in American-English and Mexican-Spanish', in A.K. Romney, R.N. Shepard and S.B. Nerlove (eds), *Multidimensional Scaling*, 2, 11-54, Seminar Press, New York City (1972).

20. Dixon, W.J. (1975) *Biomedical Computer Programs - P Series*, University of California Press, Los Angeles.

21. Dobson, R., 'Towards the analysis of attitudinal and behavioral responses to transportation system characteristics', *Transportation*, 4, pp. 267-90 (1975).

22. Dobson, R., 'Uses and limitations of attitudinal modeling', in *Behavioral Travel-Demand Models*, P.R. Stopher and A.H. Meyburg (eds), D.C. Heath, Boston (1976a).

23. Dobson, R., 'Data collection and analysis techniques for behavioral transportation planning', *Traffic Quarterly*, 30(3), 189-212 (1976b).

24. Dobson, R., 'The general linear model analysis of variance: its relevance to transportation planning and research', *Socio-Economic Planning Sciences*, 10, pp. 231-5 (1976c).

25. Dobson, R., Golob, T.F. and Gustafson, R.L., 'Multi-dimensional scaling of consumer preferences for a public transportation system: an application of two approaches', *Socio-Economic Planning Sciences*, 8, pp. 23-6 (1974).

26. Dobson, R. and Hepper, S.J. (1974a) *SINGD: A PL/1 Program for Eckart-Young Singular Decomposition*, General Motors Research Laboratories, GMT-1569, Warren, MI.

27. Dobson, R. and Hepper, S.J. (1974b) *A User's Manual for SINGD, A PL/1 Program for Singular Decomposition*, General Motors Research Laboratories, GMR-1569, Warren, MI.

28. Dobson, R. and Kehoe, 'An empirical comparison of the Tucker-Messick and INDSCAL models: measuring viewpoints about transit attributes', *Socio-Economic Planning Sciences*, 9, pp. 67-81 (1975).

29. Dobson, R. and McGarvey, W.E. (1977) *An Empirical and Theoretical Study of Trip Generation Analysis Techniques*, FHWA, US Dept. of Transportation, Washington, D.C.

30. Dobson, R. and Nicolaidis, G.C., 'Preferences for transit service by homogeneous groups of individuals', *Proceedings, Transportation Research Forum*, 15, pp. 326-35 (1974).

31. Dobson, R. and Tischer, M.L., 'Beliefs about buses, carpools, and single-occupant autos: a market segmentation approach', *Proceedings, Transportation Research Forum*, 17, pp. 200-9 (1976).

32. Eckart, C. and Young, G., 'The approximation of one matrix by another of lower rank', *Psychometrika*, 1, pp. 211-18 (1936).

33. Edlestein, P. (1976) 'Travel Behavior of an Elderly Population: An Attitudinal Model and Some Comparisons', unpublished Master's thesis, State University of New York at Buffalo, Buffalo.

34. Evans, F.B., 'Psychological and objective factors in the prediction of brand choice: Ford versus Chevrolet', *Journal of Business*, 32, pp. 340-69 (1959).

35. Federal Highway Administration, US Dept. of Transportation (1975) *Trip Generation Analysis*, US Government Printing Office, Washington, D.C.

36. Fishbein, M. and Ajzen, I. (1975) *Belief, Attitude, Intention and Behavior*, Addison-Wesley, Reading, Massachusetts.

37. Fleet, C.R. and Sosslau, A.B., 'Trip generation procedures: an improved design for today's needs', *Traffic Engineering*, 46, pp. 17-25 (1976).

38. Frank, R., Massy, W.F. and Wind, Y. (1975) *Market Segmentation*, Prentice-Hall, Inc., Englewood Cliffs, New Jersey.

39. Friedman, H.P. and Rubin, J., 'On some invariant criteria for grouping data', *Journal of the American Statistical Association*, 62, pp. 1159-78 (1967).

40. Goheen, P.G. (1970) *Victorian Toronto, 1850 to 1900: Pattern and Process of Growth*, Department of Geography, University of Chicago, Chicago.

41. Golob, T.F., 'The development of attitudinal models of travel behavior', in *Urban Travel Demand Forecasting*, D. Brand and M.L. Manheim (eds), Highway Research Board, Washington, D.C. (1973).

42. Golob, T.F., Canty, E.T. and Gustafson, R.L. (1972) *Classification of Metropolitan Areas for the Study of New Systems of Arterial Transportation*, General Motors Research Laboratories, GMR-1225, Warren, MI.

43. Golob, T.F., Canty, E.T., Gustafson, R.L. (1973) *National Studies of Urban Arterial Transportation: A Research Framework*, General Motors Research Laboratory, GMR-1274, Warren, MI.

44. Golob, T.F., Canty, E.T., Gustafson, R.L. and Vitt, J.E., 'An analysis of consumer preferences for a public transportation system', *Transportation Research*, 6, pp. 81-102 (1972).

45. Golob, T.F. and Dobson, R., 'The assessment of preferences and perceptions toward attributes of transportation alternatives', in *Behavioral Demand Modelling and Value of Travel Time*, P.R. Stopher and A.H. Meyburg (eds) Transportation Research Board, Washington, D.C. (1974).

46. Golob, T.F., Hepper, S.J. and Pershing, J.J., 'Determination of functional subregions within an urban area for transportation planning', *Transportation Research Record*, 526, pp. 16-25 (1974).

47. Grubb, E.L. and Hupp, G., 'Perception of self: generalized stereotypes and brand selection', *Journal of Marketing Research*, 5, pp. 58-63 (1968).

48. Gustafson, R.L. and Navin, F.P.D. (1972) *User Preferences for Dial-a-bus: A Comparison of Two Cities*, General Motors Research Laboratories, Research Publication GMR-1217, Warren, MI.

49. Haley, R.I., 'Benefit segmentation: a decision

oriented research tool', *Journal of Marketing*, 37, pp. 30-35 (1968).

50. Harman, H. (1967) *Modern Factor Analysis*, University of Chicago Press, Chicago.

51. Hartgen, D.T., 'Attitudinal and situational variables influencing urban mode choice: some empirical findings', *Transportation*, 3, pp. 377-92 (1974).

52. Hensher, D.A., 'Use and application of market segmentation', in *Behavioral Travel-Demand Models*, P.R. Stopher and A.H. Meyburg (eds), D.C. Heath, Boston (1976).

53. Hensher, D.A., 'Market segmentation as a mechanism in allowing for variability of traveller behavior', *Transportation*, 5, pp. 257-84 (1976b).

54. Hepper, S.J. and Gustafson, R.L. (1973) *A Multivariate Statistical Investigation of Person Movement within a Large Metropolitan Area*, General Motors Research Laboratories, GMR-1369, Warren, MI.

55. Hepper, S.J. and Pershing, J.J. (1974) *Decomposition of Flow Matrices*, General Motors Research Laboratories, GMR-1586, Warren, MI.

56. Horowitz, A.D. and Sheth, J.N. (1977) *Ridesharing to Work: A Psychosocial Analysis*, General Motors Research Publications, GMR-2216 (Revised), Warren, MI.

57. Horst, P. (1965) *Factor Analysis of Data Matrices*, Holt, Rinehart & Winston, New York.

58. Hoyt, H. (1933) *One Hundred Years of Land Values in Chicago*, University of Chicago Press, Chicago.

59. Johnson, R.M., 'Market segmentation: a strategic management tool', *Journal of Marketing Research*, 8, pp. 13-18 (1971).

60. Johnson, R.M. (1972) *How Can We Tell If Things Are 'Really' Clustered?* Presented at a joint meeting of the American Statistical Association and the American Marketing Association, Market Facts, Inc., Chicago.

61. Johnson, S.C., 'Hierarchical clustering schemes', *Psychometrika*, 32, pp. 241-54 (1967).

62. Kotler, P. (1972) *Marketing Management*, Prentice-Hall, Inc., Englewood Cliffs, New Jersey.

63. Lessig, P.V. and Tollefson, J.O., 'Market segmentation through numerical taxonomy', *Journal of Marketing Research*, 8, pp. 480-7 (1971).

64. Lichfield, N., 'Cost-benefit analysis in city planning', *The Town Planning Review*, 35, pp. 159-69 (1964).

65. Louviere, J.L., Ostresh, L.M., Henley, D.H. and Meyer, R.J., 'Travel demand segmentation: some theoretical considerations related to behavioral modelling', in *Behavioral Travel-Demand Models*, P.R. Stopher and A.H. Meyburg (eds) D.C. Heath, Boston, Massachusetts (1976).

66. Lovelock, C.H., 'A market segmentation approach to transit planning, modeling, and management', *Proceedings, Transportation Research Forum*, 16, pp. 247-58 (1975).

67. McFadden, D., 'Conditional logit analysis of qualitative consumer choice behavior', in *Frontiers in Econometrics*, P. Zarembka (ed.), Academic Press, New York (1974).

68. Mueller, E., 'Effects of consumer attitudes on purchase', *The American Economic Review*, 50, pp. 946-65 (1957).

69. Nicolaidis, G.C. and Dobson, R., 'Disaggregated perceptions and preferences in transportation planning', *Transportation Research*, 9, pp. 279-95 (1975).

70. Nicolaidis, G.C., Wachs, M. and Golob, T.F. *Evaluation of Alternative Market Segmentations For Transportation Planning*, Transportation Research Record no.649, pp. 23-31 (1977).

71. Overall, J.E. and Klett, C.J. (1972) *Applied Multivariate Analysis*, McGraw-Hill, New York.

72. Paik, U.B. and Federer, W.T., 'Analysis of non-orthogonal N-way classifications', *Annals of Statistics*, 2, pp. 1000-1021 (1974).

73. Park, C.R., 'The city: suggestions for the investigation of human behavior in the urban environment', *The American Journal of Sociology*, 20, pp. 577-612 (1965).

74. Rao, C.R. (1965) *Linear Statistical Inference and Its Applications*, Wiley, New York.

75. Rapoport, A. and Fillenbaum, S., 'An experimental study of semantic structures', in *Multidimensional Scaling*, A.K. Romney, R.N. Shepard and S.B. Nerlove (eds), 2, pp. 96-133, Seminar Press, New York (1972).

76. Recker, W.W. and Golob, T.F., 'A behavioral travel demand model incorporating choice constraints', *Proceedings, Association for Consumer Research Meeting*, 3 (1976a).

77. Recker, W.W. and Golob, T.F., 'An attitudinal mode choice model', *Transportation Research*, 10(2), 299-310 (1976b).

78. Reed, R.R. (1973) *Market Segmentation Development for Public Transportation*, Department of Industrial Engineering, Stanford University, Stanford.

79. Rubin, J., 'Optimal classification into groups: an approach for solving the taxonomy problem', *Journal of Theoretical Biology*, 15, pp. 103-44 (1966).

80. Schermer, J.H., 'Interest group impact assessment in transportation planning', *Traffic Quarterly*, 29, pp. 29-50 (1975).

81. Shepard, R.N., 'A taxonomy of some principal types of data and of multidimensional scaling for their analysis', in *Multidimensional Scaling*, R.N. Shepard, A.K. Romney and S.B. Nerlove (eds), 2, pp. 96-133, Seminar Press, New York (1972).

82. Shevsky, E. and Bell, W. (1955) *Social Area Analysis*, Stanford University Press, Stanford.

83. Simpson, R.B. and Ismart, D. (1974) *Transportation Opinion Survey*, Puget Sound Governmental Conference, Seattle.

84. Sonquist, J.A., Baker, E.L. and Morgan, J.N. (1971) *Searching for Structure*, Institute for Social Research, University of Michigan, Ann Arbor, MI.

85. Stopher, P.R. *The Development of Market Segments of Destination Choice*, Transportation Research Record no.649, pp. 14-22 (1977).

86. Stopher, P.R. and Lavender, J.O., 'Disaggregate, behavioral travel demand models: empirical tests of three hypotheses', *Proceedings, Transportation Research Forum*, 13, pp. 321-36 (1972).

87. Stopher, P.R. and Meyburg, A.H. (1975) *Urban Transportation Modeling and Planning*, Lexington Books, Lexington, MA.

88. Tischer, M.L. and Dobson, R. (1977) *An Empirical Analysis of Behavioral Intentions to Shift Ways of Traveling to Work*, presented at the 1977 Transportation Research Board meeting, FHWA, US Dept. of Transportation, Washington.

89. Tryon, R.C., 'Comparative cluster analysis of social

areas', *Multivariate Vehavioral Research*, 3, pp. 213-32 (1968).

90. Tryon, R.C. and Bailey, D.C., 'The BC try computer system of cluster and factor analysis', *Multivariate Behavioral Research*, 1, pp. 95-111 (1976).

91. Urban Mass Transportation Administration (1976) *Pricing-Transit Marketing Management Handbook*, Urban Mass Transportation Administration, US Dept. of Transportation, Washington.

92. Veldman, D.J. (1967) *Fortran Programming for the Behavioral Sciences*, Holt, Rinehart and Winston, New York.

93. Wachs, M. and Blanchard, R.D. (1975) *Life Styles and Transportation Needs of the Future Elderly*, School of Architecture and Urban Planning, University of California, Los Angeles.

94. Ward, J.H. Jr., 'Hierarchical grouping to optimize an objective function', *American Statistical Association Journal*, 58, pp. 236-44 (1963).

95. Wicker, A.W., 'Attitudes vs. actions: the relationship of verbal and overt behavioral responses to attitude objects', *Journal of Social Issues*, 25, pp. 41-78 (1969).

96. Wilkie, W.L. (1970), An Empirical Analysis of Alternative Bases for Market Segmentation', unpublished Doctoral dissertation, Stanford University, Stanford.

97. Wilkie, W.L. and Cohen, J.B. (1976) *A Behavioral Science Look at Market Segmentation Research*, College of Business Administration, University of Florida, Gainesville.

98. Wooton, H.G. and Pick, G.W., 'A model for trips generated by households', *Journal of Transport Economics and Policy*, 1, pp. 137-53 (1967).

99. Yankelovich, D., 'New criteria for market segmentation', *Harvard Business Review*, 42, pp. 83-90 (1964).

Chapter 11

MARKET SEGMENTATION IN BEHAVIOURAL TRAVEL MODELLING

Kenneth W. Heathington and David J. Barnaby

INTRODUCTION

Behavioural travel modelling as a part of the long-range
urban transport planning process has segregated travel by
trip purpose. Some urban transport planning studies have
used as little as three trip purposes while others have used
as many as twelve. Such purposes as home-to-work, shopping,
recreation, medical, etc., are normally used. Some studies,
however, may only use home-to-work, home-to-other and non-
home-based trips for stratification.

The trip generation modelling process for long-range
planning has utilised various trip purposes for over 20
years; however, in recent years there has been interest in
using market segmentation concepts that are employed by
businesses in developing forecasts for goods or services.
This new concept of segmentation is primarily oriented towards
short-range planning requirements. Many private companies
within the United States utilise a market segmentation
approach in estimating the potential demand for new or
improved products or services.

While the trip purpose is important in behavioural travel
modelling, there may be other ways to view the segmentation
of the travelling public. The concepts of market segmentation are becoming particularly useful in short-range public
transport planning and in the forecasting of the demand for
various transit services for a period of three to seven
years. There is research presently being conducted which
uses concepts of market segmentation in forecasting the
demand for transit services (see Chapter 10 of the present
volume). This chapter reviews some of the basic concepts in
market segmentation and suggests ways in which they could
apply to behavioural travel modelling. A review is made of
those concepts utilised in public transport planning, and
suggestions are made as to how these could be integrated into
the overall urban transport planning process.

THE LONG-RANGE URBAN TRANSPORT PLANNING PROCESS

Figure 11.1 shows a simplified layout of the various
activities in the long-range urban transport planning process.
To proceed with the planning process an inventory is made of
the facilities, land use, economic activity and travel
patterns in a given urban area. From the inventory data, a

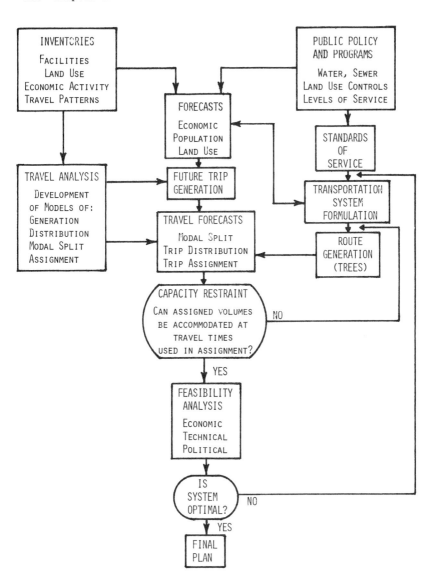

Figure 11.1   Long-range Urban Transport Planning Process

travel analysis is performed with the development of models
for trip generation, distribution, modal split and traffic
assignment.  Long-range forecasts are made on future economic
activities, population and land uses for the urban area.
These forecasts are related to public policy and programmes
of such activities as water, sewer, land-use controls and
levels of service desired for the urban area.  The desired
standards of service are to be defined for the urban area.

By taking the models that were developed from the
inventory data and using the long-range forecasts that were
made for the economic, population and land-use categories,
some expected future number of trips (i.e. perhaps 20 years
in the future) can be forecast or 'generated' for the urban
area.  The total forecasted travel is then used as input into
the other models of trip distribution, mode split and trip
assignment.  By assigning the future trips to the various
modes that would be existing in the urban area and utilising
various capacity restraints, an estimate of the adequacy of
the present system in serving the forecast volumes can be
made.  Alternative system designs can be tested in like
manner.  The results of this analysis will indicate the
expected levels of service that can be attained with various
system configurations.  Having determined the level of
transport services that can be provided, one then has to look
at the economic considerations relative to the construction
of systems providing specific levels of service and whether
or not these systems can be technically built and, indeed,
are politically feasible.  If an optimum system cannot be
attained, then one may revise either the standards for levels
of service or, perhaps, generate new transport system
alternatives for evaluation.

The trip generation phase of the travel-demand models
tends to segment the traveller by various trip-type
categories, i.e. work, shop, school, etc.  Some studies have
used as many as twelve categories while others have used as
few as three or four.  Models have been built for each
segment (trip-type) and long-range forecasts made for use in
the other phases, i.e. distribution, mode split and traffic
assignment.  Thus, there is an estimation of the total number
of work trips that will be made as well as the number of
shopping trips, school trips, etc.  These trips for each
category are then combined together in order to forecast the
total amount of trip making that will occur in the urban area
over the next, say, 20 years.

SEGMENTATION FROM MARKETING ANALYSIS

The question is now being asked as to whether the long-range
method of travel segmentation is really an appropriate one,
especially for sub-area analysis.  It is being suggested by
some that there may be better methods of segmentation for
forecasting of total travel than by trip purpose alone.
There has recently been an infusion of marketing techniques
into the transport-planning area, particularly in the public
transport field and the competitive mode context (see Chapter
10).  These marketing techniques have been used in the
private sector for many years.  The techniques concern

themselves with segmenting the market, determining the attributes of the products and/or services that are needed and testing the services and/or products.

The acceptance of a marketing orientation to transport planning seems a natural extension of traditional marketing concepts and practices to broader areas of market application other than durable and non-durable consumer products. (In fact, the marketing literature includes a wide variety of articles on the broadening of the concept of marketing and its expanded scope as 'social marketing' which reflects broader social issues than does the traditional, more visible profit motivation.) Therefore, formal integration of the various principles of marketing management into transport policy and decision-making represents one of the many seemingly natural applications of marketing that offers potential benefits to the user organisations. While some agreement exists as to the potential benefits that may accrue from implementing marketing procedures, some confusion also exists as to the exact specification of the benefits. Marketing is not a panacea; it does not promise to solve management's problems. What marketing does offer is a systematic methodology for assessing market opportunities for products and services from the perspective of the end-user or consumer. The technique of market opportunity analysis (MOA) is a sequential collection of information typologies neces- sary to meet evolving management information needs and reduce management uncertainty in the decision-making process.

The dominant conceptual foundation on which marketing strategy is based is that of market segmentation. A segmen- tation strategy is premised on a management philosophy that recognises that clusters of consumers with homogeneous needs and preferences exist in what may be viewed in the aggregate as a heterogeneous market. The marketing techniques that have been used in the private sector are not directly employable in the transport field. There must be some modifications made because of the nature of the transport field if the marketing opportunity analysis techniques are used. The Transportation Center at the University of Tennessee is engaged in a research project with the National Co-operative Highway Research Project to develop procedures for the utilisation of an MOA in the public transport field. The process that is being developed through the research programme involves procedures for segmentation that may result in differing stratifications than trip purposes. Various procedures used in the private sector are being revised for use in segmenting the market for public trans- port, defining the attributes of the transport system that would appeal to a particular market segment, and forecasting the amount of demand for systems with these attributes. The segments which result from the analysis are not necessarily related to the traditional categories used in the long-range urban transport planning process.

A basic descriptive model has been developed which indicates how the MOA can be integrated into the urban transport planning process. The procedure is discussed in this presentation with a discussion of methods of segmenta- tion.

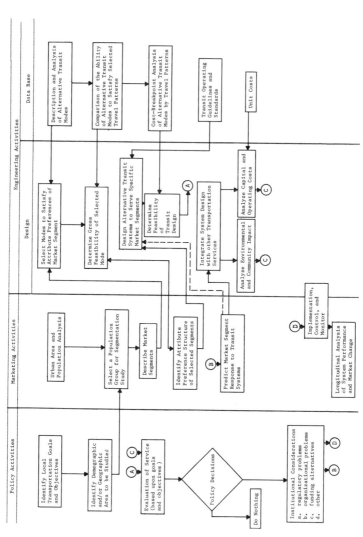

Figure 11.2   Integration of Engineering Activities with Policy and Marketing Activities

SHORT-RANGE TRANSIT PLANNING ACTIVITIES

Figure 11.2 shows the many activities of policy, marketing and engineering that should be integrated to incorporate an MOA into the short-range public transport planning process. Under the policy activities there is a need to identify local transport goals and objectives. These objectives are related to the demographic and/or geographic area to be studied. Also included in these activities is the evaluation of services based upon the goals and objectives. From this evaluation comes the policy decisions that must be made relative to short-range transit planning. These policy decisions must be viewed in the light of the institutional considerations (i.e. regulatory problems, organisational problems and funding alternatives).

In conducting the marketing activities there is an analysis of the urban area and its population characteristics. A population group is then selected for a segmentation study. Market segments are described based upon an analysis of the population. System or service attributes that are preferred by that market segment are then identified. In addition to the attribute determination there is a prediction made of the market segment's response to particular transit systems or services. If a particular system is introduced, it should be controlled and monitored by longitudinal analysis for the system's performance and market change.

There are two phases of the engineering activities - the design and data base. A portion of the engineering responsibilities requires the selection of modes that will satisfy the attribute preferences as determined from the marketing opportunity analysis. Engineering activities also include the description and analysis of alternative transit modes. An analysis is made of the ability of alternative transit modes to satisfy selected travel patterns. An evaluation is made of the user preferences of specific transit alternatives and designs are made of alternative transit systems that will be needed to serve specific market segments. Also required in the engineering area is the ability to integrate system design with other transport services which would include the analysis of environmental and community impacts as well as the capital and operating cost. It is noted that this particular descriptive model (Figure 11.2) is somewhat different from the long-range urban transport planning process although several subcategories of the activities are very similar to those found in the urban transport planning process. A more detailed discussion of the marketing activities outlined in Figure 11.2 follows.


MARKETING ACTIVITIES

The identification of consumer groupings along relevant dimensions and the subsequent analysis of matching specific segments with preferences for transport alternatives is the purpose of a market segmentation study. Recognising that a consumer orientation and a market segmentation strategy are essential for a market opportunity analysis, the paradigm in Figure 11.3 depicts the various stages of the complete

| MOA Stage | Purpose of MOA Stage | Information Needs | Output/Decisions |
|---|---|---|---|
| Aggregate Urban Trans Market Analysis | Identify and assess those urban population characteristics that are most likely to affect urban transportation usage. | Location of natural boundaries; Location of activities centers; Location and description of travel corridors; Location and population groups by selected demographic characteristics; Modal analysis of mode mixes, routes served, etc.; Travel/trip patterns (peak vs. off-peak); Travel forecasts; Population change forecasts; Activity center change forecasts | Selecting one of more broad market groups containing Transportation Demand Potential for MOA (Market Decision) |
| Market Segment Description | Identify and assess segments within broad transportation market groups to identify alternative transportation preferences. | Classificatory demographic and socio-economic characteristics; Transportation mode behavior of household members; Trip behavior of household members; Transportation mode awareness and attitudes; Media habits and use of information about transportation modes | Matching transportation mode types (e.g., bus, rail, van pools, shared-ride, taxi, etc.) with market segments. Selection of Specific Market Segment-Mode Types for further analysis (Market/Transportation Decision) |
| Attribute Preference Structure of Selected Segments | Identify and assess preferences for specific transportation attributes by mode within preselected segments. | Develop list of modal attributes from secondary literature; Classificatory characteristics based on segmentation variables; Rank of modal attribute importance; Interval scaling of attribute importance; Ratio "utility" scaling of selected attribute levels of important attributes | Selection of attribute combination required in specific modes attracting demand from matching segment (Marketing/Transportation Decision). Providing guidelines for trade-offs among inconsistent attribute expectations (e.g. short headways and frequent stops)(Market/Transportation Decision) |
| Prediction of Market Segment Response to System | Predict consumer demand response to a given transportation mode design. | Forecasting technique specific | Providing demand estimates into a transportation mode feasibility decision (Marketing/Transportation Decision) |
| Longitudinal Analysis of System Performance and Market Change | Evaluate performance of given transportation modes and to monitor market changes that may lead to mode usage shifts | Performance objective specific | Providing information to evaluate system performance |

Figure 11.3  Overview of Transport Market Opportunity Analysis

research process.  Each of the five stages presented in
Figure 11.3 will be discussed separately below.  It is
noteworthy that each stage represents distinctly different
research studies that are sequential, and the output of each
stage serves as a necessary input to the next stage.


*Aggregate Urban Transport Market Analysis*

The purpose of the initial MOA stage is to identify and
assess those urban population characteristics that are most
likely to affect urban transport usage.  The information
requirements for this stage would include assembly of
necessary data (see Figure 11.3 - Information Needs) to gain
a general appraisal of the market in order to identify one or
more broad market groups possessing sufficient transport
demand potential for further detailed market analyses.
(These types of information should, in most part, be available from secondary data sources within the urban community
under study.)  The market identification may be related to
transit usage, highway usage or specific services as may be
defined.


*Market Segment Description*

Once broad market groups are selected using information from
the first stage, a market segmentation study is necessary to
identify and assess segments within pre-selected transport
market groups and to identify alternative transport preferences within segments.  A variety of segmentation variables
offers potential bases for segments in addition to descriptor
variables necessary to characterise segments (see Figure
11.3 - Information Needs).  Analysis of segmentation data
yields segment profiles and transport mode preferences.
Selection of specific market-segment mode types provides
groups for the next stage which includes research on attribute preference idiosyncratic to specific groups.


*Attribute Preference Structure of Selected Segments*

Given specific market segments with mode preferences, the
next step is to identify and assess preferences for specific
transport attributes by mode within those segments of maximum
interests or potential.  Analysis of the attribute information provides insight into the attribute combination
requirements of specific modes necessary to attract demand
from matching segments.  At the conclusion of this stage
sufficient information should be available for initial design
of the mode(s).


*Prediction of Market Segment to System*

With a transport system outlined, a refined prediction of
consumer demand for a given transport mode design is
necessary.  A variation of concept testing provides demand
estimates for the transport mode feasibility decision.  When
sufficient funds exist for testing purposes, prototypes might

be tested to provide behavioural reaction to mode and
system design.

*Longitudinal Analysis of System Performance and Market Change*

After introduction of the new service, periodic evaluation of
performance of transport modes and monitoring of market
changes that may lead to future mode usage shifts are paramount.  Such longitudinal information will enable management
continually to appraise system performance and the extent of
the influence of external market factors on present and
predicted system usage.

SUMMARY

There are certain realisations that should be made relative
to the conduct of a marketing opportunity analysis.  It
should be readily understood that there is no guarantee that
the market segments obtained from an MOA will be different
from the trip purposes normally used in urban transport
planning.  It could well be that the market may be best
segmented by trip purpose; however, at this point this
question can only be answered through a market segmentation
study.

The attributes desired by the consumer may not be able to
be incorporated into the design of a system at acceptable
cost levels.  In order to attract individuals to the utilisation of public transport services, the public transport
system may have to have attributes that cannot be economically provided.  The automobile may well be able to provide
those attributes at the lowest possible cost.

Concept testing employs attitudinal data which may be
quite different from true human behaviour.  Personal
interviews may incorporate descriptions of transport services
to obtain attitudes towards the potential utilisation of
various types of services.  It may be difficult for respondents to comprehend fully the transport concept that is being
attitudinally tested.  Describing services or systems which
may be unfamiliar to respondents can make it difficult for
individuals to respond correctly as to their potential use of
a system.

There should be an implementation made of systems or
services having a perceived high acceptance to permit model
estimation through longitudinal analysis.  If implementation
is not made of systems perceived to be well accepted by the
public, one will never be able to determine the accuracy with
which these models forecast travel.  Since previous work in
this area has been minor, research must proceed in this
direction to determine if improvements can be made in the
urban transport planning process.  Many questions remain
unanswered, and a complete marketing opportunity analysis
with implementation of systems must be conducted in order to
ascertain the credibility of the techniques under consideration in the transport planning process.

ırkets often change in a very brief time period.   Thus, the ɪarketing opportunity analysis does not presume to solve long-range forecasting problems.   An MOA should be an integral part of a continuing process to update effectively the changes in markets on a regular basis to ensure that systems or services can be modified to meet the changing consumer needs and desires.

## ACKNOWLEDGEMENTS

The conceptual development of the Models (Figures 11.2 and 11.3) has been shared by the research team at The University of Tennessee consisting of Drs David W. Barnaby, Thomas L. Bell, Arun Chatterjee, David W. Cravens, Kenneth W. Heathington, Gerald E. Hills, Thomas C. Hood, Ray A. Mundy, Fredrick J. Wegmann and Robert B. Woodruff, and Research Assistants Carlos R. Bonilla, Grace E. Byrne and Jerrie L. McGhee.   The process has been developed through a team effort involving both marketing and transport planning personnel.   The research is part of NCHRP Project 8-16, 'Guidelines for Public Transportation Levels of Service and Evaluation'.   This paper presents interim findings from National Co-operative Highway Research Programme Project 8-16, sponsored by the American Association of State Highway and Transportation Officials.   The opinions and findings expressed or implied in this paper are those of the authors. They are not necessarily those of the Transportation Research Board, the National Academy of Sciences, the Federal Highway Administration, the American Association of State Highway and Transportation Officials nor of the individual states participating in the National Co-operative Highway Research Programme.

Chapter 12

CONSUMER SEGMENTATION

William B. Tye

CHARGE TO THE WORKSHOP

Segmentation as a pre-estimation classification procedure has
taken on increasing popularity as a potentially powerful tool
for explaining part of the variance in a modelling structure
prior to estimation.  There is a good case to be made for
identifying relatively homogeneous types of consumers with
respect to an issue under study; and there is a case for
identifying relatively homogeneous types of persons in terms
of their labour quality and spatial location (supply aspects).
Particular attention should be paid to procedures for identifying the relative similarities of persons with respect to
issues in a way that can assist in improving the predictive
reliability of our disaggregate demand and supply models.

WORKSHOP RECOMMENDATIONS AND CONCLUSIONS

'Consumer segmentation', also called 'market segmentation'
and 'segmentation', refers to any disaggregation or classification of the population which is relevant to transport
planning.  Segmentation is not confined to 'attitudinal
modelling', or marketing, but is a broad concept applicable
to some aspects of traditional transport planning and to
virtually all behavioural approaches.  Segmentation is
presently being usefully employed to improve transport
analysis in travel-demand forecasting, marketing, transport-
system design and operation, and special interest group
studies.  Segmentation concepts are practical application
tools because they emphasise behavioural methods while providing a bridge between aggregate and disaggregate approaches.
With their emphasis on marketing and relevant differences in
the population, they encourage the transport planner to think
of demand as a choice process in a highly competitive environment, much of it outside the planner's control.  Segmentation techniques are presently available which would improve
transport planning in each major application area.  However,
planners unfamiliar with segmentation concepts are encouraged
to develop small-scale projects before application at the
level of the entire metropolitan area transport system.

Segmentation approaches should be developed to meet the
goals of the analysis, the capabilities of the analyst and
the data available.  Although no hierarchy of segmentation
schemes can be specified, segmentation on the basis of
'individuals' subjective judgements' (i.e. tastes), should

not be undertaken independently of segmentation on observed
transport-system attributes (such as level of service) and
perhaps also observed variation in attributes.  By the same
token, available attributes of the choice process may not be
'exogeneous', but rather determined by individuals' subjective judgements.  The bases for segmentation may be
classified as individuals' subjective judgements, socio-
demographic variables, availability of alternatives in the
choice set, attribute levels for choices, and geography.
Analysts with an interest in probabilistic choice models tend
to place a much greater importance on the need to segment on
the basis of the heterogeneity of the choice environment than
do those with an interest in psychology, attitudinal modelling and marketing.  Highest priority implementation tasks at
the present time are to develop handbooks for planners with
case studies to illustrate how segmentation can improve
planning practice.  Research should also develop procedures
for integrating attitudinal research into probabilistic
demand models, in the hope of developing improved methods for
transferring models from one geographic area to another.


DISCUSSION OF WORKSHOP RECOMMENDATIONS

*Nomenclature and Scope*

The workshop found that considerable confusion has been
created by the extension of the terms 'market segmentation'
and 'consumer segmentation' beyond their use in marketing
studies.  In the context of transport planning they have
taken on a much broader application than marketing alone.
Although 'consumer segmentation' is preferable to 'market
segmentation' in this respect, the workshop chose to use the
simple term 'segmentation' to avoid confusion and debate
arising from the objections by marketing experts over the
less restrictive use of the term 'market segmentation'.
'Segmentation' refers to any disaggregation, stratification
or classification of the population that is relevant to
transport planning.  In this context 'planning' refers to all
relevant behavioural disciplines.  The criterion is to
achieve the maximum degree of homogeneity within groups and
the maximum degree of heterogeneity between groups with
respect to attributes relevant to the objectives of the
transport analyst.  The objectives of the analyst fall into
the following broad categories:

1  *Demand forecasting:* given a proposed transport-choice
environment, divide the population into groups that will
increase the predictive accuracy of demand-forecasting
procedures.
2  *Marketing and system design:* segment the population to
improve the marketing tasks of product development, pricing,
promotion and distribution.
3  *Evaluation:* classify the population into groups relevant
for determining the impacts of alternative policies and
evaluation of transport programmes and policies.

*Potential for Immediate Application of Segmentation Concepts*

Segmentation concepts have frequently been used in transport-planning, such as forecasts of travel demand, disaggregated by zone and use of 'cross classification' in trip generation. Increasingly sophisticated demand-forecasting techniques and more demanding requirements for project evaluation call for a greater use of more complex segmentation procedures. The segmentation techniques presently available offer substantial benefits over more aggregated procedures and can be used to improve existing procedures without requiring the planner to completely replace them. Segmentation schemes are accessible to the planner because most aggregate procedures can be disaggregated to account for important differences within the population, and segmentation is a flexible tool which can be used in a variety of ways to meet the requirements of a particular application.

The workshop concluded that the broad range of application of segmentation procedures does not permit the development of a 'hierarchy' of segmentation procedures. Rather the segmentation scheme must be developed to meet the objectives of the analyst, his level of technical sophistication and cost constraints. The workshop agreed that an unsophisticated application of elaborate segmentation schemes might possibly be counter-productive, especially if undertaken at the level of a long-range transport plan for an entire metropolitan area. A better implementation strategy would be to experiment with small-scale changes at the level of the corridor or sub-area, with less attendant risk. As the analyst developed expertise, the level of sophistication could be expanded.

*Segmentation as a Tool in Demand Forecasting*

Segmentation has traditionally been confined to the demand forecasting process in transport planning, principally in developing alternative models for different trip purposes, in disaggregating trips by geographical zone and network links, and in employing 'cross-classification' for trip generation. Recently, additional segmentation techniques have been developed from two distinct disciplines in behavioural travel-demand modelling: (1) issues raised by the requirement to aggregate and to transfer geographically non-linear individual, probabilistic, discrete choice models based on the assumption of utility maximisation (hereafter referred to as 'individual choice models'); and (2) psychological or marketing studies of 'individuals' subjective judgements' (hereafter referred to as 'attitudinal-behavioural analysis'). These two disciplines have emphasised different applications of segmentation approaches and have differed significantly in their approach to segmentation.

The interest of the individual-choice modeller in market segmentation has been generated because of the need to distinguish among trip types and persons for model specification and the need to aggregate individuals for forecasting purposes. These modelling concerns have generated segmentation schemes for two purposes: model specification to account for 'taste differences', and definition of market segments

for which separate demand forecasts can be conducted and
aggregated.

Research on non-linear disaggregate travel-demand models,
principally the multinomial logit, has identified 'market
segmentation' as one of the principal techniques for aggregating disaggregate models to produce forecasts of travel
demand for the total population (7,8,14). The requirement
for aggregation arises because, in general, a heterogeneous
group does not behave on average as if it were composed of
identical persons each having the group means for the
explanatory variables (see Charles River Associates (8) for
an explanation with examples. Prediction using group means
has been called the 'naive method', or less pejoratively,
'the direct aggregation method'). The most correct approach
is to predict the behaviour of each individual based on the
individual values of the explanatory variables and aggregate
the predictions over individuals. Because of lack of data on
the levels of the explanatory variables for each individual
in the population, the 'market segmentation' technique has
evolved as a workable compromise between the aggregate and
completely disaggregate forecasting procedures. The steps in
the procedures are:

1  Define market segments such that each segment has
relatively similar observed socio-demographic and level-of-
service attributes and measured taste attributes.  Each
'cell' groups all persons falling within a defined range of
the attributes, such as 'one-car family, within three blocks
of transit, and having two workers'.
2  Determine the relative frequency for each cell in the
population for the forecast period.
3  Forecast behaviour in each cell using a disaggregate
behavioural demand model applied to the average value of the
attributes for the individual cell.
4  Aggregate the forecast by weighting each cell forecast by
the relative cell frequency.

As a technique for aggregation of disaggregate models,
segmentation offers a 'beachhead' for introducing behavioural
approaches into transport planning because it evolves
directly from existing forecasting techniques and efforts to
improve them.  Furthermore, a planner can start with a
proposed segmentation scheme and apply either a simple or
sophisticated approach to forecasting, depending on the time
horizon for the forecast, the skills of the planner, and the
cost objectives and data available to the planner.  For
example, the planner could collect a disaggregate data set to
calibrate new models and estimate the cell frequencies, or he
could use transferred models, existing data and reasonable
approximations to provide a simplified best estimate.

The requirement to segment the population for travel
demand forecasting purposes is caused by the same considerations that led to the development of disaggregate models: the
desire to account for similarities and differences among
individuals in choices and levels of explanatory variables by
specifying a conceptually similar decision process across
groups of persons.  Viewed in this context, disaggregate
demand models should be designed to dictate appropriate

segmentation schemes, and issues in market segmentation may
be reformulated as issues in model specification and calibration.  Similarities and differences in behaviour identified
in disaggregate models should therefore produce significant
insights into appropriate market-segmentation concepts.

In addition to the aggregation problem discussed above,
the other important related issue in individual-choice modelling is the transferability of model specification and
calibration results from one geographical area to another.
The potential for transfer of models has been an important
argument for the disaggregate approach, and some encouraging
evidence is available, but the benefits and costs have not
been fully tested through careful experimentation.  The issue
of model transferability is closely related to market segmentation because an important potential cause for failure of
transferability is behavioural differences for an individual
over time (a question of individual consistency) and among
individuals (especially those residing in different geographical areas).  Failure of transferability has been ascribed to
differences in 'tastes' or 'importance weights' (coefficients
on the measured attributes in the models) and to differences
in 'decision-making criteria' (structural differences in
decision-making procedures).  While little has been
accomplished to date, integration of the results from
attitudinal-behavioural research into the structure of the
individual-choice models might enhance their transferability.
This approach would provide an excellent opportunity for
attitudinal researchers to demonstrate the value of their
methodology by meeting a perceived need of transport planners
in an actual forecasting and planning application.  The need
for improved methods of distinguishing groups with similar
tastes other than the usual use of socio-demographic
variables appears to be a priority research problem.

*Segmentation as a Tool in Marketing*

The four major functions of marketing are product development,
pricing, promotion and distribution.  The workshop concluded
that segmentation concepts, most of them highly developed in
marketing non-transportation products, have widespread
application to transportation.  Furthermore, product development, pricing and promotion must be considered together, and
the marketing function must not be reserved merely for
advertising.  Although segmentation has a role in both
operational changes to an existing system and in hardware and
total system development, the workshop shared the caution of
Heathington and Barnaby's resource paper and noted the
dangers of premature implementation of costly and sophisticated but unproven schemes.

Insights into the desirability of alternative transport
technologies or systems can be gained through 'concept-
testing' in long-range product development, but the lower-
risk strategy is to experiment with smaller-scale short-run
changes which provide information of immediate relevance to
the planner.  As an example of the insights that can be
gained through application of segmentation to psychological
market studies, research by Dobson and Nicolaidis (12)

indicated that drivers were much more fearful of automatic vehicle-control systems than non-drivers or passengers, who were accustomed to lack of control in the auto system.  Many other segmentation schemes have been developed for automatic guideway systems, such as by trip-segment type (downtown and airport distribution, line-haul, etc.).

The workshop concluded that segmentation as a marketing concept deserved more widespread application because it encourages the planner to think in terms of consumers as differentiated individuals and to recognise that these individuals may not conform to naive planning predictions that fail to consider the competitiveness of the transport-choice environment.  As indicated in the review paper by Dobson, where segmentation has been tried in transport marketing, it has generally proved to be successful.  The following caveats for application emerged from the discussions:

1  Promotion of a defective system promotes dissonance and harms credibility.  Product development, pricing and promotion must be co-ordinated to encourage successful development of a system.
2  Common-sense clustering techniques may be entirely adequate and should be tried prior to advanced clustering techniques, which may be very costly in computer time and be less valuable than more intuitive approaches.
3  The 'focus group' approach was reported to be very promising as a technique for exploring consumer preferences. The approach assembles a group with certain attributes and focuses on an issue of interest.  To probe beyond superficial replies, alternatives are mentioned and the group interacts in a loosely structured environment.  The leader prevents domination of the discussion by the most aggressive members. The procedure is not designed to produce quantitative results, but to generate useful insights and questions regarding how decisions are made.  The technique has not been well documented, especially in transport literature, and there is considerable risk of erroneous conclusions resulting from its application by untrained practitioners.
4  From the marketing perspective, undue emphasis on analysis of consumer behaviour within the range of observed experience can be highly misleading.  In developing new products, consumer reaction to poor performance may be an unreliable guide to projected response to substantially improved service, for example.

*Segmentation as a Tool in Evaluation and 'Special Interest Group Studies'*

Transport project and policy evaluation has developed a vast literature employing segmentation to discriminate among groups according to common interests, preferences and exposure to consequences of transport-system changes. Segmentation has been used both as a forecasting tool and a post-project tool of analysis.  Distinctions based on political jurisdictions and consequences for taxpayers, users and non-users have been common in transport planning.

The workshop endorsed the desirability of applying

segmentation in a broader evaluation context such as that
proposed in the resource paper by Heathington and Barnaby.
Such a planning framework would differ significantly from the
traditional approaches because it would distinguish the
preferences of different groups and it would predicate system
design on matching system performance with segment preferences.  Sensitivity to the differing preferences of population segments would have a profound impact on transport
planning.  While the workshop suggested that the technique
should first be tested on small-scale operational decisions,
the costs of implementing the procedures as designed for the
research would preclude them from full implementation except
where large investment costs are at stake.  The reviewer of
the Heathington and Barnaby discussion paper reacted favourably to the objectives of the paper, but noted that judgement
on the usefulness of the proposals must await more details on
the procedures: 'It is discouraging that a proposal to "find
out what the public wants and give it to them", rather than
something else", requires a major restructuring of the urban
transport-planning process.  The Heathington and Barnaby
proposals identify many ingredients of a good "alternatives
analysis", but they do not yet constitute a sufficiently
detailed set of procedures to be recommended for consideration as a practical approach at the present time.'


*Bases for Segmentation*

'Bases for segmentation' are the attributes of the population
which vary and thereby serve to distinguish segments of the
population.  The workshop recognised that a much broader
range of bases is available than is commonly applied.
Segmentation by trip purpose, by origin and destination, and
by income/auto ownership does not exhaust the techniques
available to the planner.  The following types of bases were
identified:
1  *'Individuals' subjective judgements'*.  This category,
also called 'behavioural segmentation', distinguishes the
attitudes, preferences, 'importance weights', tastes,
decision-making criteria, perceptions, etc., of the population.  It is frequently the primary basis employed in
marketing and psychological studies of demand.  Specification
of different models for different trip purposes or different
geographic areas is predicated on the assumption of different
underlying preference structures.  Segmentation by political
interests is also frequently based on the common preferences
of each political group.
2  *'Socio-demographic variables'*.  Attributes of the
individual, such as income, sex, age, labour force participation, auto ownership, etc., have been used as proxies for
differences in individuals' subjective judgements and are
often easier to measure.
3  *'Relevant choice set'*.  Transport demand is fundamentally
different from demand analysis for most other commodities
because all segments of the population rarely choose from the
same set of generic alternatives.  While every shopper may
face the same set of detergent alternatives, all travellers
do not choose from the same set of modes or destinations.
Some travellers are 'captive' to a mode while others are not.
The workshop recommended that the term 'relevant choice set',

taken from the literature on individual-choice models, be
replaced with 'availability of choice opportunity', to avoid
jargon.
4  *Choice attributes.*  Even when all members of a segment
choose from the same available generic alternatives, the
alternatives do not possess the same attributes, such as
level of service.
5  *Usage and observed choices.*  Actual observed choices, as
distinguished from the availability of the alternative, have
been used frequently to segment the population.  An example
is the distinction between 'users' and 'non-users'.  Segmentation on mode choice, trip purpose and time of day has been
used for this type of analysis.  Segmentation by media
exposure and habits for promotion purposes are another
example.
6  *Geography.*  Segmentation based on locational differences
has been used to segment demand by area or by network link
and as a proxy for differences in individuals' subjective
judgements.

The workshop recognised that these bases cannot be given an
order of priority.  Rather, their use depends on the study
objective and the professional training of the analyst.  A
substantial cleavage presently exists between the emphasis
given by psychologists and marketing specialists to
individuals' subjective judgements and socio-demographics as
contrasted with the emphasis given by the individual-choice
modellers to socio-demographics, the availability of opportunity and the variability of the attributes of choices.

This dichotomy of interests provides a useful point of
departure in discussing the uses of alternative segmentation
bases.  The individual-choice modeller tends to emphasise the
distinctive fact that each consumer in transport chooses from
a different choice environment while the marketing expert or
psychologist tends to emphasise the transferability of knowledge from non-transport demand studies.  The individual-
choice modeller is generally seeking to develop a demand
forecasting procedure, while the psychologist and marketing
expert (hereafter, 'attitudinal-behavioural analyst') is
frequently more interested in gaining *insights* into the
formation of preferences and habits and understanding fundamental aspects of the individual-choice process.  The
individual-choice modeller represented at the workshop
emphasised that while the problem of transferring model
parameters is common to all demand studies (i.e. different
population groups will demonstrate different tastes when
offered the same choices), the difficulties due to variation
in the attributes of choice alternatives is distinctive in
transport demand.  In transport demand, all consumers do not
choose from the same choice alternatives, and the generic
alternatives do not have the same attributes for all
consumers.  Although this heterogeneity facilitates model
calibration, it complicates forecasting.  Aggregate predictions may be erroneous if based on group means in cases where
the population is heterogeneous in its attributes, the
availability of alternatives and the attributes of alternatives.  Researchers with a background in individual-choice
model specification rely on this heterogeneity of the choice
set and levels of attributes to gain knowledge of how changes

in these attributes (time, cost, convenience, etc.) affect
choices. They are quick to criticise the direct transfer of
market segmentation approaches from non-transport markets
where heterogeneity of attributes is not a serious problem.
This view emphasises that attitudes depend both on the level
of attributes at the time the attitude is measured and the
range of variation in the attribute that has been observed.
Given that many attitudinal studies have failed to control
for the level of attribute experienced, individual-choice
modellers are sceptical of the insights inferred from such
studies. According to this perspective, marketing approaches
frequently fail to take advantage of the enormous range of
information inherent in the unstructured market experimentation caused by random variations in attributes experienced by
the population. They also neglect the aggregation problem of
forecasting demand changes in an environment where the population faces differing attributes of alternatives.

The perspective of attitudinal-behavioural analysis and
marketing, forcefully expressed at the workshop, emphasised
the insights and knowledge gained from attitudinal-
behavioural studies. Researchers with a background in
marketing and psychology generally place much less reliance
on statistical inferences from past behaviour and on observed
responses to observable attributes of the choice environment
and place much more emphasis on identifying and measuring
differences in individuals' subjective judgements. They also
are sceptical of many individual-choice models. It was
generally agreed by all that the socio-demographic variables
used as proxies for taste variations in disaggregate modelling often were inadequate to the task. Furthermore, much
individual-choice modelling fails to account for the fact
that tastes affect attributes as well as conversely, since
consumers often do not take the available choices as given.
For example, persons with taste predilections towards transit
may live near transit lines thereby affecting the observed
attributes of transit. The individual-choice models
generally assume that attributes are determined outside the
decision process, i.e. 'exogeneously'. If tastes affect
observed attribute levels there will be a tendency to calibrate model coefficients that bias the measured importance of
the observed attributes in the decision process.

In combining these different perspectives the workshop
developed a set of observations and recommendations for
implementation of segmentation and for future research.

1  While there are congruences between socio-demographics
and attitudes, the mapping function is very inexact.
Furthermore, the two types of measures correlate with
behaviour in different ways and to different degrees.
2  As the number of bases increases, the number of cells
increases multiplicatively and the requirements for data will
become voracious as the scheme becomes more complex. For
example, in applying market segmentation for forecasting from
individual-choice models, data requirements can increase
significantly as the number of cells increases, especially
when statistical dependence between the new basis variable
and other variables must be taken into account in estimating
cell frequencies. Unwieldy, complex segmentation schemes may

reduce rather than improve understanding without substantially improving forecasting accuracy.

3   Research on the level of error in aggregate forecasts based on application of non-linear individual-choice models to group means suggests that elaborate segmentation schemes should not be undertaken unless a prior analysis indicates a reasonable expectation that improved forecasting accuracy will result.  The criteria for segmentation on a variable are that: (1) it have an important effect on choice; (2) it vary significantly in the prediction group; and (3) the expected change in the transport system differentially affect segments (8).  As a general rule for many applications of individual-choice models, it appears that segmentation on the basis of taste variations may prove to be at least as powerful as segmentation on the basis of variation in level-of-service attributes.

4   Segmentation on both explanatory variables and observed choice ('choice-based sampling') can provide very useful schemes for sample stratification in data collection (see the workshop report on 'Theoretical and Conceptual Developments in Demand Modelling').

5   Segmentation on the basis of individuals' subjective judgements and attributes of alternatives should not be undertaken independently.  Rather their mutual dependence should be accounted for (e.g. by holding one variable constant and varying the other).

6   Whether segmentation should be accounted for by estimating a separate model for different groups (e.g. a shopping model and a work trip model) or by introducing continuous or discrete variables into a single-model specification is generally a matter of convenience and theoretical model considerations.  Conceptually both approaches provide grounds for segmentation in explaining and forecasting demand.

7   Segmentation techniques that do not use explicit model specifications for forecasting, such as cross classification, may be poor substitutes for a formal model (see Chapter 10). They do not provide convenient procedures for estimating the behavioural responses of relatively small changes in the transport system.  For example, they do not permit the use of elasticities.  The 'market segmentation' technique, employed with a disaggregate model, has a further advantage of flexibility to adjust the cells to suit the available data (8).

8   Research should be directed to the use of attitudinal-behavioural analysis to solve current problems of individual-choice demand modelling.  Current research into individual-choice models that explicitly allow for taste variations with regard to observed attributes might provide a point of convergence of the two approaches.

9   The value of attitudinal-behavioural methods as compared with individual-choice models depends in part on the availability of data.  Defenders of the attitudinal-behavioural approaches emphasised the relatively low cost of collecting small data sets, while proponents of the individual-choice approaches stressed the advantages of potentially transferable models and the 'bootstrap' approach to using existing transport and census data.  Both approaches require that changes in cell frequencies and the explanatory variables be forecasted for long-run estimation, and these projections may be based on dubious procedures.  The availability of data for estimation and forecasting should therefore be an important

consideration in choice of procedure.

*Recommendations of Steps to Further the Application of
Segmentation in Transport Planning*

The workshop concluded that the implementation of segmentation concepts in actual transport practice would be advanced
considerably by development of explicit procedures which
planners could adapt to their own circumstances.  The results
from NCHRP Project 8-16, 'Guidelines for Public Transportation Levels of Service and Evaluation', should be a useful
step in this direction.  The procedures should not be rigid
guidelines but should demonstrate the value of segmentation
and show how the segmentation scheme relates to the specific
objectives of the study.  The procedures should be illustrated with successful case studies, preferably where multi-
disciplinary interests in segmentation are represented.
Demonstrations of how segmentation techniques can be applied
in a multi-disciplinary framework would have much value in
providing case studies.  The procedures should specify how
the various criteria for segmentation (e.g. study objectives,
importance of the attribute, variability of the attribute,
data availability, etc.) would be used to construct the
segments.

In the area of theoretical research, the potential of
segmentation approaches in attitudinal-behavioural research
to contribute to issues in individual-choice modelling
deserves consideration.  In particular, the contribution to
the issue of transferability and procedures for adjusting the
models to account for taste variations might be considerable.
Methodology and data requirements for simplified clustering
techniques were also identified as priority research needs.


CONCLUSION

The resource papers and the workshop discussions reflect a
considerably expanded range of application of segmentation
concepts.  The workshop report and the resource papers
reflect increasing development of approaches which are
presently available for direct application.  The workshop
concluded that greater implementation of segmentation concepts will encourage greater application of behavioural
demand models.  By stressing the behavioural nature of demand
and the need to address individual differences at the
disaggregate level, segmentation offers a genuine opportunity
to improve transport planning.


REFERENCES

    1. Alpert, M.I. and Davies, C. Shane, 'Segmentation of a
market by determinant attributes', presented at 1974 American
Psychological Association meeting, Council for Advanced
Transportation Studies, University of Texas, Austin, Texas
(1974).
    2. Bass, F.M., Tigert, D.G. and Lonsdale, R.F., 'Marketing
segmentation: group versus individual behavior', *Journal of*

*Marketing Research*, 5, pp. 264-70 (1968).

3. Behnam, J. and Thomson, T.L., 'Transit service improvements as seen by non-users', *Traffic Engineering*, 46, pp. 18-21 (1976).

4. Bruno, A.V., Hustad, T.P. and Pessimier, E.A., 'Media approaches to segmentation', *Journal of Advertising Research*, pp. 35-42 (1973).

5. Byrd, J.P., 'Characteristics, attitudes, and perceptions of transit non-users in the Atlanta region', *Transportation Research Record*, 563, pp. 29-37 (1976).

6. Charles River Associates(1972) *A Disaggregated Behavioral Model of Urban Travel Demand*, Federal Highway Administration, US Department of Transportation, Washington, D.C.

7. Charles River Associates (1976) *Estimating the Effects of Urban Travel Policies*, Transportation Systems Center, US Department of Transportation, Cambridge, MA.

8. Charles River Associates (1977) *Guidelines for Using the Market Segmentation Technique to Apply Disaggregate Urban Travel Demand Models with Census Data*, unpublished working paper, National Co-operative Highway Research Program, Project 8-13, Phase II.

9. Dobson, R., 'The general linear model analysis of variance: its relevance to transportation planning and research', *Socio-Economic Planning Sciences*, 10, pp. 231-5 (1976).

10. Dobson, R., Golob, T.F. and Gustafson, R.L., 'Multidimensional scaling of consumer preferences for a public transportation system: an application of two approaches', *Socio-economic Planning Sciences*, 8, pp. 23-6 (1974).

11. Dobson, R. and Kehoe, 'An empirical comparison of the Tucker-Messick and INDSCAL models: measuring viewpoints about transit attributes', *Socio-economic Planning Sciences*, 9, pp. 67-81 (1975).

12. Dobson, R. and Nicolaidis, G.C., 'Preferences for transit service by homogeneous groups of individuals', *Proceedings, Transportation Research Forum*, 15, pp. 326-35 (1974).

13. Dobson, R. and Tischer, M.L., 'Beliefs about buses, carpools, and single-occupant autos: a market segmentation approach', *Proceedings, Transportation Research Forum*, 17, pp. 200-9 (1976).

14. Dunbar, Frederick C., 'Policy contingent travel forecasting with market segmentation', *Transportation Research Record* (1977).

15. Edlestein, P. (1976) 'Travel Behavior of an Elderly Population: An Attitudinal Model and Some Comparisons', an unpublished Master's thesis, State University of New York at Buffalo, N.Y.

16. Fleet, C.R. and Sosslau, A.B., 'Trip generation procedures: an improved design for today's needs', *Traffic Engineering*, 46, pp. 17-25 (1976).

17. Frank, R., Massy, W.F. and Wind, Y. (1975) *Market Segmentation*, Prentice-Hall, Inc., Englewood Cliffs, New Jersey.

18. Golob, T.F., Canty, E.T. and Gustafson, R.L. (1972) *Classification of Metropolitan Areas for the Study of New Systems of Arterial Transportation*, General Motors Research Laboratories, GMR-1225, Warren, MI.

19. Golob, T.F., Canty, E.T., Gustafson, R.L. and Vitt, J.E., 'An analysis of consumer preferences for a public transportation system', *Transportation Research*, 6, pp. 81-102 (1972).

20. Gustafson, R.L. and Navin, F.P.D. (1972) *User Preferences for Dial-a-bus: A Comparison of Two Cities*, General Motors Research Laboratories, Research Publication GMR-1217, Warren, MI.

21. Haley, R.I., 'Benefit segmentation: a decision oriented research tool', *Journal of Marketing*, 37, pp. 30-5 (1968).

22. Hartgen, D.T., 'Attitudinal and situational variables influencing urban mode choice: some empirical findings', *Transportation*, 3, pp. 377-92 (1974).

23. Hensher, D.A., 'Use and application of market segmentation', in *Behavioral Travel-Demand Models*, P.R. Stopher and A.H. Meyburg (eds), D.C. Heath, Boston (1976).

24. Hensher, D.A., 'Market segmentation as a mechanism in allowing for variability of traveller behavior', *Transportation*, 5, pp. 257-84 (1976).

25. Hepper, S.J. and Gustafson, R.L. (1973) *A Multivariate Statistical Investigation of Person Movement within a Large Metropolitan Area*, General Motors Research Laboratories, GMR-1369, Warren, MI.

26. Johnson, R.M. (1972) 'How can we tell if things are "really" clustered?', paper presented at a joint meeting of the American Statistical Association and the American Marketing Association, Market Facts, Inc., Chicago.

27. Johnson, R.M., 'Market segmentation: a strategic management tool', *Journal of Marketing Research*, 8, pp. 13-18 (1971).

28. Lessig, P.V. and Tollefson, J.O., 'Market segmentation through numerical taxonomy', *Journal of Marketing Research*, 8, pp. 480-7 (1971).

29. Louviere, J.L., Ostresh, L.M., Henley, D.H. and Meyer, R.J., 'Travel demand segmentation: some theoretical considerations related to behavioral modelling', in *Behavioral Travel-Demand Models*, P.R. Stopher and A.H. Meyburg (eds), D.C. Heath, Boston, Massachusetts (1976).

30. Lovelock, C.H., 'A market segmentation approach to transit planning, modeling, and management', *Proceedings, Transportation Research Forum*, 16, pp. 247-58 (1975).

31. Nicolaidis, G.C., Wachs, M. and Golob, T.F. (1977) *Evaluation of Alternative Market Segmentations For Transportation Planning*, Presented at the 1977 Transportation Research Board Meeting, General Motors Research Laboratories, Warren, MI.

32. Recker, W.W. and Golob, T.F., 'A behavioral travel demand model incorporating choice constraints', *Proceedings, Association for Consumer Research Meeting*, 3 (1976).

33. Reed, R.R. (1973) *Market Segmentation Development for Public Transportation*, Department of Industrial Engineering, Stanford University, Stanford.

34. Schermer, J.H., 'Interest group impact assessment in transportation planning', *Traffic Quarterly*, 29, pp. 29-50 (1975).

35. Sonquist, J.A., Baker, E.L. and Morgan, J.H. (1971) *Searching for Structure*, Institute for Social Research, University of Michigan, Ann Arbor, Michigan.

36. Stopher, P.R. *The Development of Market Segments of Destination Choice*, Transportation Research Record no.649, pp. 14-22 (1977).

37. Tischer, M.L. and Dobson, R. (1977) *An Empirical Analysis of Behavioral Intentions to Shift Ways of Traveling to Work*, presented at the 1977 Transportation Research Board meeting, FHWA, US Dept. of Transportation, Washington.

38. Tryon, R.C., 'Comparative cluster analysis of social areas', *Multivariate Behavioral Research*, 3, pp. 213-32 (1968).

39. Wooton, H.G. and Pick, G.W., 'A model for trips generated by households', *Journal of Transport Economics and Policy*, 1, pp. 137-53 (1967).

40. Yankelovich, D., 'New criteria for market segmentation', *Harvard Business Review*, 42, pp. 83-90 (1964).

PART FIVE

Chapter 13

QUANTITATIVE METHODS FOR ANALYSING TRAVEL BEHAVIOUR OF
INDIVIDUALS: SOME RECENT DEVELOPMENTS

Daniel McFadden

SUMMARY

This chapter is concerned with quantitative methods for the
analysis of travel behaviour of individuals.  It reviews some
of the recent developments in model specification, estimation,
model evaluation and testing, and aggregation and forecasting.
Topics in model specification include the multinomial-probit
model and its computation, and generalised-extreme value
models and their relation to sequential models.  Topics in
estimation methods include the use of choice-based samples,
sample designs and incomplete choice sets.  Model evaluation
topics include prediction-success tables and diagnostic tests
of specification.  Aggregation and forecasting topics include
aggregation by the Clark method, synthesis of the distribution of explanatory variables, and the calculus of demand
elasticities.

INTRODUCTION

This chapter is concerned with quantitative methods for the
analysis of travel behaviour of individuals.  It reviews some
recent developments in model specification, estimation, model
evaluation and testing, and aggregation and forecasting.  The
reader is assumed to be familiar with the general foundations
of disaggregate choice theory (30,31), the historical
development and properties of the multinomial logit model
(26), and the use of behavioural models in travel-demand
analysis (15,38,39).

MODEL SPECIFICATION

The choice models which have received serious consideration
in travel-demand applications are multinomial logit (MNL),
multinomial probit (MNP), and a sequential - or tree -
version of multinomial logit.

*The Multinomial-logit Model (MNL)*

A typical MNL model for joint choice of mode, destination and
auto availability is

$$P_{mda} = e^{v_{mda}} / \sum_{n,c,b} e^{v_{ncb}} \tag{1}$$

where m = mode;
       d = destination;
       a = auto availability;

and $v_{mda} = \alpha x_{mda} + \beta y_{da} + \gamma z_a$ = utility, with $\alpha$, $\beta$, $\gamma$ parameter vectors and $x_{mda}$, $y_{da}$, $z_a$ variable vectors describing the decision-maker and the alternative. Letting $P_{m|da}$ denote a conditional choice probability and $P_m$ denote a marginal choice probability, we can derive from Equation (1) the formulae:

$$P_{m|da} = e^{v_{mda}} / \sum_n e^{v_{nda}} = e^{\alpha x_{mda}} / \sum_n e^{\alpha x_{nda}} \tag{2}$$

$$P_{d|a} = \sum_n e^{v_{nda}} / \sum_{n,c} e^{v_{nca}} = \sum_n e^{\alpha x_{nda} + \beta y_{da}} / \sum_{n,c} e^{\alpha x_{nca} + \beta y_{ca}} \tag{3}$$

$$P_a = \sum_{n,c} e^{v_{nca}} / \sum_{n,c,b} e^{v_{ncb}}$$

$$= \sum_{n,c} e^{\alpha x_{nca} + \beta y_{ca} + \gamma z_a} / \sum_{n,c,b} e^{\alpha x_{ncb} + \beta y_{cb} + \gamma z_b} \tag{4}$$

Define inclusive values

$$I_{da} = \log \sum_n e^{\alpha x_{nda}} \; ; \tag{5}$$

$$J_a = \log \sum_{n,c} e^{\alpha x_{nca} + \beta y_{ca}} = \log \sum_c e^{I_{ca} + \beta y_{ca}} \; . \tag{6}$$

Then the choice probabilities can be written

$$P_{mda} = P_{m|da} P_{d|a} P_a \; ; \tag{7}$$

$$P_{m|da} = e^{\alpha x_{mda}} / e^{I_{da}} \; ; \tag{8}$$

$$P_{d|a} = e^{I_{da} + \beta y_{da}} / \sum_c e^{I_{ca} + \beta y_{ca}} = e^{I_{da} + \beta y_{da}} / e^{J_a} \; ; \tag{9}$$

$$P_a = e^{J_a + \gamma z_a} / \sum_b e^{J_b + \gamma z_b} \; . \tag{10}$$

Discussion of the historical development of the MNL model can

be found in McFadden (30); the properties of the model,
including its derivation from the theory of individual
utility maximisation, are given in McFadden (26)

*Sequential MNL Models*

Next consider the sequential or nested MNL model.  A typical
*sequential* model differs from the joint MNL model solely in
that the coefficients of inclusive values are not constrained
to equal one.  Hence, the joint MNL model is a linear
restriction on any of the sequential models.  Specifically,
a sequential model is defined by

$$P_{mda} = P_{m|da} P_{d|a} P_a \; ; \tag{11}$$

$$P_{m|da} = e^{\alpha x_{mda}} \Big/ \sum_n e^{\alpha x_{nda}} \; ; \tag{12}$$

$$I_{da} = \log \sum_n e^{\alpha x_{nda}} \; ; \tag{5'}$$

$$P_{d|a} = e^{\theta I_{da} + \beta y_{da}} \Big/ \sum_c e^{\theta I_{ca} + \beta y_{ca}} \; ; \tag{13}$$

$$J_a = \log \sum_c e^{\theta I_{ca} + \beta y_{ca}} \; ; \tag{14}$$

$$P_a = e^{\lambda J_a + \gamma z_a} \Big/ \sum_b e^{\lambda J_b + \gamma z_b} \; . \tag{15}$$

When  $\theta = \lambda = 1$, this model is identical to the joint MNL
model.  More generally, when  $\theta \neq 1$, Equations (13) and (14)
differ in the two models, and when  $\lambda \neq 1$, Equation (15)
differs in the two models.

The sequential model was introduced by Domencich and McFadden
(15) and studied by Ben-Akiva (3), and is discussed in
greater detail below.

*The Multinomial-probit Model*

A typical model for the mode-destination-auto (mda) choice
problem is obtained by assuming that each alternative mda has
a utility $u_{mda} = V_{mda} + \lambda_{mda} + \eta_{da} + \nu_a$ , where $\lambda_{mda}$ , $\eta_{da}$ and
$\nu_{da}$ summarise the influence of unobserved attributes and
taste variations, and are assumed to be jointly normally
distributed over the population.  If each individual maximises utility, the proportion of the population choosing
mda is

$$P_{mda} = \int_{\varepsilon_{mda}=-\infty}^{+\infty} \cdots \int_{\varepsilon_{ncb}=-\infty}^{V_{mda}-V_{ncb}+\varepsilon_{mda}} \cdots \int_{\varepsilon_{MDA}=-\infty}^{V_{mda}-V_{MDA}+\varepsilon_{mda}} n(\varepsilon;0,\Omega)d\varepsilon \quad , \quad (16)$$

where the number of integrals equals the number of alternatives, $n(\varepsilon;0,\Omega)$ is the multivariate normal density with mean vector 0 and covariance matrix $\Omega$, and $\varepsilon_{mda} = \lambda_{mda} + \eta_{da} + \nu_a$, with the joint normal distribution of $\lambda$, $\eta$, and $\nu$ determining $\Omega$.

The MNP model generalises a classical model of Thurstone (41) for binary choice. Bock and Jones (6) applied the model to the three-alternative case. The model was suggested for transport analysis by Domencich and McFadden (15), and first applied to transport data by Hausman and Wise (20). Further discussion is given below.

*Other Choice Models*

Several other models deserve passing note. McFadden (27) has proposed a universal, or 'mother', MNL model which can approximate an arbitrary choice model with a function of the form (1), except that $V_{mda}$ functions will depend on the attributes of all alternatives, and not solely on the attributes of mda. This model is useful for testing particular specifications, but is in general inconsistent with utility maximisation.

McLynn (37) has proposed the *fully competitive* model which is a one-parameter mapping of MNL choice probabilities into a second vector of probabilities. This model is in general inconsistent with individual utility maximisation, yet it shares with the MNL model restrictive structural properties which render it implausible in some applications. The McLynn model satisfies 'simple scalability' = 'order independence', which is closely related to the 'independence from irrelevant alternatives' property of the MNL model (see McFadden (28)).

*The Generalised Extreme-value Model*

McFadden (35) has recently proposed a family of *generalised extreme-value* (GEV) choice models which allow a general pattern of dependence among alternatives and yield a closed form for the choice probabilities. The following result characterises the family:

*THEOREM 1. Suppose* $G(y_1,\ldots,y_J)$ *is a non-negative,*
*homogeneous-of-degree-one function of* $(y_1,\ldots,y_J) \geq 0$ .
*Suppose* $\lim_{y_i \to +\infty} G(y_1,\ldots,y_J) = +\infty$ *for* $i = 1,\ldots,J$ . *Suppose for*
*any distinct* $(i_1,\ldots,i_k)$ *from* $\{1,\ldots,J\}$ , $\partial^k G/\partial y_{i_1},\ldots,\partial y_{i_k}$
*is non-negative if k is odd and non-positive if k is even.*

*Then,*

$$P_i = e^{V_i} G_i(e^{V_1},\ldots,e^{V_J}) \,/\, G(e^{V_1},\ldots,e^{V_J}) \tag{17}$$

*defines a choice model which is consistent with utility maximisation.*

The special case $G(y_1,\ldots,y_J) = \sum_{j=1}^{J} y_j$ yields the MNL model. An example of a more general $G$ function satisfying the hypotheses of the theorem is

$$G(y) = \sum_{m=1}^{M} a_m \left( \sum_{i \in B_m} y_i^{\frac{1}{1-\sigma_m}} \right)^{1-\sigma_m} , \tag{18}$$

where $B_m \subseteq \{1,\ldots,J\}$ , $\bigcup_{m=1}^{M} B_m = \{1,\ldots,J\}$ , $a_m > 0$ , and $0 \leq \sigma_m < 1$ . The parameter $_m$ is an index of the similarity of the unobserved attributes of alternatives in $B_m$. The choice probabilities for this function satisfy

$$P_i = \sum_{m=1}^{M} P(i|B_m) P(B_m) \tag{19}$$

where $P(i|B_m)$ is the conditional probability that alternative i is chosen, given the event $B_m$, with

$$P(i|B_m) = e^{\frac{V_i}{1-\sigma_m}} \,/\, \sum_{j \in B_m} e^{\frac{V_j}{1-\sigma_m}} \quad \text{if } i \in B_m ; \tag{20}$$

$$\qquad\qquad 0 \qquad\qquad\qquad \text{if } i \notin B_m ;$$

and $P(B_m)$ is the probability of the event $B_m$, with

$$P(B_m) = a_m \left\{ \sum_{j \in B_m} e^{\frac{V_j}{1-\sigma_m}} \right\}^{1-\sigma_m} \,/\, \sum_{n=1}^{M} a_n \left\{ \sum_{k \in B_n} e^{\frac{V_k}{1-\sigma_n}} \right\}^{1-\sigma_n} . \tag{21}$$

Functions of the form in Equation (18) can also be nested to yield a wider class satisfying the theorem hypotheses. For example, the function

$$G = \sum_{q=1}^{Q} a_q \left( \sum_{m \in D_q} \left( \sum_{j \in B_m} y_j \right)^{\frac{1}{1-\sigma_m}} \right)^{\frac{1-\sigma_m}{1-\delta_q}} \right)^{1-\delta_q} \tag{22}$$

where $\cup B_m = \{1,\dots,J\}$ , satisfies the hypotheses provided $1 > \sigma_m \geq \delta_q \geq 0$ for $m \in D_q$. The choice probabilities for Equation (22) and analogous functions can be written as sums of products of conditional and marginal probabilities, in a manner generalising Equation (19), with each probability element having a multinomial-logit form, and the denominator in each element equalling a representative term in the succeeding element.

Choice probabilities of the form (19) were apparently first derived, for the case of three alternatives and $B_1 = \{1\}$ , $B_2 = \{2,3\}$ by Scott Cardell [7]. For the case of disjoint $B_m$, the form (19) has been discovered, independently, by Daly and Zachary [12], Williams [42], and Ben-Akiva and Lerman [4]. The demonstration by Daly and Zachary that this choice model is consistent with random utility maximisation is particularly noteworthy in that it permits generalisation of the GEV model and provides a powerful tool for testing the consistency of choice models: suppose alternative i has a utility $U_i = w_i + y(z_i) + \varepsilon_i$ , where $V_i \equiv w_i + y(z_i)$ is the systematic component of utility and $(\varepsilon_1,\dots,\varepsilon_J)$ is a jointly distributed random vector, with a distribution function not depending on $(w_1,\dots,w_J)$, but in general depending on $(z_1,\dots,z_J)$. Suppose the choice probabilities satisfy

$$P_i \equiv P_i(V_1,\dots,V_J;z_1,\dots,z_J) = \text{Prob } [V_i + \varepsilon_i \geq V_j + \varepsilon_j \text{ for } j \neq 1] \tag{23}$$

$$\equiv \text{Prob } [V_i - V_j \geq \nu_j - \nu_i \text{ for } j \neq i] ,$$

where $\nu_i = \varepsilon_i - \varepsilon_1$ . Define the expected value of the maximum of the utilities $U_j$,

$$\bar{U}(V_1,\dots,V_J;z_1,\dots,z_J) = \text{E Max}_j \, U_j \, . \tag{24}$$

Then, the choice probabilities satisfy*

$$P_i(V_1,\ldots,V_J;z_1,\ldots,z_J) = \frac{\partial}{\partial V_i}\,\overline{U}(V_1,\ldots,V_J;z_1,\ldots,z_J) \quad , \qquad (25)$$

and the joint distribution of the differences of the random components of utility, $(v_2,\ldots,v_J)$ , satisfies

$$F(v_2,\ldots,v_J) = P_1(0,-v_2,\ldots,-v_J) \quad . \qquad (26)$$

Conversely, any choice probability functions $P_i(V_1,\ldots,V_J; z_1,\ldots,z_J)$ which satisfy the necessary and sufficient conditions for $(P_1,\ldots,P_J)$ to be the gradient of a potential** $(\overline{U})$ and for $P_1(0,-v_2,\ldots,-v_J)$ to be a distribution function,*** satisfy (23), and are consistent with stochastic utility maximisation. The key assumption, and only significant restriction, underlying this result is that the random utilities $U_i$ have linear components $w_i$ with the property that the joint distribution of the stochastic components does not depend on $(w_1,\ldots,w_J)$.****

---

*    This condition has been used by Domencich and McFadden
     (15) and Harris and Tanner (19) to establish a classical
     identity between social welfare, defined by the expected
     value (or average over the population) of the maximum
     utility for each individual, and consumer surplus,
     defined by the area under the market demand curves, or
     choice probabilities. The identity can be verified
     directly by writing out the definition of expected
     maximum utility and differentiating. The basic assump-
     tion required for the social welfare identity is a linear
     'transferable' numéraire commodity. A consequence of the
     additively separable structure of errors specified in
     (23) is that the choice probabilities are invariant with
     respect to location; i.e.,

$$P_i(V_1 + a,\ldots,V_J + a;z_1,\ldots,z_J) \equiv P_i(V_1,\ldots,V_J;z_1,\ldots,z_J) \quad . \qquad ).$$

**   Suppose $P_i(V_1,\ldots,V_J)$ is continuous and continuously
     differentiable. Then, a necessary and sufficient con-
     dition for $(P_1,\ldots,P_J)$ to be the gradient of a potential
     $(\overline{U})$ is that $\partial P_i/\partial V_j = \partial P_j/\partial V_i$ . If $P_i$ is invariant with
     respect to location, then $\overline{U}$ is homogeneous with respect
     to location; i.e. $\overline{U}(V_1 + a,\ldots,V_J + a) = \overline{U}(V_1,\ldots,V_J) + a$.

***  See Cramer (11) Sect. 8.4. The key condition for
     $F(v_2,\ldots,v_J)$ to be a distribution is that the $(J-1)$st
     difference, $\Delta^{J-1}F$, be non-negative. If F is continuous
     and almost everywhere $J-1$ times differentiable, this
     condition reduces to the requirement that the density
     $\partial^{J-1}F/\partial v_2\ldots\partial v_J$ be non-negative.

****Strictly, the condition is that the joint distribution of
     differences of the random components of utility,
     $F(v_2,\ldots,v_J)$, not depend on $(w_1,\ldots,w_J)$.

The GEV model satisfies (25) with $\overline{U} = \log G(e^{V_1}, \ldots, e^{V_J})$.

*Relation of Sequential MNL and GEV Models*

The choice probabilities corresponding to (22) can be specialised to the sequential MNL model described in Equations (11 - 15), as we shall now show. This result establishes that sequential MNL models are consistent with individual utility maximisation for appropriate parameter values, and that the coefficients of inclusive values can be used to obtain estimates of the similarity parameters $\sigma$ and $\delta$. It is hence possible to estimate some GEV models using sequential MNL models and inclusive values. Further, the GEV class provides a generalisation containing alternative sequential MNL models, and could be estimated directly to test the presence of a sequential or tree structure.

To obtain the sequential model (11 - 15) from Equation (22), index alternatives by mda for mode m, destination d, and auto availability a, and specialise (22) to the form

$$G = \sum_a \left\{ \sum_d \left( \sum_m y_{mda}^{\frac{1}{1-\sigma}} \right)^{\frac{1-\sigma}{1-\delta}} \right\}^{1-\delta} , \quad 0 \le \sigma \le \delta < 1 . \tag{27}$$

Assume $V_{mda} = (1 - \sigma)\alpha'x_{mda} + (1 - \delta)\beta'y_{da} + \gamma'z_a$ . Then (17) yields

$$P_{mda} = \frac{e^{\frac{V_{mda}}{1-\sigma}}}{\sum_n e^{\frac{V_{nda}}{1-\sigma}}} \frac{\left( \sum_n e^{\frac{V_{nda}}{1-\sigma}} \right)^{\frac{1-\sigma}{1-\delta}}}{\sum_c \left( \sum_n e^{\frac{V_{nca}}{1-\sigma}} \right)^{\frac{1-\sigma}{1-\delta}}} \frac{\left( \sum_c \left( \sum_n e^{\frac{V_{nca}}{1-\sigma}} \right)^{\frac{1-\sigma}{1-\delta}} \right)^{1-\delta}}{\sum_b \left( \sum_c \left( \sum_n e^{\frac{V_{ncb}}{1-\sigma}} \right)^{\frac{1-\sigma}{1-\delta}} \right)^{1-\delta}} \tag{28}$$

$$= \frac{e^{\alpha'x_{mda}}}{\sum_n e^{\alpha'x_{nda}}} \frac{e^{\beta'y_{da}+\frac{1-\sigma}{1-\delta}I_{da}}}{\sum_c e^{\beta'y_{ca}+\frac{1-\sigma}{1-\delta}I_{ca}}} \frac{e^{\gamma'z_a+(1-\delta)J_a}}{\sum_b e^{\gamma'z_b+(1-\delta)J_b}}$$

where

$$I_{da} = \log \sum_n e^{\alpha'x_{nda}}$$

$$J_a = \log \sum_c e^{\beta' y_{ca} + \frac{1-\sigma}{1-\delta} I_{ca}} \quad .$$

This is precisely the sequential model (11 - 15), with $\theta = (1-\sigma)/(1-\delta)$ and $\lambda = 1 - \delta$. Hence, we have established that *a sufficient* condition for a nested logit model to be consistent with individual utility maximisation is that the coefficient of each inclusive value lie between zero and one, $0 < \theta$, $\lambda \leq 1$. (The preceding demonstration for three-level trees is readily generalised to trees of any depth. The simplest proof is by induction.)  Application of the Daly-Zachary test shows that this condition is also *necessary* for consistency with random utility maximisation if the domain of $(V_2 - V_1, \ldots, V_J - V_1)$ is unrestricted.  When the necessary and sufficient condition $0 < \theta$, $\lambda \leq 1$ is satisfied, $1 - \theta$ is an index of the similarity of alternative modes, while $1 - \lambda$ is an index of the similarity of alternative destinations.

*Computation of the MNP Model*

The multinomial-probit model is an appealing conceptual model.  It allows consideration of stochastic components for tastes and unobserved attributes within an alternative, and provides a way of specifying the *structure* of dependence between alternatives.  However, MNP choice probabilities can be expressed exactly only as multivariate or iterated integrals of dimension $J - 1$, where $J$ is the number of alternatives.  Exact calculation by numerical integration is very fast for $J = 2$ or 3, moderately costly for $J = 4$, and impractical on a large scale for $J \geq 5$.  One of the more effective direct numerical integration methods, adapted for transport applications, is due to Hausman and Wise (20).

Two recent contributions have provided techniques for approximating MNP choice probabilities at moderate cost. This has made MNP a practical alternative for many transport applications.  The first method, due to Manski (25), applies a Monte Carlo procedure directly to the utilities of alternatives.  Suppose $J + 1$ alternatives, with utilities $U_i = V_i + \varepsilon_i$, where $(\varepsilon_1, \ldots, \varepsilon_{J+1})$ is multivariate normal with zero means and covariance matrix $\Sigma = (\sigma_{ij})$.  For given values of $V_i$, vectors $(\varepsilon_1, \ldots, \varepsilon_{J+1})$ from the multivariate distribution can be drawn, and the frequency with which utility is maximised at alternative i recorded.  These frequencies approximate the exact MNP probabilities when the number of Monte Carlo repetitions is large.  Because this method involves repetitive simple calculations, it can be programmed in computer assembly language to operate quite efficiently.  The approach is appealing in its generality - any joint distribution of the unobserved effects can be assumed.  In practice, the method is most effective when a relatively good initial approximation to the frequencies is available.

The second approximation method, due to Daganzo,

Bouthelier and Sheffi (13), uses a procedure suggested by
Clark (8) to approximate the maximum of bivariate normal
variables by a normal variable.   When the correlation of the
variables is non-negative, this approximation is accurate
within a few per cent.   Suppose $J + 1$ alternatives, with
utilities $U_i = V_i + \varepsilon_i$ and $(\varepsilon_1, \ldots, \varepsilon_{J+1})$ distributed multivariate normal, zero means and covariance matrix $\Sigma$.   The
probability that the first alternative is chosen is then

$$P_1 = \text{Prob } [V_1 + \varepsilon_1 > V_j + \varepsilon_j \quad \text{for} \quad j = 2, \ldots, J + 1] \qquad (29)$$

$$= \text{Prob } [V_1 - V_{J+1} + \varepsilon_1 - \varepsilon_{J+1} > V_j - V_{J+1} + \varepsilon_j - \varepsilon_{J+1}$$

$$\text{for} \quad j = 2, \ldots, J \quad \text{and} \quad V_1 - V_{J+1} + \varepsilon_1 - \varepsilon_{J+1} > 0] \quad .$$

Define $V_j = V_j - V_{J+1}$ and $y_i = \varepsilon_i - \varepsilon_{J+1}$.   Then, $(y_1, \ldots, y_J)$
is multivariate normal with mean zero and covariance matrix
$\Omega = (\omega_{ij})$ , where $\omega_{ij} = \sigma_{ij} + \sigma_{J+1,J+1} - \sigma_{i,J+1} - \sigma_{j,J+1}$.
Hence

$$P_1 = \text{Prob } [v_1 + y_1 > 0 \quad \text{and} \quad v_1 + y_1 > v_j + y_j \quad \text{for} \quad j = 2, \ldots, J]$$

$$\qquad (30)$$

$$= \int_{y_1 = -v_1}^{\infty} n_1(y_1) N_{(1)}((v_1 - v_j + y_1)|y_1) dy_1$$

where $n_{Y(X)}(y|x)$ denotes the normal density for the vector
of variables indexed by $Y$, conditioned on the vector of
variables indexed by $X$ ; $N_{Y(X)}(y|x)$ denotes the corresponding

cumulative distribution function, $N_{Y(X)}(y|x) = \int_{-\infty}^{y} n_{Y(X)}(y'|x) dy'$

and $n_Y(y)$ is the marginal density of the variables
indexed by $Y$.   As a shorthand, the set of all indices, or the
set of all indices exluding those on which a distribution is
conditioned, are omitted.   Thus, $N_{(1)}$ means $N_{2,\ldots,J(1)}$.   The
form (30), involving $J$ integrals, is the basis for exact
calculations of $P_1$.   Alternately, write

$$P_1 = \text{Prob } [v_1 + y_1 > 0 \quad \text{and} \quad v_1 + y_1 > \max_{j=2,\ldots,J} (v_j + y_j)] \quad . \qquad (31)$$

The Clark method considers trivariate normal random variables
$(X_1, X_2, X_3)$, and approximates the bivariate distribution of
$(X_1, \max (X_2, X_3))$ by a bivariate normal distribution with the
same first and second moments.   The approximation rests on
the fact that these moments for $(X_1, \max (X_2, X_3))$ can be
calculated exactly in a straightforward manner.   Applied
recursively to the expression

$$\qquad (32)$$

$$y_0 = \max(v_2 + y_2, \max(v_3 + y_3, \ldots, \max(v_{J-1} + y_{J-1}, v_J + y_J) \ldots)$$

the method allows the distribution of $(y_1, y_0)$ to be approximated by a bivariate normal distribution $n_1(y_1) n_{0(1)}(y_0|y_1)$, so that Equation (30) is approximated by the univariate integral

$$P_1 = \int_{y_1 = -v_1}^{\infty} n_1(y_1) N_{0(1)}(v_1 + y_1 | y_1) dy_1 \quad , \tag{33}$$

where $N_{0(1)}(y_0|y_1) = \int_{-\infty}^{y_0} n_{0(1)}(y_0'|y_1) dy_0'$.     Thus, an MNP

choice probability for $J + 1$ alternatives is approximated by a univariate integral involving a univariate normal density and univariate normal cumulative distribution function (which can be accurately approximated computationally by a series expansion). The approximation requires $J - 2$ applications of the Clark formula.

Manski [25] has reported good results in maximum-likelihood search methods using the approximation above, with search directions determined by numerical evaluation of derivatives. This suggests that the bias caused by the approximation is relatively stationary for evaluation of probabilities at neighbouring points. This fortuitous conclusion suggests that it is probably unnecessary to obtain analytic derivatives of $P_1$ with respect to parameters in statistical routines. On the other hand, it is possible that the use of analytic derivatives could decrease computation time. The following argument shows that the Clark procedure can be applied to yield quick approximations to analytic derivatives.

From (30)

$$\frac{\partial P_1}{\partial \theta} = n_1(-v_1) N_{(1)}((-v_j)|-v_1) \frac{\partial v_1}{\partial \theta}$$

$$+ \sum_{j=2}^{J} \frac{\partial (v_1 - v_j)}{\partial \theta} \int_{y_1 = -v_1}^{\infty} n_{1j}(y_1, v_1 - v_j + y_1) N_{(1j)}$$

$$((v_1 - v_k + y_1)|y_1, v_1 - v_j + y_1) dy_1 \tag{34}$$

The term $N_{(1)}((-v_j)|-v_1)$ can be approximated by applying the Clark procedure to the conditional distribution. The integrals in the last right-hand term of (34) each have the essential structure of (30), since $n_{1j}(y_1, v_1 - v_j + y_1)$ is proportional to a normal density for $y_1$ whose mean and variance are computed by a straightforward completion of the

square.  Then, each integral in this term can be approximated
by the corresponding analogue of (33).  We conclude that the
analytic derivative $\partial P_1/\partial\theta$ can be computed by the evaluation
of J univariate integrals, each with the generic form of (33),
and each involving J - 3 applications of the Clark procedure.
For most problems, where the number of parameters exceeds J,
this computation should be considerably faster than numerical
computation of the derivatives.

The probability $P_1$ also depends parametrically on the
covariance matrix $\Omega$.  The requirement that $\Omega$ be positive
definite can be imposed by writing $\Omega^{-1} = TT'$ , where $T = (\tau_{ij})$
is a lower triangular matrix with positive diagonal elements.
Then,  $|\Omega|^{-1/2} = \tau_{11} \cdots \tau_{JJ}$ .  Alternatively, $\Omega^{-1}$ may be
represented as an unknown non-negative linear combination,
with full rank, of known positive semi-definite matrices.
Analogues for the parameters of $\Omega^{-1}$ of the analytic derivatives above are computationally complex, and their use
appears unlikely to improve significantly on numerical
differentiation.

The key to the accuracy of the Daganzo-Bouthelier-Sheffi
approximation is the accuracy of the Clark procedure.
Because the true distribution of the maximum of two normal
variates is skewed to the right, one would expect the
procedure to tend to under-estimate small probabilities.  The
approximation will be best when the variates are positively
correlated, with widely differing means, and worse when they
are negatively correlated with similar means.  It may be
possible to adjust the Clark formulae empirically to improve
their accuracy for computation of small probabilities.
Alternately, it would be interesting to explore the possibility of adapting the Clark methodology to other trivariate
distribution.  In particular, if the generalised extreme-
value distribution were utilised, then the only point of
approximation would be the initial fit to the multivariate
normal density, since maxima of GEV distributed variates are
again GEV distributed.  This would limit approximation error
as J increases, in contrast to the Clark procedure which
becomes less accurate with large numbers of alternatives.


STATISTICAL ESTIMATION METHODS AND SAMPLING STRATEGIES

*Maximum Likelihood Estimation*

The statistical estimation of individual-choice models by the
maximum likelihood method is now well established.  For
random samples of individuals, this procedure can be shown in
general to produce estimates with good statistical properties,
at least in large samples.  The problems remaining in application of maximum likelihood estimation in this context are
primarily computational - the issues of rapid computation of
choice probabilities, the concavity or unimodality of the log
likelihood function, and the relative convergence speed of
alternative algorithms.

Estimation of sequential models, with inclusive values obtained using estimates from earlier stages of the model, has been carried out by many investigators such as Domencich and McFadden (15), and Ben-Akiva (3) treating each stage as an independent estimation problem.  This procedure neglects the fact that the use of inclusive value measures which are themselves statistics change the asymptotic distribution of the estimators, and leads to biased estimates of the standard errors of the estimators.  This problem has been pointed out by Amemiya (1), who provides the corrected asymptotic estimators for the standard errors of the estimates.

*Estimation in Choice-Based Samples*

Several recent papers have considered the problem of statistical estimation of choice models using data collected by sampling procedures other than random sampling.  Of particular interest are *choice-based* samples, utilising data collected from 'on-board' or 'destination' surveys.  Such data sources are often available to transport analysts from marketing and operations departments of operating agencies, or can be commissioned at low cost relative to random household surveys.  Manski and Lerman (23) have shown that treating choice-based samples as if they were random and calculating estimators appropriate to random samples will generally yield inconsistent estimates. (In an MNL model with alternative-specific dummy variables, the inconsistency is confined to the dummy variable coefficients.)  They introduce a weighted likelihood function whose maximisation is shown to yield consistent estimates.

Manski and McFadden* have considered more generally the problem of estimation of discrete choice models under alternative sample designs.  The discrete choice problem can be defined by a finite set C of mutually exclusive alternative responses, a space of attributes Z, assumed to be a subset of a finite-dimensional vector space, a generalised probability density, $p(z)$ $[z \in Z]$ , giving the distribution of attributes in the population, and a *response probability*, or *choice probability*, $P(i|z,\theta*)$ , specifying the conditional probability of selection of alternative $i \in C$, given attributes $z \in Z$ .  Prior knowledge of causal structure is assumed to allow the analyst to specify the response model $P(i|z,\cdot)$ up to a parameter vector $\theta*$ contained in a subset $\Theta$ of a finite-dimensional vector space.  The analyst's problem is to estimate $\theta*$ from a suitable sample of subjects and their associated responses.

The probability of $(i,z)$ pairs in the population is given by

$$f(i,z) = P(i|z,\theta*)p(z)  .  [(i,z) \in C \times Z] \tag{35}$$

The analyst can draw observations of $(i,z)$ pairs from $C \times Z$

---

*Manski, C.F. and McFadden, D., 'Alternative estimates and sample designs for discrete choice analysis', Department of Economics, University of California at Berkeley (mimeo)(1978).

according to one of various sampling rules. The problem of
interest is first, given any sampling rule, to determine how
θ* may be estimated, and second, to assess the relative
advantages of alternative sampling rules and estimation
methods.

The data layout can be visualised using a contingency
table, as illustrated in Figure 13.1. An observation (i,z)
occurs in the population with frequency f(i,z). The row sums
give the marginal distributions of attributes p(z), while the
column sums give the population shares of responses Q(i).
The joint frequency f(i,z) can be written either in terms of
the conditional probability of i given z, or choice probability, or in terms of the conditional probability of z given i,
as the formulae in the figure illustrate.

Choice Set C

| | 1 ............. i ............M | |
|---|---|---|
| z' | | p(z') |
| : | : | |
| z | ..... f(i,z) ..... | p(z) |
| : | : | |
| z'' | | p(z'') |
| | Q(1)          Q(i)        Q(M) | |

Attribute

Set  Z

$p(z) = \sum_{i \in C} f(i,z)$

$Q(i) = \int f(i,z)dz$

$f(i,z) = P(i|z,\theta^*)p(z) = q(z|i,\theta^*)Q(i)$

Figure 13.1   Contingency Table Layout of Observations

The feature of the quantal response problem which
distinguishes it from the general analysis of discrete data
is the postulate that the response probability $P(i|z,\theta^*)$
belong to a known parametric family, and reflects an underlying link from z to i which will continue to hold even if
the distribution p(z) of the explanatory variables changes.
This postulate is fundamental to the concept of 'scientific

explanation'. If the response probability function is invariant over populations with different distributions of attributes, then it defines a 'law' which transcends the character of specific sets of data. Otherwise, the model provides only a device for summarising data, and fails to provide a key ingredient of 'explanation' - predictive power. Alternately, given a population $C \times Z$ with probability distribution specified by $f(i,z)$, one might, in the absence of any knowledge of the process relating i's to z's, obtain a random sample from $C \times Z$ and directly examine the joint distribution $f(i,z)$. This exploratory data analysis approach is exemplified by the literature on associations in contingency tables, where it is assumed only that Z is finite. See, for example, Goodman and Kruskal [16], Haberman [17,18], and Bishop, Fienberg, and Holland [5].

If one believes that the elements of C index conceptually distinct populations of z values, then the natural analytical approach is to decompose $f(i,z)$ into the product $f(i,z) = q(z|i)Q(i)$, where $q(z|i)$ gives the distribution of z within the population indexed by i and $Q(i)$ is the proportion of the population with this index. This is the approach taken in discriminant analysis. There, prior knowledge allows the analyst to specify $q(z|i)$ up to a parametric family, and a sample suitable for estimating the unknown parameters is obtained from the subpopulation i [2,21].

When a well-defined process generates a value from C given any $z \in Z$, then the decomposition $f(i,z) = P(i|z,\theta^*)p(z)$ is appropriate. This decomposition, and the attending focus on the structural relation embodied in $P(i|z,\theta^*)$, is clearly the natural one for the analysis of choice data. A separate and interesting question is whether specific parametric models permit estimation of the parameter vector $\theta^*$ of $P(i|z,\theta^*)$ from convenient parameterisations of $f(i,z)$ or $q(z|i)$.

Manski and McFadden attempt to provide a general theory of estimation for quantal response models. The scope of the investigation is as follows: consider the problem of estimating $\theta^*$ from stratified samples of $(i,z)$ observations. A stratified sampling process is one in which the analyst establishes an index set B, partitions $C \times Z$ into mutually exclusive and exhaustive measurable subsets $(C \times Z)_b$, $b \in B$, and specifies a suitable probability distribution over B. To obtain an $(i,z)$ observation, he draws a subset of $C \times Z$ according to the specified distribution and then samples at random from within the drawn subset.

Within the class of all stratification rules, two symmetric types of stratification are of particular statistical and empirical interest. In 'exogenous' sampling, the analyst partitions Z into subsets $Z_b$, $b \in B$, and lets $(C \times Z)_b = C \times Z_b$. In 'endogenous' or 'choice-based' sampling, he partitions C into subsets $C_b$, $b \in B$, and lets $(C \times Z)_b = C_b \times Z$. Less formally, in exogenous sampling the analyst selects decision-makers and observes their choices while in choice-based sampling the analyst selects alternatives and

observes decision-makers choosing them.  In Figure 13.1,
exogenous sampling corresponds to stratifying on rows, and
then sampling randomly from each row, while choice-based
sampling corresponds to stratifying on columns, and then
sampling randomly from each column.

Manski and McFadden make a detailed statistical examination of maximum likelihood estimation of $\theta*$ in both exogenous
and choice-based samples.  They find that application of
maximum likelihood is wholly classical in exogenous samples.
In choice-based samples, however, the form of the maximum
likelihood estimate (MLE) depends crucially on whether the
analyst has available certain prior information, namely, the
marginal distributions $p(z)$ , $z \in Z$ , or $Q(i)$ , $i \in C$ , where
$Q(i) = \int_Z P(i|z,\theta*)p(z)dz$

The maximum likelihood estimator of $\theta$ in a choice-based
sample when p is known and Q is unknown satisfies

$$\underset{\theta \in \Theta}{\text{Max}} \sum_{n=1}^{N} \log P(i_n|z_n,\theta) - \sum_{n=1}^{N} \int_Z P(i_n|z,\theta)p(z)dz \quad . \tag{36}$$

When Q and p are both known, Equation (36) is maximised
subject to the constraints

$$Q(i) = \int_Z P(i_n|z,\theta)p(z)dz \quad . \tag{37}$$

When p is unknown, the classical conditions for maximum
likelihood estimation are not met.  However, several alternative non-classical maximum likelihood and pseudo-maximum
likelihood methods are available which yield consistent
estimators.

When Q is known and p is unknown, Cosslett (9) has shown
that the non-classical full-information maximum likelihood
estimator satisfies

$$\underset{\theta \in \Theta}{\text{Max}} \underset{\lambda \geq 0}{\text{Min}} \sum_{n=1}^{N} \log \left\{ P(i_n|z_n,\theta) \left( \sum_{j \in C} \lambda_j Q(j) \right) \Big/ \left( \sum_{j \in C} \lambda_j P(j|z_n,\theta) \right) \right\} \quad : \tag{38}$$

A second estimator, introduced by Manski and Lerman (23) and
termed WESML, satisfies

$$\underset{\theta \in \Theta}{\text{Max}} \sum_{n=1}^{N} w(i_n) \log P(i_n|z_n,\theta) \quad , \tag{39}$$

where $w(i) = Q(i)/H(i)$ and $H(i)$ is the sampling frequency for
alternative i.  Two other consistent estimators, introduced
by Manski and McFadden, satisfy

$$\underset{\theta \in \Theta}{\text{Max}} \left\{ \sum_{n=1}^{N} \log P(i_n|z_n,\theta) - \sum_{n=1}^{N} \log \sum_{j \in C} \frac{Q(j)}{N_j} \sum_{m \in N(j)} P(i_n|z_m,\theta) \right\} \quad , \tag{40}$$

where $N(j)$ is the set of observations where alternative $j$ is chosen and $N_j$ is the number of elements in $N(j)$; and

$$\text{Max}_{\theta \in \Theta} \sum_{n=1}^{N} \log \left\{ P(i_n | z_n, \theta) \frac{H(i_n)}{Q(i_n)} \middle/ \sum_{j \in C} P(j | z_n, \theta) \frac{H(j)}{Q(j)} \right\} . \tag{41}$$

If both p and Q are unknown in a choice-based sample, then provided an identification condition is satisfied, Manski and McFadden show that the non-classical full-information maximum likelihood estimator satisfies

$$\text{Max}_{\theta \in \Theta} \text{Max}_{\lambda \geq 0} \sum_{n=1}^{N} \log \left\{ P(i_n | z_n, \theta) \lambda_{i_n} \middle/ \sum_{j \in C} P(j | z_n, \theta) \lambda_j \right\} . \tag{42}$$

An important case in which the identification condition *fails* is the MNL model, where in the absence of a knowledge of Q there is a confounding of the effects of Q and alternative-specific dummies.

A second consistent estimator for this case is obtained by maximising Equation (40), with Q determined as a solution to the equations

$$Q(i) = \sum_{j \in C} Q(j) \frac{1}{N_j} \sum_{m \in N(j)} P(i | z_m, \theta) . \tag{43}$$

Note that one can, with some loss of efficiency, obtain consistent estimates for an information case by using a consistent estimator which ignores some available information. For example, the estimators (38) or (39) could be used in the case both p and Q known, and the estimator (42) could be used in any of the information cases.

*Selection of a Sample Design and Estimation Method*

Sample designs and estimation methods differ in terms of sampling and computation costs, and precision in parameter estimates and forecasts. Cost comparisons are situation-specific, and only a few general observations can be made. Comparison of the precision of alternative estimators can be made for large samples using the asymptotic covariance matrices of the estimators. In a few cases, the difference of two covariance matrices is positive semi-definite for all possible parameter vectors, and a uniform ranking can be made. More generally, rankings will depend on the true parameter vector and on the true distribution of explanatory variables. Then, rankings of designs and estimators will usually require a Bayesian approach utilising *a priori* beliefs on the distributions of parameters, perhaps based on pilot samples and previous studies. Consider sampling costs. In general, substantial economies can be achieved by stratifications designed to make it easier to locate and observe subjects. For example, exogenous cluster sampling, in which respondents are clustered geographically, reduces interviewer access time. Stratification on other exogenous

variables, such as employer, may also reduce location cost.
In many applications, choice-based sampling greatly simplifies locating subjects.  For example, subjects choosing
alternative travel modes can be sampled economically at the
site of choice.  Choice-based sampling has the greatest
potential economy in applications where some responses are
rare (e.g. choice of a seldom-used travel mode) or are
difficult to observe accurately in an exogenously drawn
sample.

Computation costs are comparable in most of the estimation
methods considered by Manski and McFadden.  The primary
component of computation costs is usually the evaluation of
response probabilities at each sample point.  For some models
(e.g. linear), this cost is minimal, for others (e.g. multinomial logit), moderate, and for some (e.g. multinomial
probit), substantial.

*Sample Designs*

Consider the precision of estimates obtained by alternative
methods from alternative sample designs.  We note first that
the level of precision, and possibly the ranking of alternatives, will depend on the prior information available on the
marginal distributions p and Q.  We shall assume the state of
this information is fixed.  However, it should be noted that
in practice the question of drawing observations on p or Q at
some cost in order to utilise more efficient estimators of
the response probability function may be an important part of
the overall design decision.

Cosslett (9) has investigated the efficiency of alternative choice-based sample designs and estimators for binary
probit, logit and arctan models with a single explanatory
variable.  All three models have form  $P(1|z,\theta) = \psi(\theta z)$ , where

$$\psi(y) = \begin{cases} \dfrac{1}{\sqrt{2\pi}} \displaystyle\int_{-\infty}^{y} e^{-x^2/2} \, dx & \text{for probit,} \\[2ex] 1/(1 + e^{-y}) & \text{for logit,} \\[2ex] \dfrac{1}{2} + \dfrac{1}{\pi} \tan^{-1} y & \text{for arctan,} \end{cases} \tag{44}$$

and z is assumed to be normally distributed with mean 2 and
variance 1/2.  Choice-based sample designs vary in the proportion of the sample H(1) drawn from the subpopulation
choosing alternative 1.  The optimal sample design for any
estimator is determined by the value of H(1) which minimises
the asymptotic variance of the estimator.

We concentrate on the case with p unknown and Q known.
The maximum likelihood estimator (38) with an optimal sample
design provides a standard against which other estimators and
sample designs can be measured.  Define the *asymptotic
efficiency* of an alternative estimator and sample design to

be the asymptotic variance of the maximum likelihood
estimator with optimal design, divided by the asymptotic
variance of the alternative estimator.  Consider as alternative estimators the WESML estimator (39), the Manski-McFadden
estimator (41), and the 'conditional' maximum likelihood
estimator (42), which does not use information on Q.  Table
13.1 gives the asymptotic efficiencies of these estimators
for each model for selected values of $\theta$.  Three sample
designs are considered: 'pseudo-random' sampling in proportion to population shares, $H(1) = Q(1)$; sampling equally from
each alternative, $H(1) = 1/2$; and sampling optimally for the
estimator.  The optimising values of $H(1)$ for these
estimators are also given in the table.

TABLE 13.1   Asymptotic efficiency of choice-based sample
designs and estimators[a]

| Probit model | Pseudo-random design $H(1)=Q(1)$ | Equal shares $H(1)=1/2$ | Optimal design | Optimal value of $H(1)$ |
|---|---|---|---|---|
|  | % | % | % | % |
| $Q(1) = 0.75$[b] |  |  |  |  |
| MLE (38) | 87.1 | 95.0 | 100.0 | 0.13 |
| MM (41) | 3.1 | 4.5 | 4.5 | 0.46 |
| WESML (39) | 3.1 | 4.4 | 4.4 | 0.47 |
| Cond ML (42) | 0.4 | 0.6 | 0.6 | 0.49 |
| $Q(1) = 0.9$ |  |  |  |  |
| MLE (38) | 62.1 | 95.2 | 100.0 | 0.30 |
| MM (41) | 6.3 | 20.5 | 21.0 | 0.42 |
| WESML (39) | 6.3 | 19.1 | 19.2 | 0.47 |
| Cond ML (42) | 1.3 | 3.6 | 3.7 | 0.45 |
| $Q(1) = 0.95$ |  |  |  |  |
| MLE (38) | 40.7 | 95.5 | 100.0 | 0.34 |
| MM (41) | 6.1 | 37.9 | 39.2 | 0.39 |
| WESML (39) | 6.1 | 32.0 | 32.0 | 0.50 |
| Cond ML (42) | 1.6 | 7.5 | 7.6 | 0.43 |
| $Q(1) = 0.99$ |  |  |  |  |
| MLE (38) | 9.5 | 96.9 | 100.0 | 0.38 |
| MM (41) | 3.4 | 78.2 | 81.2 | 0.38 |
| WESML (39) | 3.4 | 36.8 | 40.6 | 0.66 |
| Cond ML (42) | 1.4 | 17.8 | 17.8 | 0.46 |
| $Q(1) = 0.995$ |  |  |  |  |
| MLE (38) | 4.5 | 98.4 | 100.0 | 0.42 |
| MM (41) | 2.6 | 91.2 | 92.9 | 0.41 |
| WESML (39) | 2.6 | 23.4 | 30.1 | 0.77 |
| Cond ML (42) | 1.4 | 23.0 | 23.0 | 0.49 |

| | Pseudo-random design $H(1)=Q(1)$ | Equal shares $H(1)=1/2$ | Optimal design | Optimal value of $H(1)$ |
|---|---|---|---|---|
| Logit model | % | % | % | % |
| $Q(1) = 0.75$[b] | | | | |
| MLE (38) | 86.7 | 94.5 | 100.0 | 0.09 |
| MM (41) | 2.9 | 4.0 | 4.0 | 0.47 |
| WESML (39) | 2.9 | 4.0 | 4.0 | 0.48 |
| Cond ML (42) | 0.3 | 0.4 | 0.4 | 0.50 |
| $Q(1) = 0.9$ | | | | |
| MLE (38) | 62.2 | 94.3 | 100.0 | 0.26 |
| MM (41) | 5.2 | 16.1 | 16.2 | 0.44 |
| WESML (39) | 5.2 | 15.1 | 15.1 | 0.48 |
| Cond ML (42) | 0.8 | 1.8 | 1.8 | 0.50 |
| $Q(1) = 0.95$ | | | | |
| MLE (38) | 41.5 | 94.7 | 100.0 | 0.30 |
| MM (41) | 4.9 | 28.9 | 29.5 | 0.42 |
| WESML (39) | 4.9 | 24.8 | 24.8 | 0.51 |
| Cond ML (42) | 0.9 | 3.4 | 3.4 | 0.50 |
| $Q(1) = 0.99$ | | | | |
| MLE (38) | 9.0 | 95.0 | 100.0 | 0.35 |
| MM (41) | 2.7 | 66.5 | 69.3 | 0.38 |
| WESML (39) | 2.7 | 31.7 | 34.4 | 0.65 |
| Cond ML (42) | 0.9 | 8.9 | 8.9 | 0.50 |
| $Q(1) = 0.995$ | | | | |
| MLE (38) | 3.9 | 95.7 | 100.0 | 0.37 |
| MM (41) | 2.1 | 83.4 | 86.9 | 0.37 |
| WESML (39) | 2.1 | 20.3 | 25.6 | 0.76 |
| Cond ML (42) | 1.0 | 12.9 | 12.9 | 0.51 |
| Arctan model | | | | |
| $Q(1) = 0.75$[b] | | | | |
| MLE (38) | 83.5 | 91.3 | 100.0 | 0.00 |
| MM (41) | 1.8 | 2.4 | 2.4 | 0.49 |
| WESML (39) | 1.8 | 2.4 | 2.4 | 0.49 |
| Cond ML (42) | 0.08 | 0.09 | 0.09 | 0.55 |
| $Q(1) = 0.9$ | | | | |
| MLE (38) | 52.9 | 84.0 | 100.0 | 0.00 |
| MM (41) | 1.7 | 4.7 | 4.7 | 0.49 |
| WESML (39) | 1.7 | 4.6 | 4.6 | 0.51 |
| Cond ML (42) | 0.04 | 0.04 | 0.05 | 0.73 |
| $Q(1) = 0.95$ | | | | |
| MLE (38) | 27.0 | 78.1 | 100.0 | 0.00 |
| MM (41) | 1.1 | 5.9 | 5.9 | 0.49 |
| WESML (39) | 1.1 | 5.4 | 5.4 | 0.53 |
| Cond ML (42) | 0.04 | 0.03 | 0.04 | 0.83 |

| | Pseudo-random design $H(1)=Q(1)$ | Equal shares $H(1)=1/2$ | Optimal design | Optimal value of $H(1)$ |
|---|---|---|---|---|
| | % | % | % | % |
| $Q(1) = 0.99$ | | | | |
| MLE (38) | 2.2 | 64.4 | 100.0 | 0.00 |
| MM (41) | 0.5 | 11.8 | 11.9 | 0.46 |
| WESML (39) | 0.5 | 5.4 | 5.9 | 0.65 |
| Cond ML (42) | 0.04 | 0.03 | 0.05 | 0.94 |
| $Q(1) = 0.995$ | | | | |
| MLE (38) | 0.8 | 60.0 | 100.0 | 0.00 |
| MM (41) | 0.4 | 21.2 | 21.2 | 0.42 |
| WESML (39) | 0.4 | 4.1 | 5.1 | 0.74 |
| Cond ML (42) | 0.05 | 0.03 | 0.07 | 0.96 |

a. Adapted from Cosslett [9]. Asymptotic efficiency is
   defined by the ratio of asymptotic variances, with the
   optimal choice-based sample design maximum likelihood
   estimator as the standard. Note that when $Q(1)$ is not
   observed, the estimators (38), (39) and (41) are not
   available, and (42) is asymptotically efficient.
b. For the one-parameter model, knowledge of $p(z)$ and $Q(1)$
   determines $\theta$; for comparability of models, $Q(1)$ rather
   than $\theta$ has been given.

The results in Table 13.1 suggest the following conclusions:

1  Knowledge of the aggregate share $Q(1)$ is of great value
when the maximum likelihood estimator (38) is used, as indicated by the low efficiency of the conditional maximum
likelihood estimator (42) which does not utilise this
knowledge. Note however, that the information contained in
$Q(1)$ will be greatest for a one-variable model without an
alternative-specific dummy, and in general the efficiency
differential will be smaller.
2  The Manski-McFadden estimator (41) is uniformly more
efficient than the WESML estimator (39), but the differential
is small when the true parameter value is small. Both (39)
and (41) have low efficiency relative to maximum likelihood
for small parameter values, but (41) is relatively efficient
for large parameter values.
3  The equal shares sample design is generally quite
efficient for maximum likelihood estimation, and for all the
estimators yields efficiencies comparable to those for the
optimal sample designs. The behaviour of optimal $H(1)$ is
sensitive to the model and to the parameter value. Hence, in
the absence of strong prior knowledge on parameter values,
the equal shares sample design is recommended.

Table 13.2 compares the relative efficiencies of a choice-
based sample design with equal shares and an exogenous random
sample design. For $Q(1)$ known, the choice-based design is
always more efficient. For $Q(1)$ unknown, the choice-based
design is less efficient for the arctan model, and for small

parameter values in the remaining models.  Given prior
beliefs on the correct model and on the value of $Q(1)$, and
given a relative cost r of collecting an observation from an
equal shares choice-based sample compared with an exogenous
random sample, maximum efficiency subject to a fixed sampling
budget will be achieved with the choice-based design if and
only if the relative efficiency given in Table 13.2 exceeds r.

TABLE 13.2    Relative efficiency of choice-based sample design
(with equal shares) and exogenous random sample design, with
maximum likelihood estimators[a]

|  | $Q(1)$ | $Q(1)$ known | $Q(1)$ unknown |
|---|---|---|---|
| Probit | 0.75 | 1.09 | 0.19 |
|  | 0.9 | 1.53 | 0.57 |
|  | 0.95 | 2.35 | 1.23 |
|  | 0.99 | 10.23 | 5.30 |
|  | 0.995 | 21.73 | 8.83 |
| Logit | 0.75 | 1.09 | 0.16 |
|  | 0.9 | 1.52 | 0.35 |
|  | 0.95 | 2.28 | 0.69 |
|  | 0.99 | 10.54 | 3.28 |
|  | 0.995 | 24.68 | 6.26 |
| Arctan | 0.75 | 1.09 | 0.05 |
|  | 0.9 | 1.59 | 0.02 |
|  | 0.95 | 2.90 | 0.03 |
|  | 0.99 | 29.04 | 0.06 |
|  | 0.995 | 75.45 | 0.10 |

a.  Relative efficiency equals the asymptotic variance of the
    exogenous random sample maximum likelihood estimator
    divided by that of the choice-based equal share design
    maximum likelihood estimator.  A ratio exceeding one
    indicates that the choice-based design is more efficient.

*Estimation When Alternatives are Sampled Randomly from the
Full Choice Set*

A particularly advantageous use of choice-based sampling,
either in primary data collection, or in synthesising and
reducing existing data sets, is in estimation of the MNL
model from data on a strict subset of the full choice set.
This method can greatly reduce the magnitude of data to be
collected and analysed, with attendant savings in time and
cost.  The property that a choice model can be estimated
consistently using data on a strict subset of the choice set
is unique to the MNL model, and is a characterisation of the
independence from irrelevant alternatives (IIA) property of
this model.

    The following summary is drawn from McFadden (35).  Let C
denote the full choice set.  We shall assume it does not vary
over the sample; however, this is inessential and can easily
be generalised.  Let $P(i|C,z,\theta^*)$  denote the true selection
probabilities.  We assume the choice probabilities satisfy

the independence from irrelevant alternatives (IIA) assumption,

$$i \in D \subseteq C \implies P(i|C,z,\theta) = P(i|D,z,\theta) \sum_{j \in D} P(j|C,z,\theta) \qquad (45)$$

which characterises the MNL model.

Now suppose for each case, a subset D is drawn from the set C according to a probability distribution $\pi(D|i,z)$ which may, but need not, be conditioned on the observed choice i. The observed choice may be either in or out of the set D. Examples of $\pi$ distributions are: (1) choose a fixed subset D of C, independent of the observed choice; (2) choose a random subset D of C, independent of the observed choice; and (3) choose a subset D of C, consisting of the observed choice i and one or more other alternatives, selected randomly.

We give several examples of distributions of type (3):

(3.1)  Suppose D is comprised of i plus a sample of alternatives from the set $C \setminus \{i\}$, obtained by considering each element of this set independently, and including it with probability p. Then, the probability of D will depend solely on the number of elements $K = \#(D)$ it contains, and is given by the binomial formula

$$\pi(D|i,z) = p^{K-1}(1 - p)^{J-K} \quad \underline{if} \quad i \in D \text{ and } K = \#(D) , \qquad (46)$$

$$= 0 \qquad\qquad \text{if } i \notin D ,$$

where J is the number of alternatives in C.  For example, the probability that D will be any two-alternative set containing i as one alternative is $(J - 1)p(1 - p)^{J - 2}$.

(3.2)  Suppose D is always selected to be a two-element set containing i and one other alternative selected at random. If J is the number of alternatives in C, then

$$\pi(D|i,z) = \frac{1}{J-1} \quad \text{if } D = \{i,j\} \text{ and } j \neq i , \qquad (47)$$

$$= 0 \qquad \text{otherwise.}$$

(3.3)  Suppose C has four elements, and

$$\pi(\{1,4\}|4) = \pi(\{1,4\}|1) = \pi(\{2,3\}|2) = \pi(\{2,3\}|3) = 1 , \qquad (48)$$

$$\text{and } \pi(D|i) = 0 \text{ otherwise .}$$

(3.4)  Suppose C is partitioned into sets $\{C_1,...,C_M\}$ , with $J_m$ elements in $C_m$, and suppose D is formed by choosing i (from the partition set $C_n$) and one randomly selected alternative from each remaining partition set.  Then,

$$\pi(D|i,z) = J_n \Big/ \prod_{m=1}^{M} J_m \quad \text{if } i \in D , M = \#(D) \quad , \text{ and } D \cap C_m \ne \emptyset \quad (49)$$

$$\text{for } m = 1,\dots,M,$$

$$= 0 \qquad\qquad \text{otherwise.}$$

The $\pi$ distributions of the type (1), (2) and (3.1) to (3.4) all satisfy the following basic property, which guarantees that if an alternative j appears in an assigned set D, then it has the logical possibility of being an observed choice from the set D, in the sense that the assignment mechanism could assign the set D if a choice j is observed:

*Positive conditioning property:* *If* $j \in D \subseteq C$ *and*

$\pi(D|i,z) > 0$ , *then* $\pi(D|j,z) > 0$ .

The $\pi$ distributions (1), (2) and (3.1) to (3.3), but not (3.4), satisfy a stronger condition:

*Uniform conditioning property:* *If* $i,j \in D \subseteq C$ , *then*

$\pi(D|i,z) = \pi(D|j,z)$ .

A distribution with the uniform conditioning property can be written $\pi(D|i,z) = \phi(D,z)X_D(i)$ , where $X_D(i)$ equals one for $i \in D$ , and zero otherwise.

Consider a sample $n = 1,\dots,N$, with the alternative chosen on case n denoted by $i_n$, and $D_n$ denoting the choice set assigned to this case from the distribution $\pi(D|i_n,z_n)$ . Observations with an observed choice not in the assigned set of alternatives are assumed to be excluded from the sample. Write the multinomial logit model in the form

$$P(i|C,z,\theta) = e^{V_i(z,\theta)} \Big/ \sum_{j \in C} e^{V_j(z,\theta)} \quad , \tag{50}$$

where $V_i(z,\theta)$ is the strict utility of alternative i.

*THEOREM 2.* *If* $\pi(D|i,z)$ *satisfies the positive conditioning property and the choice model is multinomial logit, then maximisation of the modified likelihood function*

$$L_N = \frac{1}{N} \sum_{n=1}^{N} \log \left\{ e^{V_{i_n}(z_n,\theta) + \log \pi(D_n|i_n,z_n)} \Big/ \sum_{j \in D_n^K} \right.$$

$$\left. e^{V_j(z_n,\theta) + \log \pi(D_n|j,z_n)} \right\} \tag{51}$$

*yields, under normal regularity conditions, consistent estimates of* $\theta*$. *When* $\pi(D|i,z)$ *satisfies the uniform conditioning property, then (51) reduces to the standard likelihood function,*

$$L_N = \frac{1}{N} \sum_{n=1}^{N} \log \left\{ e^{V_{i_n}(z_n,\theta)} / \sum_{j \in D_n} e^{V_j(z_n,\theta)} \right\} \quad . \tag{52}$$

The theorem above assumes the assigned choice set for an observation may depend on the observed choice set and environment for the observation, but is independent of other observations. More generally, a set of observed choices may be used to define the assigned choice set for each observation. For example, a common procedure is to assign to all observations in a traffic analysis zone the set consisting of all the chosen alternatives observed for this zone. Assume there are N zones, with $K_n$ observations in zone n. If $K_n \to \infty$ for each n, then every alternative in C will eventually be chosen by some subject in a zone, and estimators maximising (52), with assigned sets equal to the set of observed choices for the zone, will have the same asymptotic properties as a maximum likelihood estimator for choice from the full set C. Thus, standard maximum likelihood estimation with assigned choice sets given by the set of chosen alternatives in a zone yields consistent estimates under normal regularity conditions and the usual sampling method where the number of observations in each zone becomes large when the overall sample size becomes large.

In the less common case where the number of zones N becomes large, but the number K of observations in each zone is fixed, the procedure above fails in general to yield consistent estimators. (I have benefited from discussions with Joel Horowitz on this problem.) Let $\lambda(i_n, z_n, \theta)$ denote the kernel of the 'likelihood' function for zone n, where $i_n = (i_{1n}, \ldots, i_{Kn})$ and $z_n = (z_{1n}, \ldots, z_{Kn})$ are observed in the zone. Define $D_n = D(i_n) = \{j | j = i_k$ for some $k = 1, \ldots, K\}$ and $J(D) = \{j | D = \cup_k \{j_k\}\}$. In the standard case,

$$\lambda(i_n, z_n, \theta) = \prod_{k=1}^{K} P(i_{kn} | D_n, z_{kn}, \theta)$$

$$= e^{\sum_k V_{i_{kn}}(z_{kn},\theta)} / \sum_{j \in D_n} e^{\sum_k V_{j_k}(z_{kn},\theta)} \quad .$$

Asymptotically, the likelihood function is the expectation in $z$ of terms of the form

$$\sum_{D \subseteq C} a(D) \sum_{i \in J(D)} \frac{e^{\sum_k V_{i_k}(z_k,\theta^*)}}{\sum_{j \in J(D)} e^{\sum_k V_{j_k}(z_k,\theta^*)}} \log \lambda(i,z,\theta) \quad , \tag{53}$$

where

$$a(D) = \prod_{k=1}^{K} P(D|C, z_k, \theta^*) \sum_{j \in J(D)} e^{\sum_k V_{j_k}(z_{kn}, \theta^*)}$$

Consistency requires that Equation (53) be maximised at $\theta = \theta^*$. When $J(D) \neq D^K$, the standard case fails to give this result. However, consistency can be attained by using a modified likelihood function with the kernel

$$\lambda(j_n, z_n, \theta) = e^{\sum_k V_{i_{kn}}(z_{kn}, \theta)} \Big/ \sum_{j \in J(D_n)} e^{\sum_k V_{j_k}(z_{kn}, \theta)} \quad .$$

To illustrate the impact of these results, consider a destination choice problem in which individuals face a CBD destination and a large number of suburban destinations. One is interested primarily in whether the CBD destination will be chosen. If an individual chooses the CBD destination, then he is assigned the choice set consisting of the CBD destination and one suburban destination chosen at random. (From the previous analysis, we may choose the suburban destination at random from the subset of suburban destinations chosen by some individual in the home zone of the case in question.) If an individual chooses a suburban destination, he is assigned a choice set consisting of this destination and the CBD destination. Assume J suburban destinations, with probability of selection $1/J$ for each in the case of a CBD choice. The $\pi$ distribution is then

$$\pi(\{j, CBD\}|CBD) = 1/J \qquad \text{for } j = 1, \ldots, J \; ,$$

$$\pi(\{j, CBD\}|j) = 1 \qquad \text{for } j = 1, \ldots, J \; , \qquad (54)$$

$$\pi\{D\}|j) = 0 \qquad \text{otherwise.}$$

This distribution satisfies the positive conditioning property (but not the uniform conditioning property), and hence consistent estimates can be obtained by maximising (51), which reduces to

$$\text{Max } \frac{1}{N} \sum_{n=1}^{N} \left[ V_{i_n}(z_n, \theta) - \log \left( e^{V_{CBD}(z_n, \theta) + \log J} + e^{V_{j_n}(z_n, \theta)} \right) \right] \qquad (55)$$

where $j_n$ is the suburban alternative chosen or assigned on observation n, and a term involving log J but independent of $\theta$ has been dropped. Alternately, if the model contains a CBD-specific dummy, then unweighted maximum likelihood gives consistent estimates of all parameters except the CBD-specific dummy, and gives a consistent estimate of the true CBD-specific dummy plus log J.

*Weighting and Estimation in Composite Samples*

Transport samples may be the result of a complex mixture of exogenous and choice-based sampling, or of the amalgamation of surveys conducted using various sampling procedures. The techniques of Manski and Lerman (23), and Manski and McFadden can be adapted to construct consistent estimators from these samples.

Consider first the problem of working with a composite survey, made up of subsamples collected by various procedures. Provided the subsamples are identified and the sampling procedure used for each is known, maximum likelihood estimation of parameters using the combined sample is straight-forward: the sample likelihood is the sum of the likelihoods of each of the subsamples, taking into account the sampling process used in each subsample. For example, the likelihood function for an exogenous stratified sample which is 'enriched' by a choice-based sample for minority modes is the sum of an exogenous likelihood function for the first subsample and a choice-based likelihood function for the second subsample. (Cosslett (9) has pointed out that the kernel of the composite sample likelihood will include the marginal distribution $p(z)$.) Maximisation of this composite likelihood function would require modification of most standard computer routines. An alternative consistent estimator which can be calculated using a maximum likelihood programme which allows weighting of the choice variable is the Manski-Lerman estimator (39), with $W(i) = 1$ for the exogenous subsample and $W(i) = Q(i)/H(i)$ for the choice-based subsample. Interestingly, the result that applying *unweighted* exogenous maximum likelihood estimation to an MNL model and pure choice-based sample produces inconsistency only in the alternative dummy coefficients does *not* carry over to the case of a composite sample when the exogenous subsample is stratified.

Next consider the problem of complex stratifications, such as would result from choice-based subsampling from a large exogenous stratified transport survey. The general theory of estimation from stratified samples of Manski and McFadden can be applied. In the example above, a consistent estimator would be (38), with $Q(i)$ defined to equal the marginal share of alternative i *in the exogenous stratified sample* rather than in the population.

*Non-maximum Likelihood Estimation Methods*

While maximum likelihood estimators have good asymptotic statistical properties under the conditions normally imposed in transport applications, their finite sample properties are largely unknown. There is some evidence from very limited Monte Carlo studies that maximum likelihood estimators will be unduly sensitive to observations with low calculated probabilities, and hence relatively non-robust with respect to errors in model specification or data measurement which could yield low calculated probabilities for some observed choices. These limited studies suggest that when data grouping is possible, Berkson-Theil estimators may be preferable to maximum likelihood estimators (15, p. 112). However, plausible grouping is rarely possible with transport data. An alternative approach is to develop more 'robust' estimators for individual observations. Manski and McFadden have investigated a class of such estimators, including non-linear least squares (NLLS), which satisfies (for exogenous samples)

$$\underset{\theta \in \Theta}{\text{Min}} \sum_{n=1}^{N} [S_{i_n} - P(i_n|z_n,\theta)]^2 \quad , \tag{56}$$

where $S_i$ is one if i is chosen and zero otherwise. This estimator is consistent, although not as efficient as maximum likelihood estimation, and appears in Monte Carlo studies to be less sensitive than maximum likelihood estimation to outliers caused by data measurement errors. Applications to transport data sets have not, however, resulted in significant differences between maximum likelihood and NLLS estimators.

## MODEL EVALUATION AND VALIDATION

*Model Evaluation*

The transport analyst usually has a number of alternative model specifications he considers to be *a priori* plausible, and wishes to determine empirically which alternative best fits the data. This calls for statistics which measure goodness of fit, and procedures which allow tests of hypothesised specifications.

General goodness-of-fit measures for discrete choice models which are now widely used are the log-likelihood function, the likelihood-ratio index, a multiple-correlation coefficient and a prediction-success index. The likelihood-ratio index $\rho^2$ is defined by the formula

$$\rho^2 = 1 - L/L_0 \quad , \tag{57}$$

where

$$L = \sum_{n=1}^{N} \sum_{i=1}^{J} S_{in} \log P(i|z_n,\theta) \tag{58}$$

is the log-likelihood function, with the $S_{in}$ equal to one if i is chosen, zero otherwise,

$$L_0 = \sum_{n=1}^{N} \sum_{i=1}^{J} S_{in} \log Q_i \quad , \tag{59}$$

and $Q_i$ equals the sample aggregate share of alternative i. This likelihood-ratio index is defined 'about aggregate shares', and measures the explanatory power of the model beyond that of a simple constant shares model. This index is preferable to a likelihood index 'about zero' reported by

some computer programs, which measures the power of the model
beyond that of an *equal* shares model.   A similar comment
applies to the multiple-correlation coefficient.

When the individual-choice model parameters are estimated
by non-linear least squares, an appropriate goodness-of-fit
measure is the sum of squared residuals.

$$SS = \sum_{n=1}^{N} \sum_{i=1}^{J} (S_{in} - R_n P_{in}(\hat{\theta}))^2 / R_n \quad , \tag{60}$$

where $R_n$ is the sum of $S_{in}$.   A transformation of this
statistic yields a *multiple correlation coefficient* of the
form familiar regression analysis,

$$R^2 = 1 - \frac{SS}{SS_0} \quad , \tag{61}$$

where

$$SS_0 = \sum_{n=1}^{N} \sum_{i=1}^{J} (S_{in} - R_n Q_i)^2 / R_n \tag{62}$$

with $Q_i$ the sample aggregate share of mode i as before.*

A third method of assessing the fit of an estimated model
is to examine the proportion of successful predictions, by
alternative and overall.   A *success table* can be defined as
illustrated in Table 13.3, with the entry $N_{ij}$ in row i and
column j giving the number of individuals who are observed to
choose i and predicted to choose j.**   Column sums give

---

* While the $R^2$ index is a more familiar concept to planners
  who are experienced in ordinary regression analysis, it is
  not as well behaved a statistic as the $\rho^2$ measure, for
  maximum-likelihood estimation.   Those unfamiliar with the
  $\rho^2$ index should be forewarned that its values tend to be
  considerably lower than those of the $R^2$ index and should
  not be judged by the standards for a 'good fit' in ordinary
  regression analysis.   For example, values of 0.2 to 0.4 for
  $\rho^2$ represent an excellent fit.
**The formula for $N_{ij}$ is

$$N_{ij} = \sum_{n=1}^{N} S_{in} P_{jn} \quad .$$

An alternative prediction method is to forecast that the
alternative with the highest probability will be chosen.
A dot subscript indicates summation over the corresponding index, e.g.

$$N_{i.} = \sum_{j} N_{ij} \quad .$$

predicted shares for the sample; row sums give observed shares. The proportion of alternatives successfully predicted, $N_{ii}/N_i$, indicates that fraction of individuals expected to choose an alternative who do in fact choose that alternative. An overall proportion successfully predicted, $(N_{11} + \ldots + N_{JJ})/N$, can also be calculated.

| | | Predicted Choice 1 | 2 | ... | J | Observed Count | Observed Share |
|---|---|---|---|---|---|---|---|
| Observed Choice | 1 | $N_{11}$ | $N_{12}$ | | $N_{1J}$ | $N_{1\cdot}$ | $N_{1\cdot}/N_{\cdot\cdot}$ |
| | 2 | $N_{21}$ | $N_{22}$ | | $N_{2J}$ | $N_{2\cdot}$ | $N_{2\cdot}/N_{\cdot\cdot}$ |
| | $\vdots$ | | | | | | |
| | J | $N_{J1}$ | $N_{J2}$ | | $N_{JJ}$ | $N_{J\cdot}$ | $N_{J\cdot}/N_{\cdot\cdot}$ |
| Predicted Count | | $N_{\cdot1}$ | $N_{\cdot2}$ | | $N_{\cdot J}$ | $N_{\cdot\cdot}$ | 1 |
| Predicted Share | | $\dfrac{N_{\cdot1}}{N_{\cdot\cdot}}$ | $\dfrac{N_{\cdot2}}{N_{\cdot\cdot}}$ | | $\dfrac{N_{\cdot J}}{N_{\cdot\cdot}}$ | 1 | |
| Proportion Successfully Predicted | | $\dfrac{N_{11}}{N_{\cdot1}}$ | $\dfrac{N_{22}}{N_{\cdot2}}$ | | $\dfrac{N_{JJ}}{N_{\cdot J}}$ | $\dfrac{N_{11}+\ldots+N_{JJ}}{N_{\cdot\cdot}}$ | |
| Success Index | | $\dfrac{N_{11}}{N_{\cdot1}} - \dfrac{N_{\cdot1}}{N_{\cdot\cdot}}$ | $\dfrac{N_{22}}{N_{\cdot2}} - \dfrac{N_{\cdot2}}{N_{\cdot\cdot}}$ | | $\dfrac{N_{JJ}}{N_{\cdot J}} - \dfrac{N_{\cdot J}}{N_{\cdot\cdot}}$ | $\sum\limits_{i=1}^{J} [\dfrac{N_{ii}}{N_{\cdot\cdot}} - (\dfrac{N_{\cdot i}}{N_{\cdot\cdot}})^2]$ | |
| Proportional Error in Predicted Share | | $\dfrac{N_{\cdot1}-N_{1\cdot}}{N_{\cdot\cdot}}$ | $\dfrac{N_{\cdot2}-N_{2\cdot}}{N_{\cdot\cdot}}$ | | $\dfrac{N_{\cdot J}-N_{J\cdot}}{N_{\cdot\cdot}}$ | | |

TABLE 13.3  A Prediction Success Table

Because the proportion successfully predicted for an alternative varies with the aggregate share of that alternative, a better measure of goodness of fit is the prediction-success index,

$$\sigma_i = \frac{N_{ii}}{N_{\cdot i}} - \frac{N_{\cdot i}}{N_{\cdot\cdot}} \quad , \tag{63}$$

where $N_{\cdot i}/N_{\cdot\cdot}$ is the proportion which would be successfully predicted if the choice probabilities for each sampled individual were assumed to equal the observed aggregate

shares.*  This index will usually be non-negative, with a
maximum value of $1 - N_{\cdot i}/N_{\cdot \cdot}$ .  If an index normally lying
between zero and one is desired, Equation (63) can be
normalised by $1 - N_{\cdot i}/N_{\cdot \cdot}$ .

An overall prediction success index is

$$\sigma = \sum_{i=1}^{J} \frac{N_{\cdot i}}{N_{\cdot \cdot}} \sigma_i = \sum_{i=1}^{J} \left( \frac{N_{ii}}{N_{\cdot \cdot}} - \left(\frac{N_{\cdot i}}{N_{\cdot \cdot}}\right)^2 \right) \quad . \tag{64}$$

Again, this index will usually be non-negative, with a
maximum value of

$$1 - \sum_{i=1}^{J} \left(\frac{N_{\cdot i}}{N_{\cdot \cdot}}\right)^2 \quad ,$$

and can be normalised to have a maximum value of one if
desired.

In tests of model specification, one is often concerned
with questions such as whether certain variables enter the
determination of choice, and whether certain coefficients are
equal.  For example, the question of whether in-vehicle
travel time is generic, or homogeneous-effect, can be formulated as the hypothesis that the coefficients of alternative-
specific travel times are all equal.  Such problems, where
the null hypothesis is a subset of a specified universe of
alternatives, can be tested conveniently using likelihood
ratios, as described in Theil ([40] p. 396), and McFadden
[26].  Specification tests which are less easily performed
using classical statistical methods are those in which the
model corresponding to the universe of alternatives cannot be
specified or estimated.  Examples are the question of which
of two alternative measures of travel time better explain
mode choice, and tests of a particular model specification
such as MNL against mutually exclusive alternatives such as
MNP.  Methods of statistical decision theory can be applied
to some of these problems; an exposition is beyond the scope
of this chapter.

---

*In a model with alternative-specific dummies and the
 estimation data set, estimation of parameters imposes the
 condition $N_{i \cdot} = N_{\cdot i}$.  If one predicted the choice probabilities for each individual to equal aggregate shares,
 then $N_{\cdot i}/N_{\cdot \cdot}$ would be the proportion successfully predicted
 to choose i.  This represents a 'chance' prediction rate for
 a model in which no variables other than alternative-
 specific dummies enter.  Thus, $\sigma_i$ measures the net
 contribution to prediction success of variables other than
 the alternative-specific dummies.

*Diagnostic Tests for the MNL Model*

The MNL model has significant advantages over most alternative choice models in terms of simplicity and computational
efficiency, and its independence from irrelevant alternatives
(IIA) property greatly facilitates estimation and forecasting.
On the other hand, the IIA restriction may be invalid in some
applications, resulting in erroneous forecasts.  Hence, the
validity of the IIA property should be tested in each
application.  McFadden, Tye and Train* have developed a
series of diagnostic tests for this property; the major
findings are summarised in Chapter 1 of the present volume.
One is a test of the MNL model against a 'universal' alternative, approximated by an MNL-like form in which attributes of
all alternatives can enter the 'utility' of each alternative
- this is the 'universal' logit model.  A second test is
based on the implications of the IIA property that the model
can be estimated consistently from a random sample of the set
of all available alternatives, as discussed earlier.  A third
class of tests examines residuals from the fitted MNL model,
i.e. the differences of indicators of observed choices and
the estimated probabilities of these choices.  Under the
hypothesis that the MNL specification is correct, these
residuals will have specific mean, variance and correlation
properties which can be utilised in statistical tests.


AGGREGATION AND FORECASTING

*Aggregate Forecasts*

An important use of individual-choice models is in policy
analysis of the impacts of alternative transport plans on
operating strategies.  Evaluation of these impacts usually
requires forecasts of the behaviour of the aggregate population, or of specific market segments.  Given an estimated
choice model $P(i|z,\theta)$ , the aggregate share of alternative i
satisfies

$$Q(i) = \int_Z P(i|z,\theta)p(z)dz \tag{65}$$

where $p(z)$ is the probability distribution of the explanatory
variables in the population.  For a market segment, this
formula applies, with $p(z)$ interpreted as the distribution of
explanatory variables in the segment.

A variety of methods have been proposed for the evaluation
of (65) in applications; the most practical and flexible
appears to be a 'Monte Carlo' procedure in which $Q(i)$ is
approximated by

---

*McFadden, D., Tye, W. and Train, K., 'Diagnostic tests for
the independence from irrelevant alternatives property of
the multinomial logit model', *Urban Travel Demand Forecasting Project Working Paper No. 7616*, Institute of
Transportation Studies, University of California, Berkeley
(1976).

$$Q(i) = \frac{1}{N} \sum_{i=1}^{N} P(i|z_n, \theta) \qquad , \qquad (66)$$

where $\{z_n\}$ is a sample drawn randomly from $p(z)$. The points $z_n$ may be from a representative sample of the population, or may themselves be synthesised from incomplete data sources, as described below. The formula (66) can be modified to accommodate non-uniform sampling weights. For computational purposes, it is often useful to group sample points into strata with homogeneous choice probabilities. Discussions of this and alternative aggregation procedures and their properties can be found in Koppelman [22] and McFadden [34].

*Aggregation by the Clark Method*

In general, direct evaluation of (65) requires numerical integration over the set Z, which may be of relatively high dimension. This may be impractical even if the choice probabilities are relatively easy to compute, and the problem is compounded if evaluation of the choice probabilities is expensive.

An approach which eliminates the intermediate calculation of choice probabilities has been suggested in a specific context by McFadden and Reid [36], and generalised by Manski and Daganzo. Suppose individuals maximise utility, with utility functions $u_i = \beta' z_i + \varepsilon_i$ for alternative i. Given a probability distribution $p(z)$ for $z = (z_1, \ldots, z_J)$ and a distribution of $(\varepsilon_1, \ldots, \varepsilon_J)$, one can construct the probability distribution of $(u_1, \ldots, u_J)$ resulting from *joint* variation of z and the $\varepsilon_i$. The distribution of $(u_1, \ldots, u_J)$ is obtained as a multivariate *convolution* of the probability densities of z and of $(\varepsilon_1, \ldots, \varepsilon_J)$. For some probability distributions, such as the multivariate normal case considered below, the distribution of the convolution is known. More generally, if $\phi(t_1, \ldots, t_{JK})$ is the characteristic function of the distribution of $z = (z_{11}, \ldots, z_{K1}, \ldots, z_{1J}, \ldots, z_{KJ})$, and $\psi(t_1, \ldots, t_J)$ is the characteristic function of the distribution of $(\varepsilon_1, \ldots, \varepsilon_J)$, with z and $(\varepsilon_1, \ldots, \varepsilon_J)$ assumed independent, then $(u_1, \ldots, u_J)$ with $u_i = \beta' z_i + \varepsilon_i$ has the characteristic function $\gamma(t_1, \ldots, t_J) = \psi(t_1, \ldots, t_J)\phi(t_1\beta_1, \ldots, t_1\beta_K, t_2\beta_1, \ldots, t_J\beta_1, \ldots, t_J\beta_K)$, or more compactly, $\gamma(t) = \psi(t)\phi(t \otimes \beta')$. The density of $(u_1, \ldots, u_J)$ can then be obtained from the inversion formula $h(u_1, \ldots, u_J) = (2\pi)^{-J}\int e^{-itu}\gamma(t)dt$. Using this expression in (67), one could carry out the computation of $Q(i)$ with a numerical integration of dimension at most $2J$, for an extremely broad class of distributions of z and $\varepsilon$. Application of approximation methods to the combined integral may then allow rapid computation of aggregate probabilities, even for complex choice models. Let $H(u_1, \ldots, u_J)$ denote the cumulative distribution of $(u_1, \ldots, u_J)$ and $H_i$ its derivative with respect to $u_i$,

$$Q(i) = \text{Prob } \{u_i \geq u_j \text{ for } j = 1, \ldots, J\} \tag{67}$$

$$= \int_{u=-\infty}^{+\infty} H_i(u, u, \ldots, u) du \quad.$$

Evaluation of this integral requires only a single numerical integration when $H_i$ can be obtained analytically, and at most a J-dimensional numerical integration is required to compute $Q(i)$ when the density of H is analytic.

The procedure outlined above can be applied with particular convenience to the case where z and $\varepsilon$ are assumed multivariate normal. This assumption, which yields the MNP model of individual choice, implies $(u_1, \ldots, u_J)$ is multivariate normal, with mean of $u_i$ equal to $\beta'\bar{z}_i$ where $\bar{z}$ is the mean of z, and covariances $\omega_{ij} = \beta'\Sigma_{ij}\beta + \sigma_{ij}$, where $\sigma_{ij} = \text{cov } (\varepsilon_i, \varepsilon_j)$ and $\Sigma_{ij} = \text{cov } (z_i, z_j)$. Then, $Q(1)$ can be obtained from a formula analogous to (30) for this multivariate normal distribution. The Clark approximation method discussed above then permits rapid computation of approximate aggregate shares. Further, application of the Clark formulae to the computation of analytic derivatives of $Q(i)$ with respect to $\bar{z}$, in a manner analogous to that described in (34), would allow rapid approximation of aggregate elasticities.

## The Distribution of Explanatory Variables

The computation of aggregate shares or elasticities requires knowledge of the distribution of the explanatory variables in the population, or in a market segment of the population. A random sample from the population of sufficient size, say from a major population survey, can meet this data requirement. However, it is often difficult to obtain current data of this type. Forecasts at future dates present a further problem, since the distribution of explanatory variables used in the forecasts should take into account shifts in explanatory variables over time.

Cosslett, Duguay, Jung and McFadden [10] have proposed a method of synthesising the distribution of explanatory variables at any forecast date, integrating available data sources plus information on trends. The method is particularly useful when current random survey data is unavailable, and can be applied in most urban areas using only US census data. The method utilises a classical statistical procedure for completing contingency tables, called *iterative proportional fitting*, due to Deming and Stephan [14]. This procedure allows the integration of marginal information from US census tract statistics, Public Use Samples, and the Urban Transportation Planning Package. Parametric models of some variable interactions, simple trend models for shifts in the

distribution over time, and exogenous forecasts for some
explanatory variables allow projection of the synthesised
distribution to future dates. Sampling from the constructed
distribution yields a synthesised random sample for the urban
area at the forecast date.


*Calculus for Demand Elasticities*

Demand elasticities encapsulate considerable information on
transport demand response, and are valuable tools for policy
analysis. For the multinomial logit (MNL) model, the
elasticities can be expressed in relatively simple formulae.
However, great care must be taken to avoid mechanical use of
these formulae, and to see that the computation performed
corresponds to the policy question asked. The first rules
set out below hold for any choice model. We use the notation
$P_i$ for the choice probability for alternative i, and $z_k^i$ for
the $k^{th}$ component of the vector of attributes of alternative
i.

   Rule 1 - aggregation over market segments. *Aggregate
elasticity equals the sum of segment elasticities, weighted
by segment shares of the market.* If $P_i^\ell$ is the choice
probability for segment $\ell$ , $\overline{P}_i$ is the aggregate choice
probability, and $q_\ell$ is the proportion of the population in
segment $\ell$ , then

$$\left(\frac{z_k^j}{\overline{P}_i} \quad \frac{\partial \overline{P}_i}{\partial z_k^j}\right) = \sum_\ell q_\ell \left(\frac{z_k^j}{P_i^\ell} \quad \frac{\partial P_i^\ell}{\partial z_k^j}\right) \quad . \tag{68}$$

(Note: This is a relevant elasticity only if a policy will
result in equal percentage changes for each market segment.)

   Rule 2 - aggregation over alternatives. *Elasticity for a
compound alternative equals the sum of component alternative
elasticities, weighted by the component shares of the
compound alternative.* Let $\overline{P} = \sum_i P_i$ be the choice probability
for a compound alternative (e.g. 'all transit'). Then,

$$\left(\frac{z_k^j}{\overline{P}} \quad \frac{\partial \overline{P}}{\partial z_k^j}\right) = \sum_i \left(\frac{P_i}{\overline{P}}\right)\left(\frac{z_k^j}{P_i} \quad \frac{\partial P_i}{\partial z_k^j}\right) \quad . \tag{69}$$

(Note: This is a relevant elasticity only if a policy will
result in equal percentage changes for each component
alternative.)

   Rule 3 - component effect. *The elasticity with respect to a
component of a variable equals the elasticity with respect to
the variable times the component's share in the variable.*
Suppose $z_k^j = w_k^j + y_k^j$. Then

$$\frac{y_k^j}{P_i} \frac{\partial P_i}{\partial y_k^j} = \left(\frac{y_k^j}{z_k^j}\right) \left(\frac{z_k^j}{P_i} \frac{\partial P_i}{\partial z_k^j}\right) \quad . \tag{70}$$

(Note: It is particularly important in policy analysis to look only at components influenced by a policy, such as the transit fare component of total trip cost or the bus on-vehicle time component of a multi-mode trip total on-vehicle time.)

Rule 4 - multiple effect. *The elasticity with respect to a policy that causes an equal percentage change in several variables equals the sum of the elasticities with respect to each variable.* Suppose a policy changes $z_k^j$ to $z_k^j(1 + t)$ for several j. Then,

$$\frac{1}{P_i} \frac{\partial P_i}{\partial t} = \sum_j \left(\frac{z_k^j}{P_i} \frac{\partial P_i}{\partial z_k^j}\right) \quad . \tag{71}$$

Since t is a proportional change, the left-hand side of this equation is in elasticity form. Alternately, suppose $z_k^j = w_k^j + y_k$ for several j, and a policy changes $y_k$. Then,

$$\frac{y_k}{P_i} \frac{\partial P_i}{\partial y_k} = \sum_j \left(\frac{y_k}{z_k^j}\right) \left(\frac{z_k^j}{P_i} \frac{\partial P_i}{\partial z_k^j}\right) \quad . \tag{72}$$

This formula is obtainable from a combination of Rules 3 and 4.

Often, combinations of these rules will be required to obtain the most relevant elasticity for a policy calculation. For example, suppose transit alternatives are disaggregated by access mode, and the impact of a fare increase is to be assessed. The answer is given by combining Rules 2, 3 and 4 to obtain the formula

$$\begin{pmatrix} \text{Elasticity} \\ \text{of transit} \\ \text{patronage} \\ \text{with respect} \\ \text{to fare} \end{pmatrix} = \sum_i \sum_j \begin{pmatrix} \text{Patronage on} \\ \text{mode } i \text{ as a} \\ \text{proportion of} \\ \text{total transit} \\ \text{patronage} \end{pmatrix} \cdot \begin{pmatrix} \text{Fare on} \\ \text{mode } j \text{ as a} \\ \text{proportion} \\ \text{of total} \\ \text{cost on } j \end{pmatrix} \cdot \begin{pmatrix} \text{Elasticity of} \\ \text{mode } i \text{ patronage} \\ \text{with respect to} \\ \text{total cost on} \\ \text{mode } j \end{pmatrix} \tag{73}$$

where i and j are summed over the transit access modes.

Consider elasticities for the sequential model defined earlier. Taking $\theta = \lambda = 1$, these elasticities will also hold for the joint MNL model.

$$\frac{x_{ncbk}}{P_{m|da}} \frac{\partial P_{m|da}}{\partial x_{ncbk}} = \alpha_k z_{ndak} (\delta_{mn} - P_{n|da}) \delta_{cd} \delta_{ab} \tag{74}$$

where $x_{ndak}$ is component k of $x_{nda}$ and $\delta_{mn} = 1$ if m = n, 0 otherwise, etc.

$$\frac{x_{ncbk}}{P_{d|a}} \frac{\partial P_{d|a}}{\partial x_{ncbk}} = \theta \alpha_k x_{ncak} P_{n|ca} (\delta_{cd} - P_{c|a}) \delta_{ab} \quad , \tag{75}$$

$$\frac{y_{cbk}}{P_{d|a}} \frac{\partial P_{d|a}}{\partial y_{cbk}} = \beta_k y_{cak} (\delta_{cd} - P_{c|a}) \delta_{ab} \quad , \tag{76}$$

where $y_{cak}$ is component k of $y_{ca}$.

$$\frac{x_{mdbk}}{P_a} \frac{\partial P_a}{\partial x_{mdbk}} = \lambda \theta \alpha_k x_{mdbk} P_{m|db} P_{d|b} (\delta_{ab} - P_b) \quad ; \tag{77}$$

$$\frac{y_{dbk}}{P_a} \frac{\partial P_a}{\partial y_{dbk}} = \lambda \beta_k y_{dbk} P_{d|b} (\delta_{ab} - P_b) \quad ; \tag{78}$$

$$\frac{z_{bk}}{P_a} \frac{\partial P_a}{\partial z_{bk}} = \gamma z_{bk} (\delta_{ab} - P_b) \quad . \tag{79}$$

Other elasticities are readily derived from these formulae using Rules 1 - 4 and the definitions of conditional and marginal probabilities. For example, $P_{da} = P_{d|a} P_a$ implies

$$\frac{x_{mca}}{P_{da}} \frac{\partial P_{da}}{\partial x_{mca}} = \frac{x_{mca}}{P_{d|a}} \frac{\partial P_{d|a}}{\partial x_{mca}} + \frac{x_{mca}}{P_a} \frac{\partial P_a}{\partial x_{mca}} \quad , \tag{80}$$

and the preceding formulae can be substituted in the right-hand side of this expression. Another example is the joint probability $P_{mda} = P_{m|da} P_{d|a} P_a$ , which satisfies

$$\frac{x_{ncbk}}{P_{mda}} \frac{\partial P_{mda}}{\partial x_{ncbk}} = \alpha_k x_{ncbk} \left\{ \delta_{mn} \delta_{cd} \delta_{ab} + \delta_{cd} \delta_{ab} (\theta - 1) P_{n|cb} \right.$$

$$\left. + \theta \delta_{ab} (\lambda - 1) P_{n|cb} P_{c|b} - \theta \lambda P_{ncb} \right\} \quad . \tag{81}$$

As a check, one sees that for $\theta = \lambda = 1$ , this reduces to the conventional MNL elasticity formula.


CONCLUSION

This chapter has surveyed selected recent developments in quantitative methods for travel-demand analysis. This subject has developed rapidly in the past few years on a wide spectrum of topics. As a result, the transport analyst now has available a greatly expanded 'bag of tools' with which to address policy problems. Experience makes it clear that quantitative methods are not a panacea for solving the problems of transport policy analysis. On the other hand, the use of techniques of modern mathematics and statistics relax one of the constraints which has limited the analyst in attacking a full range of transport issues.  A review of the state of the art of quantitative methods in transport demand forecasting suggests that the task of establishing a firm analytic and statistical foundation for the subject has just begun.

## REFERENCES

1. Amemiya, T. 'Specification and estimation of a multi-nomial logit model', *Institute of Mathematical Studies in the Social Sciences Technical Report 211*, Stanford University, Stanford, California (1976).
2. Anderson, T.W. (1958) *An Introduction to Multivariate Statistical Analysis,* Wiley, New York.
3. Ben-Akiva, M., 'Structure of passenger travel demand models', *Transportation Research Record No. 526* (1973).
4. Ben-Akiva, M. and Lerman, S., 'Disaggregate travel and mobility choice models and measures of accessibility', (Chapter 30 of the present book).
5. Bishop, Y., Fienberg, S. and Holland, P. (1975) *Discrete Multivariate Analysis*, Massachusetts Institute of Technology Press, Cambridge, Massachusetts.
6. Bock, R.D. and Jones, L.V. (1968) *The Measurement and Prediction of Judgement and Choice*, Holden-Day, San Francisco.
7. Cardell, S. (1975), personal communication.
8. Clark, C., 'The greatest of a finite set of random variables', *Operations Research*, 9, pp. 145-62 (1961).
9. Cosslett, S., 'Efficient estimation of choice probabilities from choice-based samples', *Alfred P. Sloan Foundation Workshop in Transportation Economics*, Department of Economics, University of California, Berkeley (1977).
10. Cosslett, S., Duguay, G.E., Jung, W.S. and McFadden, D., 'Synthesis of household transportation survey data: the SYNSAM methodology', *Urban Travel Demand Forecasting Project, Working Paper No. 7705*, Institute of Transportation Studies, University of California, Berkeley (1977).
11. Cramer, H. (1946), *Mathematical Methods of Statistics*, Princeton University Press, Princeton, New Jersey.
12. Daly, A. and Zachary, S., 'Improved multiple choice models', in *Determinants of Travel Choice*, D.A. Hensher and M.Q. Dalvi (eds), Teakfield, Farnborough, England (1978).
13. Daganzo, D., Bouthelier, F. and Sheffi, Y., 'An efficient approach to estimate and predict with multinomial probit models', Department of Civil Engineering, Massachusetts Institute of Technology (unpublished) (1976).
14. Deming, W. and Stephan, F., 'On a least square adjustment of a sampled frequency table when the expected marginal totals are known', *Annals of Mathematical Statistics*, 11, pp. 427-44 (1940).
15. Domencich, R. and McFadden, D. (1975) *Urban Travel Demand: A Behavioral Analysis*, North-Holland, Amsterdam.
16. Goodman, L. and Kruskal, W., 'Measures of association for cross-classification', *Journal of the American Statistical Association*, 49, pp. 732-64 (1954).
17. Haberman, S. (1974) *The Analysis of Frequency Data*, University of Chicago Press, Chicago, Illinois.
18. Haberman, S., 'Log-linear fit for contingency tables', *Applied Statistics*, 21, pp. 218-25 (1974).

19. Harris, A. and Tanner, J., 'Transport demand models based on personal characteristics', *Transport and Road Research Laboratory Supplementary Report 65 UC* (1974).

20. Hausman, J.A. and Wise, D.A., 'A conditional probit model for qualitative choice: discrete decisions recognizing interdependence and heterogeneous preferences', *Econometrica*, 46(2), pp. 403-26 (1978).

21. Kendall, D. and Stuart, K. (1976) *Advanced Theory of Statistics*, Vol. 3, Hafner, New York.

22. Koppelman, F. (1975) 'The Structure of Aggregated Prediction Models', Ph.D. Dissertation, Department of Civil Engineering, Northwestern University, Evanston, Illinois.

23. Manski, C. and Lerman, S., 'The estimation of choice probabilities from choice-based samples', *Econometrica*, 46(1) (1978).

24. Manski, C., 'Maximum score estimation of the stochastic utility model of choice', *Journal of Econometrics*, 3, pp. 205-28 (1975).

25. Manski, C., 'Multinomial probit model', internal memorandum, Cambridge Systematics, Inc., Cambridge, Massachusetts (1976).

26. McFadden, D., 'Conditional logit analysis of qualitative choice behavior', in *Frontiers in Econometrics*, P. Zarembka (ed.), Academic Press, New York (1974).

27. McFadden, D., 'On independence, structure, and simultaneity in transportation demand analysis', *Urban Travel Demand Forecasting Project, Working Paper No. 7511A*, Institute of Transportation Studies, University of California, Berkeley (1975).

28. McFadden, D., 'Economic applications of psychological choice models', *Urban Travel Demand Forecasting Project, Working Paper No. 7519*, Institute of Transportation Studies, University of California, Berkeley (1975).

29. McFadden, D., 'The revealed preferences of a government bureaucracy: evidence', *Bell Journal of Economics*, 7, pp. 55-72 (1976).

30. McFadden, D., 'Quantal choice analysis: a survey', *Annals of Economic and Social Measurement*, 5 (1976).

31. McFadden, D., 'A comment on discriminant "versus" logit analysis', *Annals of Economic and Social Measurement*, 5 (1976).

32. McFadden, D., 'The theory and practice of disaggregate demand forecasting for various modes of urban transportation', *Urban Travel Demand Forecasting Project, Working Paper 7623*, Institute of Transportation Studies, University of California, Berkeley (1976).

33. McFadden, D., 'Properties of the multinomial logit (MNL) model', *Urban Travel Demand Forecasting Project, Working Paper No. 7617*, Institute of Transportation Studies, University of California, Berkeley (1976).

34. McFadden, D., 'The mathematical theory of demand models', in *Behavioral Travel Demand Models*, P.R. Stopher and A.H. Meyburg (eds), Lexington Books, Lexington (1976).

35. McFadden, D., 'Modeling the choice of residential location', in *Spatial Interaction Theory and Planning Models*, A. Karlqvist, L. Lundquist, F. Snickers, J. Weibull (eds), North-Holland, Amsterdam (1978).

36. McFadden, D. and Reid, F., 'Aggregate travel demand forecasting from disaggregated behavioral models', *Transportation Research Board Record No. 534* (1975).

37. McLynn, J. (1973) *A Technical Note on a Class of Fully Competitive Modal Choice Models*, DTM Corp., Bethesda, Maryland (1973).

38. Stopher, P.R. and Meyburg, A.H. (1975) *Urban Transportation Planning and Modelling*, Lexington Books, Lexington.

39. Stopher, P.R. and Meyburg, A.H. (eds) (1976) *Behavioral Travel Demand Models*, Lexington Books, Lexington.

40. Theil, H. (1971) *Principles of Econometrics*, Wiley, New York.

41. Thurstone, L., 'A law of comparative judgment', *Psychological Review*, 34, pp. 273-86 (1927).

42. Wiliams, H.C.L., 'On the formation of travel demand models and economic evaluation measures of user benefit', *Environment and Planning A*, 9, pp. 285-344 (1977).

Chapter 14

SOME DEVELOPMENTS IN TRANSPORT DEMAND MODELLING

A.J. Daly

INTRODUCTION

In recent years more and more attention has been given to
seeking a micro-economic interpretation of travellers'
behaviour that will permit transport demand models to be
built and rigorously examined.  While this economic approach
is not the only interpretation of human behaviour from which
quantitative models of transport demand can be derived, it
has to date proved internally consistent, has yielded much
insight on model structures, and still offers a framework for
investigations beyond the current frontiers of research.  In
this chapter we discuss some of the results that have been
obtained from this approach, particularly concentrating on
those that relate to model structure.  We shall mainly report
work done in Great Britain, where the institutional framework
stresses the importance of formal evaluations using consumer
surplus measures (although other considerations are also
important).  The need to apply the economic concept of
consumer surplus tends to predispose research workers to use
an economic approach to demand models in all aspects of
transport.

    In accordance with economic theory, the traveller is
supposed to choose the journey that maximises his own benefit
or utility.  The common observation that different people
with apparently similar travel choices sometimes make different journeys means that it is necessary to allow for
individual variations of utility around the best estimate
available to the analyst.  In the first section of this
chapter we study the ways in which the structure of this
variation affects the structure of demand models and, most
particularly, the consumer surplus measures derived from
them.  In particular we find that only one class of models of
utility variation, called by Harris and Tanner (9) 'personal
difference models', permit the usual consumer surplus
measures to be applied with consistency.

    In the second section of the chapter we develop the theory
of this class of personal difference models.  For these
models we can derive a set of restrictions, including a type
of Hotelling condition, on the possible forms of demand
function.  Some examples are given of applied models both
that satisfy and that fail to satisfy these restrictions.
These conditions also make clear the arbitrary nature of any
zero of utility.

319

The third section discusses some of the practical difficulties of formulating demand models.  In the fourth section we give some examples of a particular class of demand models, which we call 'structured logit models', that are consistent with personal difference utility variation.  The intention of this section is to indicate that the scope for the model definition within even the restricted class of personal difference utility models goes well beyond currently applied demand models.  Many of the criticisms of existing demand model structures can thus be met while retaining the simplicity of the personal difference structure.  For example, the problems of defining choice sets can be greatly reduced by allowing covariance of the utilities of different choices.

In the final section, we summarise the main points of the chapter and discuss their relevance to the main streams of transport demand model theory.


MODELS OF UTILITY VARIATION

As indicated in our introductory remarks above, we intend to concentrate in this chapter on models where travel choice is represented as utility maximisation.  That is, the choice is modelled as

Choose j such that $U_j \geq U_k$ for all k, $1 \leq k \leq r$ (1)

where $U_j$ is the utility associated with choice j and r is the number of possible choices.  ('Possible' is to be interpreted as excluding choices in breach of time and money budgetary constraints on the individual as well as technical impossibilities.)  A few remarks on this representation are in order before we turn to more detailed analysis.

First, we present the choice as one of utility maximisation rather than cost minimisation.  Obviously this makes no difference to the analysis other than the reversal of signs, but the concentration on utility rather than cost helps to emphasise the arbitrariness of the scale, with respect to both multiplicative and additive constants.  For example, when 'costs' are used in choice modelling, it is a common error in application to interpret the 'cost' as a representation of the total disbenefit of a journey, as, for example, in reference to 'generalised cost elasticity'; the use of 'utility' emphasises the impossibility of measuring total disbenefit.  Further, in travel decisions more general than mode or route choice, the notion of positive utility derived from access to a destination is sometimes helpful, and indeed is the most obvious formulation when the generation of trips is included.  An attribution of 'zero utility' to the choice of not making a journey is however merely an arbitrary construct.

More generally, however, we assume without apology a cardinal utility measure.  This assumption seems inevitable in attempting to derive functions relating travellers' preferences to measured attributes of the journeys available to them.  But it is essential to realise that when such functions are calibrated by reference to revealed preferences, no absolute measure is ever available of the *degree* of preference.  Surveys in which travellers have been asked for a 'transfer price' (a change in their selected or alternative journey that would make them indifferent between the two) merely create a hypothetical additional alternative which is ranked by the respondent relative to his real choices.  The functions we obtain that relate choice to measured journey attributes are not utility functions, but somewhat indirect proxies, a fact of particular importance when extrapolations are considered.  In fact, it is not necessary for the predictive theory developed that the functions $U_j$ be utilities in the strict economic sense.  We shall, however, ignore this distinction, since our main concern is with structure rather than justification.

The choice representation as formulated is deterministic (except when two journeys have exactly equal utility).  That is, individuals are represented as certainly choosing the journey that they see as best.  The analyst, however, in attempting to model this choice, is hampered by his ignorance of the individual's exact perception of the attributes of the journeys available, the valuation of those attributes, and the way they are compared with the attributes of other journeys.  The analyst can only represent the individual's actual utility as being drawn from some distribution around the best estimate available.

In order to derive a model of choice from (1) that can be used to predict behaviour, it is necessary to postulate some relationship between the behavioural determinants $U_j$ (the actual utilities) and the analyst's best representatives of those utilities, which we shall denote by $z_j$ and call the 'measured component' of utility.  The simplest case, on which we shall subsequently be concentrating, is the 'personal difference' model:

$$U_j = z_j + X_j \qquad (1 \leqslant j \leqslant r) \tag{2}$$

where $(X_j, 1 \leqslant j \leqslant r)$ is a family of random variables with a *single* joint distribution from which all individuals are assumed to sample.  (Throughout, we use upper case symbols for random variables, lower case for non-random variables.) The vital aspect of this specification is that the X variables for each individual are drawn from the same multivariate distribution.  In particular, the parameters of the distribution are assumed not to depend on z.  This assumption seems *prima facie* implausible; for example, the variance of utility might be expected to increase with journey length, which will generally be represented in some way in the measured z variables.  But many models can be put into this form by suitable transformations.  For example, Harris and Tanner [9] give a 'personal multiplier' model with $U_j = z_j X$ where the X variables are drawn from a single multivariate distribution as before.  They then discuss the different properties that this model structure gives relative to (2).  But this model can be transformed to (2) by a simple change

to logarithmic scales, since the choice of maximum U is
equivalent to the choice of maximum log U, and the z and X
variables are defined over (so far) arbitrary scales.

In other cases, however, this reduction cannot be made.
Daly and Zachary (4) use a binary choice (r = 2) model
defined by

$$U_1 = 0, \quad U_2 = Y_1 t_1 + Y_2 t_2 + \ldots Y_n t_n$$

where $(t_1, \ldots, t_n)$ are the differences in the n measured
characteristics between the two choices and $(Y_1, \ldots, Y_n)$ are
the values attached to these characteristics by the
traveller.  In this study (4), the Y variables were assumed
to take a multivariate normal distribution with mean
$(y_1, \ldots, y_n)$ and covariance matrix V.  This model can be
written in the form

$$U_1 = 0, \quad U_2 = z_2 + X_2$$

with $z_2 = y \ t'$ (using vector notation) and the distribution
of $X_2$ is normal, with $E(X_2) = 0$, but $Var(X_2) = t \ V \ t'$ so that
the parameters of the distribution of $X_2$ depend on $z_2$,
violating the conditions required for (2).  This model cannot
be simply transformed to conform with these conditions.

One important property of the formulation (2) is its
implication for formal evaluation techniques.  For these
techniques, as applied in Great Britain, we require that
utility changes for one individual be directly comparable
with those for another individual through a measured component of utility, implemented as a negative generalised
cost $-g_j$. As outlined by Harris and Tanner (9), this
procedure requires (fairly generally) the assumption that for
all individuals

$$-dU_j = dg_j \qquad (1 \leqslant j \leqslant r) \tag{3}$$

where $g_j$ is the generalised cost measure for choice j.

Integrating with respect to $g_j$, we obtain

$$U_j = -g_j + X_j \qquad (1 \leqslant j \leqslant r). \tag{4}$$

Thus where it is required to assume that a unit change in the
measured $z_j$ (or $g_j$) is equivalent to a unit change in utility
for *any* individual, the personal difference formulation is
necessary.

It is important to note that we are not saying that only
personal difference models can give well-defined consumer
surplus measures.  An individual's consumer surplus can be
well-defined whenever his utility distributions are explicit.
But consumer surplus measures transferable between

individuals in different choice situations can be derived
easily only from personal difference models as we have
defined them.  In particular, the generalised cost evaluation
can only be derived from a personal difference model.

In the remainder of this chapter we shall concentrate on
models of this type.  This is not because we believe these
models to be the only class worth investigation, or even that
the assumptions on which they are based are particularly
plausible.  This class is, however, the simplest formulation
that contains interesting models.  It contains nearly all
transport demand models in present use, and, as we shall see,
a wide range of developments of present models.

## PROPERTIES OF PERSONAL DIFFERENCE MODELS

In this section we shall study some of the properties of
personal difference models.  For convenience we shall use
vector notation.  In this notation the model (2) is written

$$U = z + X ,  \tag{5}$$

where $X$ is a vector random variable (of r dimensions) with a
single distribution, which we shall additionally require (for
technical reasons) to exist and be finite and continuous
everywhere.  For such a model, we can determine explicitly,
for each j ($1 \leqslant j \leqslant r$), the probability

$$P_j (z) = Pr(U_j(z) \geqslant U_k(z), \text{ for all } k, 1 \leqslant k \leqslant r),  \tag{6}$$

which is the probability that an individual will make a
choice j given that the measured utilities of the choices
are z.

It is clear that an explicit statement of the distribution
of $X$ in (5) will always allow us to derive a demand model in
the form of probabilities.  In many cases, however, a demand
model is presented in the form of probabilities and we wish
to determine whether or not it is consistent with a utility
model of the form (5).  The result, partially given by Harris
and Tanner [9] and completed by Zachary [16] is that the
following conditions are both necessary and sufficient for
consistency:

(a)   $P_j(z_1+d,\ldots,z_r+d) = P_j (z_1,\ldots,z_r)$  (7)

   for all constants d and for all j;

(b)   $\lim_{z_k \to \infty} p_j(z) = 0$  (8)

   for all j and k with $j \neq k$;

(c)   $p_j(z) \geqslant 0$, for all j, and $\sum_{j=1}^{r} p_j(z) = 1$;  (9)

(d)  $(-1)^{r-1}$  $\dfrac{\partial^{r-1} p_j (z)}{\partial z_1 \cdots \partial z_{j-1} \partial z_{j+1} \cdots \partial z_r}$    (10)

exists everywhere for all j, and is non-negative and continuous;

(e)  $\dfrac{\partial p_j (z)}{\partial z_k} = \dfrac{\partial p_k (z)}{\partial z_j}$  (the 'Hotelling' condition).    (11)

This theorem can be used as a test of the acceptability of a demand model. If (and only if) the probabilities specifying the demand model satisfy conditions (a - e), then the model is a utility maximisation model of the personal difference type.

Condition (a) emphasises the arbitrary nature of the zero of the utility scale. Usually, we would relate z to measured characteristics of the choices by a calibration procedure based on choice probabilities. Condition (a) shows that identical calibration would be achieved by a model that added on an arbitrary constant to the measured component of utility.

It is interesting to note the exact correspondence of conditions (b - d) with *a priori* assumptions of McLynn and Watkins (13), which Steele and Rogers (14) suggest are 'necessarily satisfied by any reasonable model' (of mode choice). Here, we suggest that this intuitive argument can in fact be supported by the more rigorous derivation from utility maximisation. McLynn and Watkins, however, include a much more specific assumption concerning $\partial p_j / \partial z_k$ than the Hotelling condition, and thus rule out many of the more interesting models.

The Hotelling condition (e) is obviously the strong condition in this set, in the sense that models will rarely be specified that fail to satisfy other conditions, but models are often specified that do not satisfy the Hotelling condition. One example is a three-way choice model given by Langdon (11). This model is, for r = 3,

$$p_2/p_3 = Exp(c(z_2 - z_3))$$

$$p_1/(p_2 + p_3) = Exp(c(z_1 - z^*))$$

$$p_1 + p_2 + p_3 = 1$$

where $Exp(cz^*) = Exp(c.Max(z_2, z_3)) + R. \; Exp(c.Min(z_2, z_3))$ and c and R $(0 < R \leqslant 1)$ are calibration factors. This model fails to satisfy the Hotelling condition at $z_2 = z_3$ unless R = 1.

In consequence of the arguments we have outlined in this and the preceding section, models failing to satisfy the Hotelling condition cannot be derived from a utility varia-tion model of the type (5), and we cannot apply a generalised cost evaluation procedure. It is clearly difficult to

determine how significant would be the error in any particular model of an attempt to apply such an evaluation, and it
would seem reasonable not to apply such evaluations when the
demand model is not of this type.

A direct relationship exists between the function $p$ and
the distribution of the differences of the X variables.
Specifically, if we define for each j $(1 \leqslant j \leqslant r)$

$$F_j(z_1, \ldots, z_{j-1}, z_{j+1}, \ldots, z_r) =$$

$$Pr(X_1 - X_j \leqslant -z_1, \ldots, X_r - X_j \leqslant -z_r) \qquad (12)$$

as the distributions of the differences of the Xs, then it
can easily be shown that

$$\qquad\qquad\qquad\qquad\qquad\qquad\qquad\qquad\qquad (13)$$

$$F_j(z_1, \ldots, z_{j-1}, z_{j+1}, \ldots, z_r) = p_j(z_1, \ldots, z_{j-1}, 0, z_{j+1}, \ldots, z_r)$$

so that (with (a) above) the relationship is one-to-one.
Thus it is the joint distribution of the *differences* of the
Xs, rather than of the Xs themselves, that determines the
choice model. This point is frequently overlooked in the
literature. As one of its consequences we may arbitrarily
hold any one of the Xs, or even, because of (a), any one of
the Us, to zero.

A final point is that (by application of the Hotelling
condition) any *one* of the functions $p_j$ or $F_j$ may be used to
determine *all* the others. So we may conclude that, within
the context of personal difference utility models, a model $p$
of demand for r choices is exactly specified by an (r-1)
dimensional multivariate distribution F. These results will
be useful in our subsequent analysis of practical models.


PRACTICAL CONSIDERATIONS IN MODEL SPECIFICATION

In discussing the possible forms that practical demand models
might take, we shall remain within the structure of utility
variation (5). It is however necessary briefly to consider
some of the practical problems of defining the components of
the utility function, both measured and unmeasured, and the
context in which the function is maximised by the individual
traveller.

In the economic framework we have chosen for our analysis,
a travel choice is seen as maximisation of the individual's
utility subject to his budget constraints (which apply
equally to money and time) and to the technical constraints
of the travel systems available to him. As presented by
Dalvi and Daly (3) the attributes of a journey (for example,
the length of time spent in a specific journey activity) have
two conceptually separate but in practice indistinguishable
impacts on the individual: the specific (dis)utility of the
journey activity, and the opportunity cost of the time or
money spent in that activity that could have been spent on
other (possibly non-travel) activities.

In the context of the individual's perception of his
travel choice, these two disutilities run indistinguishably
into each other.  In the statistical analysis there is also a
corresponding impossibility of distinguishing disutility cost
from opportunity cost.  The separate concept of disutility
cost and the existence of technical constraints, however,
indicate a need for including in the utility function sufficient degrees of freedom to allow for the different
specific disutilities of different travel activities.  Thus
there is a strong theoretical case for increasing the
dimensions of the utility function to the limits of reasonably accurate measurement.

Much recent work has concentrated on the issues of
identifying, measuring and assessing the relevant components
of utility functions in travel decisions.  In discussing
model structure in this chapter we shall assume that as many
relevant components as possible have been measured; to derive
the complete function it is necessary to determine the
functional form in which these measured components are combined and the ways in which the unmeasured components may
vary relative to the measured ones.  The explicit form of
utility functions is inevitably a bone of contention.  To a
large extent, the different forms commonly used will make
little practical difference in most cases, and a form can be
chosen to suit the convenience of a particular situation -
e.g. to satisfy formulation (5) more nearly.  In the following development, however, we are concerned more with the
unmeasured than with the measured components.  Unmeasured
components of utility have a number of sources.  First, the
measured journey attributes may not correspond to those
perceived by the individual, either because the data is
synthesised by a model and individuals exist who are
inadequately represented by the model (e.g. they do not live
at zone centroids) or because the data is reported by the
travellers themselves and the responses are (for well known
reasons) distorted.  Second, the values attached to journey
attributes may not be identical for all individuals, and will
vary in ways that cannot be predicted exactly, even from a
knowledge of some of the individual's characteristics (e.g.
income).  Finally, many journey attributes (e.g. status,
reliability) cannot be measured at all at present, and we can
only include in the utility function a term representing
their average net impact; again it is reasonable to expect
that this net impact will vary between individuals.

Thus the overall variation of utility differences in which
we are interested is compounded from a large number of
separate variations.  The variation of some of these components is itself quite complex; for example, the waiting
time for public transport services has an unusual skewed
distribution, and little at all is known of the relative
variation of utilities of alternative destinations.  It is
therefore difficult to derive any *a priori* rationale for the
form of variation of utility differences.  What seems more
sensible is to assume some reasonably general functional
form, governed by a number of parameters which can be calibrated from observed data.  This is the approach that has
been adopted by most of the workers in this field.

For example, Cochrane (2) and Domencich and McFadden (6) assumed that the fundamental variation was that the utilities of each choice followed *equal variance independent* Weibull (or extreme value) distributions:

$$\Pr(U_j \leqslant t+z_j) = \text{Exp}(-e^{-ct}) \tag{14}$$

where c is a constant.  They then showed that this assumption led to the utility differences between choices following *equal variance independent* logistic distributions:

$$\Pr(U_j - U_k \leqslant t) = 1/(1 + \text{Exp}(c(z_j - z_k + t))), \quad j \neq k \tag{15}$$

and thence to the single-parameter multiple logit demand model:

$$P_j = \text{Exp}(cz_j) \, / \sum_{k=1}^{r} \text{Exp}(cz_k) \tag{16}$$

which has been used in so many studies.

A theoretical rationalisation of these formulae is given by Koenig (10), who deals with destination choice models in terms of the frequency of occurence of destinations with a given utility level.  He does not, however, indicate the Weibull distribution explicitly, although his model is implicitly equivalent to its use.

Andrews and Langdon (1) recommend an approach based on the assumption of normal distributions in the utilities, which of course lead to normal distributions in the utility differences and, in the binary (r=2) case, to a simple probit model.  The advantage of this specification is that the assumptions of equal variance and independence of utilities can readily be abandoned, giving easily specified distributions in utility differences *with different variances*.  Thus we might expect that the utilities of, say, journeys by two different routes to the same destination by the same mode would be more closely correlated than journeys by different modes or to different destinations.

However, Andrews and Langdon found two serious problems with the normal distribution which caused them to abandon it for practical use.  First, the estimation of the proportion of travellers who make a given choice when r > 2 involves the evaluation of an intractable multiple integral.  Second, the number of degrees of freedom in a full multivariate normal distribution is more than can be estimated from data sets of the usual size, yet useful approximations that reduce the degrees of freedom are difficult to find and to justify.  Moreover, the results obtained by numerical approximation of the normal distribution model with r=3 were very close to those obtained from a logit based model.  For practical use the logit model was therefore substituted.

Thus these approaches based on explicit assumptions of utility distributions have led either to the intractability of multiple normal integrals or the difficulties of

introducing more than one parameter to logit models.  In the
following section we introduce a class of models specified
not by the utility distributions but by the utility difference distributions, following the arguments of the preceding
section.  By allowing ourselves the freedom of not having to
derive an explicit functional form for the absolute utility
of each choice (an often useful but fundamentally unnecessary
concept), we find we can introduce new degrees of freedom in
the logit model.  These new degrees of freedom allow substantial progress to be made with several of the basic
problems of demand modelling.

## STRUCTURED LOGIT MODELS

In the previous section we discussed some of the practical
problems of demand modelling that have given difficulty
recently.  In this section we present a form of structured
logit model that solves some of these problems.  Such models
have been developed in Great Britain by Daly and Zachary (5)
and also by Williams (15).  These models are generalisations
of the well known multiple logit model:

$$P_j(z) = Exp(cz_j) \ / \ \sum_{k=1}^{r} Exp(cz_k) \ . \tag{17}$$

This model has only one degree of freedom in its covariance
structure, the scale parameter c.  This scale parameter can
be interpreted behaviourally as a measure of the sensitivity
of the choices to the measured z values.  Conditions (a) –
(e) of Section 2 are easily shown to be satisfied, so that
the model is of the individual difference type.

Following the methods of Section 2, we can investigate the
distributions of utility differences that would lead to the
model (17), in particular deriving:

$$E(U_j-U_k) = z_j-z_k, \ for \ all \ j,k \tag{18}$$

$$Var(U_j-U_k) = \pi^2/3c^2, \ for \ all \ j \neq k \ . \tag{19}$$

A further familiar property is

$$p_j/p_k = Exp(c(z_j-z_k)) \ , \tag{20}$$

i.e. the ratio of probabilities of two choices is independent
of the other zs.  This property is often termed 'independence
of irrelevant alternatives'.

It is these last two properties that we wish to generalise.
Property (19) in particular is excessively restrictive for a
model of general travel choice, where we wish to include
closely comparable alternatives (like similar routes by the
same mode) and more wide-ranging choices (such as alternative
destinations).  It would thus be desirable to allow
$Var(U_j-U_k)$ to take different values for different j,k pairs.

The model (17) can be written entirely equivalently in a hierarchical form:

$p_j = p_j$ as in (17) if j is not in S

    $qr_j$ if j is in S                            (21)

where S is some subset of the r choices, q is the probability that the choice lies in the subset S, and $r_j$ is the conditional probability of choosing j, given that we have chosen in S. If we set

$$r_j = Exp(cz_j) / \sum_{k \text{ in } S} Exp\ (cz_k) \qquad (22)$$

and

$$q = Exp(cz_S) / (Exp(cz_S) + \sum Exp(cz_k)) \qquad (23)$$

the sum being taken over k not in S, where

$$Exp(cz_S) = \sum_{k \text{ in } S} Exp(cz_k) \qquad (24)$$

then we may easily verify that the model is not changed by the hierarchical structuring. Our structuring of the mathematical formulation of the model does not, of course, imply any corresponding structuring in the mind of the traveller.

Equation (23) shows that the probability q of choosing the subset is determined in exactly the same way as the probability of making choices not in the subset. We have simply introduced a 'hybrid' or 'composite' choice, whose mean utility $z_S$ is defined by (24), and which can subsequently be treated as any other choice.

The structured logit model exploits this hitherto purely conceptual structuring in the following way. For choices outside the subset S, the model (17), with sensitivity parameter c, is retained, and the subset S appears, as in Equation (23), as a single choice in that model. For choices within the subset, however, we change the parameter c in Equation (22) to a new value c*, so allowing a different sensitivity for this group of choices:

$$r_j = Exp\ (c*z_j) / \sum_{k \text{ in } S} Exp(c*z_k)\ . \qquad (25)$$

The immediate question is whether such a demand model can remain within the framework of personal difference utility models (5). It is clear from condition (a) of Section 2 that we must also amend the specification of $z_S$ to use the new parameter c*:

$$Exp(c*z_S) = \sum_{k \text{ in } S} Exp(c*z_k)\ . \qquad (26)$$

With this amended specification, it is shown by Zachary (16) that the new model lies within the framework (5) - that is, it satisfies conditions (a-e) of Section 2 - provided only that

$$c* \geqslant c . \tag{27}$$

That is, choices within the subset must be at least as sensitive as choices external to the subset. In the equivalent terms of variance, property (19) is amended so that

$$Var(U_j - U_k) = \pi^2/3c^2, j, k \text{ not both in S, } j \neq k$$

$$\pi^2/3c*^2, \ j, k \text{ in S, } j \neq k . \tag{28}$$

The special case $c* = c$ is of course the simple multiple logit model (17) again. Again, we do not necessarily imply any structuring of the choices in the mind of the traveller; the generalisation merely allows a slightly more general structure of utility variation to be represented by the model.

Such models may be extended in two ways. First, any number of subsets may be defined, each with its own internal parameter. If required, we may without loss of generality take the whole choice set to be so divided. Second, additional levels may be introduced in the hierarchy, so that a multitude of possible structures can be considered. It is necessary for the internal sensitivity parameter of any subset, however, to be greater than the parameter at the next higher level.

The development presented here in terms of utility differences is that given by Daly and Zachary (5). The choice model they presented was used for predicting choice between car use, walking, and a number of bus routes; it was implemented in the TRANSEPT computer model for a bus study of the town of Huddersfield. Their paper gives details of calibration methods and results, showing in particular the very significant improvement in fit given by this model over the simple multiple logit model.

The development by Williams (15) is slightly different in approach. Williams considers the problem of the appropriate 'composite cost' to represent a subset of choices, to be used in the choice model at the higher level. He then shows that only the function (26) can be used in this way to give consistent prediction when choice within the subset is predicted by a logit model. Williams devotes more attention, however, to the problems of evaluation. He shows that this composite cost is equivalent to a consumer surplus measure (relative to the arbitrary zero of utility). Overall measures can be derived over the whole demand model, providing it is of the logit form throughout, by extending the process. These overall measures of surplus are both easier to extract and more accurate than the traditional 'rule of a half' approximate integration used in so many British studies.

The difficulty with such overall consumer surplus measures is that they are calculated with the measurable utility

component as used in the demand model, whereas the current
British government requirement is to apply a national
'measurable utility component', with specified coefficients
which may not be identical to those used in the demand model.
This problem is not unique to Williams' approach, however,
and seems to present a basic theoretical problem for integrating demand models and evaluation procedures when government,
not unreasonably, wishes to standardise the latter across
different studies.

A further large area of difficulty not touched on in this
presentation is that of calibration.  Williams recommends
calibration 'from the bottom up', but Daly and Zachary (5)
show that this is only an approximation, albeit an accurate
one.  The practical difficulties, however, are not with the
principles of calibration, but in reconciling the different
scales in which z is measured in practice for different
travel choices.  Unfortunately the resolution of these
problems is complicated, and greatly increases the difficulty
of presenting these models to a wider audience.


SUMMARY AND DISCUSSION

In this chapter we have outlined some of the research done by
British workers in attempts to extend the consistency and
applicability of transport demand models.  The attempt to
remain within a particular economic framework has caused
these workers to examine rather closely the theoretical basis
of their models, and a number of useful results obtained in
this process have been described.

To summarise:

transport demand models may be developed from assumptions
of individual maximisation of (cardinal) utility;

only one class of models of personal utility variation
(the 'personal difference' models) is theoretically
compatible with inter-personal evaluations using generalised cost or similar measures;

necessary and sufficient conditions that a demand model
(expressed as choice probabilities) should be compatible
with a 'personal difference' utility model may be derived;
they include a Hotelling-type condition;

these conditions emphasise the arbitrary nature of the
zero of the utility scale and concentrate our attention on
the distribution of utility differences, rather than that
of utilities.

We then discussed some of the practical difficulties in
deriving reasonable assumptions for the functional form of
utility difference distributions within the framework of
consumer choice theory.  At present it seemed there were too
many difficulties to develop these forms *a priori*, and most
workers had assumed some reasonable distributional form
(usually logistic or normal) and had attempted to calibrate
its parameters from observed data.  Problems arose in many
cases with intractable functional forms and difficulties in
specifying a useful dimensionality.

In the final section we discussed some models that avoided some of these problems.  These were the class we termed 'structured logit models', which generalise the simple multiple logit model by allowing different sensitivity parameters for different pairs of alternatives.  Ultimately the choice model may be developed as a complicated hierarchy of choice subsets, subject only to the constraint that the sensitivity parameters must increase continuously as we move down to the more detailed choices at lower levels in the hierarchy.  Consumer surplus measures may readily be extracted from these models.

It is important to distinguish the principles of these structured models from the traditional sequential transport demand model, where trip generation, distribution, mode split and assignment are performed in succession, possibly with iteration but without other interaction between the levels. This traditional model implies a sequential decision process, whereas the structured logit models represent a choice from all travel options considered simultaneously.  In terms of the hierarchy constraint $c* \geqslant c$, the traditional models represent in a sense the extreme where $c*/c \to \infty$.  Some of the debates concerning the use of the traditional models (e.g. whether mode split should be done before or after distribution) can be interpreted as a debate as to which of two extremes is nearer the truth (in the example, whether mode split is infinitely more sensitive or infinitely less sensitive than distribution).

The work described in this chapter is to a large extent parallel to and independent of recent research on journey attributes and their values.  This research has as its main objective to extend the measurable components of utility, thus reducing the variance, and hence the inaccuracy, of the models.  Some work, however, particularly by Goodwin (7), has introduced the concept of habit as a component of utility. If this extension becomes empirically established it may have consequences for model structure as well as for formulation.

The importance of the structured logit models is not in their specific form, for it may well be that other forms can achieve the same objective.  Rather, it is that they illustrate the practicability of extending the flexibility of demand models, yet remaining within a consistent economic framework.  One specific problem, that of choice set definition, which has caused so much difficulty, can be tackled by choosing appropriate subsets of reduced variability.

Throughout, we have worked with an explicit assumption of functional form for the distribution of utilities, or, more fundamentally, utility differences.  Again, however, the work presented here is largely independent of the issues of appropriateness of specific functional forms or whether calibration can be carried out without specific functional assumptions - as in Manski (12).  The problem of calibration is a serious one, and if we wish to see improvements in the models used in application it should be given priority. Recent work by Goodwin (8) suggests that the usual calibration methods may over-estimate the extent of modal choice, because of the interaction between mode and destination choice, which is of course not included in most applied

models.  In this specific instance, as so often, policy-
makers can be wrongly advised by over-simplified modelling.
The aim of this chapter is to bring forward the possibility
of improving the theoretical consistency and technical merit
of modelling, but bearing in mind the essential practical
difficulties of data availability, computational feasibility
and the need to justify recommendations to a lay audience.


ACKNOWLEDGEMENTS

I should like to acknowledge the contribution of my former
colleague, Stan Zachary, to the development of many of the
ideas presented in this chapter, and to thank him and the
referees at TRRL for their constructive comments on an
earlier draft.  I remain personally responsible for the
opinions expressed and for any errors the chapter may contain.


REFERENCES

1. Andrews, R.D. and Langdon, M.C., 'An individual cost
minimising method of determining modal split between three
travel modes', *TRRL Report LR698* (1976).
2. Cochrane, R.A., 'An economic basis for the gravity
model', *Journal of Transport Economics and Policy*, January
(1975).
3. Dalvi, M.Q. and Daly, A.J., 'The valuation of travelling time - theory and estimation', *LGORU Report T72* (1976).
4. Daly, A.J. and Zachary, S., 'Commuters' values of
time', *Local Government Operational Research Unit Report T55*
(1975).
5. Daly, A.J. and Zachary, S., 'Improved multiple choice
models', in *Determinants of Travel Choice*, D.A. Hensher and
M.Q. Dalvi (eds) Saxon House , Farnborough, England
(1978).
6. Domencich, T.A. and McFadden, D. (1975) *Urban Travel
Demand*, North-Holland/American Elsevier.
7. Goodwin, P.B. 'Habit and hysteresis in mode choice',
*Urban Studies*, 54, February (1977).
8. Goodwin, P.B., 'Travel choice and time budgets', in
*Determinants of Travel Choice*.
9. Harris, A.J. and Tanner, J.C., 'Transport demand models
based on personal characteristics', *TRRL Supplementary Report
65UC* (1974).
10. Koenig, J.G., 'A theory of urban accessibility', *PTRC
Symposium* (1975).
11. Langdon, M.C., 'Modal split models for more than two
modes', *PTRC Symposium* (1976).
12. Manski, C.F., 'Maximum score estimation of the stochastic utility model of choice', *Journal of Econometrics*,
August (1975).
13. McLynn, J.M. and Watkins, R.H. (1965) *Multimode
Assignment Model*, North East Corridor Transportation Project.
14. Steele, W.A. and Rogers, K.G., 'Predicting multi-modal
choice', *Local Government Operational Research Unit Report
C139* (1973).
15. Williams, H.W.C.L., 'The formulation of travel demand
models and user benefit measures', *Symposium on Urban
Transport Planning*, Leeds (1976).
16. Zachary, S., 'Some results on choice models', *Local
Government Operational Research Unit Transportation Working
Note 10* (1977).

Chapter 15

MEASURING THE VALUE OF TRAVEL TIME SAVINGS FROM
DEMAND FUNCTIONS

Nils Bruzelius

INTRODUCTION

The past ten years have witnessed an amazing amount of
research aimed at measuring the value of travel time.
Although empirical work in this field goes back much further,
there was a considerable boost towards the end of the 1960s
when it was shown how logit, probit and discriminant analysis
could be applied to data on mode or route choice to obtain
estimates of values of time.  Since then these econometric
models have been used in many studies, the results of which
are now widely applied, in particular in cost-benefit
analyses of investment projects in the transport sector.

   The reception given to this empirical work has not,
however, been totally positive.  A review of the literature
quickly confirms that there is little agreement about the
proper interpretation of the estimates of those parameters in
the logit, probit and discriminant models which are generally
referred to as values of time.  Some claim that these parameter estimates represent marginal values of time and others
that they are average values (23), some criticise them
because they represent what has been called 'the price of
time' (11) rather than the value of time and still others
have doubts about these estimates since they may not be
'pure' (3) or 'true' (8) values or may only represent 'curve
fitting parameters' (18).

   Some of this criticism can be interpreted as questioning
the validity of these estimates, i.e. asking whether they
really measure what we want to measure.  One consequence of
this is that although a considerable body of empirical
knowledge is available today, there is a great deal of doubt
as to the meaning of this knowledge and its relevance to
cost-benefit analysis.  This is demonstrated, for example, by
one of the papers presented at the last meeting of this
International Conference (21).

   I believe that the reason for this state of the art is
that some questions which are central to 'the value of travel
time' problem have never been asked and, of course, have even
less been given thorough answers.  These questions include:
What is meant by a value of time and why do we want a value
of time?  And given that the consumer's preferences are
revealed in terms of demand functions, how can these be used
to measure what we want to measure?  The aim of this chapter

is to try and answer these questions.  Some basic theory has
to be used in order to achieve this.  Thus I commence by
studying the question of measuring the value of a time saving
from demand functions at a general level.  The reason for
this is that it will help to reveal the special assumptions
underlying the class of demand functions to which the logit
and similar models belong and make it possible to interpret
the parameters of these models.

To begin with, let me observe that the justification of
cost-benefit analysis is to be found in welfare economic
theory.  The purpose of this normative body of theory is to
make it possible to assess states of the economy or proposed
changes in the economy.  This of course requires a ranking,
and the Pareto ranking or the Social Welfare Function (SWF)
are typically used in welfare economics.  According to the
Pareto ranking one state is considered superior to another if
everyone in this state is at least as well off, according to
their own judgements, as in the other, and if at least one
person considers himself better off.  The drawback of the
Pareto ranking is that it is only partial so that judgements
cannot be passed on a change which makes some worse and some
better off.  This can be done, however, if an SWF is available.  An SWF, which in general is assumed to contain the
Pareto ranking, also embodies value judgements as to the
distribution of welfare in society.

In welfare economic theory it is generally assumed that
both the Pareto ranking and the SWF are based on the consumers' preference orderings and that these may be
represented by utility functions.  Given a change in the
economy, this means that if we knew the utility function, it
would be possible to state immediately whether a change were
good or bad.  This is hypothetical, however, since we cannot
expect to possess such information about the consumers'
preferences.  But there is an indirect approach for obtaining
information about the utility functions.  This approach
involves the use of the Hicksian consumer's surplus concepts,
the compensating and equivalent variations (the CV and the
EV) (12), about which we can in principle obtain information,
either directly through interviews or indirectly by using
empirical demand functions.  Thus, the CV, as Arvidsson (1)
has pointed out, is directly related to the Pareto ranking,
since for a certain change it represents the amount of money
to be given to a consumer or to be confiscated from the
consumer in order to keep him at his initial indifference
level.  Indeed this is exactly what traditional cost-benefit
analysis tries to accomplish, i.e. to determine whether the
total sum to be confiscated exceeds the total sum of compensations.  If it does, the change is justified by the Pareto
ranking and thus also by the SWF, provided that the compensations are actually paid.

Often it is not possible to pay the compensations required
in order for a change to be justified by the Pareto ranking.
In this case, however, the change may be considered good or
bad in terms of an SWF.  The problem then is, how can we
measure whether a change in the economy is approved by an
SWF?  It is to solve this problem that the EV, which
represents the amount of money to be given to or taken from

the consumer to put him on the same indifference level as
would a certain change in the economy, may be used.  As has
been pointed out by Foster and Neuburger (6), the EV is the
only consumer's surplus measure which can determine unambig-
uously whether a change is acceptable according to an SWF.
This is because the EV, unlike the CV, is a strictly increasing function of the consumer's indifference level.

It may hence be concluded that there are two measures
which are of central importance from a welfare economic point
of view, the CV and the EV.  The import of this for the
subject considered here is, of course, that if for example a
road is to be built and if a consequence of this is travel
time savings, then the value of these savings should be
measured in terms of the CV or the EV.  I will now outline
how these surpluses may be measured on the basis of empirical
demand functions and how they are related to the concept of
the value of time.  I begin by reformulating traditional
consumer demand theory so that time is considered explicitly.

CONSUMER DEMAND AND TIME ALLOCATION

In order to measure the CV or the EV of a change in the
economy which affects the time duration of consumption, the
demand functions of ordinary consumer demand theory are
insufficient, since they contain only prices and income as
parameters.  Therefore this theory has to be reformulated so
that the time duration of consumption can be considered
explicitly.  One such reformulation, which appears to be
particularly relevant to transport, is a model proposed by
DeSerpa (4).  His model, as it will be used here, may be
written:

Max   $U(Y,X,L,T)$                                                     (1)

s.t.  $Y + pX - I \leqslant 0$                                         (2)

      $L + T - \overline{T} = 0$                                       (3)

and   $aX - T \leqslant 0.$                                            (4)

The utility function, which is assumed to be twice continuously differentiable and strictly quasi-concave, is defined
for two good variables, Y and X, and two time variables, L
and T.  Y is a composite commodity (13), with a price equal
to 1, and X, with the price p, is assumed to represent a
certain journey, the demand for which is to be analysed.  The
consumption of time when making the journey is given by the
time allocation constraint (4), where the time requirement,
a, indicates the minimum amount of time required per journey.
T is the actual time spent, which may exceed the minimum
amount, and L is the time spent while consuming Y or at
leisure.  Finally, I is the consumer's income and $\overline{T}$, in the
time budget constraint (3), is the total available time.

By reformulating the problem (1) - (4) in terms of a
Lagrange function, the necessary and sufficient conditions
can be derived (see Appendix 15A).  From these conditions it
is then possible to derive the demand functions

$$X = X(p,a,I) \tag{5}$$

$$Y = Y(p,a,I) \tag{6}$$

These functions differ from ordinary demand functions in that they contain the time requirement, a, as a parameter.

Some of the important properties of DeSerpa's model are (2):

1   It represents a general formulation of the time alloca-tion problem in the sense that all other models of the allocation of time may be shown to be special cases of this theory.   A precondition for this is that the time requirement need not be parametric from the point of view of the consumer but may be controlled by him.   However, this special case will not be considered here; it is assumed throughout that a is a parameter.
2   This theory does not alter any of the theorems or conclu-sions of general equilibrium analysis and welfare economics. It helps, however to identify why time savings, or time losses for that matter, constitute an economic problem. Decisions made by economic agents, such as government agencies, which affect the time requirement can thus be viewed as giving rise to effects of a public good character, i.e. affecting the consumer's utility outside of a market.
3   The demand functions (5) - (6) possess all the classical properties of the demand functions in conventional consumer demand theory.
4   The model yields a marginal value of time which is dependent on the activity for which the time is used.   The marginal value of time is represented by the ratio  / , where    is the Lagrange multiplier preceding the time allocation constraint (4) and $\lambda$ is the multiplier preceding the budget constraint (2) (see Appendix 15A).   This marginal value represents the consumer's willingness to pay to save a unit of time per journey and at a given income.   As will be demonstrated, it is also closely related to the CV and EV. Making this value a function of the activity simply means that the consumer may not be indifferent between saving a minute of time while riding a crowded bus and saving a minute riding a chauffeur-driven Rolls-Royce, which appears plausible.
5   This model is compatible with an equilibrium for the consumer where $\gamma/\lambda = 0$, i.e. a  arginal value of time equal to zero, which occurs when $T > aX$ (see Appendix 15A).   This also appears to be a valuable property, since for some journeys and in particular leisure journeys, the consumer cannot be expected to be willing to spend anything to save time; that is, the journey is demanded *per se*.

The problem that I now want to analyse is as follows. Assume that the empirical counterpart of the demand function (5) is available.   Assume further that a new road is to be built which will affect the cost of the journey from p' to p" and the time requirement from a' to a".   Given that the value of this change should be measured as the CV or the EV, how can this be achieved?

THE EXPENDITURE FUNCTION

In order to solve this problem, I use the expenditure function, which represents the minimum income necessary to put the consumer on a prescribed utility level, u (19). This function is chosen because it may be used to measure the CV and the EV and its properties may be employed to demonstrate how the expenditure function can be determined from the demand function (5).

The expenditure function may be derived by forming the dual of the problem (1) - (4), i.e.

Min   Y +pX                                                    (7)

s.t.  U(Y, X, L, T) ≥ u                                        (8)

      $\overline{T}$ - L - T = 0                               (9)

and   T - aX ≥ 0 .                                             (10)

By forming a Lagrange function the necessary and sufficient conditions for (7) to be minimised may be derived (see Appendix 15B). From these conditions, it is then possible to solve for the following so-called compensated demand functions (12).

$$X^* = X^* (p,a,u)$$                                          (11)

$$Y^* = Y^*(p,a,u).$$                                          (12)

These functions, not to be confused with the ordinary demand functions (5) and (6), indicate how much is demanded when income is continuously adjusted so that the consumer remains on the prescribed utility level, u. From these functions, the expenditure function is obtained as

$$m(p,a,u) = Y^* + pX^*.$$                                     (13)

It may be shown that m is twice continuously differentiable and strictly increasing in u (5). The expenditure function also possesses the following three properties:

$$\frac{\partial m}{\partial p} = X^*$$                        (14)

$$\frac{\partial m}{\partial a} = \pi X^*$$                    (15)

$$\frac{\partial^2 m}{\partial a \partial p} = \frac{\partial^2 m}{\partial p \partial a}$$   (16)

The first two properties are shown in Appendix 15B. The third property follows directly from the condition that m is twice differentiable. It should be pointed out that by using the two first properties, the third property may be rewritten as

$$\frac{X^{\boldsymbol{*}}}{p} = \frac{\partial X^{\boldsymbol{*}}}{\partial a}.$$

When formulated in this way, this property may be interpreted to correspond to what is sometimes referred to as the Hotelling condition for the existence of a well defined consumer's surplus measure when several prices are altered at the same time (14,20). It should also be pointed out that by rewriting the second property as

$$\frac{\partial m}{X^{\boldsymbol{*}}\partial a} = \pi,$$

the Lagrange multiplier $\pi$ may be seen to represent the willingness to pay to save a unit of time at a given level of utility. This value may therefore be called the compensated marginal value of time.

Given these properties and given that the initial utility level is u', the CV of the change in the price from p' to p" and in the time requirement from a' to a" may be written

$$CV_{\substack{p' \to p'' \\ a' \to a''}} = m(p'',a'',u') - m(p',a',u')$$

$$= \underbrace{\int_{p'}^{p''} X^{\boldsymbol{*}}(p,a'u')dp}_{A} + \underbrace{\int_{a'}^{a''} \pi X^{\boldsymbol{*}}(p'',a,u')da}_{B}. \qquad (17)$$

Assuming that the changes in p and a will put the consumer on the indifference level u", the EV is measured by

$$EV_{\substack{p' \to p'' \\ a' \to a''}} = m(p',a',u'') - m(p',a',u') = m(p',a',u'') - m(p'',a'',u'')$$

$$= \int_{p''}^{p'} X^{\boldsymbol{*}}(p,a',u'')dp + \int_{a''}^{a'} \pi X^{\boldsymbol{*}}(p'',a,u'')da. \qquad (18)$$

As can be seen, the EV and the CV are very similar, but they are not equal to each other unless the income effect is zero (12). Because of this similarity between the EV and the CV, I will henceforth refer only to the CV.

MEASURING THE CV FROM AN INDIVIDUAL DEMAND FUNCTION

In the preceding section I have shown how the expenditure function may be used to derive an expression for the CV and how this expression may be rewritten in terms of two components, A and B, both of which encompass the compensated demand function. I will now analyse how each of these components may be determined, given an empirical demand function.

The first part, A, may be measured by first determining the compensated demand curve. This can be achieved by recognising that when m=I the following must hold

$$X(p,a,m(p,a,u)) \equiv X^{\ast}(p,a,u),\qquad(19)$$

i.e. that compensated and ordinary demand coincide. If (19)
is differentiated with respect to p and the property (14) is
inserted, it is possible to obtain

$$\frac{\partial X}{\partial p} + \frac{\partial X}{\partial I}X^{\ast} = \frac{\partial X^{\ast}}{\partial p}\qquad(20)$$

which may be recognised as the Slutsky equation. It may be
shown that by using this equation as well as (14) it is
possible to derive an expression for the compensated demand
in terms of the parameter p (19). The integrability constant
of this solution may then be determined from (19). (In fact
the expenditure function is determined directly so that it
strictly speaking is not necessary to determine the compensated demand curve to calculate the CV of a change in the
price.) This can be interpreted in terms of Figure 15.1.

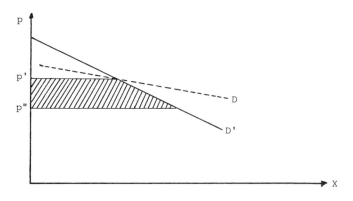

Figure 15.1   The CV of a Change in the Price

In Figure 15.1 the ordinary demand curve, X=X(p,a',I'), is
represented by the line D and the compensated demand curve,
$X^{\ast} = X^{\ast}(p,a',u')$, by the line D'. The change in price is
assumed to result in a reduction from p' to p". Given a
knowledge of the ordinary demand curve, the general shape of
the compensated demand curve can be determined on the basis
of (14) and (20). The position of this curve may then be
determined by requiring it to intersect D at the initial
price level. The value of the reduction in price is therefore, according to A, given by the shaded area in Figure 15.1.

It is more complicated to show how the second part, B in
(17), which could be called 'the value of the time saving',
may be determined. (It should be pointed out that this value
is not unambiguous since it is dependent on the path of
integration.) For simplicity I assume, however, that there
exists a finite price at which the compensated demand is

zero.  Given this assumption it may be shown that B in (17)
may be rewritten as (2)

$$\int_{p_1}^{p''} X^*(p,a'',u')dp - \int_{p_0}^{p''} X^*(p,a',u')dp \tag{21}$$

where $p_1$ and $p_0$ are price levels which make compensated
demand equal to zero when the time requirement takes on the
values a' and a'', respectively.  The justification of (21) is
explained in Figure 15.2.

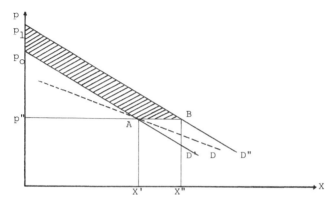

Figure 15.2  The CV of a Change in the Time Requirement

D indicates the position of the ordinary demand curve before
the change in the time requirement, $X=X(p,a',I')$, and D' the
corresponding compensated demand curve, $X^*=X^*(p,a',u')$, for
which $X(p'',a',I')=X^*(p'',a',u')$.  When a changes from a' to
a'', the compensated demand curve is assumed to shift outwards
into the position indicated by D''.  According to Equation
(21) the CV of this change is given by the shaded area.  To
see this, it may be argued that the area $p_0Ap''$ represents the
total amount of money that has to be given to the consumer so
as to make him indifferent between his initial situation of
consuming X' units of X, at p=p'' and a=a', and a situation
where X is not consumed at all.  Similarly the area $p_1Bp''$ is
the total sum to be confiscated from the consumer to make him
indifferent between not consuming X at all and consuming X''
units at p=p'' and a=a''.  Consequently the difference between
$p_1Bp''$ and $p_0Ap''$ must be taken from the consumer when a
changes from a' to a'', so as to keep him on the initial
indifference level.

MEASURING THE CV FROM AGGREGATE DEMAND FUNCTIONS

*Conditions for Consistent Aggregation*

The preceding analysis has shown how the CV of a change may be determined from an individual demand function. In real world situations, however, we cannot - due to problems of obtaining data - expect to be able to estimate individual demand functions of the general type, i.e. which give the demanded number of trips per period at different prices, etc. But several attempts have been made to estimate such functions using aggregate data. One feature of such aggregate travel demand functions is that they are expressed not in terms of individual incomes but in terms of an aggregate or an average income variable (24). This raises the question whether a theoretically correct measure of the CV can be obtained from such an aggregate function. To answer this question it is necessary to determine when consistent aggregation is possible, that is, when the properties of the individual demand functions are preserved in the aggregate function and the following relationship holds:

$$\Sigma X_i(p,a,I_i) = X(p,a,\Sigma I_i) = X(p,a,I), \tag{22}$$

where $X_i$ ($i=1,\ldots,m$) is individual demand, $X$ aggregate demand, $I_i$ individual income and $I$ aggregate income.

The conditions for consistent aggregation may be shown to be very restrictive. For the traditional consumer demand model, without any time allocation constraints, consistent aggregation is possible only when the income effect is not zero, if either all consumers always have identical utility functions and identical incomes or if all consumers have identical homothetic utility functions (10). For this model a homothetic utility function implies that all lines drawn from the origin intersect the indifference surfaces at points of equal slope, as indicated by the dotted line in Figure 15.3.

When the time duration of consumption is recognised, the first of these two alternatives must of course be strengthened to include the prerequisite that total time available, $\overline{T}$, is the same for all consumers. More important, however, is that the second alternative can no longer be expected to hold. The reason for this is that when DeSerpa's model is considered, homotheticity in Y and X will not result in points of equilibrium at different levels of income which are located on a straight line as in Figure 15.3. A necessary condition for this is that the slope at the point of equilibrium in the Y-X space is unaffected by income. Since this slope or marginal rate of substitution is given by $p+\gamma a/\lambda$ (see Appendix 15A), the marginal value of time has to be independent of income. But this can occur only when the utility function is not homothetic (2). A somewhat different conclusion is obtained if both the L and T variables in the utility function are eliminated with the time allocation constraint, $T=aX$, and the time budget constraint, $\overline{T}=T+L$. However, even in this case, homotheticity is not sufficient for consistent aggregation. This can be shown by considering

two consumers with identical utility functions, both of whom
have the same total time available but different incomes.

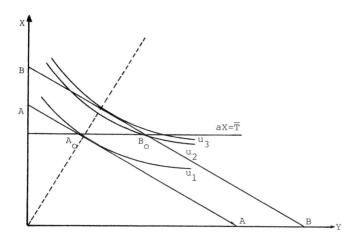

Figure 15.3   Aggregation and Homotheticity in DeSerpa's Model

These assumptions appear in Figure 15.3.   The indifference
surfaces are the same for the two consumers.   A-A is person
A's budget line and B-B person B's budget line.   The horizontal line indicates when all available time is spent consuming
X and $A_0$ and $B_0$ are the equilibrium positions of the

respective consumers.   Consistent aggregation requires that a
redistribution of income from A to B or vice versa shall not
affect the total demand for X and Y.   While this condition is
fulfilled in the absence of the time budget constraint, it
can easily be seen from Figure 15.3 that when this constraint
is present, total demand will quite possibly be affected by a
change in the distribution of a given sum of income.

In view of this and recognising that incomes vary between
consumers, there seems to be only one assumption which will
guarantee that a theoretically correct measure of the CV, as
well as of the EV, can be obtained from an aggregate demand
function.   And this assumption is that the demand for the
journey is independent of income.

*Conditions for Independence between Demand and Income*

It is well known that in order for individual demand to be
independent of income it is necessary and sufficient that (2)

$$\frac{\partial}{\partial Y}\left\{\frac{\partial U/\partial X}{\partial U/\partial Y}\right\} = 0,$$

where X is the good studied and Y a composite commodity.
This condition implies that all lines drawn parallel to the

Y-axis cut the indifference curves at points of equal slope.

This condition is still necessary when considering the time allocation model. But now, as is shown in Appendix 15A, we also have to impose the restriction that

$$\frac{\partial}{\partial Y}\left(\frac{U_3 - U_4}{U_1}\right) = 0,$$

where $U_i$ is the derivative of the utility function with respect to its $i^{th}$ argument. Now, since the marginal rate of substitution $(U_3-U_4)/U_1$ is equal to the marginal value of time $\gamma/\lambda$, this new condition for the demand of X to be independent of income requires that the marginal value of time should be independent of the consumption of Y. This in turn has a very important implication. Since any increase in income is to be spent completely on Y and $\gamma/\lambda$ is independent of Y, the marginal value of time must also be independent of income. Or, in other words, if it is assumed that the demand for the journey X is not a function of income, then this implies that when the consumer's income increases, he will not be willing to pay more to save time during this journey either.

## THE VALUE OF TIME

*A Constant Marginal Value of Time*

Up until now I have considered demand functions of a general nature. In particular it should be pointed out that no assumptions have been made concerning the marginal value of time. This value has been assumed to be a function of all the parameters of the model, p, a and I, and some limits were not imposed until the end of the preceding section. I will now go a step further and examine the consequences of assuming that the marginal value of time is a constant and not a function of any of the parameters. There are two main reasons for examining this case closely. First, practically all the empirical research concerned with the value of travel time has been carried out using econometric demand functions which are based on this assumption. When the so-called disaggregate demand models, such as the logit model, have been used for this purpose they have thus in general been based implicitly on the assumption of a constant marginal value of time. The second reason is that when considering the problems of aggregate demand functions and of obtaining data on travel demand and on the explanatory variables of this demand, one of the few options available for measuring the value of a time saving from a demand function appears to be the use of models based on constant values of time.

Assuming the marginal value of time to be independent of the parameters has some strong implications with regard to the consumer's behaviour. It may thus be shown that necessary and sufficient conditions for the marginal value of time not to be a function of a, the time requirement (2), are that (1) the marginal rate of substitution

$$\frac{U_3 - U_4}{U_1}$$

is a constant and hence not a function of Y, X, L and T; (2) the marginal rates of substitution

$$\frac{U_2}{U_1}$$

and

$$\frac{U_3 - U_4}{U_2}$$

are functions of the variable X only and independent of Y, L and T.

It may also be shown that these conditions are sufficient to make the marginal value of time independent of price. Moreover, recalling the conditions which make X independent of income, it can be seen that these two conditions mean that X is not a function of income and therefore also that the marginal value of time is independent of income. One further consequence of this is that the marginal value of time always will equal the compensated marginal value of time. This can be seen intuitively by noting that these two values are related in the same way as compensated and ordinary demand.

The conditions for the marginal value of time to be a constant mean that the utility function and therefore also the demand functions, have a very special shape. In order to analyse the latter aspect I use the equilibrium condition (see Appendix 15A)

$$\frac{U_2}{U_1} = p + \frac{\gamma}{\lambda}a, \tag{23}$$

where $\gamma/\lambda$ is a constant and the marginal rate of substitution $U_2/U_1$ is a function of X only. Since the utility function is strictly quasi-concave, this marginal rate of substitution must be monotonically decreasing. It is therefore possible to invert Equation (23) to obtain

$$X = (\frac{U_2}{U_1})^{-1} \ (p + \frac{\gamma}{\lambda}a). \tag{24}$$

In other words, when the marginal value of time is a constant, demand may be expressed in terms of one variable only. This variable may be recognised as what is often called the generalised cost of undertaking the journey X.*

To obtain a demand function of this type it may be shown that the utility function has to have the following shape (10,16):

$U(Y,X,L,T) = U(G)$

$$G = b_1 Y + F(X) + b_2 L + b_3 T + b_4 + H(L + T), \qquad (25)$$

where $b_j$ (j=1,...,4) are constants and F and H are functions of X and L+T, respectively. The utility function is thus characterised by what is known as direct additivity (9,15).

*An Example*

The implication of the utility function (25) can be illustrated by considering a concrete example. It is assumed that this function may be written

$$U = U(b_1 Y + cX - cX\log X + b_2 L + b_3 T + b_4 + e^{L+T}).$$

Maximising this utility function subject to the constraints (2) - (4) yields the following necessary and sufficient conditions:

$$\frac{\partial U}{\partial G} b_1 = \lambda \qquad (26)$$

$$\frac{\partial U}{\partial G}(c - c\log X - cX/X) = \lambda p + \gamma a \qquad (27)$$

$$\frac{\partial U}{\partial G}(b_2 + e^{L+T}) = \mu \qquad (28)$$

$$\frac{\partial U}{\partial G}(b_3 + e^{L+T}) = \mu - \gamma. \qquad (29)$$

On the basis of these conditions we can derive the following expression for the marginal value of time:

---

*It is now possible to resolve a problem encountered by McIntosh and Quarmby (17), pp. 16-17 and Goodwin (7), pp. 13-16. They observed that the demand models expressed in terms of generalised costs will forecast a decreasing demand for travel, since the value of time is an increasing function of income and since incomes are rising. They argue that this contradicts what is known about the income elasticity of the demand for travel and suggest as a remedy that demand be expressed in terms of generalised time, which is defined $(\lambda/\gamma)p+a$ instead. The answer to this of course is that they are working with the wrong model, since it is based on the premise that neither demand nor the value of time is affected by income.

$$\frac{Y}{\lambda} = \frac{b_2 - b_3}{b_1}.$$

Since $b_1$, $b_2$ and $b_3$ are constants the marginal value of time is a constant.

The relevance of this value to the measurement of the value of a time saving may be shown in the following way. The CV of a change in a from a' to a'' can, as before, be written:

$$CV_{a' \to a''} = \int_{a'}^{a''} \pi X^{\mathbf{x}}(p', a, u') da.$$

In order to measure the CV it is necessary to determine the compensated demand function. The ordinary demand function may be obtained by rewriting (27) after substituting for $\lambda, \mu$ and $\gamma$ by using (26), (28) and (29):

$$-\cdot c \log X = b_1 p + (b_2 - b_3) a,$$

which can be rewritten as the demand function

$$X = e^{-\frac{b_1}{c} - \frac{(b_2 - b_3)}{c} a}$$

An expression for $\pi X^{\mathbf{x}}$ has to be derived from this function. This can be accomplished by using the property (16), which can be rewritten as

$$\pi X^{\mathbf{x}} = \int \frac{\partial X^{\mathbf{x}}}{\partial a}(p, a, u') dp + K(a). \tag{30}$$

Since the demand for the journey X is not a function of income, the following must hold

$$\frac{\partial X^{\mathbf{x}}}{\partial a} = \frac{\partial X}{\partial a} = -\frac{(b_2 - b_3)}{c} X. \tag{31}$$

But this means that

$$\pi X^{\mathbf{x}} = \frac{(b_2 - b_3)}{b_1} X,$$

from which follows that $\pi = \gamma/\lambda$. The CV of the change may therefore be written

$$CV_{a' \to a''} = \frac{(b_2 - b_3)}{b_1} \int_{a'}^{a''} e^{-\frac{b_1}{c} - \frac{(b_2 - b_3)}{c} a} da.$$

This is the formula which in principle is used when calculating the value of time savings in cost-benefit studies of

transport investments. Of course other and even simpler
demand functions have been employed in many studies. But the
particular function considered here has also been used in
some studies.

*Relevance to Disaggregate Travel Demand Models*

The demand for only one journey has been dealt with in the
model described above. However, the analysis and results
obtained may be made much more general by expanding the
model. It can easily be reformulated to allow for several
journeys and alternative modes or routes of travel to complete a journey. (This can be done by introducing the
variables $X_j$ (j=1,...,m) to represent different alternatives
and imposing the constraint that $X=\Sigma X_j$.) It may further be
postulated that in conjunction with a given journey and a
given alternative for completing that journey, not just one,
but several different types of time may be identified. These
different types of time could represent the waiting, walking
and in-vehicle time of a particular alternative. These
different components of time can be recognised by treating a
variable such as T as a vector instead of a scalar. Of
course the time requirement, a, must then also be viewed as a
vector, giving the minimum travel time for each component in
the alternative.

When the consumer's problem is reformulated in this way
there is no longer merely one marginal value of time, but one
for each component and alternative. Now, if it is assumed
that all these marginal values of time are constants, each
alternative will be represented by a generalised cost, that
is the price of the alternative plus the sum of the time
requirements multiplied by the constant marginal values of
time. Of course, a consumer who chooses according to his
preferences will then select the alternative with the lowest
generalised cost and this alternative only for all journeys
during the period considered.

The reason for outlining this particular generalisation of
the model is that when cross-section data is available, it
constitutes the basis for deriving disaggregate travel demand
models in terms of generalised costs. Assuming that the
parameters of such an econometric model have been estimated,
the constant marginal values of time may be determined by
calculating the ratio

$$\frac{\partial X/\partial a}{\partial X/\partial p}.$$

As can be seen from (30) and (31), this ratio equals the
marginal value of time when this value is a constant. Since
values of time are determined from disaggregate models in
this way, it may be concluded that, provided the behavioural
assumptions of the model are fulfilled, then these values
represent the consumer's willingness to pay to save time
marginally during a journey. In other words, these values
are relevant to the measurement of the value of a time saving
in terms of the CV or the EV and may therefore be considered

valid. It may further be concluded that such an estimate
does not represent an average value *or* a marginal value but
an average value which is equal to the marginal value.

It should be pointed out that this constant value of time
may vary between the different components of time in the
various alternatives available to the consumer. This
reflects the assumption that the consumer is not indifferent
to what the time is used for or to what he experiences when
time is spent. These constant values of time will therefore
not represent what has been called the pure value of time.
But this is irrelevant since this value is not needed when we
want to determine whether, for example, the construction of a
new road should be considered justified in terms of the
Pareto ranking or a SWF; only the marginal values of time are
relevant for this purpose.

Moreover, it is doubtful whether it is possible to
estimate a pure value of time. In order to explain this, I
will interpret this value to represent what the consumer
would be willing to pay to have his total time available, $\bar{T}$,
marginally increased, at a given level of utility. In terms
of the model (7) - (10), it may be shown that

$$\frac{\partial m}{\partial \bar{T}} = \beta, \tag{32}$$

i.e., that this value is represented by the Lagrange
multiplier preceding the time resource constraint (19). As
is demonstrated in Appendix 15B, this value is related to the
compensated marginal value of time by the expression (33)

$$\pi = \beta - \alpha U_4. \tag{33}$$

The import of (32) is that if a pure value of time is to be
estimated from an empirical demand function, this function
must contain $\bar{T}$ as an explanatory variable. The estimation of
such a function is clearly beyond what we can hope for in
empirical research. Moreover it should be noted that when
demand is expressed in terms of generalised costs, the
function will not contain $\bar{T}$, as can be seen from (24).


SUMMARY AND CONCLUDING REMARKS

It has been argued in this chapter that the value of saving
time should be defined in terms of the compensating or
equivalent variations. It has therefore also been argued
that when examining estimates of disaggregate travel demand
models which are generally referred to as values of time,
this analysis should determine whether these estimates are of
relevance to the CV and the EV and in what way. To achieve
this I began by considering the problem of obtaining
consumer's surplus measures from empirical demand functions
of a general character. It was shown that a sufficient condition for measuring the value of saving time is that there
exist finite prices which make the demand for the journey
studied equal to zero. It was also shown that theoretically
correct measures of the CV can be obtained from aggregate

demand functions only if it is assumed that the demand will
not be affected by a change in real income.

The special class of empirical demand functions which
include disaggregate travel demand models was also considered.  It was indicated that these models, when formulated
in terms of generalised costs, may be derived by assuming
that all the marginal values of time are constants and not a
function of the time duration of a journey.  It was argued
that estimates of values of time obtained using these models
may be interpreted as estimates of the constant values of
time and that these estimates could be used to calculate the
CV of a time saving.  It was therefore concluded that such
estimates are valid.

However, an important reservation should be added to this
conclusion.  Econometric models formulated in terms of constant values of time are based on strong assumptions about
consumers' behaviour.  In particular it must be assumed that
the demand for the journey studied will not increase with a
rise in real incomes and that consumers' willingness to pay
to save time will not increase either along with a rise in
real incomes.  Since these assumptions conflict with what is
believed to be true about travel demand and the willingness
to pay to save time, some doubts arise as to the appropriateness of using these models to calculate the value of a time
saving.  Such models could probably serve as an approximation
in some instances, but how good or how bad are these
approximations?  It seems as if there is an urgent need to
provide answers to this question and that future research
concerning the value of travel time should try to seek ways
of approaching it.

## APPENDIX 15A: THE PRIMAL PROBLEM

*Necessary and Sufficient Conditions*

In terms of a Lagrange function, DeSerpa's model may be
formulated

$$L = U(Y, X, L, T) - \lambda(Y + pX - I) - \mu(L + T - \bar{T}) - \gamma(aX - T).$$

The necessary and sufficient conditions are:

$$U_1 - \lambda = 0 \tag{34}$$

$$U_2 - \lambda p - \gamma a = 0 \tag{35}$$

$$U_3 - \mu = 0 \tag{36}$$

$$U_4 - \mu + \gamma = 0 \tag{37}$$

$$Y + pX - I = 0 \tag{38}$$

$$L + T - \bar{T} = 0 \tag{39}$$

$$\gamma(aX - T) = 0, \text{ i.e. either } aX - T = 0 \text{ and/or } \gamma = 0. \tag{40}$$

From (34) and (35) it follows that

$$\frac{U_2}{U_1} = p + \frac{\gamma}{\lambda}a \qquad (23)$$

and from (34), (36) and (37) that

$$\frac{U_3 - U_4}{U_1} = \frac{\gamma}{\lambda},$$

where $\gamma/\lambda$ is the marginal value of time.

*The Income Effect*

If conditions (34) - (40) are differentiated with respect to
I it is possible to obtain

$$\frac{\partial X}{\partial I} = - \frac{H_{52}}{H} \qquad (41)$$

where H is the determinant of the bordered hessian associated
with the problem (1) - (4) and $H_{52}$ is a minor of this determinant.  In order for the income effect to be zero it is
clearly necessary and sufficient that $H_{52} = 0$.  Expansion of
this minor yields

$$H_{52} = - (aU_{41} - aU_{43} + pU_{11} - U_{21}), \qquad (42)$$

where $U_{ij}$ is the derivative of $U_i$ with respect to the $j^{th}$
argument of the utility function.  Using the conditions (34)
and (37), it is possible to write

$$p = \frac{U_2}{U_1} - \frac{a(U_3 - U_4)}{U_1} \qquad (43)$$

Using (43) to eliminate p in (42) yields $\qquad (44)$

$$H_{52} = - \{a(U_1U_{31} - U_1U_{41} - (U_3 - U_4)U_{11} + (U_2U_{11} - U_1U_{21})\}/U_1$$

$H_{52}$ is hence zero if and only if

$$a\frac{\partial}{\partial Y}(\frac{U_3 - U_4}{U_1}) = \frac{\partial}{\partial Y}(\frac{U_2}{U_1}). \qquad (45)$$

Since (45) must hold for all values of a, it is necessary and
sufficient to impose the following restrictions on the
utility function:

$$\frac{\partial}{\partial Y}(\frac{U_2}{U_1}) = 0$$

$$\frac{\partial}{\partial Y}(\frac{U_3 - U_4}{U_1}) = 0.$$

APPENDIX 15B: THE DUAL PROBLEM

*Necessary and Sufficient Conditions*

The dual of DeSerpa's model may be written

$$L = Y + pX - \alpha(U(Y,X,L,T) - u) - \beta(\overline{T} - T - L) - \pi(T - aX).$$

The necessary and sufficient conditions are:

$$1 - \alpha U_1 = 0 \tag{46}$$

$$p - \alpha U_2 + \pi a = 0 \tag{47}$$

$$- \alpha U_3 + \beta = 0 \tag{48}$$

$$- \alpha U_4 + \beta - \pi = 0 \tag{49}$$

$$U(Y,X,L,T) - u = 0 \tag{50}$$

$$\overline{T} - T - L = 0 \tag{51}$$

$$\pi(T - aX) = 0; \text{ either } T - aX = 0 \text{ and/or } \pi = 0. \tag{52}$$

From (49) it follows that

$$\pi = \beta - \alpha U_4, \tag{33}$$

where $\pi$ is the compensated marginal value of time and $\beta$ is the compensated marginal value of time as a resource or the pure value of time.

*Properties of the Expenditure Function*

If the expenditure function (13) is differentiated with respect to p it is possible to obtain

$$\frac{\partial m}{\partial p} = \frac{\partial Y^{\textbf{x}}}{\partial p} + p\frac{\partial X^{\textbf{x}}}{\partial p} + X^{\textbf{x}} \tag{53}$$

Differentiation of condition (50) with respect to p and insertion of conditions (46) - (49) and conditions (51) and (52), after the latter have also been differentiated with respect to p, yield

$$0 = \frac{1}{\alpha}(\frac{\partial Y^{\textbf{x}}}{\partial p} + p\frac{\partial X^{\textbf{x}}}{\partial p}). \tag{54}$$

It then follows that

$$\frac{\partial m}{\partial p} = X^{\textbf{x}} \tag{14}$$

Expression (15) may be shown by proceeding in the same way:

$$\frac{\partial m}{\partial a} = \frac{\partial Y^{\textbf{x}}}{\partial a} + p\frac{\partial X^{\textbf{x}}}{\partial a}. \tag{55}$$

Differentiation of (50), (51) and (52) with respect to a, and insertion of the latter two in the first expression as well as (46) - (49) yields

$$0 = \frac{1}{\alpha}(\frac{\partial Y^{x}}{\partial a} + p\frac{\partial X^{x}}{\partial a} - \pi X^{x}). \tag{56}$$

From (55) and (56) it then follows that

$$\frac{\partial m}{\partial a} = \pi X^{x}. \tag{15}$$

## REFERENCES

1. Arvidsson, G., 'On consumers' surplus and allied concepts, especially in formulations of the Pareto Criterion', *Swedish Journal of Economics* 76, pp. 285-307 (1974).

2. Bruzelius, N. (1979) *The Value of Travel Time; Theory and Measurement*, Croom Helm, London.

3. Dalvi, M.Q. and Daly, A., 'The valuation of travelling time; theory and estimation', *Local Government Operational Research Unit Report T72* (1976).

4. DeSerpa, A.C., 'A theory of the economics of time', *Economic Journal*, 81, pp. 828-46 (1971).

5. Diamond, P.A. and McFadden, D.L., 'Some uses of the expenditure function in public finance', *Journal of Public Economics*, 3, pp. 3-21 (1974).

6. Foster, C.D. and Neuburger, H.L.I., 'The ambiguity of the consumer's surplus measure of welfare changes', *Oxford Economic Papers*, 26, pp. 66-77 (1974).

7. Goodwin, P.B., 'Generalized time and the problems of equity in transport studies', *Transportation*, 3, pp. 1-24 (1974).

8. Goodwin, P.B., 'Human effort and the value of travel time', *Journal of Transport Economics and Policy*, 9, pp. 3-15 (1976).

9. Green, H.A.J., 'Direct additivity and consumers' behaviour', *Oxford Economic Papers*, 13, pp. 132-6 (1961).

10. Green, H.A.J. (1964) *Aggregation in Economic Analysis; An Introductory Survey*, Princeton University Press, Princeton.

11. Gronau, R., 'Economic approach to value of time and transportation choice', *Transportation Research Record*, 587, pp. 1-5 (1976).

12. Hicks, J.R. (1956) *A Revision of Demand Theory*, Oxford University Press, Oxford.

13. Hicks, J.R. (1968) *Value and Capital*, Oxford University Press, Oxford.

14. Hotelling, H., 'The general welfare in relation to problems of taxation and of railway and utility rates', *Econometrica*, 6, 242-69 (1938).

15. Houthakker, H.S., 'Additive preferences', *Econometrica*, 28, pp. 244-57 (1960).

16. Leontief, W., 'Introduction to a theory of the internal structure of functional relationships', *Econometrica*, 15, pp. 361-73 (1947).

17. McIntosh, P.T. and Quarmby, D.A., 'Generalised cost and the estimation of movement costs and benefits in

transport planning', *Highway Research Record*, 383, pp. 11-23 (1972).

18. Mitchell, C.G.B. and Clark, J.M., 'Modal choice in urban areas', in *Traffic Flow and Transportation*, G.F. Newall (ed.), Elsevier, New York (1972).

19. Mäler, K.-G. (1974) *Environmental Economics: A Theoretical Inquiry*, Johns Hopkins University Press, Baltimore.

20. Neuburger, H.L.I., User benefit in the evaluation of transport and land use plans', *Journal of Transport Economics and Policy*, 5, pp. 52-75 (1971).

21. Rogers, K.G., 'The use of behavioural models for deriving values of travel time', in *Behavioral Travel Demand Models*, P.R. Stopher and A. Meyburg (eds), Heath-Lexington, Lexington (1976).

22. Samuelson, P.A., 'Constancy of the marginal utility of income', in *Studies in Mathematical Economics and Econometrics*, O. Lange *et al.* (eds), University of Chicago Press, Chicago (1942).

23. Stopher, P.R., 'Derivation of values of time from travel demand models', *Transportation Research Record*, 587, pp. 12-18 (1976).

24. Talvitie, A.P., 'A direct demand model for downtown work trips', *Transportation*, 2, pp. 121-52 (1973).

Chapter 16

A MODEL BASED ON NON-HOMOGENEITY IN ALLOCATION PROBLEMS

John F. Brotchie

SUMMARY

Two basic types of activity allocation model, (1) behavioural
or predictive models of the gravity type, and (2) planning or
prescriptive models of the utility maximising type, are
initially compared. Each has application to urban land-use
allocation, transport trip distribution and modal choice, but
with different solution characteristics.

A new type of model which may be either prescriptive or
predictive is then developed in which non-homogeneity or
diversity of utility is introduced within each option. This
model is utility maximising and is shown to be related to the
gravity model and entropy maximising in which diversity is
implied. Several features of the entropy approach may be
explained in terms of diversity and utility maximisation with
this model. The essential difference between the prescriptive and predictive versions then lies in the utility itself.
In the prescriptive version, the utility of the community
including externalities is appropriate. In the behavioural
version, the utility of the decision-maker which excludes
these externalities is used. Thus the model presented may be
used for each purpose, with appropriate utilities. These
concepts are applicable to land use and travel modelling
including modal choice.

The integration of these new models in a strategic level
model of the TOPAZ type is considered. The resulting model
might be used to (1) select optimal land-use-transport plans,
(2) predict the response of the community to them, and (3)
determine policies to make the two compatible. The directions in which these studies may be extended are outlined.

INTRODUCTION

Models abstract and simplify segments of the real world.
Urban allocation models may be concerned with (1) the
distribution of land-use activities to urban spaces, (2)
linking these land uses by transport trips, or (3) assigning
these trips to routes or to transport modes. In the case of
land-use allocation, for example, a simplification made is to
assume each land use and each space or zone available for it
to be homogeneous. Variations are assumed to occur only
between activities and between zones. Thus the utility of
placing an activity in a zone is constant for all of that

355

activity in that zone.  If utilities vary between zones but
are constant over each, utility is maximised if the zones
with highest utility for the activities are filled and the
others are left empty.  However, where individual freedom of
choice exists, activities are distributed more widely than
this.  This may be due at least in part to variations in
utility for each activity in each zone caused by real or
perceived variations in the characteristics of the zone, or
of the activity, or both.  A similar situation occurs with
travel choices.  One approach to this problem is disaggregation; another is considered here.


ALLOCATION MODELS

The situation above relates to the purpose or type of model
and two basic types emerge; (1) those used to describe and
predict behaviour, and (2) those used to prescribe a policy
or plan (11).  Prescriptive models may be used to seek
maximum utility for the community as a whole.  Predictive
models on the other hand describe behavioural responses to a
particular policy or plan.  The two are complementary but
differently based, and an aim here is to relate them better
through the concepts of utility maximisation and non-
homogeneity of this utility.


*Prescriptive Models*

The Herbert-Stevens model (10) is an example of linear
land-use allocation models of the prescriptive type, and a
simplified version of it is developed here.  Utility is
interpreted here to be comprised of location benefits less
accommodation and transport costs.  Transport trips are
assumed to be between industrial activities with fixed locations, and residential activities to be located.  If all
activities, zones and trips are homogeneous, the problem can
be expressed as the allocation ($x_{ij}$) of residential activity
i to zones j such that total utility, U, is maximised where:

$$U = \sum_{ij} B_{ij} \, x_{ij} \tag{1}$$

subject to the constraints that (1) all activity is
allocated, i.e.

$$\sum_{j} x_{ij} = A_i \tag{2}$$

and (2) that no zone is over filled

$$\sum_{i} x_{ij} \leq Z_j \tag{3}$$

in which $A_i$ is the total quantity of residential activity i,

   $Z_j$ is the capacity of zone j,

      $B_{ij}$ is the utility of placing unit activity i in
      zone j, and is interpreted here as

$$B_{ij} = B'_{ij} - C_{ij} - C'_{ij} \qquad\qquad (4)$$

$B'_{ij}$ is the expected benefit or bid price for the activity in the zone,

$C_{ij}$ is the cost of accommodating and servicing the activity in the zone, and

$C'_{ij}$ is the cost of travel to other, fixed activities.

With homogeneous activities, zones, and trips, the unit net benefit $B_{ij}$ is constant over the range $0 \leq x_{ij} \leq Z_j$. (The model above is in fact a linear version of the TOPAZ model later described, and is formulated here in the notation and terminology of that model. The original Herbert-Stevens model was more disaggregated, considering housing and family types.)

Equations (1), (2) and (3) define a linear programming problem of the transportation type. Considering initially that there is only one type of residential activity (i.e. i = 1), the problem is illustrated diagrammatically in Figure 16.1. With each $B_{ij}$ constant, allocations are made to those zones with high values of $B_{ij}$, starting with the highest value and continuing downwards in order of magnitude of $B_{ij}$ filling each zone in turn until all of the activity $A_i$ has been allocated.

Some properties of this solution are evident:

(a)   zones are generally filled or left empty, except the last,

(b)   if $A_i$ and $Z_j$ are integers, $x_{ij}$ is an integer, and

(c)   if $A_i$ and $Z_j$ are unity, $x_{ij}$ is zero or one. This is then termed the assignment problem.

These three characteristics apply when there is more than one activity also. There have been a number of variations or modifications of this model (8,15).

The transport model above also applies to trip distribution, when all activity allocations are fixed. In this case, $x_{ij}$ is defined as the allocation of trips between *zones* i and j, $A_i$ and $Z_j$ define the number of *origins* and *destinations* respectively in each zone, and $B_{ij}$ (= $-C'_{ij}$) is the utility (or negative cost) of the trip between origin zone i and destination zone j. Utility $B_{ij}$ is assumed to be homogeneous for each option and the solution properties above again apply.

When more than one land-use activity is to be allocated, or where intra-activity trips are included, the travel costs in Equation (4) vary with the activity allocations, $x_{ij}$, so that substitution for $B_{ij}$ in Equation (1) gives a quadratic

or non-linear objective function.  The TOPAZ model later
discussed is one of several models of this type.

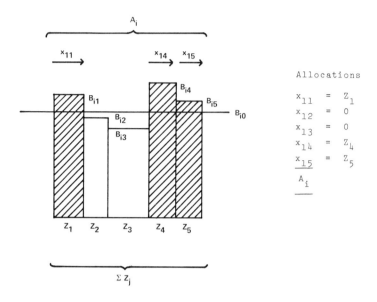

Allocations

$x_{11} = Z_1$

$x_{12} = 0$

$x_{13} = 0$

$x_{14} = Z_4$

$\underline{x_{15}} = Z_5$

$\underline{A_i}$

Fig. 1.   Allocations $x_{ij}$ of activity i of quantity $A_i$ to zones
of sizes $Z_j$, and homogeneous utilities $B_{ij}$.

*Predictive Models*

The Lowry model (13) is the best known and most widely used
(1,5,7) of the predictive land-use allocation models.  It
predicts residential and service activity allocations for
given locations of basic industry, on a gravity basis, i.e.

$$x_{ij} = C_i \, D_j \, e^{\lambda B}ij \qquad (5)$$

in which $x_{ij}$ is the allocation of activity i to zone j,

$B_{ij}$ (= $-C'_{ij}$ is the utility (negative cost) of travel
between the fixed activity and activity i in
zone j,

$C_i$  is activity size, and

$\lambda$ and $D_j$ are calibration constants.

The model is first calibrated to the existing city over two
time periods for which data are available.  It then allocates
in an iterative process, distributing work force from the
basic industry into residential areas, then allocating
service activities to cater for this work force, and residential areas for the service industry work force, on the basis
of the gravity model as a distribution function in each case.
The constants $C_i$ ensure that all activity is allocated, and
the constants $D_j$ can also be used to prevent the capacity of
zones from being exceeded.  In contrast to the previous
model, some activity is allocated to each zone.  With capacity constraints, no zone is (over) filled; no zone is empty;
although some may be almost full; and others almost empty.

The gravity model for trip distribution takes a similar
form.  In this case $x_{ij}$ is the allocation of trips between
origin i and destination j, $B_{ij}$ is the negative cost $C'_{ij}$, and
$C_i$ and $D_j$ are calibration constants which may be used to
satisfy either origin or destination constraints or both. $\lambda$
is also a calibration constant.

Wilson has shown the gravity model to be entropy maximising and therefore statistically based.  Others (15) have been
critical on the grounds that it is not utility maximising and
therefore is not behavioural in the economic sense.  Evans
(6) has shown, however, that in the limit $\lambda \to \infty$, the gravity
model solution approaches the transport model solution and is
therefore utility maximising in this case.  In the sections
following, the parameter $\lambda$ is interpreted as a measure of
non-homogeneity or diversity and this concept is used to
reduce the gap between the models above.


NON-HOMOGENEITY

In the real world, the zone j will not be homogeneous.  The
housing sites within it will vary in size, elevation and
slope, soil and vegetation, view and accessibility, and
perhaps in the services provided.  Similar variations will
occur among household units of residential activity.  They
will differ in size, income, mobility, knowledge, preferences
and needs.  (Where there is freedom of choice in site selection by households, the diversity in sites may be at least
partially matched by the diversity in needs and preferences,
so that the advantage of diversity is obtained.)  These
variations when superimposed will be reflected as a variation
in the net benefits $B_{ij}$ of the sites in zone j for the households in activity i.  Ignoring indivisibilities, if the
variation in $B_{ij}$ is normally distributed about a mean $\bar{B}_{ij}$,
with standard deviation $\sigma_{ij}$, its distribution may be illustrated as in Figure 16.2(a).  Its cumulative distribution is
shown in Figure 16.2(b).

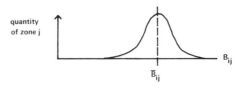

(a)

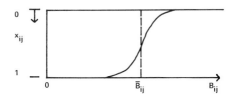

(b)

Fig. 2.   (a)   Normal distribution of site value $B_{ij}$ per unit
                area over zone j.
          (b)   Cumulative distribution over zone area $Z_j = 1$,
                from $B'_{ij} = \infty$ to $B'_{ij} = B_{ij}$.


The purely homogeneous case is represented by the single line
(line 1) of Figure 16.3.   Increasing degrees of non-
homogeneity or heterogeneity are represented by the curves 2,
3 and 4 in Figure 16.3.

    The cumulative normal distributions of Figure 16.2(b) and
16.3(b) have the form

$$y = \frac{1}{\sigma_{ij}\sqrt{2\pi}} \int_{b_{ij}}^{\infty} e^{-\frac{1}{2}x^2/\sigma_{ij}^2} \, dx, \tag{6}$$

in which $b_{ij} = B_{ij} - \bar{B}_{ij}$, and y is the proportion equal to or
greater than $b_{ij}$.   For simplicity of manipulation, Equation
(6) may be reasonably approximated by the logistic function

$$y = \frac{1}{1 + e^{\lambda b_{ij}}} \tag{7}$$

in which $\lambda$ is a calibration constant representing the degree of non-homogeneity or variation in $B_{ij}$ ($\lambda \simeq \pi/(\sqrt{3}\ \sigma_{ij})$). The range of y is $0 \le y \le 1$ for $b_{ij}$ in the range $-\infty \le b_{ij} \le \infty$. The slope of the curve is steepest at $b_{ij} = 0$, where slope = $dy/db_{ij} = -\lambda/4$. For line 1 in Figure 16.3, $\lambda = \infty$, and for curves 2, 3 and 4, $\lambda$ has finite and successively smaller values. The case of continuous uniform variation over a finite range is shown by line 5. For uniform variation over an infinite range, $\lambda = 0$. (Note that $\lambda$ should have sub-scripts i and j where $\sigma_{ij}$ varies between (activities and) zones. However, $\lambda$ will be assumed to be constant for all (activities and) zones at this stage.)

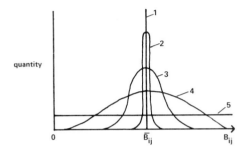

(a)

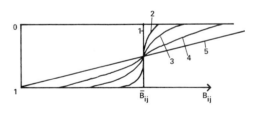

(b)

Fig. 3.  (a)  Complete homogeneity (line 1) and varying
             degrees of diversity (curves 2, 3, 4 and 5)
             in site values $B_{ij}$ about mean $\bar{B}_{ij}$.
         (b)  Corresponding cumulative distributions.

Thus $\bar{B}_{ij}$ may be the mean utility value or net benefit of locating activity i in zone j, $B_{ij}$ the value of a particular

site, and $b_{ij}$ the difference between the two.  If there is a
threshold utility value $B_{io}$ for acceptance of a site for
activity i, a site is acceptable if $B_{ij} \geq B_{io}$, and the proportion of acceptable sites for activity i in zone j is

$$P_{ij} = y = \frac{1}{1 + e^{\lambda(B_{io} - \bar{B}_{ij})}} \quad .$$

In a prescriptive or planning model, the utilities $B_{ij}$ would
be estimated utilities including external benefits and costs.
In a behavioural model, the utilities would be perceived
values of net benefit to the decision-maker, excluding
benefits and costs incurred by others from this decision.
The perceived value can contain errors or biases due to lack
of information or errors in this information.  (The proportion $P_{ij}$ can also be considered as a probability of

acceptance of a site by activity i in zone j.)

     If activity i is allocated to zone j on the criterion
$B_{ij} \geq B_{io}$, the quantity of activity allocated is given by

$$x_{ij} = Z_j \, P_{ij} = Z_j \, \frac{1}{1 + e^{\lambda(B_{io} - \bar{B}_{ij})}} = Z_j \, \frac{e^{\lambda \bar{B}_{ij}}}{e^{\lambda \bar{B}_{ij}} + e^{\lambda B_{io}}} \quad . \tag{8}$$

Similarly, if the activity is allocated among a set of zones
j = 1,2,...,N, to meet the criterion above, the allocation to
each is given by Equation (8) above.  If the constraint
$\sum x_{ij} = A_i$ is also to be satisfied, the level of acceptance
$B_{io}$ will have to be adjusted to suit.  This level of acceptance is therefore given implicitly by

$$\sum_j x_{ij} = A_i = \sum_j Z_j \, \frac{1}{1 + e^{\lambda(B_{io} - \bar{B}_{ij})}} \quad . \tag{9}$$

The allocations are therefore given by Equation (8), for all
j, in which $B_{io}$ is determined from Equation (9).  This solution is illustrated in Figure 16.4.

     When $\lambda$ is infinite, the solution reduces to that of the
transport problem and is shown in Figure 16.1, in which some
zones are filled and others remain empty.  When $\lambda$ is finite,
all zones are partly filled, but the allocations vary non-
linearly with $\bar{B}_{ij}$.

     Those which were filled when $\lambda = \infty$ will be nearly filled
when $\lambda$ is finite, and those which were empty when $\lambda = \infty$ will
be nearly empty when $\lambda$ is finite, but the extent will vary
with the degree of homogeneity $\lambda$.  With $\lambda = 0$, the activities
will be essentially uniformly distributed.

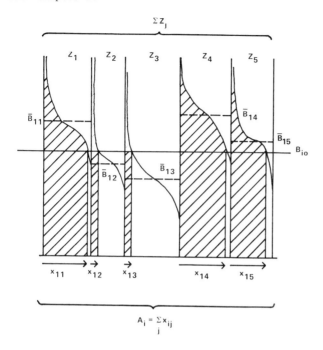

Fig. 4.   Allocation of activity i among zones j with variation
in site utility values $B_{ij}$ about mean $\bar{B}_{ij}$ in each zone
and over threshold of acceptance $B_{io}$ - selected to meet
activity constraint $\sum\limits_{j} x_{ij} = A_i$

Activities and zones would generally be selected to be as
homogeneous as possible (except where data are only available
for predetermined boundaries to zones), but the diversity can
still be substantial and will depend on the level of aggregation involved.  The allocation of trips between zones or
between modes may be treated similarly.  In the case of trip
distribution, non-homogeneity in travel utilities $B_{ij}$ again
transforms the solution of Equations (1) to (3) to the form
of Equation (8) (and Figure 16.4). The modal choice problem
may be considered similarly.

*Modal Choice*

Consider initially two modes with variable utilities $B_{i1}$ and $B_{i2}$ each with non-homogeneity factor $\lambda$.  One approach is to treat the two modes as two zones with one activity (trips) to be allocated between them (Figure 16.5).  Equation (8) gives the allocations (for $Z_1 = 1 = Z_2 = A_i$) as

$$x_{i1} = \frac{1}{1 + e^{-\lambda(\overline{B}_{i1} - B_{io})}}$$

$$x_{i2} = \frac{1}{1 + e^{-\lambda(\overline{B}_{i2} - B_{io})}}$$

and Equation (9) provides the constraint that the total activity must be allocated giving $x_{i1} + x_{i2} = 1$ and hence, from anti-symmetry,

$$B_{i1} - B_{io} = B_{io} - \overline{B}_{i2}, \quad \text{or } B_{io} = \tfrac{1}{2}(\overline{B}_{i1} + \overline{B}_{i2})$$

and substitution for $B_{io}$ in the equations above gives

$$x_{i1} = \frac{1}{1 + e^{-\lambda/2(\overline{B}_{i1} - \overline{B}_{i2})}} = \frac{1}{1 + e^{-\lambda'(\overline{B}_{i1} - \overline{B}_{i2})}}$$

$$= \frac{e^{\lambda'\overline{B}_{i1}}}{e^{\lambda'\overline{B}_{i1}} + e^{\lambda'\overline{B}_{i2}}} \tag{10}$$

which is the conventional model of modal choice (9).  The relationship $\lambda' = \lambda/2$ indicates that the net variation is twice as great when both $B_{i1}$ and $B_{i2}$ have variation $\lambda$.

Alternatively, this same result can be obtained by considering the utility of one of the modes as containing all the variation, and the utility of the other as the criterion of acceptance ($B_{io} = \overline{B}_{i2}$), i.e. mode 1 is acceptable when $B_{i1} > \overline{B}_{i2}$ again giving Equation (10).  The parameter $\lambda$ is an inverse measure of the diversity or opportunities offered en route, e.g. the number of alternative paths for each mode, the number of different submodes such as bus, train, tram, ferry, etc., that it encompasses, or the number of opportunities offered en route for a multi-purpose trip.  Thus the larger the city, the more opportunities offered en route and the values of $\lambda$ might be expected to decrease ($1/\lambda$  or  $\sigma_{ij}$ to increase) accordingly (see (3)).

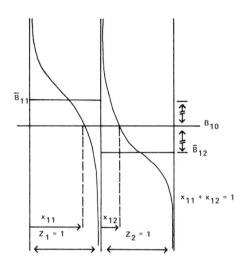

Fig. 5.   Choice between two modes where $\lambda$ represents
the degree of homogeneity in each mode.

*Multiple Activities and Zones*

Where there are multiple activities to be allocated to the
zones and these allocations have non-homogeneous utilities,
$B_{ij}$, Equation (8) may be extended to include a further calibration factor $B_{jo}$ to allow the constraints on zone capacity
to also be met,

i.e.   $$x_{ij} = \frac{1}{1 + e^{-\lambda(\bar{B}_{ij}-B_{io}-B_{jo})}} \tag{11}$$

The constraint equations that all activities are allocated

e.g.   $$\sum_j x_{ij} = \sum_j \left(\frac{1}{1 + e^{-\lambda(\bar{B}_{ij}-B_{io}-B_{jo})}}\right) = A_i \tag{12}$$

and no zone is over filled,

e.g.   $$\sum_i x_{ij} = \sum_i \left(\frac{1}{1 + e^{-\lambda(\bar{B}_{ij}-B_{io}-B_{jo})}}\right) \leq Z_j \tag{13}$$

may be satisfied simultaneously, by the calibration constants $B_{io}$ and $B_{jo}$. The effect of non-homogeneity on the objective function of Equation (1) may be considered in two parts:

(a)   The first part considers the means of the utilities, $\overline{B}_{ij}$,

i.e.   $U_1 = \sum_{ij} \overline{B}_{ij} \, x_{ij}$.

(b)   The second part evaluates the net additional utility or surplus above the mean $\overline{B}_{ij}$ (as shown in Figure 16.6 and) given by

$$U_2 = \sum_{ij} \left( \int_{b_{ij}}^{\infty} x' db' + b_{ij}(1-x_{ij}) \right) \quad \text{for } Z_j = 1,$$

in which $b_{ij} = \overline{B}_{ij} - B_{io} - B_{jo}$ and x' and b' are integration variables with finite bounds $(1-x_{ij})$ and $b_{ij}$ respectively. The integral above may be evaluated as follows utilising Equation (7), and the geometry of Figure 16.6:

$$\int_{b_{ij}}^{\infty} x' db' = \int_{b}^{\infty} \frac{e^{-\lambda b'}}{e^{-\lambda b'} + 1} db' = -\frac{1}{\lambda} \log(e^{-\lambda b'} + 1) \Big|_{b_{ij}}^{\infty}$$

$$= \frac{1}{\lambda} \log e^{-b'}/x' \Big|_{b_{ij}}^{\infty} = -\frac{1}{\lambda} \log(1-x_{ij}) - b_{ij}$$

$$\therefore \quad U_2 = -\frac{1}{\lambda} \sum_{ij} \log(1-x_{ij}) - \sum_{ij} (\overline{B}_{ij} - B_{io} - B_{jo}) \, x_{ij}$$

Thus the total utility, U, in the non-homogeneous case is

$$U = -\frac{1}{\lambda} \sum_{ij} \log(1-x_{ij}) + \sum_{ij} (B_{io} + B_{jo}) \, x_{ij} \tag{14}$$

in which $(-1/\lambda) \log(1-x_{ij})$ is the total surplus above the level of acceptance $(B_{io} + B_{jo})$ and $(B_{io} + B_{jo}) \, x_{ij}$ is the utility at this level, for each allocation.

Alternatively, by integration over the range $-b$ to $\infty$, or by combining Equation (14) with Equation (11) which may be written for this purpose in the log form

$$\overline{B}_{ij} - B_{io} - B_{jo} = -\frac{1}{\lambda} \log(1-x_{ij}) + \frac{1}{\lambda} \log x_{ij},$$

U may be expressed in the form

$$U = -\frac{1}{\lambda} \sum_{ij} \log x_{ij} + \sum_{ij} (\overline{B}_{ij} - B_{io} - B_{jo}) + \sum_{ij} (B_{io} + B_{jo}) \, x_{ij} \tag{15}$$

The allocation problem in the non-homogeneous case is then the selection of $(x_{ij})$ from Equation (11) to maximise U

(Equation (14) or (15)) subject to the constraints of Equations (12) and (13). The process has two linked steps: (1) the selection of the reduced set of parameters $(B_{io})$ and $(B_{jo})$ to maximise utility, subject to the constraints of Equations (12) and (13), and (2) the evaluation of $(x_{ij})$ from Equation (11). Iteration between these two (e.g. selection of $x_{ij}$, substituting in the objective, and evaluation of $B_{io}$ and $B_{jo}$) provides the values of $x_{ij}$ which maximise utility in the non-homogeneous case. Linearisation of the constraint equations in terms of the parameters $B_{io}$ and $B_{jo}$ by a Taylor series expansion can reduce Equations (12), (13) and (14) to a linear programming problem, but the small number of variables $B_{io}$ and $B_{jo}$ make search techniques possible also.

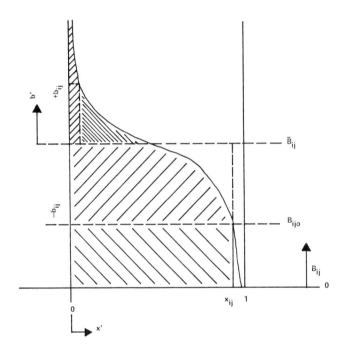

Fig. 6.   Heterogeneous utility with mean $\bar{B}_{ij}$ and level of acceptance $B_{ijo}$ resulting in allocation $x_{ij}$.

$$b_{ij} \equiv \bar{B}_{ij} - B_{ijo} \equiv \bar{B}_{ij} - B_{io} - B_{jo}$$

$$Z_j = 1$$

*Relationship with Entropy Maximising*

An alternative procedure would be to construct a Lagrangian function from the objective function (e.g. Equation (14)) and the constraint equations (assuming equality constraints on zone capacity) to give

$$L = -\frac{1}{\lambda} \sum_{ij} \log(1-x_{ij}) + \sum_{ij} (B_{io} + B_{jo}) x_{ij}$$

$$- \sum_i \gamma_i' \left( \sum_j x_{ij} - A_i \right) - \sum_j \rho_j' \left( \sum_i x_{ij} - Z_j \right)$$

The problem is then to maximise L. (The constraint implied by Equation (11) is not explicitly included here. It is not essential to the problem specification at this stage.) When L is maximised, then L' = λ L - constant is also maximised, and vice versa, where L' may be expressed in the form

$$L' = - \sum_{ij} \log(1-x_{ij}) + \lambda \left( \sum_{ij} B_{ijo} x_{ij} - U_\lambda \right)$$

$$- \sum_i \gamma_i \left( \sum_j x_{ij} - A_i \right) - \sum_j \rho_j \left( \sum_i x_{ij} - Z_j \right) \qquad (16)$$

in which $B_{ijo} \equiv B_{io} + B_{jo}$, and $U_\lambda$ is a constant utility as later discussed.

Thus maximising L' gives maximum utility in the non-homogeneous case (reintroducing the constraint of Equation (11) would only serve to reduce this utility).

A similar Lagrangian function was constructed by Senior (15) and Wilson (19) for entropy maximisation of the Herbert-Stevens model (Equations (1) - (3)) in the form

$$L'' = - \sum_{ij} x_{ij} \log x_{ij} + \lambda \left( \sum_{ij} \bar{B}_{ij} x_{ij} - U_\lambda' \right)$$

$$- \sum_i \gamma_i \left( \sum_j x_{ij} - A_i \right) - \sum_j \rho_j \left( \sum_i x_{ij} - Z_j \right). \qquad (17)$$

From the log form of Equation (11),

$$- \sum_{ij} x_{ij} \log x_{ij} = - \sum_{ij} x_{ij} \log(1-x_{ij}) - \lambda \sum_{ij} (\bar{B}_{ij} - B_{ijo}) x_{ij}$$

and substitution into Equation (17) and addition of a constant yields

$$L''' = - \sum_{ij} x_{ij} \log(1-x_{ij}) + \lambda \left( \sum_{ij} B_{ijo} x_{ij} - U_\lambda \right)$$

$$- \sum_i \gamma_i \left( \sum_j x_{ij} - A_i \right) - \sum_j \rho_j \left( \sum_i x_{ij} - Z_j \right)$$

in which only the first term on the right-hand side differs
from Equation (16). However, since $x_{ij}$ lies in the range
$0 < x_{ij} < 1$, all terms $-x_{ij} \log(1-x_{ij})$ and $-\log(1-x_{ij})$ are
finite and positive. And the nature of the solution of
Equation (11) is such that most values of $x_{ij}$ are close to 1
or 0. Thus as $x_{ij} \rightarrow 0$, $-x_{ij} \log(1-x_{ij}) \rightarrow -\log(1-x_{ij}) \rightarrow 0$,
and as $x_{ij} \rightarrow 1$, $-x_{ij} \log(1-x_{ij}) \rightarrow -\log(1-x_{ij}) \rightarrow \infty$, so that
these latter terms dominate. Hence $- \sum_{ij} x_{ij} \log(1-x_{ij})$ is
less than but reasonably close in magnitude to $- \sum_{ij} \log(1-x_{ij})$
and as $\lambda \rightarrow \infty$, $x_{ij} \rightarrow 0$ or 1, and $\sum_{ij} x_{ij} \log(1-x_{ij}) \rightarrow \sum_{ij} \log$
$(1-x_{ij})$. With a different distribution function for utility
$B_{ij}$, utility maximising and entropy maximising might be even
more closely related.

If the objective in the new model (Equations (14), (15) or
(16)) is modified to weigh the surplus ($\log x_{ij}$ or $\log(1-x_{ij})$)
by the proportion sharing it ($x_{ij}$), the entropy maximising
formulation (L" (Equation (17) or L" ') is obtained. (The
effect of this change to entropy maximising is to increase
the concentration in the nearly full zones which is somewhat
similar to increasing $\lambda$.)

The term $U_{\lambda}$ is described by Senior and Wilson as a
suboptimal value of U due to inefficiency of the allocation.
Whereas in Equation (16), in the heterogeneous case, $U_{\lambda}$ is
only a component of the maximum utility since it excludes the
surplus above the level of acceptance $B_{ijo}$. The parameter $\lambda$
is a distribution constant in the entropy approach, and
measures diversity herein.

Other interpretations are essentially the same for each,
e.g. $\rho'_j = \rho_j/\lambda$ is the (marginal) price of land in zone j -
and is related to $B_{jo}$ in the heterogeneous case; $\gamma'_i = \gamma_i/\lambda$ is
the subsidy or surplus required to ensure that all activity
is allocated - and is equivalent to $B_{io}$ in the heterogeneous
case.

Thus the entropy maximising solution of the Herbert-
Stevens model can be closely related to the utility maximising solution where heterogeneity is explicitly introduced.
The inference is that heterogeneity is implied in the entropy
maximising case. The entropy function is an approximate
measure of the component of surplus (above $B_{ijo}$) varying with
$x_{ij}$. The apparent suboptimum utility in the entropy maximising solution is the component of utility which exludes this
surplus. If the surplus is included, the total utility $U_T$ is
greater than the optimum for the homogeneous case since the
lower tails of the distribution are excluded (Figures 16.4
and 16.6), which in turn is greater than the component

$$U_\lambda \quad (= \sum_{ij} B_{ijo} \, x'_{ij})$$

i.e. $U_T > U_{\lambda \, \infty} > U_\lambda$.

Thus entropy maximising is close to utility maximising in the heterogeneous case and this heterogeneity is apparently implied by the entropy function.

The solution of the explicitly heterogeneous problem is given by Equation (11). The solution of the entropy maximising problem (Equation (17)) is derived by Senior and Wilson (16) from the optimality conditions

$$\frac{\partial L''}{\partial x_{ij}} = 0$$

in the form

$$x_{ij} = e^{-\gamma_i - \rho_j + \lambda B_{ij}} \ .$$

Thus the entropy maximising solution might be expressed as

$$x_{ij} = e^{\lambda(\bar{B}_{ij} - B_{io} - B_{jo})} \tag{18}$$

or in the normalised form (19)

$$x_{ij} = e^{\lambda(\bar{B}_{ij} - B_{io} - B_{jo})} / \sum_\ell e^{\lambda(\bar{B}_{i\ell} - B_{io} - B_{\ell o})} \tag{19}$$

whereas Equation (11) may be expressed as

$$x_{ij} = e^{\lambda(\bar{B}_{ij} - B_{io} - B_{jo})} / (1 + e^{\lambda(\bar{B}_{ij} - B_{io} - B_{jo})}) \ . \tag{20}$$

Thus the solutions (Equations (18), (19) and (20)) are essentially similar (and Equations (10) and (19) become the same in the one activity-two zone case). The constants $B_{io}$ and $B_{jo}$ will differ between the three equations for $x_{ij}$.

The coefficients $\lambda$ calibrate the models to the heterogeneity measured or implied. In the limit $\lambda \to \infty$, both solutions reduce to that of the transportation problem (Equations (1) to (3)). In the limit $\lambda \to 0$, both models give an essentially uniform distribution of activities.

Thus either form of solution might apparently be applied to prediction or prescription. An essential difference would be that in the predictive case, $B_{ij}$ is the utility of the decision-maker, whereas in the prescriptive case, $B_{ij}$ is the utility of the community including externalities.

*The Individual, the Authority and the Community*

As noted above, the difference between prescriptive and
predictive versions of the model lies in the utilities used.
In the predictive version, the utility distribution of
individuals is appropriate and will be diverse ($\lambda$ small) as
shown in Figure 16.7(a). The utility of a development
authority on the other hand will be more homogeneous
($\lambda$ large) with high initial threshold costs, and low marginal
costs on the last few lots in the zone, as shown in Figure
16.7(b). The utility of the community is equal to the
utility of individuals plus the utility of the authority
(plus any further externalities - $\lambda$ intermediate).

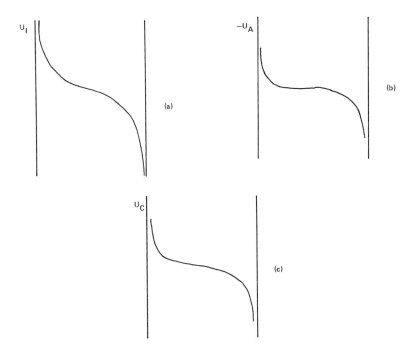

Fig. 7.   Utility distributions of (a) individuals $\lambda$ small,
          (b) authority $\lambda$ large, and (c) community $\lambda$ inter-
          mediate - $U_C = U_I + U_A$

Thus a development authority will tend to produce plans in
which zones are filled sequentially as in the homogeneous
case.   The loss of utility to individuals or the community
when the authority imposes a zone full-zone empty solution
may be obtained by integration of the peaks omitted from the

empty zones and the tails included in the filled zones, viz.

$$\Delta U_1 = \sum_{ij} (x^o_{ij} \int_{-\infty}^{-b} x'db' + (1-x^o_{ij}) \int_{b}^{\infty} x'db')$$

$$= \sum_{ij} [x^o_{ij} (-\frac{1}{\lambda} \log x_{ij}) + (1-x^o_{ij})(-\frac{1}{\lambda} \log(1-x_{ij}))] \qquad (21)$$

in which $x^o_{ij}$ is the homogeneous solution, and

$x_{ij}$ the solution with $\lambda$ finite.

The cost reduction to the authority from the solution is approximately

$$\Delta U_2 = \sum_{ij} (C_{ij}x^o_{ij} - C_{ij}x_{ij}) \qquad (22)$$

in which $C_{ij}$ is the unit infrastructure cost of activity i in zone j (assumed to be homogeneous - otherwise the solution is similar in form, but of opposite sign to that for $\Delta U_1$), and net loss in utility is then

$$\Delta U = \Delta U_1 + \Delta U_2. \qquad (23)$$

Thus the loss of utility when the authority imposes (full use of) a subset of options and withholds the remainder is given by Equations (21) - (23).

## INTEGRATION INTO STRATEGIC LEVEL STUDIES

The models above may be integrated into strategic level models of the TOPAZ type (2). The TOPAZ model (Technique for Optimal Allocation of Activities to Zones) allocates activities to zones on the basis of maximum utility. This utility has two components, (1) that of absolute location of activities in zones, i.e. the benefits of location less accommodation and service costs, and costs of travel to fixed activities as included in Equation (4) for $B_{ij}$ to give

$$U_1 = \sum_{ij} B_{ij} x_{ij}, \qquad (24)$$

and (2) that of relative location of activities in zones, e.g. the benefits less costs of travel between two sets of new activity allocations.

The utilities of these trips per unit of activity allocated are given by a trip distribution model and result in a total utility $U_2$ which is expressed in the form

$$U_2 = \sum B_{ijk\ell} x_{ij} x_{k\ell}. \qquad (25)$$

The coefficients $B_{ijk\ell}$ are evaluated by the distribution model and may be constant or vary with $x_{ij}$.

Utility $U_1$ is the linear model of Equation (1).  Utility $U_2$ results in a non-linear model.

The linear components are conventionally assumed to be homogeneous.  The non-linear components are conventionally based on a travel behavioural (gravity type) response to the activity allocations, and hence heterogeneity is implied.

The submodels defining the travel response are composed of a gravity model (Equation (5)) for trip distribution and a binary choice model of the logistic type (Equation (10)) for modal split [4].  These models evaluate the coefficients $B_{ijk\ell}$ and also the coefficients $C'_{ij}$ for trips to fixed (e.g. existing or constrained) activities which are included in the coefficients $B_{ij}$ (Equation (4)).

By introducing heterogeneity into the utilities of activities in zones, both parts of the model may be similarly based (e.g. each part may be based on Equation (11)).  Both may then be prescriptive if the utilities $B_{ij}$ and $B_{ijk\ell}$ include externalities, or predictive if they do not, and utility maximising in each case.

In the limit $\lambda \to \infty$, the model exhibits the following characteristics: (1) the modal split submodel gives a zero, one selection; (2) the gravity submodel becomes a linear transport model; and (3) the component $U_1$ of overall utility is linear, but the inclusion of transport costs $U_2$ results in a piecewise linear objective U with linear constraints (Equations (2) and (3)).  For $\lambda = 0$, a quadratic model is obtained.

By introducing heterogeneity and including externalities, optimal solutions for the community may be obtained for: (1) land-use location; and (2) travel decisions, including destination, route and mode.  By excluding the externalities, behavioural responses to these decisions may be obtained. From the differences between the two, the incentives neces- sary to implement the plan might be found.

The hierarchy of allocation decisions in the integrated model, and the way they nest together is shown in Figure 16.8.  A third level, that of route and mode choice, may be similarly included.  At the land-use allocation level, two values of $\lambda$, that of the individual and that of the community, are considered to give both the optimal solution for the community and the behavioural one.  At the trip distribution and route allocation levels, behavioural solutions only are normally considered, but the prescriptive solution could again be of interest for policy decisions.  The community optimum solution approximates that of the present TOPAZ model in which $\lambda = \infty$ at the land use level and $\lambda$ is finite (gravity solution) at the transport decision levels.

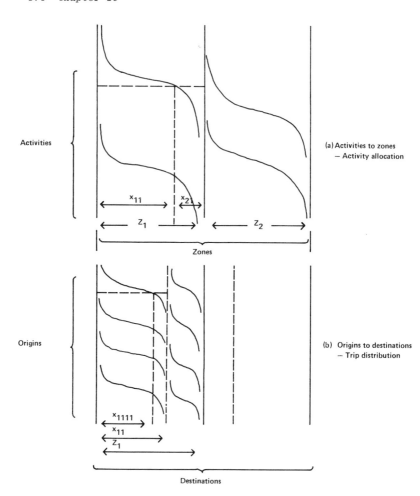

Fig. 8.   Two level model - (a) activity allocation, and (b) trip distribution and the relation-ship between them

The land-use and trip distribution models have been further integrated into a single level model in which overall utility is maximised subject to compatibility constraints between activity allocations and trip generation (as indicated in Figure 16.8).

## DIRECTIONS FOR FUTURE RESEARCH

In making a decision on a home, we in fact select a package which includes environment, accommodation and a set of home-based trips to work, school, recreational and social activities, that determines also our need for a car, and in fact defines a way of life. The urban planner, drawing lines upon a map, determines the types of packages available and their prices. The continued direction of this research is to better relate these two sets of decisions and decision-makers, using integrated planning models, to provide a better selection of packages at a better price. The steps involved include:

(a) the continued development of the heterogeneous approach, and of the relationship it provides between planning and behavioural models,

(b) development of strategic level, integrated, land-use-trip-distribution and modal choice models on this basis, which are utility maximising, heterogeneous and behavioural or prescriptive,

(c) calibration studies of the individual models – of land use, trip distribution and modal choice,

(d) studies of the differences in utilities involved and their use as incentives for implementation of plans,

(e) testing of the integrated models and then later use to develop optimal plans based on various criteria including the needs of various groups, and policies which aid the implementation of these plans,

(f) use of the models better to meet community needs and allow participation in the decisions involved,

(g) further study of the relationship with entropy maximising by relaxation of the constraint on the distribution $B_{ij}$, and by comparison of solutions, and

(h) integration with studies by others including McFadden and Reid (14), Talvitie (17), Koppelman (12), and Westin (18), on different aspects of the aggregation problem.

## CONCLUSIONS

A model based on diversity in urban allocation problems is presented. The basis of solution is utility maximisation. The utilities may be those of the community as a whole giving the optimal solution for the community, or those of the individuals making the location or travel decisions, in which case the model is behavioural. The model is related to entropy maximisation. It allows reduction of problem size by aggregation. Alternatively, heterogeneity may be reduced by disaggregation of the data such as activities and zones, but this increases costs of data acquisition and processing.

The utility parameters $\overline{B}_{ij}$ and $\lambda$ define the mean and the variation in utility for each allocation. In the case of the community, they are estimated values, including externalities. In the case of individuals, they are perceived values for those individuals, excluding externalities. The one model can thus be used to prescribe and to predict. It can also be used to indicate the pricing policies to make the two coincide, and the losses to the community if they do not.

The model is still in the initial development stage but relates closely to other prescriptive and predictive models on which wide experience has been gained.

ACKNOWLEDGEMENT

The constructive criticism of Åke Andersson, David Hensher, Paul Lesse, John Roy and Mike Taylor have added substantially to this study.

REFERENCES

1. Batty, M., 'Recent developments in activity allocation models in a British context', *Proceedings of the International Conference on Urban Development Models*, Cambridge (1975).
2. Brotchie, J.F. and Sharpe, R., 'A general land-use allocation model and its application to Australian cities', *Proceedings of the International Conference on Urban Development Models*, Cambridge (1975).
3. Davis, B.C., 'The calibration of the normal marginal utility modal split model', *Seventh Australian Road Research Board Conference* (1974).
4. Dickey, J.W. and Sharpe, R., 'Transportation and urban and regional development impacts', *High Speed Ground Transportation Journal*, 8, 2 (1974).
5. Echenique, M., 'Urban development models, fifteen years of experience', *Proceedings of the International Conference on Urban Development Models*, Cambridge (1975).
6. Evans, S., 'A relationship between the gravity model for trip distribution and the transportation problem in linear programming', *Transportation Research*, 7, pp. 39-61 (1973).
7. Goldner, W., 'The Lowry model heritage', *Journal of the American Institute of Planners*, March (1971).
8. Harris, B., 'Linear programming and the projection of land uses', *Penn Jersey Transportation Study, Paper No. 20*, Philadelphia (1962).
9. Hensher, D.A., 'A probabilistic disaggregate model of binary choice', in *Urban Travel Choice and Demand Modelling*, D. Hensher (ed.), Australian Road Research Board Special Report No. 12 (1974).
10. Herbert, D.J. and Stevens, B.H., 'A model for distribution of residential activity in urban areas', *Journal of Regional Science*, 2, pp. 11-31 (1960).
11. Kilbridge, M.K., O'Block, R.P. and Teplitz, P.V., 'A conceptual framework for urban planning models', *Management Science*, 15, 246 (1969).
12. Koppelman, P.S., 'Prediction with disaggregate models,

the aggregation issue', *Transportation Research Record*, no. 527 (1974).

13. Lowry, I.S. (1964) *A Model of Metropolis*, Rand Corporation.

14. McFadden, D. and Reid, F., 'Aggregate travel demand forecasting from disaggregate behavioural models', *Transportation Research Record*, no. 534 (1975).

15. Senior, M.L., 'Disaggregated residential location models, housing market models, and duality', *Proceedings PTRC Summer Annual Meeting*, University of Warwick (1974).

16. Senior, M.L. and Wilson, A.G., 'Exploration and synthesis of linear programming and spatial interaction models of residential location', *Geographical Analysis*, 6, pp. 209-30 (1974).

17. Talvitie, A.C., 'Aggregate travel demand analysis with disaggregation or aggregate travel demand models', *Transportation Research Forum*, 14, 1 (1973).

18. Westin, R.B., 'Predictions from binary choice models', *Journal of Econometrics*, 2, pp. 1-16 (1974).

19. Wilson, A.G., 'Some new forms of spatial interaction model, a review', *Transportation Research*, 9, pp. 167-79 (1975).

Chapter 17

THEORETICAL AND CONCEPTUAL DEVELOPMENTS IN
DEMAND MODELLING

Richard B. Westin and Charles F. Manski

## SCOPE OF THE WORKSHOP

The purpose of the workshop was to review theoretical and
mathematical developments in modelling the demand for
transport facilities.  Since the scope and flexibility of
behavioural modelling is limited by the available analytical
tools, there is a continuous need to review conceptual
developments in order to publicise important new procedures
and to focus on unresolved problem areas.  The workshop
discussion clarified which developments may be expected to
have an important impact on current modelling practices and
which developments require more research before they can be
implemented.

## SUMMARY OF RECOMMENDATIONS

In the workshop discussion, three conceptual areas were
delineated that the workshop participants felt had been
sufficiently developed and tested to warrant immediate
implementation by travel-demand modellers.  These three
areas, which were felt to hold significant promise as
practical modelling tools, are (1) choice-based sampling
procedures, (2) the structured-logit and multinomial-probit
choice models, and (3) model-transferability procedures.  In
addition, the workshop discussed several conceptual areas
which are currently being investigated or which require
further research before implementation.  In this section,
brief descriptions are given of each of these areas and the
workshop recommendations are discussed.  A detailed discussion of the three areas ready for current implementation is
then given, along with a selected bibliography.

We turn first to a summary of the three conceptual areas
that workshop participants felt are ready for current
implementation:

### Choice-based Sampling Procedures

Choice-based samples employ a sampling design that selects
individuals on the basis of their observed choice behaviour.
On-board interviews, road-side surveys, and similar
strategies are examples of choice-based sampling techniques,
as contrasted to exogenous sampling designs such as home

378

interview surveys. The workshop participants felt that choice-based survey techniques offer significant possibilities for substantial savings in data collection costs, particularly for choices which are selected by a minority of the general population (such as infrequently used modes). Estimators for choice-based sampling models have been proposed and adequately tested during the past two years, and existing computer software can be easily modified to incorporate these procedures. A useful reference for practitioners interested in using these estimators is Lerman and Manski (11).

*Alternative Choice Models: Structured Logit and Multinomial Probit*

In travel-demand analysis, it is useful to decompose the overall choice process into smaller subset decisions, such as mode choice, destination choice, route choice and car-ownership levels. Until recently, analysts lacked a consistent method of linking these models together simultaneously, and the validity of treating subset choices within a general choice model had not been established. Structured-logit analysis provides a theoretical framework for treating subset choices within a general choice model by relaxing the independence of irrelevant alternatives assumption between the subsets of alternatives. Simple estimators for structured-logit models have been proposed which incorporate some of the restrictions between subset choices imposed by the general simultaneous-choice process. These estimators can be implemented with minor changes in existing computer software. The existing literature on structured logit is difficult because of differences in approach and notation between authors, and the workshop recommended that an appropriate agency fund a project to synthesise the literature in a non-technical report for practitioners.

Multinomial probit-choice models have been developed to provide researchers with a tractable framework for incorporating departures from the independence of irrelevant alternatives assumption. Because of the convenient representation of non-independent disturbances possibly with the multivariate normal distribution, this model provides an attractive choice model when the researcher is unwilling to accept either the multinomial or structured-logit models. In addition, the multinomial probit model can incorporate taste variations between decision-makers, allowing researchers to relax the assumption of a fixed utility function for all individuals. Recent advances in computational techniques have made the multinomial probit model feasible for a wide range of choice modelling problems.

*Model Transferability Procedures*

Developing adequate behavioural models is relatively expensive in data requirements and analyst's time. Transferring existing models to new situations and areas provides a low-cost alternative to the development of new models and should be considered as a possible option for forecasting

purposes.  There is a growing body of empirical literature on
transferability that indicates that the estimated coefficients of behavioural models show differential sensitivity to
changes in population groups and geographical areas.  Fairly
simple procedures are available for adjusting and updating
the coefficients of transferred models based on available
data.  In addition, aggregation procedures have been devised
for building group, zonal and area-wide forecasts from the
predictions of individual choice probabilities given by
behavioural models.

   In addition to the above recommendations, the workshop
participants felt that several conceptual areas hold significant promise as modelling procedures but require more
research before implementation.  Other areas were felt to
require additional research to clarify outstanding problem
areas.  The workshop recommendations for further research
were as follows:

1  Consumer surplus evaluations drawn from choice models
require better theoretical justification and sensitivity
evaluation.  Assumptions necessary for consistency with
individual utility maximisation and for consistent aggregation across consumers must be clarified (see the chapters in
this volume by Bruzelius and by Daly).
2  Models incorporating taste variations between consumers
should be further investigated.  The multinomial-probit model
provides a possibility for incorporating taste variations in
choice models, and taste variations in other transport submodels should also be investigated (see the chapter in this
volume by Brotchie).
3  Generalised choice models that relax the assumption of
independence of irrelevant alternatives should be more fully
investigated.  Both the generalised extreme value and the
multinomial probit model offer potentially attractive
alternative procedures when specification tests for i.i.a.
fail (see the chapter in this volume by McFadden).
4  Robust estimation procedures are required to reduce the
sensitivity of conventional choice models to data coding
errors and data outliers.  Re-weighting and truncation
procedures that are sometimes used require theoretical
justification.  Additional research on different sample
designs is also necessary (see Manski and McFadden (13)).
5  Models incorporating inter-related discrete and continuous decisions deserve further development.  Particularly
important is the ability of these procedures to model choice
situations where the decision-maker has some degree of
control on the attributes of the alternatives.  Additionally,
these models are capable of handling choice situations where
data is systematically missing as a function of the choice
process (see Westin and Gillen (23)).
6  The inter-temporal aspects of travel behaviour require
further investigation.  This information is necessary in
order to assess the rate at which travellers will react to
changes in the attributes of transport networks.  The workshop participants suggested that retrospective questioning
might provide a low-cost alternative to panel studies for
gathering inter-temporal data for behavioural models.
7  There is a continuing need for more research on the
definition of choice sets.

This concludes the summary of the major recommendations of the workshop. We turn next to a detailed discussion of the three conceptual areas identified as ready for current implementation by practitioners.


DATA COLLECTION METHODS: CHOICE-BASED SAMPLING

The prevailing 'revealed-preference' approach to parameter estimation in models of travel behaviour presupposes the availability of a suitable data sample. Specifically, it is assumed that for each of a set of individuals, the analyst observes the set of travel alternatives available, the actual travel choice made and a vector of attributes characterising the decision-maker and his alternatives. It is important to recognise that there exist many distinct ways to draw such a sample of choice data and that the statistical properties of a given estimation method will in general depend on the sampling process used.

The traditional sampling process, exemplified by the home interview survey, is to stratify the decision-making population based on observed attribute dimensions, determine a sample size to be drawn from each stratum and then, within each stratum, to draw the required number of decision-makers at random. Such a process, termed 'exogeneous sampling', is assumed in the statistical development of such familiar estimators as the McFadden (14) maximum likelihood estimator for the multinomial logit model and the Theil (20) regression estimator for the same model.

Recently an important alternative to exogeneous sampling has been posed and appropriate estimation methods developed. In this approach, exemplified by transit on-board surveys, one stratifies the decision-making population based on the observed chosen travel alternative, determines a sample size to be drawn from each stratum and then within each stratum draws the required number of decision-makers at random. Termed 'choice-based sampling' the new approach appears to have great practical usefulness.

In Section 1 below we summarise the available theoretical results on estimation in choice-based samples. In Section 2 we discuss the potential role of choice-based sampling in empirical practice. Section 3 considers the general question of sample design for choice-model estimation and poses a number of sampling processes as yet unexplored.


1. *Estimation Theory for Choice-based Samples*

The existing estimation theory for choice-based samples has been developed in papers by Manski and Lerman (12) and Manski and McFadden (13). Less technical presentations of some of the relevant results are contained in Lerman and Manski (11) and in the chapter by McFadden in this volume.

Let us first express the likelihood of an observation under choice-based sampling. Let C be the choice set, T the population of decision-makers and for each $i \in C$, $t \in T$, let

$z_{ti} \in R^k$ be an attribute vector. Let $z_t = (z_{ti}, i \in C)$ and define the attribute space $Z = (z_t, t \in T)$. Let $p(z)$, $z \in Z$ be a generalised probability density function giving the marginal distribution of attribute vectors in the population of decision-makers and let $P(i|z,\theta^*)$ be the conditional probability that a decision-maker with choice set attributes z would select alternative i. We assume that the analyst has specified the choice model $P(i|z,)^*)$ up to the unknown parameter vector $\theta^* \in R^k$. Finally, for each $i \in C$, let $H(i)$ be the frequency, determined by the sampler, with which decision-makers are drawn from the stratum selecting alternative i. Then it can be shown that the choice-based sampling likelihood is

$$\lambda_c(i,z) = \frac{P(i|z,\theta^*)\ p(z)}{\int_Z P(i|y,\theta^*)\ p(y)\ dy} \cdot H(i). \tag{1}$$

The form (1) may be contrasted with the likelihood obtained under exogeneous sampling. There we have

$$\lambda_e(i,z) = P(i|z,\theta^*)\ g(z) \tag{2}$$

where $g(z)$ is a generalised density function on Z describing the attribute-based stratification selected by the sampler.

Because of the appearance of the integral $\int_Z P(i\ y,\theta^*)\ p(y)\ dy$ in the denominator of (1), maximum likelihood estimation is quite cumbersome in choice-based samples. The first practical estimator with desirable statistical properties was developed in Manski and Lerman [12]. Given a sample $(i_n, z_n)$, $n=1,\ldots,N$ where $i_n$ is the chosen alternative and $z_n$ the attribute vector for decision-maker n, their 'weighted' exogenous sampling maximum likelihood estimator is

$$\underset{\theta \in R^k}{\text{Max}} \sum_{n=1}^{N} \frac{Q(i_n)}{H(i_n)}\ \log P(i_n|z_n,\theta) \tag{3}$$

where, for each $i \in C$, $Q(i) \equiv \int_Z P(i|y,\theta^*)\ p(y)\ dy$ is the marginal probability with which the decision-makers in T select alternative i. In applying the estimator, it is assumed that the marginal probabilities $Q(i)$, $i \in C$ are known to the analyst.

Manski and McFadden [13] undertake a comprehensive study of estimation in choice-based samples. Their work demonstrates that prior knowledge of the marginal distributions $Q(i)$, $i \in C$ and $p(z)$, $z \in Z$ is useful but not necessary for estimation in such samples. Of particular practical interest is their 'conditional' maximum-likelihood estimator

$$\underset{\substack{\theta \in R^{k} \\ \tilde{Q} \in \Omega}}{\text{Max}_{k}} \quad \sum_{n=1}^{N} \quad \frac{P(i_n | z_n, \theta) \dfrac{H(i_n)}{\tilde{Q}(i_n)}}{\sum_{j \in C} P(j | z_n, \theta) \dfrac{H(j)}{\tilde{Q}(j)}} \tag{4}$$

where $\tilde{Q} = (\tilde{Q}(i), \; i \in C)$ and $\Omega$ is the unit simplex. Use of this estimator requires prior knowledge of neither Q nor p. Under fairly general conditions the estimates obtained for $\theta*$ and Q are consistent and asymptotically normal.

Finally, an important negative result obtained by Manski and Lerman (12) is that in the context of a choice-based sample, the unweighted exogenous sampling estimator

$$\underset{\theta \in R^{k}}{\text{Max}_{k}} \quad \sum_{n=1}^{N} \quad \log P(i_n | z_n, \theta) \tag{5}$$

is generally inconsistent. This result highlights the need for care in ensuring that the estimation method used in practice is appropriate under the sampling process used.

## 2. Choice-based Sampling in Practice

The theoretical basis for estimation in choice-based samples is now as firm as that for exogenous-sample estimation. Moreover, a variety of computationally and statistically appealing choice-based sampling estimation methods are presently available. Practical experience with choice-based sampling has not however yet accumulated. In this section we discuss various issues that will have to be faced as travel-demand researchers begin considering the use of the new sampling method.

Perhaps the broadest issue concerns the proper domain of choice-based sampling vis-à-vis exogenous sampling. The great practical advantage of the former approach over the latter one is its relative cost. Whenever decision-makers are physically clustered according to the alternatives they select, choice-based sampling can achieve economies of scale in surveying not available in exogenous sampling. For example, in studying choice of mode for work trips, it is often less expensive to survey transit users at the station and car users at the parking lot than to interview commuters at their homes. Similarly, in examining choice of destination for shopping trips, surveys conducted at various shopping centres offer significant cost savings relative to home interviews.

While choice-based sampling has a cost advantage, exogenous sampling often permits the interviewer to obtain more detailed information from respondents. It is more likely that individuals will agree to an in-depth interview in their homes than while engaged in a trip. We might therefore speculate that choice-based sampling will find its most fruitful applications in situations where decision-makers cluster according to their chosen alternatives and detailed interviewing is not necessary.

In comparing choice-based and exogeneous sampling approaches to data collection, the empirical researcher should attempt to assess the relative statistical precision of estimation available from each approach. Unfortunately, as indicated in McFadden's chapter in this volume, there presently exists almost no theoretical basis for conducting such a comparison. The absence of theoretical results derives primarily from the fact that in general, the probability distribution of choice-model estimates is a function not only of the sampling process and estimation method used but also of the unknown true parameter value $\theta*$. Moreover, this dependency on $\theta*$ is in some respects quite complex. It appears that few general results on comparative estimation precision are possible. Empirical researchers will, it seems, have to approach this issue in a case-by-case manner, using whatever prior knowledge regarding $\theta*$ that they have available.

We have earlier stated that estimation in choice-based samples can be facilitated by prior knowledge of the marginal distributions $p(z)$, $z \in Z$ and/or $Q(i)$, $i \in C$. It should be recognised that existing estimation theory for choice-based samples presumes that these distributions, when known at all, are known perfectly. In practice only estimates of the attribute distribution $p$ or aggregate shares $Q$ are likely to be available. The impact of errors in estimation of $p$ and $Q$ on the statistical properties of existing estimators is not known yet. Care should therefore be taken in using $p$ and $Q$ estimates in practice. On a related point, care should also be taken to specify correctly the population of interest. For example, if one is modelling mode choice in work trips, decision-makers not making such trips should be censored from the sample and the relevant $p$ and $Q$ distributions will be those for work-trip makers, not those for all trip makers.

*3. A Remark on the General Question of Sample Design*

In addition to its direct value, the recent work on choice-based sampling is important because it frees travel-demand research from exclusive reliance on the home interview survey and because it demonstrates that estimation methods must be suited to the sampling process used.

Beyond exogeneous and choice-based sampling, there exist numerous other sampling processes which may have practical value in choice modelling but whose uses remain largely unexplored. One such process is the mixed exogeneous-sampling-choice-based sampling approach in which a primary exogeneous sample is collected and then 'enriched' with a secondary choice-based sample composed of decision-makers selecting alternatives chosen rarely in the original sample. A second process, related to choice-based sampling, is the one in which an interviewer surveys all trip makers passing through a location (such as a transit station or shopping centre) within a given time interval. In such sampling, the ultimate size of the sample is random as well as its

composition.  Third, travel-demand researchers might
consider using sequential sampling procedures in which the
decision to collect further observations is conditioned on
the data collected thus far.  Clearly, there exist many
possibilities for imaginative sampling which the profession
may want to consider.

## CHOICE MODEL SPECIFICATION: THE STRUCTURED-LOGIT AND MULTINOMIAL-PROBIT MODELS

In the individual choice modelling approach to travel-demand
analysis, the multinomial-logit model has come to occupy a
dominant position for empirical work.  The logit model's
advantages are well known.  The logit-choice probability
form is theoretically grounded in that it can be derived from
a plausible behavioural model, namely that of individual
utility maximisation.  The form is computationally attractive
in that the logit probability is easy to evaluate and is
analytically differentiable.  A further reason for the
profession's acceptance of the logit model is its simplicity
- the ease with which its essential features can be grasped.
See McFadden (14) for the seminal paper on the model.

   The primary shortcoming of the multinomial-logit model is
its lack of flexibility as a representation of behaviour.
Viewed as a random utility model, the logit structure implies
that all individuals have the same tastes.  Moreover, it
implies that the unobserved attributes of all alternatives
can be construed as independent drawings from a single
probability distribution, the Weibull distribution.

   Not surprisingly the restrictive properties of logit have,
over the past five years, led numerous researchers to seek to
develop theoretically sound, computationally tractable choice
models having greater inherent flexibility.  Recently,
efforts to improve on the logit model have resulted in two
breakthroughs.  First, the 'structured' or 'sequential' logit
models, which previously lacked a theoretical basis, have now
been shown to fit within the random-utility framework.
Second, the multinomial probit models, which previously
lacked computational feasibility, have now been made tractable.  Because both structured logit and multinomial probit
are substantially more flexible than the multinomial logit
model, these developments may have important impacts on
travel demand forecasting practice.

   In Sections 1 and 2 below we introduce the structured-
logit and multinomial-probit models and summarise the recent
relevant developments.  Section 3 briefly discusses the
respective roles that we see for these two models and for the
multinomial logit model over the next few years.

## 1. Structured-logit Models

Sometimes termed 'sequential' or 'hierarchical' logit, the
class of structured logit models was introduced in Domencich
and McFadden (7) and studied by Ben-Akiva (2) and by Daly and
Zachary (6) among others.  Notational differences have often

obscured the relations among the various authors' efforts. Within the present volume, superficially distinct but essentially equivalent definitions are given in the chapters by Ben-Akiva and Lerman, by Daly, and by McFadden. The reader is referred to those chapters for the alternative definitions in use.

To many, the structured logit models have offered an appealing framework for the modelling of behaviour in multidimensional choice situations, for example joint destination, mode and route choice. This class of models is flexible enough to depict choice processes ranging from a true sequential process, in which the dimensions of choice are considered in some order, to semi-sequential processes, in which a hierarchy of dimensions exists but some interdependencies are allowed, to the simultaneous process in which no dimensional hierarchy exists. The relative computational simplicity of structured-logit models and the fact that multinomial logit is a special case within the class have added to the models' attractiveness. On the other hand, these models have often seemed *ad hoc* and perhaps behaviourally unreasonable in some unknown way.

Now, independent work by McFadden (17) and by Daly and Zachary (6) gives the structured-logit models a firm theoretical foundation. Specifically, McFadden demonstrates that these models fall within the class of 'generalised extreme value' (GEV) models, a set of random-utility models where additive disturbances have a fixed multivariate extreme value distribution. Daly and Zachary give a distinct proof that structured logit models are 'personal difference' random-utility models, this being the class of models whose additive disturbances have some fixed multivariate distribution. An important finding by both sets of authors is that the random utility interpretation of structured logit holds only if certain very intuitive restrictions on the coefficients of 'inclusive price' variables are satisfied.

The McFadden and Daly and Zachary results clarify a good deal about the basic nature of structured logit models and provide a firm basis for their use in appropriate empirical contexts. Further investigation of the GEV class within which structured logit falls should prove quite fruitful. In particular, initial formulation of a choice model in the GEV form can allow the analyst to determine which among alternative structured-logit models best represents observed choice behaviour.

## 2. The Multinomial-probit Model

The multinomial-probit models are simply random-utility models whose additive disturbances have, for each decision-maker, a multivariate normal distribution across alternatives. In the more interesting forms of the model, this distribution varies systematically across decision-makers rather than being fixed. For example, McFadden (15) and Albright, Lerman and Manski (1) have worked with a random coefficients version allowing a distribution of tastes across the population of decision-makers while Daganzo, Bouthelier and Sheffi (5) and

Hausman and Wise (8) have developed other versions in which
the disturbance distribution is a function of decision-maker
and alternative attributes.  The multinomial probit class
clearly offers enormous flexibility in modelling-choice
behaviour.

   Until recently, computational difficulties in evaluating
multinomial-probit choice probabilities limited applications
of the model to quite small problems.  Now however a sequence
of developments has significantly reduced the computational
problem.  First Hausman and Wise (8), using a series approximation to the bivariate normal distribution function, wrote
a programme capable of inexpensively handling choice sets
having three alternatives.  Then Albright, Lerman and Manski
(1), using a simulation approach, developed a general purpose
but fairly costly probit estimation procedure.  Of greatest
practical importance has been the adaptation by Daganzo,
Bouthelier and Sheffi (5) of Clark's (4) approximation to the
distribution of the maximum component of a multivariate-
normal vector.  Independent tests by Daganzo, Bouthelier and
Sheffi (5) and by Albright, Lerman and Manski (1) verify that
the approximation is generally accurate and quite fast.  The
Clark approximation makes multinomial-probit estimation
computationally feasible for a wide range of choice modelling
problems.

## 3. The Choice Among Choice Models

The empirical researcher now may select from a much richer
set of choice models than in the past.  Some guidance in
making this choice may be useful.

   Where its assumptions are deemed acceptable, multinomial
logit should remain the preferred choice model.  The analytical and computational simplicity of the model and its
ease of interpretation cannot be matched.  Where the
researcher is unwilling to assume that the random-utility
disturbances are independent and identically distributed but
is willing to assert that their distribution is fixed across
decision-makers, the structured logit models and, more
generally, the GEV models should prove useful.  Multinomial
probit can also be used in this context.  Where the disturbance distribution must vary across decision-makers as a
function of attribute values, the multinomial-probit models
should provide the modelling framework.  It is in principle
possible to extend the GEV models to have varying disturbance
distributions also.  However, the structure of these models
would apparently make such an extension difficult in practice.

## MODEL TRANSFERABILITY PROCEDURES

Model transferability involves the use of an existing
estimated model to predict travel behaviour in a different
situation.  In this sense, all uses of travel-demand models
involve transferability to some degree.  Demand models are
used to predict in different time periods than the interview
date, for different or larger groups of travellers, and
perhaps for different choice situations.  Transferability

essentially therefore involves the question of whether demand models capture traveller behaviour that is predictable in new situations or whether the models are a statistical artefact of the existing sample.

In practice, there is no such thing as a perfect model. Because of measurement and specification difficulties and the wide range of determinants of human behaviour, no model can contain all the attributes of general choice situations.  As a practical matter, transferability involves the question of how different choice settings can be before existing models lose useful predictive power.  Of particular interest in practice is whether models can be transferred to geographic locations other than the area from which the original sample of travellers was drawn.  Since the initial development of behavioural models is relatively expensive and time-consuming, the ability to transfer existing models to new situations and geographic areas would provide a low-cost and quick forecasting alternative for practitioners.

The discussion of model transferability can be usefully organised under two sub-headings: (1) tests of transferability as a test of model specification; and (2) procedures for updating and modifying existing models to improve transferability.  The first of these sub-headings views model transferability as the ultimate test of the adequacy of a model specification and attempts to identify empirical conditions under which existing model specifications are transferable.  The second sub-heading attempts to extend the range of model transferability by the inclusion of additional information from the area to which the model is to be transferred.

## 1. Tests of Transferability as a Test of Model Specification

Travel-demand analysts generally accept the principle that no single model specification is adequate for all purposes. Indeed, the common practice of estimating separate submodels for work trips and non-work trips can be viewed as an attempt to circumvent problems of model specification by a process of market segmentation.  In general, the negative finding that a particular demand model is not transferable can be viewed as implying that the model specification used is not adequate to explain differences in observed choice behaviour.  The analyst's options are either to improve the model's specification or to segment the travel choices to eliminate the sources of mis-specification.

Several tests of the transferability of particular model specifications have been reported in the literature (see papers by Ben-Akiva and Atherton (3), Kirshner and Talvitie (9), Pecknold and Suhrbier (19), and Watson and Westin (21). Criteria for judging the transferability of a particular model are: (a) similarity of the coefficients for models estimated on separate samples; (b) ability of a transferred model to replicate individual-choice behaviour within specified tolerance limits; and (c) ability of transferred models to replicate observed aggregate shares within specified tolerance limits.  In addition, Watson and Westin

provide information on the differences in the distribution of
the independent variables between the transferred areas.
This latter information is useful in order to judge whether
the model specification is only transferable for small
changes in the distribution of the independent variables or
whether a particular model specification is valid for
significantly different choice situations.

The workshop discussion dealt at some length on the
relevant criteria for judging model transferability.  The
ultimate purpose of travel-demand forecasting is to predict
changes in the aggregate shares of the choices.  Many workshop participants were of the opinion that performing tests
of statistical differences between coefficients estimated in
different areas was too strict a criterion for judging model
transferability.  Model coefficients that may be judged
different on the basis of statistical significance may be
similar enough to provide forecasts of aggregate shares
within reasonable error limits.  Indeed, because of errors in
model specification, sufficient data will almost always show
that the model coefficients are statistically significantly
different between different areas.  The view of the workshop
participants was that measures of statistical differences in
model coefficients are only indicative of possible problems
of model specification and must be judged against the ability
of the models to predict changes in aggregate shares.
Particularly useful are results that demonstrate conditions
under which the model specification shows persistent large
differences in estimated coefficients and a poor ability to
predict market shares.  For example, two studies (9,21)
identified the condition of whether or not the trip was CBD-
oriented as significantly affecting the transferability of
the model in its coefficients and in its predictive ability.
This result can be used as a basis for looking for improvements in the model specification to reflect attributes of the
trip particularly dependent on CBD-orientation.

Further research is also needed on the sensitivity of
predictions of aggregate shares to variations in estimated
coefficients.  This research would clarify the question of
what differences in estimated coefficients are tolerable
before the transferability of the model for predicting
aggregate shares is affected.  Koppelman (10) has proposed a
general framework for analysing errors in prediction, and
further work in this area is required.

Furthermore, model transferability should not be viewed as
an either-or proposition but rather should examine the transferability of specific components of the models for use in
particular forecasting situations.  McFadden (16) has
distinguished three types of structural taste changes that
may affect transferability: (1) shifts in alternative-
specific parameters; (2) shifts in the scale of parameters;
and (3) shifts in relative coefficient values, such as those
used to estimate model values of time savings.  Empirical
evidence indicates that the first type of structural taste
change may be common and is often severe while the third
type is relatively unimportant and will generally not cause
problems in transferability.  The severity of the second type
of structural change has not been established (in effect, our

previous recommendation regards examining this problem), and additional evidence regarding changes in individual co-efficients is required.

The implication of viewing model transferability in a partial sense is important for predictive purposes. Alternative-specific effects generally capture the mean levels of unobserved variables and therefore can reasonably be expected to be non-transferable to geographic regions with different transport networks and urban configurations. The inability to transfer alternative-specific effects will generally mean that transferred models will predict absolute levels of aggregate shares poorly. (Note the similarity of this situation to the use of choice-based samples with the unweighted multinomial-logit model containing alternative-specific variables. In this case, only the alternative specific coefficients and not the other structural coeffic-ients will be biased.) Predictions of *changes* in aggregate shares due to changes in measured system characteristics will be dependent on the generic coefficients of the model, and these parts of the model are much more likely to be trans-ferable. Because of the non-linearity of the choice probabilities to changes in system characteristics, predicted changes in aggregate shares will depend on both estimated structural coefficients and the predicted base shares (see McFadden and Reid (18) and Westin (22)). For this reason, updating procedures for model coefficients will be important in obtaining accurate predictions of base shares.

Therefore, parts of existing models, especially if combined with the updating procedures for model coefficients to be discussed next, can be transferred for particular fore-casting problems at substantial cost savings as compared to the development of new models.

*2. Procedures for Updating and Modifying Existing Models to Improve Transferability*

Up until this point, transferability has been discussed in terms of using information from one particular geographic area or group of travellers to predict in other situations. Updating procedures attempt to use information available from the area to which the model is to be transferred in order to improve the model's predictive ability. Updating procedures are discussed by Ben-Akiva and Atherton (3) and McFadden (16) and will only be summarised here.

As stated above, alternative specific taste variations are likely to be the most important cause of non-transferability, and these variations will be most important in predicting absolute levels of aggregate shares. If information on base line aggregate shares is available in the area to which the model is to be transferred, the alternative specific co-efficients can be changed to replicate these aggregate shares as closely as possible. For the multinomial logit model, this alteration can be performed to replicate existing aggregate shares exactly, and the required adjustments can be computed fairly readily using existing logit software. (The correction procedure is the same as that used to correct multinomial-logit models estimated on choice-based samples.)

If additional information is available in the area to which the model is to be transferred (such as a small compatible data set), additional updating procedures can be employed.  If models can be estimated in both areas, Ben-Akiva and Atherton (3) suggest a Bayesian updating technique based on the asymptotic normality of the estimators to combine the information of the two samples.  This procedure should be most useful when the prior beliefs of the analyst are combined with estimates of sampling variability of the coefficients in the Bayesian weighting procedure.

In summary, the workshop participants believe that transferring existing models to new situations and geographical areas provides a low-cost and flexible alternative to the development of new models.  Existing empirical evidence indicates that the model coefficients show differential sensitivity to transferability, and easily-implemented updating procedures are available to modify model coefficients.  Further research is required on transferability and the sensitivity of predictions of changes in aggregate shares to changes in the value of coefficients.

REFERENCES

1. Albright, R., Lerman, S. and Manski, C., 'The multinomial probit estimation package: features and operational tests', Cambridge Systematics, Inc., Boston (1977).
2. Ben-Akiva, M. (1973) 'Structure of passenger travel demand models', unpublished Ph.D. Dissertation, Department of Civil Engineering, MIT.
3. Ben-Akiva, M. and Atherton, T., 'Transferability and updating of disaggregate travel demand models', *Transportation Research Record*, no.610 (1976).
4. Clark, C., 'The greatest of a finite set of random variables', *Operations Research*, 9, pp. 145-62 (1961).
5. Daganzo, C., Bouthelier, F. and Sheffi, Y., 'Multinomial probit and qualitative choice: a computationally efficient algorithm', Department of Economics, MIT (1977).
6. Daly, A. and Zachary, S., 'Improved multiple choice models', in *Determinants of Travel Choice*, Hensher, D.A. and Dalvi, M.Q. (eds), Saxon House, Farnborough, England (1978).
7. Domencich, T. and McFadden, D. (1975) *Urban Travel Demand: A Behavioral Analysis*, Amsterdam, North-Holland.
8. Hausman, J. and Wise, D., 'A conditional probit model for qualitative choice', *Econometrica*, 46(2), pp. 403-26 (1978).
9. Kirshner, D. and Talvitie, A., 'Specification, transferability and the effect of data outliers in modelling the choice of mode in urban travel', *Transportation*, 7(3), pp. 311-32 (1978).
10. Koppelman, F., 'Methodology for analyzing errors in prediction with disaggregate choice models', *Transportation Research Record*, 592, pp. 17-23 (1976).
11. Lerman, S. and Manski, C., 'Alternative sampling procedures for calibrating disaggregate choice models', *Transportation Research Record*, 592, pp. 24-8 (1976).

12. Manski, C. and Lerman, S., 'The estimation of choice probabilities from choice based samples', *Econometrica*, 45 (1977).

13. Manski, C. and McFadden, D., 'Alternative estimators and sample designs for discrete choice analysis', unpublished (1977).

14. McFadden, D., 'Conditional logit analysis of qualitative choice behavior', in *Frontiers of Econometrics*, Zarembka, P. (ed.), Academic Press (1974).

15. McFadden, D., 'Quantal choice analysis: a survey', *Annals of Economic and Social Measurement*, 5, pp. 363-90 (1976).

16. McFadden, D., 'Properties of the multinomial logit (MNL) model', *Urban Travel Demand Forecasting Project Working Paper No. 7617*, University of California, Berkeley (1976).

17. McFadden, D., 'A closed form multinomial choice model without the independence of irrelevant alternatives restriction', *Urban Travel Demand Forecasting Project Working Paper No. 77*, University of California, Berkeley (1977).

18. McFadden, D. and Reid, F., 'Aggregate travel demand forecasting from disaggregated behavioral models', *Transportation Research Record*, 534, pp. 24-37 (1975).

19. Pecknold, W. and Suhrbier, J., 'Tests of transferability and validation of disaggregate behavioral demand models for evaluating the energy conservation potential of alternative transportation policies in nine US cities', Cambridge Systematics, Inc., Boston (1977).

20. Theil, H., 'A multinomial extension of the linear logit model', *International Economic Review*, 10, pp. 251-9 (1969).

21. Watson, P. and Westin, R., 'Transferability of disaggregate mode choice models', *Regional Science and Urban Economics*, 5, pp. 227-49 (1975).

22. Westin, R., 'Predictions from binary choice models', *Journal of Econometrics*, 2, pp. 1-16 (1974).

23. Westin, R. and Gillen, D., 'Parking location and transit demand: a case study of endogeneous attributes in disaggregate mode choice models', *Journal of Econometrics*, 8(1), pp. 75-101 (1978).

PART SIX

Chapter 18

THE ROLE OF DISAGGREGATE MODELLING TOOLS IN THE POLICY ARENA

Stein Hansen and Ken E. Rogers

INTRODUCTION

The purpose of transport planning is to provide the executing
agencies with realistic advice on the potential and desirability of various transport developments or operations.
This implies acquisition of technical knowledge concerning
what can be done and how people would react to it, as well as
a sound understanding of the limits of the techniques
available in the policy decision-making process concerning
transport.

In the preparation of transport plans and policies, models
with little, if any, foundation in behavioural theory have
served as the basis for the major parts of impact analysis,
forecasts, marketing studies and policy planning. Such
models have traditionally been composed of equations inspired
by analogy to simple Newtonian mechanics. As the requirements of transport models have been increased in response to
the development towards more comprehensive transport policy
planning, the shortcomings of these models have become
increasingly apparent.

Substantial resources have been devoted over the past
decade to develop disaggregate models of travel behaviour,
i.e. demand, choice and attitudes, to replace important parts
of the traditional transport planning packages that have had
such a major influence on urban transport policies and
planning. The rationale behind this research orientation has
among other things been:

1  If travellers' behaviour can be adequately modelled, more
permanent structural relationships will be developed and
impact analysis and forecasting will become more accurate.
Results from one urban area may under certain conditions be
valid for other urban areas thus reducing the cost of studies
needed.
2  The transport system can be described in terms of system
attributes (cost, travel time, number of transfers, etc.).
Such attributes are not only crucial to the travellers who
base their choice on them, they also constitute very
important parameters in the hands of the policy-makers since
they are under their control and also provide a basis for
evaluating alternative policies.
3  Since the ultimate goal of transport policy decisions is
to satisfy the travelling (and non-travelling) populations, a
better understanding of individual reaction to, and

appreciation of, changes in transport system variables
become a central issue to transport policy-makers as well as
to suppliers and designers of transport systems.

It is against such a background that a critical evaluation of
the relevance or otherwise of disaggregate modelling tools in
the policy arena must be undertaken.


THE TRANSPORT POLICY ARENA

The policy arena in which disaggregate travel behaviour
models should be evaluated against alternative model
approaches is heterogenous and comprehensive.  Many of what
we now consider to be elements of the transport policy arena
would not have been so considered fifteen or twenty years
ago.  The most obvious illustration of this is the development in the area of land-use transport modelling.  The bulk
of such models developed over the past two decades consists
of two independent submodels; one for transport which
considers the land use as predetermined, and another for land
use which assumes a prefixed transport system and transport
attributes.  Only recently have simultaneous land-use transport interaction models for policy evaluation purposes been
developed (5).

Experiences with such models have clearly revealed that a
number of so-called non-transport policies may have substantial impacts on the travel market.  Such policies should
therefore be included in a comprehensive definition of the
transport policy arena.  However, since almost everything is
more or less interdependent, practical considerations call
for some limitation.  This chapter will exclude general
income policies from consideration although it is known that
several important categories of trip-making are highly income
elastic.

The arena for possible application of disaggregate
behavioural models therefore consists of user - and supplier
- related policies that affect travel decisions directly, as
well as those that affect travel decisions indirectly through
changes in land use.  These categories can be further subdivided into policies related to the use of existing stocks
of transport systems and other land uses, and policies
related to alterations in these stocks.  Furthermore, a
subdivision can be made between the use of the models for
pure prediction purposes and for evaluation purposes.

Alternatively one can subdivide the policy arena into
short- and long-term policies.  It is felt, however, that
this classification is rather ambiguous and therefore less
efficient unless one has a clear notion of what is short term
and what is long term, and one can assume that a given policy
element can be uniquely classified as either short- or long-
term even under very different political situations.  Neither
of these conditions are normally satisfied.  A well known
dilemma facing economists whenever attempting to apply text-
book theories on real-world situations is the distinction
between short-term and long-term marginal costs.  Policies
classified as long-term oriented under one type of political

setting may well become short term under a different political setting.


## THE EVALUATION CONTEXT

The problems facing transport and land-use planners today are
substantially more comprehensive than those considered only
ten years ago.   The new demands require that a transport
model contains elements that make it capable of:

1  Explaining the existing travel-market structure in terms
of travel attributes and socio-economic characteristics.
2  Detecting structural changes and producing reliable
demand and choice predictions that are responsive to
transport- and land-use policy changes.
3  Revealing the travellers' expected reactions to the
introduction of hitherto unused policies, modes and systems
of travel.

Disaggregate-behavioural travel models have been developed in
response to these new requirements.   The researchers who have
been involved in this process have put forward a number of
convincing arguments in favour of disaggregate approaches.
However, proper evaluation requires careful selection of the
alternative modelling approaches to be subjected to comparison, and the evaluation of disaggregate-behavioural
models in the transport policy arena is presently complicated
for the following reasons:

1  Two different models may require two different data sets
in order to perform efficiently.   Comparisons carried out by
applying different models to the same data set may therefore
be misleading.
2  Comparisons are complicated due to the multiple objective
nature of the policy-evaluation problem.
3  Theoretical model properties pertain to the present state
of the art and may well improve substantially and at different rates for the alternative modelling approaches with
further research.

In evaluating the relative merits of disaggregate travel-
behaviour models, several criteria must be considered, and
severe difficulties are encountered in weighting and trading
off among these.

    The major areas in which disaggregate travel-behaviour
analysis has been applied as an alternative to other methods
are outlined below in order that some evaluation of relative
merits can be made:

1  Estimation of parameters for use in policy-impact studies.
2  Revealing tastes and preferences as well as choice and
demand reactions for use in design and marketing.
3  Estimation of parameter values for use in project evaluations of benefits and costs.

ACHIEVEMENTS IN POLICY IMPACT ANALYSIS

Urban transport policy decisions deal with aggregates.  These
may be geographical zones or socio-economic groups.  The
conventional aggregate-transport models deal with spatial
zones, and travel demand as well as supply variables are
therefore zonal aggregates.  A basic premise for disaggregate
models to be successful in a policy planning context is that
efficient aggregate results from disaggregate models can be
derived.  This issue has received surprisingly little attention until quite recently.  However, intensive and highly
qualified research has clarified the aggregation problems
faced when interzonal flows are to be forecasted on the basis
of correctly calibrated disaggregate models.

Recent research has demonstrated that the aggregate and
disaggregate models have a common form and that it may be
possible to use a synthesis of the models to facilitate calibration and improve forecasting accuracy (8).  Biases
resulting from direct calibration of an aggregate model when
the individual behaviour conforms to the disaggregate model
have been derived, and their implications for forecasting
have been illuminated.  Furthermore the biases resulting from
using some zonal-average variables along with disaggregate
variables in estimating behavioural parameters have been
identified and correcting formulae derived (8).

Disaggregate travel-choice models have been developed and
improved rapidly over the last ten years both in terms of the
theoretical foundations of the models, their computational
efficiency and their ability to provide inputs into the
policy planning process (3).  Early disaggregate-transport
models contained two explanatory variables; door-to-door time
and cost.  Today travel survey questionnaires will explicitly
address such travel attributes as: line haul time; walk time;
wait time at first stop; number (and perhaps length) of
transfers; seat availability; fare; parking charge; operating
cost, plus a number of socio-economic attributes of the
individual traveller.

Travel demand in conventional aggregate models is usually
measured as a function of one impedance variable only.  This
variable therefore expresses the so-called generalised cost
of travel.  This is composed of a weighted sum of all the
travel attributes assumed to affect the trip-making decisions
of the travellers.  For the decision-maker, however, it is
important to know the relative importance of each of these
attributes because (1) travel demand is more sensitive to
changes in some attributes than others, and (2) the resource
cost of changing travel demand by a *given* amount depends on
which travel attribute is chosen to be altered.  Empirical
estimates of such detailed demand and choice parameters make
it possible for decision-makers to choose transport systems
more closely tailored to the preferences of the traveller
within the severe budgeting constraints normally imposed upon
the urban transport authorities.

The present disaggregate approach takes explicit account
of the observation that travellers seem to dislike long waiting and transfers very much.  Rather than using door-to-door

time as the key system design and travel-demand variable,
three separate effects are identified and their relative
impacts on, for example, choice of mode determined.  The
approach also makes it possible to calculate the implicit
trade-off values, such as how much extra money or line-haul
time an average traveller is willing to sacrifice in order to
avoid a transfer or in order to reduce waiting times by five
minutes.

The introduction of disaggregate-behavioural mode-choice
models has thus made the entire urban transport modelling
approach more compatible with economics which suggests the
following approach to policy evaluation:

1  Identify the technical *efficiency frontier* in the existing system and relate the actual system performance to this
level.
2  Perform a proper economic supply and demand study in
terms of resource costs to identify *economic efficiency* in
system performance.
3  Evaluate system expansion on the basis of the results
derived under (1) and (2) and on equity considerations.

In conventional transport models it is normally implicitly
assumed that the system is performing efficiently in technical terms, and that system attribute values reflect resource
costs in such a way that economically efficient travel-demand
levels have been identified.

These assumptions are not unrealistic when modelling
inter-city traffic in well organised and highly developed
countries.  However, when modelling intra-city traffic in
developing countries, assumption (1) is normally severely
violated.  The second evaluation model assumption is violated
in virtually every city experiencing congestion (Singapore
may come closest to being an exception to this statement).
Optimal use of the transport system calls for careful monitoring of user charges (petrol costs, parking charges, public
transport fares, etc.).  The travel volumes resulting from
efficient pricing policies are those relevant for analysis if
and when expanding the physical infrastructure is appropriate.

Obviously it is extremely difficult and costly to identify
the economically optimal solution to co-ordinated operation
of the private and public components of the urban transport
system.  However, the analysis techniques of disaggregate-
choice models have opened up substantial new ground for
information-gathering that will help the planners and
decision-makers to identify more efficient uses of the
existing infrastructure and vehicle fleets before embarking
upon the appraisal of expanding the road and rail infrastructure.

The major emphasis has been on the analysis of the
journeys to/from work.  This has been a natural starting
point for three reasons:

1  Work journeys are peak-period trips with the most severe
external effects (congestion, pollution).
2  From a modelling point of view it is convenient that not

too much realism is lost in introducing the simplifying
assumptions of completely inelastic overall-work journey
demand and zero-trip substitution between peak- and off-peak
periods.
3 Origin and destination of such journeys are prefixed for
fairly long periods.

In a number of metropolitan areas in USA, Europe and
Australia empirical disaggregate-behavioural mode-choice
studies have been conducted for work journeys. (See (4) for
the most complete survey available.) These studies have
revealed very useful information about the travel choice
impacts of changing the various travel attributes as well as
the values of journey attributes. This has paved the way for
much sounder analysis of choice between policy-options,
because it introduces an explicit impact comparison of different low-cost early-action alternatives. Such alternatives
to immediate infrastructural expansion are often ignored or
forgotten because the conventional aggregate models have no
way of identifying them. The results of such studies have
not only been of considerable value to decision-makers they
have also led to the development of more general disaggregate
travel models.

The trip categories to receive attention next were shopping, recreation and personal business. Such trips have been
rapidly increasing in importance with the development of
large shopping and service centres outside downtown areas.
Such locational patterns result in circumferential rather
than radial trips and therefore in a high percentage of car
journeys, since circumferential public-transport supply is
always relatively poor.

The two simplifying assumptions that made work-journey
mode choice a relatively simple modelling exercise cannot be
maintained for these non-work journeys without deviating too
much from the real world picture. In attempting to model
non-work trips one has to assume that the number of trips
will depend on the travel conditions, the time of day for
shopping can be altered, and that the place of shopping can
be changed. The models developed have therefore addressed
some or all of these travel choices in the same model (1,2,3).
This has called for an additional set of explanatory variables to describe the attractiveness of the destinations
available to the travellers.

The following comments are common to all the disaggregate
models, i.e. the partial-mode choice and the multiple-choice
models:

1 Standard impact measures such as the partial elasticity
of choice must be carefully interpreted because the elasticity value in itself may be rather sensitive to both the
initial probability and to the initial attribute level.
Typically, the partial elasticities of choice of car may be
very low because virtually everyone uses a car anyway.
2 However, the partial car-choice elasticities may be very
high when the probability of choosing car is 0.5 for the same
group of people.
3 What is needed for proper policy impact analysis is

sufficient variability in the data to estimate accurately the variation in the partial-choice elasticities with the initial situation.  Only then will the models be useful to the planners in providing information on the expected reaction to policy changes and on the proper dosage of each policy from a given initial situation.

Specific to multiple-choice models are the following features:

1  Multiple-choice models can be joint or sequential (1). Joint models yield estimates on the partial choice elasticities that are different from those of sequential models applied to the same data set.  Typically the values of line-haul and excess time which are derived as the ratio of time and cost coefficients become approximately twice as high when a joint model is chosen.
2  A typical sequential model differs from the joint model solely in that the coefficients of inclusive prices are not constrained to equal one.  Hence, the joint model is a linear restriction on any of the sequential models.  The sequential model is better in the sense that it is more likely to be valid, but worse in the sense that if the joint model is valid, it can be calibrated and tested with more powerful methods than the sequential models (7).

Furthermore, multiple-choice models that include destination choice will yield information of value to land-use policy decisions with transport-sector implications.  A typical example is the use of attraction variables in the destination choice equation.  Experience has shown that floorspace is a good attraction proxy in a model for choice of shopping destination (2).  Obviously, the initial market share affects the partial destination-choice elasticity, but for a given initial market share, disaggregate models yield valuable information on what percentage increase in market share for a given shopping location can be expected if the floor area is increased 1 per cent.

Experience from Sweden (2) indicates that the elasticity increases with floorspace, and with threshold values (i.e. from where the market share elasticity with respect to floor-space surpasses one) established for each initial market-share level.  One conclusion to be drawn from such models is that as shopping centres already above the threshold level expand and attract proportionally more trips, shopping centres below the threshold value will contract but at a less than proportionate rate.  Thus the total outcome will be net shopping trip generation in addition to the redistribution of the initial trips.  Thus disaggregate multiple-choice travel models can reveal policy impacts of, and thereby provide input into, the simultaneous planning of land use and transport.

The other direction of model improvement has been in the area of clarifying the underlying behavioural assumptions of the disaggregate models.  The models came about as statistical curve fitting models and for many years the search for a logical behavioural underpinning was rather unsystematic. Faced with several equally good statistical models (probit,

logit, discriminant, etc.) attempts were made to single out
their relative merits from a behavioural theory point of
view.   So far the logit-model has been chosen due to pioneering research by McFadden (3).  However, the theoretical
clarifications have also revealed the weakness of the logit-
model, i.e. the situations in which it is likely that the
logit assumptions will be violated and the model predictions
misleading.  Methods have recently been developed to test
these assumptions (7).

   Inter-city transport presents much less of a modelling
challenge than intra-city transport.  Unfortunately, models
developed for inter-city planning have dominated the intra-
city planning arena as well.  The crucial difference between
inter- and intra-city transport is the role of transport
supply.  In inter-city models little reality is lost by
exogenising supply and concentrating all model development on
the demand side.  Intra-city transport is, however, strongly
dominated by severely congested peak periods.  The main
characteristic is thus a supply which is just as endogenous
as the demand side.  This introduces the scope of much more
complex economic models and with the refinements achieved on
the demand side by introducing disaggregate models, an
imbalance within the overall model is felt unless the supply
side is also given more attention.

   In recent years supply models have been developed on the
basis of disaggregate data in order to provide more appropriate access and egress times for zones (6).  Such supply
models contain travel-volume and transport-system attributes
as explanatory variables, and by proper use of the disaggregate data through aggregation, unbiased and accurate zonal
time estimates are derived which will improve the accuracy of
the demand side of the model.

   In such simultaneous supply/demand models of the urban
travel market the travel volumes are endogenous and so is
travel time.  Other transport-system attributes of a more
traffic engineering nature could well enter as exogenous
aggregate variables in the supply submodel in order to
improve the accuracy of the supply forecasts.

   Such variables have for the most part been ignored in
supply models developed up to now, or alternatively they have
been accounted for by segmenting the supply side.  Even then,
it is important to distinguish between actual performance and
technically efficient performance.  Typical examples are the
speed/flow relationships of which there is one for each type
of street category.  In well-organised cities with an
experienced traffic police and a carefully monitored traffic-
engineering programme, actual performance comes very close to
what can be expected theoretically.  In rapidly developing
cities, however, this assumption is normally severely
violated and it is not unusual to find for a well-defined
street category that maximum traffic flow for a given speed
is only one half of what is achieved on a similar street in
efficiently organised cities.  The logical but unfortunately
very often forgotten remedy is to strengthen the corps of
traffic police and to improve the enforcement of traffic
laws, to improve and integrate traffic signals, and to work

out a comprehensive traffic-engineering programme to be
implemented in the system.  Studies carried out in Tehran (5)
indicate that a 50-100 per cent increase in performance on
existing streets can be achieved, thus postponing what
otherwise seems to be the optimal implementation time for
capital-intensive system expansion programmes and thereby
saving substantial sums for alternative uses.

By introducing disaggregate-behavioural travel models
nothing is really gained with respect to the particular
assumption of technically-efficient system performance.
Neither the conventional nor the disaggregate-behavioural
models address this particular transport policy issue.  The
main reason it is neglected is - we believe - that the models
have been developed and tested in countries where this
technically-efficient assumption is *not* severely violated.
Unfortunately, many consultants have adopted these models
uncritically and applied them in developing countries as
well, thus often recommending authorities to rush into
expensive system-expansion programmes too rapidly.

Although we strongly criticise transport model developers
for having given this important supply aspect of transport
planning a cinderella treatment, we do not believe that this
is an area where the introduction of conventional disaggregate-behavioural travel models will contribute much.  What is
needed here is a more thorough analysis of the partial
impacts of the various traffic-engineering measures on
overall flows and speeds for a given intrastructural system.
The individual system user will, however, perceive these
policy impacts in terms of the system attributes upon which
*his* choices are made (cost, time, safety, convenience,
comfort, etc.) and to trace these impacts, the disaggregate
approach seems very promising.


DIRECTIONS FOR FURTHER DEVELOPMENT

This analysis has reviewed the contributions of disaggregate-
behavioural travel-choice models developed and used over the
past ten years.  Achievements have been substantial, yet it
is felt that one can by now refer to these models as the
conventional techniques.  The major reason for this being
that the increased knowledge and awareness of how people
behave provided by these techniques has raised questions
which are, necessarily, unanswerable by the methodology that
provided the awareness in the first place.

Typically the conventional-disaggregate techniques have
been designed to provide answers to the urban transport
policy and planning problems of cities with well-developed
alternatives to the private car for radial trips to/from the
city centre.  One has therefore concentrated research and
experimental efforts on the study of travellers who are
facing such a choice.  This direction of research has implied
a restriction of studies to real-choice situations which, due
to the limited number of alternatives, has the unfortunate
effect of fixing the parameters of models, in large part, by
the relatively small number of people in marginal-choice
situations.

Data have been collected from questionnaires designed to provide a proper field for studying how people do behave in relation to what they think their choices are.  The key word has been choice analysis and its area of application has been rather static.  Perhaps more appropriate when trying to extend the use of disaggregate-behavioural study techniques to the transport problem of rapidly growing and severely malfunctioning cities is a distinction between change and selection.

In order to increase productivity of further developments in disaggregate techniques one should turn to the Third World and the problems faced in their cities.  The conventional choice question seems less relevant there since alternatives are virtually non-existent - at least in a Lancasterian sense.  This is certainly true for the car users who will find the present public transport system inferior in every respect.  It also holds true for the public transport users who view car availability as something completely out of reach.  This being the case, a conventional disaggregate study has little chance of success since hardly anybody is in a competitive choice situation.  Thus the creation of a proper sample becomes a major obstacle to the study.

In this situation the policy issue of immediate concern is one of altering the relative performance of the available modes, so that choice becomes a meaningful concept to the travellers.  Their choice may then be to change mode.  In a developing country the traveller is likely to have much less - if any - prior knowledge about the alternative than we have found to be the (prerequisite) case for proper disaggregate choice analysis in well-developed cities.

Based on experience from a number of such change studies one may embark upon the second choice aspect which is more related to the long-run equilibrium issue of *selection* of alternatives.  Obviously it is onerous to change back and forth between modes for repetitive trips.  The individual will be much more at ease when following a routine pattern in this effort-minimising activity.  He is therefore concerned with selecting a way of performing this activity that will remain stable for long periods at a time.  To the policy-maker as the ultimate system supplier it is therefore important to convince the travellers who are in the transition stage where they are facing day-to-day changes, to select the public mode as their more permanent means of travel for repetitive journeys.  Surely, change and selection are both choice issues, but in making the distinction the approach becomes more compatible with the dynamics of the Third World cities with growth rates often between 5 and 10 per cent per annum.

In the absence of explicit options for people to choose from, conventional choice-analysis methods become somewhat inefficient.  Insufficient observations for proper estimation of the parameters needed for impact and evaluation analysis becomes a major obstacle.  These studies now have to be supplemented by studies of *how people think they would behave* in relation to explicitly defined options.  Such studies provide for the examination of a wide range of choice

options, the use of an extended range of analytical techniques, a much closer examination of individual choice
mechanisms and a broader basis for the evaluation of
alternative disaggregate methodologies.  Thereby the general
applicability and attractiveness of such modelling tools to
those operating in the policy arena should become more convincing.  At the same time, a somewhat more appropriate basis
for the comparative evaluation of alternative methodologies
emerges.

The feasibility of this type of approach has recently been
tried in Tehran (9).  The experimental design which is suggested for a comprehensive and generally applicable disaggregate travel-behaviour study provides data on *both how people
do behave in relation to what they think their choice options
are, and how people think they would behave in relation to
explicitly defined options.*  By collecting both types of data,
a proper examination of transferability is possible, and the
potentials for future savings in data collection efforts can
be estimated.  The approach allows for checks on the consistency of each individual's responses.

In preliminary discussions concerning the survey, it was
common to hear comments like, 'Iranians aren't used to
surveys, they won't be able to answer your questions', 'You
can't expect to find a systematic explanation of Tehrani
travel behaviour', 'You can't expect sensible answers to
hypothetical questions', and so on.  All such comments and
doubts can now be discounted.  It is clear that when people'
take the trouble to answer the questionnaire they do make
sensible responses.  A rational explanation was found for the
behaviour of all individuals who completed the questionnaire.

Although the approach adopted was due to the peculiar
nature of travel in Tehran, it is, of course, generally
valid.  Since it affords an opportunity for the study of
individual travel behaviour at a much greater level of detail
than is possible with conventional disaggregate methodology,
the range of application of the technique extends far beyond
Tehran, Iran, or even developing countries.

The questionnaire contained ten sections.  The first six
sections provide - along with socio-economic background (also
included in Section 10) - what is needed for conventional
disaggregate mode-choice analysis, assuming that the conditions for such an approach are met (for example that the
traveller has proper information about alternatives *not* used).
The remaining sections provide complementary attitudinal
data, for example, information on the individual's perception
of his probable behaviour in relation to a number of
explicitly defined alternatives.  These answers (Section 7)
can be used in conjunction with the attribute values of
Section 9 to give for each individual a set of corresponding
values for the probability that the individual will change
mode and utility (or generalised cost) level.  The data can
be analysed to enable both parameter estimates and attribute
values to be obtained.  Both can then be compared to the
parameter estimates and attribute values obtained on the
basis of the actual choice information contained in Sections
1 to 6 if the survey is undertaken in an environment where

actual choice situations can be identified.

In brief, the issues addressed in this 'double' approach
are as follows:

Section 1.  What is your usual mode of travel to work?
            (car, taxi, bus, other)
            What is your usual mode of travel from work?

Section 2.  What is your alternative mode of travel to work?
            What is your alternative mode of travel from work?

Section 3.  What factors cause you to choose the mode(s) of
            travel given in Section 1?
            Please give due weight to these factors in answer-
            ing the remainder of this questionnaire.

Section 4.  Give details of your usual journeys to and from
            work, i.e. walking time, waiting time, in-vehicle
            time, if you have a seat, if you have to transfer,
            cost or, of car users, distance.

Section 5.  For bus users; give details of the car journey you
            would have undertaken if you owned a car and used
            it for your work journeys.

Section 6.  For non-bus users; give details of the bus
            journeys you would have undertaken if you used the
            bus for your work journeys.

Section 7.  Respondees are asked to indicate, on a continuous
            scale from 0 to 1, how likely they would be to use
            the bus for 22 bus journeys defined in terms of
            walking time, waiting time, having a seat or not,
            having to transfer or not, cost, and difference in
            in-vehicle time from present chosen mode of
            travel.  Bus users are asked to suppose they have
            a car.

Section 8.  Eleven journey factors are identified, six of
            which are quantifiable.  Respondees are asked to
            indicate, by a number between 0 and 20, the
            importance they attach to improvement ir these
            factors.

Section 9.  For the six quantifiable variables of Section 8,
            respondees are asked to answer a number of
            questions of the form:

            'How much longer (in minutes) would you be pre-
            pared to walk to be certain of having a seat?'

            'How much longer would you be prepared to wait to
            avoid 10 minutes of in-vehicle time?'

Section 10. Respondees are asked to indicate their income.

Each section of the questionnaire constitutes a design
element in an integrated framework for the examination of observed behaviour and the evaluation of alternative techniques
and methodologies.  This framework is shown schematically in
Figure 18.1.

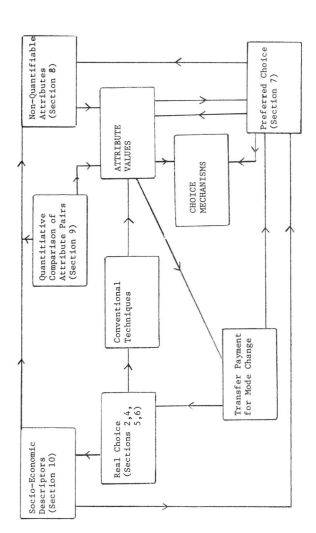

Figure 18.1   Interactions Between Design Elements

The pilot survey was undertaken in Tehran to test the questionnaire and to evaluate the validity of modelling individual-choice mechanisms and attribute evaluation with extensive use of attitudinal data. It was not therefore possible to examine all the effects which this approach allows. The survey is described in detail elsewhere (9); here, it is only necessary to summarise the major results. These are given below:

1   The travel choice of all individuals (n = 38) was rational.
2   The distributions of individuals' values for six travel attributes were positively skewed.
3   The correlation matrix of individuals' attribute values showed significant positive correlation between the values of attributes which related to the length or difficulty of the journey.
4   Individuals' attribute values are uncorrelated with income.
5   Model change behaviour is specific to the mode of travel used implying that, for example, the same modal-choice model should not, in general, be used for both bus users and car users.
6   The previous result may be explained by the further result that the values of some non-quantifiable attributes (i.e. reliability and being able to make other trips during the working day) are specific to the mode of travel selected by individuals.

The Tehran survey was restricted to the consideration of the journey to work, probably the choice situation most studied by conventional methodology and best understood. Nevertheless, conventional methodology cannot give results at this level of detail. These results, if confirmed by further work, add significantly to our knowledge of individual behaviour; they should be relevant to policy-makers in developing both transport systems and marketing strategies. Further sophistication of transport planning models may be required if the differences in behaviour at the individual level lead to differences at the aggregated level.

Certainly substantial improvements with this approach are possible, but if this is found to be the path to follow in extending the disaggregate approaches from their conventional areas of application, improvements will come about very rapidly and much existing research with direct policy implications can be expected.

REFERENCES

1. Ben-Akiva, M. (1973) 'Structure of passenger travel demand models', unpublished Ph.D. thesis, Dept of Civil Engineering, MIT, Boston, Mass.
2. Berglund, C.-O., Tegner, G. and Widlert, S. (1976) *Val av resmal och fardsatt vid inkøpsresor - en individorientert beteendestudie* (Choice of destination and mode for shopping trips - a disaggregate behavioural study) Almanna Ingeniørsbyran AB, Stockholm.
3. Domencich, T.A. and McFadden, D. (1975) *Urban Travel Demand - A Behavioural Analysis*, North Holland Publishing Co., Amsterdam.

4. Hansen, S., 'An analysis of low cost early action local transport improvements', *Technical Report No. 5*, Tehran Development Council Secretariat, Tehran (1976).

5. Hensher, D.A. and Dalvi, M.Q. (eds) (1978) *Determinants of Travel Choice*, Saxon House, Farnborough, England.

6. Hilsen, N. and Talvitie, A., 'An aggregate access supply model', *Proceedings*, 15th Annual Meeting Transportation Research Forum, San Francisco (1974).

7. McFadden, D., 'Properties of the multinomial model', *Working Paper No. 7616*, from Urban Travel Demand Forecasting Project Institute of Transportation Studies, University of California, Berkeley (1976).

8. McFadden, D. and Reid, F., 'Aggregate travel demand forecasting from disaggregated behavioural models', *Transportation Research Record No. 534*, Washington, D.C. (1975).

9. Rogers, K. and Hansen, S., 'Analysing the travel behaviour of individuals', Norconsult, Stockholm, Sweden (unpublished).

Chapter 19

ROLES OF BEHAVIOURAL TRAVELLER MODELS IN URBAN
POLICY ANALYSIS

Fred A. Reid

INTRODUCTION

An investigation of the policy role of an analysis tool must
necessarily be based on a description of the tool and an
identification of the activity of policy analysis.  This
chapter begins with a general descriptive definition of the
type of models considered, discusses their features, reviews
examples of their applications, identifies considerations in
policy applications, and assesses the status and potential
for behavioural models in policy analysis and planning.  The
behavioural scope and choice structure of the models for
which policy analysis roles are considered are identified;
and some examples of calibrated models are used to review
their features and limits.

     A particularly important part of breaking from the 'solu-
tion looking for a problem' status into a standard tool of
the trade is to understand the trade.  Hence, considerable
space is taken in reviewing applications examples from the
point of view of how far they have extended into the policy
realm as well as how much they use behavioural methods.
Attempts are made to define a policy analysis 'role space' in
planning activity objectives.  This is not merely a realm of
issues or of social goals, though these in turn are examples
of the dimensions of the space.  Other dimensions include the
behavioural generality and the resources required of
analyses.  The definition of a policy role space will necessarily be based on the realm with which the author is
familiar - urban region passenger transport planning.  Even
in this limited applications context the attempt independently to identify the *role* separately from the *tool* is risky,
as there may be analytic potential of models not apparent
from the policy analysis point of view.  However, the reverse
process - the development of a tool from a methodology - is
less likely to define useful applications.  Also, it is
expected that readers, being predominantly methodologists,
will have little problem imagining new analytic applications
of the models.

     We also assess and speculate on extensions of the current
policy analysis package categories with the aid of the role
space dimensions.  The standard and evolving analysis
packages are criticised and some new categories are suggested
which more effectively meet planning objectives and in most
cases employ behavioural methods.  Within these analysis

systems as well as in other policy roles, both promising and less worthwhile applications of the models are identified.

Finally, suggestions are separately made for research and for demonstration project support necessary to realise the potentials for behavioural models in effective policy analysis.

## DESCRIPTIVE DEFINITION OF BEHAVIOURAL DEMAND MODELS

The tools for which the role is investigated here are disaggregate travel behaviour models.   These are models expressing the transport-related choices of a decision unit (usually an individual) as a function of the characteristics and travel environment of the unit.  Though the discussion will be in terms of individual or household decisions, such as choice of mode of passenger travel, the framework is similar for discrete decision models of individuals or organisations in economics and psychology.  Examples are decisions on shopping frequency, purchase of goods or services, job selection, family size, and so forth.

These discrete choice decision unit models are necessarily probabilistic as they attempt to express individual choices without complete information on the chooser or his travel situation.   Common forms of these models are multinomial logit, probit and discriminant analysis functions, where the explanatory variables enter as linear sums (5,24,31,49).   In order to qualify further the models for which a policy role is being investigated, a possible set of the choices being considered will be broadly defined.

Multiple alternatives might be involved in each of the following *choice* dimensions:

L residential locations;
A auto-ownership levels;
F trip frequencies;
D trip destinations;
M modes of travel.

The set of choices considered in a particular analysis may be smaller or greater than this depending on the planning objectives.   The general mathematical framework for these choices is described further below to form a consistent behavioural base for revealing what scope of modelling is necessary to satisfy different analysis roles.

A general expression for the behaviour of a decision-maker facing the choice dimension set above is given by the function g representing the joint probability of his choosing one alternative in each dimension:

$$P_{mdfa\ell} = g(\beta, x_{mdfa\ell}, s) \; ; \tag{1}$$

where

$\ell$, a, f, d, and m index the five choice dimensions listed above;

x is a vector of attributes for each of the alternatives of each choice dimension;
s is a vector of socio-economic characteristics of the decision-maker;
is a vector of parameters relating choice to the attribute and characteristics vectors.

This form is usually termed the *disaggregate direct* or *economic demand model* and presumes no ordering of the decision process (7,16). Domencich and McFadden (11) discuss the properties of this framework, pointing out that it may, without loss of generality, be decomposed into a product of conditional probabilities, that is, separate demand functions, more feasible for estimation and forecasting. Thus the probability of the previous five-dimensional joint choice can, for example, be expressed as:

$$P_{mdfa\ell} = P_{m|dfa\ell} \cdot P_{d|fa\ell} \cdot P_{f|a\ell} \cdot P_{a|\ell} \cdot P_\ell \, , \tag{2}$$

or

$$= P_{md|fa\ell} \cdot P_{f|a\ell} \cdot P_{a\ell} \, , \tag{3}$$

or any other series of products of single or joint conditional probabilities.

This will most generally consider interdependencies between the joint choices when:

1   the functions expressing the conditional probabilities are actually conditional on all of the choices to the right in the product series of conditional probabilities (the more marginal probabilities), and
2   the expressions for the more marginal probabilities are in terms of attributes of all the choice dimensions to the left in the conditional series, i.e. for all those choices not entering conditionally.

Thus, equation (3) could be expressed as

$$P_{mdfa\ell} = g_1(x_{md}, s) \Big|_{fa\ell} \cdot g_2(x_{mdf}, s) \Big|_{a\ell} \cdot g_3(x_{mdfa\ell}, s)$$

where the gs are the choice functions for the conditional or marginal probabilities, and the subscripts on the xs show which choice dimensions the attributes must cover for simultaneous choice generality.

For example, $g_2$ could be a different expression for each $a\ell$ combination or merely contain the variables car ownership and location, at which levels it could be evaluated and multiplied by the marginal probability $g_3$.

Generality may be maintained while dealing with less than the large number of conditional variables and attributes suggested above. Some conditions on some of the choice probabilities may be found insignificant by testing, or judged to be independent. The marginal probabilities can be

estimated without the attributes of all the conditional
choices to the left if a representative set of the mix of
those choices is used to evaluate the attributes of the
marginal choice dimensions. This considerably reduces the
number of variables. However, the resulting model will only
accurately forecast for a group with the same set of unspeci-
fied variables. The traditional four-step *sequential*
transport models are an example of a modelling system with
this latter limitation.

Few disaggregate models have been estimated which express
behaviour in as many as two joint-choice dimensions. Many
have considered multiple alternatives within a choice
dimension, such as that of mode of travel. Three models -
one of two-dimensional choice, a conditional joint-choice
structure version of it, and an elaborately specified mode-
choice model - are outlined below to illustrate the type of
model structures available to form a basis for policy
analyses.

Lerman and Ben-Akiva [18] developed a behaviourally
general (simultaneous) joint-choice model of car ownership
and mode to work. Choices were expressed in terms of the
probabilities that a household would have any of the combinations of A (three) ownership levels and M (two) modes of
travel to work by the principal worker. The probabilities
were the alternatives of a single logit function of variables
that were expected to influence either mode or ownership
choice. The model has the form

$$p_{ma} = \frac{\exp \left[ x_m' x_a' x_{ma}' \right] \beta}{\sum\limits_{v}^{A} \sum\limits_{u}^{M} \exp \left[ x_u' x_v' x_{uv}' \right] \beta} ,$$

where $p_{ma}$ is the joint probability of the two choice sets;

$\quad$ $x_m$ is a vector of attributes of the work mode-choice;

$\quad$ $x_a$ is a vector of attributes of the car ownership
$\quad\quad$ choices;

$\quad$ $x_{ma}$ is a vector of attributes which vary jointly over
$\quad\quad$ both choices;

$\quad$ s $\quad$ is a vector of socio-economic characteristics of
$\quad\quad$ the household decision unit;

$\quad$ $\beta$ $\quad$ is a vector of model coefficients with length equal
$\quad\quad$ to the sum of the elements in the x and s vectors
$\quad\quad$ plus choice-specific constants;

$\quad$ and the primes on the vectors represent their transpose.

Included in $x_a$ were attributes of non-work-trip impedance
alternatives. Work-trip travel times by mode were included
in $x_m$. Total household costs is an example of the elements
of $x_{ma}$. Household income is an example of the elements of s.

This joint-probability model is equivalent in assumptions
and results to estimates of two simultaneous models of mode
and car-ownership choice. Using the conditional joint-choice
decomposition described in the last section, these joint

choices can and have been estimated by their authors as
separate simpler models of car ownership and mode-choice
conditioned on ownership (6).  This allows more feasible
estimates of larger alternative sets.  To be generally consistent with simultaneous choice behaviour, the car-ownership
model must include all of the significant $x_m$, $x_a$, and $x_{ma}$
attributes.

The residential and work locations were assumed to be
given in this model.  If these choices were modelled separately, honouring the above framework, they too could be
incorporated without loss of generality.  Attempts to develop
behavioural location-choice models are rare.  Aggregate
models are currently the only viable location forecasting
alternatives.

The Urban Travel Demand Forecasting Project at the
University of California, Berkeley, has developed models of
work mode-choice, conditioned on car ownership, which predict
selection probabilities for seven mode-submode alternatives.
They include policy-responsive variables not previously
related to the choice function (46).  The model expresses the
probabilities of choice in the following form:

$$p_m = \frac{\exp [x_m' s']\beta}{\sum\limits_{u}^{M} [x_u' s']\beta} .$$

The M (seven) choice alternatives are:

> car driver, alone;
> carpool or passenger;
> bus, with walk access;
> bus, with drive access;
> rail, with walk access;
> rail, with drive access;
> rail, with bus access.

Examples of elements of the vector $x_m$ of mode-choice
attributes are:

> variable cost of trip alternatives;
> on-vehicle travel times;
> walking times for transit trips;
> first headways for transit trips;
> transfer headways for transit trips.

Examples of elements of the vector s of socio-economic
characteristics of the household from which the trip is
being made are:

> income of household;
> number of household drivers;
> residential tenure;
> density at workplace;
> head of household boolean.

The model also includes the variable 'number of cars per
driver in the household'.  Thus it can be used as a component

of a conditional joint-choice prediction structure as the
previous models.  Alternatively, for policies which do not
greatly effect car ownership, mode-choices can be predicted
with base-line values of the number of cars.

Behavioural variation across socio-economic groups was
accounted for by dividing the cost variables by wage rates
and allowing the on-vehicle time coefficients to be different
for urban and suburban groups.  These coefficients were also
found to be significantly different for car and transit modes.
These models, calibrated in a bus and car mode environment,
were validated against the behaviour observed after the
introduction of a rail mode in the same region.  The sum of
absolute errors in the aggregate shares of the seven modes,
expressed as a percentage of all travellers, was 16.7 per
cent for predictions of the calibrated model in the new
environment.

The tests also show that the magnitude of error due to
assuming independence among seven alternatives when using the
multinomial-logit choice function is small compared to other
sources of error such as data uncertainty and misspecifica-
tion.

This mode-choice model includes car-ownership variables
allowing (and implying) conditional joint-choice formulation
to be made with a car-ownership model for forecasting.  Since
work trip-end locations are assumed to be fixed, this and all
of the model examples are not sufficiently general to forecast joint-choice with location.

These models represent the range of choice dimension,
choice alternatives, explanatory variables and choice
structure generality which have been well tested for policy
analyses.  The conspicuous shortcomings are in location
choices.  The trip-timing decision, though not mentioned in
the list of choice dimensions, is another element of behaviour not well understood but potentially important to policy
analysis.

Other approaches to choice modelling not mentioned include
the calibration of probit-choice functions and attitudinal
models.  Probit functions are potentially valuable to account
for the distribution of preferences across travellers but
have not reached a practical level in estimation theory or
aggregate prediction methods (see Chapter 17 of this volume).

Models based on attitudinal variables are more useful for
guiding research on behaviour than in forecasting because of
the difficulty of predicting shifts in attitudes as a first
step to modelling behaviour in terms of attitudes.  The
premise of this investigation is that policy analysis must
relate demand and its impacts to the quantifiable parameters
of transport operations and the socio-economic characteristics of the population.


EXAMPLES OF THE USE OF BEHAVIOURAL MODELS

A number of examples of the *use* of behavioural models by
public agencies are reviewed below.  These applications are

grouped in the typical planning categories: conventional
urban-transport planning (UTP), generalised short-run policy
analysis and special purpose or project studies.  (Generalised short-range policy analysis is used as a category here
rather than sketch planning models since behavioural applications in this role have tended to be short-range but
otherwise more general policy tools than sketch planning
models.)  Key examples are discussed in more detail with
special attention to their policy role.  A recent review by
Spear (40) of the US Federal Highway Administration discusses
some important disaggregate model applications.  His examples
will only be given brief comment here.


*Conventional UTP Systems*

Several metropolitan planning organisations (MPOs) have
incorporated disaggregate-behavioural models into their urban
transport forecasting packages.  Wigner (50) reported on one
of the first such efforts by the Chicago Area Transport
Study.  Work and non-work binary mode-choice models were
calibrated on individual traveller choice variables and
aggregate explanatory variables.  The motives for developing
a behavioural model for their forecasting system was to
increase the policy sensitivity of their analysis and to
update the results.  The results can be considered behavioural in their sensitivity to more policy variables.  They fail
to capture individual sensitivities to explanatory variables
if the aggregate observations used do not have the *same
correlations between variables* as fully individual observations would.  Insertion of the mode-choice model into the
traditional UTP choice sequence did not increase the
behavioural scope because of its hierarchical choice
structure.  Aggregation factors were ignored in this work.
Hence, expected improvement in accuracy from a disaggregate
calibration may have been overshadowed by aggregation error.
Aggregate calibrations would probably have been more
accurate.  The calibrations were used separately for project
policy analysis.  These will be discussed later.

The New York State Department of Transport incorporated a
mode-choice model, calibrated on individual behaviour and
explanatory variables, into their UTP system (19).  This work
is discussed in detail in the Spear review.  Its main result
was to show that disaggregate methods can lower the cost
(the data and calibration requirements) of model development
while predicting at least as accurately as models calibrated
on aggregate data.  It achieved no increase in policy sensitivity or behavioural scope.

The Metropolitan Council of the Twin Cities Area
(Minneapolis-St Paul) incorporated a multinomial mode-choice
model in their UTP system (14).  The major achievements in
this effort were to predict several different levels of
carpool occupancy as well as driver and transit modes.  It
increased the policy responsiveness of the system by estimating the model's sensitivity to additional behavioural
variables.  It also demonstrated the data efficiency and the
feasibility of disaggregate calibrations on readily available
UTPS software (48).

The most ambitious effort to incorporate behavioural
methods into a UTP system was accomplished for the Metropolitan Transportation Commission (MTC), a San Francisco MPO, by
Cambridge Systematics, Inc. [36]. Most of the twenty-one
models used in this travel forecasting system were calibrated
with disaggregate data.  The component models were predominantly of the logit form.  Significantly, the car-ownership
decisions were expressly considered and endogenously
forecasted.  The traditional assumption of mode-choice
following trip distribution was relaxed by calibrating
separate conditional choice model decompositions of these
joint choices for work trips as discussed in Section 2.  For
non-work trips, distribution and mode-choice were jointly
estimated and generation was separately estimated and applied
in the conditional joint choice structure.  Figure 19.1 is a
block diagram of the model components and behavioural
linkages used in the MTC UTP system.  Each small block is a
component choice model.  The solid arrows show the sequence
of choices assumed and the flow of data.  However, the dotted
arrows show the passage of variables important in simultaneous choices *back* to any model of joint choice, thus not
constraining choices to be made sequentially.  This, along
with the conditional choice structure, yields an equivalent
simultaneous choice decision for these subsets of the model
system.  Since the forecasts are done with zonal aggregated
data there is a limit to the behavioural accuracy of the
structure.  However, the zones are disaggregated into three
income groups explicitly included in each of these joint
choices.

The passage of results of work mode choices to the non-
work mode choices means that car availability for these trips
is dependent on any use in the work mode choice.  These
conditional joint-choice relationships make this the most
behaviourally general of UTP framework implementations.
Together with its inclusion of many policy responsive
explanatory variables it should be expected to forecast
travel accurately over a wider range of policies and demographic changes than traditional UTP systems at some increase
in cost.

The MTC package accepts inputs from an urban growth
forecasting system and produces forecasts of changes in
levels-of-service which are inputs to the growth forecasts.
It is thus possible, though expensive, to iterate the two
model systems to forecast land use-transport interactions.
This is not planned at MTC.

*Short-range General Policy Analysis Applications*

This category of applications is emerging from needs for
quick response, low-cost analyses.  It is often associated
with sketch-planning or short-range analyses.  Two studies in
this category which have exploited behavioural models quite
broadly are the carpool incentives policy analysis done by
Cambridge Systematics, Inc. (CSI) for the Federal Energy
Administration (FEA) and the evaluation of pollution control
strategies done by Charles River Associates, Inc. (CRA) for
the Environmental Protection Agency (EPA) [2,10].  These are

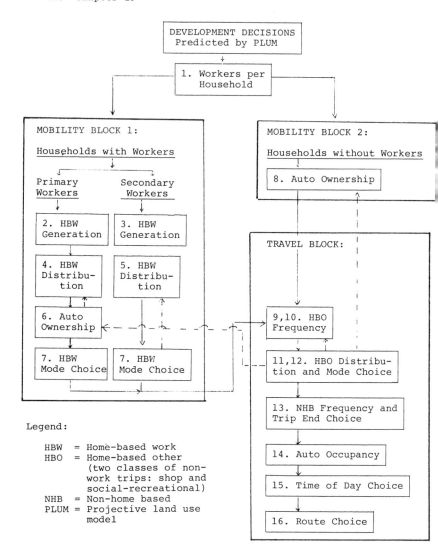

Figure 19.1  The MTC Travel Demand Model (UTP) System

categorised as short-range since they do not contain elements
for generating work trips from future year demographic inputs
nor consider land-use-transport service interactions.  The
CSI effort, however, does consider joint car ownership and
work mode choice decisions and joint generation, distribution
and mode-choice of non-work trips.  It draws from the same
conditional joint choice framework used in the MTC UTP type
forecasting system discussed previously (Figure 19.1).

These are broad policy analysis models.  They consider
across-the-board pricing and/or service changes, differentiated by travel time component, trip distance or subregion
of incidence, not by service on specific links.  Their
outputs are broken down by the number of travellers in large
geographical and socio-economic markets.

On the other hand, this broad approach has allowed the
analysis of the largest number of policy alternatives of any
behaviourally-based studies.  They have done this at low cost
(compared to conventional UTP packages).  Their analysis of
new policies requires no transport network recoding, although
the methodologies do allow prediction with a newly coded
network.  However, they may analyse all but major additions
to the capacity or routes of the transport system without
recoding networks.  Even these changes may be crudely predicted using only transformations of the base data.

Both of these studies are distinguished by their forecasts
of what different policies would do to the aggregate behaviour of a small but *representative sample* of travellers from
a region.  Disaggregate models of work and non-work trip
mode-split plus non-work trip destination and frequency are
used to predict the behaviour of the individuals in the
sample.  Regional aggregate travel by mode is then obtained
by scaling up the sum of the choice probabilities in each
segment by the inverse of the original sampling rate.

Transport policy alternatives are input to these analyses
by linearly transforming the base set of trip attributes of
the representative sample of travellers to produce the trip
values expected under the new policy.  Thus increased petrol
taxes, for example, would be represented by a per-mile
scaling up of the car costs of each trip in the sample.
Neither the trip impedance data nor the methodologies of
these studies allow precise prediction of demand under
capacity constraints.  Hence analysis of policies which may
effect these constraints will usually result in over-
estimates of the demand changes.

Some noteworthy aspects of the CRA analysis are mentioned
below, followed by a discussion of a recent study at the
University of California which applies an extended version of
the CSI models to regional policy analysis.

From a policy analysis point of view one of the distinctions of the CRA study is the range of evaluative
interpretations of the basic forecasts.  In addition to
showing the shifts in shares of trips using four modes,
changes in user costs, car and total vehicle miles travelled
(VMT), and resource costs are shown for the various policies.

The policies considered in this study were all on car-mode
pricing: taxes on petrol, emissions and parking. Other
aspects of note were the dependence of the results on data
for only 187 of the conventional 11,664 inter-zonal trip
interchanges characterising the region. Yet validation tests
carried out for the base forecasts showed this sample to
replicate regional mode shares and VMT well enough for broad
policy analysis. Thus the costs and data requirements for
this method are moderate for a large number of policy
alternatives.

The referenced short-run policy analysis study at CSI has
been extended on two fronts. CSI has extended its work on
the system (now called the Short-range Generalised Policy
programme (SRGP)), for the FEA and the MTC. The MTC version
has been extended and applied by the University of California
to San Francisco area data (35). All of the features were,
in principle, included in the original FEA version. The
behavioural scope thus includes all of the short- and medium-
range choices identified above. Only the long-range
decisions on residential and job locations as well as
employment and population growth are assumed given.

Useful analysis also requires predictions sensitive to
policies of interest. The extended version of the CSI model
can separately resolve behaviour due to the various components of trip travel times and costs as well as to different
socio-economic groups. Effects of changes in transit access
distance, headways, car parking, operating costs and other
policy variables may be separately resolved. With these a
range of differentiated pricing and service policies are
being assessed for MTC. These include changes in full taxes,
tolls, parking costs and traffic controls (including those
differential across modes), and transit fares and headways of
line-haul and feeder buses. In addition to predicting the
travel changes by mode and car ownership level, these models
have or can compute the VMT, fuel usage and revenue consequences of the policies.

Since the analysis is based on a sample of individuals for
which many socio-economic and travel characteristics are
known, the results may be output for any subgroups identifiable by these characteristics. The current studies have
broken down their outputs by income, subregional political
groups, and revenue districts. The study with the MTC has
provided forecasts for the Commission's decisions on the
setting of San Francisco Bay Bridge tolls. The models have
allowed planners to progress from the idealistic position of
maximising transit usage, subject to funding constraints, to
one of finding equitable service-to-subsidy ratios and
evaluating their associated impacts. These generalised
behavioural model systems allow the revenue versus service
predictions by mode and sub-area necessary for goal definition and assessments.

A third broad policy analysis model is of interest more
because of its impact on decision-makers than as an application of behavioural models. The study of transport impacts
for Southern California Association of Governments' (SCAG)
Short-range Program was done by RAND. It used the

n-dimensional logit mode-choice models developed by Peat,
Marwick, and Mitchell (3,27).  It forecasts work mode-choice
and non-work generation separately for 400 *markets* of
travellers (100 distance or trip attribute segments and four
income groups) representing the region.  The principal
relevance of this study is that the methods used treated the
policies of interest to decision-makers, was comprehensible
to them, integrated the many issues in a consistent framework
and, most significantly, had a major impact on Southern
California transport policies.  (This information was contained in a personal communication from Frances Bannerjee,
Los Angeles Community Redevelopment Association (formerly of
SCAG).)

There are different opinions on the value of this study.
In the implementation stage there was difficulty in making
the study's recommendations specific and criticism of the
analysis assumptions.  One part of the implementation - the
Santa Monica Freeway exclusive bus and carpool 'diamond lane'
was notorious in its political failure.  However, several
other preferential freeway operations such as the San
Bernardino Freeway busway are successful.  Previous studies
by the EPA and the region had proposed more notorious car-
control plans or unacceptably expensive transit plans.  The
RAND study is credited by SCAG staff as having cleared the
air of isolated solutions.  It integrated multi-modal service
pricing and control policies in a framework intuitively
sensible even to the politicians on the SCAG board.  The
model displayed simple regional outputs on costs, revenues,
mobility, energy use and environmental impacts.  They felt
they had not only the basis for, but found the consensus to
act on, some of the alternative policies.

Behavioural theory contributed to this directly through
the use of models sensitive to important policy variables and
the ability to disaggregate over subgroups of interest.  The
behavioural spirit of the approach contributed to the conceptual appeal of the model, even for the layperson.  Some of
the weaknesses of the study, such as its very generalised
treatment of policy descriptions or simplified behavioural
assumptions could be improved in future studies by a more
complete application of behavioural models.  This will be
explored further below in the discussion of the frontier role
of policy models.

*Special Purpose or Project Applications*

Though there has been much work on project level policy
applications of behavioural models, the literature is disappointing in the number of fully developed examples of their
impact on plans and designs.  Only three studies were found
where a range of real policy decision alternatives were
evaluated and only one where the project was completely
evaluated.  Most purported applications are actually *calibrations* of models on a limited behavioural setting (21,23).
Others are principally exercises of such models on very
idealised alternatives (26,37).

Lisco and Tahir's analysis of the impact of parking taxes in Chicago gave simple but directly useful analyses in the form of plots of the changes in mode shares versus these taxes (22). It was based on the behavioural model calibrated by Wigner and applied to the specific Chicago downtown area (50). This parking study is noteworthy because it showed that a small effort can give useful results on a current limited issue. Further, it showed how results could be extended to analyse further potential pollution, revenue and congestion impacts.

A thorough design and evaluation of a feeder bus system for a suburban rail station was done by Tahir and Hovind (42). Though this study was discussed in detail in the Spear review, some of its positive and negative aspects are worth noting here. It is the most complete application found by this author in its carry-through to designs for implementation. The things that are most important to implementation - revenue analysis and the fare policy - make good use of its demand model. It stops short of estimating any optimum policy, leaving the decision-makers to decide between the fare-subsidy trade-offs revealed. It is disappointing in its acceptance of external and *a priori* definitions of the service levels even though the model was capable of checking these trade-offs in the design.

From a policy role point of view, probably the boldest use of the behavioural model results for special projects is the analysis by Tahir of the relative advantages of feeder buses and parking construction at rapid transit stations (41). In addition to computing an elaborate measure of user and social benefits for the policy alternatives based on value-of-time results, it used an available behavioural model for demand predictions. The benefit measures were intended for evaluating capital grant requests for these types of improvements throughout the state of Illinois. As such it used relevant behavioural tools and concluded with clear and simple recommendations. The simplicity of the study (not the basic method nor its ability to reach simple conclusions) depended on a key assumption on the constancy of feeder bus costs with distance and a coincidence of benefits measures on two of the modes. Other assumptions in the analysis might also require relaxation to achieve generality of the analysis. Walk and wait service times were related in a crude constraint. Bus capacities were not considered. No total trip variables relating to generation rates were considered. Removing these limitations would not have unduly complicated the analysis and would have resulted in a policy tool with broad evaluation of social benefits.

This concludes the examples of the policy applications of behavioural models. Together with the next section, categorising the dimensions of the role space for these tools, the examples will form a base for the last section, which attempts to identify the most promising policy roles for these models.

GENERAL DIMENSIONS OF THE POLICY ROLE

In the previous meetings of the International Conference on Behavioural Travel Modelling, the workshops and papers on applications have stressed prediction methods for multi-dimensional choices, aggregation problems, system equilibration necessities, model theory limitations and support requirements for implementation (20,29,47). An applications guide manual for the UTPS software, titled *Introduction to Urban Travel Demand Forecasting*, additionally stresses the importance of being able to handle many policy issue inputs and providing a variety of evaluative outputs (4). Two earlier papers on the role of transport and location models by Harris and by Lee defined typologies of models on dimensions such as comprehensiveness, descriptive-analytic, holistic-partial, level of abstraction, evaluative-normative, and so forth (12,17).

Drawing from the framework suggested in these two papers, but also including the aspects of applications pointed out by the others, a set of dimensions of a *policy role* space are defined and discussed below to provide a perspective from which to compare and guide models in the policy arena. These attempt to express the planner's point of view of the analysis realm as opposed to the modeller's perspective represented by Harris and Lee. However, these dimensions are put in *terms* used by the technical analyst. Note also that these policy realm dimensions are distinct from the dimensions of *behaviour* previously discussed. Behaviour dimensions are one breakdown of the fifth role dimension - level of generality. The role dimensions should apply to policy analysis in general, though behavioural theory motivates the inclusion of some of them and dominates much of their discussion.

The role dimensions are:

1 Policy issues scope.
2 Resources required.
3 Scale of plan.
4 Level of detail.
5 Level of generality.
6 Level of evaluation.

There are many other aspects of policy planning that are not specifically included because they are not related to traveller or consumer behaviour. Most of these are related to the role of modelling through their evaluative dimension. They include transport system costing, finance, system design, operations management, the level of economic activity, the political decision process and sociological impacts. A dynamic-static dimension is not included since most planning is concerned with the steady-state outcome of policies.

The first dimension - policy issues scope - refers to the inputs to which an analysis is sensitive. This contrasts with the behavioural breadth or resolution of outputs of the analysis which are included in other dimensions. The scope of issues can include the analysis of changes in highway and

transit operations, pricing and capacities.  Of particular
interest is the differentiation of these policies by
implementation categories and for all travel modes.
Behavioural models make possible the accurate resolution of
the effects of policies such as parking versus fuel costs and
transit frequencies versus line-haul time.  They can better
distinguish specific modal policies such as those for car-
poolers or feeder submodes.

For policy issues to be analysed and the results to be
useful, planners' or politicians' policy concepts must be
translated into realistic transport service terms and thence
to data inputs to the models.  This has traditionally been
done by recoding network files and transforming them to trip
alternative data by operating path selection algorithms.
Other methods can more simply realise the issue scope
potential of behavioural models.  These include scaled
transformations of base policy trip attribute files such as
described in the short-range general policy examples and the
new work on analytic supply policy models (35,44).  The
policy issues dimension is especially important because it
points to the difficult interface between the analyst and the
decision-maker in clarifying policies in such a way that they
can be made specific to a transport system and represented as
data for input to analysis systems.

The resource dimension refers to the initial and running
costs of the analysis method, including required data, skills,
training, computing facilities and turn-around time.  In
keeping with the planning point of view, the feasibility or
the accuracy of a prediction was not separately mentioned as
a role dimension.  Some level of feasibility is presumed.
However, the accuracy is a resource issue in terms of the
investment desired in data and the level of behavioural
generality desired.  This association of accuracy with
resources points out the need for the planner and analyst to
evaluate and decide the level of effort on these aspects of
an analysis to achieve an economical balance.  Expensive
precision may not be required.  There may also be planning
institution compatibility issues, beyond the initial cost and
data factors which require further evaluation of analytical
results to effect policy.

The scale dimension is simply the size of the area or
population being analysed.  It is included here mainly to
distinguish this factor from the level of detail.  This is
an important difference, especially where quick-response
analysis of a large area is of interest but detail (not
necessarily accuracy) may be sacrificed.  Thus large-scale
models need not be either detailed or comprehensive (in Lee's
sense of including many system or behavioural interactions).

The level of detail dimension attempts to encompass
Harris's conceptual-specific scale.  This is distinguished
from the general versus special purpose scale which is
discussed later.  It is possible to have a model that is
very specific on the spatial distribution of travel effects
which is either general or special purpose.  For example, a
general model would not depend on behavioural or system
assumptions, limiting it to the particular situation

analysed, while the special purpose model may do so for
simplicity.  This dimension is the inverse of Lee's level of
abstraction measure.  It is loosely related to the level of
disaggregation (in that conceptual or coarse analysis will
usually employ highly aggregated outputs) but does not
necessarily associate aggregate outputs with an aggregate or
non-behavioural model choice structure.  It is possible and
often desirable to have a behaviourally structured analysis
operating on a small data set with highly aggregated *outputs*,
accurate only at a coarse level.  This is appropriate when
the underlying theory is not very detailed as well as when
only coarse or sketch planning results are needed.  The CRA
analysis for EPA and the RAND analysis for SCAG are such
examples.

Decisions about where an analysis should aim on the level
of detail dimension are probably the key to the success or
failure of many studies.  This level can control whether the
results will be understandable by the decision-makers, timely,
or even within a feasible cost.  A different level of detail
may be appropriate at a different stage in the planning
process or for different audiences.  The common error is
usually to include too much detail.  The planner is often
oriented towards detailed spatially differentiated analysis.
Neither he nor the behavioural modeller is easily able to
make the extensions from geographic aggregate predictions to
other traveller market impacts such as those by socio-
economic groups.

The level of generality dimension guages most closely what
behavioural modelling is about.  As used here it not only
means how many traveller choices and transport alternatives
are considered in the analysis but how much of their behavioural and system interactions are included within the model.
Thus a model which includes different choice dimensions such
as those of car ownership and travel mode, but constrains the
choices to be sequential or independent by its structure, has
a less general scope than one which allows these choices to
be interdependently predicted.  Of course, from an applications point of view this is only true if there is empirical
evidence that the choices are significantly interdependent
under some conditions of policy change.  Consideration of
dependence between traveller choices and system characteristics would similarly be measured on this dimension.  This
might include supply-demand equilibration of travel or
location choices.

The generality dimension is arbitrarily separated from
that of the issues scope because they are viewed separately
in practice.  Theoretically, analysis of more issues simply
requires inclusion of a greater scope of causes and effects
(generality).  The issues scope of a traditional four-stage
sequential planning method is greatly expanded by substituting a mode-choice model sensitive to more system and socio-
economic variables.  This is of great interest to the planner
in determining the role of the model for short-run or special
purpose applications.  It does not increase greatly the
generality which might be important to analyse trip genera-
tion-mode choice interactions or longer-term effects
involving car ownership or location choices.

The time horizon of an analysis is considered here to be a natural aspect of the generality dimension. Obviously, longer-range predictions need consider more general behaviour such as decisions on car ownership, trip generation and land-use changes.

The presumption, from a policy relevance point of view, is that more general models are best. Some studies have been found useless because they stressed a limited behaviour scope, such as choice of mode, only to find that trip generation rates or land-use changes were more important. However, from the standpoint of the costs or even political or institutional constraints on an analysis, it is important to limit the generality only to that necessary for the situation. If generation and distribution are constrained such as for work trips, mode-choice analysis may be adequate. However, car ownership and trip timing may be important to consider. If there is *no* way policies on land use may change it is futile to consider them. The other side of the coin, though, is that failure to consider inviolate policies may obscure social benefits that should cause the fixed policies to be re-examined.

The last dimension of the policy role of behavioural models is that of evaluation level. This scale runs from simple predictions of the different population demand dimensions through various impact analysis to normative or optimal assessments. Philosophical and practical arguments can be made for the appropriate role of modelling, or of the analyst on this scale, but the object here is to identify the possibilities. At a minimum, policy analysis identifies the demand level of the population for each alternative of each choice dimension considered. This may be broken down by different sub groups or markets (geography, socio-economic, choice category, supply categories, and so on). The ability for this segmentation is a function of the generality of the analysis; its degree is a function of its level of detail. A next level of evaluation is impact predictions. Most impacts are direct derivatives of demand outputs and the impact rate coefficients such as energy consumed per trip-mile. Thus net operating revenues/costs, vehicle-miles travelled, location densities, pollution impacts and energy consumption are all derivatives of the basic demand outputs. Exceptions are accessibility measures which may, to a first order, be derived directly from transport system design without demand analysis and economic activity measures.

A next level of evaluation may be subsystem benefit analysis. User benefits employing value-of-time (VOT) figures or consumer surplus theories may be compared to user and/or total social costs. Subsystem design optimisation may be considered by looking at benefits of different system design components at constant costs. Decisions might thus be made between proposals to increase transit headways or route miles.

Finally, analysis on this dimension may be carried to identification and quantification of one or more general social-benefit measures. This is usually thought of as a single socially-weighted account of costs and benefits. It

might differentially account for benefits and costs for
different economic groups and weigh these equally and
linearly or otherwise, as desired.

This speculative subject is worth exploring because of the
value it has in clarifying the requirements of an analysis
and the true objectives of a public plan.  In a recent study
for the MTC, the Urban Travel Demand Forecasting Project at
the University of California, Berkeley, discovered, late in
the process, that one objective of the agency was to predict
the transit service levels corresponding to a given rate-of-
return of revenues to costs.  This could only be very
inefficiently done with the modelling system being developed
because it required iterations with the output revenues.  If
the rate-of-return objective had been explicit initially, a
more efficient analysis system could have been devised.

More will be said on the role of models on this evaluative
dimension in the concluding section.  Identifying the
emphasis of modelling on this role dimension is one of the
most challenging and fruitful for analysts and decision-
makers.

FRONTIER POLICY ROLES AND IMPLEMENTATION PROBLEMS

With the definitions and examples of the previous sections,
the chapter concludes by reviewing what is and is not being
accomplished from the policy planning point of view by
behavioural methods.  It attempts to identify which of these
analytical roles are most worth stressing or developing.  The
major problems inhibiting useful analyses and the actions
necessary to open this potential are also indicated.

The section begins by stating which examples of applications of behavioural methods are believed to be most
promising.  Other planning tasks are mentioned where they
should not be expected to contribute.  Some frontier roles
within the conventional demand-forecasting frameworks are
reviewed.  The section then speculates on new categories,
similar to the conventional and emerging prediction packages,
that may better fit the policy roles space defined before.
Recommendations are made for many of the weaknesses in the
applications.  Finally, a number of applications and research
topics where support is needed are mentioned.

The most promising way that behavioural models are
penetrating policy analysis is in the development of generalised quick-response prediction packages such as the work for
FEA by CSI.  These prediction packages are based on general
multi-dimensional behaviour and are sensitive to policy-
relevant socio-economic and trip data.  They satisfy many
current *planning* concepts of where analysis should fit on the
role space dimensions - issue sensitivity, low effort,
freedom from restrictive behavioural assumptions and ability
to predict impacts over varied socio-economic or market
groups.

Though most promising applications are in demand -
derivative analyses, rather than in system design or economic

feasibility studies – some contributions can be made in these latter areas.  To the extent that design or financing are equivalent to optimisation, or even making a trade-off which involves traveller demand, behavioural methods can be used. For example, the use of the relative elasticities of demand to prices and service levels, can help the realistic achievement of design objectives.  However, the limitations of elasticities in representing varied populations and transport systems requires recognition of their over-simplified averaging properties or resignation to a process of iteration of design and prediction.  Possibly the development of much simpler prediction models can lead to a more efficient use of the breadth of behavioural information available to transport system design and administration.

Other important policy areas that are *not* likely to be impacted by behavioural methods are source financing (apart from revenue analysis), system costing, economic activity impacts and freight planning.  All these are generally outside of the individual behaviour realm.  Impacts of transport policy on the economy, though in the limit involving individuals, are too complex for any but aggregate modelling, if possible on that level.  Freight demand prediction in terms of individual-firm decisions is theoretically possible, though only sparse advances have been made in this area (33).

Other policy areas often thought to be separate from the trip prediction realm, such as marketing, operations management, benefit analyses and conceptual models of transport, are considered here to be prediction or prediction-based. Though marketing and operations management have many aspects outside of behavioural modelling, such as promotion, behavioural intention surveys and experience assessment, their key element is reliable prediction of the dependence of usage on service attributes.  Thus a test of value of prediction models is their ability to serve operations and marketing needs.  This fits well into project or small-scale scope of analysis and planning.

The conceptual understanding of the behaviour of people and the flow of benefits to them under transport policy change is especially important to the decision-maker and manager.  These have often been based on past experience, rules-of-thumb or idealistic promotions.  Elasticity figures from behavioural models are an improved substitute for rules-of-thumb.  They have broader application than system prediction in that they can be included in many transport analyses which cannot afford the complexity of incorporating a prediction model.  Use in a system design trade-off is an example.  However, their serious limitations in revealing behaviour due to multiple policies or the variation of impacts over an area requires qualification of the limits in their use.  Much could be added to the concepts of past experience and idealistic models by developing very simple conceptual prediction methods demonstrating, for example, the trade-offs in people's behaviour between transport prices and service levels or between trip-making rates and congestion.

Within the prediction package realm there are aspects
which can and those which cannot be well-served by behavioural, rather than aggregate or correlative, models. The
decision between these approaches should be based on how much
of individual behaviour is well known and how much disaggregate data is available. Generally the behavioural models and
data are preferred as they have a better chance of capturing
the demand interactions and the range of underlying explanatory characteristics. They also can output more distributions of impacts. However, if the behavioural theory is weak
or the models untested against experience, such as with
present individual location models, they may fail to include
some important factors embodied in aggregate or summary
variables which merely show a correlation to demand. The
latter are more likely to pick up unknown effects. (Of
course, it is possible to include any aggregate variables in
a disaggregate model, but this defeats its behavioural and
transferable generality.) Similarly, if the data for
*prediction* is not as disaggregated (for availability or
efficiency reasons) as that used for the calibration of the
model being used, prediction error could be large. No
aggregation method can compensate for data not available to
recover the original disaggregate interactions. Thus if no
disaggregate data will be available for forecasting, models
calibrated on aggregate data will be no more inaccurate than
disaggregate models used with aggregate data. Efficiency is
another argument for aggregate data *forecasting*, which is
easily overlooked in the rush for behavioural precision. In
large-scale forecasting, the number of individual observations can imply very expensive data collection and
computation costs, possibly avoidable if accuracy is not
critical.

Computation efficiencies exceeding those of the traditional large-scale aggregate-forecasting systems are possible
with the use of new disaggregate data classification methods
(34). The effects of the mix of individual household
characteristics and local service attributes cannot be
resolved unless some disaggregate data is available or the
cost of collecting it is faced. Without some of this data
behavioural models are useless in applications. With it,
aggregation methods based on the classification procedures
show promise of lower costs of predictive calculations than
those based on geographic aggregate models. More will be
discussed about these procedures in the section entitled
'Conclusions on Applications and Data Support Needed'.

*Behavioural Model Effectiveness in Conventional*
*Forecasting Frameworks*

Before speculating on some new emphases in prediction models
for policy analysis, experiences with behavioural model
applications in the existing categories of applications will
be evaluated in terms of the role dimensions defined above.
These analysis categories were: conventional UTP, short-range
general and special purpose analyses.

The categories are not simply emphases of one or two of
the role dimensions. There is, however, a tendency to

identify UTP analyses as large scale, short-range general
analyses as multi-issued, and special purpose analyses as
small scale.  The UTP systems, in addition to forecasting
over large urban regions, are expected to produce outputs
sufficiently detailed for network link-loading and to include
enough phenomena to be pertinent for long-range forecasting.
A rising concern has also been the time and money costs of
these analyses as the need for short-run forecasting has
increased.

Three types of questions on the traditional UTP category
are instructive.  Are they presently meeting expectations?
Can behavioural models help?  Is the present form of the UTP
analysis category the most useful?  On the first, it appears
the resources to use these models are excessive.  Original
set-up time, updating and operating costs, and turn-around
have been motives for changes such as calibrations by
behavioural methods.  Though not always apparent to the
planner, the long-range applications of these systems are
straining their assumptions on behavioural and system interactions.  Many long-range issues involve the land development
and traffic-generating implications of transport policy.
These are only analysed by the traditional models at great
expense or with assumptions causing over-estimates of the
changes; i.e. those which yield only partial equilibrium
solutions.  Behavioural models can contribute and have
contributed to reducing these problems.

An expanded UTP category is represented by the example of
the MTC system developed by CSI.  It has increased both the
issue sensitivity and the behavioural scope of this framework.  Operating with other traditional land-use, assignment
and evaluation packages, it would rank well on all of the
role dimensions except resource use, on which it would be a
high burden.  Still, this is a feasible direction in planning
and one where behavioural methods could contribute to issue
breadth and long-range behavioural consistency with efficient
calibration procedures.

A trend in this direction, however, may be unfortunate.
Increasing the policy issue sensitivity of the conventional
UTP packages further increases their costs unless they are
relieved of their more general features of generation and
distribution.  Even so, they are expensive packages for
multi-issue analysis.  Hence it is probably wise to encourage
the development of multi-issue breadth in the short-range
generalised (i.e. sketch planning) models rather than in UTP
type frameworks to meet overall policy analysis objectives.
Once policy alternatives have been narrowed by the less
expensive packages, their distributional effects can be
revealed by a more spatially detailed specification of the
major policies to the UTP package.

Before exploring these potential revisions to the
traditional UTP forecasting category the emerging general
purpose analysis category will be compared to the role
dimensions to reveal if this category provides alternate
possibilities for the emphasis of behavioural methods.  For
example, if the long-range needs can be more easily satisfied
with generalised systems, less behavioural complexity may be

appropriate for the detailed UTP-type analyses.

The new generalised analyses reviewed above, as well as
sketch-planning systems, have arisen principally from the
requirement for lower resource use, the consideration of
numerous policy issues, and the provision of broad evaluative
outputs. They are meeting the resource dimension requirements but are only slowly affecting decisions since they are
new to the institutional framework and are difficult to
relate to some specific implementations because of the
generalisations of policies in the models. They are meeting
the role objectives of responsiveness to many policy issues
because of their basis in behavioural methods. These methods
have also allowed the analyses to expand the phenomena considered to generation and distribution choices without
constrained choice hierarchy assumptions. In two new
generalised studies, supply-demand equilibration has also
been incorporated at low cost (27,45). These analyses in
addition provide evaluative outputs such as derived environmental impacts.

Behavioural methods are the foundation of these generalised analysis systems and will find one of their clearest
roles here. Development of their issue sensitivity will
depend on the ability of researchers to reveal and validate
model responsiveness to basic transport service attributes.
Their resource costs will depend in part on discovering the
correct balance between detail required by the policies and
simplicity of the models and evaluative outputs. Aggregation
method research can contribute to the lowering of their costs.

Behavioural methods contribute indirectly to the
evaluative capability of these policy analysis systems by
allowing outputs to be expressed over most any market segment
of the region without errors resulting from homogeneity
assumptions. The segmentation also facilitates calculation
of impacts such as those proportional to vehicle-miles of
travel. Benefit measures and social objectives can use the
segmentation capability for equity and welfare considerations.
The value-of-time conclusions from behavioural modelling can
be an important aid to constructing benefit measures over
differences in user and socio-economic groups.

As short-range tools these methods are meeting all the
policy space objectives well, with the possible exception of
providing clear links between generalised policies and
specific implementation. They have not been developed for
long-range analyses. However, considering the resources
necessary to achieve behaviourally consistent forecasts,
including supply and land-use equilibration, it seems at
least as feasible to put efforts into generalised systems for
long-range planning than into detailed UTP systems.

The category of existing systems termed special-purpose or
small-area analyses was the first where behavioural models
had significant use. The ability to calibrate the models on
small data sets or to transfer them from another area made
the models available to these low budget situations.
Prediction was similarly possible at low costs because of the
small data sets needed. These have been and will continue to

be strong reasons for the applicability of behavioural models to small area analyses.  Because of the low budget and *ad hoc* nature of most of these analyses, they do not explore the range of issues or evaluative impacts that are possible with behavioural tools.  In the case of two of the study examples discussed, conspicuous alternatives between the frequency and density of service were untested though it would have been straightforward to do so.  Thus superior design trade-offs may have been missed.  These studies also stop at relatively narrow limits of behavioural complexity or supply equilibration.  Transit access studies usually avoid overall trip mode-choice interactions and capacity interactions, much less considering demand generation.

There are two problems in these weaknesses worth discussing for understanding how these applications can be developed.  Strong beliefs or vested positions are often inherent in any activity with sufficient interest to generate support for an analysis.  This is true at all scales of analysis, but particularly at the local or project level where fewer key persons are involved and the problem is credibly handled by intuition.  Thus analysis may only be done to support *a priori* policies.  Second, the models usually have neither the precise accuracy at small scale nor the credibility from any past experience or familiarity.  Thus there are probably large obstacles to overcome in their fully effective use in small-scale planning.

*Departures from the Conventional Forecasting Frameworks*

It was suggested in the last section that the evolution of the UTP analysis category using behavioural models might best relegate their long-range choice dimensions to other analysis tools as they increase their complexity to achieve behavioural generality and issue sensitivity.  Correspondingly it was suggested that the short-range general analysis category could evolve into a long-range form at much lower cost than UTP systems, filling the gap with an even more behaviourally general forecasting tool.

Speculations on these and other possibilities for the evolution of planning and policy analysis categories are illustrated in Table 19.1.  It shows the ratings of eight analysis categories on the role space dimensions.  The ratings are H, M and L for high, medium and low on each of the dimensions.  High does not necessarily imply superior but depends on the objectives.  For example, low resources and high generality are usually desired but the ranking on the scale and detail dimensions depend entirely on the planning objectives.  The first four categories are identical to the three conventional categories discussed in the last section, except that the behaviourally general expansion of the UTP framework is shown as a separate (the second) category.

The first category is the traditional UTP approach.  If it has incorporated behavioural models it is medium in ability to consider too many policy issues but expensive due to its geographic scale and detail.  The generality dimension is ranked medium, even though many choice dimensions are

TABLE 19.1  Rating of categories of transport analysis systems on the policy role dimensions

| Transport analysis category | Policy role dimensions | | | | | | Package examples (bibliographical reference) |
|---|---|---|---|---|---|---|---|
| | Issues scope | Resources required | Scale | Detail | Generality | Evaluative outputs | |
| UTP (four-step, sequential) | M | H | H | H | M | – | 14, 19, 48, 50 |
| Expanded UTP | M | H | H | H | H | – | 36 |
| Short-range General | H | L | H | L | M | M | 2, 10 |
| Short-range Special Purpose | L | L | L | H | L | L | 41, 42 |
| Market Classification | H | L | H | L | M | – | 11, 35 |
| Long-range General | M | M | any | L | H | M | 3 |
| Short-range UTP | M | M | H | H | L | – | subsystems of 14, 19 |
| Medium-range Programme | M | M | M | any | M | M | |

H = high level
M = medium level
L = low level

modelled, because of the sequential choice constraints and
the difficulty and expense of equilibrating with supply and
land use.  Yet this is the most used analysis category
because of its development and availability.  The second
category differs from the first in its incorporation, as in
the MTC example, of a conditional joint-choice structure and
associated explanatory variables across choice-dimension
functions to overcome sequential-behaviour constraints.  This
type of development achieves a high level of generality but
an even higher cost than the first category.

The third category is the short-range general category
discussed in detail above.  Though the specific packages used
as examples for this category are new, they are rapidly
becoming established in the roles of the conventional sketch-
planning analysis.  They are the highest of all in issues
scope.  Their costs are low - in the tens of dollars per
policy analysis after initial set-up - but are low in detail
of outputs.  Their lack of consideration of work-trip generation or capacity constraints gives them only a medium
generality rating.  Their integral provision of derived
impacts rates them medium in the level of evaluative outputs.

The fourth category emphasises that special-purpose or
small-area analyses have all been for a short-range scope.
Together with their usual limitation to single dimensions of
choice, for example, of mode, they are only rated low in
generality.  Data and computation costs, and lack of necessity, limit them to a low issues scope.  They are by
definition low in geographic scale.  They are usually
detailed, but seldom have evaluative outputs.

The remaining categories, though all but one have prototypes in research or applications, are speculative from the
standpoint of their becoming established in the planning
repertoire.  The fifth category in Table 19.1 is actually a
simplification of the third, short-range category, based on
improvements discovered for aggregate forecasting with
disaggregate models.  Classification of travellers into
homogeneous situational or choice-utility markets allows
great reductions in the number of computations necessary for
accurate predictions compared to those required for geographically classified or individually enumerated prediction
(11,15,34).  Thus these methods can achieve the same levels
on the role dimensions as the short-range general category,
but at considerably lower resource costs.  Policy analyses
have been shown feasible by these methods with hand
calculators.

The balance between model complexity and efficiency has
another aspect pulling towards the simplicity potential of
this market classification category.  Often the goals of a
transport plan are the optimisation of some benefit or the
achievement of a state of the system which is indeterminant
from a one-step prediction of demand levels for a set of
policy inputs.  Such a system goal might be determining the
service and pricing levels that will yield a given rate of
return of revenues to costs (the subsidy rate).  Finding such
a level with models requires the availability of transport
cost and supply models and *iterative* methods to reach revenue

equilibrium.  There are two important implications here.  The
development of behavioural methods requires an associated
balance in the development of supply (cost versus performance
and capacity) models.  Second, the complexity and operating
costs of the demand predictors should not be so great that it
is not reasonable to run them in multiple iterations.  This
reinforces earlier suggestions of low costs for possible
iterations to achieve equilibrium with capacity constraints
or land usage.  Supply models need development for the
generation of a range of policy alternatives commensurate
with the power of these demand analyses.

The sixth category of Table 19.1, long-range general
analysis, has been implied above in the extensions of the
third short-range category.  This is a speculative one - the
closest example being the RAND transport impacts model for
Southern California Association of Governments.  Moderately
long-range sketch planning models such as CAPM and TRIMS are
also patterns (28,38).  The development of this category -
especially around behavioural models - is motivated by the
success of the short-run generalised packages and the need
for efficient analyses considering travel generation as well
as interactions with population growth and supply constraints.
This could relieve UTP systems of an expensive part of this
role.  Some of the previously mentioned needs to predict
supply-demand balance and to optimise policy goals may be
long-range issues.  This category is seen as a natural
extension of the short-range general analyses which satisfies
many of the concerns expressed over the costs and behavioural
approximations of UTP analyses.

The ratings for this category in the table were assigned
with two major considerations in mind with respect to the
short-run generalised category.  The behavioural scope must
increase to make this a generally useful forecasting tool,
potentially replacing the traditional UTP systems.  The
emphasis suggested is more towards broad behavioural scope
than issue sensitivity or geographical and system detail as
these can be covered by other categories.  The set-up costs
of this approach should be no higher than implementation of
its precursor - the short-range generalised models.  Its
eventual turn-around costs should be only moderate even with
equilibration.

The role of behavioural methods in this category is
central to its long-range scope but may be limited by cost
and feasibility considerations.  It is appealing to attempt
to develop behaviourally unconstrained predictions of all the
choice dimensions including residential location, possible in
the conditional joint choice framework.  Long-range, large-
aggregate population and employment predictions, distinguished, for example, between major urban and suburban areas,
could be used in an iterative framework with essentially
short-range packages to extend their generality beyond the
rarely attempted iteration of UTP and location models, but at
a fraction of the cost.  Yet this would answer a major
question planners are asking.  That is: what will be the subregional population and travel growth implications of a new
policy?  By contrast, the UTP approach to this issue, while
trying to retain the detail of its impact on particular

transport links, has failed either to accurately give long-range link-loadings or to control analysis costs.

The seventh, a proposed short-range UTP category, *presumes* the development of more spatially aggregated long-range models such as the sixth category in the table.  With generation inputs from a general model, efforts in this category could be on distribution and mode forecasting and the detail of the geographical and market breakdowns of the results.  This would reduce their complexity and costs. Short-range issue sensitivity might also be more compatible with this model scope.  The applications examples of behavioural models at CATS, NYSDOT and Twin-Cities have moved in this direction as they have attempted only mode-choice improvements and calibration efficiency.  Regardless of whether UTP analyses continue in the traditional role or move towards shorter-range methods, behavioural methods can contribute efficiency in updating and issue breadth.

The eighth category is the extension of the short-range special purpose analyses into at least a medium-range scope. Many project studies suffer from failure to consider long-range effects.  There will continue to be a need for and potent application of behavioural models in low-budget, special-purpose or project planning because of their potential transferability and generality properties which can circumvent the extra costs of local model estimation efforts.

The major distinctions made in defining these categories have been on the relative emphasis of issue sensitivity, detail and behavioural generality.  The shortcomings in the existing applications examples above, clarified by the policy role dimensions outlined earlier, have guided the categorisation to match the range of perceived planning goals.  The key to effective analysis in a particular context is the emphasis of the important role dimensions, for example, generality and evaluative outputs, not detail in medium- or long-range sketch planning and emphasis of detail, not generality in a comprehensive analysis.  Behavioural models are a valuable aid to this orientation because of their flexibility in decision structure and level of prediction aggregates.  They can be formulated and applied for single or multi-dimensional behaviour choices without choice hierarchy assumptions.  They can be aggregated over familiar urban zones or any market or geographic segment up to the total group of analysis.  This also gives considerable flexibility in the data and analysis costs.  It opens the possibilities of low cost but still general analysis and simpler economical models for use with other systems such as land-use or economic-activity models.

There are many categories in Table 19.1, not all of which may have great promise or will receive adequate support.  At the outset of this section it was stressed that the short-range general analysis category and its extensions have the greatest promise for aiding policy decisions, at least at the urban regional level and above.  This short-range category is getting good research attention as the examples attest.  It is suggested here that the frontier lies in the further development of the longer-range generality of this type of analysis (the sixth category), especially for considering the coarse grain growth distribution impacts of transport policies, and in the development of the evaluative outputs of these models.  The role of the UTP type and project level analysis systems can largely be found with reference to examples of a clear role developing for the general (or sketch planning) categories.

The necessity for development of these models on the evaluative dimension has been only lightly discussed.  It has been shown that behavioural models facilitate expression of travel, cost and environmental impacts over markets or traveller groups of interest.  The motive for considering alternative transport policies is often to handle an expected increase in demand, in costs or the imposition of a new social constraint.  To the extent that alternative actions are possible, their rankings towards meeting the basic motives, much less other potential social impacts, cannot be assessed without evaluative measures beyond the forecasts of demand by mode.  It was pointed out that though evaluation is not strictly demand modelling, it is the primary requirement of policy analysis.  The role of behavioural models should be shaped by this need.  Attention to the type and structure of models needed to differentiate transport service or environmental impacts at the decision level may be preferable to breadth on the issue scope or generality scale.  For example, if decisions are to be made between rail and bus investments, the models should surely distinguish the separate usage, revenue and impacts of these modes in preference to emphasising the pricing sensitivity of a single general transit alternative.  If the primary impacts of interest are on land use, it is more important to simplify the detail of the model to render its feasible iterative use with location models, than to refine its geographical or issue sensitivity at an expense too great for joint model solutions.

A policy-maker is often dangerously unable to keep a broad perspective in a complex system.  He is likely simply to consider maximum transit usage or a single popular environmental impact to the the social-benefit measure.  Ideally, the politician could identify complete social-benefit measures, help the analyst to quantify them, and the model would rank the alternatives on the benefit scale.  This is fraught with problems.  There is no consensus benefit scale.  If there were one it would not easily be quantified.  The incomplete knowledge in these areas tends to make any attempt very narrow.  However, politicians are, or at least should be, faced with making discrete decisions between alternative policies.  Granted, some decisions are 'pork barrel' compromises among conflicting interests.  But to the extent that there are some generally superior consensus policies possible, they require ranking for decision.  Relegating as many benefits and impacts as clearly as possible to a rating scale would at least organise a part of the decision and give a framework in which to consider evaluation.

Since transport may be only a part of what a politician is dealing with, a rating may at least give him a handle for comparing transport alternatives with other issues.  It is

better that a transport alternative be compared with a
housing programme with a more complete social measure than,
for example, on the basis of a currently popular measure such
as the energy consumption rate or transit subsidy level.

*Conclusions on Applications and Data Support Needed*

The author concurs with the conclusions of the two previous
Behavioural Travel-demand Modelling Conferences that demonstration programmes of on-line policy models and development
of standardised analysis packages are needed to promote their
use.  This is especially true in their application to project
level and small-area analyses where resources are limited.
Many of these restrictions in the use and power of project or
small-scale analyses by local agencies could be relaxed if
there were more standardised analysis systems or examples and
aids available from central institutions.  These could
incorporate more general behavioural and system features and
give examples to aid both the initiation and breadth of the
analyses.

Most planning agencies either have vested interests in a
current travel-prediction system or have limited resources to
develop new ones from scratch.  Demonstration support for use
of packages such as short-range generalised analyses,
especially in a 'before and after' policy setting, would do
much to help determine the best emphases of their features as
well as overcome the credibility obstacles mentioned earlier.

The major aspects needing resources or demonstration
funding support in this respect are: (1) data collection
support - base-line socio-economic, level-of-service and
longitudinal validation data; (2) model estimation software,
especially if transferability research shows negative
results; (3) system theory tutorials and guidelines on
determining the appropriate level of behavioural and system
equilibrium complexity for particular planning goals; (4)
prediction software support - probably in the form of
standardised packages; and (5) evaluation measures support -
numerous impact rates such as pollution and energy per VMT:
system cost data; examples of benefit assessments of various
alternative policies in other areas, including costs, time
savings, accessibilities; environmental impacts, and so on.

It is important that these demonstrations be an integral
part of the institutional structure of the planning or
operating organisations making the decisions.  Hence, support
funding should go to them.  However, since there are many
pressures toward expedient analysis in real-time planning it
is expected that generally useful demonstrations require a
definite give and take between planners and behavioural
analysts.  This might be facilitated by joint funding of
consulting or research groups with planning agencies for the
study projects, especially in areas where quantitative planning staffs are limited.

Regarding the support necessary for policy analysis data,
much of the past practice in data collection, even for
behavioural modelling, is not serving policy objectives

well.  Two things are needed.  The first is more frequent
representative household level surveys of planning areas.
These need *not* contain either as many questions nor as many
respondents as the comprehensive surveys of the 1960s.  Even
research studies do not need hundreds of survey questions.
Very good predictions may be done with less than a dozen
traveller variables.  Coarse grain analyses may be done on
less than 0.1 per cent of area households, not the 1 per cent
samples often collected in the past.  These small surveys can
be conducted at moderate cost by random-digit dialing
telephone methods.

This level of periodic surveying is judged necessary
because disaggregate data is needed to resolve accurately
travellers' behaviour.  Individual characteristics cannot be
reliably forecast over time, even if location models were
reliable.  Where fine grain (UTP-type) travel predictions are
required, only rarely will large surveys also be necessary.
Assuming that minor socio-economic trends do not quickly
become major ones, the low density surveys may be used to
periodically re-weight the larger samples for current
conditions.

An alternate method of deriving household data, updated at
least as frequently as the US census, has been developed by
the Urban Travel Demand Forecasting Project and the
University of California, Berkeley (9).  It is based on the
US census 'public use sample' and census tract data, and on
urban-location model growth rates.  It has shown good
aggregate and fair sub-area validation results.  It has yet
to find a situation where disaggregate data is available for
validation.  The availability of low-density representative
surveys would at least aid in this type of validation.  If
validation results are negative these would provide the
alternative data base for policy analysis.  This need for at
least low-density, moderate-length surveys *disagrees* with the
recommendations of Liou and Hartgen in the Applications
Workshop of the Second International Conference on Traveller
Behavioural Modelling (20).

Another area of data collection needed is field observations for validation and assessment of the transport system.
Much of this need has traditionally been filled by the large
comprehensive surveys or, rarely, by operating agencies.  But
with the fiscal crises at local levels, these data are less
and less and less available.  It should be possible to
collect them at small cost, compared to the large household
surveys, even including moderate-scale representative telephone surveys.  Special economies and benefits to operating
agencies should be possible if these efforts are conducted by
the operators, funded in part by central government agencies,
under the condition that the collection and storage methods
are uniform and the data available to all.  This would be a
valuable combination of field experience data for operators
as well as for before-and-after checks of policy forecasts.

A final comment on the need for disaggregate data is
necessary since two of the forecasting developments mentioned
earlier may be misleading in this respect.  Where behavioural
models are judged worthwhile, disaggregate data is necessary

or the advantage of the models will be lost.  This has been
found true even with traditional traffic zone aggregates in
the case of socio-economic and transit access data.  It is
even more important for coarse aggregate forecasting.  The
development of the expanded UTP category of forecasting on
aggregate data *plus* income subclass information does not
contradict this.  These subclassifications were found necessary to make accurate predictions.  They could only be
derived from disaggregate income data.

   Similarly, the development of market-classification
methods of aggregate forecasting does not mean that old
geographic class data can be factored to or replaced by
collection of data on the basis of a new kind of sampling
class.  Choice market classes are not even fairly well
defined by geography.  Thus they cannot be derived from
existing area aggregate planning or census data.  They are
not even definable on single socio-economic classifiers such
as income.  Behavioural models have shown that choice determinants cut across a wide mix of travellers' characteristics
and trip attributes.  Thus individual data must be collected
to reveal this mix.  The savings that are possible in forecasting are *in the amount of processing* necessary on this
data to make accurate forecasts.  Classification research has
shown that a relatively small number of market types or
classes, defined on this mix of individual data, are necessary for accurate prediction (11,15,34,35).  Hence,
disaggregate observations are needed.  Once reduced to choice
market classes, only a dramatically smaller amount of data
need be carried through and computed upon for forecasting.
There are trade-offs between the degree of this compression
of data into classes and the resolution of alternative
policies possible in forecasting.  The references have shown
that considerable policy analysis is possible with aggregate
forecasting by classification.


*Conclusions on Research Support Needed*

There are at least three areas of relatively basic research
needing attention to support the analysis goals already
mentioned.  These are development of some of the basic choice
theories, development and testing of prediction theories, and
supply models.

   The most severe basic model limitations seen for effective
moderate or long-range analysis is the lack of any practical
behavioural models of work generation and distribution; i.e.
residential and job location models.  Were either models of
joint location and mode-choice or separate location models
available, simpler than the large scale zonal location of the
Lowry type, then far more general transport impact analysis
could be done at modest costs compared, for example, to
iterative computations between UTP and zonal location models
such as by Putnam (30) at the University of Pennsylvania.

   Though a few behavioural non-work generation-distribution
and mode-choice models exist, their accuracy has been only
fair under testing.  They do not have the ability to represent all trip types.  Research is necessary to determine

which trip types can be modelled together and to develop
accurate models on reasonable data requirements.  The
question of accuracy is an issue even in the relatively well-
developed modal-choice models.  Seldom do the models predict
the modal demand due to a major policy change with better
than 10 per cent accuracy, even for large aggregates.  Much
of this error may be due to the common assumption of uniformity in travellers' tastes in most of the models.  Research is
necessary to relax this assumption and determine the resulting effect on prediction structure (for example, on the
aggregation issue).  Research on multinomial-probit models is
a step in this direction (6,13).

Needed research in prediction methods refers to prediction
error theory, validation tests, aggregation theory, transferability tests and equilibration methods.  No measures are
available to show the total ranges of error in a prediction.
This should certainly be expressed for planners and modellers
alike.  It should separately identify cross-section and
longitudinal model error (the former from estimation statistics, the latter from before and after validation tests),
aggregation error and data error components.  Very few
before-after validation tests have been performed.  This is
the key to evaluating the value and direction of modelling
efforts.  It requires support of timely data set collections
and applications-oriented research efforts.

The development of aggregation theory has important
implications for both accuracy and costs of analyses.  Little
is presently known of the optimum trade-off between the level
of disaggregation of various types of data (for example,
between socio-economic and level-of-service components and
their breakdowns) necessary for a given level of prediction
accuracy.  Since there are fairly large sources of error in
naive use of aggregate data *and* large economies in using it
where possible, research in this area has a great return.

Much of the hope of the value of behavioural methods lies
in the expectation of the *general nature* of the findings –
such as values of time, elasticities and the transferability
of the models, especially for low-cost application such as in
small area or project analyses.  Yet very little testing of
this expectation has been done.  Some tests have been in
conflict and shown some disturbing results on the stability
of the results on different samples (1,43).  More research
and testing are definitely necessary to determine the conditions over which models might be transferable, even over
different data sets in the same region.

Most development of equilibration theory such as between
supply and demand or travel service and location has either
been at the micro (network link) or the macro (full model
iteration, e.g. Putnam) scale.  Talvitie and Hasan's work
(45) on equilibration over only the 'supply markets' (links),
where significant capacity constraints exist, is a pattern
for an area of research that could produce very cost-
effective models, still considering complex system interactions.

The third area of research needed is in supply models, including network methods and analytic supply models. The typical regional network data files are too complex for analysis of many policy alternatives. They are useful for refinement of major policy impacts. Network data, combined with the supply-policy transformation capabilities of the generalised short-range analyses discussed before, are an efficient combination, overcoming the traditional recoding expense for new alternatives. This potential has been weakened and has limited the ability of generalised model analyses to be related to specific implementations due to the failure of most network data development groups to categorise the different types of links in the network. The type of service on different links has been categorised by some into freeways, arterial and local streets on highway networks, and express and feeder links on transit networks - categories subject to separate policy manipulation. Some increase in basic network software systems capability may be needed to do this. This small increase in complexity of network data can be offset for generalised analyses by cutting back on the number of nodes of the network, as has been done in some of the existing sketch planning packages. The research in aggregation methods can give the guidelines on the limits of these simplifications.

Analytic supply models have intriguing promise in their simplicity and for readily handling supply-demand equilibration and policy optimisation. Talvitie's work (44) is a pattern for this. They face a tough challenge in equalling the ability of network files to represent the unique physical distribution of services, especially with complex geography. However, the networks have been shown to be of questionable accuracy themselves. The return on power and simplicity of successful analytic models would be high. Research is needed to both further develop and validate analytic supply models.


ACKNOWLEDGEMENTS

Research was supported in part by National Science Foundation grants GI-43740 and APR74-20392-AO2-AO3-AO4, Research Applied to National Needs Program, to the University of California, Berkeley.


REFERENCES

1. Atherton, T.J. and Ben-Akiva, M., 'Transferability and updating of disaggregate travel demand models', *Transportation Research Record No. 610* (1976).
2. Atherton, T.J., Surhbier, J.H. and Jessiman, W.A., 'The use of disaggregate travel demand models to analyze carpool incentives', *Transportation Research Record No. 599* (1976).
3. Bigelow, J.H., Goeller, B.F. and Petruschell, R.L. (1973) *A Policy Oriented Urban Transportation Model: the San Diego Version*, Report No. R-1366-SD, Appendix 4, RAND Corp., Santa Monica, California.
4. Cambridge Systematics, Inc. (1974) *Introduction to Urban Travel Forecasting*, two volumes, Software Systems

Development Program, Urban Mass Transit Administration, US Department of Transportation, Washington, D.C.

5. Charles River Associates, Inc. (1972) *A Disaggregate Behavioural Model of Urban Travel Demand*, National Technical Information Service, Springfield, VA, Charles River Associates, Inc., Report 156-2, Cambridge, Massachusetts.

6. Daganzo, C.F., Bouthelier, F. and Sheffi, Y., 'An efficient approach to estimated multinomial probit models', unpublished paper, Department of Civil Engineering, Massachusetts Institute of Technology, Cambridge, Massachusetts (1976).

7. Domencich, T.A., Kraft, G. and Valette, J., 'Estimation of urban passenger travel behaviour: an economic demand model', *Highway Research Board Record No. 238* (1968).

8. Domencich, T.A. and McFadden, D. (1975) *Urban Travel Demand: A Behavioural Analysis*, North-Holland, Amsterdam.

9. Duguay, G., Jung, W. and McFadden, D., 'SYNSAM: A methodology for synthesizing household transportation survey data', *Urban Travel Demand Forecasting Project, Working Paper No. 7618*, Institute of Transportation Studies, University of California, Berkeley (1976).

10. Dunbar, F.C., 'Evaluation of the effectiveness of pollution control strategies on travel: an application of disaggregated behavioral demand models', *Proceedings of the Transportation Research Forum* (1975).

11. Dunbar, F.C., 'Policy-contingent forecasting with market segmentation', *Transportation Research Record* (in press) (1978).

12. Harris, B., 'Quantitative models of urban developments: their role in metropolitan policy making', in *Issues in Urban Economics*, H. Perloff and L. Wingo, (eds) Johns Hopkins University Press, Baltimore (1968).

13. Hausman, J.A. and Wise, D.A., 'A conditional probit model for qualitative choice: discrete decisions recognizing interdependence and heterogeneous preferences', presented at the joint session of the NSF/NBER Conference on Decision Rules and Uncertainty and the MSSB Conference on the Theory and Measurement of Economic Choice Behavior, University of California, Berkeley (1976), *Econometrica*, 46(2), 1978.

14. Hoffmeister, J., McLynn, J.M. and Schultz, G., 'The Twin Cities modal choice model: development and calibration', unpublished paper, The Metropolitan Council of the Twin Cities Area, DTM, Inc., and R.H. Pratt Associates, Inc. (1976).

15. Koppelman, F.S., 'Guidelines for aggregate travel prediction using disaggregate travel choice models', *Transportation Research Record No. 610* (1976).

16. Kraft, G. and Wohl, M., 'New directions for passenger demand analysis and forecasting', *Transportation Research*, 1, pp. 205-20 (1967).

17. Lee, D.B., Jr. (1968) *Models and Techniques for Urban Planning*, Report UY27474-G-1, Cornell Aeronautical Laboratories.

18. Lerman, S.R. and Ben-Akiva, M., 'A disaggregate behavioral model of auto ownership', *Transportation Research Record No. 569* (1976).

19. Liou, P.S., Cohen, G.S. and Hartgen, D.T., 'Application of disaggregate modal choice models to travel demand forecasting for urban transit systems', *Transportation Research Record, No. 534* (1975).

444  Notes to chapter 19

20. Liou, P.S. and Hartgen, D.T., 'Issues in the implementation of disaggregate, behavioural travel demand forecasting
technique', in *Behavioural Travel-Demand Models*, Stopher, P.R.
and Meyburg, A.H. (eds), Lexington Books, Lexington (1976).
21. Liou, P.S., 'Comparative demand estimation models for
peripheral park-and-ride service', *Preliminary Research
Report No. 71*, New York State Department of Transportation
(1974).
22. Liou, P.S. and Talvitie, A.P., 'Disaggregate access
mode and station selection models for rail trips', *Preliminary Research Report No. 53*, New York State Department of
Transportation (1973).
23. Lisco, T.E. and Tahir, N., 'Travel mode choice impact
of potential parking taxes in downtown Chicago', *Technical
Papers and Notes Series No. 12*, Office of Research and
Development, Illinois State Department of Transportation
(1974).
24. Lisco, T.E. (1967) 'The value of commuter's travel
time: a study in urban transportation', unpublished Ph.D.
thesis, Department of Economics, University of Chicago,
Chicago.
25. McFadden, D., 'Measurement of urban travel demand',
*Journal of Public Economics*, 3 (1974).
26. McLynn, J.M. and Goodman, K.M. (1973) *Mode Choice and
the Shirley Highway Experiment*, prepared by DTM, Inc., for
the Office of Transit Planning, Urban Mass Transit Administration, US Department of Transportation.
27. Mikolowsky, W.T., *et al.* (1974) *The Regional Impact of
Near Term Transportation Alternatives: A Case Study of Los
Angeles*, Report No. R-1524-SCAG, RAND Corp., Santa Monica,
California.
28. National Capitol Region Transportation Board, 'Small
area analysis - a tool for local decision-making: TRIMS
model', *Technical Report No. 9*, Metropolitan Washington
Council of Governments, Washington, D.C. (1973).
29. Perle, E.D., 'Policy issues: workshop report',
*Behavioural Demand Modelling and Value of Time*, Transportation Research Board Special Report No. 149 (1974).
30. Putnam, S.H., 'Preliminary results from an integrated
transportation and land-use models package', *Transportation*,
3, 3 (1974).
31. Rassam, P., Ellis, R.H. and Bennett, J., 'The
N-dimensional logit model: development and application',
*Highway Research Record No. 369*, pp. 135-47 (1971).
32. Reid, F.A. (1974) 'A set of models for optimizing the
benefits of a transportation plan', unpublished M.S. dissertation, University of California, Berkeley.
33. Roberts, P.O., 'Forecasting freight flows using a
disaggregate freight demand model', *Transportation Research
Record* (1978).
34. Reid, F.A., 'Aggregation methods and tests', *Urban
Travel Demand Forecasting Project, Final Report Series,
Volume VII*, Institute of Transportation Studies, University
of California, Berkeley (1977).
35. Reid, F.A. and Harvey, G., 'Regional policy analysis
case study', *Urban Travel Demand Forecasting Project, Final
Report Series, Volume IX*, Institute of Transportation
Studies, University of California, Berkeley (1977).
36. Ruiter, E.R. and Ben-Akiva, M., 'A system of disaggregate travel demand models: structure, component models,

and applications procedures', *Proceedings of the USA Transportation Research Forum*, XVIII (1977).

37. Schindel, S.E., 'Impact of station choice and access mode choice due to the establishment of a commuter rail station at Arlington Park, Illinois', *Technical Papers and Notes Series No. 14*, Illinois State Department of Transportation, Illinois (1973).

38. Schleifer, H., Zimmerman, S.L. and Gendell, D.S., 'Community aggregate planning model: an urban transportation sketch planning procedure', *Transportation Research Record* (1976).

39. Small, K., 'Bus priority, differential pricing, and investments in urban highways', *Urban Travel Demand Forecasting Project, Working Paper No. 7613*, Institute of Transportation Studies, University of California, Berkeley (1976).

40. Spear, B. (1977) *Applications of New Travel Demand Forecasting Techniques to Transportation Planning: A Study of Individual Choice Models*, Federal Highway Administration, Office of Highway Planning, US Department of Transportation, Washington, D.C.

41. Tahir, N., 'Feeder buses as an alternative to commuter parking: an analysis of economic trade-offs', *Technical Papers and Notes Series No. 15*, Office of Research and Development, Illinois State Department of Transportation (1974).

42. Tahir, N. and Hovind, M., 'A feasibility study for potential feeder bus service for Homewood, Illinois', *Proceedings of the USA Transportation Research Forum*, XIV (1973).

43. Talvitie, A. and Kirshner, D., 'Specification, transferability, and the effect of data outliers in modelling the choice of mode in urban travel', *Transportation*, 7(3), pp.311-32 (1978).

44. Talvitie, A. and Deghani, Y., 'Supply model for transit access and linehaul', *Urban Travel Demand Forecasting Project Working Paper No. 7614*, Institute of Transportation Studies, University of California, Berkeley (1976).

45. Talvitie, A. and Hasan, I., 'An equilibrium mode-split model of work trips along a transportation corridor', *Urban Travel Demand Forecasting Project, Working Paper No. 7621*, Institute of Transportation Studies, University of California, Berkeley (1976).

46. Train, K., 'A validation test of disaggregate travel demand models', *Urban Travel Demand Forecasting Project, Working Paper No. 7619*, Institute of Transportation Studies, University of California, Berkeley (1976).

47. Tye, W.B., 'The application of behavioural travel demand models: workshop report', in *Behavioural Travel-Demand Models*, P.R. Stopher and A.H. Meyburg (eds), Lexington Books, Lexington (1976).

48. Urban Mass Transit Administration (1975) *Urban Transportation Planning System (UTPS)*, US Department of Transportation, Washington, D.C.

49. Warner, S.L. (1962) *Stochastic Choice of Mode in Urban Travel: A Study of Binary Choice*, Northwestern University Press, Evanston.

50. Wigner, M.F., 'Disaggregated modal-choice models of downtown trips in the Chicago region', *Highway Research Record, No. 446* (1973).

Chapter 20

THE ROLE OF DISAGGREGATE TRAVEL MODELS IN TRANSPORT
POLICY ANALYSIS

Alistair Sherret

WORKSHOP SCOPE

The workshop was charged with exploring the justifications
for pursuing improved approaches to travel modelling by
addressing the following questions.  To date, what uses have
been made of analysis and prediction procedures based on
disaggregate travel models?  What types of transport policy
issues have disaggregate approaches been most effective in
analysing compared to more conventional aggregate travel-
analysis procedures?  What further types of issues appear
amenable to analysis using available disaggregate modelling
procedures?  What steps need to be taken to ensure wider use
of appropriate techniques in analysing these issues?
Finally, what are the shortcomings of existing models and
procedures in addressing current policy issues, and where is
further development work most needed?

   In addressing these questions, the workshop first defined
the range of disaggregate-travel models and prediction
procedures which have been used or are currently available
for use in analysing transport options.  Second, we discussed
the types of current and future transport policy issues for
which responsive analysis tools are most required.  Third,
by comparing the outcomes of the first two stages of the
discussion, we attempted to assess the strengths and weaknesses of existing travel models and procedures in analysing
transport policy issues.  The workshop drew heavily on
information and ideas contained in the resource paper prepared for the workshop by Reid (see previous chapter).

   Discussions of travel modelling are frequently confused
by a failure to distinguish clearly between the estimation
(or 'calibration') and application (or 'prediction') stages
of the modelling process.  The principal concern of this
workshop was with the second of these stages, i.e., with
travel-prediction procedures.

   The models discussed in the remainder of this report are
all 'disaggregate' in that their parameters are estimated
from data on individual traveller or household travel
choices.  The advantages and efficiencies of travel models
estimated with disaggregate data are well-documented elsewhere (for example (4)),and are not dwelt on here.  However,
although based on disaggregate estimated travel models, not
all the procedures discussed here predict travel at a

446

disaggregate level.   Indeed, one of the principal topics of
the workshop's discussion was to compare the roles of prediction procedures which apply disaggregate travel models in
a disaggregate manner with more commonly used procedures
(which apply disaggregate models in an aggregate zone-based
framework).


SUMMARY OF WORKSHOP DISCUSSIONS

*Travel Modelling Procedures*

The workshop's discussions concentrated on disaggregate
travel modelling procedures such as those developed and
applied by Charles River Associates (CRA) for the Environmental Protection Agency (3), and by Cambridge Systematics,
Inc. (CSI), for the Federal Energy Administration (2).
Both studies used a set of individual traveller-choice models
as the basis for evaluating the aggregate travel impacts of
alternative transport control strategies.   (The CRA study
assessed various strategies to reduce automobile-caused air
pollution; the CSI study evaluated the effectiveness of
carpooling incentives in reducing overall fuel consumption.)
Recent work by CSI and by the Travel Demand Forecasting
Project at the University of California, Berkeley, has
extended the procedure developed for the Federal Energy
Administration study.   The result has been referred to as
Short-run Generalised Policy (SRGP) Analysis Procedure.

   The SRGP procedure is characterised by a conditional
joint-choice model structure in which individuals' decisions
about automobile ownership and choice of work-journey mode
and their decisions about the frequency, destination and
modal choice for non-work trips are modelled as joint decisions.   A further important feature of the SRGP procedure is
that it applies the models in a disaggregate way by random-
sample enumeration.   The sample enumeration procedure selects
a representative sample of households from the study area,
and then predicts the effects of implementing a candidate
policy separately on the travel-choice probabilities of each
household in the sample.   (This sample will typically be the
sample used in the original model calibration.)   Overall
travel impacts may then be computed for any desired geographic or socio-economic grouping by summing the effects on
the separate households (1).

   The sample-enumeration prediction method employed by the
SRGP procedure is in contrast to previous disaggregate
modelling procedures which, although using individual-choice
data to estimate the parameters of the choice models, apply
the estimated relationships to aggregates of individuals in
prediction, so introducing problems of 'aggregation bias' in
the forecasts.

   The travel-model system currently being developed for the
San Francisco Bay Area's Metropolitan Transportation Commission (MTC), described in Ref. (4), represents the most
comprehensive (and conceptually probably the most convincing)
example of an attempt to apply disaggregate models to predicting travel in the context of a conventional traffic-zone

system and network representation of the transport system.
Among other features, the model is intended to include many
policy-responsive variables, forecast car ownership endogenously, account for the dependence of non-work car availability  on work-travel modal choice, and account for the
simultaneous nature of modal choices and destination choices.
The MTC model system shares its basic, conditional joint-
choice structure and many of its component 'submodels' with
those used in the SRGP prediction procedure.  Although
successful implementation of the MTC model system has yet
to be demonstrated, the workshop took the MTC model system
as the 'standard' of aggregate prediction procedures for
comparison with disaggregate SRGP procedures.

*Models for Policy Formulation, Prediction and Evaluation*

The workshop focused on modelling procedures intended to
predict the travel impacts of implementing alternative transport policies or plans.  However, participants recognised
that predicting the volumes and patterns of travel associated
with a particular proposal is not a useful end in itself;
rather, the ultimate objective is evaluating the benefits
and costs of implementing alternatives.  Estimating travel
volumes and their distribution among modes is a key component
of the evaluation process, but it is not the only one.  The
travel modeller must also share a responsibility for design-
ing or formulating policy, not simply predicting or assessing
the consequences of policies proposed by others.

    The workshop agreed that current predictive models such as
those referred to above are effective and ready for implementation in many situations.  However, good predictive models
are not necessarily good models for policy formulation or
evaluation.  For example, models which include only 'conventional' time and cost variables to describe the attributes
of alternative modes may well give acceptable predictions of
travel volumes.  At the same time, by omitting other 'qualitative' variables considered in traveller choices, they
provide an incomplete understanding of travel behaviour and
the value or importance of different transport-system
attributes in determining choices, and so may be inadequate
tools for evaluation.  In other words. for a model to be an
effective evaluative tool, we need to know the composition
of the 'constant term' in the predictive equation.

    Conversely, models which provide an acceptably complete
understanding of travel behaviour for purposes of evaluation
are unlikely to be practical in prediction, largely because
of the difficulties of predicting changes in the qualitative
attributes of travel modes and quantifying traveller perceptions and attitudes towards these changes. However,
'attitudinal' models exist which provide a considerably
fuller understanding of travel behaviour than is possible
from 'practical' prediction models alone.  Although the workshop did not discuss attitudinal models in detail, their
importance as policy-relevant tools was acknowledged.  It
was felt that development of attitudinal approaches should be
continued and ways found to incorporate them in the travel
modelling process as supplements to 'pure' prediction models,

to enhance the ability of the analyst both to identify
options and evaluate their consequences.

An advantage of disaggregate models in evaluation,
especially in current UK/Australian transport planning practice, is that they are readily amenable to specification in
terms that allow consistent and comprehendible measures of
consumer surplus (and related values-of-time).  However, the
requirement that a model provide consistent monetary values
of consumer surplus implies constraints on the way in which
the model is specified (especially how cost variables are
entered).  These constraints may well give rise to models
which are less efficient predictors than would otherwise be
the case.  Prediction procedures such as the SRGP model which
allow individual-by-individual prediction have the advantage
that travel impacts and benefits can be aggregated according
to any socio-economic or geographic grouping of the population.  This allows the incidence of user costs and benefits
to be assessed in detail and thereby allows the equity
impacts of transport policies and plans to be evaluated more
easily and flexibly than can be accomplished with aggregate-
prediction procedures.

*Modelling Requirements of Policy Analysis*

Reid's resource paper suggests a list of 'dimensions' along
which it is appropriate to assess the suitability of travel
modelling procedures to the requirements of policy analysis.
They may be summarised as follows.

*Responsiveness to issues*.  This dimension describes the
ability of the model to estimate the outcome of implementing
specific policy options and to distinguish between the
effects of changing the supply and price of transport service
in different ways.  For example, can the model predict the
effects of changing the costs of driving; and can it distinguish between the effects of doing this by increasing parking
charges as one policy option or increasing fuel costs as
another?  The dimension essentially describes the number and
types of variables to which the model outputs are sensitive
and the way in which they are specified in the model.  An
important aspect of the dimension is the ease with which
particular policy options (as specified by 'the decision-
makers') can be translated into terms which the model can
accept as input.

*Comprehensiveness*.  This dimension describes the extent to
which the model comprehensively describes the outcome of a
particular policy option.  Does the model consider effects on
traveller choices about trip frequency, destination, mode,
route, trip complexity and time of travel?  Is the simultaneous nature of these decisions taken into account?  Are the
interactions between the demand for, and supply (or capacity)
of, transport services considered?  Are both short-run effects on travel patterns and long-run effects on travel and
land-use activity accounted for?  In short, the 'comprehensiveness' dimension describes the basic scope and structure
of the model and its ability to represent the complex of
responses to changes in the characteristics of the transport
system.

*Evaluation measures*.  This dimension describes the types of model output which are required by the analyst or decision-maker in evaluating the relative costs and measures of effectiveness for alternative policy options.  The range includes simple aggregate measures of ridership or travel volume, measures of travel time savings, more complex consumer-surplus benefit measures, the impact of changes in transport service on the mobility of particular population groups, the social costs of traffic congestion, air pollution and energy consumption, and measures of environmental impact on non-users.

*Geographic scale*.  This concerns the extent of the area for which model outputs are required.  The area may range from the entire urban area to corridor, local area or specific facilities or links of the transport network.

*Geographic detail*.  This concerns the level of detail required in the model outputs, and the fineness of the population segments or transport system components for which summaries of the outputs are required.

*Resources required*.  The cost of using a particular modelling procedure clearly has a bearing on its applicability to a particular policy analysis requirement.  The costs involved are made up of several components; the most important of these are the costs of collecting data to estimate, validate and apply the models.  These data costs, in turn, depend on the complexity of the data required, the survey methods necessary to collect the data, and the sample sizes required. A second cost component is the computer time cost required to estimate and apply the models.  A third and frequently understated cost is the manpower required to estimate and apply the models.  The technical sophistication, skill and experience required in the analyst is an important determinant of this cost component.  A fourth component is the time required to set up the modelling procedure, and a fifth, the 'turnaround' time required for the procedure to respond to a particular enquiry.

*Feasibility*.  Closely related to the 'resources required' dimension is the possibility of model estimation and application given the inevitable constraints of limited resources. Perhaps the most common of these constraints is the requirement that models be estimated and analyses conducted within the confines of an existing data set.  (Although the limitations of our knowledge about travel processes and how to explain them are themselves perhaps the most serious constraints of all.)

*Comprehendibility*.  Finally, one of the most important criteria for assessing the applicability of a modelling procedure to a particular policy analysis is how understandable the procedure and its outputs are to those who have to interpret and make decisions on the basis of the model outputs.  More will be said on this subject later.

Judgements about the usefulness of a particular modelling procedure to a policy analysis problem, or conversely, the requirement of a particular policy analysis for a modelling

procedure, must be made on the basis of all the above dimensions. This will necessarily involve making several trade-
offs, some of them fairly clear-cut, others less so. The
remainder of the workshop's discussions focused on these
trade-offs, and particularly the advantages and disadvantages
of the SRGP class of disaggregate models described earlier.
The following sections record some of the conclusions reached.


*Issues Amenable to Analysis by Disaggregate Modelling
Approaches*

The elements of traveller behaviour which have so far been
modelled most convincingly by disaggregate approaches are
modal choice, and to a lesser extent, trip frequency and
trip-length choices (for non-work travel at least). These
may be regarded as short-run elements of travel choice (in
contrast to the long-run elements associated with residence
and work-place location choices). As has been demonstrated
to some extent by the CRA and CSI studies referred to earlier
(2,3), existing disaggregate modelling approaches are capable
of addressing these short-run issues effectively. Although
there is clearly room for developing and improving them,
existing models and procedures are ready to be applied to
these short-run planning concerns. (On the other hand, long-
run concerns of residence location, work-trip lengths and
interactions between transport demand and supply are not
adequately addressed by existing procedures.) The short-run
applicability of existing models is also reflected by their
emphasis on analysing low-cost and marginal improvements in
transport system service and pricing levels.

Policy-makers are currently concerned with making more
efficient use of existing urban-transport facilities (as
reflected, for example, by the recently-introduced requirement for a Transportation System Management (TSM) element of
analysis in US transport-planning practice), rather than
with making large investments in constructing additional
transport capacity. The marginal analysis approach of the
SRGP family of models is very much in tune with these current
policy issues.

A fundamental difference between the disaggregate and
conventional 'UTP' modelling structures is that the latter
requires a network representation of the transport system
at both the calibration and application stages. The expense
and time consumed in modifying the network tends to be a
major limitation of the UTP structure, particularly in the
application stage where the modeller desires to screen or
conduct broad sketch-planning analyses of a large number of
alternatives. In these situations, the disaggregate SRGP
models show clear advantages of flexibility. Of course, a
price is paid for this advantage in the detail of output
which is provided. The disaggregate models provide coarse-
scale results which are inappropriate for many applications,
for example, where measures of travel volume on specific
transport facilities or network links are required. Nevertheless, the disaggregate models are capable of providing
estimates of VMT or VMT-derived evaluation measures such as
fuel consumption at aggregate or corridor levels.

These comments are addressed primarily to the question of
how modelling procedures can be applied to already defined
issues - the assumption being that 'the decision-maker' has
the necessary knowledge to define the relevant issues.   But
the modeller must share in the responsibility for identifying
relevant policy options.   The 'behavioural' logic of disaggregate travel models, to the extent that it allows an
improved understanding of the causes of individual travel
behaviour, enhances the modeller's ability to assist in this
process of generating policy options.

*Flexibility of Output Measures*

Perhaps the greatest appeal of the SRGP disaggregate modelling
approach is the flexibility it offers for summarising output
evaluation measures.   The explicit sample-enumeration method
of model application permits outputs to be aggregated on the
basis of any desired geographic or socio-economic stratification (providing only, of course, that the required stratification data are contained in the respondent data files).
This is important to both the decision-maker and the evaluation process in that it allows the incidence of user impacts
and distributional effects of a given transport policy to be
considered explicitly.   Thus, 'equity' considerations can be
taken into account by aggregating outputs by low- and high-
income groups; and 'political' considerations can be taken
into account by aggregating according to various constituen-
cies of interest to the decision-maker.   (Of course, the
output of a travel model, whether disaggregate or otherwise,
addresses only user-related impacts and benefits; it does
not allow non-user and other external impacts to be assessed.)
The ability to address these political concerns explicitly
is an advantage which transport analysts have possibly overlooked too often.

*Data Needs of Aggregate and Disaggregate Procedures*

At the estimation stage, disaggregate modelling procedures
offer well-documented advantages in data efficiency, because
they: (1)  use every trip-record observation as a sample
point in the estimation;  (2)  incorporate within-sample
variance; and (3)  eliminate biasing 'ecological correlations' among aggregations of individual observations.   However, for data needs at the application stage, the advantages
of disaggregate over aggregate approaches are not so clear-
cut.   For a given required level of detail in model outputs,
equally detailed information is needed as input to the forecasting stage for both disaggregate and aggregate approaches.
The advantage and potential economy of disaggregate models
instead lies in the flexibility they offer for allowing
various levels of coarseness in the outputs.

RECOMMENDATIONS AND CONCLUSIONS

*Areas Needing Further Research and Development*

As has been pointed out, disaggregate modelling approaches

have been most effective in short-run planning contexts where changes in travel demand (as evidenced by traveller mode choices) resulting from marginal changes in transport service are assessed. The obvious area for extending this modelling capability is to long-run applications where (1) further dimensions of traveller decisions such as trip timing, residence and work-place location decisions are taken into account; and (2) interactions between the demand for and supply of transport facilities are effectively taken into account. However, as evidenced by concerns voiced by work-shop members, there is room for further debate about the most appropriate form of models to address these longer-term policy concerns.

One major concern is that the more rigorous and more policy-sensitive the disaggregate model, the more information is required to calibrate and apply it. Forecasting basic information such as zonal averages of employment, number of households and income levels is a difficult enough task. Forecasting more detailed information such as the number of licensed drivers, duration of residence and other specific life-cycle information some 20 years ahead is clearly much more difficult. On the other hand, if a set of long-range forecasting disaggregate models rely only on those variables which are commonly used by existing 'UTP' models, then their relative advantages over existing models in terms of fore-casting accuracy, ease of application and policy sensitivity may be negligible. The advantages of disaggregate procedures have yet to be clearly demonstrated in these respects.

A related concern is the 'aggregation problem' which arises when zonal data are used as input to a disaggregate model at the forecasting stage. It has not yet been demon-strated that market segmentation or other available methods effectively take care of this problem. Such a demonstration is especially important when the market segmentation is based on a single variable. Some planners feel that aggregated models may have an advantage in long-range forecasting since most long-range planning is focused on districts, zones and other aggregates of the population.

Another difficulty is in identifying alternatives to define the choice sets faced by travellers. As a model be-comes more completely descriptive of travel behaviour, more choice sets and more complicated sets must be defined. How choice sets are defined is currently a subject of much interest among researchers and difficult enough to resolve for short-run situations. Identifying alternative choices faced by travellers for a long-range future system is an even more demanding task.

In summary, the benefits of disaggregate modelling procedures over existing aggregate procedures are at the present not as apparent in the long-range as in the short-range planning context. Clearer evidence and demonstration of the advantages of disaggregate models are needed to encourage transport planners to switch from the conventional to disaggregate modelling structures.

*Considering the Decision-making Process*

Of the eight dimensions listed earlier in this report, one of the most important, but also most frequently overlooked, is the need for modelling procedures and their outputs to be understandable to decision-makers who use them. Disaggregate-travel models have an advantage in this regard because their underlying logic - relating the decisions of individual travellers to the options facing them - is simple, obviously behavioural and intuitively easy for non-modellers to understand and accept. As a result, disaggregate models and their outputs have the potential for acceptance once their scope and applicability can be demonstrated.

The most effective (and perhaps only) way to convince decision-makers of the usefulness and applicability of disaggregate models is by example. The tools will gain acceptance if they can be successfully applied to specific problems currently facing decision-makers. The problem then becomes one of identifying key issues to which existing models can be applied and persuading decision-makers that disaggregate-modelling approaches should be given a chance. Clearly, this must be a co-operative venture between the modeller and the decision-maker. This raises the important point that 'comprehendibility' is a two-way problem. It is necessary for the modeller to show the decision-maker the inside of the 'black box' of his model, and so give him confidence that the model is indeed helpful in making his decisions. But if the modeller is to be effective in developing useful models and convincing the decision-maker of their usefulness, he needs to understand the workings of the decision-making 'black box'. The problem is not simply one of matching models to issues; it is also necessary for the modeller to know the process by which decisions are made about the issues, and the way in which information provided by his models is used by the decision-maker. To achieve this, a two-way programme of education, training and discussion will be necessary between modeller and decision-maker. The workshop's discussions were valuable to the participants in allowing some part of this education to take place. Similar forums should be encouraged to allow those responsible for transport policy and decision-making to interact with those responsible for analysis. For the moment, perhaps the first priority is educating the modeller to understand the needs of the decision-maker and appreciate the constraints under which he operates.

At this Third International Conference, the workshop on the 'Role of Disaggregate Models in Transport Policy Analysis' was the only workshop devoted principally to the problems of modelling from the decision-maker's point of view; the other nine focused on the modeller's view of modelling problems. There is a need to change this emphasis. At the Fourth International Conference, attended one hopes by many more planning officials and decision-makers, a similarly-named workshop might be only one of several devoted to discussing the various aspects of the problems of policy analysis, its information requirements and the decision-making process.

REFERENCES

1. Ben-Akiva, M. and Atherton, T.J., 'Choice model predictions of carpool demand: methods and results', *Transportation Research Record* (1977).
2. Cambridge Systematics, Inc. (1975) *A Behavioral Model of Automobile Ownership and Mode of Travel* (report to the Office of the Secretary of Transportation and the Federal Highway Administration) Cambridge Systematics, Inc., Boston.
3. Charles River Associates, Inc. (1976) *Policies for Controlling Automotive Air Pollution in Los Angeles* (report to the Office of the Federal Energy Administration), Charles River Associates, Inc., Boston.
4. Spear, B.D. (1977) *Applications of New Travel Demand Forecasting Techniques to Transportation Planning: A Study of Individual Choice Models*, Federal Highway Administration, Urban Planning Division, Washington, D.C.

PART SEVEN

Chapter 21

VALUES OF TIME, MODAL SPLIT AND FORECASTING

M.E. Beesley

INTRODUCTION

We are interested in this chapter in forecasting situations
in which the passage of time is foreseen to be essential to
the decisions which forecasting is to serve. Thus, we are
not mainly concerned with 'forecasting' in the sense of
ability to simulate, from limited information, the current
behaviour of transport users, or to transfer from one con-
temporaneous set of data about users notions how another
might now behave. These problems yield issues relevant to
our concern; but the focus is on situations in which the
forecasting structure involves the transport adviser in predictions for fairly remote dates, during which considerable
changes in behaviour of both demand and supply must be
anticipated.

This definition implies that there will inevitably be a
division of function between a 'transport adviser', who makes
the forecast, and the provider of key variables deemed
exogenous to, but necessary for, the prediction. Factors
such as the growth of real income over time, changes in
population and its distribution, typically used in transport
forecasting, are normally the output of other sources or
specialist advisers. The transport adviser has to be
especially aware of the implications of accepting others'
forecasts; but because of the necessary division of labour,
he will have to be sparing in his own modifications of these
inputs. Hence, our focus is on what the transport adviser
can realistically do within his own area of special competence. Another implication of the way we have defined the
topic is that forecasts are only 'good' if they are used, and
this user will be a 'client' other than the transport adviser.
There arise also, then, questions of acceptability – the
adviser's work must be capable of being demonstrated to be
legitimate in terms understandable to the client. This can,
and should, influence forecasting strategy.

A natural starting point is issues arising where most work
has been done, namely in the evaluation of inter-city
infrastructure and that connected with large-scale urban
transport studies. Current practice in these areas typically
separates issues of predicting the quantity of transport
demanded, and its distribution between modes, from the
evaluation of options, which characterise expected or possible supplies of transport. This division is highly likely
to persist, because of the inherent difficulty of specifying

459

and measuring the relationships between demand and supply,
the sheer complexity of models having the required interactions, and the capital invested in the present approach.
The purpose of this chapter is to suggest some developments
in research in time values which may help to improve valuation as it is now performed, and to discuss developments in
strategies for forecasting which may usefully complement what
is done now, thus challenging present strategies, hopefully
in a constructive way. We start with values of time savings
and then move to consider modal-split forecasting. Forecasting *what* modes will be used, and how much, raises the more
searching questions about the organisation and strategy for
improvement of forecasts. The discussions later in the
chapter accordingly set modal-split issues - more correctly
labelled 'modal-use issues' - in the practical forecasting
context.

## VALUES OF TIME SAVINGS

Interest in empirical work on time valuation began when the
need was felt to narrow down the range of values to be
imputed as benefits to options. Benefits were treated as
cost savings, for the simple reason that the quantities
demanded were derived in an indirect manner, from sources
exogenous to the model. Recent developments in time valuation open the possibility of making valuations and modal
split formally consistent. Since it is now understood that
measurements of time values must consider both the implications of the allocation of time amongst competing uses and
the quality of the experience represented by time spent on or
between modes, one has the possibility simply to proceed by
imputing modal-time values to options, and decide modal
splits by assuming only that users act to minimise costs.
But there are practical difficulties in pursuing this option.

For one thing, growing empirical evidence in the last ten
years has not increased confidence in values of time, at
least so far as these are applied to issues of modal disutility or leisure time. In the most comprehensive recent
survey (of over 50 studies) Hensher was only able to say
'even though there is a diversity of single values, the commuter time values tend to show some semblance of consistency
when converted to international units' (10). The study by
Guttman (9), however, illustrates the difficulty of arriving
at a permanent conclusion. Guttman focused on a neglected
element in other studies, namely the variation of values of
time by time of day of trip, disclosing a large and, on the
evidence, highly significant difference, most probably connected with the differing car-driving conditions encountered.
No one has yet been able to fit a model which will satisfactorily test the continually growing number of attributes
seemingly worthy of consideration. In these circumstances,
only the more general regularities in value of time findings
are likely to appeal to forecasters in practical work. Such
regularities Hensher described thus - the value of time
savings 'tends to range from a high for working time through
business travel time and commuting time to a low for non-work,
non-commuting time' (10).

Another impediment to integrated treatment of modal-split behaviour and evaluation has been rival modelling objectives. For example, by far the most influential source for applying time values in the UK is the advice given by the Department of the Environment. It has available advice, occasionally updated, on 'Standard generalised cost parameters for modelling inter-urban traffic and evaluating inter-urban road schemes' (5). The advice not only summarises 'best practice', but also has the great advantage, from a client's point of view, of defensibility in public. An updated note on appropriate parameter levels has recently been issued (October 1977). (This is in practice widely followed for urban and other modal schemes too.) It advocates another approach to consistency between modal modelling and evaluation - the use of 'equity values' of time in the steps before evaluation. The values referred to concern leisure time, and 'equity' values represent a long-standing policy position that 'behavioural' values (i.e. those derived from revealed behaviour) for each individual should not be used for evaluation. 'Equity values' - a form of average of behavioural values - are set at 19 per cent of average household income, and are applied to individuals. This perhaps is a revealing commentary on much recent endeavour to pinpoint 'behavioural values'. Modelling consistency is achieved at the cost of abandoning information on behaviour. In the DoE advice, the number of values of leisure time is now (1976) as it was in 1967 - namely two, representing in-vehicle, and walk and wait time. The less contentious area of working-time values has seen considerable proliferation. There are now ten, for different occupants of drivers of vehicles.

The use of disaggregated values of time is also now discouraged by the fact that their application must always be in a situation where local values are really at issue. As the DoE advice remarks, 'the use of local values of time implies a much greater knowledge about the valuation of time than we could claim'. Behavioural values are also related in some way to income, and this causes difficulties when, as is usual, very poor independent estimates of local income exist. Thus, in current UK practice, income estimates for zones needed for trip distribution are usually derived *from* car-ownership observations. So, in effect, when these local estimates are used, the level of car ownership will determine the leisure-time values to be imputed. If these values are to affect predicted modal choice, obvious possibilities of bias arise.

While, therefore, work towards integration must obviously continue, and will involve better accounts of behavioural time values, a separation of modal modelling and evaluation will continue. The most commonly used type of model forecasts local population and its distribution, employment and its distribution, national car ownership and national growth in income, as modified in the light of local data, and, as a consequence, travel demands. Over the first four and the sixth elements, the transport adviser will have little influence. He can hope to improve forecasting performances in the fourth and seventh elements, and in the subsequent separate evaluation phase. What extra work *would* be most likely to improve estimates?

National car-ownership forecasts are perhaps the most developed as a form of transport forecasting, and have a good track record.  Travel demands are fairly well understood too. If we wish to improve weakness, the most promising area for support seems to be evaluation.  Within this, there is one outstanding and little rationalised area - the *trend* value to be imputed to leisure time.  This is important for these reasons: leisure-time trips are growing much faster than 'in-work' trips, and will increase their dominance: leisure-time value forecasts depend on GDP per head forecasts, notoriously liable to change and often politically sensitive, so mistakes, if any, in leisure-time values will tend to be magnified; and recent work in the principles of valuation can provide a framework for criticism of present practice and, hopefully, suggest research which may reduce the uncertainties.

Practice in treating trends varies.  In the UK, the DoE's recommended values run from 1974 as base year through 1981 and five-year intervals to 1996.  Of the major components of valuation, time and other resource costs associated with the use of transport, the value of time is predicted to grow the most quickly.  Both working and leisure time, of whatever category, is deemed to rise at the predicted rate of growth of GDP (2-3 per cent per annum), giving an increase between 1974 and 1996 of 65 per cent.  Resource costs of operation vary with the type of vehicle and assumed speeds, but have a much lower rate of increase - yielding a 15 per cent growth for cars to 1996, and 19 per cent for public service vehicles (including buses).  (The latter are taken net of indirect taxation, and reflect a combination of the slowly rising real cost of vehicle purchase and use of recent years.)  Thus, use of the recommended forecast values involves accepting not only that leisure-time saving itself is the leading component in valuation - always a characteristic of valuation in most transport cases - but also that it will become sharply more important in the total evaluation.  This contrasts, for example, with Australian practice, in which the trend value is set at 0 (information from J.K. Stanley, Australian Commonwealth Bureau of Roads).

True, there has been a tendency over the last decade to focus less on the long-run, large-scale infrastructure problem and more upon issues of small-scale modification and management of existing networks, reflecting a more realistic time-scale for effective action, and budget stringency. Nevertheless, decisions for the now more usual 10-15-year view are much influenced by the trend assumed.  If one has reason to change an assumed trend value for leisure time, the following effects at least would ensue: the weight attached to net benefits of transport investment as a whole, vis-à-vis those of other sectors will be affected.  Choice between transport options would be affected also.  These vary markedly in their capital intensity.  Higher rates for time-savings, together with relatively lower rates for operating costs favours more capital-intensive options (e.g. adaptations to, rather than on, existing networks).  They favour new systems rather than schemes for changes in vehicle use, etc. because returns are usually relatively slower in larger works.  Raising leisure-time values relative to other resource costs will tend to penalise projects with high proportions of commercial vehicles.   So the issue is of some importance.

For forecasting purposes, we are interested not in the rather arbitrary starting point for leisure-time values (the postulated 19 per cent of household income for example in the DoE advice) which represents an estimate for today, but in how it may change over time.   The most developed theoretical apparatus we have to guide us is the neo-classical account of time values, much elaborated over the last decade (see Chapter 15 in the present book).   All that we can usefully hope for from this account is some expectation of the direction of change associated with expected changes in the elements entering into that account, viewed as partial adjustments.   This should, however, be a framework from which to form judgements and suggest further clarifactory work.

If then, we take the elements in the account, we find (to paraphrase a recent summary by Dalvi and Daly (4)) that a person is assumed to maximise his utility from allocating his time in conditions where:

1   He has a complete, consistent preference ordering among alternative commodity bundles and exhibits rational behaviour.
2   Utility is a function not only of commodities and income but also of the time allocated to them.
3   Choice is subject to constraints, of which time is the first.

But activities such as work are paid for, and generate income. Some other activities will require payment.   Once a 'work' decision is made, therefore, the person's decision is subject to two resource constraints, money and time.   The decision to consume a specified amount of any commodity or to work, requires that some time be allocated to it, which can be varied by the individual.   Since one is thinking of a long-run equilibrium, an assumption that work can be freely varied seems justified.

Now, the present measures of 'leisure' time values observe the trade-offs between leisure time and money derived principally from work.   Using the foregoing account, we can list possible sources of change in the measured values.   We are not concerned here to enquire whether the measures are 'correctly' done: we are interested in the forces operating to change them.   We are also working in 'real' terms, i.e. we neglect the effect of inflation on prices.

The possible sources seem to be that:

1   the time budget available may change;
2   the disutilities attached to production or utilities attached to consumption may change (e.g. onerousness of work per unit time, or the pleasure to be derived from a given leisure activity);
3   the price at which the commodities can be produced or bought can change (more pay per hour or cheaper cinema seats for example); and
4   the time inputs to these activities necessary for a given level of disutility or utility might change.

These may alter the trade-offs facing the rational consumer, deemed to be interested in allocating his time efficiently in a system where utility diminishes at the margin.

In this set of ideas, the decision to allocate time is the fundamental one; thus, for example, income earned is a result of the individual's decision to spend time at work rather than, say, leisure activities costing no money or those that do.  The relevant changes are of prices and quantities he faces.  Other changes relevant to forecasting may of course also emerge, most notably that individual consumers' tastes may change and that consumers themselves will be replaced, in time, by different consumers.

It is obvious that not only are the relationships between the elements potentially extremely complicated, but that statements about how they may individually change are very speculative.  But one must start somewhere, and since the question of the likely sign of the individual changes has not, to my knowledge, been canvassed, some progress in shaping ideas may at least emerge.  We must also note the problem that household members, where present, must at least strongly influence individual behaviour and hence measured values. (Some statement of the relation between individual and house-hold income and time values must be adopted in practical modelling.)  We should enquire whether systematic change is likely here also.  We should make clear too that the specula-tions are about the UK; other countries may well have quite different experiences with respect to some elements.

To begin with the *time budget*:  24 hours a day cannot be changed, as often, and redundantly, pointed out.  Neither is change in life expectancy, i.e. the life-time budget, likely to be a factor.  If a person becomes convinced that he would live longer, he might alter his allocation of time over time. But we probably can ignore this for practical purposes.  The main source of an effective (or 'operational') change in time budgets is systematic effects on the time inputs necessary for available or potential sets of activities.  Thus, if the time required to consume or to go to work generally falls, this in effect increases total available time to be allocated to more activities, or not, as the case may be.  This is indeed what transport investment in particular is meant to accomplish.  Transport investment and vehicles of all types will go on accumulating in the future as in the past. Increasing availability of labour-saving devices of all kinds have like effects.  It seems probably that the independent (partial) effect of this is to reduce the value of time savings over the forecasting period.

It is perhaps a fair guess that the average *disutility of work* will fall over the years.  Working conditions and standards (e.g. in offices) have been rising fast and will continue to do so.  Automation will, on balance, increasingly remove the worst forms of tedium.  The partial effect of this on the measured value of time is to raise the value of time, for two reasons.  For a given set of workers in an initial position, decreasing disutility of work should, *ceteris paribus*, lead to an increased willingness to work, and a shift into remunerated time.  Employment will also expand,

given some elasticity of demand of labour.  More people will
enter the work force, and since the chief reservoir is among
females, presumably the value of time on average to be
imputed to females will rise.  Future changes in tastes
probably reinforce this.  Occupational demands are changing
in favour of white-collar rather than blue-collar workers,
e.g. as GDP per head grows, the proportion of the work force
devoted to manufacturing falls.  We should remember here that
in the neo-classical model individuals act in their own
interests, and their level of welfare does not depend on
others' consumption, etc.  This, amongst other things, dis-
allows interpersonal effects within households, a point to be
taken up later.

The outlook seems to be towards increasing the *utility of
leisure*, mainly because of innovation in ways of using
leisure time and the effect of increasing educational
resources of individuals.  The innovation referred to concerns the quality of the experience on offer, not mere
ingenuity in making the use of leisure time more efficient,
an effort already taken into account in describing the likely
course of the overall time budget.  In terms of the growth
and extension of the package holiday market, for example,
this means extending the amount of sun (or other desirable
features of holidays) available per unit of time, not simply
reducing the time needed to consume a holiday, e.g. by better
organisation of package tours.  One hesitates to predict that
there will not come over the horizon an innovation comparable
to television, but it does seem likely that this effect,
though positive in its effect on time value, will perhaps not
be strong.  But, as remarked earlier, the prospect is that
complements to leisure - i.e. ways to use it pleasurably -
will proliferate, and this is also a force raising leisure-
time values.

These leisure pursuits ('commodities') referred to in the
last paragraph, are ones with which payment is associated.
There are certain leisure pursuits which do not require payment.  It is possible that a switch to these may modify the
effects just noted.  One's feeling about this, however, is
that paid-for leisure pursuits, have been displacing non-paid
in the past, and that this reflects underlying preferences.

As to the real *prices offered for work*, and the real *price
of leisure pursuits*, the expected trends may be rather
different.  The chief problems here seem to be - first, that
the price of labour depends very much on the movements of
relative productivity of a given country, say the UK, versus
other countries with which, increasingly, it has to trade;
second, leisure time pursuits are, one expects, relatively
more labour-intensive in the direct inputs needed to sustain
them than are other activities.  Over the long run, the real
price of labour may rise at a rate different from that suggested by post-war UK economic performance, and which is
reflected strongly in the 'recommended' figures noted earlier.
But it will doubtless rise.  And the real price of leisure
pursuits, because labour-intensive, will rise even faster.
The effect of these factors is to reduce the value of leisure
time.

There is next the question of the trend in tastes for
leisure.  There seems little doubt that a shift towards
leisure in this respect has taken place in recent years in
the UK at least,  though the evidence is highly circumstantial.  Not only has there been a widely observed decline in
what might be called the individual Puritan work ethic, but
there has also been growth of social approval for not working,
as evidenced in the rise in unemployment pay as a proportion
of the average wage or salary.  This will probably continue,
and its effect, *ceteris paribus*, is to raise the measured
value of leisure time.

To summarise, we have noted the principal elements that
seem likely to influence the future trend of time values,
focusing on leisure time as the most important time element.
Viewed independently, time budgets and the net effect of the
prices of leisure and work activities will reduce the
measured values.  Falling disutility of work, and innovation
in leisure activities will increase them.  Though partial,
these are the directions of change to which the neo-classical
formulation points us.  Changes in tastes in individuals seem
likely to raise values.

As mentioned earlier, one important modification of that
formulation should be noted - the likely effect of the household's influence on individual behaviour, for which the
neo-classical account is drawn up.  Households represent a
constraint on individual behaviour, reflected in present
measured values of time.  The observations are for individuals, and moreover very often for the wage- and salary-
earners in the household.  The practice in valuation is to
assign positive values to time savings by household
dependents (e.g. to assign a housewife the same leisure time
value per hour as her husband, and, sometimes, to assign a
proportion to children also).  This accords with the common-
sense notion that wives' and children's time must have some
positive value, and with the observation that even children
will often pay to save time, or vice versa.  Nevertheless,
the assignments are *ad hoc* and lack precise rationalisation.
This is probably inevitable; bringing in values for housewives and children is to recognise that evaluation should do
more than reflect that part of the economy concerned with
cash transactions only.

We assume, therefore, that such practice will continue, as
seems sensible.  The task here is not to give it a rationalisation, but to enquire whether we may expect systematic
changes to occur which could affect the assigned values.
Presumably the main factors are the incidence of dependants
in the household, and attitudes within the household.  The
first is important in that a trend value in the size of
household is implicit in all demographic forecasts on which
transport forecasts themselves rest.  A fair guess for the UK
is possibly that household size will fall.  If so, this will
show up in the populations to whom values of time are to be
assigned (less children, for example).  But the independent
effect of fewer dependants on measured values of leisure time
is, one expects, to raise that value so far as the wage- or
salary-earners in the household are concerned.  Less of the
workers' budget has to be spent on them, and more will be

spent on time saving. So far as the new wage- or salary-
earner - the housewife - is concerned, the opposite will be
true. Less time has to be spent on serving children.  As to
attitudes towards dependants, it is anyone's guess about how
the intra-family ties will change.  If the desire to spend
time with the family increases, so will leisure-time values.
If the opposite occurs, and family life becomes more a matter
of economic convenience to members, the values will, *ceteris
paribus*, fall.

From the foregoing, it seems clear that the factors to
which neo-classical accounts of consumer behaviour direct us,
and which are our only reasonably well-established framework,
do not permit an unequivocal prediction of the direction of
change in measured values of time, let alone the adoption of
a value strongly increasing in line with GDP or other
measures of real income change.  For the most directly relevant element for which the trend in GDP is a defensible proxy
- the price of work - the expectation is for a decline in
leisure-time values.  There are, nevertheless, quite strong
forces working the other way.  And one can easily point to
other sources of uncertainty.  For example, changes in tax,
and especially income tax, will affect the leisure/work
trade-off and thus measured values.  If past tax trends continue, one might expect, *ceteris paribus*, an increase in
values of leisure time.

In a situation in which there is considerable uncertainty
about the signs of the expected partial changes, as here, and
given that *some* account of future values must be adopted, a
sensible attitude would seem to be to adopt a zero trend
value.  (Were leisure time not so important an element, one
might let it assume the same relative value as others.)  The
appropriate reaction to uncertainty in forecasts is two-fold
- first, to attempt to reduce it by further research, to be
considered in the next section, and to pay closer attention
to the organisational conditions in which the forecast is
made.  The latter raises a point made familiar by writers on
technological forecasting, another area in which the time-
scale is long and the inherent uncertainties great.  They
emphasise the need to get agreement by forecasters on the
approaches to forecasting to be adopted and the active interaction between forecasters and users.  This issue will be
discussed later.  In the meanwhile, were the zero trend in
values to be adopted, this would presumably induce considerable shifts in choice between options evaluated, as fore-
shadowed earlier.  This is itself an hypothesis worth testing
before resources for research are committed.


INDICATIONS FOR RESEARCH ON VALUES OF TIME

The most obvious implication is that the passage of time
between observations should be noticed in attempted explanations of values of time.  Studies of values of time have been
wholly cross-sectional and one-period in focus.  So much
attention has been given to the problem of 'improving' cross-
section estimates, mainly to derive modal disutilities, that
little or no attention has been paid to the forecasting need
to get a feel for the impact of change over time.  Critiques

of value-of-time studies (e.g. (1)) have failed to suggest
that inconsistencies in measured values as between studies
may to some extent be due to real change in values in a trend
sense.  Now that repeated cross-sections of behaviour from
transport studies are becoming more frequently available, it
is clear that some attention should be given to discovering
and explaining shifts in implied values over time.

Moving to the specific factors enumerated earlier, one can
formulate hypotheses which, were they tested, could help to
build judgement in forecasts.  The first concerned time
budgets.  It was suggested that the main source for an
effective increase in time available might well be changing
accessibility to points of consumption.  The hypothesis to be
tested is that, as accessibility improves, more time is
generated to be used for all activities, whether for work,
leisure or sleeping, and that observed values of time fall,
*ceteris paribus*.  The principal opportunity to observe such a
phenomenon is probably the acquisition of a car.  Having a
car brings most points of consumption much nearer in time.
How then will measured time values be affected?  Holding
*ceteris paribus* in the likely experimental condition would be
quite difficult, of course.  Amongst other things, points of
consumption and work have to be held constant, and acquiring
a car has effects on disposable income which also require
standardisation.

Perhaps the most difficult link to be established in such
work is the change in the measured value of time.  One wants
to observe both it and the change in time availability.  For
good reasons, most research workers in this field are now
apprehensive about inferring values from choices involving
cars as a competing mode.  To maximise likely success with
the present objective, therefore, one should observe car
owners, before and after acquisition, in situations where
they choose between public transport alternatives, or indeed
alternatives not involving transport.  As has been often
pointed out, value-of-time trade-offs are not confined to the
transport area; it simply happens that some of the better
experimental conditions can be found in transport, and the
demand for values has principally arisen in transport.

A closely allied question is the use of the time that is
generated.  This belongs to an area of growing interest - of
the portion and extent of time budgets that are allocated to
purposes involving transport.  This is of interest to forecasting for the light it may throw on the appropriate ways to
constrain the estimated future consumption of transport.
Characteristically, forecasts of this future consumption are
made only implicitly.  They arise from forecasted car ownership, other vehicle ownership, and of vehicles' respective
use.  These elements are unlikely to be displaced.  Rather
they will be refined, by closer attention to the effects of
real-price changes on ownership, the ways in which local
variance in ownership deviates from national projections, etc.
Good forecasting, however, requires attention to the implications of these forecasts for total time (and money) deemed to
be devoted to transport in the future.  Implied money budgets
for transport are relatively easily calculated for future
dates and compared to household expenditure implied by

forecasts of GNP per head, though this seldom seems to be
done in practice. The implied expenditure of time is more
important, because it is the more fundamental constraint.

Goodwin (8) has recently and usefully suggested the notion
of a movement towards a postulated saturation level for
expenditure of time in travelling, which is parallel to the
notions already used for vehicle-ownership forecasting. The
evidence is now, it seems, largely confined to cross-section
comparisons of income groups, in which the leading effect
seems to be a strong positive relationship of time spent and
income, but with some evidence of tailing off at the highest
income levels, which itself prompted the notion of a saturation level. Since the explicit or implicit collection of
time budget data is relatively new, we cannot yet expect
enlightenment from time series or repeated cross-sections,
and will have to be content with single-period observations.
More investigation of the effect of income on time spent in
transport of all types, including holiday travel, should
prove useful. Essential features, from the point of view of
the present discussion, are that, apart from the obvious
standardisation requirements for age, sex, family size, etc.,
accessibility to different points of consumption and to work,
the price of travel and possibly also the speed of travel, as
a proxy for its disutility, should be allowed for. Probably
also one would have to take more care than is usual in transport studies with the definitions of income and its relation
to assets or wealth.

Research workers in transport can do relatively little to
throw light on issues about the disutility of work, or the
postulated likely growth of complements to leisure. These
are essentially the province of industrial- and leisure-
market forecasting. However, transport expertise could
usefully be brought more to bear on the question of trends in
the price of transport inputs to leisure pursuits, and that
of travel in work, respectively. In the former, work has
usually concentrated on the expected price of car ownership
and use, stimulated partly of course by recent oil price
changes. It has always been an active research concern
because of the coincidence of the interests of large-scale
users - manufacturers and road planners. Perhaps the time is
now ripe to switch some concern to likely trends in the real
price of all other forms of transport, and in particular, for
leisure-time trip evaluation, of public passenger transport.
(Within work-trip evaluation, the most important sector is
haulage, and there is the same need to take a view of trends
in real prices here.)

What will happen to real prices of inputs in these sectors
and to productivity? Over the long term in the UK, the
supply of labour to both sectors should rise, principally
because the chief skill, driving, is getting ever more
common. In passenger transport - but not in road haulage, in
which there is easy entry in the UK - the current price of
labour is heavily influenced by restriction of entry, via
licensing and other institutional barriers, and by trade
union control. Fundamentally, the latter depends on the
former in industries which have virtually constant returns to
scale, as does most transport operation. Hence predictions

depend on the view taken of the economics of institutional
restrictions in transport, which used to be a much more
fashionable area for research (when rail-road competition was
itself an important issue) than it is now.

On productivity, the two sectors - public passenger and
road haulage - seem to provide a strong contrast.  Productivity depends most on increasing the capital-labour ratio.
For haulage, predictions are relatively straightforward,
concerning the likely growth in average lorry size, as
modified by such institutional factors as the effect of EEC
regulations.  For passengers, many more options, to substitute different forms of capital, and to explore ways in
which labour may be combined with it, seem to be available in
the long run - e.g. to cope with the characteristically
peaked nature of public passenger demand.  Here, a comparative study of the productivity of buses, conventional or
unconventional, in different cities, within and between
countries, might well give interesting indications of the
possibilities.

Likely shifts in tastes will, predictably, remain
unattackable by transport research, not only because of their
inherent elusiveness and dependence on social forces at large,
but because, in principle, such shifts can only be detected
in transport models as some form of unexplained trend
variable, with satisfactory specification of changes in
relevant groups of individuals for whom tastes are postulated
to change.  But one might make useful progress in one respect.
Classification of individuals in transport models tends to be
highly conventional (by income, socio-economic group, age,
etc.).  There is certainly no *a priori* reason to believe that
these, though in themselves highly useful and necessary in
forecasting practice because they relate to principal available exogenous variables, are especially revealing of
similarities in attitudes to transport variables.  Testing
for *a priori* grouping of individuals with respect to tastes
is a proper aim for such techniques as non-metric scaling,
using individuals drawn from the populations now recognised.
If one can derive what seems from these experimental situations to be more appropriate groupings, we should have a
firmer basis for subsequent work.

The prospects for clarifying the issues of the household's
effects on time values is perhaps better, though it requires
a careful research strategy.  One should probably first
attempt to test for systematic relationships between values
of time of wage- or salary-earners in the family and other
members, chiefly to test the hypothesis that within households variance of values of time is less than that between
households.  This itself might well be conducted by means not
requiring observations of cash-time trade-offs in the real
world, e.g. by attitude surveys, testing with simulated
budgets and choice situations, etc.  Again, since the ultimate focus is on valuing time at the leisure/work margin, if
transport examples are used these should minimise differences
involving varying modal disutility.

One can then proceed to test, on a cross-section of heads
of households, and other household earners, for the effect of

dependants on time values, holding cash available for transport expenditure constant.  This would have to involve real-world choices, not simply because of experimental doubts about devising usable values from situations where actual sacrifices and gains are not made, but also because the client usually needs to show that the results being used appeal to the non-expert 'common-sense'.

Useful speculation could also be made about the inter-dependence of members of families in their leisure time, with the help of family time budgets.  The degree to which one person's utility is dependent on another household member's joining in the same activity is of some interest to the fore-caster.  One exogenously given trend in the UK is the increasing participation of women in the labour force.  To the extent that the earner's - say the husband's - utility of leisure is dependent on wifely leisure activity, and that is lessened by her joining the formal labour market, we might suppose a diminution in the attractiveness of leisure time, and a fall in its value.  What happens to the amount of joint-leisure activity in the face of a big household change, like adding to the list of family earners, is one of several ways in which interdependence may be investigated.  Another set of questions concerns the total amount of time in the family devoted to travel when substantial changes of access-ibility occur for one member - e.g. when the wife learns to drive the car, or a second car is acquired.  How far the expected increase in total time for her is offset by less time spent by other members of the family, including the husband, is possibly of significance in keeping forecasts within plausible totals.  These are sensitive, explicitly or implicitly, to assumed increases of numbers of drivers in the family and second car ownership.

These, then, are some ways in which research within the competence of the transport specialist might assist in the variety of issues brought up by the problem of forecasting values of time.  The plausibility of forecasts can also be helped, it has been suggested, by work directed at setting total constraints on time and money expenditure on travel. Arguably the most difficult forecasting task, however, is to give weights to present and future modes for reasons that owe as much to the forecasting context as to the technical problems.  We turn to the question of modal use in the next section.


MODAL-USE FORECASTING

The required output of a forecasting model will vary with the 'client'.  As examples, those responsible for decisions on infrastructure will require forecasts centred on numbers of 'vehicle miles' (performed in cars, trains, etc.) using alternative investments among which they must choose.  Those responsible for producing the 'vehicles' require a focus on numbers of vehicles of different types.  Managing a given infrastructure is usually a task separated from that of decisions about it.  The interest for such managers is on how the vehicle miles respond to instrumental variables forecast to be available (e.g. road pricing, existing and new forms of

traffic management, etc.).  The 'clients' are subject to
varying requirements for public accountability, i.e. the need
to justify their decisions.  For example, in managing public
transport the natural focus is the passenger miles potentially achieved by alternative schemes, and their distribution among the members of the public.  Efficient forecasting
procedures consist of comparing required outputs with the
feasibility and costs of the effort involved.  What is
efficient for one 'client' is not necessarily so for another,
because of the varying specific requirements.  Keeping this
in mind, we attempt here to point out what seem to be major
needs to improve the present state of the art.

Reduced to its essentials, an economic account of future
passenger miles, modal miles and vehicle numbers for a given
geographical area will contain:

1  The scale and distribution of expected population.
2  The scale and distribution of expected employment.
3  Expected income per head.
4  Expected prices of modes, in terms of time and cash
outlays required to use them.
5  Expected prices of other goods and services.

Since travel is largely a derived demand, the function of the
first two points above is to indicate the market potential,
consisting of locations between which travel can take place.
The function of the last three points is to predict how far
this potential will be realised.  The output of these elements is actual demand for person-trip miles, by mode.  From
this, the facts of importance to 'clients' can be derived -
such as passenger miles, vehicle modes and numbers.  But this,
in turn, requires at least the prior specification of possible future infrastructure and modes, which enter into the
'prices' of modes.  These can be set up with, or without, the
information derivable from the model.  In so far as innovation is involved, and it almost always is, the model must be
contingent, and exploratory.  Clients wish to use the model
to accept or reject ideas for improvement.

A forecasting strategy requires attention to what is to be
considered exogenous or endogenous.  In general, the more
that is made endogenous, the higher the forecasting cost.
This is a matter of judgement - of the client's interest and
of the likely success in incorporating factors within the
model.  So for example, in principle, the modal changes can
not only affect prices of modes, but also population and
employment distribution.  This is the familiar problem of the
interaction of transport and land use within a region.  Even
more, the transport outputs can affect incomes and the level
of population and employment.  This is the problem of inter-
regional competition.  The current state of the art severely
limits the possibilities of dealing successfully with these
latter two types of feedback, and a particular client may be
more or less vulnerable to this fact.  This inherent uncertainty, for the client, is a function both of his interest
and of the state of information.

This can be dealt with in several ways, but will, in any
case, require client-forecaster interaction.  A 'client' must

learn and react in order to convey his attitudes to uncertainty to the forecaster.  Thus, for example, a 'client' may
be shown the sensitivity of modal outputs to changes in the
model structure or inputs.   These can very easily be complex
and involve selection.   The client explores the assumed consequences of his choice.   Exploration of the implications of
truncating the model by limiting feedback and interactions
within the model will simplify and lower the cost of the
forecaster's task.   Thus, taking the elements described
earlier, it may be advantageous to combine the third and last
in a specification of disposable income per head available
for transport purposes, or to regard car ownership as
exogenously determined.   Car *use* is then defined as endogenous, to be determined by location, competition between modes,
disposable income, etc.

Forecasting strategy requires not only attention to what
is technically possible now, but it must also be open to
possible progress in the state of knowledge about the
economic facts about travelling.   Thus, the neo-classical
account of time valuation referred to earlier now stresses
the concept of a time budget.   This, as suggested, might take
a place analogous to that of a monetary budget (income)
already incorporated in forecasting models, so long as the
prospective gain in output relevant for the client's purposes
outweighs the cost of incorporation, including the greater
complexity and uncertainty generated.   The present state of
the art probably militates against attempts formally to
include projections of time budgets, but this may well change
as knowledge of time budgets increases.

This view of the forecasting task immediately suggests the
outline of a strategy to improve forecasts.   Starting with
what is available now (mostly the models associated with
defining future use of inter- or intra-urban infrastructure),
one would seek to rank the model components in terms of their
predicted impacts upon the individual client's interest as
identified in model outputs - passenger miles, vehicles, etc.
Comparison would require a series of common percentage
variations to be applied to each constituent, and their testing in combination.   One would then systematically consider
the feasibility and cost of improving the constituents.   This
contrasts with what is usually done now.   Sensitivity tests
typically performed refer to what the forecasters believe to
be the least plausible items in their evaluation of alternatives.   Conspicuous among these is the assumed value of time.
The principle of selection of items is thus their supposed
individual defensibility, not a search for model improvement.
Defensibility is indeed an important attribute from a
client's point of view, but it is only one of the aspects of
devising a better forecasting strategy.

A corollary of the client's responsibility for proposing
alternative projects and investments for testing, and of the
interaction between forecaster and client, is that one object
of testing for model sensitivity should be a critique of the
projects and instruments initially set up for testing.   The
question might be - are there conspicuous gaps in these which
can be inferred by tests performed within the model?   At what
cost can these be filled?   Clearly, it is not feasible here

to work out all or even most of the implications of specific
research strategies along these lines, as many clients and
objectives are involved.  However, it may be useful to consider likely pay-offs to improving existing modal split
procedures in the inter-urban or intra-urban contexts,
proceeding by way of important UK examples.


INTER-URBAN MODAL-USE MODELLING

Modal-use modelling is most critical when the issue is one of
choice between future modes.  Perhaps the most recent occasion in which a major effort was made in the inter-city
forecasting area to deal explicitly with this was in
'Comparative Assessment of New Forms of Inter-City Transport',
by an interdepartmental working party in which the Department
of the Environment, Department of Trade and Industry and the
Ministry of Aviation Supply were concerned (13).  The modes
compared were car, bus, advanced passenger train (APT), conventional take-off and landing aircraft (CTOL), tracked
hovercraft (TH), vertical take-off and landing aircraft
(VTOL), and (later) short take-off and landing aircraft
(STOL).  The test inter-city pairs were London-Glasgow and
London-Manchester, and the time-scale from 1970-85.  The
treatment of modal split, the part it played in the assessment and possible directions of improvement are all of
interest.

The concerns of the client departments in the assessment
were not made clear by the published information.  We may
reasonably assume that chief among them were the Department
of the Environment's interest in the growing commitment of
British Rail to the APT, following a major effort to build up
inter-city passenger traffic in conventional trains, and the
other departments' interests in backing technological
development.  In particular, at the time, the question of
tracked hovercraft's future was of great moment.

The working party included two of the main modal pro-
tagonists, ensuring that the exercise was at least partly one
of negotiation and compromise.  For example, a critical
assumption about vehicle loading - settled at 60 per cent for
*all* modes - was explicitly a compromise.  Passenger wait-for-
service times were set at 30 minutes for *each* mode, penalis-
ing the faster modes.  Both of these might have been expected
to be determined endogenously, as a part of the model
processes, because of their obvious implications for realised
costs of modes and marketability.  Again, apparently, an
appreciable amount of time was spent on the issue of how to
treat the future of sleeper passengers on rail.  They were
eventually allowed to retain current levels throughout the
forecast period - in a situation in which the whole rationale
of the exercise was to illumine the impact of the possible
introduction of super-fast modes rendering, one might have
supposed, overnight travel in the UK obsolete!

The expected modal competition - the 'behavioural' part of
the model - was predicted by first calibrating modal split on
existing modes, for origin and destination pairs relevant for
the London-Glasgow corridor.  The time values emerging from

this, plus assumed fares for future modes, made cash-equivalent options facing the future travellers, who were deemed to choose on this basis. Because of the geographical possibilities of access to the future modes (all 'new' modes were deemed to require loading at specific single points at each end of, or limited points along, the corridor), the populations of zones, and varying assumed income levels and trip purposes, each future mode could expect to be assigned some captive custom. Attempts to make fares and service levels endogenous to the model failed, because of a tendency for the simulated market competition to price out modes with fixed costs. Thus where, for example, as an arbitrary starting point, a future mode was assigned a fare and volume at which costs happened to be above fares, it was assumed that the mode's reaction would be to raise the price. This price would tend to lower the mode's market share, leading to a further price rise, and so on. A corollary of these difficulties was to yield a necessarily rather artificial account of expected financial performance for each public mode, according to assumptions about the share of the total market (thought to be capable of realising between 2.5-6.0 million passengers a year).

Nevertheless, the report was able to come to definite conclusions. At one end of the spectrum, APT was likely, it was thought, to make a profit at all likely shares of the market, and was vulnerable (because of its fixed costs) only if the total market was actually very small. At the other end, TH's revenues failed to cover costs at any postulated market share or fare. The other modes performed variously around their break-even points.

In the summary, the report remarks: 'It has proved difficult to develop a method of assessment', and goes on to list desirable developments. They include: 'to extend the models to networks, rather than single London-based routes'; 'to include new travel generated by improvement in transport and by longer term regional developments'; and the 'further refinement of the behavioural model'. The authors go on to stress the last point. Whereas the model described above is 'the best means that exist for making projections into the future for travel on many modes', its 'central place in the assessment is a clear indication of the need to develop and refine it still further' ((13), p. 74).

There has been little such development of the original models since, probably for the reason that the conclusions were quite sufficient to satisfy the clients - rail's APT could go ahead, and tracked hovercraft should not be supported further. Actually, it seems that, contrary to the report's summary, a strategy to improve the usefulness of forecasts in this case called for rather different emphases. First, the outstanding fact about the envisaged market was the potential entry of very fast new modes. But they were, as we have seen, also potentially fixed, with few points of entry. Such modes are markedly sensitive to the geographical market around these points. Speed advantage gained on the line haul diminishes rapidly away from the point of entry. It follows that a crucial problem is search for the correct points of entry, and the cost of such access. Such a search, and its strategic position, are not mentioned in the report.

'APT appears to be assured of success', says the report. If the client in this case had needed more reassurance, there were obvious points to seek it. Conspicuous among these was the lack of even elementary economic caution in accepting the range of rail-cost estimates, specifically of the initial capital outlays. The scale of capital outlays is a critical assumption both for estimated break-even points and risk. On a normal financial risk analysis, i.e. of expected *variance* of profitability (or loss-making) with possible volumes, APT, on the report's own analysis, is the next most vulnerable to TH itself, because of its ratio of fixed to variable costs. Although the discussion of the rail capital outlays displays extremely dubious features, no sensitivity analysis about assumptions is suggested ((13), pp. 17-18). For example, costs are allocated to APT on a ton-mile basis; costs specific to APT are put in this cost base, so that freight is assigned some of these costs; the range tested is produced by postulating a separate two-track route for APT without, apparently, considering the effect on total costs to all traffic; and the range also depends on assuming 'proportional increases in all other traffic' - an unbelievable reversal of previous railway experience. The lack of sensitivity analysis is probably due to the position of railways in the organisation of the study; they were not only the most experienced in arguments about costs but they were most obviously likely to be regarded as the best informed of the modal representatives, and APT was in a far more advanced state of development than the other modes.

Improvement in the behavioural model was indeed possible. It is relevant to note that in the model as developed, more knowledge of modal-time values would hardly have helped. Even if it were possible to identify some specific modal attributes other than elapsed time (which was not contemplated, it seems, simply because no one could really specify market differentiation in terms other than in relative speeds), the production of specific modal disutilities, such as comfort, or in-vehicle service, would probably have shifted relative cost/volume predictions but slightly. What would have made a considerable difference, within the description of competition adopted, was to have made service frequency for each mode endogenous to the model. This would not only have required more detailed modelling of competitive time savings, but, equally important, it would greatly have affected predicted modal costs. The report recognised the inadequacy of its assumption of invariant service times. But, as we have seen, modelling future costs was not a strength of the study; and a strategy for improving the forecasts should certainly have stressed this.

One should perhaps stress again the forecasting situation. An exercise like that described must, for participating operators, be a defensive exercise. If they individually wish to forecast *their* business, their best efforts are likely to be along well-established lines of business forecasting, stressing opportunities and threats in the environment, related to the business's own assets, and its strengths and weaknesses. Thus railways would properly think

of the markets it could attack, the probability of competitors eroding markets (for example, how far will it be
possible to assume, as in the past in the UK, that inter-city
bus services would be highly constrained by licensing laws of
1930s vintage?).* They might logically think of potential
use of railway assets - a collection of properties and
locations as well as traditional transport equipment.  The
focus should be on corporate strategy as a whole.  In this
context, exercises like that described are not to be regarded
as inputs to a strategy; rather they would be treated as part
of strategy implementation.  In other words, the way in which
the work was organised did not maximise the likelihood that
researchers' time would be well spent, or that unbiased
results would ensue.

The limitations of conventional modal-split modelling when
faced with a need to predict the outcome of competition
between modes was also clearly illustrated in the early 1970s
by another important example - the Channel Tunnel studies.
These were essentially concerned with three actual or
potential modes - ships, hovercraft and the proposed tunnel.
Again, the approach was to predict total cross-channel
passenger volumes and then attempt to allocate them to the
competitors.  No plausible account of competitive actions as
they might be in the event emerged.  Improvement of prediction in this case turns on explicitly modelling an 'industry'
consisting of a differentiated oligopoly, perhaps the
trickiest of all market structures to analyse.  As the
Monopolies Commission's report on the cross-channel car ferry
services in 1974 implied (11), critical to understanding here
is a view of international cartel operations in shipping,
development of the conditions for entry into the industry
(e.g. port and land facilities and their control), technological developments in hovercraft greatly affecting their
range, and many other elements.  A differentiated oligopoly
might well, in the event, succeed in expanding its total
market considerably.  An important question facing the forecaster here is how the total demand will behave when a fixed
single-point, but always available, competitor (the tunnel),
can be used alongside less reliable but often much more conveniently located and flexible competition, one at least of
whom, the hovercraft, might well offer substantially improved
local access characteristics, and which could materially
influence effective journey times.  In such a market structure, the pricing strategy of the least flexible, but highest
capacity 'firm' - the tunnel - would be complex and critical
to its financial success.  As Glaister has shown, it is
possible to throw more light on this than was achieved by the
official enquiries (7).  But prices would be only one
dimension of competition; to make modal-split models simply

---

*One extraordinary feature of the report under discussion was
the assumption that buses would take, in 1985, 7.00 hours to
get from London to Manchester (an average of less than 30
mph), whereas cars would take 3 hrs 10 mins.  The 1968 base
year time was 8.00 hours, which no doubt influenced the
assumption.  With freedom to arrange schedules, buses, at
best, cannot be far behind car times between two centres
served by motorways.

more capable of simulating this aspect is an intrinsically
limited ambition.

Conventional modal-split modelling is, as these examples
show, designed in effect to reduce the problem of modal competition to one of competing for custom where market areas
basically defined by location of customers may overlap.
Where this formulation is appropriate to the client's objective, the approach is likely to be efficient; but one wonders
in that case whether much development is needed. For example,
a most prominent recent application of such a model formed
part of the Roskill Commission's Enquiry into the Third
London Airport (3). This indeed was a problem of competing
locations (the rival airport sites). The airports were
deemed to serve a similar air market; what chiefly differentiated the sites was their marginal effect on drawing this
custom. The ground-side model was adequate for this purpose.
With hindsight, to refine it further would have been far less
important to the outcome of the exercise than more attention
to the question of the predicted number of aircraft movements
and their loading. (The latter predictions proved to have
under-estimated the impact of big jets. It turned out that
not only would the 'need' for extra capacity be much greater
than assumed, on which Roskill was asked for an opinion, but
that the problem of airport capacity turned more into the
question of handling passenger numbers.)

None of these examples, nor the conventional modal-split
approach in general, is addressed to the important problem of
deciding what modes should be in the modelling set. These
are typically very constrained and often arbitrarily arrived
at. To make progress is partly a question of organising the
nomination of modes, and partly how far modal-split modelling
can be made to contribute towards a search process, to
indicate ways of serving markets and filling market gaps.
The former is obviously important where there are, a priori,
drastically different ways of serving distinct markets, as in
our inter-city example. The latter is of great potential
interest in urban areas especially. It asks the question how
far can we specify a means to feed back accounts of modal
split to criticise the client's choice of options, including
not only investment alternatives but also policy options,
such as means to regulate modes, traffic management, etc. In
both, client objectives and relations to the forecasts are
inevitably involved.

The nomination of modes is a problem akin to that facing
large firms who wish to anticipate long-term market developments. What are the chances of new products appearing, as
threats or opportunities, in the future? There seems no
reason apart from custom and inertia why clients in the
transport area should not learn from best practice in the
private sector here. Thus, for example, one well regarded
survey of 18 forecasting techniques singled out 7 appropriate
to investigating 'new product possibilities' (2). These
were: 'product life-cycle analysis', 'trend projections',
'historical analogy', 'visionary forecast', 'panel consensus',
'market research' (enquiring of present customers about
future developments), and the 'Delphi' technique. Interestingly, other techniques such as regression models,

econometric models, and input-output models are not considered apt for new product forecasting.  Four of the
approved techniques, namely product life-cycle analysis,
trend projections, market research and historical analogy are
really projections for products recently introduced, so that
for our purposes, the three other techniques, which are
essentially qualitative, are most relevant.  The authors have
little doubt that of these, the Delphi technique is superior
- 'visionary forecast' is a prophecy, which uses personal
insights and when possible, facts about different scenarios
of the future.  'Panel consensus' is based on the assumption
that several experts can arrive at a better forecast than one
person.  There is no secrecy and communication is encouraged.
As is well known, Delphi involves a panel of experts, inter-
rogated by a sequence of questionnaires in which the
responses to one questionnaire are used to produce the next
one.  Information is thus passed to all experts.  Use of the
(anonymous) questionnaire 'eliminates the bandwagon effect of
majority opinion'.

Further insights might well be gained by considering the
private sector's attitudes to technological forecasting in
general.  Here again, as an example, a well-regarded and
typical work is Quinn's *Technological Forecasting*, still much
used in business schools despite its present rather advanced
age (12).  As a review of techniques it has been overtaken by
such contributions as that just noted, but its discussion on
forecasting the organisational problems they face is still
valid.  Noting that it is 'difficult to convince managers
unaccustomed to using the technique that it is valuable and
perhaps essential in making critical decisions', he offers
four suggestions: Forecasts should help this year's decisions;
they must not only look ahead but also lead to present
action; the forecast must be communicable to managers; and
the forecaster must be prepared to 'teach' its key elements.
Forecasts should be fitted in with regular cycles of
executives' decisions, and more especially, with budgeting
decisions, so that managers become used to long-term thinking,
at the appropriate time.  Finally 'promising executives
should, wherever possible, be exposed to planning and forecasting activities as a routine part of their training'.
This is an ambitious agenda, especially when added to the
forecasting requirements themselves, and transmitted to the
public sector organisations typical for transport decisions.
Nevertheless, it clearly suggests that, with new modes in
particular, forecasting improvement is as much a matter of
attention to organisational arrangements as of improvement in
techniques.


URBAN MODAL-USE MODELLING

It was suggested earlier that the need for improving the
search process was a potentially strong purpose for development of modal-use modelling, and that this potential was
greatest for urban areas.  This is not to say that urban-area
forecasting does not have other important potentials for
improvement.

Forecasting for urban areas indeed raises a great number of threats to plausibility. As things stand, probably the most sensitive areas are the links between transport infrastructure provision, the urban core and suburbs. The core typically contains the most potential for modal competition since cars are ubiquitous and public transport a function of density. Infrastructure costs are higher, hence the projects to be tested will be sensitive to the ratio of core to suburban provision. So also for instruments for urban traffic policy, which are highly correlated with the provision, or not, of core capacity. Core areas typically experience larger changes in employment and population relative to suburban areas, and so the forecasts will be most affected by the core-suburb relation, which can generate much potential variance in demand for passenger miles. Cores are more liable to be atypical of the area as a whole, because of itinerant populations, visitors, etc. All this suggests not an attempt to solve problems of the interaction of land uses and transport in the urban area as a whole, but a relative concentration on the core as a transport market, part of which is concerned with movement to and from the suburbs, especially where the urban area is large. London Transport Executive, for example, is developing a model of its markets in which geographical separation of this core from suburban markets, with a commuter market straddling the centre and suburbs, is a leading feature.

In deciding forecasting strategy as it concerns modes, the first question must always be: are all possible and potentially important modes accounted for? This raises the problems dealt with in the inter-urban context. Second, an economic account requires that the potential 'prices' of modes have to be foreseen. This is a function of what must be regarded as exogenous variables like the real cost of labour and material inputs, and of partly endogenous factors which will determine the numbers of person-miles per vehicle, and hence vehicle miles. The important question here is the terms on which different public modes are supplied. Their loading and frequency critically affect both their demands relative in particular to cars, the dominant mode, and their costs.

In practical forecasting contexts, these relationships have not been well explored in the past, partly because 'clients' have been predisposed to set up options for testing, highly constrained by the conventional modes important for public transport at the time. Part of the mutual task of the forecaster and client is to enrich these connections. One predictable result of this would be more attention to the market area between cars and buses. Where they are unencumbered by regulation, taxis and hire cars are the growing areas of public transport, indicating one reaction to the market opportunities opening up between the conventional bus and rail and the car. So also for less formal provision of car-like substitutes (lifts, carpooling, jitneys, etc.) now becoming known collectively, with taxis and hire cars, as 'para-transit'.

Potential modes, loadings, prices in terms of time and outlays therefore interact, and in total will certainly

affect the use of private cars and, on the evidence of
Fairhurst (6) possibly car ownership.  The probability is
that a high degree of differentiation between mode-types
would emerge, given the opportunity, now conspicuously constrained in Western urban areas by regulation, monopoly of
bus supply, subsidy commitments, etc.  Part of the forecasting task is to foresee the institutional changes involved and
to judge their timing.  This, again, might usefully involve
the tapping of independent expert opinion as described
earlier.

But how may the present typical outputs of modal-split
modelling and its procedures be turned to the problem of
systematic search for modal opportunities?  The process may
be of varying sophistication.  Thus in searching for the
potential role for forms of public transport, one might seek
to take advantage of discoveries about values set on comfort,
speed, variance in service characteristics, etc., emerging
from modal split and value-of-time work.  The unit of analysis for the search would be person journeys, classified
probably by length and certainly by elapsed time.  Certain
types of journeys viewed as sums of attributes - e.g. to
work - might be shown to imply a different aggregate disutility from others - e.g. for shopping.  Journeys thus are
given implicit 'time' prices.

Now, the chief problem in deciding whether there is
potential for a public mode is how it performs vis-à-vis the
car.  Assuming that journeys can always be made by car, a set
of time-price differentials from the journey by car could be
developed.  The least unfavourable of these (or, exception-
ally, the favourable ones) would indicate the existence of
market potential for a public mode.  The problem then becomes
the terms on which a 'mode' can be supplied.  Potential
journeys must be aggregated to form potential modes.  These
aggregations will vary in size (numbers of persons involved
per unit time), service frequency per unit time and, correspondingly, cost in terms of cash outlays of operation.  There
may well be a need to investigate here ways in which labour
costs could be reduced by using part-timers.  Once a list of
potential services is identified, the 'markets' could be
extended to potential users from non-car-owning households.
Promising public 'modes' could then be inserted into a conventional distribution and modal-split model.  Surviving
'modes' could then be further tested for feasible operations,
a test which would have to include an appreciation of the
overall scale across an urban area which each might achieve,
because of the likely influence on required outlays.

A less ambitious version would be a systematic search for
possibilities to save time spent in a fixed-trip matrix
devised from a conventional model, including conventional
mode options.  The time potentially saved to travellers by
providing opportunities would be valued, where appropriate,
by known attribute disutilities, e.g. walking and waiting.
The variance of these opportunities would indicate the preferred area for possible supply.  For example, the aggregate
time spent in interchanges on the existing underground system
might well indicate potential for new pedestrian aids -
travellators, elevators, escalators, or new walk-ways.  Major

barriers to existing trip-making such as rivers could also be investigated for potential in this way. Some preliminary investigations on these lines have been done at the London Graduate School of Business Studies.


CONCLUSION

It will be seen that the thrust of the argument of the last few paragraphs is for developments running rather against the grain of most efforts to improve modelling strategy for urban areas. There is no emphasis on combining modal split with distribution and making the *whole* supply-responsive, i.e. to deal with generation, in a 'comprehensive' way. For one thing, too little is known now about future real prices of present or potential modes successfully to derive future-demand curves for urban travel. Generating the supply curves from which to fit them needs much research in itself. Relatively little emphasis has been put on using time values in the sense of measures of representing modal disutilities. The endeavour rather would be to disclose what one might call 'shadow services', not present, but contingently realisable. The further exploration of these shadow services, the conditions for viability, the necessary conditions of labour supply, other factor costs, vehicle characteristics, etc., involves the transport analyst in an alliance with more general economic analytical skills. This need to keep the demand and supply sides of modal forecasting in explicit relation to each other, and to allocate effort accordingly, applies equally to inter-urban problems.

Shadow services also implicitly question present institutional and legal constraints, and often imply criticisms of present management of public transport systems and infrastructure. As has also been implied in the discussion on inter-urban modelling, the interests of many potential clients may not run with this endeavour. To be responsive, and thus admit innovation potential in his choice of projects or instruments, a client has to be able to withstand the arguments, and initially far superior information, of existing modes, organisational arrangements and their representatives. Yet the knowledge of current operation and administration is, of course, essential to success, and, as argued, forecasting must emanate in current and practical decisions. This indeed amounts to a considerable challenge to central and local government to review not only the content and purpose, but also the organisation of urban and inter-urban transport planning. This might, among other things, consider building alternative sources for knowledge, including the systematic use of independent experts at the critical stage of definition of future modes.


REFERENCES

1. Beesley, M.E., 'Conditions for successful measurement in time valuation studies', *Behavioural Travel-Demand Modelling and Valuation of Travel Time*, Transportation Research Board Special Report 149 (1974).
2. Chambers, J.C., Mullick, S.K. and Smith, D.D., 'How to

choose the right forecasting technique', *Harvard Business Review No. 71403*, Jul-Aug. (1971).

3. Commission on the Third London Airport, *Report*, Her Majesty's Stationery Office, London (1971).

4. Dalvi, M.Q. and Daly, A.J., 'The valuation of travelling time: theory and estimation', *Transport and Road Research Laboratory Report No. T72* (1976).

5. Department of the Environment, 'Standardised generalised cost parameters for modelling inter-urban traffic and evaluating inter-urban road schemes', *Economic Directorate Note*, Department of the Environment, London (1975).

6. Fairhurst, M.H., 'Influence of public transport on car ownership', *Journal of Transport Economics and Policy*, IX, 3 (1975).

7. Glaister, S., 'Peak load pricing and the Channel Tunnel: a case study', *Journal of Transport Economics and Policy*, X, 2 (1976).

8. Goodwin, P.B., 'Travel choice and time budgets', in *Determinants of Travel Choices*, D.A. Hensher and M.Q. Dalvi (eds), Teakfield, Farnborough, UK (1978).

9. Guttman, J., 'Avoiding specification errors in estimating the value of time', *Transportation*, 4 (1975).

10. Hensher, D.A., 'Valuation of journey attributes: existing empirical evidence', in *Determinants of Travel Choices*, D.A. Hensher and M.Q. Dalvi (eds), Teakfield, Farnborough, UK (1978).

11. Monopolies Commission (1974) *Cross-channel Car Ferry Services: A Report on the Supply of Certain Cross-channel Car Ferry Services*, Her Majesty's Stationery Office, London (1974).

12. Quinn, J.B., 'Technological forecasting', *Harvard Business Review*, Sept.-Oct. (1967).

13. Transport and Road Research Laboratory, 'Comparative assessment of new forms of inter-city transport', *Transport and Road Research Laboratory Report SR 1* (1973).

## Chapter 22

BEHAVIOURAL MODELLING: AN EVALUATOR'S PERSPECTIVE

John K. Stanley

SUMMARY

This chapter discusses the nature of policy evaluation, stressing its value basis. The evaluation links between behaviour, preferences and welfare are considered, to highlight the importance of behavioural modelling. Against this background a number of issues which determine the bounds of application of behavioural modelling in policy evaluation are considered. These issues include both ethical and practical considerations.

INTRODUCTION

Evaluation can be described as the process of identifying and weighing up the advantages and disadvantages of particular courses of action. These courses of action, which in the public sector may alternatively be termed 'policies', may have already been taken or they may be under consideration for the future. Traditionally the courses of action subject to most formal evaluation in the public domain have been investment options but there is no conceptual reason why one cannot evaluate anything that involves choice from among available alternatives.

The purpose of evaluation in any sector is to improve the quality of decision-making. For public-sector decisions it is generally argued that improvement should be judged in terms of what society prefers. Society's preferences are then usually defined as some aggregation of the preferences of individuals in society. In other words, the value judgement that in general terms individual preferences should be normative for social choice decisions is adopted. Nash, Pearce and Stanley (29,30) have recently drawn attention to the role of value judgements in evaluation. Any evaluation technique must reflect one or more value judgements which describe the ethical basis for the technique. Such value judgements determine which effects of particular policies are to be counted as advantages and which as disadvantages (benefits and costs) and how they are to be traded off against each other, both at the level of the individual and across groups of individuals.

Social customs will tend to limit the range of value structures existing in particular communities, but within this broad confine several such sets of values will exist in heterogeneous societies. Since there is no way that the unique superiority of one set of values over another can be demonstrated there can be no uniquely correct way to evaluate public policies. Against this background, Nash, Pearce and Stanley [30] then ask a series of questions related to application of the individual preferences value judgement in policy evaluation. Some of these questions are basic to any assessment of the role of behavioural modelling in evaluation of alternative policies. In particular: (1) *whose* preferences should count in policy evaluation? (2) *which* preferences should count? (3) *when* should individual preferences not count?

This chapter considers behavioural modelling in policy evaluation in the light of the author's value judgements and practical experiences on these particular questions.

The general policy evaluation framework in which behavioural modelling is considered is of the cost-benefit analysis (CBA) variety. CBA seeks to investigate individual preferences in terms of willingness-to-pay, usually based on Hicks' [20] compensating variation for measuring individual costs and benefits. Assuming no capital rationing (which may be handled by shadow-pricing techniques), the conventional cost-benefit decision rule may be expressed as

$$\underset{ti}{\Sigma\Sigma}CV_{Gi}(1+r)^{-t} > \underset{tj}{\Sigma\Sigma}CV_{Lj}(1-r)^{-t} \quad \text{implies desirability} \qquad (1)$$

where $CV_{Gi}$ = compensating variation of $i^{th}$ gainer

$\quad\quad CV_{Lj}$ = compensating variation of $j^{th}$ loser

$\quad\quad r$ = social rate of discount

$\quad\quad t$ = time period of analysis.

As Nash, Pearce and Stanley [29] point out, this decision-rule implies the following main value judgements:

$J_1$ = 'the decision criterion shall reflect individuals' preferences';
$J_2$ = 'these preferences shall be weighted by market power'.

In practice one can add, $J_3$ = 'preferences of future generations should be ignored', largely for reasons of choice of discount rate. If $J_1$ were modified to 'shall reflect expressed individual preferences', $J_3$ would not be needed if one's purpose was to exclude preferences of future generations. The present author would certainly not wish to do this.

One can generalise the conventional monetary cost-benefit decision-rule by inserting a set of 'equity' weights into that rule. Then one would have

$$\underset{ti}{\Sigma\Sigma}e_i CV_{Gi}(1+r)^{-t} > \underset{tj}{\Sigma\Sigma}e_j CV_{Lj}(1+r)^{-t} \quad \text{implies desirability.} \qquad (2)$$

Nash, Pearce and Stanley argue in favour of use of 'equity'
weights which serve to remove the influence of market power
on the measure of strength of preference used in policy
evaluation.  Instead they favour weighting compensating
variations to treat all persons effectively as if they had
equal income and wealth.  Note that, in addition, in rule (2)
one might seek to attach extra weight to gains to particular
groups (e.g. the poor).  This approach formally rejects value
judgement $J_2$.  Nash, Pearce and Stanley also argue in favour
of using a variety of equity weights in policy evaluation, as
a sensitivity test of evaluation results to choice of value
judgements on how individual measures of strength of preference should be aggregated.

In the discussion which follows, the *policy evaluation*
framework into which the outputs of behavioural modelling
are fed is a model as in decision-rule (2) above. Prime
emphasis, however, is on inputs to the compensating variation
measure at the individual level. One need not, therefore, use
support for decision-rule (2) over rule (1) as a test of
whether or not to read further.  It is taken as read that the
need for *some* policy evaluation rules need not be explained.
The use of money measures in such a rule is not misplaced
materialism but a convenient means of aggregation.  Nash,
Pearce and Stanley (30) discuss ethical and practical limitations on cost-benefit type decision-rules.


THE ROLE OF BEHAVIOUR IN POLICY EVALUATION

The basic value premise that individual preferences should
generally be normative for social choice questions (policy
evaluation) by itself says nothing about how one identifies,
records or aggregates preferences.  Further value judgements
and evaluation conventions are required.

One of the most important *conventions* that has operated in
economic evaluation of public policy has been the convention
that behaviour is the ultimate evidence of preference.  Thus
one could investigate an individual's attitude towards some
particular object but, if the individual's behaviour was out
of line with this expressed attitude, behaviour has been
conventionally taken as the guide to 'real' preferences.

Behaviour is equated with choice and this is taken to be
determined by an individual's preferences.  Preferences also
play a central role in economics as the basis of welfare
judgements, the convention being that a preferred position
involves a higher level of individual welfare (28).  Preference thus provides the link between behaviour (choice) and
welfare.  Any consideration of the role of behavioural
modelling in relation to policy (welfare) evaluation needs to
consider various aspects of this link.  Sen (35) has critically reviewed the choice-preference-welfare link.  Some of
his work is referred to in the chapter.  The chapter seeks to
do this and also to consider alternative approaches to
assembling data for policy evaluations if behavioural data is
not available.

Because of the choice (behaviour)-preference-welfare basis for the economists' approach to policy evaluation, the conceptual importance of behavioural modelling to policy evaluation scarcely needs to be argued. Instead, the chapter seeks to develop an interpretation of what is required of behavioural modelling for evaluation and to identify qualifications to an uncompromising application of the behavioural convention in such evaluation.

Throughout, 'behavioural' modelling is understood to be any modelling activity that seeks to incorporate explicit influences on human behaviour as independent explanatory variables, the dependent variable being some associated behavioural response (e.g. a demand for housing, travel demand). This includes both aggregate and disaggregate type models. Whilst most examples used are drawn from the transport field, they are not narrowly restricted to behavioural travel-demand models. Behavioural modelling as an input to evaluation of particular non-user consequences of transport policies/projects is also given attention. Statistical properties and the functional form of models are not discussed. Instead, the emphasis is heavily on evaluation requirements.


WHOSE PREFERENCES SHOULD COUNT?

It is quite clear that behavioural modelling techniques, including those used in the travel-demand field, can only measure the preferences of those who actually do 'behave' in relation to particular choice objects. It needs to be stressed, however, that such an approach can shed little light on a potentially important class of preferences, the *non-participant* preferences. Behavioural modelling of demand for trips to wilderness areas or for trips on a new railway (say) in an area previously unserved by public transport are perhaps the best examples with which to illustrate this problem. Such modelling can be used to estimate user travel demand in both cases and to evaluate user benefits. However, whether or not such benefits constitute the total benefits derived from the existence of the wilderness area and railway line respectively depends on whether these facilities have: (1) any option value; and/or (2) any existence value. More formally, of course, any biases in estimation of actual user benefits would also lead to incorrect estimation. This is particularly relevant in estimation of benefits of wilderness areas. (See, for example, Clawson and Knetsch [7], Cichetti and Smith [6].) This matter is discussed further on p. 491.

The notion of *option value* is primarily due to Weisbrod [41], who pointed out that an individual may value a particular commodity (e.g. a railway line or wilderness area) even though he may not intend to use it. This value is a kind of insurance value; that is, the individual may value the existence of the commodity just in case he wishes to use it one day. In conventional cost-benefit terms, the individual would be willing to pay something to preserve the right to exercise this usage option. Cichetti and Freeman [5], have shown that option value exceeds consumers' surplus for risk-averse individuals confronted with a decision on whether to

buy an option to consume a good in the future at a predetermined price. The 'free-rider' or public good problem makes
investigation of option value difficult but behavioural
modelling alone will miss it. That is, if sufficient other
persons are willing to pay to preserve the existence of the
railway (for example), this individual can benefit by understating his preferences and enjoying a free ride (no pun
intended) at their expense.

*Existence value* is an even more detached notion than
option value, since it has nothing to do with use or possible
use. The central idea of existence value is that an individual may consider the existence of a particular item (e.g. a
wilderness area or threatened species) as worthwhile in its
own right, irrespective of any use considerations. This may
be due to a desire to preserve ecosystem stability as a basic
precondition for mankind's survival or it may be that the
individual ascribes rights to the existence of other species.
When one recognises that most perception and learning is via
the media, the potentially powerful nature of existence value
is apparent. Thus blue whales might enter my utility function because I know they are there and may not survive. I
may also know what they look like, even though I've never
actually seen one. Such reasons are not central to the
argument. What is central is the fact that some individual
does consider these things of value and that, therefore, in
line with the individual-preferences value judgement, they
are of relevance to policy evaluation concerning the items
in question.

This discussion is intended to help define some broad
limits of usefulness of behavioural modelling for policy
analysis. Such usefulness lies in predicting usage consequences of particular policy changes and evaluating them
from the user viewpoint. This is clearly basic to any
evaluation if one accepts the individual-preferences value
judgement. However, user preferences are not necessarily
exhaustive of all preferences of relevance to policy
evaluation and one needs to be aware of the possible existence of option value and existence value of particular items
which may be the subject of policy analysis. Behavioural
techniques do *not* illuminate such values.

User-benefit calculations based on behavioural modelling
may provide sufficient benefit estimates in some cases to
provide a basis for policy decisions. Thus Krutilla and
Fisher [24], examining possible hydroelectric and recreation
alternatives in Hells Canyon, found that the recreation
alternative could be supported over development alternatives
even without quantifying the less tangible recreation values
(i.e. option and existence values). However, this is
certainly not a situation one can depend on and increased
research into quantification of possible option and existence
values is needed for evaluation of conservation versus
development-type policy matters. This work is not instead of
but, rather, in addition to work on estimating benefits
through behavioural modelling exercises.

# WHICH PREFERENCES SHOULD COUNT?

*The Common Property Resource Problem*

A common property resource is a resource used, if not necessarily owned, in common by all members of the community. Neither exclusion nor discrimination is permitted with respect to its access. Sen (35) has shown how use of revealed preferences (i.e. behaviour) as a basis for making normative social judgements about resource allocation can be misleading in situations of welfare interdependencies. Such situations are defined as those where one economic agent engages in an activity that has a direct effect on the welfare or productivity of one or more other agents (i.e. externalities are created). Common property resources (e.g. roads, wilderness areas) are prone to such difficulties. Table 22.1, based on some of Sen's ideas, illustrates some of the problems.

TABLE 22.1   An environmental 'prisoner's dilemma': strategy rankings

|  |  | Person 2 | |
|  |  | Visit | Do not visit |
|---|---|---|---|
| Person 1 | Visit | 3, 3[a] | 1, 4 |
|  | Do Not Visit | 4, 1 | 2, 2 |

a. The first entry in the cell is Person 1's ranking, the second Person 2's ranking.

| Person 1 strategies | Preference ranking | Comments |
|---|---|---|
| He alone visits the wilderness area | 1 | He loves wilderness areas |
| Nobody visits the area | 2 | He cares more for the existence and preserva-tion of wilderness than he does for the pleasure he personally derives from a visit to such an area, given that 1 is not achievable. |
| Everybody visits the area | 3 | If all except him visit the area degradation will be virtually the same as it would be if he joined in, so he might as well get something out of it |
| Everybody but him visits | 4 | The area is degraded and he has missed out on any pleasure he might have derived from a visit |

Table 22.1 and the associated strategy assessments reflect
the common property resources problems common to many
environmental questions.  Wilderness areas are used for the
example, but a wide range of other equally suitable examples
would be found (e.g. air pollution, residential area improvement).  The essence of the problem is that consumption of the
particular commodity (in this case the wilderness area) by
one individual (or at least a small number of individuals)
affects the pleasure which other persons can derive from it.
Increased visitor pressure on a wilderness area can thus be
expected, after a time, to cause destruction of the very
qualities which many persons visiting such areas are seeking,
especially solitude (see Cichetti and Smith [6]).

Person 1 in considering whether or not to visit the
wilderness area, may initially decide to go ahead with the
visit, on the basis that he cannot really affect the behaviour of others.  If a number of others who have the same
ranking of strategies as Person 1 react in the same way, all
will visit and a socially suboptimal choice results.  From
inspection of Table 22.1 it can be seen that a (3,3) ranking
results.  If no wilderness lovers had visited the area a
(2,2) ranking could have resulted, all persons being better
off (a Pareto improvement).

If Person 1 thinks about the implications of his behaviour
he may or may not decide to stop visiting the area.  He may
consider that his own possibility for influencing the
behaviour of others is so small as to ignore, in which case
he is unlikely to change his behaviour (i.e. he prefers ranking 3 to ranking 4).  On the other hand he may decide to
abstain in the hope that sufficient others do likewise (i.e.
2 is preferred to 3).  However, if he follows this course of
action and nobody else abstains, he ends up in the worst
possible position (4).

The wilderness problem is one of going from behaviour to
welfare.  All the possible courses of behaviour for Person 1
reflect his preferences.  The latter have not changed.
However, the interdependence of welfares that exists means
that one cannot simply observe behaviour and draw welfare
conclusions.  The individual in this common property resource
situation cannot alone be sure of attaining his own most
preferred, or even second most preferred, situation, and
resource allocations resulting from individual behaviour can
be socially suboptimal when welfare interdependencies are
present.  Estimating the benefits of the wilderness recreation site (in Table 22.1) could thus be very misleading.  If
all persons visited the area, the (3,3) rankings would be
used for estimating willingness-to-pay measures.  However,
all visitors can envisage situations where they would be
willing to pay more than this (e.g. the (2,2) situation where
*no one* visits, underlining again the need for consideration
of existence value).  In the example given, behavioural
analysis of the (3,3) situation will under-estimate the
willingness-to-pay for wilderness preservation (2,2 situation).  The latter may be better investigated using the
'Search Conference' type of participatory approach, where the
common property resource issue can be considered by people
acting as part of a group.  The 'free rider' public goods

problem then needs to be considered.   This is discussed on
p.492.

If one abstracts from the problem of existence value,
behavioural modelling may still lead to over-estimates of the
benefits of developing common property (wilderness) resources.
In other words, preservation will be under-valued.   This
conclusion is implied by the work of Cichetti and Smith (6),
who have developed a methodology for incorporating the
effects of congestion in wilderness areas on the users'
willingness-to-pay for the wilderness experience.   Congestion
in their study relates primarily to use effects where the
solitude which users experience is diminished because of
encounters with other users.   Such encounters reduce the
willingness of users to pay for the wilderness experience,
moving the relevant demand curve to the left.

Where such congestion effects exist, behavioural demand
modelling along elementary Clawson-Knetsch lines may produce
misleading results.   Such modelling is generally based on
cross-sectional usage data and will define demand for the
area, given its quality at that time.   If a major increase in
the area's accessibility take place, willingness-to-pay
measures based on the previous demand curve will over-
estimate benefits of increased access if congestion effects
are now experienced at the area.   In Figure 22.1, DD is the
Clawson-Knetsch demand curve based on use level $0Q_0$.   Access
costs fall from $S_0$ to $S_1$, and use of demand curve DD gives
estimated benefits of ABCE.   However, increased visitor
levels beyond $0Q_0$ cause congestion and reduce the willingness
of users to pay for the wilderness experience.   The demand
curve moves to D'D', usage increasing to $0Q_2$.   Benefits of
this increased usage are not ABCE, however, but D'FE *less* DBA.

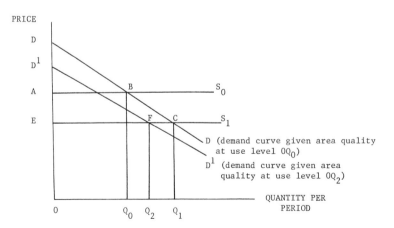

Figure 22.1   Demand for, and Benefits of, Wilderness Recreation

Behavioural modelling, as a basis for benefit estimation
in policy matters related to common property resources needs
modification if congestion effects are present. Attitudinal
adjustments which investigate the solitude (encounter)
willingness-to-pay relationship can provide a useful sup-
portive tool in such situations. Option values and existence
values also need to be considered. Group type participatory
techniques seem likely to be required in the latter regard,
for the reasons discussed on p.491.   In both these situations the familiar 'free-rider' problem will need to be faced.

It has long been recognised that individuals have an
incentive to misrepresent their willingness-to-pay for common
property resources: (1) downwards if they are likely to have
to pay but can enjoy a free or cheap ride at somebody else's
expense by understating willingness-to-pay; (2) upwards if
they do not have to pay anything (see, for example, Bowen
(4), and Kurz (25)). Bohm (2,3) has argued, however, that if
the method of paying for a particular project is uncertain
and respondents know that several payment alternatives are
possible, the incentives to distort statements of willingness-
to-pay are not unambiguous. Bohm tested this by using five
different approaches to investigate willingness-to-pay, varying the need for payments to be made. None of his approaches
revealed significantly different average maximum willingness-
to-pay, when applied to a random sample of consumers.
Attitudinal type approaches thus appear rather more promising
in the investigation of willingness-to-pay for public goods
than has been allowed in the conventional wisdom.

*Attitudinal or Behavioural Preferences?*

The discussion on behavioural modelling and evaluation where
common property resources are being considered led to a
conclusion that study of attitudes might be necessary to
augment behavioural data for policy evaluation. Practical
problems of identifying behavioural preferences can lead to
similar conclusions.

From the evaluation viewpoint, a limited range of price-
quantity observations for marketed commodities will generally
be sufficient for one to derive willingness-to-pay measures
for evaluation of price changes from the consumer's viewpoint.
These observations, in the case of marketed commodities, can
usually be behaviourally based and there seems little reason
in general to argue against use of such data. However, aside
from common property resource problems discussed above there
are some important situations where this position might not
obtain: (1) where a new commodity is being introduced; and
(2) where the 'commodity' in question is not sold directly in
the market place.

By definition one cannot use revealed preference data in
relation to existing commodities to assess consumer behaviour
in relation to a new commodity. Instead, however, one may
follow the Lancaster (26) approach to consumer theory and
deal in terms of the attributes of commodities, rather than
directly in terms of commodities. If one can then identify
and physically measure the relevant attribute set, it is

conceptually possible to obtain behavioural data (revealed preference variety) relating to such measured attributes. Definition of the attribute measures for the new commodity and use of the existing behavioural data set then opens the way for one to estimate consumption of the new commodity and consumer willingness-to-pay.

Stanley [37] has argued that policy evaluation in willingness-to-pay terms, based on an attribute approach, requires measurement of: (1) the individual marginal rates of substitution (MRS) between the relevant affected attributes and money; (2) the relevant attribute changes to apply to these MRSs; (3) a social choice rule to aggregate individual measures of willingness to pay. In this section we concentrate on the first of these requirements.

Two main difficulties in determining marginal rates of substitution between attributes and money, for use in policy evaluation, are: (1) not all commodity attributes will be physically measurable; (2) it is not always easy to gather and quantify consumer behavioural data in relation to commodity attributes which are readily measurable in physical terms.

*Attributes non-measurable.* Consequences of particular public policies might be conveniently classified according to whether they are: (1) direct market impacts (e.g. operating cost savings); (2) physically measurable non-market impacts (e.g. time savings); (3) intangible consequences having no obvious physical or monetary dimensions for measurement purposes (e.g. destruction of a view). In policy evaluation using willingness-to-pay measures, one is seeking to impute 'market' measures to consequences of types (2) and (3), so they can be readily added to consequences of type (1). Behavioural modelling work for evaluations relates primarily to consequences of type (2), although one may carry out behavioural modelling investigations into direct market impacts of a particular policy. Where there are no physically measurable attributes to which behaviour can be related, there can be no behavioural modelling.

Recent work by Hensher, McLeod and Stanley [17], Hensher and McLeod [19] and the Commonwealth Bureau of Roads [11] has sought to understand traveller mode-choice decision processes for category (3) type attributes in a way that one can go on to identify physical attributes which help to explain behaviour. This work initially used a probabilistic attitudinal choice model to investigate mode choice. The choice model used was based on the underlying assumption that the perceived utility difference between two modes (A and B) can be expressed as follows for the i[th] individual:

$$\Delta U^i = \sum_{j=1}^{m} I_j^i (S_{A_j}^i - S_{B_j}^i) , \qquad (3)$$

where $I_j^i$ = a measure of the importance of attribute j to the i[th] individual in the given choice situation

$S_{A_j}^i$ = a scaled measure of the satisfaction the $i^{th}$
individual derives from the $j^{th}$ attribute of Mode A.

Attitudinal measures of the importance of particular modal attributes to an individual in a given mode-choice situation were obtained, along with satisfaction scores on each attribute, for the usual and alternative modes. With the linear additive specification of the individual's utility function

$$if \quad \sum_{j=1}^{m} I_j^i (S_{A_j}^i - S_{B_J}^i) > 0, \tag{4}$$

the individual should choose mode A. In a mode choice setting, the attitudinal importance and satisfaction scores can be obtained and used deterministically as in (4) to predict mode choice. Alternatively, one can use the IS scores in a probabilistic (e.g. logit) model to predict choice. Behaviour is known, so one can check predicted behaviour with actual behaviour. This provides a useful check on the attitudinal data. Poor correspondence between predicted and observed choices may mean, *inter alia*, the linear additive specification is inappropriate, the attitudinal data is unreliable, or both.

The advantage of this attitudinal approach in mode-choice work is that one can handle modal attributes that are not obviously measurable on a physical dimension (e.g. comfort and convenience) on a comparable basis with attributes that are readily measurable (e.g. time and cost). This enables an appreciation of the possible influence of less quantifiable attributes. The difficulty is, of course, that unless physical attribute measures for the less quantifiable attributes can be derived, one cannot determine how *changes* in such attributes will affect modal patronage levels or how to evaluate such change. Thus one may know that comfort is very important in mode choice, but this by itself is not very useful for policy analysis if one does not know the relevant dimensions of comfort.

In an attempt to overcome this difficulty, Hensher, McLeod and Stanley (17) investigated the perceived component dimensions of comfort and convenience. Hensher (16) used open-ended questions to identify component dimensions of comfort and convenience. Importance-satisfaction scores for these dimensions were then used in a regression analysis by Hensher, McLeod and Stanley (17) to seek to explain variations in the perceived overall level of comfort and convenience of particular modes. The latter analysis identified the major component dimensions of *comfort* as being: (1) seating quantity; (2) relaxing, quiet, read; and (3) ease of travel (crowds). Major dimensions of *convenience* were identified as: (1) ease of access/egress; (2) availability; and (3) time/speed. It is interesting to note the time-relatedness of 'convenience'.

Following on from these studies, the Commonwealth Bureau of Roads (11) has sought to obtain physical measures which most nearly approximate some of these component dimensions. Successful identification of such measures will enable one to extend the range of physically measurable variables which are used in mode-choice modelling. One can then assess sensitivity of mode split to changes in a greater range of explicit policy variables. Methods similar to those used to evaluate time savings in money terms can then be used to evaluate possible policy changes in the 'new' (i.e. newly physically measurable) attributes.

The Commonwealth Bureau of Roads work is based on the following implicit model:

Choice    = f (Utility)

Utility   = g (Attitudes)

Attitudes = h (Attributes).

The second, or attitudinal, step has been introduced as a means towards extending the scope of behavioural mode-choice models which, in the past, have relied on measures of just time and cost. Once the attitudinal measures can be related back to attributes, the attitudinal step can be dropped out and one has an extended range of physically measurable attributes. The attitudinal work is thus being used to complement and extend more traditional behavioural models. The extended models will provide for both predictive and evaluative work with a greater range of policy sensitive variables.

A second stream of work at the Bureau of Roads has been concerned with development and evaluation of street improvement programmes in residential areas. This may involve such projects as street closures and conversion of street space to active or passive recreational space of various sizes. One such project in Middle Park (see Stanley (38)), involved an intensive programme of local involvement to identify (1) the issues locals saw as relevant to the use of street space in the area, and (2) relative preferences of locals for various possible rearrangements of the use of local street space.

The issue-identification stage of the study produced the following evaluative dimensions of concern to locals: (1) traffic levels on streets; (2) provision of small areas of open space; (3) provision of large areas of open space; (4) ease of vehicular access within the area; and (5) ease of vehicular access/egress to/from the area. Three 'improvement' plans were developed for the Middle Park area, involving varying levels of these dimensions for the area as a whole and in particular locations. A do-nothing plan was also used, making four alternatives for evaluation.

From an evaluation viewpoint, no external behavioural data on relevant individual marginal rates of substitution between these evaluative issues (dimensions or attributes) and money or between the issues themselves was available. Furthermore, the problems (inc. cost) of collecting such behavioural data over the relevant issue 'levels' that would be relevant in the study area were considered prohibitive, particularly given the likely cost of any improvement works (up to about $100,000). Basic preference data thus needed to be collected

for the study and behavioural data was ruled out.   Instead,
attitudinal techniques were used.

Various attitudinal approaches were used for evaluation,
for two reasons: first, different approaches reflect different value judgements, which Nash, Pearce and Stanley (29,30)
argue increases the public accountability of the evaluation
process; and, because of the element of doubt that always
surrounds attitudinal data, alternative approaches provide
consistency checks on one another.   The alternative evaluation approaches used were:

1   a simple ranking of alternatives by each individual,
rankings being then 'scored' and summed across respondents to
produce the 'best' plan;
2   importance-satisfaction scoring of attributes (dimensions) and alternatives on each attribute, individual scores
being summed to provide an overall measure of preferences;
3   the same as in (2), with respondents' scores constrained
so that each individual has the same number of 'votes' to
allocate across attributes and plans;
4   an attitudinal willingness-to-pay question related to
each improvement plan as a whole, phrased so as to endeavour
to minimise the free-rider problem; and,
5   as in (4), with stated willingness-to-pay of the $i^{th}$
individual, whose income is $Y_i$, being factored by the ratio
of the mean income in the area to his income (i.e. $\overline{Y}/Y_i$).

Thus if an individual has half the mean income of the area,
his willingness to pay is doubled.   This 'market voting'
approach is discussed in Nash, Pearce and Stanley (30), being
an explicit equity-weighted version of rule (4).   (This
general approach to evaluation was discussed earlier, see
p.486.)

Evaluation approaches (1), (2) and (3) avoided any questions
of willingness-to-pay, which entered separately in approach
(4) and its equity-weighted equivalent (5).   Plan rankings
based on the five approaches revealed a high degree of correspondence, providing support for the attitudinal approach.
Based on results of the importance-satisfaction questions,
moreover, it was possible to identify which attributes
merited further attention so that a plan superior to all
those evaluated could be developed from the residents'
viewpoints.

An attitudinal questionnaire provided the basis for the
Middle Park evaluation.   It included point allocation
importance-satisfaction questions, which sought to place the
individual in a situation of thinking in trade-off terms.
The alternative evaluation questions provided consistency
checks on one another.   All things considered the exercise
was an encouraging use of attitudinal preference data,
achieved at low cost.   Behavioural preference data in relation to the evaluative dimensions was not available and would
have been both difficult and expensive to collect.   The case
against collecting original behavioural data for small-scale
(e.g. $100,000) area improvement type programmes seems
strong, since carefully designed attitudinal approaches can
produce adequate evaluative data with low cost and effort.

*Attributes measurable*. The latter lesson was also learnt in
the mode-choice study discussed previously and in Commonwealth Bureau of Roads (11). That study sought to develop,
in line with consumer preferences, improvements in road-based
public-transport services to particular groups identified as
being 'transport disadvantaged' for the work trip. The steps
in the methodology (provided by David Hensher) used in that
study were:

1   *Issue* identification.
2   Given (1), *identification of the location* of the groups
meriting attention. This step involves:
    (i)    selection of criteria which best define the issue;
    (ii)   empirical measurement of criteria;
    (iii) decision rules to combine criteria to identify
         groups;
    (iv)   selection of zones (and groups) for further study.
3   *Project identification*, to serve the zones/groups
identified:
    (i)    select appropriate models that will be able to
         identify potential project types, including a role
         in assessment of benefits to groups in line with
         the criteria used for location identification;
    (ii)   data collection for input to these demand models;
    (iii) estimation of demand models and use of *attributes*
         identified as significant influences on choice, to
         identify project types likely to produce signifi-
         cant benefits. Model estimated on individual
         traveller observations;
    (iv)   identify travel corridors from zone, and for
         groups, where these improvements should be con-
         centrated;
    (v)    combine modelling output and corridor identifica-
         tion with non-modelling considerations (e.g.
         institutional, regulatory constraints) to produce
         project types for further consideration.
4   *Project evaluation*, looking at cost-benefit considerations and financial implications.
5   *Choice of preferred solutions*.

In stage (3), disaggregate mode-choice models, with extended
variables (i.e. additional to various time components and
cost) were used to identify those modal attributes which, if
changed, would produce worthwhile improvements in the level
of public-transport service to travellers, inducing some
switching. In other words, project generation came from the
demand side not the supply side as is more usual. However, a
number of problems were encountered with this approach.

The attribute identification stage (3(iii)), which used
the disaggregate models to identify the important modal
variables if mode switching was the goal, proved to be
expensive when compared to the cost of typical local projects
which might emerge from such investigation. Whether or not
the use of such modelling techniques can be considered
efficient in cost-benefit terms for particular project investigations depends primarily on how transferable the results
are between different areas. If one believes in the
economist's conventional indifference map, one could not
expect much transferability, necessary preference data being

marginal and thus depending generally on specific circumstances.

Where projects are expected to cost several millions of
dollars a modelling phase as part of project identification
would be more feasible.  For lower cost exercises, however,
separate modelling exercises would be hard to justify,
although if it can be shown that results are reasonably
transferable (an issue for the segmentation workshops), outputs of other studies may be used.  Instead, attention should
be given to the use of low-cost participatory approaches.
Thus, for example, one might use a series of in-depth discussions with members of the client group(s) to identify key
attributes and possible improvement projects.  The Middle
Park Study discussed above was carried out in accord with
this philosophy.  This approach was also used in the Bureau
study, similar projects being suggested as were implied by
the modelling techniques.

Low-cost discussion techniques do not produce quantitative
estimates of the demand implications of particular policies,
so an *ex ante* evaluation will not be possible if mode switching is a goal.  However, the cost saved by avoiding the
individual data-collection and model-estimation stage (where
models are used to search for key attributes) would finance
smaller demonstration projects on a trial basis.  These may
provide a better basis for evaluation than the behavioural
models, provided the projects can be given a fair trial
period.*  This issue turns largely on project scale and cost,
which influence feasible planning overheads, and transferability of results from modelling exercises.  Too little
attention has been devoted in the literature to date on this
approach to evaluating behavioural models.

The mode-choice study raises questions of the extent of
original behavioural data collection and modelling work for
local applications, for reasons of cost.  A separate set of
difficulties for behavioural modelling arise when the attributes to be measured are clear and unambiguous, but one is
not able to establish statistically significant and meaningful results.  The example of traffic noise is perhaps the
best example of this problem.

A number of authors have examined the question of
behavioural evaluation of traffic noise.  See, for example,
Pearce (32,33) and Harrison (14) for some discussion of such
work.  *A priori* one would expect noise costs to show up in
house-price differentials and that one could, therefore,
obtain a guide to the cost of noise nuisance by examining
behaviour in the housing market.  Two major problems exist
here, however: first, the relationship between house prices
and noise levels has not been satisfactorily demonstrated for
road traffic; and, second, the relevance of house-price

---

*In all demonstration projects, to a greater or lesser
extent, long-run consequences may be quite different to
short-run consequences.  Demonstration projects will
probably be most useful where the choice being investigated
is between alternative public-transport services.

depreciation to the cost of noise is the subject of theoretical debate.  The house-price market has proved to be by no means easy to analyse for the purposes of imputing monetary costs to noise nuisance.  A large number of different environmental factors (among other things) can be expected to influence house prices.  Since many of these are likely to be correlated one with another, the influence of particular factors is difficult to disentangle.  Furthermore, as Harrison (14) has noted, comprehensive and consistent data on house prices and on possible influencing factors is difficult to assemble.  Nevertheless a relationship between house prices and aircraft noise was established by the Roskill Commission Research Team on the Third London Airport (10), but similar relationships for road traffic have proved more elusive.

The Roskill approach to estimating property-price depreciation was attitudinal rather than strictly behavioural. Estate agents were *asked* what the relative house-price depreciation was between noisy and non-noisy areas (9).  Agents' responses tended to be quite variable for houses of particular price ranges but data from the Inland Revenue tended to confirm the average depreciation rates used for properties in particular price ranges.  Diffey's (13) work on traffic noise and house prices concluded no negative effects could be distinguished.  Similar results were also reported by Towne (39) and Colony (8).  Abelson (1) has established some relationship between house prices and traffic noise in a Sydney suburb, but the gross nature of his treatment of noise levels makes application of his findings of limited usefulness - that is, marginal valuations are not really derivable from his work.  For a second suburb Abelson studied in Sydney, no relationship between traffic-noise level and house price could be established.

Alternative approaches to valuation of noise nuisance have been used.  For example, Starkie and Johnson (36) examined cases where people incurred a cost of insulation to avoid or reduce noise.  This alternative cost behavioural approach, if the cost is incurred, will certainly provide some part of an estimate of noise cost for the persons concerned.  However, the problem of valuing residual noise, both inside and outside, at places of employment, recreation, etc. and for non-insulators remains.  Hoinville's (21) priority evaluator approach has also been used in the pursuit of noise cost. However, as Harrison (14) notes, the significance of results obtained using such simulation techniques has yet to be properly assessed.  As one of the faithful, however, one holds good hopes for this attitudinal work.

Similar problems to this noise cost problem arise in respect of behavioural modelling applications in the travel-demand field.  The logic of behavioural modelling in evaluation is that one uses the model to predict behavioural response to particular policy initiatives and then uses individual valuations implied by the model to evaluate the policy from the viewpoint of the client group.  The practical reality, however, is that the significant variables in terms of influencing behaviour (as determined by the model) are not always all the variables which one would expect people to

value.  Thus, for example, relative modal costs may not be a
significant variable in terms of a particular choice model
but one would expect people to value cheaper travel.  How
should this be handled?

From an evaluation viewpoint, one can envisage two
situations where benefits from a particular policy intiative
might be credited to the project (policy) even though the
behavioural model implies that the particular item is not
valued (i.e. is not a significant influence on group
behaviour): (1) where the nature of the 'benefit' is such
that it can be accumulated by the individual and will be used
in a way that improves his well-being, even though he may not
be able to attribute such benefit (conversely for costs); and
(2) where one adopts the standpoint that behavioural modelling cannot really be looked at to identify all the relevant
behavioural influences in a given choice situation and that
such modelling should always be interpreted in the light of
exogenous knowledge.

If behavioural modelling does indicate a statistically
significant relationship between house prices and traffic
noise (or any other environmental nuisance), how does one use
the output in policy evaluation?  One might expect that if
introduction of a new facility into the urban area (say) was
going to cause x houses to fall in value $y each, because of
noise effects, the noise cost of the project would by $xy.
Pearce (33), however, shows that the answer is not quite so
easy.  Using a simple model, Pearce shows that the appropriate noise cost is equal to the difference between the
initial and new prices of houses that are to be noise-
affected (i.e. $xy) *plus* the change in consumers' surplus
between the initial 'no noise' and the new noise situation.
Costs emerging from behavioural analysis thus need a sound
theoretical basis if they are to be used in policy evaluation.

Walters (40) has suggested an approach which uses market
price data alone to measure the economic cost of noise.
However, his model assumes a very restrictive constant returns Cobb-Douglas utility function, giving rise to a unitary
elastic demand curve for quiet with respect to price and
income.  As Walters himself notes, this is inconsistent with
an income elasticity of demand for quiet of 2 implied in the
house-price depreciation work of Roskill.

*The Appropriate Marginal Valuations*

Discussion of the appropriate 'cost' of noise for input to
policy evaluation raises the more general question of what
costs or values behavioural demand-modelling exercises have
sought to measure and how these relate to the marginal
valuations needed for policy analysis.

Consider an individual who is faced with making a trip,
the timing, origin and destination and purpose of which have
been decided.  All that remains is for him to choose the most
suitable mode for the trip.  This choice is assumed to depend
only on the trip time and cost of the (only) two alternative
modes available, A and B.  (This assumption does not limit

the generality of the methodological conclusions reached.)
In Figure 22.2, $T_A$, $T_B$ represent trip times by modes A and B
respectively and $C_A$, $C_B$ trip costs.  It is assumed that both
these attributes are regarded onerously by the individual and
thus generate disutility (DU) for him.  It is assumed the
individual chooses the travel mode to use by minimising
disutility.  If the individual chooses mode B rather than A,
it is thus implied that

DU(B) < DU(A)

i.e.

$DU(T_B, C_B)$ < $DU(T_A, C_A)$.

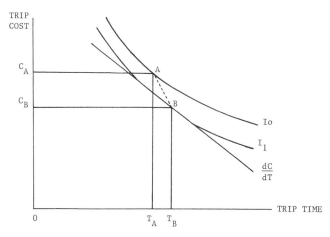

Figure 22.2   Individual Choice of Travel Mode

Given that the individual chooses B, a small change in $T_B$
would be valued by dC/dT, the marginal rate of substitution
between C and T, at point B.  If a small improvement is made
to the travel time by mode B, then the value of dC/dT is the
relevant marginal evaluation of the saving by this individual.

By choosing mode B rather than A, the individual is giving
up time to save cost.  He could have spent $(C_A - C_B)$ to save
$(T_B - T_A)$ time but did not.  Therefore,

Value of Time Saving at B $< \dfrac{C_A - C_B}{T_B - T_A}$ (= slope of AB)

i.e. dC/dT measured at B < slope of AB.

This result will always hold, provided mode times and costs

are not identical.  In other words, investigation of the
amount an individual would have to pay to save time will
generally *over-estimate* that persons valuation of a time
saving on the chosen mode if that person does not choose to
pay such an amount.  Conversely, for an individual who does
choose to buy time savings, the amount paid to save time will
generally *under-estimate* the individual's evaluation of a
small saving in time by the chosen mode.

The purpose of probabilistic mode-choice models is
generally to estimate implicit weights which a group of
travellers who choose between modes A and B place on trip
time and cost.  By seeking to maximise the power of the model
to classify users by choice of mode, a single value of dC/dT
across the group is sought.  This value will under-estimate
the value of time savings to users who do buy time savings if
the trip time of their chosen mode is improved by a small
amount.  This is usually all that is possible.  If travellers
who do not choose the faster mode receive a faster trip by
their chosen mode, the use of dC/dT derived from investigating the mode-choice decision across all travellers will be an
over-estimate.  Because fast modes tend to be used by those
with a relatively high value of time and slow modes by those
with a low one, use of a common value will tend to bias
investment into accelerating slow modes too much relative to
fast ones.  The need thus exists for mode-specific estimates
of values of travel time.

The value of dC/dT estimated from probabilistic mode-
choice models may be interpreted as a marginal valuation
typical for a group of buyers and non-buyers of time savings
(usually) across various amounts of time saved.  Figure 22.2
and its associated argument indicated the theoretical error
involved in using such valuation to evaluate time savings
resulting from improvements to a particular mode.  Further
error may also be introduced if the value of dC/dT also
depends on such things as: (1) characteristics of the group
of travellers (e.g. their income levels); and (2) the amount
of time saved and/or the trip length.

A number of studies (e.g. Hensher [18]) have attempted to
get around such problems by introducing these factors as
independent variables in explaining the value of dC/dT or
else by stratifying the data set to allow for broad variations between (say) income levels, amount of time saved and
trip length.  Such refinements help to ensure that valuation
of dC/dT is appropriately *marginal* with respect to such
variables.  However, they do not get around the problem that
separate values are needed for buyers and non-buyers of time
savings if 'biased' evaluation is not to result (as discussed
in the preceding paragraph).

One 'saving grace' so far as this valuation difficulty is
concerned lies in the 'equity' approach to time valuation.
This approach, largely due to the UK Department of the
Environment, seeks to apply a common value of time savings to
all persons for evaluation purposes (see, for example,
Harrison [14]).  Nash, Pearce and Stanley [30] have argued
more generally for evaluation of public policies in this way,
marginal mean valuations of consequences across the

population being estimated and applied to all groups regardless of their income.  Use of an equity value of time can
thus be defended in terms of particular evaluative value
judgements, avoiding the need to estimate mode specific
values for different income groups.  The problems of differing values by trip length, time of day, amount of time saved,
etc. remain.

Perhaps the main problem the equity value approach
introduces for evaluation is that if certain (e.g. poor)
sectors of the community value time at below the mean rate,
this equity weighting procedure will place undue weight on
reducing their travel time rather than their money cost of
travel.

One means of deriving separate attribute values for buyers
and non-buyers of time saving, should such separate values be
required, is the 'diversion price' approach (Lee and Dalvi
(27)), refined by Hensher (18) as the 'transfer price'
approach.  Hensher did not calculate values of time by mode
but his approach allows such a calculation to be made.  These
approaches allow one to calculate separate time values for
traders and non-traders by investigating the change in time
or cost that would be required on either the chosen mode or
alternative mode to move an individual to a position of
indifference.  Valuation can thus conceptually occur along an
indifference curve for each individual, facilitating the
derivation of separate values for traders and non-traders.
The validity of such calculation clearly depends on the
ability and/or willingness of respondents to answer hypothetical questions concerning transfer prices/times.  If such
values are then to be applied in evaluation of particular
modal time savings, the attribute range over which valuation
took place needs to approximate the range over which the
attribute of the given mode will change.  This applies
equally to values derived from the more conventional probabilistic approach.  Figure 22.3 illustrates these propositions.

As in Figure 22.2, the individual initially chooses mode
B, having attributes $(C_B, T_B)$ rather than mode A $(C_A, T_A)$.
Under the transfer price/time approach one looks for changes
in $C_B, T_B, C_A, T_A$ which will move the individual to a position
of indifference between choosing modes A or B.  In Figure
22.3, if the travel time by mode A was reduced to $T_{A'}$, the
individual would then be indifferent between mode B and mode
A', both leaving him on indifference (disutility) curve $I_1$.
Under the transfer price/time approach, the transfer time in
this instance is $(T_A - T_{A'})$.  Using this, an estimate of the
individual's value of time saving can be derived based on
indifference curve $I_1$.

$$\text{VOTS} = \frac{C_A - C_B}{(T_B - T_A) + (T_A - T_{A'})} = \frac{C_A - C_B}{T_B - T_{A'}} \ .$$

This value of time saving is the individual's value over the
time range $T_B$ to $T_{A'}$.  If the travel time by the individual's

existing mode (B) were to be reduced from $T_B$ to $T_{A'}$ or thereabouts, this value would be an appropriate estimate of the unit worth of such time saving to the individual.

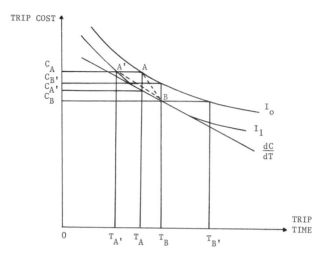

Figure 22.3   The Many Individual 'Values' of Time Saving

Alternatively, in Figure 22.3, one can derive different values of time saved by examining:

1   the fall in cost of mode A required to move the individual to a position of indifference between modes B and A. This would be achieved by a fall in cost of mode A to $C_{A'}$. The VOTS derived using knowledge of this transfer price is

$$VOTS = \frac{(C_A - C_B) - (C_A - C_{A'})}{T_B - T_A} = \frac{C_{A'} - C_B}{T_B - T_A} \ .$$

This value would be appropriate for valuing a time reduction on the existing mode B from $T_B$ to $T_A$;

2   the increase in trip time by mode B required to induce indifference.  If time by mode B were to increase to $T_{B'}$, indifference between modes A and B would result and a VOTS could be derived:

$$VOTS = \frac{C_A - C_B}{T_{B'} - T_A} \ ;$$

3   the increase in cost of the existing mode required to bring indifference

$$\text{VOTS} = \frac{C_A - C_{B'}}{T_B - T_A} \; .$$

Using the transfer price/time approach one can thus derive
several different values of time saving for the individual,
depending on where one locates in the individual's preference
system.  If there is little perceived difference between
alternatives in the initial situation, all four values may be
quite close.  In this situation, choice of value for use in
evaluation of a particular modal improvement is no problem,
provided the range of time differences used in derivation of
such values is approximately the same range which will result
from the particular modal improvement being evaluated.

If the modal alternatives have quite different attributes
in the initial situation, time values derived from time and
cost changes to the alternative mode may be more appropriate
for use in evaluation since they will be values based on the
individual's existing indifference surface.  Most projects
involving time savings will primarily benefit current users
of the mode concerned, as distinct from persons who will be
induced to change mode, so values based on the initial
indifference surface seem more appropriate.  Again time
changes valued need to approximate those from which values of
time savings were derived.

Perhaps the major difficulty in practice with this
approach concerns the individual's knowledge of the attributes of the alternative mode and the role of those
attributes in the individual's decision process.  If the
individual's behaviour is largely habitual in relation to a
particular modal 'choice', do the attributes of his chosen
mode have to change to induce a search for superior alternatives or can a change in the attributes of the alternatives
themselves induce such a search?  A priori one would expect
that a sufficiently large change in either set of attributes
could induce a search for new alternatives, but that for an
individual choosing out of habit such change would need to be
larger for the alternative than for the chosen mode.  Does
this mean that asking a transfer price/time question based on
the alternative mode will produce meaningless answers?

In mode-choice applications one can always check the
individual's knowledge of the alternative mode by asking
perceptions of times/costs, etc.  If these are reasonable,
one can have some confidence that the individual is making a
conscious choice decision in which preferences and trade-off
evaluations can be investigated.  If knowledge of the alternative is poor, such confidence will be eroded.  Even then,
however, the individual may be making a conscious choice from
which his preferences and trade-off rates can be implied,
albeit that such preferences are based on poor information.
Thus the individual may be choosing what he would regard as
an inferior mode had he known all the relevant attribute data
accurately.  This does not mean we cannot learn something
about his preferences by examining his perceptions of the
relevant attributes of alternatives.  Thus his perceived rate
of trading-off time for cost or vice-versa need not necessarily be ruled out as invalid just because the attribute

*magnitudes* on which such trade-off is based happen to be wrong. One therefore has no major qualms about using trade-off data based on perceptions of an alternative mode, provided the individual is able to perform a conscious trade-off. Questions should be included to aid an assessment of this problem. This conclusion applies for both the transfer-price technique and disaggregate probabilistic techniques which require attribute inputs for two or more modes. Mis-perceptions may, of course, lead to a case for the government seeking to change mode split. This is a separate issue.

## WHEN SHOULD INDIVIDUAL PREFERENCES NOT COUNT?

There is no inconsistency in arguing that certain types of policy consequences should be evaluated with respect to individual preferences (measured through behavioural response) but that others should not. Individuals may possess the knowledge and experience to choose between certain types of alternatives but not others. Given incomplete consumer information, restriction of the sovereignty of individual preferences and imposition of consumption choice might thus be defended, for example, because of long-term regret which may follow the choice decision. Such a restriction would be seen as part of a learning process, to permit more informed free choice thereafter. Economists typically refer to the goods which are the subject of such imposed preferences as 'merit goods'. Increasing incidence of standards and regulatory devices related to environmental matters serves as evidence that one can also have 'merit bads'.

Some pollutants generate perceived costs only after a lengthy delay and in such a way as to be irreversible when they actually are perceived. Nobbs and Pearce [31] give the example of the ingestion and inhalation of cadmium. If individual preferences are to be investigated in relation to such a commodity, where general ignorance will be the norm, behavioural analysis would be inappropriate. Rather, one would probably need to investigate such preferences in an experimental situation where risks are made clear to respondents.

A further example of the ignorance problem related to individual preferences concerns fatal accident costs. Economists at various times have used four major techniques for costing the value of human life: (1) the human capital approach (e.g. Reynolds [34], Dawson [12]); (2) implied values from human behaviour, i.e. risky jobs, life assurance, safety expenditures and implied values by government; (3) social valuations: court awards; (4) individual preferred valuation. Approach (2), provided it is the individual at risk making the choice, is consistent with the requirements of cost-benefit analysis. Generally, however, implied values have been limited to values implied by others in making decisions on the behalf of those at risk. Approach (4) is also consistent with the cost-benefit framework for policy evaluation. Jones-Lee [22,23] has sought to investigate willingness to pay for small changes in the risk of death, from the viewpoint of the person at risk. To such evaluation one would need to add the valuation by others of changes in

the degree of risk to this person.  Ignorance of degrees of
risk and the imperceptibility of small changes therein (e.g.
because of a new road being opened) make behavioural analysis
of doubtful validity for evaluating such risk changes.
Instead, Jones-Lee has used hypothetical questions involving
gambling prizes, income levels and risk of death.  This
overcomes the ignorance problem but still raises questions of
whether respondents can give sensible answers for small
changes in risk, such changes as policy evaluation is usually
concerned to assess.  Threshold problems pose major difficulties for such work.

     The question of when individual preferences should not
count for policy evaluation extends further than questions of
ignorance.  In particular, environmental irreversibilities
and their implications for future generations need to be
considered.  With future generations, as Pearce (33) has
pointed out, the issue is not one of always taking their
(unknown) wants into account but of preserving for them the
options to make choices based on their wants.  Zero demand
for many environmental goods today will mean their disappear-
ance with no option of reintroduction (because of irreversibilities).  This is unlike the situation implicit with most
commodities where, provided production technology is known or
knowable, markets can be opened and closed as desired.  With
these irreversible environmental commodities, then, one might
set preservation of a minimum viable unit for future expansion as a constraint on existing economic activity.  This
constraint is not costless.  However, here one is implicitly
valuing freedom to choose, which is quite different to asking
what value people place on commodities they are able to
choose (33).


PUBLIC ACCOUNTABILITY

Nash, Pearce and Stanley's (30) second criterion for selecting policy evaluation techniques is 'public accountability'.
If respect for individual preferences is thought to be
desirable not just in the evaluation of particular effects
but also in relation to the use to which such evaluation is
put in the decision-making process, it is important that the
evaluation process be publicly accountable.  By this, one
means that interested persons can see what has been done in
an evaluation, what value judgements have been made and how
various models have been used to produce outputs.  Such
persons should, by this criterion, be able to insert their
own values and/or make their own assumptions on key issues,
as a part of the process of reaching their own views of
project/policy desirability.

     *Behavioural modelling*, including its data collection
stages, is a high energy activity.  'Energy' is used here to
characterise the whole range of inputs to an activity,
including, time, money, etc.  Energy requirements of such
modelling include expert training, expert model construction,
calibration/estimation of models, data collection, computer
operation, analysis, etc.  The implication is that the
activity is exclusivist, amenable to comprehension by few.
If one is a supporter of participatory democracy - the

accountability criterion implies such support - behavioural
modelling can be seen as very questionable, more in the
nature of a legitimating device for professionals than an
indispensible research tool (15).

The increasing importance of participatory democracy will
mean an increasing devolution of decision-making power to
those who are affected by the decisions.  One aspect of this
will be an increase in absolute and relative importance of
local small-scale issues over regional large-scale issues.
Cost-effectiveness considerations in relation to planning of
the local level will dictate increased emphasis on low energy
methods for guiding such decision-making.  If transferability
of behavioural modelling results at the micro level proves
feasible, modelling will have a better future at such level.

Behavioural modelling work is likely to be limited
primarily to guiding policy of a broader strategic nature -
for example, assessments at a regional (urban) level of the
travel effects of increasing fuel prices.  Large-scale,
capital-intensive projects (e.g. a major urban freeway) can
also carry the cost of such work, but there is a great need
to improve the means of communicating the methods and
implications of such analysis to interested parties if
accountability of the policy-evaluation process is to be
served.


CONCLUSION

The days when evaluation of public policy was seen primarily
as a technical problem which could be left to the experts are
largely gone - but not completely, since macro-economic
policy is still characterised by the expert syndrome.  Policy
evaluation is now recognised as an essentially political
activity, where the prior role of value judgements over
technique is recognised.

The role of policy evaluation is seen as the identification of: (1) the *needs* which a particular policy (project) is
to address; (2) the *issues* which particular affected/inter-
ested groups see as relevant to an assessment of whether or
not the policy is desirable; (3) the *value positions* of these
groups on such issues.

Behavioural modelling is a potentially powerful and useful
device in such evaluation because of the behaviour (choice)-
preference-welfare relationship which underlies much of the
economist's approach to evaluation.*  Its usefulness has two
aspects, prediction of behavioural response to policy and as
input to the evaluation of that response from the 'behaver's'
viewpoint.**  However, the value basis of evaluation into

---

\* Not all of the approach, however, as argued in this chapter
  (viz. non-participant preferences).
\*\*Achievements in this latter regard have not been great over
  the past decade.  Apart from more refined values of travel-
  time savings, behavioural models have given us few implied
  monetary evaluations of particular consequences of public
  policies.

which the outputs of behavioural modelling are fed should not be forgotten. A consideration of that value basis and a consideration of some practical problems with using behavioural models has suggested some of the bounds to use of such models for policy analysis. In summary, the following issues have been covered.

1   Behavioural models provide no guidance to non-participant preferences (i.e. 'option' and 'existence' preferences), a much-neglected aspect of policy evaluation.
2   The welfare signals given by behavioural models may be misleading where a congested, common-property resource is under consideration.
3   For some consequences of public policy, there are no readily measurable response units against which behaviour can be assessed (e.g. comfort of a travel mode).
4   Even if measurable responses can be identified, the cost of behavioural modelling makes application at the micro level questionable unless transferability of results can be established.
5   In some situations, behavioural measurement has not identified a behavioural response, even when *a priori* the latter can be expected (e.g. road traffic noise).
6   When a behavioural response can be identified, one needs a well-founded theoretical model to translate this response into a cost (or benefit) for use in willingness-to-pay type evaluations (e.g. noise-cost and house-price depreciation).
7   Given such a theoretical model, it is important to ensure that the right marginal valuations are used (e.g. time-savings valuation).
8   One has reservations about obtaining evaluation inputs from behavioural modelling if the activity to which response is being modelled is one about which consumers have little knowledge and this lack of knowledge could lead them to experience considerable regret.
9   Similar reservations are held about behavioural response in relation to environmental irreversibilities, where future generations may lose the right to exercise any choice, and where the common property resource difficulty arises for analysing preferences of the current generation.
10   The ethics of participatory democracy mean that a considerable communications effort will be needed where modelling is used as part of a participatory evaluation process.

In a number of these situations, direct preference assessment has a role to play (e.g. attitudinal techniques; group search conferences). In some instances this role is complementary to that of behavioural modelling (e.g. cases (1), (2) and (3)), whilst in other situations it may be as a replacement for such modelling (e.g. case (4)).

REFERENCES

1. Abelson, P.W., 'Environmental factors and house prices', *Bureau of Transport Economics Research Paper* (1977).
2. Bohm, P., 'An approach to the problem of estimating demand for public goods', *Swedish Journal of Economics*, 73 (1971).
3. Bohm, P., 'Estimating demand for public goods: an experiment', *European Economic Review*, 3 (1972).

4. Bowen, H.R., 'The interpretation of voting in the allocation of economic resources', *Quarterly Journal of Economics*, 58, pp. 27-48 (1943).

5. Cichetti, C.J. and Freeman III, A.M., 'Option demand and consumer surplus: further comment', *Quarterly Journal of Economics*, 85, pp. 528-39 (1971).

6. Cichetti, C.J. and Smith, V.J. (1976) *The Costs of Congestion: An Econometric Analysis of Wilderness Recreation*, Ballinger, Cambridge, Mass.

7. Clawson, M. and Knetsch, J. (1966) *Economics of Outdoor Recreation*, Johns Hopkins Press, Baltimore.

8. Colony, D.T. (1967) *Expressway Traffic Noise and Residential Properties*, US Department of Transportation, Bureau of Public Roads.

9. Commission on the Third London Airport (CTLA) *Consumer Surplus in Housing: Report of Survey Work*, Further Research Team Work, CTLA, London (1970).

10. Commission on the Third London Airport (CTLA) (1971), *Report*, Chapter 7 and Appendices 22 and 23.

11. Commonwealth Bureau of Roads (Hensher, D.A., Smith, R.A. and Hooper, P.G.), 'An approach to developing transport improvement proposals', *Occasional Paper No. 2*, Commonwealth Bureau of Roads, Melbourne (1976).

12. Dawson, R.F., 'Cost of road accidents in Great Britain', *TRRL Report LR 79* (1967).

13. Diffey, J., 'Investigation into the effect of high traffic noise on house prices in a homogeneous sub-market', *Seminar on House Prices and the Micro-economics of Housing*, Centre for Environmental Studies, London (1971).

14. Harrison, A.J. (1974) *The Economics of Transport Appraisal*, Croom Helm, London.

15. Healey, P. and Stanley, J.K., 'Transportation studies and normative social planning', *Royal Australian Planning Institute Journal*, July/October, pp. 82-7 (1975).

16. Hensher, D.A. (1974) *Consumer Preferences in Urban Trip Making* (5 Vols.), Melbourne, Commonwealth Bureau of Roads.

17. Hensher, D.A., McLeod, P.B. and Stanley, J.K., 'Usefulness of attitudinal measures in investigating the choice of travel mode', *International Journal of Transport Economics*, 2(1), pp. 51-75 (1975).

18. Hensher, D.A., 'Valuation of commuter travel time savings', *Journal of Transport Economics and Policy*, 10(2), pp. 1-10 (1976).

19. Hensher, D.A. and McLeod, P.B., 'Towards an integrated approach to the identification and evaluation of the transport determinants of travel choices', *Transportation Research*, 11(2), pp. 77-93 (1977).

20. Hicks, J.R. (1956) *A Revision of Demand Theory*, Clarendon Press, Oxford.

21. Hoinville, G., 'Evaluating community preferences', *Environment and Planning*, 3, pp. 33-50 (1971).

22. Jones-Lee, M., 'Valuation of reduction in probability of death by road accident', *Journal of Transport Economics and Policy*, 3(1) (1969).

23. Jones-Lee, M. (1976) *The Value of Human Life*, Martin Robertson, London.

24. Krutilla, J.V. and Fisher, A.C. (1975) *The Economics of Natural Resources*, Johns Hopkins University Press, Baltimore.

25. Kurz, M., 'Experimental approach to the determination of the demand for public goods', *Journal of Public Economics*, 3, pp. 329-48 (1974).

26. Lancaster, K.J., 'A new approach to consumer theory', *Journal of Political Economy*, 74, pp. 132-57 (1966).

27. Lee, N. and Dalvi, M.Q., 'Variations in the value of travel time', *Manchester School*, 37(3) (1969).

28. Little, I.M.D. (1950) *A Critique of Welfare Economics*, Clarendon Press, Oxford.

29. Nash, C.A., Pearce, D.W. and Stanley, J.K., 'An analysis of cost-benefit analysis criteria', *Scottish Journal of Political Economy*, 22(2), pp. 121-34 (1975).

30. Nash, C.A., Pearce, D.W. and Stanley, J.K., 'Criteria for evaluating project evaluation techniques', *Journal of the American Institute of Planners*, 41(2), pp. 83-9 (1975).

31. Nobbs, C. and Pearce, D.W., 'The economics of stock pollutants: the example of cadmium', *International Journal of Environmental Studies*, 8, pp. 245-55 (1976).

32. Pearce, D.W., 'The economic evaluation of noise generating and noise-abatement projects', in *Problems of Environmental Economics*, OECD, Paris (1972).

33. Pearce, D.W. (1976) *Environmental Economics*, Longman, London.

34. Reynolds, D.H., 'The cost of road accidents', *Journal of the Royal Statistical Society*, Series A (General), 119(1), pp. 393-408 (1956).

35. Sen, A.K., 'Behaviour and the concept of preference', *Economica*, 40, pp. 241-59 (1973).

36. Starkie, D. and Johnson, D., 'Exclusion facilities and the valuation of environmental goods', in *Papers from the 1973 Urban Economic Conference*, vol. 1, Centre for Environmental Studies, London (1975).

37. Stanley, J.K., 'Cardinal utility approach for project evaluation', *Socio-Economic Planning Sciences*, 8, pp. 329-38 (1974).

38. Stanley, J.K., 'An evaluation of residential area improvement strategies from the residents' viewpoint', *Socio-Economic Planning Sciences*, 11(2) (1977).

39. Towne, R., 'An investigation of the effect of freeway noise on apartment rents', R. Towne and Associates, Seattle (1968).

40. Walters, A. (1975) *Noise and Prices*, Oxford University Press, London.

41. Weisbrod, B., 'Collective consumption services of individual consumption goods', *Quarterly Journal of Economics*, 78, pp. 471-7 (1964).

Chapter 23

THE RELATIONSHIP BETWEEN BEHAVIOURAL MODELS,
EVALUATION, FORECASTING AND POLICY

Robin C.Carruthers

STATEMENT OF SCOPE

The scope of the workshop was to review the policy framework
in which the evaluation and other advice emmanating from
behavioural travel models is used. When modellers, transport
planners and policy advisers meet, the complaint is frequently heard that the modellers do not produce the information
required by the planners, and the planners do not produce
the analyses that the policy advisers feel that they need.
To understand the role of the modeller in the general policy
process, the basic structure of decision-making in various
circumstances requires understanding. There is little point
in modellers producing models with increasing and varied
levels of sophistication, if they still fail to provide the
inputs to the analyses necessary to the adviser. Similarly,
the adviser needs to understand something of the nature of
the models that are being used on his behalf. Without an
understanding of the capabilities, and strengths and weaknesses of the models, the adviser is liable to become frustrated in his attempts to formulate policy advice, or worse,
to misapply the information provided.

SUMMARY OF WORKSHOP CONCLUSIONS AND RECOMMENDATIONS

*Introduction*

The workshop was fortunate in having two resource papers
that raised most of the issues in the sphere of interaction
between behavioural models, forecasting and evaluation, and
policy that were of concern to the other members of the
workshop. Together with the review paper, those formed the
basis of the workshop discussion.

It was necessary at an early stage to identify the role
of behavioural models and the results they produce, in the
policy-making process. The varied background and experience
of the workshop participants provided an insight into most
stages of the process of involving behavioural models in
policy-making, from the more conventional applications in
large-scale investment decisions, to the rapidly developing
sphere of local area policy proposals. In the introduction
to Guttman's review paper, he observed that both resource
paper authors were sceptical of the performance of conventional behavioural models, a scepticism that was typical of
members of the workshop.

512

*Models and Policy Formulation*

The interface between modellers and decision-makers is not
being operated successfully, resulting in a potential under-
utilisation of model output. In addition it appears that
the models currently available are not applicable to the
problems being faced by decision-makers.

*Transferability of Data*

The problems now being faced by decision-makers are generally
of a smaller scale than those of the period when the current
generation of models were developed. A greater emphasis on
transferability of data between models and modelling situations would reduce study costs and increase the usefulness
of models.

*Value of Time over Time*

Investigation of an essential modelling assumption, that
concerning the value of time over time, has been largely
ignored by transport modellers. The two assumptions most
usually made can lead to a difference in benefits of the
order of 50 per cent. This one item, being frequently the
major if not the only benefit of many transport projects,
merits more serious consideration.

*Goods Movements*

Goods movement modelling has not attracted the attention
justified by the importance of freight in both urban and
inter-urban situations. The adaption of person–trip models
to goods movements is indicative of this neglect, and a new
approach to goods movements modelling, perhaps stimulated by
the goods movement workshop, is required.

*Transport Supply*

Current transport models are generally concerned with demand
changes as a consequence of investment changes and are not
suitable for the analysis of changes in transport supply
conditions. In particular, transport models should be capable of analysing changes in regulatory policy, which too
often is assumed to remain unchanged over time.

*Evaluation Framework*

The conventional use of user benefit and operator surplus as
measures of social benefit were recognised as being insufficient. More use should be made of the results of attitudinal studies in formulating the evaluation framework of
transport models.

DISCUSSION OF CONCLUSIONS AND RECOMMENDATIONS

*Models and Policy Formulation*

The results produced by behavioural models are rarely seen directly by the politicians responsible for making decisions, there being an effective filter in the form of the policy adviser.  The adviser interprets the advice produced as a result of technical studies in the light of political constraints - the adviser does not act in the altruistic manner often assumed by producers of technical reports.  Although a concept of the 'public interest' may be an important principle for both the technician and adviser, the adviser will use his own interpretation of the political environment to present the technical advice in what he judges to be the most appropriate way.

To be successful in the first instance, the technician should direct the advice at the policy adviser and not the decision-maker himself.  A useful guide to the presentation of advice was that politicians - the successful ones at least - often have a mission, and are looking for projects and policies which will have a recognisable impact within their own tenure of office, typically less than three years. If technical advice can be offered in a way which fulfils this sense of 'mission', then it will be more appealing to the adviser and stand a better chance of ultimate acceptability.

There was a divergence of opinion as to whether the role of the modeller should include this political aspect or whether the modeller should attempt to be impartial, and whether the modelling work was designed to be, in some abstract sense, in the public interest.  This contrasted with the view of others, just as strongly held, that it was impossible for the modeller to provide impartial, independent advice however sincere the attempt.  The modeller would be held in a better position to influence policy decisions if this fact of life was recognised.  The partiality of the modeller could intervene at almost any stage of the modelling process, from the selection of input data, through the methods of analyses chosen to the final evaluation advice offered. Models and advice where this partiality were made explicit would be of greater use to policy advisers than those where it was not.

Concern was expressed on the level of confidence that should be expressed in the model results.  The policy adviser is not really interested in the results of exhaustive sensitivity tests of models, whether to alternative data sets, assumptions or parameters.  The level of confidence of the technical modeller should be conveyed to the policy adviser, and the modeller himself should be satisfied that the advice being offered was reliable.  This appeared to indicate that the modeller should investigate the limits within which the technical advice offered would stand uncontradicted, and that the interpretation of the reasonableness of these limits was largely the prerogative of the modeller rather than the adviser.

As would be expected, the interface between modeller and decision-maker is more direct on the smaller scale and at a more local political decision level. The policy adviser is still involved in the process but operates more as a co-ordinator between the modeller and decision-maker rather than as a filter on the advice offered. As a consequence, modelling advice at a local level needs to be more closely related to political consequences than on the more traditional problems addressed. The closer integration in the political process requires a different type of model, one that can provide information on more specific impacts of policy than previously.

It is only comparatively recently, and then only in a few isolated instances, that behavioural models have been used to identify problems and generate potential solutions, rather than simply to evaluate a predetermined set of solutions to an already perceived problem. This historically limited use of behavioural models represented a serious loss of opportunity on the part of the modeller in making the best use of the tools that he has available and a lack of appreciation on the part of the policy adviser on the assistance that can be provided.

However, the form of the models used for problem recognition and the generation of potential solutions needs to be rather different to the usual policy or project evaluation model. For example, in problem recognition there is generally little use for a predictive modal-split model, more for a model that produces measures of accessibility, given a current modal split. Relatively simple analytical models can be used to identify social groups that have significantly different characteristics to the general population, or instances where public transport services do not follow the trip patterns of the people they are designed to serve. The use of models to identify real problems would reduce the occurence of studies being designed to evaluate remedies to imagined social problems, but only demonstrate at the end of an exhaustive evaluation process that no real problem exists.

*Transferability of Data*

It was apparent that the current range of available behavioural models was not meeting the needs of policy advisers. Although many reasons were advanced for this lack of coincidence of interest, the changing nature of transport problems and types of potential solution were consistently the most dominant in the discussion. Whereas until recently the size of the problem was sufficient to justify the generation of new models specific to the problem, with the associated survey costs for data collection, estimation and/or verification of the model parameters, the problems now requiring behavioural analyses are not sufficient to justify the cost of this approach. All members of the workshop were able to cite examples where the cost of setting up the analytical model was of the same order of magnitude as the cost of the project itself, or the likely cost advantage between the best and worst solutions. If we hope to

continue to apply behavioural models to such problems, then
the nature of the models themselves must change.

One particular respect in which modellers appear to have
been extravagant is in the need for problem-specific data
and for models estimated for each application. There was a
strong feeling in the workshop that the level of transferability of data between models and applications could be
much greater than at present. Experience in different
countries supported this intuitive feeling. For example,
in Australia and the United States there are no generally
accepted or recommended values for such basic model inputs
as unit values of time, whereas in the United Kingdom, the
relevant government department produces a set of intermitt-
ently revised values of time for different journey purposes.
This precludes the need to estimate time values each time
they are required.

Although there was not general support for the principle
of disaggregation by trip purpose as being sufficient, there
was support for the principle of geographically constant
time values, once certain other key variables had been
controlled for. The workshop was particularly concerned to
suggest a minimum set of variables to be controlled, which
would allow reasonable reliance to be placed on the application of values of travel time to models other than those
from which they were derived. The most important of these
key variables was thought to be time of day, with a sufficient distinction being between peak and off-peak. Reported
experience from the United States was that time values could
vary by a factor of up to 10 between peak and off-peak, the
major difference being accounted for by a disutility element
of peak travel rather than by a difference in opportunity
cost. It was reported that this disutility difference was so
great that it obviated any need to control for journey
purpose. Although there was an obvious change in the purpose
split between business and non-business trips from peak to
off-peak periods, this was not sufficient to explain the large
change in time values. Other important variables to control
would be income, trip frequency and mode of travel. It was
also important to ensure that the models where the data
transfer was being applied were of a similar structure. For
example, the change in time values between peak and off-peak
would not be applied to a model where various components of
travel disutility were being accounted for separately.

Controlling for the relevant variables might have a small
cost penalty on the model used to derive the parameter
values, but this extra cost is likely to be small compared
to the cost savings in avoiding unnecessary data collection
and model estimation in subsequent applications.

There are some instances where transferred data may be
preferable to directly observed data on grounds other than
cost. Where the population being modelled does not have
direct experience of the particular range of a variable
being tested - very high levels of public transport service
for example - it might be preferable to transfer cost-demand
elasticities derived from populations who had experienced
that particular level of service. Extrapolation of

elasticities derived from the direct experience of the modelled population could be misleading.

Directly observed data is difficult to apply where
behavioural evidence is not necessarily a reasonable reflection of individual preferences, where for example there is a
significant degree of personal integration of utilities,
typically in 'prisoner's dilemma' situations.  It may be
necessary to transfer data not from similar behavioural
models where revealed preferences may be just as misleading,
but from attitudinal surveys designed specifically to overcome such problems.

There is an increasing acceptance of attitudinal data,
both to supplement and to replace behavioural data.  This
represents a change in attitudes to attitudinal data itself,
which until recently tended to be disregarded as inferior in
all respects to behavioural data.  But behavioural data
itself has shown in practice to have problems ot its own,
particularly its inability to produce consistent results
when applied to similar situations.  These practical problems
are in addition to the more conceptual problems raised in
Stanley's paper with respect to some of the evaluation consequences of behavioural models.  Techniques of attitudinal
survey in the transport sector have improved considerably in
recent years, possibly through the application of more
experience from psychological attitudinal tests where similar
techniques are long established.

Attitudinal data also has an important role in assessing
the impact of transport changes that go outside the realm
of current experience.  Behavioural modellers have approached
the problem by assessing the impact of attributes, but come
up against problems where there is no simple unit of measure
of the attribute.  It is hoped that attitudinal methods can
help in defining some of the components of attributes, particularly for public transport modes.  Some work has already
been undertaken, but insufficient resources are being devoted
to the method for imminent general application of the results.
It is believed that attitudinal survey data requires much
smaller sample sizes than those conventionally used in
providing data bases for disaggregate behavioural models.
In addition to the reduced cost at the initial survey stage,
there is a higher probability that the results will be
applicable in other studies.

*The Value of Time over Time*

There was little general discussion of the forecasting of
exogenous input data for behavioural models.  Beesley's main
point, taken up by Guttman, related to the forecasting of
values of time, particularly changes in values over time.
The value of time and its changes over time have a significant
effect on the evaluation of transport proposals, since most
of them are designed to reduce travel times in one way or
another.  If the UK practice of assuming that the value of
time increases over time at a rate equal to the change in
GDP per head, assumed to be 3.25 per cent per year, then the
total value of time savings over the life of a 20-year project

will be 47 per cent higher than if the value of time were
assumed constant.   For a 30-year project, the increase would
be 65 per cent.

Although the value of time is assumed to increase over time
in almost all UK transport investment appraisals, it is not
common practice elsewhere.   The USA and Australia have no
government recommendation for such an assumption as exists
in the UK, and the value of time is frequently held to be
constant over time.   The difference in practice would be even
more marked in developing countries where the expected rate
of growth of per capita income is greater, and particularly
important where the projects being considered may themselves
have an effect on per capita income.

This therefore is an area of potential exploration by
behavioural modellers, since they have available the necessary
expertise to test the hypothesis that the value of time is
related to per capita income and does change over time.
Although travel time savings do not assume the importance of
a few years ago, now being supplemented by measures of comfort and convenience, they are still often the largest item
in the evaluation of transport proposals.


*Goods Movements*

An examination of the resource papers presented at the
various conference workshops was supporting evidence for the
complaint that goods movement analysis was being neglected
through an exaggerated attention to person trips.   This
current balance of interest between the two sources of demand
for transport services was perhaps more justified in the days
when our principal concern was with peak-period, radial movements.   But with a changed emphasis now being more related
to basic production functions and resource allocation, the
costs of distribution in both urban and rural areas merits
more of the modellers attention.

Until relatively recently any serious attempt to integrate
an understanding of the behavioural aspects of goods movements in the structure of transport models foundered on the
sheer complexity of the task and the lack of adequate, reliable data related to the problems being considered.   Even
recent attempts to construct freight models as in the New
Zealand Transport Policy Study (12)or the South Western
Australia Transport Study have relied too heavily on the
experience gained in person-trip modelling and the application of over-simplified economic theory.   The workshop wel-
comed the papers prepared for the goods movement workshop,
as a sign that this neglect might at last have been
recognised, but felt that little in the way of constructive
effort had yet been achieved and that we were still in the
stage of trying to define what the relevant models should
look like.   There was still a long way to go before any
useful models might emerge, and the process would benefit
from more attention from those currently engrossed in the
detail of person-travel modelling.

*Transport Supply Considerations*

Few of the currently developed behavioural models relate to
the supply side of the transport industry.  Our understanding
of how the supply of transport services reacts to changes in
the transport infrastructure, to changes in regulatory
policy, to technical innovation or to changes in its own
internally imposed constraints is inadequate and unexplored
by behavioural modellers.

A major criticism of the behavioural models currently
used, is that in too many instances they make no attempt to
relate basic characteristics, such as car-ownership rates or
trip rates, to proposed changes in the transport infrastructure or other supply variables.  A better understanding of
such relationships, even if they were to show that the basic
characteristics were largely unaffected by supply changes,
would go far in demonstrating that the modellers have some
appreciation of the existence of demand-supply interactions.

Too often models assume that regulatory policy will
continue unchanged, or overlook the possibility that changes
in regulatory policy may provide a more socially acceptable
solution to a problem than an investment in further infrastructure facilities.  Perhaps this preference for capital-
intensive solutions has been unconsciously encouraged by
politicians, or political advisers to whom regulatory change
has traditionally been a less attractive option.  This
preference may have been emphasised by engineering-based
consultants whose principle interest has been capital development rather than administrative reform.  However, this is
one area where the changing nature of the problems to which
the models are being applied is rapidly forcing a change in
the models themselves.  They are increasingly being used in
situations where regulatory change is part of one of the
solutions being examined, and in some cases, such as the
New Zealand Transport Study or the South Western Australia
Transport Study, they are addressed to problems created by
slowness of regulatory change and where regulatory/adminis-
trative measures form the entire solution set.

However, where regulatory change is involved in the
solutions examined, we generally suffer from a lack of
comprehension and understanding of how individuals and institutions in the transport sector react to such changes.
There is a general recognition of some of the more obvious
constraints on regulatory change, such as labour agreements
which prohibit the introduction of new forms of transport
and the entrenched interests of existing transport operators,
who are well attuned to the mechanisms of the policy process
and able to resist changes which would not appear to be in
their interests.  Despite ample evidence of how binding
these and similar constraints might be, our lack of understanding of the root causes tempts us not to investigate
the circumstances that might lead to their being relaxed.
But whether such investigations are necessarily the prerogative of the behavioural modeller is debatable.  The
behavioural modeller should be able to incorporate the
experience of industrial physchologists and others.  But
where such experience appears to be non-existent or incapable

of being translated into the structure necessary for a transport model, then the modeller inevitably becomes involved in
attempting to analyse the industrial and social relationships
involved.

*The Evaluation Framework*

Stanley's contention that measurement of user surplus and
operator benefit does not tell the whole story was generally
accepted.  Even for users there is the problem of option and
existence values, but these are likely to be relatively
small.  Behavioural evidence does not necessarily lead to a
measure of consumer surplus.

   Experience in the application of user-benefit evaluation
was mixed.  In Australia there was limited use of user-
benefit as a means of assessing the justification of schemes,
whereas in the UK this appeared to be the sole measure of
benefit for justification of many road schemes at least.
Experience from the USA indicated that transport user-benefit
evaluation was a relatively insignificant part of transport
studies, practical operating effects and financial results
assuming much greater importance.  The essential points of
existence and option values would seem to be relatively
insignificant in the Australian and USA context, but could
be more relevant where user benefit itself assumes importance.

   Given the acceptance of many of the principles raised by
Stanley, we are then faced with an assessment of their
importance and relevance in particular situations.  By their
nature, many of the arguments are incapable of quantification
or assessment in terms more specific than used by Stanley.
Where there are alternative evaluation methods available,
they tend to suffer from similar if not identical conceptual
problems.  It is perhaps in the measurement of attributes
and the use of choice-utility-attitude attribute models as
described by Stanley that there is some prospect of improving
the performance of current evaluation models.

   The scepticism revealed in Stanley's paper, and echoed in
the views of some other workshop participants appeared to be
related more to the incompleteness of consumer benefit as a
measure of the benefit of a proposal than to significant
errors in its application or measurement.  Sympathy with the
proposal to supplement consumer benefit measures with
evidence from attitudinal studies perhaps indicated more a
hope that attitudinal information could be relied upon rather
than a judgement resulting from a critical examination of its
performance.

REFERENCES

   1. Affleck, F (1976) *Transport Regulation in Western
Australia*, a report prepared for the South Western
Australia Transport Study by F. Affleck and Associates, Perth,
Western Australia.
   2. Barrel, D.W.F., 'Equity and incidence in transport
evaluation', *Proceedings of 1974 Summer Annual Meeting,*

*Planning and Transportation Research and Computation Co.*, London (1974).

3. Beesley, M.E., 'Economic criteria for the maintenance, modification or creation of public urban and suburban transport services', *Report of the 24th Round Table on Transport Economics*, ECMT, Paris (1974).

4. Frost, M.J. (1971) *Values for Money*, Gower Press, London.

5. Hall, P.G., Batty, M.J. and Starkie, D.N.M., 'The impact of free public transport upon land-use and activity patterns', *Transport and Road Research Laboratory, Supplementary Report 37 UC*, Crowthorne, England (1973).

6. Harrison, A.J. (1974) *The Economics of Transport Appraisal*, Croom Helm, London.

7. Hoinville, G. and Johnson, E. (1971) *The Importance and Values Commuters Attach to Time Savings*, Social and Community Planning Research, London.

8. Huckfield, L., 'Transport policy: the report of a study group', *Socialist Commentary*, London (1975).

9. Lichfield, N., Kettie, P. and Whitbread, M. (1975) *Evaluation in the Planning Process*, Pergamon Press, Oxford.

10. Perrod, P. *et al.*, 'Application of modern methods (with special reference to planning, programming, budgeting techniques) to the choice of investment projects', *Report of the 10th Round Table on Transport Economics*, ECMT, Paris (1971).

11. Plowden, S.P.C., 'The future of transportation studies', *Proceedings*, PTRC Summer Meeting, Warwick, England (1974).

12. Smith, Wilbur and Associates (1973) *New Zealand Transport Policy Study*, Ministry of Transport, New Zealand.

13. Starkie, D.N.M. (1973) *Transportation Planning and Public Policy*, Pergamon Press, Oxford.

14. Starkie, D.N.M., 'The policy-modelling interface', *Southern Western Australia Transport Study*, Perth, Western Australia (1976).

15. Starkie, D.N.M. (1976) *Transportation Planning, Policy and Analysis*, Pergamon Press, Oxford.

16. Stopher, P.R. and Meyburg, A.H. (1975) *Urban Transportation Modelling and Planning*, Lexington Books, Lexington.

PART EIGHT

Chapter 24

URBAN GOODS MOVEMENT: PROCESS, PLANNING APPROACH AND POLICY

Peter J. Rimmer and Stuart K. Hicks

> The elements that...should concern
> us most...are the conflicts that
> exist between vehicles and other
> vehicles, between vehicles and
> persons, and between vehicles and
> the environment. (Watson (29)p.8)

A STUDY IN CONFLICT

Conflicts between people and goods movement in urban areas
have jolted decision-makers and transport planners. Yet,
their past preoccupation with urban passenger movements has
provided them with few strategies for coping with a newly-
recognised set of stresses involving people *and* goods
movement.

Rather than grapple directly with these stresses the
immediate reaction of the transport planner has been to
attempt to incorporate urban goods movement in the urban
(passenger) transport planning process. As Watson (29)
indicates, the main research thrust has been to model urban
goods movement in an analogous fashion to urban people movement. A series of parallel submodels are used to represent
transaction (trip generation), flow (trip distribution),
means change (modal split) and network assignment; an
industrial location model is added to make the links between
industrial activities and commercial vehicle movements
explicit. A refinement to the basic framework by Meyburg and
Stopher (15) shifts attention from commercial vehicles to
consignments and adds a vehicle loading submodel to follow
the modal-split stage in the urban transport planning process
which converts consignments into vehicle movements.

The use of the consignment as the unit of analysis has
strong parallels with developments in people-trip modelling
where the unit of analysis has switched from zones to
individuals following the adoption of a disaggregative
behavioural modelling approach. According to Hensher and
McLeod (7), this procedure assumes that an individual's
travel behaviour stems from his subjective perception of
various spatial movement alternatives and his individual
choice process remains constant despite changes in his choice
behaviour pattern. While the need for a disaggregative
approach is accepted, this chapter does not pursue the
passenger analogy further. As Watson ((29)p.101) stresses,
there are few problems to which the revamped model can be
addressed. Instead, this chapter highlights the need to
understand the urban goods *process*, not only as a means of
obtaining better information on urban goods and their

interaction with people but as the foundation of some future
behavioural model.

An understanding of the urban goods process depends on an
appreciation of the role conflicts between a large number of
actors involved in making a large number of decisions.   Three
types of role conflict are recognised.

1   *Activity conflict* involving friction between shipper and
goods vehicle operator.
2   *Technological conflict* involving friction between goods-
vehicle driver and car driver/pedestrian.
3   *Land-use (activity structure) conflict* involving friction
between goods-vehicle owner and urban residents as reflected
in the emotive problem of the 'Big Truck' discussed by Smith
(26).

Many political decision-makers operating at varying scales
(local, regional and national) are anxious to intervene in
these conflicts to lessen stresses.   However, interference
may increase the benefits accruing to one group at the
expense of another.   Before such implications can be appreciated we need: (1) to deepen our knowledge of the pattern of
urban goods movement, the functions of interacting role
players and the task of the decision-makers and transport
planners; (2) to identify the variables that policy-makers
can change; and (3) to indicate how they can be used to
develop new policies.   Thus, this chapter provides a framework for examining the urban goods process.   As it is
important that the analytical problem (or modelling procedure) should not be the sole issue in studying urban goods
movement the chapter also outlines the associated planning
problem, and examines the relative feasibility of policy
options and the nature of policy instruments available for
resolving conflicts.   The emphasis on putting the analytical
problem into its proper context within a transport planning
framework highlights that any model (behavioural or otherwise) should be a constructive means of investigating the
relationships on which sound policies are based.   Even if the
model is not practically operational it must make a positive
contribution towards understanding the urban goods process.

URBAN GOODS-MOVEMENT PROCESS

The past preoccupation with modelling commercial vehicle
movements within the traditional transport study framework
has provided little information on either the purpose of
urban goods movement or on the conditions under which goods
move.   Data collection on commercial vehicles within transport studies is invariably subjugated to the requirements of
a pre-existing model.   While the modelling approach is
informative, in that it provides a picture of the spatial
pattern of commercial vehicle movements, it does not supply
the reasons as to why goods move.   The key issue of this
chapter, therefore, is the need to understand the urban
goods-movement process before outlining a planning approach
and examining policy options and their likelihood of attainment ((21)p.33).

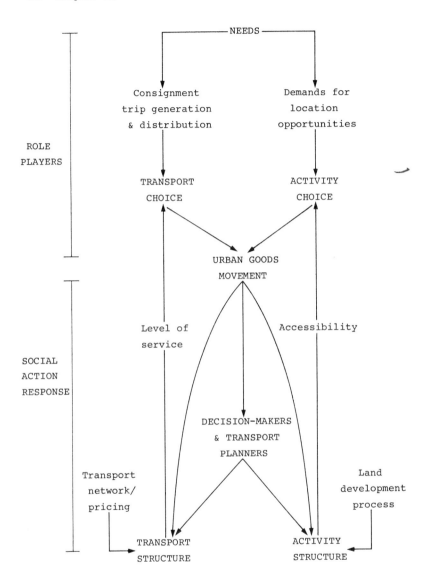

Figure 24.1   A General Framework of the Urban Goods—Movement Process

In comprehending the urban goods-movement process within
the general framework depicted in Figure 24.1, attention is
directed initially at the *pattern* of urban goods movement.
Then the *roles* of the actors involved in the urban goods
process are studied. Finally, the functions of the political
*decision-makers* and *transport planners* are identified and
discussed. As existing modelling procedures can only handle
relatively trivial urban goods-movement problems an emphasis
on understanding the process through an examination of pattern, roles and decision makers is critical.

*Pattern*

A deeper understanding of the urban goods process depends on
shifting interest from inter-zonal spider networks and inter-
zonal commercial vehicle movements and commodity flows to two
different types of connections recognised by Smith and Hay
(26) - the linkage and the strand.

1 *Linkage* is used to describe the spatial connection
between receiving and despatching units or nodes within urban
areas.
2 *Strand* is employed to encompass the number of transactions that occur between receiving and despatching units
(i.e. consignor and consignee).

In making these distinctions it is assumed that the strands
can be combined for all links over a given inter-zonal route
to obtain total commodity flow. However, before these
distinctions can be appreciated the linkages and strands have
to be described in more detail.

*Linkages*

Interest in examining linkages is centred on the network of
connections for moving goods between the shipper and ultimate
consumer as indicated in Figure 24.2. We are concerned in
particular with the connections between the locations of
transport terminal operator, manufacturer, wholesaler,
retailer, householder, institution and waste disposer. An
understanding of the linkages between these nodes is facilitated by adapting and extending Hutchinson's (10) and Meyburg
and Stopher's (15) external and internal linkages and
illustrating them in more abstract terms in Figure 24.3 (23).

1 *External linkages* are those that *either* originate *or*
terminate outside the urban area. These external linkages
can be further subdivided into:

(a) *direct linkages* that pass to or from a transport terminal,
manufacturer, wholesaler, retailer, householder, institution
or waste disposer without contact with an intermediate unit
located in the defined urban area,

(b) *through linkages* which facilitate goods movement through
the urban area en route to other destinations without contact
with an economic unit.

2 *Internal linkages* are those that *both* originate *and*
terminate inside the urban area. These internal linkages

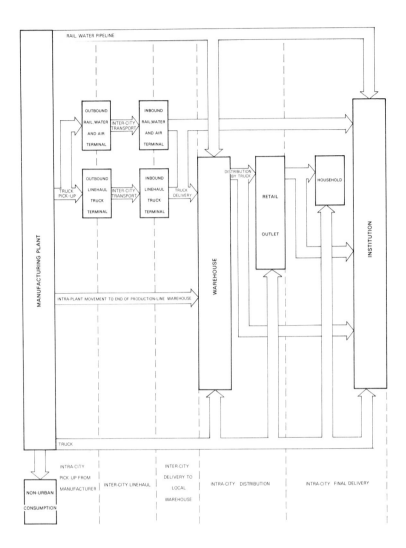

Figure 24.2  Generalised Distribution Pattern Adapted from
Kearney and Voorhees (12)

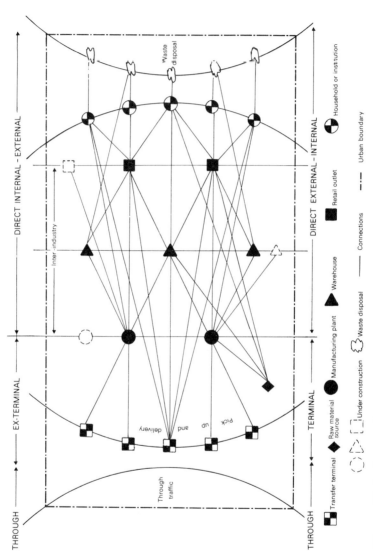

Figure 24.3   External and Internal Linkages within an Urban Area derived from Rimmer (23)

can be subdivided into:

(a) *inter-unit linkages* incorporate those that permit movements of finished and semi-finished goods between terminal
operator, manufacturer, wholesaler, retailer, householder,
institution and waste disposer.

(b) *intra-unit linkages* embrace those that allow goods movement within the confines of a transport terminal, factory,
warehouse, retail outlet, house, office or waste disposal
site.

These distinctions allow us to concentrate on a particular
set of linkages. While through linkages are a contributor to
total commodity flow and traffic and intra-unit linkages are
an under-estimated topic of research, attention is focused on
direct external linkages and internal inter-unit linkages.
Any analysis of the urban goods-movement process must try to
quantify these linkages in terms of the *thinness* and *thick-
ness* of the strands.


*Strands*

The accounting unit for determining the strength of these
strands is the consignment. Such a unit is chosen because
the ultimate objective of the urban goods-movement process is
to position the consignment over time and in space so as to
maximise its potential contribution to productivity and
utility subject to owner, user, distributor or governmental
constraints. Productivity is defined in terms of the cost of
the unit of output per unit of input and utility in terms of
the price an individual or organisation is willing to pay to
receive the consignment (5).

   The essential characteristics of the consignment are
described by Meyburg and Stopher (15) as its physical state
(i.e. solid, liquid or gaseous), mass (weight), volume
(average density), shape, value, perishability, time and date
of despatch and origin and destination. Such characteristics,
in part a response to the nature of existing and potential
methods of transport, constitute some of the determinants of
direct demand and provide a basis for developing supply-
oriented strategies such as road and terminal pricing and
regulations about working hours because it is assumed that
initially the provision of transport will respond to demand.

   Where any characteristic is dominant, it requires, as
emphasised in Figure 24.4, a special type of vehicle. For
example, tankers are needed for gases and liquids, refrig-
erated or insulated vans for perishables, custom-built
equipment for fragile items, armoured vehicles for high-value
articles, high-powered vehicles for express services,
specially designed vehicles for high-density freight and low
loaders for heavy haulage. If we eliminate consignments with
these special characteristics from further consideration we
can concentrate on solid, non-perishable, durable, low-to—
medium value, regular-shaped, non-urgent, low-to-medium
density and low-to-medium mass goods that are usually
designated as general cargo.

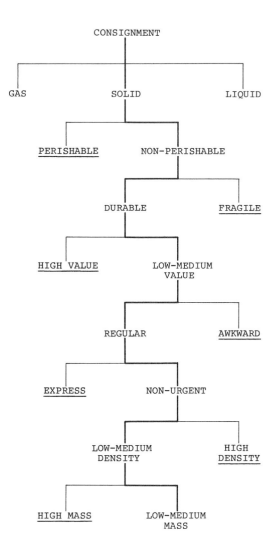

Figure 24.4   The Different Characteristics of a Consignment

*Consolidation and Distribution*

In examining general freight it is helpful to follow Smith and Hay (26) and recognise two distinctive components of the urban goods-movement process: consolidation and distribution. *Consolidation* embodies the accumulation and transfer (beginning with the initial shipper) of consignments into a successively larger load until some *limiting consignment size* (i.e. carrying unit) is reached. At this limit the load is transferred between the largest shipper and largest receiver without any reduction in the size of shipment. Then, *distribution*, the reverse of consolidation, occurs with the large shipment undergoing repeated division within the distributive chain until it reaches the ultimate consumer (consignee).

Consolidation and distribution can be elaborated further by indicating the two main directions of the transaction - manufacturer to consolidating wholesaler (or freight forwarder) whose activities produce a maximum consignment size (i.e. full unit load) and from this wholesaler (or one like him) to the final consumer in the distributive chain. The distinction between the two directions is highlighted by reference to Figure 24.5 in which the smallest manufacturer and smallest consumer are located near the right angle (X) of the triangle.

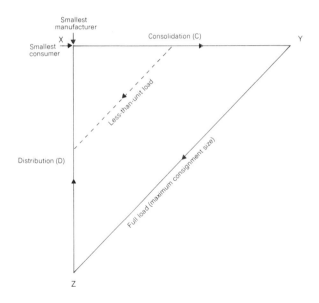

Figure 24.5  Diagrammatic Representation of Bulking, Distribution and Maximum Consignment Size

Movements from X to Y involve consignments of increasingly
greater size and represent bulking. Progression from Z to
X reflects distribution and implies successive breaking of
bulk. The limiting (consignment size is) represented by
the side YZ of the triangle ... Thus, given stages of
bulking and distribution may be characterised by identical
consignment sizes, but may differ in direction. Further,
(urban goods movement) does not necessarily involve this
progression up to the limiting (consignment size), then
down to the smallest consumer; lateral or near-lateral
exchanges (of less-than-unit-loads) may occur in the
bulking and distribution processes between participants
who' operate at the same scale. Diagrammatically, the
(less-than-unit-load) moves across the triangle ((26)pp.
122-3).

Smith and Hay have extended their schema with the addition
of imports, exports and large-scale producers and translated
it into a graphical model which, in turn, has been tailored
in Figure 24.6 to serve as the basis for discussing the
urban goods-movement process. Features of the model include
the small manufacturer Ms, the large manufacturer Ml, the
importing terminal Ti and the range $C_1$ to Cn which refers to
consolidation activities. A complementary scale $D_1$ to Dm
covers distribution activities and reference is also made to
the exporting terminal Te, retailer R and the household (or
institution) H. The transaction Cn to Dm is the limiting
consignment. These values of n and m, reflecting the number
of links in the consolidation and distribution chain, are of
critical importance in determining the nature of transport
services as instanced in the three hypotheses that can be
derived from the graphical model.

1. *Frequency of exchanges.* As trade occurs along $C_1$ to Cn
and Dm to $D_1$ or from any Ci to Dj it is expected that the
frequency of transactions between Ci and Dj will be a
function of the proximity to the diagonal as suggested for
one instance in Figure 24.7(a).

2. *Length of strands.* Assuming that the frequency distribution of different sized shippers is positively skewed and
that they are located in a fairly uniform fashion it is
anticipated that the distance from Cn to Dm will be longer
than that between $C_1$ and $D_1$ as indicated in Figure 24.7(b).

3. *Size of strands.* If it is assumed that the size of
consignment progressively increases from $C_1$ to Cn and progressively decreases from Dm to $D_1$ and the nature of any
transfer between Ci and Dj is dictated by their comparative
size-ranking it could be postulated that the limiting consignment size or full consignment load would occur in the
lower right-hand corner as shown in Figure 24.7(c).

Other hypotheses could be developed about the hierarchy of
shippers and receivers and strands but the size of consignment is particularly important as it has a direct bearing
on the type of transport required.

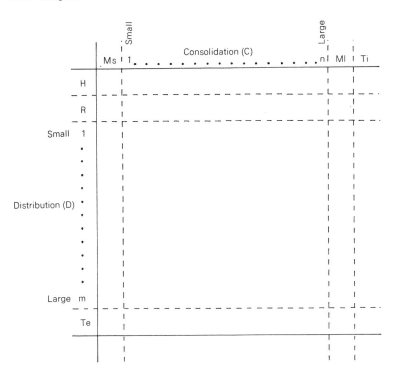

M = manufacturer; Ms = small manufacturer; Ml = large manufacturer

C = consolidating wholesale chain

T = terminal; Ti = importing terminal  Te = exporting terminal

D = distributive wholesale chain

R = retailer

H = household or institution

Cell entries are shipments of various sizes; small in upper left quadrant; large in lower right quadrant.

Figure 24.6  A Graphical Model of Urban Goods Movement
Adapted from Smith and Hay (26)

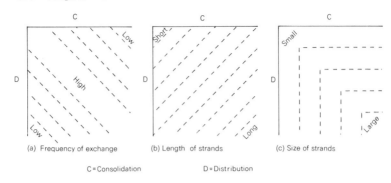

(a) Frequency of exchange      (b) Length of strands      (c) Size of strands

C = Consolidation            D = Distribution

Figure 24.7  Hypotheses Concerning Attributes of Strands
Developed from the Graphical Model derived from Smith and
Hay ((26)pp.125-6)

*Consignments*

It is helpful to identify four different categories of
consignment (24).

1  *Full unit loads*, averaging 15,000 kg in mass (weight) and
62m$^3$ in density, requiring fork lifts but no consolidation
facilities or intermediate drops, would occupy the capacity
of the carrying unit.  Such units may be either:

(a) semi-trailer (capacity 20 tonnes mass or density of 68m$^3$),

(b) jumbo road/rail containers (capacity 20 tonnes mass and
density of 74m$^3$),

(c) ISO (International Standards Organisation) type containers (20 ft x 8 ft x 8 ft) with a capacity of 17 tonnes mass
and density of 31m$^3$).

2  *Large part loads* are those consignments that use 50 per
cent or more of the capacity of the line-haul unit.  Averaging 10,000 kg in mass and 41m$^3$ in density, these loads would
be forklifted or manhandled onto a line-haul unit (with gates
and tarpaulins where necessary) and then returned to the
originating terminal to be topped up, or alternatively
proceed to another half-load pick-up.  Such loads generally
signify the use of a semi-trailer or jumbo road-rail container rather than an ISO container.

3  *Small part loads* or 'smalls' are those between one and
ten tonnes (i.e. less than half the carrying capacity of the
unit) and are normally carried from a consignor's premises
to a consolidating depot by table-top feeder trucks.

4  *Parcels* refer to loads less than one tonne in mass with
the majority of very small consignments (under 35 kg) being
handled by the specialist express companies.  These parcels,
averaging 35 kg and 0.1m$^3$ are manhandled, collected by parcel
van and require dock labour to sort them in terminals.

A shipper, therefore, with fifteen tonnes for pick-up for an

external destination is offering a full load.  However, a
shipper with fifteen tonnes for pick-up for three different
external locations is not offering a full load.  Where a consignment of ten or more tonnes of dense freight (e.g. lead
ingots) is involved, but is less than full mass capacity of
the line-haul unit, it will be routed back to the consolidation terminal to be topped up with high cubic freight (e.g.
tennis balls).  The marriage of these types of part-load
consignments is the essence of the consolidator's operation.
Small-part loads always incur handling costs at both despatch
and receiving terminals and are married in the consolidator's
terminal with other general cargo.

Once the type of consignment is known, therefore, it can
be matched against the fleet of line-haul trucks and trailers,
table (flat) tops and parcel vans.  When this information is
augmented by details of the origin and destination of the
consignment and the nominated line-haul mode the consignment
can be programmed into the pick-up and delivery schedule of
the vehicle fleet.

*Pick-up and Delivery*

As instanced by Jordan (11) and demonstrated in Figure 24.8
it is possible to recognise four basic patterns in the pick-
up and delivery of full load, part load and parcel consignments by line-haul trucks and trailers, table tops and parcel
vans.

1  *Diffuse location to discrete location* involving the
consolidation of consignments into large shipments as exemplified by the progressive collection of goods for delivery to
a transport terminal.
2  *Discrete location to discrete location* embracing a unique
movement of a consignment from one point to another as
reflected in the transfer of goods between a manufacturer and
an institution.
3  *Discrete location to diffuse location* encompassing the
distribution of consignments from a single receiver to
multiple receivers as exemplified by the delivery of goods
from a transport terminal to a warehouse.
4  *Diffuse locations to diffuse locations* covering a composite case involving both the consolidation of consignments
and their distribution.

An investigation of these activity sequences in urban goods
movement should complement Hensher's (6) investigation of the
urban travel patterns of individual people and assist us in
explaining where, when, how and why such behaviour occurs.
Yet, these sequences have been neglected because the preoccupation with modelling the flow of commercial-vehicle
movements on road networks has resulted in each link in the
chain being considered as a one-way trip from an origin zone
to a destination zone for a single purpose.  However, a
concern with activity sequences involving node visitations
(i.e. land-use configurations) and paths through the network
emphasise that an urban area provides a set of constraints
(and opportunities) to the movement of consignments through
space and time.  The constraints are either *spatial*, such as

those occasioned by the location of nodes and the friction of
distance between them, or *temporal*, such as those reflected,
for instance, in the varying operating hours of institutions.
Given these constraints the sequencing of consignments is
conceived as occurring within a three-dimensional space-time
framework in which vehicle activity is structured by the
obligation to call at certain nodes at particular times (4).

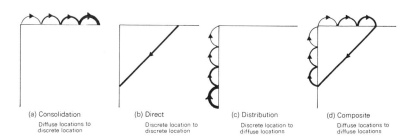

(a) Consolidation
Diffuse locations to
discrete location

(b) Direct
Discrete location to
discrete location

(c) Distribution
Discrete location to
diffuse locations

(d) Composite
Diffuse locations to
diffuse locations

Figure 24.8   Activity Sequences in the Pick-up and Delivery
of Urban Goods

The compulsory calls circumscribe the vehicle's potential
area of operation and are reflected in the division of table-
top trucks and parcel vans into *run trucks* and *floaters*.
Run trucks performing regular (daily/weekly) calls operate in
defined territories (with a driving stem linking terminal and
zone).   In contrast, floaters are used to mop up any overload
within the defined territories and accommodate late calls
that cannot be fitted into the run trucks schedule without
backtracking.   Floaters are also used to handle seasonal
peaks such as Christmas and Easter.   Thus, floaters permit a
measure of limited flexibility in the routine operation of
run trucks.   When changes in land use happen through, for
example, the opening of a new industrial estate or alterations occur in transport policy via, for instance, mass
(weight) limitations on trucks operating on particular
routes, adjustments to the territories of run trucks will be
necessary.

The recognition of the four basic activity sequences in
urban goods movement should, at least, lead to a more
sensitive modelling procedure in matching the load system
(consignments and vehicles) with the capacity of the road
system if this is still to be the main task in studying urban
goods movement.   Perhaps more importantly, if an interven-
tionist policy is sought, the set of activity sequences
should trigger an interest in the importance of allocating
time in urban goods movement and the need to comprehend the
spatial and temporal repercussions that it generates (cf.
(1)).

Before identifying and effecting a list of candidate
improvements, however, it is important to recognise that
their ultimate effectiveness depends on decision-makers and

planners becoming aware of the viewpoints of the communities
of interest from which the changes will be considered.  The
task of devising a system which ensures that the consignments
demanded by the consumer arrive at the right place at the
right time has to wait, therefore, until we further our
understanding of the urban goods process by specifying the
expected behaviour (or plans of action) associated with the
actors or communities of interest.  Attention is confined to
the task of identifying the actors and defining their usual
roles.

*Roles*

The task of identifying communities of interest within the
urban goods-movement process is difficult because socio-
economic organisations differ in the range of functions they
perform.  Railway authorities, for example, are responsible
for both vehicles and track whereas these items in road
transport come under different ownership.  This dilemma is
resolved by recognising that the urban goods-movement process
consists of a number of interlocking subsystems each revolv-
ing around the role performed by an actor at a particular
location or within a bounded area.  As a result the set of
interrelated, partly inconsistent functions performed by
individuals and organisations in the urban goods process are
divisible and can be allocated to one of six distinctive
roles: shipper/receiver, prime contractor, vehicle owner,
transport-terminal operator, highway authority, traveller,
and impactee.

    A glass manufacturing plant, therefore, with an ancillary
fleet of pick-up and delivery vehicles would be recognised as
fulfilling three roles - shipper/receiver, prime contractor
and vehicle owner.  Similarly, the functions performed by a
conglomerate transport company would have to be assigned to
prime contractor, vehicle owner, transport-terminal operator
and traveller.  It would be possible to examine intra-agency
conflict stemming from an organisation's involvement in a
multiplicity of roles and explore the substructure of roles
by examining the degree of co-operation between management
and employees in fulfilling a role's mission.  However,
attention is confined to the use of the role concept in
understanding the urban goods-movement process.  Hence,
interest is centred, with the aid of Table 24.1, on highlighting the distinctive attributes, objectives, constraints and
options of each of the six roles (cf. (20,22)).

1  *The shipper/receiver's* objective is to minimise perceived
costs in despatching/receiving goods by tailoring loading/un-
loading facilities and fostering labour relations to speed
pick-up/delivery of goods - a schizophrenic role in which
attention to loading often outstrips interest in unloading
because of the widespread belief that delays occasioned by
these inadequacies are not part of the cost of purchasing the
goods.  Inefficiencies can be resolved either by changes in
existing premises or by relocation.
2  *The prime contractor's* objective, functioning as a
transport broker in organising a range of different priced
services, is to maximise profits from a set of urban and

inter-urban services.

3  *The vehicle owner's* objective, operating either as a hire
and reward carrier, ancillary operator or owner driver, is to
maximise vehicle earnings by varying the scheduling and routing of line-haul and pick-up and delivery units in the short
term or the composition of the fleet in the long term within
the constraints imposed by the prime contractor.

4  *The transport-terminal operator's* objective, whether
public or private owner of rail, road, sea and air facilities,
is to maximise throughput; periodic congestion may be overcome either through software solutions such as scheduling the
arrival of vehicles, or hardware solutions such as building
new terminals or relocating activities in less congested
areas.

5  *The highway authority's* objective, encompassing national,
provincial and local variants, is not really specified but
could probably be couched in terms of net social benefits
although it is usually more concerned with capabilities than
objectives which can be changed by the use of traffic
management or extensions to the road system.

6  *The traveller's* objective is to move at the least perceived cost from one place to another to engage in trip-end
activity - a journey that may be frustrated by congestion.
While the action of the traveller is not unlike that of the
shipper in choosing between alternative modes to overcome the
separation of distance, the role is identified separately to
avoid discussing the urban goods-movement process in isolation from that involving the transport of people: pedestrians,
cyclists, car drivers, bus drivers, truck drivers and
passengers.

7  *The impactee's* objective, as representative of the community at large, is either to minimise disruption or maximise
benefits from changes to the transport system; an activity
associated with lobbying or demonstrations to emphasise a
sectional interest.

None of the plans of action or expected behaviour associated
with the specified roles in the urban goods-movement process
are sufficiently embracing to weld the disparate series of
participants into coherent organisation in such a way that
they co-operate in initiatives to improve economic efficiency
or promote equity.  Such action is the task of the decision-
makers.

*Political Decision-makers and the Transport Planner*

The decision-makers fulfil a dual function.  This embraces
the allocation of resources to different sectors of the
economy including transport and the establishment of operational norms (what, how, when and where) governing the
urban goods process in such a way that the participants are
sensitive to changing resources and standards for judging
services.  In fulfilling this task the decision-makers have,
according to Vickers ((27)pp.23-5), to promote a level of
satisfaction, either in terms valuable to themselves or by
those to whom they are accountable, while simultaneously
maintaining a dynamic balance in a budgetary sense between
other claims on scarce resources.  Where the state of affairs
in urban goods movement is out of line with expressed

TABLE 1  Roles in Urban Goods Movement Process

| | Shipper/receiver | Prime contractor | Vehicle owner | Terminal operator | Highway authority | Traveller | Impactee |
|---|---|---|---|---|---|---|---|
| *Variants* | Long haul<br>Short haul | Government<br>Private (including freight forwarder) | Ancillary<br>Hire and reward<br>Owner driver | Government<br>Private | National,<br>Provincial,<br>Local Governments | | |
| *Objectives* | Minimise perceived costs | Maximise volume | Maximise vehicle earnings | Maximise throughput | Maximise net social benefits | Minimise perceived costs | Minimise disruption<br>maximise benefits |
| *Constraints* | Labour relations<br>Loading/unloading facilities<br>Vehicle suitability<br>Access problems<br>Handling equipment | Nature of consignments | Spatial/temporal access<br>Terminal congestion<br>Highway congestion<br>Vehicle use<br>Security<br>Documentation<br>Regulation | Terminal congestion<br>Terminal site<br>Terminal location | Capacity<br>Desing of infrastructure<br>Location | Truck induced high congestion<br>Air pollution | Noise<br>Air pollution<br>Property severance<br>Community disruption<br>Safety |
| *Options* | Improvements *in situ*<br>Relocation | Type of service offered<br>Pricing | Vehicle scheduling and routing<br>Size and type of fleet | Scheduling vehicle arrival<br>Re-siting or relocation of buildings<br>New and different sized buildings | Change control signals<br>Widen roads<br>Construct new roads | Change time/route of journey<br>Change mode<br>Change destination | Adjustment *in situ* (e.g. modify perception insulation)<br>Appeals lobbying publicity<br>Migration |

*Source:*  Ref. 22.

objectives, the decision-maker performing the satisficing-
balancing role can intervene in the process to vary opportunities available to participants.  Such intervention
involves, according to Watson ((29)pp.7-8), finding policies
to eliminate or internalise negative externalities by transferring the costs back to the individual or organisation
creating the problem.  In fulfilling this task much emphasis
is placed by the decision-maker on advice from the transport
planner.

Traditionally, the transport planner, as a technical
adviser, has provided the decision-makers with value-loaded
recommendations from which courses of action have been taken
to which participants in the urban goods-movement process
have had to respond (a task often obscured by urban transport
planning having no government at a corresponding scale).
Reactions to the traditional system have prompted attempts,
as instanced by Voorhees (28), to recast the relationships
between the transport planning team, decision-makers and
participants in the urban goods-movement process.  In the new
system the transport planning team's responsibility is to
provide information on the cost, feasibility and impacts of
these alternatives both to participants in the urban goods
process and to decision-makers.  Particular emphasis is
placed on mutual assistance between actors and the transport-
planning team in specifying problems, alternative plans and
relevant evaluation issues.  The function of decision-makers
in the revised planning process is to develop policies and
take action on the basis of recommendations from role players
in the urban goods-movement process (which reflect their
respective value positions) and technical advice from the
transport-planning team.

The need for a forum to resolve apparently conflicting
interests in the urban goods-movement process could be met by
an activity structure/transport gaming-simulation as illustrated in a not too dissimilar context by the Planning
Research Centre (18).  Urban goods movement could be generated by the interaction of role players and the effect of the
lack of adequate information and intervention by political
decision-makers could be traced.  However, it is more
pertinent to take the suggested recasting of the function of
the transport-planning team a stage further by recommending
a planning approach that is more attuned to our level of
understanding of the urban goods-movement process.


PLANNING APPROACH

As it is difficult, given the present state of the art, to
unravel the complexity of the urban goods process, it is more
realistic to adopt a planning approach that investigates an
urban goods issue incrementally by building from limited
knowledge in a sub-optimisation process.  The incremental
planning approach is conceived by Etzioni (3) as being pre-
occupied with the search for realisable, short-range and
low-capital projects as a counter to the conventional planning approach, epitomised by the traditional transport study
framework, which generally restricts the policy options to
capital-intensive long-term recommendations.  In developing

a planning approach tailored to examining the urban goods-
movement process it seems appropriate to adopt a minimum-
regret strategy which follows Hensher (5) in abandoning
Etzioni's short-term/long-term dichotomy and assesses policy
options in terms of their certainty of 'success'.

Under the proposed planning approach options are graded
from the relatively certain, non-capital, flexible, short
gestation period projects through to the relatively uncertain,
capital-intensive, inflexible, long gestation period projects.
This grading system permits the adoption of a planning
approach which begins at the certain end of the uncertainty-
certainty continuum and works towards more general projects
involving greater uncertainty.  The application of such a
planning approach in a situation of uncertainty will emphasise realisable projects which make more efficient use of
existing technologies and promise early improvements rather
than major capital investments in plant, equipment or
technology.  However, the latter options are not excluded
provided they meet the criterion of relative certainty.

A framework incorporating the revised incremental approach
is adapted from the work of the Commonwealth Bureau of Roads
(2) in employing market-segmentation procedures to urban
travel and illustrated with reference to goods movement by
the target issue of relieving the congestion of transport
terminals (the costs of attempting to rectify these inadequacies should not be high and would almost certainly be
outweighed by the resulting benefits).  As outlined in
Figure 24.9 the task, after selecting the decision unit (in
this case the size of consignment), is to use the operational
unit (the consignment or where this is unavailable the commercial vehicle) as the basis for defining and measuring the
terminal's level of performance.  After the performance data
is collected decision rules have to be applied to rank order
terminals and identify the location of inefficient facilities.
Attention can then be given on the one hand to the cost of
operating and improving terminals, the collection of data
from interviews with terminal operators and the relation of
costs to a range of potential transport and non-transport
policy options and on the other hand to selecting shippers/
receivers and vehicle owners for interview, collecting data
from them and developing models to gain an insight into their
behavioural values.  From these twin sources an evaluation
framework has to be developed in which potential projects can
be identified in terms of issues raised by shippers/receivers,
prime contractors and vehicle owners, different types of
improvement and constraints.  Once the potential projects are
recognised they can be evaluated in terms of their economic,
marketing, social and practical applicability.  The final
project (or projects) can then be selected and, after
implementation, monitored.

Such a systematic approach is particularly apposite for
examining key issues in urban goods movement as it provides a
measure of flexibility in accordance with changing needs.  As
complete information on changing needs is difficult to obtain
in advance there is a high level of risk in allocating
resources to improvements on the basis of hazy planning and
implementation horizons.  The risk is minimised if we build

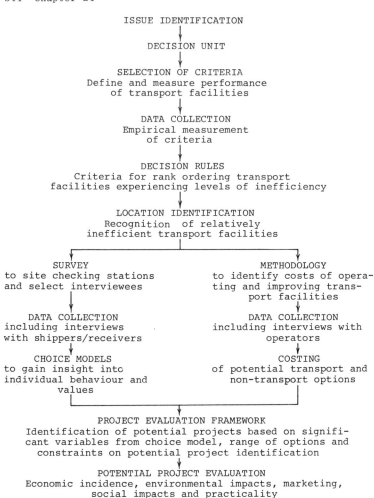

ISSUE IDENTIFICATION
↓
DECISION UNIT
↓
SELECTION OF CRITERIA
Define and measure performance
of transport facilities
↓
DATA COLLECTION
Empirical measurement
of criteria
↓
DECISION RULES
Criteria for rank ordering transport
facilities experiencing levels of inefficiency
↓
LOCATION IDENTIFICATION
Recognition  of relatively
inefficient transport facilities

SURVEY
to site checking stations
and select interviewees
↓
DATA COLLECTION
including interviews
with shippers/receivers
↓
CHOICE MODELS
to gain insight into
individual behaviour and
values

METHODOLOGY
to identify costs of operating and improving trans-
port facilities
↓
DATA COLLECTION
including interviews with
operators
↓
COSTING
of potential transport and
non-transport options

↓
PROJECT EVALUATION FRAMEWORK
Identification of potential projects based on signifi-
cant variables from choice model, range of options and
constraints on potential project identification
↓
POTENTIAL PROJECT EVALUATION
Economic incidence, environmental impacts, marketing,
social impacts and practicality
↓
SELECTION OF FINAL PROJECT(S)
↓
MONITORING

Figure 24.9  A Framework for Identifying Projects to Improve
the Efficiency of Transport Facilities adapted from Commonwealth Bureau of Roads (2)

up an understanding of the urban goods process by concentrating on using undertakings at the certain end of the spectrum
as demonstration projects.  Such experimentation enables us
to comprehend the pattern of changing needs over time as part
of the determination of long-term objectives.  Once the
nature of the deficiencies inherent in the urban goods
process are known we will, within the context of the revised
incremental approach, be in a better position to assess
policy options.

POLICY

As policies are conditioned by the objectives of decision-
makers it is important to state what should be their aims.
When these ends are known attention can be given to the
elements that can be controlled by decision-makers in the
urban goods-movement process.  Once the possible range of
changes has been canvassed, interest can be focused on the
set of instruments available to decision-makers for effecting
policy ends.

*Objectives*

The main objectives of an urban goods-movement policy are
directed towards the reduction of generalised social cost.
According to Hicks (8,9) generalised social cost in this
context incorporates transport operation costs, external
costs, community costs and urban structure costs.

    A reduction in *transport operation costs*, which are
essentially private costs incurred in either consolidating or
distributing goods, can be realised through the lowering of
either running costs or overheads (which affect the 'price'
to the user), the lessening of other distribution costs such
as packaging, storing and insuring goods for transport, and
the abating of 'quality' costs (or non-'quality' costs)
incurred by goods being lost, damaged or delayed in transit.
These reductions may be achieved through a more efficient use
of such existing resources as terminals, road space and
vehicles, particular attention being given to the selection
of more appropriate vehicle types, the consolidation of
shipments and routes, the management of modal interchanges
and the organisation of passenger-goods interfaces.  Such an
investigation is warranted because a principal share of
benefits in most major road investments is attributed to
goods vehicles.  Whether the high value placed on commercial
vehicle time (usually the driver's wage plus an additional
amount) should continue to be applied across all organisations, on the assumption that they can capture all possible
marginal productivity gains, must be questioned.

    If the aim is to reduce generalised social costs the
questioning must go beyond the private costs incurred in
transport costs and encompass a set of *external costs*.  These
costs include noise, air pollution and vibration stemming
from the actions and characteristics of trucks (17).  There
are, in addition, delays caused by trucks to other vehicles
and pedestrians, personal and property damage occasioned by

road accidents and psycho-social disturbances triggered in individuals through the close proximity of trucks. While the quantification of such indirect external costs is difficult compared with direct transport operation costs, any policy designed to reduce generalised social cost must endeavour to encompass the task.

The policy must also include *community costs* incurred by governments and their agencies in assuming responsibility for urban goods activities that are best performed by public authorities. These costs to the community are related to the provision and administration of rail and road facilities, the provision and operation of terminal facilities such as sea-ports and airports, the operation of government goods services such as waste disposal and post office mail services, and the functioning of regulatory and planning bodies. The list is not exhaustive but covers the most important community costs that have to be examined in reducing generalised social costs; these community costs are relatively easy to identify compared with urban structure costs.

Just as *urban structure costs* influence the nature of urban goods movement the converse also holds in that the costs of moving goods influences urban structure. The latter is reflected in the siting of factories adjacent to major arterial roads in the outer suburbs of cities; a phenomenon that may attract or repel residential development. Although our knowledge of the relationship between transport costs and urban structure is not very great the attempts to alter the nature of such costs by varying the range of urban structures does promise, at least, to reduce generalised social cost.

Whilst the isolation of transport operation costs, external costs, community costs and urban structure costs assists in revealing the character of generalised social cost these components overlap and are interrelated. The system effects of the urban goods-movement process are such that considerable care has to be undertaken to ensure, for instance, that a reduction in transport operation costs in one part of the system is not counteracted by an increase in another part. A consolidation terminal in the Central Business District (CBD) may, for example, make more efficient use of available truck capacity but may also necessitate shippers having larger warehouse space to offset less frequent deliveries. Such complex interrelationships prompt caution but should not divert decision-makers and planners from the challenge of discovering and implementing measures which reduce the total generalised cost of urban goods movement to the lowest possible level. The possibility of reducing the total amount of resources devoted to the urban goods task prompts an investigation of the elements that can be changed by decision-makers.

*Controllable Elements*

Policy-makers in the urban goods-movement process are able to alter the *variable constraints* in the decision-making environment. Some examples of this set of constraints are discussed with reference to three elements ((4)p.17).

1 *Socio-economic organisation (or activities)* which, for example, includes the variable conditions for entry into particular nodes (transport terminal, manufacturer, wholesaler, retailer, institution, household or waste disposal site).
2 *Technical aids for communication and physical movement* that bind the location of nodes together in space and over time.
3 *Locational structure (or activity structure)* which embraces the spatial pattern of nodes.

In discussing these elements it is important, as instanced in Table 24.2, to make a distinction between changes decision-makers can effect in the urban goods process itself and changes they can make in the urban goods environment. It should be highlighted that the response time necessary to effect changes in urban goods movement is less in the urban goods-movement process itself than in its environment. Within both sets of elements the response time increases progressively from activities, through technical aids, to activity structure. As a result the two sets of elements provide decision-makers with a hierarchy of variables.

The quickest way that decision-makers can influence the urban goods-movement process is by changing socio-economic activities by, for example, banning vehicles from designated residential streets. Changes via technical aids take longer to introduce as it may take several months, for instance, to supply a new vehicle type. Alteration to the physical structure by the introduction of such innovations as consolidation terminals for goods originating or terminating in a central business district involve an even longer gestation period and are a decidedly more risky undertaking according to de Neufville *et al.* (16) despite favourable studies to the contrary (see, for example, (13)).

TABLE 24.2  Elements amenable to varying degrees of control by decision-makers

| Element | Response | Changes in urban goods-movement process itself | Changes in the urban goods-movement environment |
|---|---|---|---|
| Activities | Low | e.g. vehicle bans | e.g. shoppers collecting and delivering own parcels |
| Technological aid (*inter alia*) transport | Medium | e.g. new vehicle types | e.g. substitution of telecommunications for goods movement |
| Activity structure | High | e.g. consolidation terminals | e.g. urban location |

Decision-makers intent on making changes to the urban goods-movement process via variations to its environment, say to reduce the need for urban goods movement, will find such policies take longer to implement and are less easy to control (8). The alteration of activities to induce the replacement of urban goods movement by passenger movement (i.e. moving people to the goods rather than *vice versa*) can only be achieved in a relatively few instances such as persuading shoppers to take their own parcels home (and may even increase the passenger problem). Opportunities for the replacement of urban goods movement by technical aids such as telecommunications are also limited. The reductions of the distances commodities have to travel by changing the activity structure of the urban area (i.e. by locating shippers/ receivers more optimally) may have promise for new cities as illustrated by the LUTSANC (Land Use and Transport Systems Alternatives for New Cities) Study (14). Although such policies may be feasible for new cities the investment and disruption involved in redesigning existing urban areas restrict their opportunities for anything other than minor or long-term changes. Such a conclusion prompts a review of the policy instruments that decision-makers have at their dis-posal for effecting more realisable changes.

*Policy Instruments*

Four main policy instruments for reducing the generalised cost of urban goods movement are available to decision-makers.

1 *Education programmes* which can be designed to disseminate information from research and development and alert all participants to the generalised social costs of urban goods movement.

2 *Financial incentives (or disincentives) embodied in taxes* which can be used to ensure that community costs are passed on to the participants in the urban goods-movement process or in fees which can be used in urban road pricing to offset congestion arising from the joint use of roads with passenger vehicles.*

3 *Regulations* that cover vehicle mass and design charac-teristics, land-use zoning, traffic control, noise, air pollution, building codes, safety and public ownership.

4 *Investment* which can range from expenditure on small demonstration projects through to spending on new cities.

These instruments will rarely involve a reduction in all elements of social cost simultaneously. There must almost always be a compromise in which certain costs are increased to gain an overall reduction in cost (8). Such trade-offs illustrate the basic nature of the conflict which has to be resolved by government policy.

In resolving the conflict it can be assumed that each role player in the urban goods-movement process is capable of more

---

*J.A. Gouldstone (Pers.comm.) is against taxing the vehicle operators. He refers to a speech by Sir Garfield Barwick made at a Liberal Convention in Brisbane 1976 which argues that the transporter uses the road 'to serve the people who want to buy or sell and it is their business to provide the road'.

or less assessing his needs and capabilities and implementing
activities to achieve his objectives.  At least there can be
no doubt that each actor *tries* to do so.  The actual goods-
movement process we observe in operation in any particular
urban area is simply the combined result of their efforts.
Only by the greatest of coincidences will these unco-
ordinated efforts result in an urban goods process which is
'desirable' from the wider community point of view.

It is this omnipresent conflict of interests and activities
which presents the major government policy problem.  Virtually whatever action the government takes, some actor
perceives his own interests to be harmed.  The temptation
exists for government to leave well enough alone, to treat
the problem as a 'private enterprise' issue and to refrain
from the equity judgements which intervention entails.  In
many cases this abstention may be perfectly defensible.  Yet,
without further knowledge, we cannot be sure how many areas
there are in which government activity would yield useful
results.  However, it is apparent that conflict between the
truck and the car, the truck and the environment, and the
truck and the pedestrian will be among them.

The natural hesitancy of government to intervene in the
urban goods-movement process indicates a risk of a weak-
handed policies approach, preferring to rely on education
programmes or present tax structures.  The most distasteful
weapons for improvement of the process are:

1 *Directives*.  Government edicts - rules, laws and
regulations - specifying that a part of the process should be
performed in a particular prescribed manner, or that it
should not be performed in a particular proscribed manner.
2 *Spending*.  Major or marginal improvements requiring some
government outlay.

The reason that the second weapon should be disliked is clear.
Few governments like spending money; fewer like wasting it,
and the policy advisers are not certain that the desired
results of the expenditure would be achieved - hence, the
focus on demonstration projects.

The reason for the dislike of the first weapon is probably
similarly related to the governmental fear of commitment in
an unchartered area of urban life.  We believe, however, that
it may be the most useful weapon in the armoury, and that
assistance in drawing up such directives should be a prime
aim of researchers in the area of urban goods movement.  Why,
for example, cannot researchers provide greater assistance in
promulgation of building codes and traffic laws which relate
to urban goods movement?

Regulatory devices can be most refined and accurate
weapons.  Greater knowledge of the urban freight process will
enable us to gauge their side-effects and, in particular, to
assess the magnitude and location of those elements of the
generalised social cost which will increase with the introduction of a new regulation.  Present-day technology is
entirely capable of producing freight vehicles which are
safer, more manoeuvrable, quieter or less pollutant.  It is

not difficult to formulate regulations which would compel
vehicle manufacturers to take greater advantage of this
technology.  The difficulty lies instead in assessing whether
the increased *transport operation costs* are outweighed by the
decreases in *external* and other costs.  Equally, the full
effects of the more adventurous weapons like road pricing and
public ownership are proportionally more difficult to
measure (19).

CONCLUSION

This study has abstracted urban goods movement from total
urban movement.  Even in isolation from movements of people
the internal physical and socio-economic organisation of the
urban goods-movement process is a highly complex activity.

We have seen that a better understanding of the physical
situation can be obtained by going beyond inter-zonal flows,
zonal origin and destination patterns, vehicle types, vehicle
numbers and vehicle utilisation.  This has been achieved
through the use of such concepts as linkages, strands and
activity sequences which relate land-use configurations and
consignment movements.  In discussing socio-economic organisation we can also go further than the recognition of public
ownership of road and rail track and private ownership of the
majority of road vehicles operating in an essentially free
market with firms of varying size, a fair degree of vertical
integration and some degree of horizontal integration between
urban and line-haul activities.  This has been attained by
examining interactions between different role players.  In
addition, we have identified the function of the political
decision-makers and transport planners, an appropriate
planning framework, a range of policy variables and a list of
policy instruments.  One way of integrating this catalogue of
ingredients would be to absorb them into a gaming-simulation
framework to further deepen our understanding of policy
trade-offs.

While the degree of aggregation of roles to operate gaming
simulation successfully may be too unrealistic for our needs
the mere attempt at integration should, at least, emphasise
that urban goods movement is not optimally organised from the
viewpoint of society as a whole.  Perfect competition within
the urban goods-movement process is hampered by inadequate
information flows between role players and the failure to
reflect noise, pollution and congestion in pricing.

At this stage we may have to accept much of the existing
organisation of the urban goods process with its lack of
perfect competition and absence of mechanisms for pricing
externalities and attempt to increase the efficiency of
operation through improved information flows.  In this way we
should, in time, be able to provide the information for making decisions on policies dependent on the pricing of road
use and external effects, regulations and public ownership.

ACKNOWLEDGEMENTS

The authors acknowledge comments on the original draft by
John Black (University of New South Wales), Joe Gouldstone
(TNT Management Services) and David Hensher (Macquarie
University). They are also grateful to Manlio Pancino,
Cartographic Section, Department of Human Geography, Research
School of Pacific Studies, The Australian National
University, Canberra, who drew the accompanying figures.

REFERENCES

1. Chapin, F.S. (1974) *Human Activity Patterns in the
City: Things People Do in Time and Space*, Wiley, New York.
2. Commonwealth Bureau of Roads (Hensher, D.A., R.A. Smith
and P.G. Hooper), 'An approach to developing transport
improvement proposals', *Commonwealth Bureau of Roads
Occasional Paper No. 2*, Commonwealth Bureau of Roads,
Melbourne (1976).
3. Etzioni, A., 'Mixed-scanning: a "third" approach to
decision-making', in *A Reader in Planning Theory*, A. Faludi
(ed.), Pergamon Press, Oxford, pp. 217-29 (1976).
4. Hägerstrand, T., 'The impact of transport on the
quality of life', in OECD, *Transport in the 1980-1990 Decade*,
Fifth International Symposium on Theory and Practice in
Transport Economics, Organisation for Economic Co-operation
and Development, Paris, pp. 1-51 (1974).
5. Hensher, D.A., 'Incremental planning and uncertainty',
in *Goods Movement and Goods Vehicles in Urban Areas*, K.W.
Ogden and S.K. Hicks (eds), Commonwealth Bureau of Roads,
Melbourne, pp. 100-5 (1975).
6. Hensher, D.A., 'The structure of journeys and nature of
travel patterns', *Environment and Planning A*, 14, pp. 65-72
(1976).
7. Hensher, D.A. and McLeod, P.B., 'Approaches to modelling consumer preferences and demand in transportation',
*Australian Transport Research Forum: Forum Papers Sydney
1975*, Australian Government Publishing Service, Canberra,
pp. 307-53 (1976).
8. Hicks, S.K., 'Urban goods movement: a political
economist's viewpoint', in *Goods Movement and Goods Vehicles
in Urban Areas*, K.W.Ogden and S.K. Hicks (eds), Commonwealth
Bureau of Roads, Melbourne, pp. 26-50 (1975).
9. Hicks, S.K., 'Urban freight', in *Urban Transport
Economics*, D.A. Hensher (ed.), Cambridge University Press,
Cambridge (1977).
10. Hutchinson, B.G., 'Estimating urban goods-movement
demands', *Transportation Research Record*, 496, pp. 68-79
(1974).
11. Jordan, F., 'Freight study progress report: background
to Newcastle and Wollongong freight studies', unpublished
note, Urban Transport Study Group of New South Wales, Sydney
(1975).
12. Kearney Management Consultants in association with
Voorhees, A.M. and Associates, Inc. (1975) 'Urban goods-
movement demonstration-project design: final report on phases
I and II - a primer on urban goods movement', unpublished
report to Urban Mass Transportation Administration, US
Department of Transportation, Washington.

13. McDermott, D.R., 'A simulation study examining the benefits of an urban consolidation terminal', *Traffic Engineering*, 46, pp. 26-32 (1976).

14. Maunsell and Partners Pty Ltd. (1975) *New Structures for Australian Cities: Main Report*, Cities Commission, Canberra.

15. Meyburg, A.H. and Stopher, P.R., 'A framework for the analysis of demand for urban goods movements', *Transportation Research Record*, 496, pp. 68-79 (1974).

16. de Neufville, R. Wilson, N.H.M. and Fuertes, L., 'Consolidation for urban goods movements: a critical analysis', in *Transportation Facilities Workshop: Passenger, Freight and Parking*, W.H. Kraft and J.J. Fruin (eds), American Society of Civil Engineers, pp. 426-51 (1975).

17. Organisation for Economic Co-operation and Development (1973) *Road Research: Effects of Traffic and Roads on the Environment in Urban Areas*, Road Research Group, Paris.

18. Planning Research Centre (1974) *An Interactive Land Use/Transport Gaming Simulation: Mate:2 Operational Manual*, Department of Town and Country Planning, University of Sydney, Sydney.

19. Reid, K., 'The economics of urban goods movement', in *Goods Movement and Goods Vehicles in Urban Areas*, K.L. Ogden and S.K. Hicks (eds), Commonwealth Bureau of Roads, Melbourne, pp. 106-10 (1975).

20. Rimmer, P.J., 'Transport decision-making and its spatial repercussions: shooting an arrow at a moving target', *Monash Publications in Geography*, 11, Melbourne (1974); revised version in *Urban Transport Economics*, D.A. Hensher (ed.), Cambridge University Press, Cambridge (1977).

21. Rimmer, P.J. (1975) *Urban Goods Movement: A Review of the Magnitude of the Urban Freight Issue, Recommendations on Very Certain Policy Options and the Way Ahead*, Report to the Commonwealth Bureau of Roads, Commonwealth Bureau of Roads, Melbourne.

22. Rimmer, P.J., 'A conceptual framework for examining urban goods movement', in *Goods Movements and Goods Vehicles in Urban Areas*, K.W. Ogden and S.K. Hicks (eds), Commonwealth Bureau of Roads, Melbourne, pp. 15-25 (1975).

23. Rimmer, P.J., 'Urban goods and commercial vehicle movements in Sydney', unpublished paper delivered at the Seminar on Future Transport Strategies for Metropolitan Sydney, School of Transportation and Traffic Engineering, The University of New South Wales, 15-16 June 1976.

24. Rimmer, P.J., 'Freight forwarding: changes in structure, conduct and performance', in *Economics of the Australian Service Sector*, K.A. Tucker (ed.), Croom Helm, London (1977).

25. Smith, K.J.G., 'Urban retail deliveries and the problem of heavy goods vehicles in towns - a review of some of the literature', *Transport Operations Research Group Working Paper No. 12*, Department of Civil Engineering, Division of Transport Engineering, School of Advanced Studies in Applied Science, University of Newcastle-upon-Tyne (1975).

26. Smith, R.H.T. and Hay, A.M., 'A theory of the spatial structure of internal trade in underdeveloped countries', *Geographical Analysis*, 1, pp. 121-35 (1969).

27. Vickers, Sir G. (1965) *The Art of Judgement: A Study of Policy Making*, Chapman and Hall, London.

28. Voorhees, A.M. and Associates Inc., 'A prototypical study design', unpublished report to the *Commonwealth Bureau of Roads*, Melbourne (1975).

29. Watson, P.L. (1975) *Urban Goods Movement: A Disaggregate Approach*, Lexington Books, Lexington.

Chapter 25

URBAN GOODS MOVEMENT: BEHAVIOURAL DEMAND-FORECASTING
PROCEDURES

Paul O. Roberts and Brian C. Kullman

INTRODUCTION

Knowledge of the specific industry structure and commodity
flows in a region is a first big step towards a full understanding of a regional economy and its freight-transport
requirements. However, transport planners are necessarily
concerned with the choice of transport mode as well as with
total volumes of commodity flow. In consequence of this, it
is necessary to extend traditional industry/commodity
analysis one step further to include the capability of predicting a shipper's or receiver's mode choice. This in turn
requires that models of freight mode choice be developed and
combined with knowledge of commodity flows to and from the
region.

The model of freight demand described here represents a
disaggregate approach to the problem of predicting shipper
behaviour. The choice of a disaggregate model is based on
many researchers' experience with alternative approaches (8).
Use of aggregate models, that is, models that draw on
aggregations of data for estimation and forecasting, have not
been successful in practice, and theory has shown that by
suppressing deviations from the mean of an aggregate data
point, models will be biased. Therefore, it is necessary to
develop demand-forecasting models that are at least as
disaggregate as the industry/commodity statistics described
above. Recently disaggregate freight-demand models have been
specified that take advantage of (1) detailed information on
commodities, industries, and carriers' services, and (2)
advances in probabilistic-choice modelling in the urban
passenger-transport field. These new disaggregate modelling
approaches predict the joint origin, shipment size, frequency,
and mode-choice behaviour of the shipper of specific commodities and are used as the basis of the approach described
here.

This chapter has three broad goals. First, it develops a
methodology that can be used to establish the industry/com-
modity data base that is essential to all commodity-based
freight-transport analysis. Second, it develops procedures
that can be used to specify and estimate disaggregate demand
models of freight mode choice, using the industry/commodity
data developed above as well as carrier attributes, i.e.
transport levels of service for different modes. Third, it
describes how the industry/commodity data base can be

553

combined with the probabilistic choice models to create a
powerful tool for analysis.


## INDUSTRY/COMMODITY ANALYSIS OF INPUTS TO THE REGION

Determining the volume of cargo that will flow in a given
market is the starting point for any quantitative analysis of
freight policy.  This is true whether the issues being
addressed are those of a carrier that would like to consider
changes in a particular pricing policy or a government
attempting to justify the capital expenditure for a new
facility.  The flow of cargo in a given market is simply the
sum of individual shipper decisions.  The transport decisions
of individual shippers are, in turn, conditioned by the
specific inputs and outputs of each of the production processes involved.  Understanding the consequences of a
proposed policy change or investment in the transport system
thus depends upon knowledge of the industry (and population)
structure of the region.

This section focuses on the *inputs* of intermediate goods
to producing industries and the *inputs* of both intermediate
and final consumption products to the wholesale and retail
industries.  The reason for focusing on inputs as opposed to
outputs is three-fold: (1) the choice of the origin and
transport mode for a shipment is most frequently controlled
by the buyer (receiver) of the goods; (2) it is more difficult to obtain information on consumption than production
of commodities since government statistics are compiled on
the basis of production; and (3) the process described can be
applied to the outputs of industry in a very simple manner.
In short, if inputs can be properly modelled, output modelling is a much easier task and will be straightforward.

There are two issues generally involved in predicting the
annual inputs or usage rates of commodities.  The first issue
is the development of a methodology and the supporting
information to make prediction possible.  The second issue is
how to develop, handle and store the information required in
the prediction process.  The second issue, although it is
straightforward, is a non-trivial task.  There are, after
all, a large number of industries and firm sizes and they use
a large variety of goods as inputs to their production
processes.

This section presents and demonstrates a method for
representing the urban-industrial structure and predicting
the annual usage rate of commodities by industry and by firm
size.  It also describes a scheme for grouping industries in
sectors and treating the inter-industry linkages between
these sectors logically.  Predictions are made in such a way
that they will be consistent between commodities and between
industries.  Computations are simple and feasible using data
which exist for any area in the USA.  The approach provides a
powerful tool for attacking the problems of urban and regional structure, industry interaction and various other areas as
well as serving as a basic input to freight-demand forecasting.

*The Basic Approach*

The basic approach advanced here starts by estimating the
output level of a firm, then, by using input/output relationships, the inputs required to produce a given output are
developed. The input/output coefficients represent the
dollar volume of inputs purchased from a particular industry,
i, required to produce one dollar's worth of output of
industry, j. When multiplied by the dollar output of an
industry, this coefficient gives the dollar volume of purchases of inputs required from each industry, i.

Within each of these buying industries, there are any
number of different size firms. In view of the fact that
the most important characteristic of a firm's logistic
process in a particular commodity is the usage rate, the
determination of the size distribution of buying firms
becomes important. Obtaining firm-size distributions is not
difficult to do in the United States since it is comparatively easy to determine the number of firms of a given
industry in each of several size categories working with
statistics from the County Business Patterns published by
the US Bureau of the Census.

*The Urban Industrial Structure*

A simple way to represent the industrial structure of a
given geographical unit, a region, an SMSA, or a city, is to
use an industry/firm-size distribution matrix (1) as illustrated in Figure 25.1. This matrix shows the number of firms
in each industry by each size category. Firms are classified
into eight size categories based on employment:

| Firm-size class | Number of employees |
|---|---|
| 1 | 1-3 |
| 2 | 4-7 |
| 3 | 8-19 |
| 4 | 20-49 |
| 5 | 50-99 |
| 6 | 100-249 |
| 7 | 250-499 |
| 8 | 500+ |

Firm-size information is taken from the County Business
Patterns. These data are given for each 4-digit SIC code.
To simplify processing, it is necessary to aggregate and
convert the 4-digit SIC code to a new industry code with
69 classifications to match available input-output information. Basically, the new 69 industry code is a modified
2-digit input-output code, with 62 processing industries and
7 final-demand sectors. See Table 25.1 for the list of
these codes.

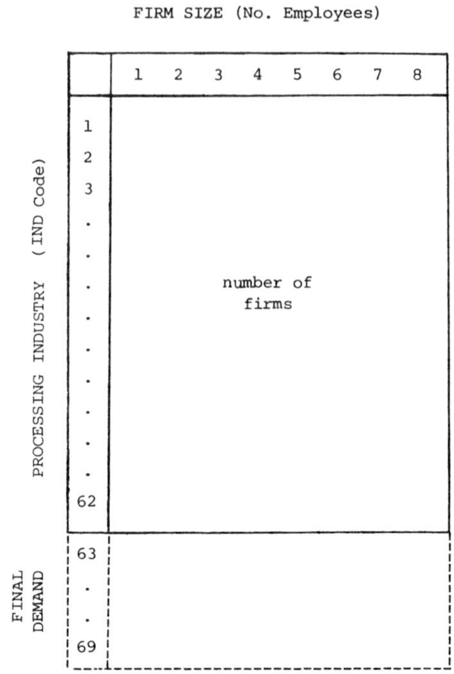

FIRM SIZE (No. Employees)

Figure 25.1   Urban Industry/Firm-size Distribution Matrix

TABLE 25.1   Industry codes

1 Agricultural, forestry and fisheries services

2 Construction

3 Ordnance and accessories

4 Food and kindred products

5 Tobacco manufactures

6 Fabrics

7 Textile products

8 Apparel

9 Miscellaneous textile products

10 Lumber and wood products

11 Wooden containers

12 Household furniture

13 Other furniture

14 Paper and allied products

15 Paperboard containers

16 Printing and publishing

17 Chemicals and selected products

18 Plastics and synthetics

19 Drugs and cosmetics

20 Paint and allied products

21 Petroleum and related industries

22 Rubber and miscellaneous plastics

23 Leather tanning products

24 Footwear and leather products

25 Glass and glass products

26 Stone and clay products

27 Primary iron and steel products

28 Primary non-ferrous manufactures

29 Metal containers

30 Fabricated metal products

31 Screw machine products, etc.

32 Other fabricated metal products

33 Engines and turbines

34 Farm machines and equipment

35 Construction and material handling machines and equipment

36 Metalwork machinery

37 Special machines and equipment

38 General machines and equipment

39 Machine shop products

40 Office and computation machines

41 Service industry machines

42 Electric transmission equipment

43 Household appliances

44 Electric lighting equipment

45 Radio, TV, etc. equipment

46 Electronic components

47 Miscellaneous electrical machines

48 Motor vehicles equipment

49 Aircraft and parts

50 Other transport equipment

51 Professional and science equipment

52 Medical and photographic equipment

53 Miscellaneous manufactures

54 Agricultures, forestry and fisheries

55 Mining

56 Retail trade

57 Wholesale trade

58 Transportation and warehousing

59 Communications and public utilities

60 Service industries, excluding auto repair

61 Auto repair and services

62 Government enterprises

63 Personal consumption expenditures

64 Government purchases

65 Gross private fixed capital formation

66 Net inventory change

67 Imports

68 Net exports

69 Other final demand sectors

---

With the industry/firm-size matrix in hand, the composition and level of output of the region represented can be estimated by multiplying information on the productivity per employee from the Census of Manufactures for each category: the number of firms in an industry of a given size times the average number of employees in that size category times the average productivity per employee in that industry becomes the estimated total dollar output of all the firms in the given industry/size category. This can be stated mathematically as:

$$\hat{X}_{IND,SIZE} = {}^{NFS}IND,SIZE * {}^{AVEMPL}SIZE * {}^{PVEM}IND \qquad (1)$$

where:

$\hat{X}_{IND,SIZE}$ = the product value of the output for all firms in an industry/firm-size category

$NFS_{IND,SIZE}$ = the number of firms in the given industry/firm-size category

$AVEMPL_{SIZE}$ = the number of employees per firm of the given size category

$PVEM_{IND}$ = the productivity per employee in the given industry

In estimating the average employment for each size firm using the County Business Patterns, special attention must be paid to the size category with an employment of 500 or greater. Firms in this category are big firms in which the number of employees could go from 500 to several thousand. One way to solve the problem is to estimate the employment size of this category backward, that is, to estimate the employees of the first seven categories, then subtract this subtotal from the total number of employees in the industry. The Census of Manufactures provides additional detail for firms with 500+ employees (500-999, 1000-2499, 2500+).

An argument could be raised against using the average productivity per employee for all size categories within an industry. To use average productivity per employee for all size categories assumes constant economies of scale in the industry. Economies of scale can be reflected in the model, if more detailed information on productivity by firm-size categories can be obtained and manipulated. The Census of Manufactures has the productivity information by different firm sizes; the classification by size is very similar to, but not identical with, the County Business Patterns.

*Input-Output Relationships*

The input-output table shows how the output of each industry is distributed among other industries and sectors of the economy, and how much input to that industry from other industries is required to produce that output. Before going into detail on how the coefficients of an input-output table are used to predict the usage of various commodities, it is worthwhile to undertake a quick review of the make-up of the table and the interpretations involved (3). An input-output table is usually a square column by row matrix, each element representing the sales from one industry or sector of the economy to another as shown in Figure 25.2. As it is usually organised, the upper left-hand corner of the table, labelled the processing sector, is the portion of the input-output table that contains transactions between the industries producing goods and services. Among them are agriculture, various manufacturing industries, transport, communications and other utilities, wholesale and retail trade, and other service industries. The rows at the bottom of the table are set off under the heading 'payments sector'. This sector includes gross inventory depletion, depreciation allowances, households, etc. The right-hand side of the table is called

the final-demand sector.  It consists of personal consumption,
investment, inventory and export.

| | Agri. | Mnfg. | Trans. | Comm. | Util. | Whole. | Retail | | Pers. | Invs. | Inv. | Export |
|---|---|---|---|---|---|---|---|---|---|---|---|---|
| Agri. | | | | | | | | | | | | |
| Mnfg. | | | | | | | | | | | | |
| Trans. | | The Processing Sector | | | | | | | Final Demand Sector | | | |
| Comm. | | | | | | | | | | | | |
| Util. | | | | | | | | | | | | |
| Whole. | | | | | | | | | | | | |
| Retail | | | | | | | | | | | | |
| Inv. Dep. | | | | | | | | | | | | |
| Dep. Allow. | | Payment Sector | | | | | | | | | | |
| Households | | | | | | | | | | | | |

Figure 25.2   Input-Output Transactions Table

The entries in an input-output transactions matrix show
commodity flows between industries in monetary units.  It is
possible to construct other forms of this table which are
more useful for present purposes.  By dividing the inputs of
a given industry by its total output, a matrix of *technical
coefficients* can be obtained.  An input-output technical-
coefficient table shows the amount of input required from the
industries in each row to produce one dollar's worth of the
output of an industry in a given column.  However, one must
be careful in interpreting the coefficients, especially those
for service industries.  They do not show the total product
associated with the industry in question, but merely the
amount used in the production process.  For the service
industries, this can be misleading.

For the purpose of illustration, let us pick two columns
and two rows of a technical-coefficient table, one row and
column representing a manufacturing industry and the other a
service industry as given in Figure 25.3.  The technical
coefficient $C_1$ simply shows the amount of lumber (in dollars)
required to produce one dollar's worth of plywood.  Coefficient $C_2$ gives the amount spent on transport of inputs
required to produce one dollar's worth of plywood.  The
transport costs of shipping the lumber to the plywood factory
are part of the transport mentioned above.  Coefficient $C_3$
shows the amount of lumber required by the retail industry to
produce a dollar's worth of retail output.  The consumption
of lumber by the retail industry for its own use is *not* the

amount of lumber carried by the retailer for the purpose of
selling to customers.  That portion which is sold in retail
does not appear in the technical-coefficients matrix at all,
but is found in the final demand portion of the I/O table.
In studying the transport decision of a retailer or wholesaler, the total demand both for internal use and for selling
to customers are of interest.  For many products, the retail
demand which is not shown in the input-output table at all is
even more important than intermediate sales to industry.  A
strategy has been devised to develop these retail and wholesale demands.  Since this is a key point of the model, it
will be discussed in detail later.

Processing Industries

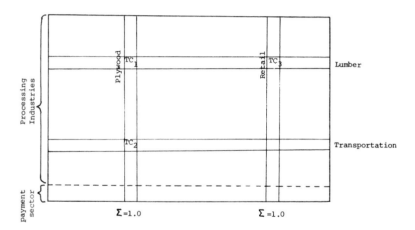

Figure 25.3   Input-Output Technical-coefficient Table

The matrix of technical coefficients thus treats the service
industries, transport and wholesale and retail trade as
separate industries with commodities valued free-on-board
(f.o.b.) at the factory.  If delivered prices were used
instead, with transport, wholesale and retail costs included
in the cost of the good rather than as separate industries,
an entirely different input-output matrix would be produced.
The element-by-element difference between the 'producer'
matrix as the first is known, and the 'delivered-price'
matrix, produces a matrix of 'mark-ups'.  The vectors of
transport mark-up as well as wholesale and retail mark-up can
be tabulated separately in individual matrices.  The matrices,
called the transport, wholesale and retail mark-up matrix
respectively, are in the same format as the coefficient table,
and show the transport (wholesale, retail) mark-up required
for shipping (selling) the commodity from a row industry to

a column industry. As shown later, these mark-up matrices are useful in investigating the spatial flow of the commodities.

The input-output information described here is based on the 1976 Bureau of Economic Analysis (BEA) input-output table, which has 484 rows and 494 columns. For use by urban planners, the original 'square' input-output table can be reconstructed to convert it to a 'rectangular' table, such that each column industry represents a 2-digit level of aggregation of the buying industries as described earlier, while the row industries (which represent the sellers in an urban area) are kept as disaggregate as possible. The original 485 rows are similarly collapsed to 355 rows and matched to STCC 4-digit commodities. All told, five separate compatible 355   69 input-output tables can be developed from the BEA 485   494 matrix, namely:

1   Direct transactions
2   Technical coefficients
3   Transport mark-ups
4   Wholesale mark-ups
5   Retail mark-ups

*Industry Demand Prediction*

With these data bases, the annual demand for all firms in an area can be predicted by multiplying the industry/firm-size output matrix of the production area by the input-output technical-coefficient matrix to get the demands for 4-digit STCC commodities by the firms in each industry/firm-size category. Mathematically, this is

$$A^K_{IND,SIZE} = \hat{X}_{IND,SIZE} * a_{K,IND} \tag{2}$$

where:

$A^K_{IND,SIZE}$ = demand for commodity K of all firms in an industry/firm-size category

$K$ = 4-digit STCC commodities

$a_{K,IND}$ = technical coefficients

The demand by an individual firm in a firm-size class is derived simply by dividing the estimated total demand for that firm-size class by the number of firms in the industry/firm-size category:

$$DEMD^K_{IND,SIZE} = A^K_{IND,SIZE}/NFS_{IND,SIZE} \tag{3}$$

where:

$DEMD^K_{IND,SIZE}$ = annual demands for commodity K by an individual firm in the industry/firm-size category IND,SIZE

The result of the calculation is a three-dimensional
62 × 8 × 351 table of information on the demand of each
commodity by each firm size in each industry.

*Personal and Government Consumption*

As indicated above, however, the work is not yet finished.
The goods consumed by processing industries usually represent
only a portion of total consumption. The proportion of the
consumption between processing industries and final demand
sectors varies from commodity to commodity, depending on
whether the commodity being addressed is ordinarily used as
an intermediate good or as a final good. Overall, commodities consumed by processing industries amount to only about
40 per cent of total consumption.

In predicting the consumption by final demand of a
production area, two most important sectors are included, i.e.
expenditures for personal consumption and government purchases. These two sectors represent more than 90 per cent of
total final-demand consumption. Export is also one of the
final-demand sectors. However, it is not of interest here
since it is the transport decision about the *input* to a firm,
not the *output*, that is of interest. Investment and inventory are not included either, since they are not only small,
but they are aggregations over all firms and all industries
and there is no way to classify them by firm size or
industry. They are not significant in terms of quantity, if
one compares them to personal and/or government consumption.

The personal and government consumption of a production
area is estimated by applying a regional-share ratio to
national final demands, or by computing a per capita figure.
This per capita figure for personal consumption is then
applied to the population of the area on the assumption that
consumption per capita is the same for all urban areas. The
per capita consumption could be different from urban to
rural; however, the computations can be elaborated further to
take these factors into consideration, provided that the data
are available. In addition, a ratio based on regional
factors such as government employment, etc., can also be
derived to estimate the regional government consumption.

The estimates of final personal and government demand,
once available, can be added to the intermediate demands
required by industry to produce their output. The result is
the *total input of each commodity used by population, government, and industry within the urban area.*

## FORMULATION OF A DISAGGREGATE FREIGHT-DEMAND MODEL

The disaggregate data base of inputs and outputs developed
from the industry/firm-size analysis requires a comparable
disaggregate model of shipper behaviour. This section
develops a general specification for a disaggregate freight-
demand model and then specialises for the case of the
urban-regional planner.

The impetus for a disaggregate approach to freight-demand modelling originated in the urban passenger-transport field, where recent breakthroughs in the modelling of individual travel behaviour have occurred. Urban transport researchers, after years of frustration with aggregate models of travel behaviour, concluded that the choice of transport mode is intertwined with choices of work place, residential location and car ownership, as well as household characteristics such as family size, income and number of workers. The response to this condition was to develop models (or model systems) that would make, for the individual household, probability estimates of choosing each of the possible combinations of mode to work, work place, residential location, etc., based on household characteristics and transport levels of service.

In the freight field, a similar situation exists. The shipper must make three simultaneous choices: where to buy, how much to ship, and by what mode (or carrier), based on annual requirements for a commodity, storage, ordering and other costs, the levels of service offered by competing modes, the price of the commodity quoted at different origins, and characteristics of the commodity (shelf life, value, packaging, special handling requirements, etc.). As in the passenger transport field, it is conceptually appropriate to formulate a choice model which assigns probabilities to combinations of shipper alternatives (origin, shipment size and mode) based on shipper, transport-mode and commodity attributes [6].

The mathematical form most frequently specified for disaggregate choice models is the multinomial-logit function of the form:

$$P(\underline{X}|\underline{A}) = \frac{e^{U(\underline{A})}}{\sum_{\ell} e^{U(\underline{A})}} \qquad (4)$$

$X$ = vector of choice combinations
$A$ = vector of attributes
$P$ = probability of choosing a particular combination $X*$
$U$ = utility function based on all the attributes

For the case of freight-demand prediction, this general form specialises as follows:

$$p^k(i,mq|ALTS) = \frac{e^{U(T,C,M,R)}}{\sum_{\ell} e^{U(T,C,M,R)}} \qquad (5)$$

where:

$U(T,C,M,R)$ = the utility function of the receiver
$k$ = commodity index
$i$ = supply (origin) point
$mq$ = mode/shipment-size combination
$ALTS$ = alternatives available to the receiver
$U$ = utility function
$T$ = transport attributes
$C$ = commodity attributes
$M$ = market attributes
$R$ = receiver attributes

Figure 25.4 defines the T,C,M and R variables that could enter the utility function.

Transport Attributes

W = wait time (days)

T = transit time (days)

R = reliability (days)

L = loss and damage (unitless, $0 \leq L \leq 1$)

$ = freight rate ($/lb)

C = special charges ($/lb)

Commodity Attributes

V = value ($/lb)

D = density (lb/ft$^3$)

S = shelf life (days)

Market Attributes

P = relative price (unitless)

O = ownership (binary 0-1)

Receiver Attributes

A = annual use rate (lbs/yr)

M = mixed order (unitless $0 \leq M \leq 1$)

$S^1$ = seasonal purchase (unitless, $0 \leq S^1 \leq 1$)

Q = shipment size (lbs)

U = reliability-of-use rate (days)

G = guarantee of availability (unitless %)

Figure 25.4   Variables that can Enter a Utility Function

The task of the freight-transport analyst is quite clearly the *specification* and *estimation* of the utility function U(T,C,M,R). While several specifications for U(T,C,M,R) are possible, they are all estimated using maximum-likelihood techniques.

Specifying the utility function of the receiver can be done in light of the logistics process he/she is trying to manage. Basically, total costs consist of purchase costs plus logistics costs, as follows (4):

Total Costs = purchase cost + order cost and handling cost + transport cost + capital carrying and storage cost + stockout cost

The utility function of the shipper is developed by combining the variables previously specified with appropriate parameters as shown in Figure 25.5. By careful specification, the parameters can be presented in such a way that they can be interpreted as constants, interest rates, elasticities or dimensionless. This would allow the estimated models to be checked for reasonableness and extended to other environments where estimation is not practical. It would even allow the model to be used without estimation should that be necessary.

The utility function in Figure 25.5 is based on the 'classic' calculation of logistics costs, with parameters taking on values that represent various aspects of a shipper's/receiver's cost structure. The equation is constructed to yield a disutility measure in units of $/lb, i.e. the *total* cost of the commodity from time of purchase until the time of consumption. Each cost element in Figure 25.5 is discussed below.

$$U(R,C,M,R) = \text{purchase cost} + \text{order and handling cost} +$$
$$\text{transport cost} + \text{capital carrying and storage}$$
$$\text{cost} + \text{stockout cost}$$

## Purchase Cost

cost to buyer

$$\alpha_1 P \cdot V$$

## Order and Handling Cost

set up charge

$$\alpha_2 \, 1/Q \cdot M$$

## Transport Cost

| capital cost in transit | loss & damage | perishability |
|---|---|---|

$$\alpha_3 \frac{W+T}{365} \cdot P \cdot V \cdot 0 + \alpha_4 (L \cdot P \cdot V) 0 + \alpha_5 (\frac{W+T}{S})^n \cdot P \cdot V \cdot 0$$

transport charges

$$+ \alpha_6 (\$+C) \, 0 \cdot M/Q$$

## Capital Carrying Cost and Storage Cost

$$\alpha_7 (\frac{Q}{2A} + \frac{R+U}{365}) \cdot V \cdot P \cdot S^1 + \alpha_8 (\frac{Q}{2A} + \frac{R+U}{365}) \cdot S^1/D$$

## Stockout Cost

stockout

$$\alpha_9 (1-G) \cdot (1/Q)$$

Figure 25.5  Specification of the Utility Function

*Purchase cost.* P, the relative price at each origin, is multiplied by V, the value-per-pound delivered, to obtain a local price. Both P and $\alpha_1$ are dimensionless, and $\alpha_1$ should be equal to unity.

*Order and handling cost.* The cost of the personnel and paperwork required to order each shipment, where $1/Q = F/A$. The term M varies from 0 to 1.0 and is the per cent by weight of this commodity in a mixed shipment. Mixed shipments in which more than one commodity is involved but where the sum

of all commodities is less than a full truckload, are transported under the freight rate applicable to the highest-rated
commodity applied to the combined shipment weight. Both
transport costs (see $\alpha_6$ below) and order costs must be
apportioned over all items in the shipment. The term $\alpha_2$ will
be the cost of placing each order.

*Transport cost.* The first term represents the capital
carrying costs of the goods while in transit; O is a 0-1
variable which signifies ownership. If the receiver is
buying the goods free-on-board at the origin, O takes on the
value of 1. If the receiver is buying f.o.b. at the destination, the shipper bears these charges and O = 0. The
parameter $\alpha_3$ is simply the cost of capital for the receiver.

The loss and damage term hinges on the transport attribute
L which is derived from a separate model and represents the
fraction of units totally destroyed. The cost of damaged
units is subsumed in L. The parameter $\alpha_3$ should calibrate
to unity.

The perishability term has the units of \$/lb, and $\alpha_5$ is
consequently expected to calibrate to unity. The parameter
n on the term $[(W+T/S)^n$ is designed to modify the influence
of this term and must be specified exogenously. For example,
if the commodity is fresh fruit, n may lie in the range $0 \leq n \leq 1$
to reflect the fact that as W + T approaches S, there will be
a significant loss due to spoilage. Alternatively, if the
commodity does have a finite shelf life but does not lose
value until W + T is very close to S, then n will take on a
positive value greater than 1. The determination of a proper
value for n will be left to the judgement of the analyst and
the experience of the traffic managers who assist in data
preparation.

Transport charges are the freight rate per pound times the
binary variables Q to indicate who pays the freight and the
mixed shipment variable M described above. Dividing by Q
distributes the freight charges on a per pound basis. The
parameter $\alpha_6$ should calibrate to unity.

*Capital carrying and storage cost.* The first term is the
cost of the merchandise while in the receiver's warehouse
prior to consumption. The term Q/2 is the average level of
stock on hand (exclusive of safety stock) and the term
(R + U)/365 represents the safety stock required to protect
from transit-time unreliability and usage-rate unreliability.
The variable R is the number of days beyond the mean transit
time in which there is a probability G that the shipment will
arrive, given a constant rate of daily commodity use. The
variable U represents the variation in the rate of use of
stock and is measured by the standard deviation of the
(presumed normally distributed) use rate. The term $S^1$
represents the influence of seasonality and the fraction of
the year that the item is held in inventory. Hence $0 \leq S^1 \leq 1$.
Note that for an item which is specially ordered as needed
(i.e. not held in inventory) $S^1$ = 0 as would be expected.
For an item used for a production run which lasts only four
months, $S^1$ = 0.33. The parameter $\alpha_7$ should calibrate to
the cost of capital.

The second term is an expression for the costs of warehousing and represents the average amount of goods on hand $(Q/2A + (R+U)/365) \cdot S^1$ times the reciprocal of the density. This expression is valid for either bulk or packaged commodities. The term $\alpha_8$ will be the cost of storage per *cubic* foot. (Note that warehouse costs are normally calculated on a per square foot basis so that stacking height of cartons becomes key. However, it is too difficult to generalise about this variable, so it was not put in the model.) Thus, the parameter $\alpha_8$ for packaged commodities will pick up the influence of both warehouse costs per square foot and stacking height which will increase the variance of this parameter estimate.

*Stockout costs.* The final term is a representation of the cost per pound of stocking out, which occurs with probability $1 - G$ each time there is a shipment (F times per year) and is distributed over the annual use A, where $F/A = 1/Q$. The parameter $\alpha_9$ is a measure of the cost of each incidence of stockout. The variable G is measured in terms of the probability that the shipment arrives in time to prevent a stockout and will typically range from 0.90 to 1.0.

This brief discussion of logistics costs is in no way intended to substitute for a careful reading of any of the available texts on physical-distribution management. It is intended only to provide an overview of the way that logistics terms might enter the model. In a given situation, the regional planner would probably have to add, delete or modify terms to meet local data requirements.

The four attribute vectors (T,C,M,R) can be developed in the following way. *Receiver attributes* are determined through exercise of the industry/firm-size table and input-output model coefficients.

$$A^K_{IND,SIZE} = a_{K,IND} \cdot \hat{X}_{IND,SIZE} \qquad (6)$$

where:

$A^K_{IND,SIZE}$ = annual usage rate of commodity K by firm of size class SIZE in industry IND

$a_{K,IND}$ = input-output coefficients

$\hat{X}_{IND,SIZE}$ = average output of all firms of industry IND and size class SIZE

Note that the annual usage rate developed by this procedure will only be as detailed as the input-output table used. It is possible that regional planners may have access to input-output tables that differ in degree of detail from national tables which are typically at the 3-4-digit SIC. Also, the result is in dollars. To convert to physical units, the result may be divided by the value per pound, using the commodity attribute described below. Other receiver attributes, such as available facilities, whether the commodity is used as an intermediate or final good, and whether the receiver uses mixed orders are a function of the receiving industry. Once the receiving industry is known, these inputs can be determined quickly.

*Transport level-of-service variables* can be developed for a given situation using three separate classes of level-of-service models by mode.  These models, developed by the Freight Transport Group at MIT are (5):

1  Waiting and transit time and reliability models
2  Loss and damage models
3  Freight-rate estimation models

Each is described in detail in the references; however, a brief description is given here.

The waiting, transit time and reliability models predict time distributions for waiting plus transit from origin to destination as a function of number and type of terminals and the line-haul distance, speed and frequency of service between terminals for each of the modes as given in Figure 25.6.  Since the principal cause of delay in the system is that which occurs at terminals, this approach has produced very good comparisons with observed travel-time distributions measured in the real world (2).  The probability of delay at a given type of terminal is represented by a cumulative function of the time available between arrival and the next regularly scheduled departure.

Loss and damage by mode is a function of the commodity attributes and the particular transport mode under consideration.  The models used are simple regression models based on the experience record of the mode.

Freight rate estimation using a model is essential because of the complexity of the commodity rate structure and its huge size (more than 3 trillion separate commodity tariff rates are on file in the US).  The model uses the various commodity attributes as input (i.e. density, value per pound, shelf life, etc.) and the distance and shipment size by mode. The regression models have been developed using actual waybill or freight-bill information for the various modes. Though point estimates of the freight rate are produced, it is possible, using the error distribution produced by the regression, to predict the distribution of likely freight rates if this becomes desirable.

*The commodity attributes*, the third class of variables, are available for 1200 5-digit STCC commodities from the MIT Commodity Attribute File (7).  This file uses the Standard Transport Commodity Classification (STCC) code at the five-digit level to record the following information in machine readable form:

1  STCC code number
2  35 digits of description
3  Wholesale value per pound ($/lb)
4  Density (lbs/ft$^3$)
5  Shelf life (weeks)
6  State (solid, liquid, gas, particulate)
7  Environmental protection required (frozen, temperature, pressure, shock)

Given the commodity, this information can be made available quickly.

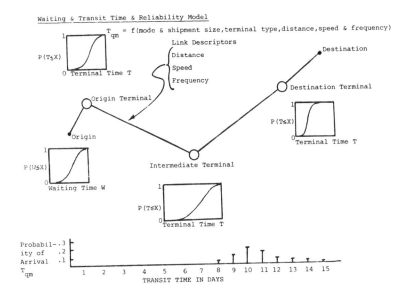

Waiting & Transit Time & Reliability Model

$T_{qm}$ = f(mode & shipment size, terminal type, distance, speed & frequency)

Loss and Damage Model

$L^k_{mq}$ = f(Mode & Shipment Size, Commodity Attributes)

Tariff Rate Estimation Model

$R^k_{mq}$ = f(Transport Attributes, Commodity Attributes, Mode & Shipment Size, Distance)

Fig. 6  FEATURES OF THE LEVEL OF SERVICE MODELS

*Market attributes* are more difficult to secure. The most
important is price. The price variable is designed as a
*relative price*, which when multiplied by the wholsesale value
per pound becomes a local price. Using relative price
enables differential model prices to reflect the spatial
distribution in prices in the input. Even wholesale and
retail mark-ups can be simulated. Price and availability
information can be obtained from the Office of Business
Economics, the Agricultural Marketing Service, and the Bureau
of Mines of the US Government. Or, this data may be fur-
nished by a macro-economic model used in conjunction with the
study. Data on ownership and facilities is normally a
function of the industry from which the commodity of interest
is drawn. For example, the food industry normally sells its
products with cost, insurance and freight (CIF) delivered
prices and has rail sidings available for loading rail cars.

## Model Outputs

The output of the model is the probability that a particular
receiving firm located in the region will secure its input
from origin i in shipment size q by mode m. When this
probability is multiplied by the annual use rate, $A^k_{IND,SIZE}$
calculated from the industry/firm-size analysis previously
described, the result is the commodity k moving from each of
the known producing regions to the mode choice/shipment size
for each region for the firm under consideration.

$$V^K_{i,qm,IND,SIZE} = p^k(qm|i) \cdot A^k_{IND,SIZE} \tag{7}$$

where:

$p^k(qm|i)$ = probability of shipping commodity k by mode m
at shipment size q from origin i

$A^k_{IND,SIZE}$ = annual use rate of commodity k by industry IND
of size SIZE

When summed over all firms and firm sizes in the region, the
results can be stored in a single three-dimensional table
(Figure 25.7). If the commodity moving is of no particular
interest, then the result should be presented using only two
dimensions.

$$V_{i,qm} = \sum_k V^k_{i,qm} . \tag{8}$$

A third possible approach is to summarise flows by major
commodity grouping or segment. In this case:

$$V^{kseg}_{i,qm} = \sum_{k \in kseg} V^i_{i,qm} \tag{9}$$

kseg = aggregation of individual commodities k.

This minimises the extent of the third analysis (k at 5
digits would be approximately 1,200).

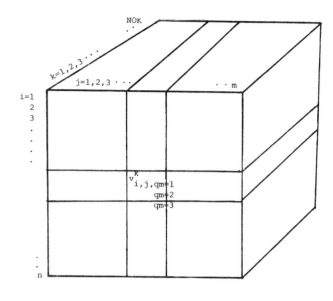

$$v^k_{i,j,qm} = P^k(i,qm \mid j) \cdot A^k_{IND}$$

Figure 25.7  The Output of the Model

Obviously, the computations to be performed are voluminous if
the number of commodities is large and the number of origin
regions is extensive.  The process described above is, in
fact, merely an enumeration.  The best dimensions for sampling appear to be those concerning industry, IND, and firm
size, SIZE.  However, the commodity k, also looks as though
it should be sampled, particularly because of its size.  On
the other hand, the dimension i, covering each origin region
is a good choice for enumeration since there is considerable
reason to preserve spatial detail.

A random sampling process can be set up for selecting the
representative sample so that the commodity k, to be used is
selected based on the relative size of the $K^{th}$ row of the
input-output table.  Finally, a firm size is selected based
on the relative size of the appropriate row in the industry/
firm-size matrix.  Each such randomly-selected point is used
for computation until a sufficiently large sample is available that the total, by mode and shipment size from the
different supply points, can be adjusted until it equals the
total volume flowing.

COLLECTION OF ESTIMATION AND FORECASTING DATA

As stated in the previous section, it is recommended that a
highly disaggregate data set of shipper behaviour be collected for the purpose of estimating a disaggregate freight-
demand model even though comparably disaggregate data will
not generally be available for exercising of the model.  The
reasons previously cited concern: (1) the ability to check
the importance of market attributes on mode choice and choice
of origin; (2) the ability to perform more detailed analyses
if desired; and (3) the capability of utilising generally-
available market-attribute data, if such became available in
the future.

Given the need for a highly-detailed calibration data set,
four options can be identified for its collection:

1  use of trade associations
2  use of a broadly-distributed, single-observation
   questionnaire
3  use of individual establishment interviews
4  use of carrier waybills with establishment interviews

Each needs some elaboration.

*Option 1: Use of trade associations.*  Many manufacturers have
state or regional trade associations that can provide information that may be valuable.  For example, state associations
of manufacturers frequently publish directories that give the
names, addresses, SIC code and sales volume or work force
size of every manufacturer in a state.  Beyond this, however,
such associations can provide little in the way of hard data.

*Option 2: Use of a broadly-distributed, single-observation
questionnaire.*  This approach is designed to overcome two
basic concerns of many managers: (1) businesses do not like
to reveal the names of their customers; and (2) the development of data takes time and effort.  The use of a broadly-
distributed, single-observation questionnaire would reduce
the resistance of managers to contributing data since it
deals with suppliers rather than customers and since it asks
for only *one* observation ... clearly, not enough to reveal
any secrets of company operations.  Since one observation
would also not require the commitment of company resources,
it should therefore not require high-level approval.

The basic approach would be to send a carefully designed
questionnaire that is interesting and easy to complete to an
individual such as the traffic manager, or purchasing agent.
By using traffic or purchasing associations, a basic file of
individual names could be developed easily and with the
backing of the association, a cover letter designed to appeal
to the individual's association loyalties.  The questionnaire
would request information on the firm's attributes, the
annual use of the product shipped according to the waybill
sampled, the places where the product could be obtained and
the relative price at each location.  The questionnaire would
be followed up with a telephone call to improve the response
rate.

*Option 3: Use of individual establishment interviews.* The individual establishment interviews to be performed would be selected at random from a list of approximately 20 cities and towns chosen to be representative of the variety of modes, products and conditions in the US as a whole. By grouping the observations in specific destinations, the cost of the field interviews and the difficulty of developing transport level-of-service information for each observation would be reduced. Interviews would be arranged ahead of time and the information recorded on forms much like the questionnaires used in Option 2, but without the need for an approved questionnaire.

Physically going out and talking to a sample of traffic managers or purchasing agents is a potential vehicle for collecting disaggregate data. The choice of firms to be interviewed would be based on the known or presumed contribution of the firm to the regional economy and its contribution to freight demand. To develop a full data set in this manner would require up to 100 interviews, with each interview yielding approximately 20 data points. The interviews themselves would take at least one day each and could go on longer if the party interviewed is interrupted (which is likely) or does not have needed information readily available. The interview itself would be structured so that a minimum amount of information would be required for each observation. The process would naturally generate much additional information about a firm's purchasing or shipping practices that may be important but which may not be generalised readily for the purposes of modelling.

*Option 4: Use of shipper/carrier waybills and limited establishment interviews.* A more discriminating approach to the establishment—interview process is a two-stage approach for the securing of a large sample of waybills and a follow-up interview on a subset of the waybills to obtain additional data. The concept derives from the desire to collect a broadly-based data set on origin, destination, freight rates, mode choice and commodity, while digging deeper into a smaller data set to obtain market, shipper and receiver attributes. Selecting waybills as the basis for interviews is expected to make the interview process much more efficient and could conceivably eliminate the need for site visits. Since the option has as its second stage an interview (by phone or in person) with manufacturing establishments, it is convenient to begin the collection of waybill data there as well. Shippers may also feel more secure about disclosing information on their own operations than carriers will feel about disclosing their customers' activities.

*Recommended option.* Without further analysis it is possible to recommend Option 4 as the preferred course of action for obtaining a calibration data set. It provides the most complete information and yields two data sets that are independently valuable. It also provides for a much more efficient use of both the regional planner's time and the manufacturer's time.

The information that is desired for estimation consists of the following:

1   annual volume of commodity purchased (at 4- or 5-digit
    level)
2   freight rates of competing modes from chosen origin
3   freight-service levels of competing modes from chosen
    origin
4   why a particular carrier/routing was chosen
5   estimate of the cost to the firm of a stockout of the
    commodity
6   what other options were available for the acquisition of
    the desired item, in particular, availability in other
    cities or from local wholesaler
7   indication of whether the point of purchase was condi-
    tioned by levels of service of available transport modes
    (as well as price of commodity)
8   whether commodity in question is typically purchased
    f.o.b. origin or destination or whether this is nego-
    tiated case by case.

The above information, when added to the data already known
from the waybill or freight bill and supplemented by commodity and transport attributes as described above, provides
the necessary data set to estimate a disaggregate demand
model.

*Collection of forecasting data.* As stated in the previous
section, a disaggregate freight-demand model can be used by a
regional planner in a restricted fashion by assuming that the
choice of an inbound shipment origin is already given. This
precludes for general forecasting purposes the need to gather
commodity origin, but requires other assumptions as to the
mix of commodities flowing in from each origin. Similarly,
outbound shipments can have an assumed destination, leaving
the shipment-size/mode-choice combination up to the model.

The first part of this chapter described in detail a
process that could be used to analyse the annual commodity
inputs and outputs, which are key pieces of information
required for forecasting purposes. The next part described
how estimates for the commodity and transport-attribute
vectors could be made and how probability estimates and
annual industry-use rates could be combined to forecast the
volume of commodity k shipped from an (known) origin i to the
region in question by mode m and shipment size q:

$$V^k_{i,qm} = \sum_{IND} \sum_{SIZE} P^k(qm \mid i) \cdot A^k_{IND,SIZE} \tag{10}$$

Note that the forecasting data base is by design specialised
to the case of shipments to and from a specific region.
However, the estimated model is not region specific and is *in
theory* transportable to new environments. Whether the
estimated model is *in practice* transportable is a function of
how completely the utility function describes all factors
that influence freight demand. Missing variables could bias
the model or otherwise reduce its transportability.

*Application of the technique.* The methodologies for the
estimation of disaggregate freight-demand models and for the
development of a forecasting data base have many applications
for urban and regional planning. Presented here are several

hypothetical cases where planners could use the data bases
and models to answer questions that arise in urban/regional
planning.

*Example 1: Estimate commodity flows to and from a proposed
industrial park.*  By obtaining from the developer the kind of
firms expected as tenants and the number of employees, the
industry/firm-size technique could be used to determine what
annual freight movements by commodity would be.  If it
appeared that the commodities were subject to intermodal
competition, it would be further possible to assume the
likely points of purchasing and sales and use the probability
choice model to determine the expected mode/shipment-size
choice for competitive commodities.  The results could be
compared to the capacity/utilisation of present modal facilities to determine whether additional investment would be
required.

A variant of this same technique could be used as an aid
in zoning.  The capacity of existing transport infrastructure
could be matched against the freight (and passenger) requirements imposed by various levels and kinds of development on
commercially/industrially-zoned property.  This information
could be used as a check on the property of various land-use
classifications or as the basis for forecasting transport-
investment requirements.

*Example 2: Analyse regional freight flows.*  The industry/
firm-size analysis alone will allow a regional planner to
describe total freight-transport requirements for a region,
with commodity detail available.  Projections of future
regional growth by SIC category could be translated into
commodity-specific freight requirements which, when combined
with estimates of passenger-transport requirements and
through traffic, would provide an aggregate analysis of
regional-transport requirements.

*Example 3: Analyse O-D or corridor flows.*  If a planner
wanted more detailed information on the freight traffic
moving along a corridor or between the region and another
SMSA, the choice model would allow the analysis to proceed at
the level of detail for which data are available.  For
example, if the commodity mix is important, using the two-
digit control totals to develop O-D or corridor estimates is
possible.  If questions of market attributes or transport
policy/level of service are most important, the analyst would
be required to develop market-attribute information for
competing origins or destinations and apply the full
(unrestricted) choice model.  For example, if the impact of
allowing double trailers on a highway system were of interest,
an O-D analysis of particular transport-cost sensitive commodities could be used to determine whether new points of
supply (or sales) would emerge in response to an altered
transport-cost structure.  The same approach could be used to
determine the consequences of improving modal service levels
through changes in operating policies or through new investment.  A case in point would be the impending investment in
railroad roadbeds and yard facilities.

*Example 4: Application to intra-regional flows.* The methodology developed could easily be applied to analyse certain
traffic flows that fall under the heading of urban goods
movement. The marketing/firm-size analysis, which has
previously always been based on the county level, could be
refined to some smaller geographic unit through use of more
detailed information on firm location. This detail could be
obtained from trade associations, industrial directories, or
Dun and Bradstreet market directories. Geocoding could be
based on zip codes or street addresses. Care would have to
be taken to ensure that the location of the plan is the same
as the corporate address. One company may have multiple
establishments within an SMSA.

Freight moving in or out of the region must, of course,
have a component of its route within the region where it
mixes with purely local traffic. Knowledge of the specific
commodity inputs and outputs at the traffic-zone level could
be matched to O-D flows to determine what local requirements
for inter-regional trips exist.

Although the model and analysis framework described here
have only been applied in a limited way to date, the expectation is that data for estimation and testing of the disaggregate demand model will soon be available. A US
Department of Transport sponsored data collection and
estimation effort currently being undertaken by the MIT
Center for Transport Studies has already begun and preliminary results are expected to be available by late 1977.

REFERENCES

1. Chiang, Y.S. and Roberts, P.O. (1976) *Representing
Industry Population Structure for Estimating Freight Flows*,
Center for Transportation Studies Report 76-8, Massachusetts
Institute of Technology, Cambridge, Massachusetts.
2. Martland, C.D., 'Improving railroad reliability: a case
study of the southern railway', unpublished paper,
Massachusetts Institute of Technology, Cambridge,
Massachusetts (1975).
3. Miernyk, H. (1965) *The Elements of Input-Output
Analysis*, Random House, New York.
4. Roberts, P.O., 'The logistics management process as a
model of freight traffic demand', *International Symposium on
Freight Traffic Models*, Amsterdam, Holland (1971).
5. Roberts, P.O. (1975) *Transport in Intercity Markets: An
Overview of the Physical Distribution System*, Center for
Transportation Studies Report 75-17 (Task II Report),
Massachusetts Institute of Technology, Cambridge,
Massachusetts.
6. Roberts, P.O. (1976) *Forecasting Freight Flows using a
Disaggregate Freight Demand Model*, Center for Transportation
Studies Report 76-1, Massachusetts Institute of Technology,
Cambridge, Massachusetts.
7. Samuelson, R.D. and Roberts, P.O. (1975) *A Commodity
Attribute Data File for Area Freight Transportation Studies*,
Center for Transportation Studies Report 75-20, Massachusetts
Institute of Technology, Cambridge, Massachusetts.
8. Terzier, M.N., Ben-Akiva, M. and Roberts, P.O. (1975)
*Freight Demand Modelling: A Policy Sensitive Approach*, Center
for Transportation Studies Report 75-6, Massachusetts
Institute of Technology, Cambridge, Massachusetts.

Chapter 26

INDICATORS FOR URBAN COMMODITY MOVEMENTS

Marc R. Wigan

SUMMARY

Behavioural or modelling analyses of freight issues require
considerable care in specification and testing, as the complexity of freight issues is such that the links between the
observable consequences and the policy or operational changes
require stringent control of the model or behavioural
analysis *ab initio*.

The number of different dimensions of urban commodity
movements and impacts is such that cross-sectional surveys
are inadequate to keep track of geographical, spatial and
demand shifts, and the employment, environmental, economic
and social impacts. Links between data sources, policy
issues and indicators for monitoring purposes are set up.

The limitations of cross-sectional methods are covered in
detail, and land-use influences and changes in operational
use are also pinpointed through the comprehensive and diverse
data available for the Greater London Council area.

INTRODUCTION

The objective of this chapter is to identify a useful set of
indicators to represent the balance and changes in level of
urban commodity movements in a manner appropriate to monitoring local and national transport policy. Commodities move
only as a result of economic activity and their movement is a
basic function of commerce and industry. The flexibility of
the freight transport system raises numerous difficulties in
setting up simple indicators, as the range of possible
reactions to changes in the economic or the regulatory
climate is so extensive and complex. Transport is an integral part of the production, distribution and sales operations of all kinds of enterprise, and the accommodation to
economic or regulatory changes may be reached by adjustments
to all parts of the system. As an extremely diverse range of
data sources is fundamental to this issue, the Greater London
Council area of the UK, covering the physical (if not the
full economic) area of London, has been used due to the good
statistical bases available.

In this chapter we start by defining the different types
of variable involved in commodity movements, and discuss some
standardised systems for describing commodities and commodity

movements.  Some ambiguities inherent in specifying commodity
movements are discussed using data from the 1971-2 GLTS Road
Freight Survey (3), and by comparisons between Swindon (9)
and GLC commodity tonnages.  The different interacting
interests of mode of transport, land use, commodity and
vehicle used, each offer pertinent monitoring statistics
which are appropriate to different aspects of freight-
movement policy.  At a more detailed level, the road vehicle
used, the commodity concerned, the drop sizes made, and the
characteristics of the vehicle trips provides the framework
for a similar but more detailed monitoring.

The types of data required for such monitoring indicators
are not widely available at a suitable level of detail, or
with sufficient coverage to provide any trends over time.  By
considering the most probable policy-sensitive areas for
different sizes of urban area it is possible to set up
appropriate indicators at different levels of utility and
detail, and link these to the types of resources needed to
service them.

THE COMPONENTS OF COMMODITY-MOVEMENT SYSTEMS

Commodity movements are an integral part of the production,
distribution, storage and sales elements of manufacturing,
servicing and retailing industries.  Movements arise only
when the sources of supply and the points of use are
physically separated.  While transport costs are an important
element in the location of production and processing plants
with respect to their markets, for most industries labour
availability and capital-investment factors substantially
outweigh the transport element.  For those industries wherein
transport costs dominate the industry, and demand for the
commodities produced is widely dispersed, planning controls
on the production point (i.e. detailed land-use controls) can
have a considerable effect on commodity movements.

Aggregate extraction in the London green belt, for
example, is becoming very much more difficult to obtain
agreement with planning authorities over, with the result
that aggregates are progressively being trucked in from
greater and greater distances, as the demand at the point of
consumption is not directly affected by regulation of the
points of production.  Commodity tonnages delivered are a
function of the land use, and not of the freight system.  It
is therefore a question of monitoring production and
consumption-activity levels, rather than freight-traffic
movements.  The manner in which these goods are delivered,
and the distances from which they come, are separate issues
which are directly related to freight-traffic movements.

Monitoring commodity tonnages delivered will not necessarily give an accurate view of the tonnages moved, as a
single consignment of goods may be handled several times
through warehouses and wholesalers of different kinds before
consumption.  If packaging, part-processing and assembly are
included as 'handling', the tonnage of goods may appear
several times in the form of deliveries and ton km.  It is
clear that such processes are best regarded as points where

the 'commodity' changes its nature, and its category.  The
problems of specifying a suitable commodity-classification
system are considered later, but it is clear that the
boundaries between commodity classifications may be crossed
several times before a single shipment of raw materials
reaches the consumer.

Movements through warehousing and open-storage areas
require special care in interpretation.  Depending on the
styles of distribution and production involved, the same
final tonnage delivered may give rise to several times this
amount delivered at different commercial land uses en route.
This will not only inflate the 'tonnage delivered' figures
without change of the final tonnage required for consumption
solely as a result of distribution system changes, but also
alter the ton km generated by this same tonnage.  Depending
on the efficiency of the distribution system, ton km could
rise or fall, but it is clear that ton km is affected by
distribution and freight system changes even in the absence
of any alteration in demand level or location.  Commodity
movements can therefore be affected by warehousing, wholesaling, depot and modal interchange alterations: these
changes will alter both ton km and tonnage-delivered figures,
without necessarily affecting the quantity consumed or
delivered at the consumption or production point.

In so far as production and consumption are linked locally,
commodity movement could be affected by industrial location
policies, but as a substantial proportion of production
plants forms part of a national pattern of specialised
processing and manufacturing even within a single firm, the
distribution system itself - with its depots and trunking
infrastructure - is of greater practical significance.  The
functional elements in distribution which can be affected by
planning are warehousing and depot function, scale and
location.  The whole nature of distribution industries - such
as those at wholesaling, steel stockholding, specialist
distribution companies, cash and carry and other warehouse
operators - is to link manufacturer and consumer by buffering
the stocks inevitably held in transit between production and
consumption, and to introduce economies of scale by concentration into the distribution process by channelling trade.
It is, therefore, of special importance to planning authorities to understand the nature and function of these channels
of trade, as they are the most sensitive to planning
authority intervention in the complex business of freight
movement.

Table 26.1 summarises the parameters describing commodity
movement as the commodity moves from production to consumption point.  These parameters, used to monitor commodity
movements, require further refinement before they could
usefully be used to monitor policies or complement other
transport and economic indicators.  The wide variety of
changes which can occur, without actually influencing con-
sumption-demand level - or even location - significantly,
means that the policies to be monitored must be brought into
the discussion before an adequate set of indicators can be
proposed.

TABLE 26.1  Parameters of commodity movement systems

| Consumption | Distribution | | Production |
| | Movement | Wholesale transfer storage | |
| --- | --- | --- | --- |
| Deliveries | Vehicles | Vehicles (In) | Deliveries (In) |
| Tonnage | Ton km | Vehicles (Out) | Deliveries (Out) |
| Commodities (In) | Commodities | Commidities | Commodities (In) |
| Commodities (Out) | Mode | Mode (In) | Commodities (Out) |
| Land Use Activity | System | Mode (Out) | Vehicles (In) |
| Vehicles (In) | (e.g. container) | Processing | Vehicles (Out) |
| | | | Land Use Activity |

POLICIES AND PARAMETERS

General indicators can be drawn from the elements of Table
26.1, each bearing on a different policy area.

1  Commodity consumption        area measure of economic
   (demand) levels:             activity.

2  Commodity movement levels:   measure the level of
                                activity of distribution.

3  Land-use activity levels:    measure economic activity
                                and consumption, production,
                                distribution balance.

4  Vehicle-activity levels      measure environmental impact
   (movement):                  potential of commodity
                                movement.

5  Vehicles activity levels     measure activity and
   (location):                  environmental levels.

6  Vehicle activity levels      measure parking and loading
   (stopping):                  conflict levels and the
                                balance between dropsize,
                                environment and vehicle
                                presence.

                                measure potential conflict
                                between freight and other
                                road users.

It is immediately evident that any indicators suitable for a
rural area will also be useful for urban areas, and that the
rural indicators would most appropriately be drawn from
groups 2,4,5 and 6.  Urban areas contain a greater concentration and variety of activities, but the smaller the area, in
general the simpler the freight system and the easier it is
to obtain information on it.  For the larger, metropolitan
districts, the system is substantially more complicated and
very much more difficult to monitor.  This might appear to
suggest that useful indicators for small urban areas would
present fewer problems in their establishment and maintenance
than for large cities.  This is misleading.  Although the
monitoring of activities within a smaller urban area is more

easily undertaken, the regional role of major parts of the
freight distribution system - i.e. the structural background
to any planned modification of commodity servicing - is
obscured. In a large metropolitan area this regional distribution function is more readily apparent, while for London
and a few other places a certain national role is superimposed. The boundaries of the economic region defined by
even a metropolitan county are unlikely to coincide with the
administrative boundary. This is certainly true of London as
an economic area and London as defined by the GLC administrative boundaries.

This mismatch is likely to cause few fresh monitoring
problems, as the data for monitoring commodity and freight
activities is generally tied to similarly restrictive
administrative boundaries. The real problems arise when
policies of regional significance are undertaken, when the
administrative boundaries provide little constraint on the
industrial and locational reactions and trends.

The geographical scale and the variety of freight
activities has meant that the only co-ordinated sources of
data are major surveys carried out at a particular time, and
with few continuing sources of data to support any monitoring
function.

At the national level, the UK DoE Road Goods Transport
Survey (RGTS) [4] has in the past provided a useful source of
commodity-movement data. Unfortunately, after the comprehensive coverage achieved in the 1968 survey, drastic
simplifications were made, thereby sharply limiting the value
of this annual survey of commodity movements. The details
will be considered later, but RGTS remains the only present
basis for any form of annual geographical monitoring of
commodity movements by road, and publishes data only at a
very coarse level of disaggregation. This was found to be
the major bar to National Freight Model syntheses by several
workers, notably Heyman [8]. Some land-use and transport
surveys included limited freight data-collection [2] but the
only really comprehensive UK cross-sectional survey of
commodity movements in land-use and transport surveys (LUTS)
was in London during 1971-2 [3]: even this survey stopped
short at a 3 per cent sample of vehicles, and thereby limited
the degree to which reliable, detailed and geographically-
specific results could be extracted. Considerable use is
made of this data later in this chapter.

The intimate and disparate relationships between commodity
movement, load and delivery size, vehicle use, and land uses
of different types made it quite clear that a better appreciation of the specialised behaviour of different types of
operators, receivers and despatchers of goods was needed to
underpin policies which could modify the operation of the
freight industry. Three such local and specialised studies
have been commissioned: in Swindon (TRRL, Atkins Planning),
Hull (TRRL, Wilbur Smith and PA Consulting Group) and
Lewisham/Greenwich (GLC/TRRL: Metra, SCPR,* RSGB**). These

_____
* Social Community and Planning Research Ltd, London, UK.
**Research Services; Great Britain, London, UK.

surveys required progressively greater effort in line with
the progression from small towns to part of a major city.
The surveys were of three types:

1  interviews with businesses and organisations to probe
their operation in depth;

2  interviews with visiting vehicles and the premises
visited;

3  cordon counts of moving vehicles with a very brief
questionnaire.

In view of the scale of effort involved in these cross-
sectional surveys, it is doubtful if more than a very limited
version of the types (2) and (3) surveys could be sustained
for monitoring purposes.  Certainly the whole apparatus of
the GLTS 1971-2 Freight Survey could not possibly be used
more frequently than every three to five years, even if
financially justified.

   It is therefore likely that any commodity-flow monitoring
process would have to be based either on modified forms of
the surveys listed above, or on the very few existent data
sources.  Consequently a more searching appraisal of both
cross-sectional (static) indicators, and shift (i.e. change
monitors) indicators is needed, in order to define indicators
suitable for the policies likely to be considered in the
future.  Table 26.2 is a brief summary of descriptive
(static) and indicative (shift) variables of potential
utility.  The top row is split into questions of mode, land
use, vehicle and commodity: the question of just how vehicles
are used is treated later.

   Column 1, row 1 in Table 26.2 emphasises the need to
consider modes other than road: not so much for the overall
picture, but to assess the changes in the traffic carried
or not carried by rail or water.  A further complexity not
treated in Table 26.2 is that of distribution *system*.
Container transport is such an issue: the commodity is of
little interest and may not even be known by the driver of
the vehicle: multi-modal systems such as Freightliner* make
good use of containers, and systems such as container, piggy
back, ro-ro, kangaroo and pipeline should profitably be
considered as extra - if complex - 'modes'.

   Table 26.3 outlines policy issues pertinent to the
indicators listed in the corresponding element of Table 26.2.
The presentation of issues and data in terms of interaction
between land use and commodity (etc.) emphasises how policy
interests arise primarily where two or more of these factors
are simultaneously involved.  Commodity movements *per se* are
of dubious practical value, although when taken in conjunction with commodity *deliveries* or some other factor, a
knowledge of tonnage moved can be of considerable utility.

---

*Freightliners Ltd: A UK firm operating line haul by rail
for ISO containers, with cartage also available.

TABLE 26.2  Relevant specification data - broad elements

| Mode | Land use | Vehicle | Commodity | |
|---|---|---|---|---|
| Technical through puts<br><br>Facilities in use<br>Modal transfers | Progress of rail/water feeds; line sections and rail heads Port handling, depot, ICD[a] and container rates | Fleet, wagon and special vehicle system useage | Market shares and commodity types | Mode |
| Industrial activity shifts | Degree of linkage between land uses<br><br>Warehousing and depot changes | Pattern of fleets used to service land uses Intensity of activity | Tonnage demands and production (by commodity) | Land Use    s<br><br>    t |
| Fleet composition trends | Trends in vehicle size Servicing given Land uses | Flow on roads by type and location noise/air pollution levels<br><br>Scrapping rates<br><br>Fleet composition by rate of change of type | Types of vehicle/mode used for different commodities<br><br>Drop size character- istics Consolida- tion usage |     a<br><br>    t<br><br>    i<br><br>    c<br>Vehicle |
| Relative tonnage shifts | Consumption trends land use Distribution system changes | Vehicle size trends | Multiple handling part pro- cessing and warehouse handling<br><br>Changes in balance production- consumption | Commodity |

s  h  i  f  t  s

a. Inland Customs Clearance Depots

TABLE 26.3   Policy factors for which indicators may be useful - broad factors

| | Modal Network | Land Use | Vehicle | Commodity |
|---|---|---|---|---|
| **Modal Network** | Depot or location<br><br>Section 8 rail applications[a] wharf safeguarding | Balance of road/rail water investments and regulation (Infra-structure Investment and inter modal regulation | Balance of economic regulation, or fiscal controls | Locational strategies for these commodities depots for them |
| **Land Use** | Locational strategy for consumers of commodities | Monitoring level of IDCs[b] for warehousing and distribution<br><br>Planning application trends | Effects of bans and other restrictions and regulations on fleet composition | Levels of activity linked to employment as a basis for a balance between economic and environmental expectations |
| **Vehicle** | Matching railway and waterway capacity to possible mode shift support | Locational and economic effects of regulating different land uses | Levels location and trends in truck movement and activity location<br><br>The rate at which policies can work through fleet purchases | Levels of vehicle activity linked to given scales of economic activity against balancing environmental effects |
| **Commodity** | Modal shifts policies and trends in market shares to anticipate amelioratory action | Levels of economic activity balanced against truck activity levels | Economic impact of new regulations on vehicle size | Control of vehicle and ton-km carried<br><br>The city's role as a net producer or consumer of goods and services |

Static

Dynamic

a. A special UK legislative provision for 50 per cent of capital grants to rail freight facilities formally supported by local government
b. Industrial Development Certificates: i.e. permits for (industrial) use of land.

The questions of vehicle routing, of vehicle origin, of access need, and the relationships between ton mileage and vehicle mileage are all part of a quite separate grid of interactions.  Vehicle mileage, vehicle size and commodity tonnage are closely related to most of the powers in the UK Heavy Commercial Vehicle Act (7): the location of these activities and their concentration are of direct environmental concern.  It would have been tempting to concentrate on these interactions alone were it not for the highly integrated nature of the service functions discharged by goods movement and the freight industry.  The policy areas outlined in Table 26.3 are no less important, just as the issues raised in Tables 26.4 and 26.5 (where the vehicle system is analysed) are primarily environmental in object with economic influences, those of Tables 26.2 and 26.3 are primarily economic, with environmental consequences. Balanced policies could not ignore either aspect.

TABLE 26.4   Relevant specification data - vehicle elements

|  | Vehicle | Commodity | Drop | Vehicle trip |
|---|---|---|---|---|
| Vehicle (V) | Size distribution: Base location | - | - | - |
| Commodity | Tonnages<br>- delivered<br>- collected<br>- multiple handling % | Raw materials<br>- semi-finished<br>- manufactures<br>- consumption<br>- sales | - | - |
| Drop | - | Throughput of depots warehouse and other concentration or multiple handling points | Drop size<br>Time of day<br>Variations in deliveries and delivery sizes | - |
| Vehicle Trip | Vehicle km<br>Vehicle hours<br>Vehicle routes | Ton km<br>commodity<br>Length of haul | Drops/trips<br>Dropsize destination<br>Time of day variations<br>(by size of vehicles) | Ton km by V/type<br>km by V/type<br>km by V/type<br>(Noise etc.) |

The indicators listed in Table 26.4 are probably more readily obtained than those in Table 26.2, but in so far as they are concerned with land uses and vehicle stops they too require some new and continuing basis for monitoring; moving-vehicle cordon counts with stoplight interviews are fairly commonplace and vehicle-log methods have been used for RGTS, GLTS and the special Swindon/Hull/GLC freight studies with some success.  It would be reasonable to look to these methods again.  The regional - and even national - basis for much of freight distribution is emphasised by the considerable number of vehicles in the Swindon/Hull/GLC studies which visited the survey area from over 100 miles away. Externally based vehicles deliver a third of all tonnage delivered in London. Clearly, a sampling frame of vehicles resident in even an area as large as London is not going to be adequate.

TABLE 26.5 Policy factors for which indicators may be useful for monitoring and control - vehicle specific factors

|  | Vehicle | Commodity | Drop | Vehicle/Trip |
|---|---|---|---|---|
| Vehicle (V) | The level of the city's own fleet. The degree to which it handles the city's goods | – | – | – |
| Commodity | This in conjunction with consumption figures will permit the monitoring of the distribution industry, and aid planning wholesale market, warehouses, IDCs[a] | Shifts in manufacture to regional distribution may be detected early | – | – |
| Drop | Effects on trades and the public of vehicle use, routing, loading and access restraints | Planning control and locating warehouses and termini | Effectiveness and impact of cumulative economic and regulation reactions to restricted delivery access | – |
| Vehicle Trip | Vehicle fleet usage effectiveness trend, when taken with commodity tonnages | Trends to altered distribution patterns which may require adjustment | (as above) also bus lane interactions | Balance between V/km, V/trips and location in respect of economic and regulation pressures. |

a. Industrial Development Certificates: a formal approach means of land-use control in UK.

DATA CLASSIFICATION

The prime problem is that of classifying commodities. If
commodity groups are not sufficiently numerous, the value of
the data declines sharply, as misclassification anomalies
increase. The GLC experience has been that the 172-category
CST code ('CST' Standard Commodity Codes for Trade: predates
NST for EEC usage) and the 99-category system used for the
GLTS in 1972, are both substantially adequate. Both systems
aggregate and disaggregate commodities slightly differently
but both are fine enough to permit a reasonable degree of
compatibility between CST surveys (such as the DoE RGTS) and
GLTS categories. For international compatibility the CST
code had real advantages, but for UK use the adoption of the
NST code for the continuing RGTS is a more persuasive argument. On the other hand, the Swindon and Hull TRRL surveys
are only roughly compatible with either scheme, although the
Lewisham study uses the GLTS code with the addition of a
load-compatibility index.

Classification schemes are therefore of considerable
importance in international comparisons, providing some
perspective on this question, Figure 26.1 shows the daily
tonnages delivered by road in Swindon and in the GLC. By a
reasonable regrouping of the GLTS codes, the two urban areas
have been made to appear very similar, with only haulage
depots in London substantially different from Swindon values.
This figure is misleading as both land-use categories and
commodity categories differ between the two areas and the
apparent agreement depends on numerous minor variations in
classification. Swindon is a small, rapidly-developing town
at a major crossroads with a motorway, and an ideal distribution centre. There was heavy building activity at the centre
at the time of the survey, while in London the population
has been declining, and building activity was at a low ebb
at the time of the GLTS survey.

When the broader policy issues of Table 26.2 are admitted,
the classification of land-use *activities* becomes important.
Table 26.6 is collated from several reports on freight and
industrial location (e.g. (10)) and illustrates the limited
correspondence of SIC (Standard Industrial Classification)
and GLTS categories. This table also graphically reaffirms
the quite different character of London (a metropolitan area
par excellence) and the other quite substantial centres in
the south-east, in terms of major manufacturing centres.
(If head offices are also included, then London-based
industry is clearly still of major national significance.)

Industrial premises are of special interest in the monitoring of commodity movements as the commodity inputs and
outputs are different, although directly connected. Figure
26.2 shows how commodity inputs and outputs are related, and
is a single stage via a series of matrices of transformations
which allow the quantity of tonnage moving within London to
be determined by convolution, and the frequency of handling
and multiple counting to be estimated from the then identifiable material throughputs. Judicious use of such transformation percentages may extend the effectiveness of limited
monitoring surveys. The quantity of goods outbound by road

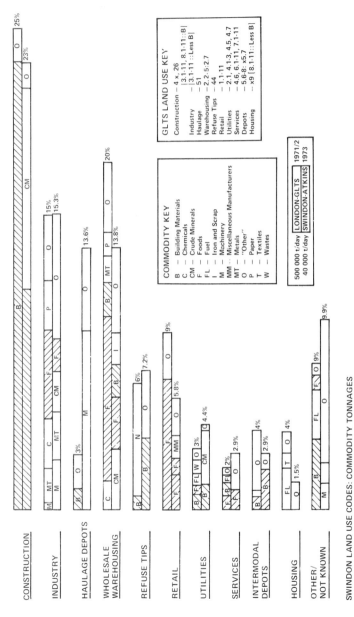

Figure 26.1 Daily Tonnages Delivered by Road to Specified Land Uses in London (1971) and Swindon (1973)

TABLE VI   Industrial Premises

| BOT (Board of Trade) | GLTS Industry | BOT 1945-65 Moves with origins in the South East | | GLTS - 1971 Goods delivered by ROAD | | Industry Stocks (over 100 employees) | | | | | | | |
|---|---|---|---|---|---|---|---|---|---|---|---|---|---|
| | | | | | | 1971 South East | | | | | Planning Studies | | ALL |
| | | 1963 Stocks | 20 yr Movement * per 100 plants | Millions tons/day delivered to | Delivered in GLC from | Inner | NE | SE | SW | NW | Outer Met. Area | Outer SE | |
| III Food | 31 Food drink tobacco | 1402† | 4.7% | 3.23 | 2.81 | 97 | 40 | 6 | 7 | 35 | 63 | 56 | 304 |
| IV Chemicals | 32 Chemicals | 1239 | 9.6 | 1.01 | 0.75 | 41 | 37 | 8 | 15 | 39 | 68 | 35 | 243 |
| V Metal manufacturers | 33 Metal manufacture and goods | 556 | 6.3 | 0.58 | 0.43 | 15 | 13 | - | 3 | 8 | 24 | 10 | 73 |
| IX Metal goods (N.C.S.) | | 3165 | 4.0 | | | 40 | 28 | 1 | 16 | 30 | 51 | 26 | 192 |
| VI 1 Mech.Engineering | 34 Mech.Eng.& Ships | 7267 | 7.4 | 0.61 | 0.43 | 90 | 56 | 15 | 48 | 90 | 236 | 108 | 543 |
| 2 Elec. | | | | | | 46 | 46 | 18 | 36 | 67 | 142 | 67 | 422 |
| VII Shipbuilding | 35 Elec.Eng.& goods | 419 | 1.4 | 0.70 | 0.23 | 2 | 4 | - | - | - | 4 | 14 | 24 |
| VIII Vehicles | 36 Vehicles | 650 | 11.7 | 0.31 | 0.06 | 12 | 3 | - | 5 | 24 | 67 | 40 | 151 |
| X Textiles | 37 Textiles,clothing | 590 | 4.1 | | | 7 | 9 | 1 | 2 | 4 | 14 | 13 | 50 |
| XI Leather | leather, laundry | 826 | 2.3 | 0.44 | 0.31 | 7 | 3 | 1 | 1 | 1 | 2 | 4 | 20 |
| XII Clothing | | 4146 | 3.3 | | | 82 | 27 | 1 | 2 | 3 | 37 | 35 | 187 |
| XIII Bricks | 38 Bricks, cement pottery, glass | 1444 | 5.3 | 2.26 | 1.37 | 13 | 9 | 4 | 2 | 14 | 58 | 32 | 132 |
| XIV Timber furniture | 39 Timber, paper, etc | 3588 | 2.5 | 1.12 | 0.58 | 43 | 37 | 4 | 7 | 13 | 50 | 36 | 190 |
| XV Paper, print pub | | 4740 | 2.3 | | | 137 | 22 | 6 | 30 | 38 | 156 | 58 | 447 |
| XVI Other m/f | 30 Other m/f | 1947 | 7.1 | 0.33 | 0.26 | 31 | 22 | 6 | 15 | 24 | 67 | 27 | 132 |
| ALL ESTABLISHMENTS | | 31979 | 4.9 | 8.6 | 4.54 | 663 | 356 | 71 | 190 | 390 | 1039 | 561 | 3270 |

* Moves within an area (GLC) are excluded from BOT Statistics, as are completely new firms.

† All firms exceeding 10 employees at some stage.

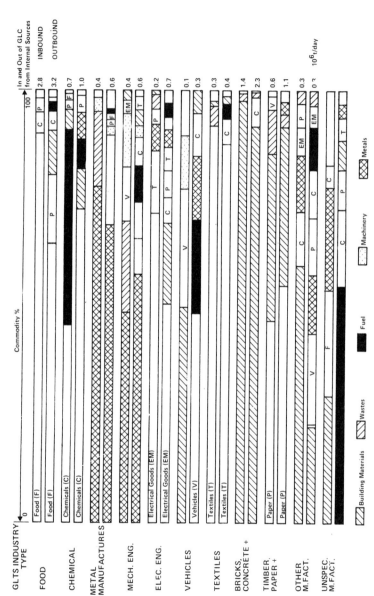

Figure 26.2   Commodity Transformations In and Out of Industrial Land Uses (London, 1971)

to each industry exceeds the inbound flow in every case: the
leakages due to other modes, and movements by vehicles other
than goods vehicle would contribute to a 'loss' while the
inclusion of water and other 'non-freight' inputs are
primarily responsible for this 'gain'. In view of the very
heavy demands on resources which would be posed by any comprehensive monitoring of commodity movements, devices such as
this should be developed to make maximum use of the few
cross-sectional surveys currently available.

Retail outlets are also of special interest, as most goods
'disappear' from the commodity movement pattern at this point,
as comparatively few HGV (Heavy Goods Vehicles) or other
goods vehicles are used to remove goods from shops, with the
exception of home deliveries of larger goods, and servicing
calls, by goods vehicles. Tonnages delivered and vehicles
making individual calls have no direct relationship: many
goods-vehicle movements do not involve goods movements *per
se*, however the environmental impact of goods-vehicle movements remains irrespective of the purpose or economic
importance of the visit. Shops account for more deliveries
than any other land use although they are not quite so important in terms of delivered tonnage. A rapid decline in food
outlets has complemented the increasing role of supermarkets,
with their bulk delivery systems. Deliveries of food per
head in Swindon and London are very similar (Table 26.7).

TABLE 26.7  Food deliveries/head in kg/day

|  | Swindon | London |
|---|---|---|
| Fresh fruit and vegetables (retail) | 0.87 | 0.44 |
| Other food for retail sale | 2.06 | 2.75 |
| Total retail food deliveries | 2.97 | 3.19 |
| Non-retail food deliveries | 10.06 | 10.31 |
| All food | 12.99 | 13.50 |

This is not due solely to the consumption of food, as 27 per
cent of all goods delivered in London is to warehouses,
leading to substantial double counting of the same materials.
31 per cent of the total tonnage of all commodities delivered
by road from points within London originates in warehouses
(or wholesale markets) and 23 per cent of food deliveries in
London arise from warehouses in London. The correspondence
between the two towns would therefore indicate that similar
distributions and consumption patterns applied in both towns,
as such a large proportion delivered through warehouse
buffers would otherwise distort the comparison significantly.

The national trend to fewer food outlets might therefore
be expected to have similar effects on deliveries and tonnage
moved in each town, and an appropriate indicator with which
to assess the deliveries and tonnage delivered to food
outlets would be given by the decline in food outlets in
Table 26.8, which shows the pace of concentration of outlets.
This has three effects,

1  use of fewer, larger vehicles through supermarket chains'
distribution depots;

2  use of fewer, larger vehicles direct from manufacturers
to large food outlets;
3  greater throughput via cash and carry warehouses to
sustain the independent shop.

Similar trends are evident in many countries.

TABLE 26.8  UK food outlets 1957-71 census of distribution
and production in (1000s)

|                               | 1957 | 1965 | 1971 |
|-------------------------------|------|------|------|
| Grocers                       | 151  |      | 108  |
| Other food retailers          | 123  |      | 100  |
| Dairies                       | 7.5  | 4    |      |
| Butchers and dairies combined | 49   |      | 39   |
| Fish shops                    | 8    |      | 6    |
| Greengrocers                  | 40   |      | 29   |
| Bread/flour shops             | 18   |      | 18   |
| Off licences                  | 9    |      | 9    |

The examples in this section illustrate the interaction
between patterns of commodity movements and the actual
effective quantities of the commodities moved.  The two
points enphasised are:

1  classifications of land uses and commodities need to be
detailed so that aggregation for comparison purposes is
possible;
2  commodity movements do not necessarily have a simple
relationship to the tonnages involved: industrial premises
transform commodities, retail outlets 'lose' them to cars,
warehousing causes a large amount of double counting, and the
key commodity in terms of urban deliveries (food) is subject
to rapid concentration of outlets, and in consequence to
increasingly sophisticated distribution arrangements for
what has remained a stable total quantity of food: a dominant
material in terms of tonnage and vehicle km.  Vehicle types
carrying different commodities must therefore be identified.

The classification scheme used by RGTS in 1968 was adequate
to cover all commodities, but although trips were also
covered, no information on vehicle types was reported.
Consequently the physical appearance on the road of the
commodity movements could not be traced.  After 1968 the 178-
commodity code was collapsed to a 20-commodity system, and
the three-digit national geographical coding of movements
collapsed to the 11 planning regions to match a much reduced
survey sample.  Although the sampling fraction was raised
once more in 1973, these limited categories remained, thereby
vitiating the value of the RGTS as an annual monitoring
scheme with any detailed geographical utility.  The recent
commitment of the NST coding system (adopted by the EEC for
internal trade from 1975) of 58 commodities adopted by the

EEC Statistical Office has led to a further change, requiring dual coding for the 1975 cycle preparatory to the 1976 changeover to NST and geographical coding revised to reflect the new county structures. In view of the slight national decline in both tons and ton/km 1974 (w.r.t. 1973) it is clear that shifts in the manner of transport and changes in locational emphasis are of rapidly increasing·importance – not least because such changes between coding systems could well have 'caused' such an apparent effect. This would strongly suggest that RGTS be refined to 3-digit geographical level once more, and the CST (178 commodity) code be used. NST is compatible with CST at the two digit level, thereby satisfying the formal EEC requirements, while retaining an adequate level of detail for commodity movement monitoring.

As stressed in the 1972 European Conference of Ministers of Transport (ECMT) Round Table (5) on these subjects, the simultaneous use of tons *and* ton/km is essential: i.e. in the climate of present public interest vehicle/km by vehicle type and ton/km are both imperative additions.

At a national level of aggregation the lack of any land-use coding of origin or destination of goods is of no great importance, but as certain areas are the regions wherein loading, consumption and production are concentrated, a land-use coding is needed to permit the policies of Tables 26.3, 26.5 to be viewed in conjunction with commodity-movement patterns and transport and commercial land uses have a flexibility which is of considerable potential importance in monitoring. Table 26.9 shows the key land uses which act as sources and links for commodity movements in London, and the 15 per cent of all tonnage delivered in London from the few transport land uses starred account for only the road-borne movements from these sources. The tonnages of direct road deliveries from sources outside London are very substantial, but still amount only to one third of all delivered tonnage in London. 15 per cent of tonnage delivered to warehouses is only the input: 9 per cent is redelivered in London, the importance of double counting of delivered tonnages demonstrated. This imbalance in warehouse tonnage once again illustrates the regional nature of distribution systems, as 40 per cent of tonnage delivered by road to GLC warehousing is redelivered outside the GLC administrative area itself.

TABLE 26.9 Destinations and sources of GLC commodity tonnage (1971/2 data)

| | | | | | |
|---|---|---|---|---|---|
| FOOD | Warehouses | 37% | Warehouses | 23% | (ES[a] 37%) |
| | Food/drink industry | 19% | | | |
| | Grocers shops | 12% | | | |
| | Other food shops | 10% | Other food shops | 23% | |
| | Goods vehicle depots | 4% | Goods vehicle depots | 4% | |
| CHEMICAL | Chemical industry | 28% | Chemical industry | 23% | (ES 56%) |
| | Warehouses | 25% | Warehouses | 12% | |

| | | | | | |
|---|---|---|---|---|---|
| | Food, drink industries | 6% | Food, drink tobacco industry | 5% | |
| METALS | Warehousing | 22% | Warehousing | 6% | (ES 41%) |
| | Mechanical industry | 14% | Mechanical industry | 8% | |
| | Chemical industry | 16% | | | |
| MACHINERY | Plant hire/ Builders yard | 24% | | | (ES 18%) |
| | Private Homes | 11% | Private Homes | 12% | |
| | Wharves | 11% | | | |
| | Roads | 9% | Roads | 9% | |
| ELECTRICAL | Electrical eng. industry | 34% | Electrical eng. industry | 24% | (ES 40%) |
| | Warehousing | 25% | | | |
| VEHICLES | Open storage | 15% | | | (ES 73%) |
| | Warehousing | 12% | | | |
| TEXTILES | Clothing industry | 32% | Clothing industry | 24% | (ES 18%) |
| | Warehousing | 15% | Warehousing | 15% | |
| | Hotels | 13% | Hotels | 13% | |
| BUILDING | Building construction | 34% | Sand, gravel workings | 7% | (ES 42%) |
| | Bricks, cement | 12% | | | |
| | Building yards | 10% | | | |
| PAPER | Warehousing | 22% | Warehousing | 19% | (ES 28%) |
| | Paper industry | 20% | Paper industry | 9% | |
| | Docks, wharves | 8% | Docks, wharves | 11% | |
| | Goods vehicle depots | 6% | Goods vehicle depots | | |
| WASTES | Refuse tips | 75% | Building construction | 34% | (ES 3%) |
| | | | Private homes | 11% | |
| | Local authority depots | 6% | Local authority depots | 7% | |
| FUELS | Petrol stations | 19% | Petrol, oil storage | 32% | (ES 35%) |
| | Private homes | 16% | | | |
| | Petrol, oil storage | 6% | | | |
| ALL | Warehousing | 15% | Warehousing | 9% | (ES 36%) |
| | Building construction | 12% | | | |
| | Food, drink industry | 6% | Food, drink industry | 6% | |
| | Refuse tips | 6% | Brick, cement industry | 3% | |
| | Private homes | 5% | Clay, sand gravel workings | 2% | |
| | | | Roads | 4%* | |

| | |
|---|---|
| Goods vehicle depots | 3%* |
| Petrol, oil depots | 3%* |
| Docks, wharves | 3%* |
| Railway land | 2%* |

*Source:* Wigan (12)
a. ES = External sources or vehicles

The imbalances in Table 26.9 draw attention to the need to identify the role and scale of other modes of commodity transport. Pipeline deliveries of oil and petrol products might appear to account for much of the fuel imbalance, and while pipelines play a major movement role inside London, river transport is of greater importance.

Commodity flows by other modes are generally limited to building materials, bulk foods and similar materials, by rail; and refuse, paper, timber, steel by water. Although some of these commodities are delivered directly to manufacturing centres, most are carried on to their final consumption point by road. The potential double counting of goods carried by rail and water is a further point of interest and emphasises the primary role of such modes as truck-feeders. Freight policies may affect the precise terminals used, and the quantities trunked by rail/water, but in most cases will not affect the tonnage carried by road, merely the ton km.

If it is accepted that commodity movements are of greatest planning interest not in their quantity but in the manner and location of their movement, then the classification of road vehicles used is very important, as to serve the consumption and production needs of a city, the only degrees of freedom left are the location of freight generators and handling points and the types of vehicle (and mode) used to serve them.

The RGTS surveys of DoE provide no published data on vehicles, but the GLTS 1971 surveys use two different categories. For vehicles based (registered) in London, the unladen weight is recorded in five steps. For vehicles not based in London, a general vehicle-type code was used:

1  Goods vehicles with two axles and four tyres (approximately 1.5 ton, unladen weight)
2  Goods vehicles with two axles and six tyres
3  Rigid vehicles with three or more axles (approximately 4.5 ton, unladen weight)
4  Articulated vehicles.

While these categories readily identify vehicles in a traffic stream, there is a wide variety of vehicle weights in each category, the main anomalies being large specialised vehicles with six tyres and small, highly manoeuvrable articulated vehicles designed specially for urban use. Regulations on parking, movement and access of lorries are based on unladen or laden weight, but until a clearer form of standard sizing

of vehicles is agreed by DoE there is no reliable direct
visual link to the type of vehicle.  For environmental
protection, the vehicle, by type of vehicle, and by area, are
the major factors.  The times at which given types of vehicles
operate also have environmental significance, even when both
route and type of commodity remain unchanged.

Any categorisation of vehicles must be both easy to apply
for enforcement and for survey purposes, and also be directly
linked to the specifications to be used for regulation.

Table 26.10 shows the total commodity tonnages delivered
by road to the GLC in terms of the visual vehicle type.  Some
idea of the visual importance of these movements is given by
Table 26.11, where the deliveries made by the same vehicles
are shown.  The vastly different roles of internally and
externally based vehicles and 4-tyre and articulated/rigid
3+ axle classes is evident.

TABLE 26.10   Daily GLC tonnage by vehicle type

| Vehicle | Internally based | | Externally based | | All | |
|---|---|---|---|---|---|---|
| | $10^3$ | % | $10^3$ | % | $10^3$ | % |
| 4 tyre/2 axle | 29 | 9 | 4 | 2 | 33 | 7 |
| 6 tyre/2 axle | 168 | 52 | 65 | 36 | 233 | 41 |
| Rigid/3+ axle | 76 | 23 | 41 | 23 | 117 | 23 |
| Articulated | 47 | 15 | 70 | 39 | 117 | 23 |
| All | 320 | 100 | 180 | 100 | 501 | 100 |
| | 64% | | 36% | | 100% | |

TABLE 26.11   Daily GLC deliveries by vehicle type

| Vehicle | Internally based | | Externally based | | All | |
|---|---|---|---|---|---|---|
| | $10^3$ | % | $10^3$ | % | $10^3$ | % |
| 4 tyre/2 axle | 276 | 49 | 14 | 26 | 290 | 47 |
| 6 tyre/2 axle | 247 | 44 | 26 | 47 | 273 | 44 |
| Rigid/3+ axle | 27 | 5 | 6 | 11 | 32 | 5 |
| Articulated | 14 | 2 | 9 | 18 | 23 | 4 |
| | 564 | 100 | 55 | | 619 | |
| | 91% | | 9% | | 100% | |

It is also possible to relate the type of vehicle most
commonly used to service given land uses, and the commodities
involved.  This, and the types of vehicles used to link different land uses, are of the most direct application for
locational planning purposes.

EMPLOYMENT DENSITIES AND FREIGHT LAND USE

The land-use and employment aspects of freight activities
provide a point of conflict between freight and other land
uses.  The common currency in which this can be expressed is
in terms of employment densities on either a site or a floor-
space standard.

Industrial land uses are commonly supposed to have higher
employment densities than commercial land uses.  The steady
shift in UK from labour-intensive industries to capital-
intensive methods of production has introduced a steady
overall trend to lower employment densities even on purely
industrial sites.  This trend is most clearly discerned in
new towns, where modern industries form a higher proportion
of the whole than in older areas.  New towns have also
attracted many of the new plants set up by older industries
cramped for redevelopment space and capital in their
historical urban locations.

Regarding each industrial group as having a single average
employment density/ha, the overall average of the densities
of employment per ha in UK new towns has declined from 196
(standard deviation $\sigma$ = 120) in 1960, to 132 ($\sigma$=73) in 1967,
and with a further decline to 125 ($\sigma$=71) in 1974.  At the
same time, floorspace/employee moved from 20 m  ($\sigma$=9) in 1960
to 24 $m^2$  ($\sigma$=8) in 1967 and 29 $\overline{m}^2$  ($\sigma$=9) in 1974, as the
productivity of new plants reduces the use made of labour.

Against this background we can marshal recent data on
employment densities of freight land uses.  Several such
sources are available, and the most important of these is the
Greenwich and Lewisham Freight Sector Study, carried out in
1975.  The major freight land uses of interest are, (1)
warehousing, (2) depots.  Each of these land uses may be
further subdivided by function into:

1   cash and carry warehouses (outside the use classes order),
2   open storage,
3   covered storage;

and:

1   distribution depots (generally 'own account' operations),
2   hauliers depots.

A finer classification could be made, but would not be
justified on the basis of present data.

The employment in depots and warehouses must be related to
the areas of land that they cover, and the floor areas in
use.  Table 26.12 summarises the areas in use in 22 boroughs,
and the changes which have taken place since 1966.  Clearly
such land uses are only a very small part of the whole, and
the 1 to 3 per cent that they require might reasonably be
accommodated in almost every borough.  That freight land-use
changes of up to 70 per cent can occur in a mere five years
in an area as established as Greater London shows how
important such location shifts can be in areas of greater
volatility such as Melbourne.

TABLE 26.12   Time shifts in freight land use

| Borough | 1971 total area (ha) | 1971 area as % of total (ha (%)) | | | Changes from 1966 to 1971 | | |
|---|---|---|---|---|---|---|---|
| | | Depots site area (ha) | Warehousing covered storage site area (ha) | Warehousing occupied floorspace m² | Depots site acres (ha) | Warehousing and covered storage | |
| | | | | | | Site (ha) | Occupied floorspace m² |
| Barking | 3,400 | 28(0.8%) | 91(2.6%) | 480 | 6(21%) | 23(26%) | 200(40%) |
| Barnet | 9,000 | 18(0.2) | 38(0.4) | 270 | 3(18) | -2(-8) | 9(4) |
| Croydon | 8,700 | 25(0.2) | 31(0.3) | 240 | 2(8) | 6(21) | 60(23) |
| Ealing | 5,500 | 26(0.4) | 97(2.1) | 810 | 4(14) | 36(37) | 330(41) |
| Enfield | 8,100 | 23(0.2) | 87(1.1) | 530 | 3(12) | 2(2) | 30(5) |
| Greenwich | 4,700 | 57(0.4) | 45(0.9) | 320 | 7(30) | 3(7) | 70(24) |
| Hackney | 1,900 | 17(0.8) | 35(1.8) | 590 | 2(12) | 1(4) | 30(5) |
| Hammersmith | 1,600 | 10(0.5) | 22(1.3) | 250 | -1(-8) | -.4(-2) | -5(-2) |
| Haringey | 3,000 | 21(0.7) | 40(1.3) | 340 | 1(6) | -.4(-1) | 7(2) |
| Havering | 11,800 | 30(0.2) | 46(0.4) | 230 | -4(-12) | 5(13) | 30(12) |
| Hillingdon | 11,100 | 23(0.2) | 46(0.4) | 240 | -1(-3) | 4(9) | 30(19) |
| Hounslow | 5,900 | 40(0.6) | 109(1.9) | 690 | 2(5) | 17(15) | 120(18) |
| Kingston | 3,800 | 15(0.4) | 12(0.3) | 90 | 6(41) | -3(-23) | -30(-30) |
| Lambeth | 2,700 | 23(0.9) | 18(1.2) | 430 | 8(33) | -3(-8) | -80(-20) |
| Merton | 3,800 | 11(0.2) | 42(1.1) | 270 | -1(-7) | 9(22) | 80(31) |
| Newham | 3,600 | 29(0.8) | 71(1.9) | 560 | -.4(-1) | 15(22) | 9(15) |
| Redbridge | 5,700 | 17(0.3) | 11(0.2) | 75 | 5(28) | .4(4) | 30(13) |
| Richmond | 5,500 | 12(0.2) | 18(0.3) | 130 | 1(7) | 5(27) | 28(21) |
| Sutton | 4,300 | 9(0.2) | 19(0.4) | 90 | 0(0) | 6(35) | 40(40) |
| Tower Hamlets | 2,000 | 27(1.2) | 77(3.9) | - | 2(6) | 4(5) | - |
| Waltham Forest | 4,000 | 11(0.25) | 30(0.2) | 240 | .4(4) | 2(9) | 40(15) |
| Wandsworth | 3,500 | 18(0.2) | 23(0.7) | 230 | 4(20) | -2(-10) | -30(-12) |

GLC Land-use Surveys: data for both 1966 and 1971 available only for 22 of 33 boroughs.
: each borough contains on average 0.25 million population.

L.H. Baker (1)has obtained some approximate relationships between employment and floor-area from data of this kind. These are currently being re-estimated for Sydney by the Urban Transport Study Group, who have set up Baker's Planning Parameters Allocation Model. Those specially relevant to freight are:

1   Wholesale operatives (non-office staff) / m   floor area =
                          6.4 + 0.3
                          (Area of general storage)
                          + 6.6
    $R^2$ = 0.57          (Area of wholesale markets)

2   Retailing operatives in general storage/unit floor area =
                          7.8 + 0.1
                          (Area of general storage)
                          + 2.1
                          (Area of department stores)
                          + 0.9
                          (Area of food shops)
                          + 1.2
                          (Area of service shops)
                          + 2.8
    $R^2$ = 0.89          (Area of non-food shops)
                          + 0.7
                          (Area of public houses)

These two equations show how freight employment is concentrated.

    Baker also obtained overall average ratios of floorspace to site area of:

(a) 0.664 General storage
(b) 0.749 Markets
(c) 0.668 All commerce
(d)(0.867 All industries)

These ratios are examined in more detail later, but a Freight Transport Association Survey carried out in 1971 of own-account depots suggests that covered depot areas are on the average:

1   Warehousing    46%
2   Garaging       12%
3   Load/In Load   11%
4   Parking         8%
5   Offices        16%

and that 26 per cent of the site area is covered.  This enables us to add:

(e) 0.74 Distribution depots.

A special survey of Southwark depots carried out in 1970 suggested that

for road-haulage depots   : (f) 0.809
and for own-account depots: (g) 0.554.

The disagreements between these different values emphasise
the need to use data for individual sites of a specified type
if reliable employment-density data is to be obtained: it
might also be concluded that the disagreements between road
haulage and own-account figures suggest Baker's overall
average figures are in themselves misleading.

The Greenwich and Lewisham Freight Sector Study carried
out in 1975 provided such data, for three categories of land
use:

1   Road Hauliers
2   Distributors
3   Cash and Carry warehouses

The results follow in summary form:

For the 29 haulier sites; average employment density =
104/site ha ($\sigma=76$)

$$\text{Employment} = 23.5 + .0037 \text{ (Site area m}^2) \qquad R^2 = 0.33$$

$$\text{Employees/Site ha} = 54.15 - 14.7 \text{ (Site area ha)} \qquad R^2 = 0.11$$

$$M^2 \text{ of site/Employee} = 107 + 60 \text{ (Site area ha)} \qquad R^2 = 0.29$$

For the 16 distributors' sites; average employment density
= 141/site ha ($\sigma=69$)

$$\text{Employment} = -3.93 + 0.0157 \text{ (Site area m}^2) \qquad R^2 = 0.65$$

$$\text{Employees/Site ha} = 45.35 + 16.9 \text{ (Site area ha)} \qquad R^2 = 0.17$$

$$M^2 \text{ of site/Employee} = 115 - 34.5 \text{ (Site area ha)} \qquad R^2 = 0.25$$

For the 7 cash and carry warehouses; average employment
density = ha ($\sigma=64$)

$$\text{Employment} = 20.23 + 0.0272 \text{ (Floor area m}^2) \qquad R^2 = 0.99$$

$$\text{Employment/Floor ha} = 88.3 + 20.54 \text{ (Floor Area/ha)} \qquad R^2 = 0.79$$

$$M^2 \text{ floor area/Employee} = 114 - 54 \text{ (Floor Area/ha)} \quad R^2 = 0.42$$

The average site employment densities of the two types of
depot per ha are therefore 104 ($\sigma=77$) and 141 ($\sigma=69$), and the
equivalent figure for cash and carry warehousing is (using
the Freight Transport Association (FTA) result of 1974) 168
($\sigma=86$). These values should be compared with the 1974 new
town value of 126 ($\sigma=67$). It is therefore probable that
freight land uses such as haulage depots and distribution
depots are not significantly different from modern low-
density industry in their employment intensity - and that
cash and carry warehousing is significantly higher. This
question of the age of the industry concerned is underlined
by a 1965 Industrial Premises survey in Middlesex of 4,839
firms which gave a value of 212 employees/site ha at that
date.

The monitoring of employment and land-use shifts in
freight land uses is therefore closely connected with the
more general question of land-use monitoring as a whole for
employment purposes, but is a major issue in its own right
for freight-forecasting purposes. The swift location shifts

possible - and essential - to freight operations moves the
point of environmental impact and behavioural studies for
urban freight very close to land-use forecasting: a finding
very closely in tune with the current study of a possible
Outer Ring Freeway for Melbourne, where the use of Lowry
models and direct-demand estimation procedures shows signs
of being sensitive to these factors.

THE SPECIAL ROLE OF WAREHOUSING

The importance of distribution between different levels of
warehousing or bulk storage, and the major differences
between the characteristics of vehicles and drop sizes for
delivery to and deliveries from commercial land uses, leads
directly to the question of shifts in operational style on
freight land uses.

The problem is neatly summarised in Table 26.13 drawn from
Talbot (11), for a single day in one warehouse handling one
group of homogeneous commodities.

TABLE 26.13  Inbound and outbound movements from one warehouse

|  | Number of consignments | Average weight (kg) | Total weight involved (kg) |
|---|---|---|---|
| IN | 4 | 8 810 | 8 080 |
| OUT | 185 | 270 | 50 070 |

This gives little indication as to where the goods are going,
but one may deduce a 'break-bulk factor' of 32 as the ratio
of consignment size *in* and *out* bound. Drawing once again on
the GLTS 1971 data of the GLC, we can obtain Table 26.14 for
deliveries from commerce (i.e. warehousing) to shops alone.

TABLE 26.14  Break-bulk factors for commerce to shop
deliveries (GLC 1971)

| Commodity | GLTS code | No. of daily deliveries | Break bulk factor |
|---|---|---|---|
| Meat, fish, eggs | 13 | 9 100 | 58 |
| Fresh fruit, vegetables | 16 | 7 800 | 9 |
| Groceries | 18 | 7 500 | 7 |
| Drinks | 19 | 8 500 | 7 |
| Sugar, confectionery | 15 | 3 700 | 4 |
| Bread, etc. | 12 | 3 100 | 13 |
| Tobacco | 1X | 2 700 | 1 |
| Milk | 14 | 2 000 | 20 |
| Frozen fruit, vegetables | 17 | 1 700 | 6 |
| Newspaper | 14 | 1 600 | 20 |
| Haberdashery | 73 | 1 400 | 13 |
| Wallpaper, cardboard | 92 | 1 000 | 11 |
| Metal furniture | 30 | 1 000 | 6 |
| Batteries, bulbs, etc. | 5X | 900 | 2 |

This variation in break-bulk factors underlines the
vulnerability of freight vehicle-movement forecasting to
swiftly changing operational factors in the overall physical
distribution management system (PDM, as it is known, is a
discipline combining stockholding, warehousing, distribution
and delivery functions treated as a single system). The 3
per cent vehicle sample on which this GLTS data was based
also contributes to this variability.

The vehicles serving this two-stage operation can be
described by a selective analysis of cross-sectional data
from a conventional Land Use and Transport Survey (LUTS):
Tables 26.15 and 26.16 are also drawn from Wigan (12), and
also serve to show how far the cross-sectional sample of 3
per cent of London's vehicles has to be stretched for these
results to be extracted. In the first column 'E, I' refer to
the percentage of all the tonnage (Table 26.15) or deliveries
(Table 26.16) that comes from vehicles based outside or
inside the GLC area. Fryer *et al*. (6) gives a more comprehensive picture of movement, vehicle and commodity patterns.

INDICATORS AND EFFECTS OF SCALE

Freight movement is a regional function and for this purpose
GLC London is only part of the London economic region. This
is emphasised by the large quantity of outbound shipments
from GLC warehouses, amongst other indicators. The policies
which apply to a whole region are therefore of great interest
in such large metropolitan areas, but in smaller districts
would not be of any great concern other than as a considered
reaction to the backwash effects of broader policies.
Indicators of commodity tonnage movement are essentially
measures of the economic activity and population of the area
concerned. These are useful, but more at a regional or
national level of concern. The balance between through
traffic and visiting traffic is an aspect of commodity
movement which increases as the size of the area decreases.
Only 10 per cent of Swindon's freight traffic passed through
without a stop. The use of cordon survey checks on freight
traffic is therefore of increasing value as the size of the
urban area diminishes.

The diversity of land uses and the internal servicing
function of warehouses and depot sites both increase with
the size of the urban area: small towns will have a few
clearly-defined freight generating areas or sites, comparatively concentrated shopping facilities, and a substantial
amount of deliveries from a considerable distance away from
regional distribution points in retailer and manufacturer
distribution systems. Smaller towns and cities also tend
to have very few rail or other mode terminals and few
major manufacturing centres. A direct monitoring process
is therefore feasible in such cases, although of less
value than in a larger metropolitan region. The special
problem of London and similar concentrations of employment
are shown by Table 26.6, where the manufacturing centres
are listed. It is therefore reasonable to list suitable
indicators by the scale of the urban area concerned
(Table 26.17).

TABLE 26.15   GLTS 1971: Tonnage Analysis: Vehicle Types Used to Serve Different Land Uses

| Commodity | Vehicle types | External vehicle split | Internal vehicle split | Key vehicle – land use linkages: External | | Key vehicle – land use linkages: Internal | % of all (Group) (I or E) tonnage |
|---|---|---|---|---|---|---|---|
| FOODS<br>E : Commerce 46%<br>I : Commerce 36% | 4 tyre/2 axle<br>6 tyre/2 axle<br>Multi axle rigid<br>Artic | -<br>26%<br>24%<br>50% | 10%<br>46%<br>19%<br>12% | Artic/Commerce<br>Artic/Industry<br>6 tyre/Commerce<br>Rigid/Industry | 29%<br>16%<br>10%<br>10% | 6 tyre/shops<br>Artic/Commerce<br>6 tyre/Commerce<br>Rigid/Commerce | 22%<br>14%<br>14%<br>7% |
| CHEMICALS<br>E : Industry 45%<br>I : Industry 54% | 4 tyre<br>6 tyre<br>Rigid<br>Artic | 1%<br>22%<br>21%<br>55% | 12%<br>71%<br>3%<br>14% | Artic/Commerce<br>Artic/Industry<br>Rigid/Industry<br>6 tyre/Commerce | 30%<br>21%<br>14%<br>10% | 6 tyre/Industry<br>6 tyre/Commerce<br>6 tyre/Open areas<br>Artic/Industry | 37%<br>15%<br>10%<br>10% |
| METALS<br>E : Commerce 54%<br>I : Industry 48% | 4 tyre<br>6 tyre<br>Rigid<br>Artic. | 10%<br>43%<br>7%<br>50% | 11%<br>59%<br>28%<br>2% | Artic/Commerce<br>6 tyre/Industry<br>6 tyre/Commerce<br>6 tyre/Open areas | | 6 tyre/Industry<br>Artic/Industry<br>6 tyre/Commerce<br>Artic/Utility | 25%<br>15%<br>13%<br>10% |
| MACHINERY<br>E : Open 63%<br>I : Commerce 38% | 4 tyre<br>6 tyre<br>Rigid<br>Artic | 12%<br>14%<br>5%<br>72% | 16%<br>33%<br>36%<br>15% | Artic/Open areas<br>6 tyre/Commerce<br>Artic/Utility<br>4 tyre/Residen-tial | 63%<br>9%<br>6%<br>6% | Rigid/Commerce<br>6 tyre/Commerce<br>Artic/Transport<br>6 tyre/Industry | 24%<br>12%<br>9%<br>8% |
| ELECTRICAL<br>E : Industry 50%<br>I : Industry 35% | 4 tyre<br>6 tyre<br>Rigid<br>Artic | 3%<br>30%<br>5%<br>62% | 26%<br>51%<br>3%<br>20% | Artic/Commerce<br>Artic/Industry<br>6 tyre/Commerce<br>6 tyre/Commerce | 33%<br>29%<br>19%<br>9% | 6 tyre/Commerce<br>6 tyre/Industry<br>Artic/Industry<br>6 tyre/Transport | 16%<br>15%<br>12%<br>6% |
| VEHICULAR<br>E : Commerce 32%<br>I : Industry 39% | 4 tyre<br>6 tyre<br>Rigid<br>Artic | 1%<br>24%<br>-<br>64% | 28%<br>35%<br>-<br>37% | Artic/Commerce<br>Artic/Transport<br>6 tyre/Transport<br>Artic/Industry | 31%<br>16%<br>12%<br>8% | Artic/Industry<br>Artic/Open areas<br>6 tyre/Industry<br>6 tyre/Transport | 17%<br>17%<br>17%<br>13% |

Table 26.15 (cont'd)

| Commodity | Vehicle types | External vehicle split | Internal vehicle split | Key vehicle – land use linkages: External | | Key vehicle – land use linkages: Internal | % of all (Group) (I or E) tonnage |
|---|---|---|---|---|---|---|---|
| **TEXTILES** | | | | | | | |
| E : Industry 28% | 4 tyre | 17% | 18% | 6 tyre/Industry | 28% | 6 tyre/Industry | 33% |
| | 6 tyre | 59% | 80% | 6 tyre/Residential | 20% | 6 tyre/Residential | 15% |
| I : Industry 42% | Rigid | 23% | – | Rigid/Commerce | 11% | 6 tyre/Commerce | 15% |
| | Artic | 11% | 2% | Rigid/Shops | 11% | 4 tyre/Industry | 8% |
| **BUILDING** | | | | | | | |
| E : Commerce 29% | 4 tyre | 1% | 6% | 6 tyre/Commerce | 18% | Rigid/Utility | 25% |
| | 6 tyre | 50% | 52% | 6 tyre/Utility | 13% | 6 tyre/Utility | 21% |
| I : Utility 47% | Rigid | 29% | 38% | 6 tyre/Utility | 12% | Rigid/Industry | 9% |
| | Artic | 20% | 4% | Rigid/Utility | 11% | 6 tyre/Residential | 8% |
| **PAPER** | | | | | | | |
| E : Commerce 35% | 4 tyre | 9% | 18% | Artic/Open | 19% | Artic/Industry | 25% |
| | 6 tyre | 33% | 39% | Rigid/Industry | 16% | 6 tyre/Industry | 14% |
| I : Industry 45% | Rigid | 19% | 6% | 6 tyre/Commerce | 15% | 6 tyre/Commerce | 10% |
| | Artic | 39% | 37% | Artic/Commerce | 15% | 6 tyre/Utility | 8% |
| **WASTES** | | | | | | | |
| E : Utility 100% | 4 tyre | – | 7% | 6 tyre/Utility | 100% | 6 tyre/Utility | 52% |
| | 6 tyre | 100% | 70% | – | – | Rigid/Utility | 22% |
| I : Utility 79% | Rigid | – | 23% | – | – | 6 tyre/Commerce | 6% |
| | Artic | – | – | – | – | 6 tyre/Open areas | 3% |
| **FUELS** | | | | | | | |
| E : Commerce 40% | 4 tyre | – | 2% | Artic/Commerce | 28% | 6 tyre/Residential | 23% |
| | 6 tyre | 19% | 55% | Artic/Transport | 18% | Rigid/Transport | 8% |
| I : Residential 33% | Rigid | 34% | 33% | Rigid/Commerce | 12% | Artic/Transport | 8% |
| | Artic | 47% | 10% | Rigid/Transport | 12% | Rigid/Residential | 8% |
| **TOTAL** | | | | | | | |
| E : Commerce 37% | 4 tyre | 2% | 9% | Artic/Shops | 20% | 6 tyre/Utility | 14% |
| | 6 tyre | 36% | 52% | 6 tyre/Commerce | 12% | Rigid/Utility | 10% |
| I : Utility 26% | Rigid | 23% | 23% | 6 tyre/Industry | 9% | 6 tyre/Commerce | 9% |
| | Artic | 39% | 15% | Rigid/Industry | 9% | 6 tyre/Shops | 6% |
| | | | | Artic/Industry | 7% | 6 tyre/Residential | 6% |
| | | | | Artic/Transport | 6% | Rigid/Industry | 5% |
| | | | | 6 tyre/Utility | 5% | Artic/Industry | 5% |

TABLE 26.16  GLTS 1971: Deliveries Analysis: Vehicle Types used to Serve Different Land Uses

| Commodity | Vehicle types | External vehicle split | Internal vehicle split | Key vehicle – land use linkages: External | % | Key vehicle – land use linkages: Internal | % | % of all (Group) (I or E) tonnage |
|---|---|---|---|---|---|---|---|---|
| FOODS<br>E : Shops 47%<br>I : Shops 62% | 4 tyre/2 axle<br>6 tyre/2 axle<br>Multi axle rigid<br>Articulated | 15%<br>58%<br>13%<br>15% | 41%<br>50%<br>7%<br>3% | 6 tyre/Shops<br>6 tyre/Commerce<br>4 tyre/Shops<br>Artic/Commerce | 32%<br>12%<br>8%<br>7% | 6 tyre/Shops<br>4 tyre/Shops<br>4 tyre/Residential<br>Rigid/Shops | | 37%<br>19%<br>13%<br>5% |
| CHEMICALS<br>E : Industry 34%<br>I : Shops 41% | 4 tyre<br>6 tyre<br>Rigid<br>Artic | 29%<br>38%<br>15%<br>16% | 59%<br>36%<br>4%<br>1% | 6 tyre/Industry<br>6 tyre/Commerce<br>4 tyre/Commerce<br>4 tyre/Shops | 18%<br>12%<br>12%<br>11% | 4 tyre/Shops<br>6 tyre/Shops<br>4 tyre/Industry<br>6 tyre/Industry | | 24%<br>16%<br>10%<br>9% |
| METALS<br>E : Industry 52%<br>I : Industry 34% | 4 tyre<br>6 tyre<br>Rigid<br>Artic | 45%<br>34%<br>9%<br>12% | 44%<br>50%<br>3%<br>4% | 4 tyre/Industry<br>6 tyre/Industry<br>4 tyre/Residential<br>6 tyre/Commerce | 29%<br>15%<br>9%<br>8% | 4 tyre/Industry<br>6 tyre/Industry<br>6 tyre/Residential<br>4 tyre/Residential | | 18%<br>14%<br>11%<br>10% |
| MACHINERY<br>E : Commerce 28%<br>I : Industry 24% | 4 tyre<br>6 tyre<br>Rigid<br>Artic | 38%<br>46%<br>3%<br>13% | 63%<br>26%<br>7%<br>4% | 6 tyre/Industry<br>4 tyre/Commerce<br>6 tyre/Commerce<br>6 tyre/Utility | 16%<br>16%<br>12%<br>11% | 4 tyre/Industry<br>4 tyre/Offices<br>6 tyre/Industry<br>4 tyre/Commerce | | 16%<br>13%<br>8%<br>7% |
| ELECTRICAL<br>E : Industry 36%<br>I : Shops 26% | 4 tyre<br>6 tyre<br>Rigid<br>Artic | 33%<br>45%<br>9%<br>14% | 66%<br>32%<br>–<br>1% | 6 tyre/Industry<br>6 tyre/Shops<br>4 tyre/Industry<br>6 tyre/Commerce | 20%<br>12%<br>11%<br>11% | 4 tyre/Residential<br>4 tyre/Shops<br>4 tyre/Industry<br>6 tyre/Shops | | 19%<br>15%<br>10%<br>10% |
| VEHICULAR<br>E : Transport 32%<br>I : Transport 45% | 4 tyre<br>6 tyre<br>Rigid<br>Artic | 31%<br>30%<br>–<br>39% | 91%<br>9%<br>–<br>1% | 4 tyre/Transport<br>Artic/Shops<br>Artic/Commerce<br>6 tyre/Open & Water | 21%<br>20%<br>9%<br>7% | 4 tyre/Transport<br>4 tyre/Shops<br>4 tyre/Industry<br>4 tyre/Residential | | 42%<br>13%<br>11%<br>10% |

Table 26.16 (cont'd)

| Commodity | Vehicle types | External vehicle split | Internal vehicle split | Key vehicle – land use linkages External | % | Key vehicle – land use linkages Internal | % of all (I or E) tonnage |
|---|---|---|---|---|---|---|---|
| **TEXTILES** E : Shops 35% I : Residential 27% | 4 tyre | 30% | 46% | 6 tyre/Industry | 19% | 4 tyre/Shops | 16% |
| | 6 tyre | 50% | 51% | 6 tyre/Shops | 17% | 4 tyre/Residential | 15% |
| | Rigid | 14% | 1% | 4 tyre/Residential | 11% | 6 tyre/Residential | 12% |
| | Artic | 5% | 3% | 4 tyre/Shops | 10% | 4 tyre/Industry | 11% |
| **BUILDING** E : Commerce 25% I : Residential 38% | 4 tyre | 17% | 37% | 6 tyre/Commerce | 16% | 6 tyre/Residential | 20% |
| | 6 tyre | 38% | 51% | 6 tyre/Industry | 10% | 4 tyre/Residential | 18% |
| | Rigid | 4% | 9% | 6 tyre/Utility | 9% | 6 tyre/Utility | 7% |
| | Artic | 32% | 28% | 4 tyre/Shops | 7% | 6 tyre/Shops | 6% |
| **PAPER** E : Industry 22% I : Shops 27% | 4 tyre | 37% | 61% | 6 tyre/Shops | 12% | 4 tyre/Shops | 20% |
| | 6 tyre | 51% | 36% | 6 tyre/Industry | 12% | 4 tyre/Offices | 13% |
| | Rigid | 3% | 13% | 6 tyre/Transport | 11% | 6 tyre/Offices | 8% |
| | Artic | 5% | 13% | 6 tyre/Commerce | 11% | 6 tyre/Residential | 8% |
| **WASTES** E : Utility 100% I : Utility 66% | 4 tyre | – | 13% | 6 tyre/Utility | 100% | 6 tyre/Utility | 48% |
| | 6 tyre | 100% | 73% | – | – | 6 tyre/Commerce | 14% |
| | Rigid | – | 14% | – | – | Rigid/Utility | 13% |
| | Artic | – | – | – | – | 4 tyre/Utility | 5% |
| **FUELS** E : Transport 45% I : Residential 66% | 4 tyre | – | 19% | 6 tyre/Transport | 27% | 6 tyre/Residential | 43% |
| | 6 tyre | 46% | 63% | Artic/Commerce | 15% | 4 tyre/Residential | 12% |
| | Rigid | 22% | 17% | Artic/Transport | 12% | Rigid/Residential | 11% |
| | Artic | 32% | 1% | Rigid/Commerce | 7% | 6 tyre/Industry | 6% |
| **TOTAL** E : Commerce 23% I : Shops 34% | 4 tyre | 26% | 49% | 6 tyre/Shops | 12% | 6 tyre/Shops | 16% |
| | 6 tyre | 47% | 44% | 6 tyre/Commerce | 11% | 4 tyre/Shops | 15% |
| | Rigid | 11% | 5% | 6 tyre/Industry | 11% | 4 tyre/Residential | 12% |
| | Artic | 16% | 2% | 4 tyre/Shops | 6% | 6 tyre/Residential | 8% |
| | | | | Artic/Commerce | 6% | 4 tyre/Industry | 6% |

TABLE 26.17   Indicators by scale of area

| | Small town | City | Metropolitan area (county scale) | National |
|---|---|---|---|---|
| Tons delivered | ✓ | ✓ | ✓ | ✓ |
| by land use | × | ✓ | ✓ | × |
| by commodity | × | ✓ | ✓ | ✓ |
| by vehicle | ✓ | ✓ | ✓ | × |
| by mode | × | ✓ | ✓ | × |
| | | | | |
| Ton/km | × | ✓ | ✓ | ✓ |
| by vehicle | × | ✓ | ✓ | ✓ |
| by commodity | × | × | ✓ | ✓ |
| by mode | × | × | ✓ | ✓ |
| | | | | |
| V/km | ✓ | ✓ | ✓ | ✓ |
| by commodity | ✓ | ✓ | ✓ | ✓ |
| by type | ✓ | ✓ | ✓ | ✓ |
| | | | | |
| Deliveries made | | | | |
| by size | × | ✓ | ✓ | × |
| by land use | × | ✓ | ✓ | ✓ |
| by vehicle | ✓ | ✓ | ✓ | × |
| by mode | × | × | ✓ | ✓ |
| by time of day | ✓ | ✓ | ✓ | × |
| | | | | |
| Vehicles | | | | |
| registered | × | × | ✓ | ✓ |
| licensed | × | × | ✓ | ✓ |
| scrapped | × | × | ✓ | ✓ |
| through/stopping ratio | ✓ | ✓ | ✓ | × |

None of these indicators cover the structural parameters of the commodity movement which require links between these major factors with the addition of land use (Tables 26.2, 26.3). Presently available data sources are not sufficient to monitor commodity movements below the county (or London borough) level, and commodity movement data at this level will soon not become generally available again. Vehicle-movement data is much more readily available, and at national, county and metropolitan level classified counts are regularly obtained. At present commodity data is not collected, as the data-collection process does not cater for an interview. There is little reason to doubt that this could not be done on a regular basis, given the survey resources required. As there is a substantial amount of empty running of goods vehicles, interview methods will be needed to monitor commodity movement.

For smaller urban areas (or parts of metropolitan districts) the balance between 'through' and 'stopping' traffic is of considerable importance, as commodity movements and commodity needs in the area concerned could differ substantially.

Licensing and registration data are considerably more useful and pertinent to the larger areas, and traffic commissioners offices could become an effective source of vehicle-monitoring data for such regions.  In smaller urban areas the value of such information for monitoring commodity movement is far less direct.  Data on vehicles registered, licensed and scrapped will not give any direct information on the degree of *use* made of vehicle fleets, which may have many vehicles laid up when commodity demand falls off.  Monitoring changes in concentration, commercial and retail activity will provide a direct indication of the demand for urban-commodity movement other than through-vehicle traffic.

DATA SOURCES

The data sources on which to base indicators for urban commodity movement are of different kinds, and relate to tonnage delivered, tonnage handled, ton/km.  These are listed in Table 26.18 in the same format as in Table 26.17: it is unlikely that any other city or county will have any greater regular statistical coverage.

TABLE 26.18   UK data sources for indicators of urban commodity movement: regularly available

| | Small town | City | County/Metro. | National |
|---|---|---|---|---|
| Tons delivered | × | × | RGTS | RGTS |
| by land use | × | × | × | × |
| by commodity | × | × | RGTS | RGTS |
| by vehicle | × | × | × | × |
| by mode | × | × | NPC,BR,BWB,RGTS | NPC,BR,BWB,RGTS |
| Ton/km | × | × | RGTS | RGTS |
| by commodity | × | × | RGTS | RGTS |
| by vehicle | × | × | × | DOE |
| by mode | × | × | BR,BWB,RGTS,AMLBO | RGTS, BR, BWB |
| Vehicle k/m | × | × | DOE 1300pt survey | DOE 1300pt survey |
| by commodity | × | × | × | × |
| by type | × | × | DOE/LA surveys | DOE/LA surveys |
| Deliveries made | × | × | × | × |
| by size | × | × | × | × |
| by land use | × | × | × | × |
| by vehicle | × | × | × | × |
| by mode | × | × | × | × |
| by time of day | × | × | × | × |
| Vehicles registered | DOE/LA | DOE/LA | DOE/LA | DOE/LA |
| licensed to operate | TC | TC | TC | TC |
| scrapped | DOE/LA | DOE/LA | DOE/LA | DOE/LA |
| through/ stopping ratio | × | × | × | × |

Key: RGTS  - DOE Road Goods Transport Survey
     NPC   - National Ports Council
     BR    - British Rail

BWB   - British Waterways Board
AMLBO - (London) Association of Master Lightermen and Barge Owners
TC    - Traffic Commissioners
DOE   - UK Department of the Environment
LA    - Local Authorities: second tier Government in the UK

Table 26.18 shows how limited is the regularly available
information.  There are a number of local freight surveys
which pick up one or more of these indicators on an *ad hoc*
basis, and as many of the characteristics of commodity movements remain stable from year to year, it is clear that
annual monitoring would not be necessary for many indicators.
If, however, planning and economic policies are introduced to
alter the location of freight generators, or to constrain the
manner of commodity movement, then the need for an annual or
biennial sample survey becomes apparent.  Consumption and
demand for commodity movement will continue to be linked to
economic growth, but the manner and location of commodity
movement and demand can be altered irrespective of the comparative stability of the final demands.

Generally, most of the structural interactions discussed
in this paper can be adequately traced by the national
research programme under way, or by slight changes in its
emphasis.  The effects on the ground in specific urban areas
will, however, demand specific data to be collected.  A basic
set of monitoring surveys would fall into the categories:

1  origin-destination questions    (through traffic: visiting
   on cordons:                      tonnage/vehicles)
2  commodity questions on screen   (ton km by vehicles and
   lines and sample points:         commodity type)
3  delivery/collection activity    (double handling, double
   or throughput on a sample of     counting, and the links
   land uses/freight generators:    between vehicle/km and
                                     tonnage registered
                                     deliveries).

Surveys of type 1 would be suitable for small towns and
cities, where types 1,2 would be essentially a single
exercise.  For small cities, local transport planning would
also require type 3 but this would be comparatively simple
to obtain.

Metropolitan areas would need to set up a bi- or tri-
ennial sample survey of type 3, and an annual survey of type
2, wherein type 1 would form a subsidiary part.  In big city
or planning regions the RGTS enhanced by greater vehicle and
geographical detail, would serve as a basic movement monitoring tool in the background.  Attention should be paid to
collating and utilising commodity consumption and production
estimates in urban areas as a basis for monitoring the degree
of freight movement actually taking place.  Census of Production and Distribution frameworks might be the least complex
approach to this end.

A regular sample enquiry based on licensing records would
cover over half of commodity movement, but the remainder is
based elsewhere - even in as large a region as London - and
would still require coverage of freight movements on the
ground.

The main policy areas of interest for public policy
impacts are those affecting movement patterns, types and
locations: movement patterns require local monitoring on a
regular basis.  Movement types do also, while movement
between locations is also subject to land-use and location
policies and will best be monitored accordingly.


CONCLUSIONS

Indicators of urban commodity movement are not readily
available and would require complementary indicators of
vehicle and land-use activity to be effective for the monitoring of local and regional transport planning.  The influence
of demand factors and the structure of the distribution
industries on patterns of service will both be more clearly
understood when current national and local investigations are
further advanced.  Beyond a brief list of vehicle-oriented
and survey-based indicators, it is not yet clear which
structural factors should be included in any monitoring
process.  Until then, vehicle-based survey methods will
suffice to monitor the movement levels and impacts as directly perceived by the public and by other road users and
frontagers.

In view of the fairly static levels of demand and movement
needs serviced by the freight industry, annual national
surveys (such as RGTS) could fulfil the major needs of larger
areas, if the surveys were to be slightly modified.  To
monitor changes in the manner in which this basic demand
affects a specific area will need local surveys.  The
frequency and scale of them will depend on the types of
policy envisaged for the districts concerned.

This chapter has not covered the complementary issues of
environmental and economic impacts, as the requirements for
environmental indicators are less directly involved with the
data needs to monitor transport policies and programmes than
the specific policy balance which might be sought through
them.  It is arguable that indices of transport cost should
form part of any set of monitoring processes, but the existence of several scales of transport costs (e.g. commercial
motor tables), and the ambiguity of the 'transport cost'
component of the Census of Production justifies separate
investigation in order to separate out the costs and transfer
costs inherent in existing and potential available series of
data.

The vulnerability to rapid change of the patterns of
vehicle and commodity movements - which at first sight appear
to be more stable than passenger travel behaviour as described by cross-sectional LUTS - has been demonstrated.  The
need for analytical methods designed to forecast the short-
run responses of the freight industry and its uses is clearly
greater than that for further refinements of trip generation
and distribution functions based on cross-sectional data.
The intimate links between land use, consumption patterns,
PDM factors of stockholding, service level and distribution,
and vehicle fleet interactions with drop sizes and service
levels, strongly indicate that an efficient and selective

technique for monitoring of operations and impacts is likely
to be of considerable practical value.

REFERENCES

1. Baker, L.L.H., 'A planning parameter allocation model
for Greater London', *Greater London Council, Research
Memorandum 460* (1975).
2. Barber, J., 'Appraisal of existing freight models',
*Mathematical Advisory Unit Note No. 204*, Department of the
Environment, London (1971).
3. Berthoud, R., Kenyon, G., Tyrell, W., Walker, D.,
Weald, D.E. and Yates, L.B. (1975) *The Greater London
Transportation Survey, Goods Vehicle Survey Report 3, Goods
Vehicle Survey Technical Manual 6*, Greater London Council,
London.
4. Department of the Environment (1968) *The Road Goods
Transport Survey*, Department of the Environment, London.
5. European Conference of Ministers of Transport (1976)
*Freight Collection and Delivery in Urban Areas, 31st Round
Table on Transport Economics*, European Conference of
Ministers of Transport, Paris.
6. Fryer, J.A., Hasell, B.B. and Wigan, M.R., 'Goods
vehicle activity in Greater London', *Greater London Council
Research Memorandum 491* (1977).
7. Her Majesty's Stationery Office (1973) *Heavy Commercial
Vehicles (Control and Regulations) Act 1973* (The Dykes Act),
London.
8. Heyman, R., 'Initial attempts to model freight flows',
*Mathematical Advisory Unit Note No. 211*, Department of the
Environment, London (1971).
9. Hitchcock, A.J.M., Christie, A.W. and Cundhill, M.A.,
'Urban freight: preliminary results from the Swindon freight
survey', *Transport and Road Research Laboratory Special
Report 126UC* (1974).
10. Howard, R.A. (1968) *The Movement of Manufacturing
Industry in the United Kingdom 1945-65*, Her Majesty's
Stationery Office, London.
11. Talbot, M.F., 'Warehouse traffic generation study',
*Greater London Council Research Memorandum 400* (1974).
12. Wigan, M.R., 'The scope for transshipment and consolidation operations in London: an analysis of available goods
vehicle and commodity movement data', *London Freight Conference Background Paper 8*, Greater London Council, London
(1975).

Chapter 27

URBAN GOODS MOVEMENT: RESEARCH REVIEW[*]

Keith J.G.Smith

INTRODUCTION

In recent years new areas of research have often encumbered themselves with a complex language derived largely from American sociological studies of the 1950s and 1960s. Some urban goods-movement (UGM) researchers at the moment seem to be set on the same course. It is imperative that UGM research does not become an esoteric movement, understood only by the initiated. Its purpose is to deal with real problems in the real world and therefore both its framework and its language should be as simple, clear and realistic as possible. How quickly a new area of research can become divorced from reality can be seen in the related field of physical distribution management where, in a short time, researchers lost touch with practitioners. This led to a situation in which the managing director of one of the largest total distribution companies in Great Britain had recently to call for a return to the use of four letter words (such as 'vans' and 'shed' let it be added!) in order to restore a little sanity to the subject. The plea is therefore for us to avoid the necessity of being called down from our ivory towers.

There is, however, another tendency to which UGM researchers could all too easily succumb (some have already done so), and that is to carry out research for its own sake. Whilst this can sometimes be justified in areas such as history or astronomy, it is surely not to be encouraged in UGM. It is noticeable that many researchers pay lip service to the idea of conducting meaningful and realistic research, and yet become so interested in the *process* of conceptualisation or modelling for instance, that they lose touch with the very problems they are supposed to be solving. It is surely difficult to keep to a straight and narrow path, but the attempt must certainly be made.

NEED FOR RESEARCH INTO URBAN GOODS MOVEMENT

Turning to the question of why we are conducting research into UGM, let us first of all admit that there are those

*This chapter is a non-modellers viewpoint of research requirements in urban goods movement. It is designed to provide an additional direction to that emphasised in Chapters 24, 25 and 26.

amongst us to whom it is as a new-found toy.  It is, at the
moment, fashionable to study UGM.  This applies particularly
to some of those people who treat it as an extension to the
studies that have taken place in the field of people movement.  Many of them appear to see it as an interesting parallel to their own studies.  This is a negative approach.  It
encourages modelling almost for its own sake, when more
attention should be directed towards problems and solutions
in the political context in which they arise.  UGM has become
an important area of research precisely because it has become
a political issue.

In Great Britain for instance, the early 1970s saw the
heyday of various environmental pressure groups.  Great concern was expressed by the public, through them, at the environmental damage being caused by heavy goods vehicles (HGVs)
and public demand for action led to a political response.
Particularly notable in this respect was the passing of the
Dykes Act in 1973* which charged local authorities with undertaking a comprehensive review of the movement of HGVs in their
areas, and with drawing up measures to control such movements
in and through environmentally sensitive areas.  The government was also rushed into planning a national lorry route
network.  The need for action led to the need for research,
and thus there was instituted a government-funded programme
of research, the general objective of which could be said to
be the identification of areas in which government could make
a contribution to the reduction of the generalised social
costs of UGM,and the assessment of the benefits and disbenefits of intervention.

It is interesting to follow through the example of what
took place in Great Britain a little further at this point.
At first local authorities, confronted with the problem of
deciding upon a goods-movement policy, envisaged that their
actions would be regulatory and almost wholly negative as far
as the freight industry was concerned.  Indeed, this point of
view was encouraged by central government for political
reasons.  However, as they came to have a better understanding
of goods movement, its complexity and its cost, they came to
realise that most of the controls they had considered would
produce considerably more monetary disbenefits than benefits.
In the context of an inflationary economy, the government
was particularly anxious not to introduce any measures which
would add to the inflationary spiral, and so reversed its
earlier exhortations for comprehensive regulation of goods
movements by local authorities in favour of minimal action,
and a policy of 'scratch where it itches'.  This in itself
was not such a negative policy as it sounds, for central
government and some local authorities perceived that the real
need now was for positive measures designed to help the
freight industry - to increase its efficiency and to mitigate
its social effects.  Therefore a new phase of research was
instituted, partly via a joint government/industry forum (the

*The Heavy Goods Vehicles (Controls and Regulations) Act
1973, generally known as the Dykes Act because of its
sponsor Hugh Dykes, MP.  It defined an HGV as a goods
vehicle of over 3 tons unladen weight.

Lorries and Environment Committee), which had positive
objectives such as the design of better vehicles and the
elimination of empty running when possible (2).

   This description of the political processes at work in
Great Britain over the past few years has introduced two
main reasons for studying UGM: first, to enable control of
goods-vehicle movements when environmental reasons dictate
that this should be necessary; secondly, to enable, if
possible, an increase in the efficiency of the UGM industry.
Note that it should *not* be an objective of studies to model
the UGM process as a whole.  Modelling may or may not be a
technique to be employed, but is not an objective in itself.
Some recent, and regrettably some current, research does,
however, appear to have modelling as its sole objective.


SOME APPROACHES

Work carried out so far has indicated that UGM is much too
complex a process to model as a whole.  Indeed, there is
little point in attempting to do so when the implications of
our reasons for studying it are considered.  The type of
model envisaged by its proponents has the purpose of predicting such variables as the probability of a given receiver
selecting a certain shipper, the probability of a shipper
selecting a certain level of service, the amount of goods
likely to be generated by any particular shipper, and so on.
The efficient functioning of the model would ostensibly
enable an understanding of the UGM process, and would allow
for the testing of various options or constraints.

   This type of model could never truthfully represent the
UGM process.  It cannot encompass all the quantifiable
variables let alone those that are unquantifiable or difficult
to quantify.  And, in many situations, it is the latter
variables that are decisive as far as the shipper is concerned.  It is unrealistic to make generalisations about, say, a
shipper's probable course of action because what he does is
dependent upon what type of commodity he is shipping, where
he is sending it to, what other goods are being sent to the
same area, what facilities are available, what service levels
are laid down by his firm, whether he has special arrangements
with certain hauliers, and so on.  The variables are so
diverse and so numerous that each shipper represents a
special case, and indeed each shipper's decisions on each day
can represent a special case.

   Given that the first objective is to find a method which
enables assessment of the effects of various controls on
goods movements, the basic information required is the form
goods movements take at the moment, and the nature of
vehicle operators' reactions to the various controls.
Because UGM is so complex, its characteristics will vary from
country to country, region to region, town to town.  This
means that it is necessary to examine, town by town, the
effects of the policy options under consideration.  What is
needed, therefore, is a model limited to representing the
existing patterns of a town's goods movements such that any
one town's activities can be an input.  The next stage would

be to interview the operators of vehicles to assess how they
would react to the controls posited. This information would
then be fed back into the model, and changes in the pattern
of activities could then be measured for their effect on both
private and public costs. This limited model would achieve
our first objective, without modelling the UGM process as a
whole. It merely examines an existing set of circumstances
in one location, posits a certain set of constraints, and
measures their effect. It is not concerned with such devices
as the probability of a certain amount of goods being
generated by a certain type of shipper, because it deals with
actual vehicle movements (see for example Figure 27.1).

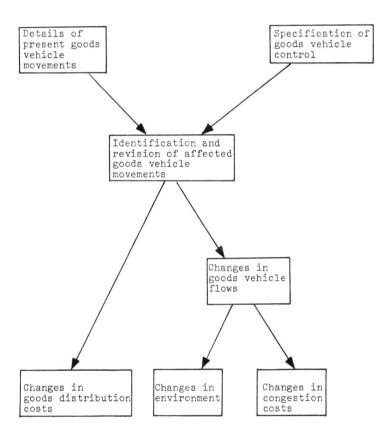

Figure 27.1 Schematic Representation of the Swindon Freight
Model

Such a model has in fact been developed by the Transport
and Road Research Laboratory (TRRL) in England, using data
for the town of Swindon (1,3,4,5). Its principles are also
being applied to two other urban areas, Hull and a sector of
London. The modelling process described does have its weaknesses. A substantial data collection exercise must take
place, although the realism of the model more than outweighs
this. But, more important, the operators' stated attitudes
towards a hypothetical list of controls do not of course
guarantee an identical reaction in reality. It is true that,
because of this, flexibility was built into the model by
allowing two alternative inputs other than the operators'
stated attitudes. These were based on a critical analysis of
transport objectives, and on preferences decided by the model
user himself. Neither of these is, for obvious reasons,
entirely satisfactory. Another weakness of the model is
inherent in the difficulty of trying to measure environmental
and congestion costs. TRRL has developed a submodel to
assess these costs, but its reliability is dependent on the
extent to which it is held possible to quantify such costs.

Nevertheless, the model uses 'live' data and its development does allow the assessment of controls on HGVs in particular urban areas. It is unlikely that further research
will substantially improve on it, except perhaps by refinement
of some of the measures used in it. At the time of writing,
researchers in other parts of the world appear to be unaware
of its existence, which is unfortunate because it does seem
to make many of their attempts to conceptualise or model the
UGM process passé. The developers of the model have surely
recognised that there is no easily identifiable UGM process -
thus their decision for each town to be investigated individually.

Of course, having run the model it is possible to make
certain generalisations about the type of town, whether it
has a good system of primary and secondary routes, whether it
has a by-pass, and whether environmentally sensitive areas
and non-sensitive areas are on the whole separated, for
instance, and then relate the findings to other towns of a
similar size and nature. But this will be merely a pointer
and not a definite indicator, although such a process would
still give more realistic results than other more abstract
models.

The second objective stated was to find ways of increasing
the efficiency of the UGM industry. Again, the trend of
researchers has been to discuss UGM as one process. This is
misleading. It is a series of processes, some related, some
not, and in connection with the second objective it can only
be examined at the level of the individual process. What is
this level? It is really the activities of a single firm,
although firms handling one particular commodity can be
grouped together usefully provided that the differences between firms owning their own vehicles and those not doing so,
firms engaged solely in manufacturing and firms involved in
other sectors of the economy, large firms and small firms,
and many other distinctions are remembered.

If we are to examine ways of making the UGM industry more
efficient - such as the improvement of vehicles or the
introduction of new operating methods - it is essential to
investigate the freight industry commodity by commodity, and
when investigating each commodity to cover as broad a range
of firms as possible in as disaggregate a way as possible.
Only then will the researcher or planner be able to see the
effects on the ground of various solutions he puts forward.
It is academic arrogance to discuss concepts such as transshipment depots or consolidation of routes, or relocation of
freight facilities, in the abstract, although this has often
been done.  The way forward must be for planners and the
freight industry jointly to set the objectives of research,
and for them jointly to carry out the research at a disaggregate level.


PROPOSED MODELLING APPROACHES

The papers presented at the UGM workshop (Chapters 24,25 and
26) appear to form a cohesive trilogy.  That by Rimmer and
Hicks sets out to provide a framework within which the urban
goods movement process can be considered.  That by Wigan draws
links between possible indicators and land use, economic,
environmental and social impacts and shows the relation
between data sources, policy issues and indicators for monitoring purposes.  Finally, that by Roberts and Kullman outlines the development of a behavioural demand model for UGM.
Closer investigation, however, reveals that they do not hang
together as nicely as could be wished.

    The title of the paper by Rimmer and Hicks, 'Urban Goods
Movement: Process, Planning Approach and Policy' indicates
at once that they see UGM as a cohesive, albeit complex
process, which I have argued it is not.  In their introduction
they seem unaware of the work undertaken in Great Britain and
they state that the main research thrust has been to model
UGM in an analogous fashion to urban people movement, and
this lack of information has influenced their approach.
Their aim is to provide a framework, in other words to conceptualise the UGM process, and they correctly point to the
need to understand UGM before outlining a planning approach
and examining policy options and their likelihood of attainment.  However, the crucial point is *how* we can come to
understand UGM, and here the development of a conceptual
model by the authors is a red herring.  It promulgates the
myth of a cohesive UGM process and tends to hide the complexity of movements involved, each of which needs separate
analysis.

    Considering their avowal to 'break away from the model as
the central issue in studying UGM', it is strange that they
adapt the terminology and methodology of Smith and Hay.  The
build-up and description of their graphical model of UGM
does not appear to be very helpful in the context of this
paper, and Rimmer and Hicks seem to be on surer ground when
they turn to a discussion of consignments and pick-up and
delivery patterns, although here the categorisations are a
little idealised; for instance the division of lorries into
'run trucks' and 'floaters' could not be applied to many
businesses.

Rimmer and Hicks state that the recognition of four basic
activity sequences in UGM (consolidation, direct, distribution and composite) should lead to a more sensitive modelling
procedure 'in matching the load system (consignments and
vehicles) with the capacity of the road system *if this is
still to be the main task in studying UGM*'.  The emphasis is
mine, and draws attention to that lack of clarity concerning
objectives which conceptualisation of the process of UGM
leads to.  It is as if the researcher is trapped in a self-
generating educative process in which the original aims
become blurred.

In discussing the role of the political decision-maker and
the transport planner, a new system is emphasised in which
there is mutual assistance between all participants in
specifying problems, alternative plans, and relevant
evaluation issues.  This relates very closely to the function
of the British Lorries and Environment Committee, but instead
of developing how this sort of forum could work and how it
could best effect a programme of research, the authors revert
to modelling - in this context, an attempt 'to resolve
apparently conflicting interests in the UGM process could be
provided by an activity/transport gaming-simulation'.

Under 'Planning Approach' Rimmer and Hicks adopt Hensher's
schema with its assessment of policy options in terms of
their certainty of success.  This incremental planning
approach is certainly wise, but surely it must also be
mirrored  in our development of an understanding of UGM?  In
other words, rather than attempting to conceptualise the UGM
process, we must build incrementally towards our understanding of UGM by examining features of it.

It has been accepted by some researchers that the government's role in UGM is to attempt to minimise its generalised
social costs.  Rimmer and Hicks conclude that to do this more
use should be made of regulations and directives.  Indeed,
they suggest that a prime aim of researchers should be to
assist in drawing up directives.  Regulatory devices can,
they say, be most refined and accurate weapons.  British experience shows that the *differential* effects of such measures
as control of HGVs by time or weight category for instance,
can be disastrous.  The lesson is that such devices must be
used with great care and only when absolutely necessary.
This does not detract from the conclusion that researchers
should assist in drawing up directives, but only emphasises
that it should not be their prime role.  The researchers'
role must surely be to investigate the present patterns of
UGM, to discover why such patterns exist, and then to suggest
pragmatic improvements.  Many of these improvements will be
positive rather than regulatory.

The paper by Rimmer and Hicks reveals the fundamental
dichotomy in current research between the search for
practical solutions in specific problems, and the treatment
of UGM as an academic problem requiring an academic solution.
It is interesting that in the introductory paragraphs they
state that 'The key issue of this paper ...is the need to
understand the UGM process before outlining a planning
approach and examining policy options and their likelihood

of attainment' and yet later on they state that 'As it is
difficult, given the present state of the art, to unravel the
complexity of the urban goods process, it is more realistic
to adopt a planning approach ...'. This appears to be a hint
that the conceptualisation of the UGM process *is* a red
herring, but their conclusion does not bear this out.

Wigan's chapter, 'Indicators for Urban Commodity Movements'
is valuable reading for anyone wishing to conduct research
into UGM. It raises a number of extremely valid questions
and goes some way towards setting out an appropriate course
of action, although it would have been very useful if a
little more space had been devoted to the latter purpose.
The objective is, however, clear enough, 'to find a useful
set of indicators to represent the balance and changes in
level of urban commodity movements in a manner appropriate
to monitoring transport policy local and national objectives'.
This is important because conventional land-use transport
studies are a mere camera shot, giving us a picture of one
moment in time. We do need rather more than this if we are
to see the effects of policy and planning changes.

Wigan considers the different types of variables involved
in commodity movements and concludes that the types of data
required for monitoring indicators are not widely available
at a suitable level of detail, or with sufficient coverage,
to provide trends over time. Indeed 'beyond a brief list of
vehicle-oriented and survey-based indicators, it is not yet
clear which structural factors should be included in any
monitoring process'. Therefore, vehicle-based survey methods
will have to suffice for the moment because they do measure
both movement levels and impacts 'as directly perceived by
the public and by other road users and frontagers'. Bearing
in mind survey costs, perhaps they are effective enough to
suffice. Wigan, however, suggests that modified annual
national surveys (such as RGTS) could fulfil the major needs
of larger areas, as long as they are coupled to local surveys
to monitor changes on the ground. Local surveys would be
necessary in smaller urban areas. The cost of such surveys
is certainly a worrying factor, although the author hedges
against this by stating that the frequency and scale of them
would depend on the types of policy envisaged for the
districts concerned.

This raises a very important issue. For, unlike much of
the current research in UGM, Wigan proposes a policy-based
programme of research. Considering the complexity of UGM,
this is realistic and deserves serious consideration. It is
the sort of programme currently being carried out in Great
Britain. Wigan breaks through a lot of the mumbo-jumbo
surrounding the problems of conceptualising and modelling
UGM, and realises that there are a set of problems associated
with UGM and there are a set of solutions. Given that, all
we have to do is to discover a means of testing and measuring
policy effects - that should be the major aim of the UGM
researcher. Rimmer and Hicks moved towards this view during
their paper, but insisted that it is vital to conceptualise
the UGM process prior to policy evaluation. They stated
that it is essential to do this in order that any data to be
collected 'reflects the essential nature of the process.'

What the data ought to reflect, as Wigan proposes, is the
nature and pattern of movements before a given solution, and
afterwards, rather than the essential nature of the process
itself.  This may seem to be a subtle difference, but it
represents a difference in emphasis.  Rimmer and Hicks are
more concerned with the UGM process itself, Wigan with the
problems to which it gives rise.  It is a question of at what
point the researcher should break into the study area.

It is a pity that Wigan did not examine the issue of the
environmental impacts of UGM, as this is central to our
reasons for research.  Also, the requirement for suitable
environmental indicators is linked up inextricably with the
data needs for monitoring transport policies and programmes.
To a large extent, data collection will be constrained by the
sort of environmental indicators in use.  The importance of
the subject of data classification is demonstrated neatly in
the comparison of daily tonnages delivered by road in Swindon
and the GLC whereby the regrouping of data classifications
makes two dissimilar urban areas appear to have similar
characteristics.  The section devoted to an analysis of
employment densities and freight land use is particularly
relevant in view of the assumptions underlying much of the
current programmes of land-use incentives, and it ought to be
compulsory reading for all the local authority staff
concerned.  Because of its importance in causing a large
amount of double counting, and because it is vulnerable to
swiftly changing operational factors, warehousing is picked
out for special mention.  The relevance of Wigan's findings
here cannot be disputed, but a little more depth to the data
used in the examples would have been welcome.

The most significant sentence in the paper, however, is
to be found in the last paragraph of all, 'The need for
analytical methods designed to forecast the short-run
responses of the freight industry and its uses is clearly
greater than that for further refinements of trip generation
and distribution functions based on cross-sectional data'.
This is central to the whole basis of our further study of
UGM, and Wigan in his paper has set us firmly on the right
sort of path to follow, a path based very much on his
delineation of what is possible and what is practicable in
the way of monitoring procedures.

Before analysing the paper by Roberts and Kullman, 'Urban
Goods Movement: Behavioural Demand Forecasting Procedures',
one has to ask the fundamental question 'why model?'  Meyburg
elsewhere in this book (Chapter 28) provides an answer -
'Modelling has proved to be a useful and sometimes necessary
aid to understanding complex phenomena and processes.  It can
serve as a means of investigating relationships on which
policies can be based.  Behavioural modelling as a specific
subcategory of modelling is concerned with decision processes
which involved the evaluation of perceived alternatives.'
But it is still necessary to discriminate carefully what a
model will be used for.  Modelling can be an aid in the
taking of strategic decisions, or it can be used at the tactical level.  It is my contention that if we want to use the
model as a strategic aid, in planning road networks or lorry
routes for instance, then a simple model using aggregate data

will be robust enough for most purposes. Disaggregation of data and procedures will serve no useful purpose and indeed may be positively harmful and misleading. Experience has shown that disaggregate models only make sense in UGM when they deal with limited policy issues such as 'What are the benefits of a "No entry except for access" policy in such and such a town?' In such cases it is essential to get down to a fine level of detail in order to discern policy effects on the ground.

Roberts and Kullman attempt to develop a model of freight mode choice which can be combined with knowledge of commodity flows to and from a region,

> The output of the model is the probability that a particular receiving firm located in the region will secure its input from origin i in shipment size q by mode m. When this probability is multiplied by the annual use rate $A^k_{IND,SIZE}$ calculated from the industry/firm-size analysis previously described, the result is the commodity k moving from each of the known producing regions to the mode-choice/shipment size for each region for the firm under consideration.

In essence, their model should be used for strategic planning purposes and their attempt to describe its use at a tactical level is half-hearted and even a little presumptuous. The relevance of this paper to a discussion of UGM is suspect because their type of modelling procedure can only be used at a strategic level. Any attempt to apply it to, say, a single industrial estate would be unrealistic because its inherent defects, its use of averages, its disregard of non-tangibles, and all its other weaknesses, would make its use at such a level woefully inadequate. Irrespective of its relevance to UGM, how robust, how effective is the model? The biggest question mark against it is the very possibility of predicting mode choice. Anyone with inside knowledge of the freight industry would immediately suggest that there are far too *many* variables to enable any valid generalisations to be made. There are also many intangible and unquantifiable reasons for mode choice - collusion and 'under the counter' payments between traffic clerk and haulier to give but one common example.

Even if it is accepted that prediction of mode choice is possible using certain variables, the validity of Robert and Kullman's measurement of each variable is in doubt. In their introduction they state that one of their reasons for the choice of a disaggregate model was that 'theory has shown that by suppressing deviations from the mean of an aggregate data point, (aggregate) models will be biased'. It is extraordinary, therefore, to find that the majority of their variable measurements and other inputs to their model are based on simple means, in such instances as the calculation of per capita personal consumption figures for an urban area by applying national final demands to the population of the area on the assumption that consumption per capita is the same for all urban areas. Also the calculation of a freight rate applicable to a mixed shipment by using the highest rated

commodity and the combined shipment weight.  Such examples,
and there are many, grossly distort reality, and this
manipulation of a set of averages belies the title
'disaggregate modelling'.

    Roberts and Kullman merely paint their model with a
spurious reliability when they increase the number of variables they include over other models.  But there is little
point in thus including stockout costs, for instance, when
they are to be so generalised as to be meaningless, or for
that matter any of the other logistics terms they mention.
The calculation of such costs is a highly specialised procedure and both the method and the results will vary from
firm to firm.  It is not possible to generalise from a firm
handling a particular commodity to all firms of the same
size handling that commodity as would be necessary in the
authors' scheme.  Goods movement is much too complex to
generalise in this way: sometimes logistics variables will be
commodity dependent, sometimes firm-size dependent, sometimes
market-policy dependent.  There can be no one rule for a
group of companies.  For this reason, it is vital to recognise that goods movement is totally different from people
movement, and that the methodologies employed in analysis
of the latter just will not do for investigation of the
former.

    There appear to be few, if any, merits in the approach
adopted by Roberts and Kullman over more simple predictive
procedures.  There are certainly no gains in accuracy - the
reverse is more likely since each additional variable builds
in a greater chance of error.  Even the necessity for a mode-
choice model in UGM can be called into question.  In most
urban areas there is *no* modal choice: virtually all goods
(except for those using an isolated and more-or-less fixed
supply of private railway sidings) are constrained to
complete their journeys in urban areas by road.  To change
existing patterns of modal split at all significantly would
demand levels of capital expenditure and physical and
economic reorganisation that make it inconceivable.  Thus
it is that we return to Wigan's statement that 'The need for
analytical methods designed to forecast the short-run
responses of the freight industry and its uses is clearly
greater than that for further refinements of trip generation
and distribution functions based on cross-sectional data'.

    It is to be hoped that research into UGM will eventually
gel by a process of cross-fertilisation between researchers
and operators in different countries, and their different
approaches and experiences.  Here the value of an international conference such as that at Tanunda is immeasurable.
However, there is a long way to go yet, and it would
certainly help if researchers kept each other informed of
developments in their work and that of their colleagues.  A
lack of co-ordination has been a major constriction on
progress so far, and it will be a major achievement if this
situation can be righted.

REFERENCES

1. Christie, A.W. *et al.*, 'The Swindon freight study: assessment of goods vehicle controls', *Traffic Engineering & Control*, May, 252 (1977).

2. Lorries and the Environment Committee (1977) *The First Two Years*, Her Majesty's Stationery Office, London.

3. Purcell, R.M. *et al.*, 'The Swindon freight study 1. objectives, surveys and model-building', *Traffic Engineering & Control*, April, 162 (1977).

4. Transport and Road Research Laboratory, 'Urban freight: preliminary results from the Swindon freight survey', *Transport and Road Research Laboratory Supplementary Report 126 UC* (1974).

5. Transport and Road Research Laboratory, 'Swindon freight study: collection of data and construction of computer model', *Transport and Road Research Laboratory Supplementary Report 158 UC* (1975).

Chapter 28

THE APPLICABILITY OF BEHAVIOURAL MODELLING TO THE
ANALYSIS OF GOODS MOVEMENTS

Arnim H. Meyburg

STATEMENT OF SCOPE

The analysis of goods movements traditionally has been neglected in the transport-planning process. Consequently,
very little progress has been made in the development of
tools and techniques suitable for goods-movement analysis.
Yet goods movements constitute a substantial element of the
transport sector, both in terms of volume and in terms of
economic, social, environmental and traffic impacts.

The workshop investigated whether new approaches, as
presently applied to an increasing degree in the passenger
travel-modelling area, are applicable and useful in the
goods-movement context. Techniques are needed to permit
successful modelling of movement of goods and to evaluate the
consequences of alternative policies affecting freight and
passenger movements. Due to the absence of appropriate work,
the workshop was forced to concentrate on the goods-movement
process in its entirety rather than start with the investigation of specific sub-topics.

SUMMARY OF CONCLUSIONS AND RECOMMENDATIONS

*Introduction*

Through a detailed analysis of the process involved in goods
movement, the workshop developed an understanding of the
actors and their decision processes which would constitute
the elements of a behavioural-modelling approach to goods-
movement analysis. It was concluded that the only meaningful
basis for modelling any components of the goods-movement
process would be first to establish the process and its
composition in its entirety. From this basic vantage point
a number of recommendations were developed. These recommendations, out of necessity, have to be of a rather general
nature, given the scope of the workshop. Underlying the
concern with understanding the goods-movement *process* is the
conviction that modelling for modelling's sake is of no
interest to the problem-solver. Also, data collection and
modelling exercises should be determined and delimited by a
thorough understanding of the ultimate purpose of those
efforts in the context of solving goods-movement problems.
The lesson was learnt in the early large-scale passenger
transport-planning studies when data collection and modelling

took place without proper understanding of purpose and
direction.  The analysis of the travel decisions based on
human behaviour is a relatively recent development, and it
illustrates the need for completely different types of data
information and modelling techniques than those traditionally
used.

*Planning Versus Efficiency Concerns*

The freight transport sector can be viewed from two major
concerns, namely from the planning or regulatory context of
social efficiency and from a private efficiency point of view.
The former constitutes a somewhat more comprehensive outlook
with societal concern and welfare being the over-riding
issues, while the latter represents the freight industry's
viewpoint which is focused on the provision of service and
profit maximisation.

It was concluded that, in the short run, research efforts
should concentrate on those decision elements that determine
the physical flow of goods, with shippers, receivers and
operators being the prime actors whose behaviour and decisions are to be analysed.  It was felt that the effects of
decisions made by planners and regulators can only be
evaluated after the decision process directly affecting the
physical goods-movement process is well understood.  As a
consequence land-use planning and policy analyses will need
increased attention after the characteristics of the physical
distribution process have been analysed and understood.

*Urban Versus Inter-urban Goods Movements*

After careful scrutiny of the goods-movement process, it was
concluded that for analysis and modelling purposes, a
distinction between intra- and inter-urban is not meaningful
or useful.  The characteristics of the choice situations
faced by the actors in the goods-movement process are
virtually identical for the two transport contexts.  Also,
the vast majority of inter-urban freight movements have an
urban component at either end of the trip (11).

*Applicability of Behavioural Modelling*

The workshop concluded that behavioural modelling constitutes
a useful, if not the only useful approach to analyse the
goods-movement process.  This conclusion is based on the
premise that the purpose of modelling the process is its
description as well as its understanding and prediction.
Behavioural modelling is appropriate since the goods-movement
process is characterised by a number of actors facing choice
situations whose outcome affects the volume of flow, its
origin and destination, its costs, the transport mode by
which it moves, and the regulatory and planning framework
which affects the whole process.

Unlike in the passenger context, the decision-maker in the goods-movement process is not the moving element. Nevertheless, the decision-making process concerning origin, destination, volume, transport mode and route is very much akin to that in the passenger sector. The workshop concluded that research efforts should be initiated to develop appropriate behavioural models analysing the various decision stages in the goods-movement process.

*Development of Individual-choice Models*

The goods-movement process has been identified as a complex process. Modelling of the entire process is impractical at this stage. Models of individual choices faced by the actors involved in the physical-distribution process should be developed. Specifically, research should be initiated to develop shipper level-of-service choice models, operator-choice models, and receiver-choice models. The other individual choice-process models, namely shipper-generation and receiver-attraction models should be developed concurrently.

DISCUSSION OF CONCLUSIONS AND RECOMMENDATIONS

*Social Versus Private Efficiency Concerns*

The objective of the goods-movement process can be defined as the efficient movement of supplies consistent with minimised generalised activity, community and transport costs. This definition implies the involvement of actors representing governmental agencies on one side and actors representing the freight industry on the other (12). The concerns of the former can be summarised as follows:

1  Trade-off between passenger and goods-movement sectors.
2  Employment, location and economic viability.
3  Community and environmental impacts.
4  Level-of-service.

These concerns are typically realised by means of regulatory guidance and constraints. In a broad sense they could be labelled planning issues.

The actors representing the freight industry operate within the constraints set by governmental agencies. Their primary concerns focus on:

1  Level-of-service.
2  Operational and cost considerations.
3  Overall transport-system efficiency.

The freight-industry actors are obviously primarily interested in issues of industry efficiency.

This dichotomy of primary concerns is quite significant for purposes of identifying the actors whose choices are the object of a proposed behavioural analysis. The physical-distribution process lies at the heart of goods-movement

activities while the regulatory and planning concerns
surround this central physical process and interact with it.
The workshop members reached agreement that it would be
meaningful initially to concentrate research efforts on the
core process whose actors are guided by efficiency considerations.

Figure 28.1 illustrates a comprehensive view of the
freight movement process as it was discussed in the preceding
paragraphs. The 'Enterprise' side corresponds to the
industry efficiency argument developed earlier while the
'Public Sector' side corresponds to planning and regulatory
concerns. The inter-action among the four main elements in
the figure comprise both physical and informational flows.

*Urban Versus Inter-urban Goods Movements*

Transport decisions concerning goods movements affect intra-
and inter-urban trips equally. Furthermore, the two segments
of the goods-movement sector are very tightly intertwined in
terms of the physical facilities and the vehicle technologies
used as well as the actual goods-movement operations. The
decisions faced by the actors involved in the goods-movement
process are very similar for both contexts. A number of
references can be cited which offer a more detailed
discussion of relationships between urban and inter-urban
freight movements (e.g.(11)).

*Applicability of Behavioural Modelling*

Modelling has proved to be a useful and sometimes necessary
aid to understanding complex phenomena and processes. It
can serve as a means of investigating relationships on which
policies can be based. Behavioural modelling as a specific
subcategory of modelling is concerned with decision processes
which involve the evaluation of perceived alternatives.

The goods-movement process constitutes a significantly
complex process. Within this process a number of decision-
makers (actors) are faced with possible alternative courses.
of action. The evaluation of the consequences of different
courses of action would be the object of behavioural
modelling in the goods-movement area. The workshop, after
agreeing on the general definition of modelling and its
purposes, decided that the goods-movement process, or elements thereof, would be useful targets for modelling purposes.
The justification for attempting behavioural-modelling
efforts is based on the identification of actors and choice
situations within the goods-movement process. Figure 28.1
shows the actors in the goods-movement process. The systems
boundaries are defined such that a most comprehensive
picture of the physical and regulatory processes is developed.
The identifiable actors in the process are the shipper,
transport operator (vehicle and terminal operator or forwarder), receiver and government (in a regulatory, control and/or
planning function).

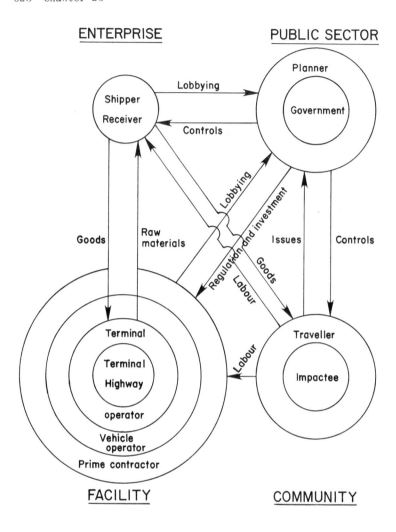

Figure 28.1   Components of the Goods-movement Process
(Adapted from P.J. Rimmer and A. Tsipouras (18))

The workshop members confined the discussion to the
physical collection and distribution process involving
shippers, transport operators and receivers only.  A good
argument can also be made for an understanding of the
physical-distribution process involving shippers, operators
and receivers as a prerequisite for generating meaningful
policy objectives to be pursued by regulatory, control and
planning agencies.  The evaluation of the consequences of
alternative policies on the physical collection and distribution element would constitute the logical follow-up to
the analysis of the physical collection and distribution
process.

Figure 28.2 constitutes a simplified illustration of the
physical goods-movement process.  The actors identified in
the figure are shippers, transport operators, and receivers.
Also indicated on the diagram are those points in the
process where decisions by these actors affect the flow of
commodities.  At each of these decision points, the actors
in the goods-movement process face specific choice situations.

*Development of Individual-choice Models*

The identifications of the decision-makers and decision
points within the goods-movement process are illustrated in
the preceding section.  In order to proceed to the specification of individual-choice models to be developed in the
course of future research activities, it is necessary to
identify the determinants of choices confronted by shippers,
operators and receivers.

Model development requires three prerequisites, assuming
that the basic objective of the modelling effort has been
agreed upon; the specification of appropriate variables;
quantifiable measures for these variables; and the stage in
the goods-movement process at which the modellers enter the
system.  The workshop concluded that the only reasonable way
of analysing the goods-movement process was to break it down
into its component elements, namely the individual choices
of the decision-makers involved in the process.

The determinants of choice which could be considered by
the shipper, transport operator or receiver, respectively are
shown in Table 28.1.  The shipper who can either be a
producer or a supplier makes his decision with respect to the
type of transport service he selects.  His decision as to
whether to use his own vehicles, if he is a private transport
operator, or whether to choose a freight forwarder or a
transport-service operator directly may be affected by the
variables listed under 'Shipper' in Table 28.1.  The importance and relevance of any one of these variables depends on
the shipper's perception of the characteristics of the
transport-service alternatives available to him.  The
appropriate measures for the decision variables are identified in the columns adjacent to the respective variables.

The transport operator deals with the choice of route and
with the question of what equipment to use for the specific

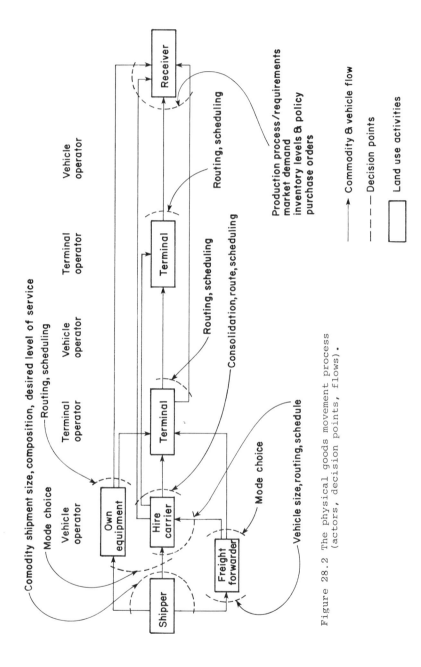

Figure 28.2 The physical goods movement process (actors, decision points, flows).

TABLE 28.1  Determinants of transport choice in the goods-movement process

| SHIPPER (Choice of transport service) | | OPERATOR (Choice of route and equipment) | | RECEIVER (Choice of supplier and volume) | |
|---|---|---|---|---|---|
| Determinants of choice | Measures | Determinants of choice | Measures | Determinants of choice | Measures |
| Rate | Money | Cost | Money | Commodity price | Money |
| Time | Hours | Charges (rates) | Money | | |
| Loss propensity | Insurance costs | Volume | Weight, dimensions, density | Perceived service quality | Ranking |
| Probability of delay | Probabilities | | | Potential delay | Hours |
| Shipper control | Yes/No | | | | |
| Availability of appropriate vehicles and capacities | Yes/No | | | | |
| Consignment characteristics | Weight, volume, density, perishability | | | | |

shipping task on hand.  As indicated in Table 28.1, his
decision is determined by consideration of costs, the rate
structure set by government or through industry competition
and by the volume of commodities involved in the shipment to
be transported.  The volume to be carried can further be
influenced by consolidation efforts on the carrier's part,
where he holds back shipments in order to build up a volume
whose transport would be more advantageous to him from a
profit point of view.  The receiver is confronted with the
choice of supplier and the choice of volume to be ordered.
The two main determinants affecting these decisions are the
commodity price and the perceived transport-service quality
or potential delay experienced in the delivery.

In their basic structure the proposed probability models
have a substantial amount of resemblance to those models used
in behavioural passenger-travel analysis.  Of course, in the
goods-movement context more than one decision-maker affects
the movement process, requiring several individual decision-
making models as exemplified by the following.

The *shipper level-of-service choice* model would analyse
the probability of a shipper choosing a certain level-of-
service for a given consignment to be shipped to a specific
receiver.  The *operator-choice model* would determine the
probability of an operator choosing a certain transport means
for transporting a given consignment from a certain shipper
to a certain receiver.  The *receiver-choice model* would
specify the probability of a receiver choosing to purchase a
consignment or set of consignments from a certain shipper
(producer or supplier).

The workshop recommends strongly that available data
sources be used and new data sets generated which would
allow researchers in the goods-movement area to pursue the
actual specification and estimation of the proposed models.
The interaction of the complete set of individual models
representing the physical goods-movement process is illustrated in Figure 28.3.  The long-range purpose of this model
development would be to test the policy sensitivity of these
models with respect to planning and regulatory interference
in the goods-movement process.

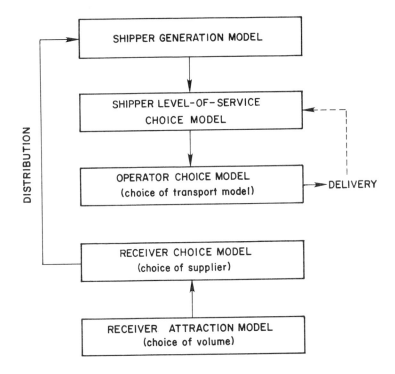

Figure 28.3   Model Components of the Physical Goods-movement Process

REFERENCES

1. Baumol, W.J. and Vinod, H.D., 'An inventory theoretic model of freight transport demand', *Management Science*, 16, 7 (1970).
2. Bayliss, B., 'Modal split in freight transport', *Proceedings, Conference on Freight Traffic Models*, Planning and Transport Research and Computation Company Ltd, London (1971).
3. Black, W.R., 'Inter-regional commodity flows: some experiments with the gravity model', *Journal of Regional Science*, 12, 1 (1972).
4. Fisher, G.P. (ed.) (1973) *Goods Transportation in Urban Areas*, Conference Proceedings, US Department of Transportation, Washington D.C.
5. Fisher, G.P. (ed.) (1976) *Goods Transportation in Urban Areas*, Santa Barbara, California, US Department of Transportation, Washington, D.C.
6. Hartwig, J.C. and Linton, W.E., 'Disaggregate mode

choice models of inter-city freight movements', unpublished Report, The Transportation Center, Northwestern University, Evanston, Ill. (June 1974).

7. Hutchinson, B.G., 'Estimating urban goods movement demands', *Transportation Research Record 496*, 1-15 (1974).

8. Kullman, B. (1973) 'A model of rail truck competition in the inter-city freight market', Ph.D. Dissertation, Massachusetts Institute of Technology, Cambridge, Massachusetts.

9. Mathematica (1967) *Studies in the Demand for Freight Transportation*, vol. I, Princeton, N.J.

10. Mera, K., 'An evaluation of gravity and linear programming transportation models for predicting interregional commodity flows', *Techniques of Transport Planning*, 1, Appendix A, Brookings Institute, Washington, D.C. (1971).

11. Meyburg, A.H. and Stopher, P.R., 'A framework for the analysis of demand for urban goods movement', *Transportation Research Record*, 496, pp. 68-79 (1974).

12. Meyburg, A.H., Diewald, W. and Smith, G.P., 'An urban goods movement planning methodology', ASCE, *Journal of Transportation Engineering*, 100, TE4, 791-800 (November 1974).

13. Meyburg, A.H., 'Modelling in the context of urban goods movement problems', in *Proceedings of Engineering Foundation Conference, Goods Movement in Urban Areas*, G.P. Fisher (ed.), Santa Barbara, California (September 7-12, 1975); US Department of Transportation (Report DOT-OS-60099) 127-168 (May 1976).

14. Miklius, W., 'Estimating freight traffic of competing transportation modes: an application of the linear discriminant function', *Land Economics*, 43 (1969).

15. Ogden, K.W. and Hicks, S.K. (eds) (1975) *Goods Movement and Goods Vehicles in Urban Areas*, Commonwealth Bureau of Roads, Melbourne.

16. Rimmer, P.J. and Hicks, S.K., 'Urban goods movement: process, planning approach and policy', Resource Paper, 3rd International Conference of Behavioural Travel Modelling (Chapter 24 of present volume).

17. Rimmer, P.J., 'A conceptual framework for examining urban goods movements', in *Goods Movements and Goods Vehicles in Urban Areas*, K.W. Ogden and S.K. Hicks (eds), Commonwealth Bureau of Roads, Melbourne, 15-25 (1975).

18. Rimmer, P.J. and Tsipouras, A., 'Ports and urban systems: framework and research needs in resolution of port-generated conflicts', *Australian Transport Research Forum*, Forum Papers, Melbourne (1977).

19. Roberts, P.O. and Kullman, B., 'Urban goods movement: behavioural demand forecasting procedures', Resource Paper, 3rd International Conference of Behavioural Travel Modelling (Chapter 25 of present volume).

20. Roberts, P.R., 'Predicting freight modal choice and its role in national transportation policy', *A Report on the 1970 Conference on Mass Transportation*, Popular Library, 238-255 (1970).

21. Schwerdtfeger, W., 'Staedtischer Lieferverkehr' *Veroeffentlichungen des Instituts fuer Stadtbauwesen*, Heft 20, Technische Universitaet Braunschweig, Federal Republic of Germany (1976).

22. Slavin, H.L., 'Demand for urban goods vehicle trips', *Transportation Research Record 591*, pp. 32-7 (1976).

23. Starkie, D.N.M., 'Intensity of commercial traffic

generation by industry', *Traffic Engineering and Control*,
pp. 558-60 (January 1967).
    24. Starkie, D.N.M., 'Traffic and industry', *Department of
Geography Research Paper No. 3*, London School of Economics
and Political Science, Chapter 6 (1969).
    25. Surti, V.H. and Ebrahimi, A., 'Modal split of freight
traffic', *Traffic Quarterly* (October 1972).
    26. Traffic Research Corporation Ltd (1969) *The Develop-
ment of a Goods Vehicle Forecasting Procedure*, Merseyside
Area Land Use/Transportation Study, Technical Report No.7.
    27. Watson, P.L. (1975) *Urban Goods Movement*, Lexington
Books, D.C. Heath and Company, Lexington, Massachusetts.
    28. Wigan, M.R., 'Urban goods movement: the British
experience', Resource Paper, 3rd International Conference of
Behavioural Travel Modelling (Chapter 26 of present volume).
    29. Zavattero, D.A., 'Suggested approach to urban goods
movement and transportation planning', *Transportation
Research Record 591*, pp. 41-3 (1976).

PART NINE

Chapter 29

BEHAVIOURAL MODELLING, ACCESSIBILITY, MOBILITY AND NEED:
CONCEPTS AND MEASUREMENT*

M. Quasim Dalvi

INTRODUCTION

Concepts of accessibility, mobility and travel need have been
used recently in the transport literature. There is some
confusion over terminology and the precise role these terms
have been assigned to play in modelling transport behaviour.
The aim of this chapter is first to trace the developments
which led to the present interest in the subject, then to
clarify the concepts and finally to examine the role which
accessibility plays in the theory of travel choice.

THE BACKGROUND

In tracing the background to the present interest in the
concepts of mobility, accessibility and travel need, it is
useful to begin by considering the recent developments in the
objectives of transport policy. In earlier decades when by
present standards there was an abundance of road space for
the traffic being carried, transport planning was the chief
concern of the highway engineer, whose primary concern was to
ensure maximum vehicular mobility for the anticipated
increase in traffic. It was assumed that every household
would soon have a car and would be able to use it, so that
emphasis was placed on movement of vehicles rather than of
people, and the evaluation of alternative schemes was in
terms of network capacity and speeds. As the number of
vehicles on the road rose and the scale of the problem was
increased, it soon became apparent that full car use could
not be accommodated in most cities, so a switch had to be
made - from attempting to accommodate the unrestricted use of
vehicles to regulating and restraining their use and encouraging travel by public transport. This change in emphasis
necessitated a switch of interest from vehicular mobility to
personal mobility with the person trip becoming the basic
unit of analysis. However, as pointed out by Jones (19) and
Hillman (15) in the British context, the philosophy of the
earlier approach was still apparent: transport models were
still concerned with motorised trips (to the neglect of walking trips and pedestrian movement in the city centre) and
evaluation was primarily in terms of travel-time savings
attributed to motorised investments.

---

*I am grateful to Peter Jones, Oxford University, for his
useful comments on the earlier version of this paper.

This is substantially still the way in which urban transport studies are carried out in the UK, but the recent emphasis on personal mobility and the consequent need to improve public transport have relegated to the background the narrow point of view of the highway engineer, and instead brought into prominence the contributions of other disciplines, notably those of mathematics, economics, geography, etc. As a result, transport studies have evolved a more sophisticated approach to traffic forecasting with the use of mathematical models and computer-based estimation techniques, and transport planners have begun to perceive transport problems rather differently.

One aspect of this change in the approach of transport planners is that instead of worrying about 'traffic congestion', they have begun to concern themselves with 'accessibility provision' (28). The emphasis on accessibility provision as a transport policy objective may be attributed to two different motivations. One was the realisation, arising primarily from working on gravity-type trip-distribution models, that in order to optimise a transport decision, it is not enough to focus simply on the characteristics of the transport system. It is equally necessary to consider the spatial distribution of opportunities, so that transport policies might be evaluated not only in terms of moving the people to the opportunities but also moving the opportunities to the people. This means that land uses and locational planning are as vital to the efficiency of resource use in transport as the management of transport services and the determination of modal split.

Another motive for emphasising the objective of accessibility provision stemmed from the concern shown for importing income-distributional considerations in the cost-benefit appraisal of transport alternatives. Thus, Wilson (28) for example, has argued in favour of calculating an accessibility index for different types of people in different locations, and even for allowing for a variation for a family in which a member has a car available and uses it, and one in which a member does not. However, the idea of incorporating the accessibility objective on income–distributional grounds is open to criticism, for apart from introducing essentially a political judgement in the selection of transport policies, there is the question of developing an operational criterion with an appropriate weighting system which would attribute high values to chosen income, locational or trip-making categories. In the absence of such specific weightings it is difficult to see accessibility as differing significantly from the usual user–benefit appraisal (11). Moreover, if the advantages of improved accessibility and mobility are to be presented independent of user benefit and differentiated on equity grounds, there is the danger that conventional assessments may result in double counting and thereby inflating the rate of return on proposed investments.

ACCESSIBILITY AND TRIP GENERATION

Whilst the debate on accessibility and mobility as objectives
of transport still remains unsolved, there is another area
where the role of accessibility has received considerable
attention, viz. the impact of accessibility on urban trip
generation. For example, in Britain, in the consideration of
the motorway network proposals in the Greater London Development Plan (GLDP), following criticism that accessibility had
not been adequately allowed for, the Greater London Council
(GLC), argued that, on the basis of either time series or
cross-section analysis, the motorway network could not be
shown to have a significant impact on trip making. Although
the GLC's argument was rejected at the GLDP inquiry as it was
based on inadequate empirical evidence, the debate raised
important issues for the forecasting of traffic demand and
the evaluation of transport investment proposals.

Of particular relevance in this context was the question
about the treatment of benefits to generated traffic, which
have invariably accounted for a sizeable portion of the
measured benefits in the cost-benefit appraisals of transport
proposals in Britain. If the GLC's hypothesis that changes
in accessibility have no effect on travel generation were to
be accepted as a valid basis for obtaining traffic forecasts,
this would call into question the conventional economic
evaluation scheme which devoted considerable time to estimating the net benefit from generated traffic. Indeed, the
whole issue boils down to testing a fundamental substitution
hypothesis of the micro-economic theory of demand: since in
the last resort, accessibility is a cost which must be borne
by the community, both directly by individuals and, beyond
that, socially by the entire community, the question is
whether or not an improvement in accessibility on account of
a proposed investment in transport would induce individuals
to make more journeys or otherwise change the pattern of
their transport behaviour by substituting expenditure from
other items of consumption to transport. Thus, it is not
simply the question of measuring the diversion of existing
journeys from one destination to another, or one mode to
another as a result of changing relative accessibilities -
the point underlying the GLC's argument: it is essentially
the question of considering what happens to journeys which
presently are not just made, but would be made in response to
a slight improvement in overall accessibility.

At present the problems of trip generation are being
examined from the old established point of view. In the UK,
for example, while price-substitution effects have been
introduced in modelling the distribution and modal-split submodels, no allowance is made for the effect of the predicted
change in travel cost in estimating the trip-generation
submodel. Indeed, the model most commonly used to obtain
trip-end forecasts, viz. the category-analysis model origin-
ally developed by Wootton and Pick, is hardly geared to
incorporate elasticity into trip generation, let alone to
test the validity of the substitution law of the classical
demand model.

It must be emphasised here that, in modelling travel
demand, elasticity considerations not only refer to the
effect of transport system characteristics; they also refer
to the effect of land-use characteristics and the locational
changes of the firms and households affect on individual's
destination choice.  In the long run, and particularly in the
recent development of the interactive, simultaneous demand
models, these latter affects are reflected in the form of
changes in trip lengths.  However, in the short run, and
especially in the conventional sequential-demand models, the
trip-production version of trip forecasts accounts only for
the effects of changes in transport—system characteristics.
The effects of land use and locational changes are either
ignored or shown indirectly as measures of the relative
attractiveness of destination zones in the trip-distribution
submodel.

The remainder of this chapter is concerned with the
definition and measurement of accessibility and with the
question of how accessibility effects have been incorporated
in the modelling of both short-term and intermediate travel
demand.  However, before discussing these problems, it is
necessary to clear some confusion arising from a mistaken
emphasis laid on mobility and travel need as criteria of
transport planning.

## MISTAKEN EMPHASIS ON MOBILITY AND TRAVEL NEED

Whilst attempts have been made, as we shall see below, to
define and measure the term 'accessibility' in operational
terms, so that it may be used as a criterion in evaluating
alternative transport systems, very little attention has been
given to define the terms 'mobility' and 'travel need' in
precise terms.  The term mobility has been used rather
indiscriminately to refer both to the supply function and the
demand function of transport services.  In an attempt to
clear the confusion, Burkhardt [3] suggested that mobility
should be used to refer to the supply constraint facing an
individual, whilst travel need may be used to refer to his
actual behaviour when he uses transport services.  While this
distinction is conceptually appealing, the difficult problem
arises when one tries to convert the terms into quantifiable
magnitude.  Clark et al. [6] have defined personal mobility
as 'the number of annual kilometers of personal travel (by
all modes, including walking and cycling) per head of population', while Hillman et al. [16] define mobility as 'the
capacity that a person possesses for getting around in his
daily life'.  The Independent Commission on Transport [17]
found that the term mobility refers to 'the ease with which a
person can move about or the amount of movement he performs'.
More precisely the term mobility can be defined in two
distinct senses: first as a concept and then in terms of some
physical measure of that concept.  As a concept, mobility
refers to 'the degree to which a person is free of ties in
the home or elsewhere': in this respect a mother with young
children would be considered less mobile than someone without,
although this constraint might be translated in terms of the
scarcity of time resources, etc.  As a physical measure,
however, mobility indicates the ease with which a person can

move about or the amount of movement he performs, measured in
terms of the number of trips made by all modes, the availability of different modes, and presumably some indication of
the resources he can expend on travel.   In all these measures,
unfortunately, as Jones (19) has pointed out, attention
focuses on the amount of travel undertaken, rather than the
potential for travel and this, in turn, leads to emphasis on
raising mobility standards by encouraging an increase in trip
rates, or the annual mileage travelled.   The current emphasis
on mobility is, in fact, a hangover from the days when
engineers were in charge of transport planning and management,
and thus misses the whole point that travel is a derived
demand, a means to an end.

Much the same problems arise with respect to the concept
of travel need.   The prescription of travel need as a criterion for transport planning is based on an implicit
rejection of the income distributional implications of the
existing pattern of resource allocation in the transport
sector.   For example, recent emphasis of the British transport policy on the 'rural transport problem' is motivated by
the concern to bring the rural poor to some national level of
transport adequacy.   In the urban context, too, the criterion
of 'need' has sometimes been mentioned with reference to the
transport plight of disadvantaged groups, locked in the
declining inner areas of cities, whose welfare is particularly sought.

The problem with the criterion of travel need, however,
lies not in accepting the policy (which after all is essentially a political value judgement beyond the competence of a
transport planner) but in defining it in operational terms
for evaluating alternative transport projects.   Well-
intentioned as they are, all previous attempts to define
travel need are ambiguous and arbitrary.   Although there are
many possible definitions of need, they usually depend on:
(a) the collective opinion of a body of experts as to the
amount of travel people ought to undertake to meet some
standard of living; (b) the actual amount of travel undertaken by some group whose behaviour is taken as a norm; or
(c) the target group's own perception of what it would like
to have.   Burkhardt (3), for example, has made an ingenious
attempt to compute the travel needs of households in five
selected areas of rural poverty in the United States.   The
average number of trips made by all persons more than five
years of age in the United States was chosen as a norm of
travel for his computation.   This figure, according to the
Nationwide Personal Transportation Survey of the Federal
Highway Administration, was 807 trips per person per year.
Against this norm of travel behaviour, the actual number of
trips taken by households in the five study areas were compared, and the difference was taken as the number of trips
needed by each study group.   The total number of trips needed
per area was then the total number of households in need
weighted by the number of trips needed.   Burkhardt found that
a poor person in the five study areas only travelled one-
sixth to one-fifth as much as the average American.   The
futility of such an exercise is immediately obvious when one
realises the phenomenal increase in the amount of tripmaking
that would be necessary to bring the rural poor to the

average tripmaking behaviour of the American nation. Note
that Burkhardt's calculations include only the vehicular
trips and do not account for walk trips.  Assuming that walking is done more by the poor than by the rich, the inclusion
of these latter trips in the norm would still render the
needs criterion undesirable on economic grounds.

The trouble with this approach, as with mobility, is that
it views trip-making in isolation from the individual's other
activities for which transport is needed as a means to an end.
As Jones has argued, this emphasis on moving people has
prevented local authorities in Britain from considering why
the people need to travel and whether some simpler, cheaper
solution does not exist.  Moreover, the use to which transport service is put is closely tied to the price that will be
paid.  Thus, the recent concern for the transport plight of
the disadvantaged is a perfectly valid objective to pursue on
equity grounds, but at the same time it is important to bear
in mind the cost implications of this for the urban community
as a whole.  Transport already accounts for a bulk of national
investment almost everywhere.  Therefore, at a time when
practically every nation is confronted with the shortage of
investible resources, it seems rather pointless, indeed
reckless, to emphasise mobility and travel need as criteria
for evaluating transport policy.  What is needed is a totally
new approach in which the necessity for travel is traded at
each stage against the cost of travel, including the possibility of avoiding travel altogether by moving opportunities
to people.  Such an approach leads to the consideration of
accessibility to which we shall now turn.

ACCESSIBILITY: DEFINITION, MEASUREMENT AND INTERPRETATION

Accessibility denotes the ease with which any land-use
activity can be reached from a location using a particular
transport system.  This definition suggests that there are
two elements in the term accessibility: the land-use activities or opportunities desired by individuals to satisfy their
wants, and the service provided by transport to reach the
opportunities.  These elements have been measured and combined in a variety of ways and so there is no unique and
unambiguous measure of accessibility.  In general, accessibility measures have taken two forms: opportunities
weighted by a decreasing function of the interaction costs in
time or money of reaching those opportunities (e.g. (14)), or
cumulative functions of the opportunities which can be
rendered within a specified travel time (10,25,26).  These
measures, however, have little or no strong underlying theory
from which causality in transport decision-making can be
inferred.

*Accessibility Measures Based on Average Distance*

Accessibility measures have been used in human geography to
describe the growth of towns; the location of facilities and
functions; and the juxtaposition of land uses.  These
measures have one thing in common: they are concerned with
overcoming some form of spatially operating source of

friction (for example, distance and/or time). The starting point is the assumption of a planar surface and a network in the form of a planar graph so that the mathematical theories of network can be used (12). Points where links of the network cross are called nodes or vertices; all links are counted equally, and on this basis simple measures descriptive of a network's form can be developed. The first attempt to develop network measures of this kind was by Kansky (20), but his measures were essentially alternative ways of describing the network in terms of the number of links and the number of nodes, and did not constitute an index of accessibility. To measure accessibility along a given network it is necessary to establish a common reference point on *a priori* grounds as an appropriate centre of attraction. Location relative to a central area is such a measure.

To clarify these points and to develop operational forms for accessibility, Ingram (18) has suggested a distinction between (a) relative accessibility and (b) integral accessibility. Relative accessibility has been defined as the degree to which two places (or points) on the same surface are connected, while integral accessibility is defined as the degree of inter-connection for a given point with all other points in the same surface. The integral accessibility of the $i^{th}$ point is thus a scalar point function of the relative accessibilities at that point, i.e.

$$A_i = \sum_{j=i}^{n} a_{ij}$$

where $A_i$ is the integral accessibility at the $i^{th}$ point and $a_{ij}$ is the relative accessibility of point $j$ at $i$. Since such an index would be sensitive to network size, the average accessibility to all relevant points is more appropriate if it is necessary to make comparisons between points on different networks:

$$A_i = \frac{1}{n} \sum_{j=1}^{n} a_{ij}$$

The simplest surrogate of relative accessibility $a_{ij}$ is the straight line distance between two points $d_{ij}$, but the assumption of a linear response to distance is not compatible with most real-world situations. One further modification is thus necessary in order to introduce a behavioural element into the formulation of accessibility - that the perception of accessibility declines increasingly rapidly as distance increases. In terms of relative accessibility, the simplest gravity formulation is:

$$a_{ij} = d_{ij}^{-\alpha}$$

where $\alpha$ is a constant of the system to be estimated. An alternative formulation is an exponential function:

$$a_{ij} = \exp(-\beta d_{ij});$$

where $\beta$ is a parameter of the model.  The exponential
function, which places more emphasis on accessibility over
short distances, is derived from Wilson's entropy-maximisation model as we shall see below.

Ingram has shown that both of these functions tend to
decay too rapidly in comparison with the empirical evidence.
He suggests that a modified Gaussian form is superior:

$$a_{ij} = K \, \exp(-d_{ij}^{2} \, \upsilon^{-1})$$

where K is a scaling factor and $\upsilon$ is a constant determined
for a given network.  This derives from a trip-generating
function developed on the basis of the opportunity and
potential models in which the probability of a given trip
being made to a destination j is given by:

$$P_r(j) = g_i d_{ij} \, \exp(-d_{ij}^{\theta} \, \emptyset)$$

where $g_i$ measures the attractive of j, and $\theta$ and $\emptyset$ are constants of the system.

*Behavioural Basis of Accessibility*

If accessibility is to be used as a factor influencing travel
demand or as a criterion for evaluating alternative transport
options, its formulation must incorporate more behavioural
content than is implicit in the form developed in urban
geography.  There are two ways in which accessibility can be
linked with behavioural theories: (a) first via the choice of
an appropriate impedance function to reflect the perceived
cost of transport; and (b) second via the choice of appropriate attractiveness variables to reflect the availability
of opportunities at destinations to satisfy the particular
wants of travellers.

The impedance function selected must, in principle,
reflect the excess surplus accruing to a traveller when he
selects a given destination over all other destinations
within the range of his choice to satisfy his particular
demand.  Ideally the impedance function most appropriate to
this end is a generalised cost of travel which includes all
the relevant elements as its argument, including the separate
components of travelling time, and in each case it is the
perceived values that matter.  Since different modes of
transport have different characteristics of comfort, convenience, crowding, etc., these values will be perceived
differently for different modes.  Thus, an appraisal in
behavioural terms of even a very simple definition of
accessibility reveals a series of different accessibilities
depending on the means of transport and its characteristics.
This connects with Vickerman's argument for separate models
of trip generation by each mode, since no single index of
accessibility can be defined (24).

However, a more pressing argument for differentiating the
accessibility index arises from the requirement of linking

mobility patterns to the spatial distribution of opportunities. The first step in this respect is to relate integral forms of accessibility to urban land-use models incorporating *a priori* hypotheses about the distribution of opportunities in urban space. The earliest attempt in this regard is the work of Harris (13) in the United States and Clark (5) in Britain, who developed the concept of economic potential which is inversely linked to accessibility. If each node on the surface is given a weight $w_i$ and the transport costs between each i and j are given as $c_{ij}$ then the economic potential of j is given by:

$$P_j = \sum_i W_f \, f(c_{ij})$$

where $f(c_{ij})$ usually takes some inverse powered or exponential form. The weights can be in terms of employment or income if the potential for industrial or retail location is needed.

If the potential of residential locations is calculated by using weights, such as residential population or retail turnover of each area, this effectively defines an index expressing the total attraction felt by individuals at each point i in terms of the weighting of all possible destinations, j, modified by the travel costs of reaching those destinations:

$$A_i = \sum_j W_j \, f(c_{ij}).$$

This, in effect, is Hansen's accessibility index which is comparable to the balancing factor in single-constrained trip-distribution models of the form:

$$T_{ij} = a_i \, O_i \, W_j \, f(c_{ij}),$$

where

$$a_i = \frac{1}{\sum_j W_j \, f(c_{ij})}$$

$O_i$ = number of trips originating at i

$W_j$ = measure of attractiveness at j.

Wilson (27) has in fact shown that all the traditionally constructed accessibility measures can be related directly to spatial—interaction models. A doubly-constrained interaction model can be written as:

$$T_{ij} = a_i \, b_j \, O_i \, D_j \, f(c_{ij}),$$

where

$$a_i = \frac{1}{\sum_j b_j \, D_j \, f(c_{ij})}$$

and

$$a_j = \frac{1}{\sum_i a_i \, O_i \, f(c_{ij})}$$

Subject to the total trip constraints

$$\sum_i T_{ij} = D_j$$

$$\sum_i T_{ij} = O_i.$$

In this, $a_i$ and $b_j$ are the balancing factors which are solved iteratively.

It is well known that if an additional constraint equation is introduced on $T_{ij}$ in the form of travel-cost expenditure,

$$\sum_i \sum_j T_{ij} \, C_{ij} = C,$$

the solution to the entropy-maximisation problem yields a trip-density function and balancing factors all containing inverse exponential form for the travel cost function. The reciprocal of the balancing factor for origins

$$\frac{1}{a_i} = \sum_j b_j \, D_j \, \exp(-\beta c_{ij})$$

is interpreted as the relative accessibility to job opportunities for residents at location i. Similarly, the reciprocal of the balancing factor for destinations

$$\frac{1}{b_j} = \sum_i a_i \, O_i \, \exp(-\beta c_{ij})$$

is a measure of the relative accessibility to residential opportunities for workers at a given work place j [1]. It can be seen from the above equations that since in a doubly-constrained model the accessibility index $1/a_i$ is proportional to the other balancing factor, its value is larger when there are many potential jobs in the vicinity, and not many residents competing for them. By a similar argument, the accessibility to residential opportunities, $1/b_j$, will be large where there are many residences and few work places.

*Measures of Attraction*

The key point remaining for discussion is the selection of appropriate attraction measures reflecting the importance of alternative destinations for the satisfaction of particular travel demands. Attempts have been made in Britain [7,8,24] and in the United States [4] to construct purpose-specific as well as mode-specific measures of accessibility to explain travel behaviour. Unfortunately, there is no strong

underlying theory defining the selection of attraction
factors for specific journey purposes, although at the operational level it is customary to use total employment or
residential population to define journeys to work, aggregate
floor space, retail turnover or retail employment to define
shopping journeys, and total population to define social
journeys, etc. Empirically, however, accessibility profiles
developed for London by Dalvi and Martin (7), using four
attraction factors - total employment, retail employment,
households and population - indicated high correlation
between opportunities of different types at equal distance
from residential zones, although the correlation was lowest
between measures based on total employment and population.
Similarly, high correlations were also found in the purpose-
specific accessibility measures developed by Burns and Golob
to explain car-ownership decisions. Because of this correlation, Burns and Golob found that simple attraction measures,
such as relative population or relative total employment in a
zone, served equally as well in model goodness-of-fit as more
complicated measures of relative land area devoted to various
uses or relative employment by specific industries. These
findings are, however, by no means conclusive and further
research is necessary to relate satisfactorily the accessibility profiles to the spatial distribution of opportunities. For one thing, the accessibility indices constructed
to date are mostly at a zonal or district level of aggregation and hence are likely to conceal patterns that are more
apparent at a lower level of aggregation. For another, apart
from the effects of attraction measures, these indices are
also sensitive to the choice of parameters of travel-cost
function and the type of zoning system used at model-building
stage (7). All these difficulties together with the non-
homogeneous distribution of opportunities destroy the
usefulness of accessibility indices as explanatory variables
unless extreme care is exercised in constructing them.


ACCESSIBILITY AND TRAVEL CHOICE

In this section, I comment briefly on some of the recent
empirical studies on the role of accessibility in urban
households' travel decisions. In Britain, these studies have
focused mainly on the impact of accessibility on trip generation (7,8,9,24), but in the United States, Burns and Golob (4)
have also investigated the role of accessibility in household
car-ownership decisions. All these studies have been carried
out within the conventional sequential framework using a
single-equation estimation procedure to estimate model
parameters.

Vickerman has constructed attraction-specific accessibility indices as well as a combined attraction-cum-accessibility
index incorporating expenditure weights to test the impact of
accessibility on travel generation. He modelled separately
for different journey purposes and also for different modal
choices, such as car, bus and walk, and used both a multivariate regression model and factor model to estimate his
model parameters. His regression results were found
unsatisfactory but factor-model estimates showed that whilst
aggregate attraction had an important effect on leisure

travel, accessibility *per se* was a dominating influence on
shopping travel. Dalvi and Martin (8) also used attraction-
specific accessibility measures to investigate the impact of
accessibility on non-work trip making in inner London areas,
but unlike Vickerman did not find accessibility to be a
particularly significant determinant of trip generation. On
the other hand, disposable income and certain household
characteristics proved to be the main explanatory variables
in their model. The work of Doubleday (9) also confirmed the
ineffectiveness of accessibility in explaining travel generation, although his analysis lacked any goodness-of-fit tests,
as he employed conventional category analysis to test his
hypothesis.

Burns and Golob (4) developed a utility-maximising model
of travel behaviour to explain the role of accessibility in
households' car-ownership decisions. In this theory, direct
utility functions were formulated for households in various
states of car ownership. Through determination of utility-
maximising trip frequencies and distributions among
destinations, indirect utility formulations which express the
value of travel by automobile or public transport were
developed in terms of accessibility to opportunities rather
than in terms of actual trips. The results of their multinomial logit calibrations showed that accessibility measures
(the indirect utility terms of the maximising solutions)
explained a significant proportion of the residual variance
not accounted for by the consumption term. However, as
mentioned earlier, the purpose-specific accessibility was
found to be as good a determinant of car-ownership behaviour
as the general attraction accessibility measures, thereby
indicating that such complications were not necessary at
least in their particular study.

Obviously, these results are tentative because of the need
to use data in a zonally aggregated form and also because the
data were not specifically collected to explain the role of
accessibility. Besides, there is the more basic limitation,
namely that, as noted above, these studies have used a
single-equation approach to explain what is otherwise an
essentially interactive decision process. Thus, the Greater
London Council had argued during the GLDP enquiry that the
really important effects of accessibility are reflected, not
in terms of a single decision, viz. the frequency of trips
made, but in terms of the whole set of inter-related land-use
and transport decisions, affecting journey lengths and mode
choice. All this indicates that on both theoretical and
practical estimation grounds a simultaneous-equation framework is more sensible to assess the role of accessibility.
For in a simultaneous-equation model, the individual's trip
decision is evaluated, within a common framework, along with
his decisions on car ownership, residential space consumption,
choice of transport modes and length of journey to work, etc.
Hence, if it is true that accessibility, being a measure of
spatial relationship, affects all the stages of travel
decision, including the choice of residential/work place
locations, then it would be possible to identify, in the
simultaneous structure, not only its short-run effect in
terms of variation in trip frequencies but also the long-run
effects in terms of changes in trip lengths and modal split.

To evolve a balanced transport plan a knowledge about trip-
length changes and model splits is as vital as a knowledge
about trip frequencies.

## ACCESSIBILITY AND LAND USE

In discussing its operational forms, we have seen that the
concept of accessibility has been linked inexorably with land
uses and the distribution of opportunities in urban space.
This link has a two-way causation.  There is the impact of
exogenous changes in land use and the distribution of
journey-specific opportunities on the pattern of travel
behaviour which has been considered in the previous sections.
There is also the impact of accessibility on land use and the
development of urban patterns which has traditionally been
the focus of attention amongst urban geographers.  This
aspect has figured recently in the theories constructed to
explain the rural urban land-use conversions and the optimal
growth of urban areas (21,22).  Here accessibility has been
defined, in a rather conventional sense, as a reduction in
the generalised cost of transport or as an improvement in the
degree of mobility desired by society (21), and the effect of
this has been worked out in terms of changes in land use
usually by assuming a monocentric form of urban structure,
more or less on the lines of the Von Thunen model of urban
growth.  The limitation of the monocentric assumption of
urban development has been pointed out by Beesley and Dalvi
(2) in the context of explaining long-run changes in trip
lengths.  The limitations of this assumption in explaining
land-use changes, and the standard and size of urban area as
a function of accessibility changes is also evident in the
work of Stone and Richardson.  The difficulty of this
approach arises from the fact that changes in land use have
much wider connotation than the mere urban/rural land-use
conversions assumed by the theory; for example, they also
indicate decentralisation of activities formerly located in
the CBD areas or the relocation of activities to decongest
the inner areas of the cities, or more significantly, the
growth of 'new towns' around the older metropolis where some
of its functions have been transferred in the wake of improved accessibility.  Attempts have been made recently to
quantify some of these effects by constructing an integrated
land-use/transport forecasting model - for example, modifications have been made in the Lowry model to account for
changes in accessibility on urban development.  However,
there is still a need to develop a satisfactory model of
urban growth in which changes in land-use pattern can be
meaningfully analysed in relation to changes in transport
accessibility, particularly in the context of multi-concentric
urban pattern.  Needless to say, this problem is beyond the
purview of this chapter.

## CONCLUSION

The recent interest in the role of accessibility has arisen
partly due to a shift in the objectives of transport policy
and partly due to the concern of model builders to investigate its role in travel decision-making.  In the earlier

decades, transport policy was focused primarily on relieving traffic bottlenecks faced by car users in urban areas, but lately policy-makers have given up this narrow approach to transport solutions and instead have concerned themselves with accessibility provision to all sections of the community. In this new approach, the emphasis is no longer on improvements to a particular mode of travel; indeed, it is not even on the transport system as a whole, but instead the emphasis is now on transport as well as optimal location of opportunities. From this point of view, the interest of model builders has now centred on investigating the role of accessibility, which has been motivated partly by their concern to introduce price elasticity into trip generation, but partly also with a view to understanding how the spatial distribution of opportunities affects travel-demand decisions. We have also argued that the emphasis on mobility and need, which appears in the writings of some authors, is completely misguided. Mobility refers to the supply side of transport, and emphasises the role of trip-making *per se* without regard to the resource constraint facing the society. Need, on the other hand, focuses on the transport demand of the disadvantaged, but it is essentially a normative concept and so it is difficult to define and operationalise as a planning criterion without regard to political value judgement. Hence, if any concepts are required as working criteria to evaluate transport decisions, then it is accessibility to which we must return. Although the measurement of accessibility needs more attention and thought, it is a readily quantifiable concept based on objective measures and plausible hypotheses about travel behaviour.

REFERENCES

1. Angel, S. and Hyman, C.M., 'Urban transport expenditures', *Papers of the Regional Science Association*, 29 (1972).
2. Beesley, M.E. and Dalvi, M.Q., 'Spatial equilibrium and journey to work', *Journal of Transport Economics and Policy*, VIII (1974).
3. Burkhardt, J.E., 'Need as a criterion for transportation planning', *Highway Research Record No. 435* (1972).
4. Burns, D. and Golob, T.F., 'The role of accessibility in basic transportation choice behaviour', *Transportation*, 5, 2 (1976).
5. Clark, C., 'Industrial location and economic potentials', *Lloyds Bank Review No. 82* (1966).
6. Clark, N., Lee, J.A. and Ogden, K.W., 'The use of energy for personal mobility', *Transportation Research*, 8, 4/5 (1974).
7. Dalvi, M.Q. and Martin, K., 'The measurement of accessibility: some preliminary results', *Transportation*, 5, 1 (1976).
8. Dalvi, M.Q. and Martin, K., 'Estimate of non-work trip demand: a disaggregated approach', paper presented to the Leeds Conference on Urban Transport Planning (1976).
9. Doubleday, C., 'Spatial mobility and trip generation', paper presented to the Leeds Conference on Urban Transport Planning (1976).
10. Falcocchio, J.C., Pignataro, L.J. and McShane, W.R., 'Measuring the effect of transportation accessibility on

inter-city unemployment', paper presented at 52nd Annual Meeting of Highway Research Board (1975).
11. Gwilliam, K.M., 'Economic evaluation of urban transport projects: the state of the art', *Transportation Planning and Technology*, 1 (1972).
12. Haggett, P., Chorley, J.R. (1969) *Network Analysis in Geography*, Edward Arnold, London.
13. Harris, C.D., 'The market as a factor in the localisation of industry on the United States', *Annals of the Association of American Geographers*, 44 (1954).
14. Harris, B., 'Notes on accessibility', discussion paper, *Institute for Environmental Studies*, University of Pennsylvania (1966).
15. Hillman, M., 'Social aspects of transport planning', Memorandum in Second Report from the Expenditure Committee HC 59 session, 1972-73, *Urban Transport Planning*, vol. 2, Her Majesty's Stationery Office, London (1972).
16. Hillman, M., Henderson, I. and Wholley, A., 'Personal mobility and transport policy', *PEP Broadsheet No. 542* (1973).
17. Independent Commission on Transport (1974) *Changing Directions*, Coronet Books, London.
18. Ingram, D.R., 'The concept of accessibility: a search for an operational form', *Regional Studies*, 5 (1971).
19. Jones, P.M., 'Accessibility, mobility and travel need: some problems of definition and measurement', paper presented to the IBC Transport Geography Study Group Conference (1975).
20. Kansky, K.J., 'Structure of transportation networks', *Research Paper 84*, Department of Geography, University of Chicago (1963).
21. Richardson, W.H. (1973) *The Economics of Urban Size*, D.C. Heath, Saxon House, Farnborough.
22. Stone, P.A. (1973) *Structure, Size and Costs of Urban Settlements*, Cambridge University Press, Cambridge.
23. Vickerman, R.W., 'A demand model for leisure travel', *Environment and Planning, A*, 6 (1974).
24. Vickerman, R.W., 'Accessibility, attraction and potential: a review of some concepts and their use in determining mobility', *Environment and Planning, A*, 6 (1974).
25. Wachs, M. and Kumagal, T.G., 'Physical accessibility as a social indicator', *Socio-Economic Planning Sciences*, 7 (1973).
26. Wickstrom, G.V., 'Defining balanced transportation- a question of opportunity', *Traffic Quarterly*, 25 (1971).
27. Wilson, A.G., 'A family of spatial interaction models', *Environment and Planning*, 3 (1971).
28. Wilson, A.G., 'Developing issues in urban transport planning', Memorandum in Second Report from the Expenditure Committee, HC 57 session 1972-3, *Urban Transport Planning*, vol. 2, Her Majesty's Stationery Office, London (1972).

Chapter 30

DISAGGREGATE TRAVEL AND MOBILITY-CHOICE MODELS
AND MEASURES OF ACCESSIBILITY

Moshe Ben-Akiva and Steven R. Lerman

SUMMARY

Existing measures of accessibility are not based on an
explicit behavioural theory. This chapter proposes an
accessibility measure which is consistent with the application of random-utility models to individuals' decision
processes. The proposed measure is the expected maximum
utility that a consumer derives from a given situation. The
chapter presents the properties and advantages of this
measure and its derivation for the special cases of the
multinomial logit and probit choice models.

Given the multidimensional nature of the travel and
mobility choices, this measure is also shown to provide a
logically consistent linkage between component models in a
complete model system of travel and mobility choices. The
use of this measure is demonstrated for joint and sequential
logit models. Finally, the chapter addresses key problem
areas in the state-of-the-art of spatial-choice modelling
which directly affect the development of measures of accessibility defined over a large set of spatial alternatives.

INTRODUCTION

As Dalvi and Martin (10) point out, the term accessibility is
widely used but rarely defined in a rigorous and satisfactory
fashion. The existing literature on accessibility is often
focused on the development of measures of accessibility,
without sufficient thought or insight into precisely what
accessibility means. In this chapter, we will define accessibility in a manner directly related to individuals'
travel-decision processes. Furthermore, we shall demonstrate
that under most commonly encountered conditions, our definition of accessibility can be easily operationalised and used
in empirical studies.

The term accessibility, as used in this chapter, refers to
some composite measure which describes the characteristics of
a group of travel alternatives as they are perceived by a
particular individual. In a sense which shall be formalised
later in this chapter, accessibility logically depends on the
group of alternatives being evaluated and the individual
traveller for whom accessibility is being measured. When
considered from this perspective, existing measures of

accessibility have a distinctly *ad hoc* character.

Consider, for example, the most widely used measure of accessibility proposed by Hansen [15]:

$$A_{ik} = \frac{\sum\limits_{j=1}^{n} W_{jk}\, e^{-\beta c_{ij}}}{\sum\limits_{j=1}^{n} W_{jk}}$$

where:

$A_{ik}$ = the accessibility of zone i to opportunities of type k in zones j=1,...,n;

$W_{jk}$ = a measure of attractiveness of zone j to activities of type k;

$c_{ij}$ = the generalised cost of travel from i to j; and

$\beta$ = a scale coefficient.

In this measure, there is no consideration of the individual for whom the accessibility is being computed; all individuals in the same zone i have the same level of accessibility, despite the fact that they may perceive the set of travel alternatives (in this case, the destinations j=1,...,n) quite differently. Furthermore, the causal justification for the functional form is relatively arbitrary, and the measurement of attraction is usually some population or employment level.

The underlying justification for these (and virtually all other) measures of accessibility is that the term $W_{jk}$ is in a crude sense a proxy representing how valuable a visit to destination j would be for fulfilling a specific trip purpose k, and the term $e^{-\beta c_{ij}}$ represents the perceived cost of getting to j. While there is nothing intrinsically wrong with this general line of thought, it is clear that the concepts of attraction and impedance as used in these measures are rooted in a relatively naive perspective on how travellers make decisions. This perspective has been largely abandoned in more recent work in travel-demand modelling, yet for reasons which are difficult to explain, the crude measures of accessibility persist.

Other types of accessibility measures such as those used by Dunphy [13] and Sherman, Barber and Kondo [30] rely on isochron analysis. For example, one might use the fraction of regional employment within x minutes by car as a measure of highway accessibility. Such measures have obvious deficiencies, including the arbitrary selection of x and their lack of differentiation between opportunities which are adjacent to the origin and those just within the x minute isochron.

A further shortcoming of existing measures of accessibility is their simplistic underlying assumption about how people travel. All measures of accessibility developed in the past use only the level of service from the origin to potential destinations; they ignore the fact that travellers quite frequently chain together destinations into trip tours. Ideally, the concept of accessibility should incorporate the trip-chaining potential of the available destinations. Some destinations may contribute very little to total accessibility when the home is used as the origin but may in reality be visited quite frequently because they are near other, non-home sites.

The remainder of this chapter will explore the implications of a definition of accessibility which is directly tied to travel demand decisions, and will show how this concept of accessibility can be operationalised. The next section will set forth the formal definition of accessibility and discuss some of the general properties of this measure. Then we will discuss how this definition can be applied in conjunction with existing, disaggregate behavioural travel-demand models. In the following three sections, an example of the use of this measure to construct hierarchical models of choice will be presented. Finally, we will describe some of the unresolved issues implicit in the use of the proposed accessibility measure.


DEFINITION OF ACCESSIBILITY

As discussed above, accessibility can be viewed as the outcome of an operation on a set of travel alternatives. In order to formalise this concept, we shall denote some set of feasible travel alternatives as $C_t$, where the subscript t denotes the individual for whom the accessibility is being determined. The members of the set $C_t$ are by definition mutually exclusive and collectively exhaustive, and one and only one member can be chosen by t in any single decision. We will assume that there exists a utility that individual t associates with each alternative in $C_t$, and that the individual selects the member of $C_t$ which maximises that utility. Suppose furthermore that the utilities are random variables, which we shall denote as a vector $U_t = (U_{it}, \ldots, U_{nt})$ where n is the number of alternatives in $C_t$. For the purposes of this chapter it is irrelevant whether one adopts the perspective of the econometrician that the underlying utilities are deterministic but imperfectly observed or the perspective of the mathematical psychologist that the utilities individuals associate with the alternatives are actually random in nature. Manski [24] develops random utility theory from the former perspective, while Luce and Suppes [22] work from the latter.

Under these assumptions, a natural definition of accessibility is simply the utility of the choice situation to the individual t. For any single decision, the individual will select the alternative which maximises his/her utility; thus, a simple definition of accessibility is

$$\underset{i \varepsilon C_t}{\text{Max}} \ U_{it}$$

However, since the vector $U_t$ is random, this value is not directly measurable. Given the inherent uncertainty in the outcome of the choice process, a reasonable alternative to this measure is *the expected value of the maximum of the entries in* $U_t$, or $E(\underset{i \varepsilon C_t}{Max} U_{it})$. This is the value which we shall define as the accessibility of set $C_t$ to individual t.

It is worth noting that under a set of restrictions on the distribution of $U_t$ the expected value of the maximum utility has a direct interpretation as a consumer surplus measure. Several researchers including Harris and Tanner (16) and Williams (34), have used disaggregate choice models and derived this value for specific cases. Neuburger (26) derived the appropriate benefit measure of a gravity model which is the same expression derived for the expected maximum utility in a disaggregate logit model. Lerman (21) used this formulation directly to derive an approximation for the utility of a group of nearly homogeneous alternatives; in this case, the multinomial-logit model was used. Sheffi and Daganzo (29) have explored the properties of this function in the context of the network assignment problem.

While later sections of this chapter will discuss this measure under specific assumptions about $C_t$ and the distribution of $U_t$, it is useful to explore two properties of this measure and the conditions on the distribution of $U_t$ under which they hold. Consider the following properties:

1    *monotonicity with respect to choice-set size*

$$E\,[Max\,(U_{1t},\ldots,U_{nt})] \leq E\,[Max\,(U_{1t},\ldots,U_{nt},\,U_{n+1,t})]$$

In words, this property implies that the measure does not decrease if a new alternative is added to the choice set.

2    *monotonicity with respect to mean utilities*

If $V_t = (V_{1t},\ldots,V_{nt})$ denotes $E[U_t]$, then

$$\frac{\partial}{\partial V_{it}}\; \underset{i \varepsilon C_t}{E\,[Max\, U_{it}]} \geq 0, \text{ for all } i \varepsilon C_t$$

In words, this property implies that the measure of accessibility does not decrease as the mean of any one utility increases, all else held equal.

As will be demonstrated, the first property, monotonicity with respect to choice-set size, holds for all situations in which $E(\underset{i \varepsilon C_t}{Max} U_{it})$ exists with and without the n + 1 alternative. To prove this, define the random variable k as follows

$$k = Max\,(U_{1t},\ldots,U_{nt}), \tag{1}$$

and let f(k) denote the probability density function of k. Williams (34) offers an independently developed proof of this proposition. Furthermore, let U* be the utility of some additional alternative, and let f(U*, k) denote the joint probability density function of U* and k. Consider first

$$E[k] = \int_{-\infty}^{\infty} f(k)k \, dk. \tag{2}$$

For reasons that will be clear shortly, this expression will be written as

$$E[k] = \int_{-\infty}^{\infty} k \int_{-\infty}^{\infty} f(U^*,k) \, dU^* \, dk$$

$$= \int_{-\infty}^{\infty} \int_{-\infty}^{k} k \, f(U^*,k) \, dU^* \, dk + \int_{-\infty}^{\infty} \int_{k}^{\infty} kf(U^*,k) dU^* dk. \tag{3}$$

The two integrals in the last expression correspond to the ranges of k and U* such that $U^* < k$ and $U^* \geq k$, respectively. These are the areas in $(U^*,k)$ space for which the additional alternative is not chosen $(U^* < k)$ and the additional alternative is chosen $(U^* \geq k)$.

Now consider

$$E[\text{Max } (U^*,k)] = \int_{-\infty}^{\infty} \int_{-\infty}^{k} k \, f(U^*,k) \, dU^* dk + \int_{-\infty}^{\infty} \int_{k}^{\infty} U^* f(U^*,k) dU^* dk. \tag{4}$$

Comparing expressions (3) with (4), we note that

$$E[\text{Max } (U^*,k)] - E[k] = \int_{-\infty}^{\infty} \int_{k}^{\infty} (U^*-k) \, f(U^*,k) dU^* dk. \tag{5}$$

Furthermore, since over the range of the integral U* is greater than or equal to k (and since $f(U^*,k)$ is by definition greater than or equal to zero), the value of the integral must be non-negative. Thus, the addition of an extra alternative cannot decrease the value of the accessibility measure.

A corollary of the above proof is that the expected value of the maximum is greater than or equal to the maximum of the means of the utilities, i.e.

$$E[\text{Max } U_{it}] \geq \text{Max } (E[U_{it}]) \, . \tag{6}$$
$$\quad i\varepsilon C_t \qquad \quad i\varepsilon C_t$$

A sufficient condition for (6) is that

$$E[\text{Max } (U_{1t},\ldots,U_{nt})] \geq E[U_{1t}], \tag{7}$$

since the order in which utilities are numbered is totally arbitrary. To prove that (7) holds let us define $k_j$ as

$$\text{Max}(U_{1t},U_{2t},\ldots,U_{jt}).$$

We note that

$$E(Max(U_{1t}, \ldots, U_{nt})) = E(Max \ (k_{n-1}, U_{nt})). \tag{8}$$

By the property of monotonicity of choice set size,

$$E[Max \ (k_{n-1}, U_{nt})] \geq E[k_{n-1}]. \tag{9}$$

By recursively using this property, we can show

$$E[k_{n-1}] \geq E[k_{n-2}] \geq \ldots \geq E[k_1], \tag{10}$$

where

$$E(k_1) = E(U_{1t}),$$

thereby proving Equation (6). A simpler proof of (6) was suggested by Charles Manski. Since maximisation is a convex function, Jensen's inequality holds. Expression (6) follows immediately ((27)p. 58).

One might logically expect that the property of monotonicity with respect to mean utilities would also hold in general. Oddly, this is not the case. Consider the binary choice model in which

$$U_{it} = V_{it} + V_{it}\delta_{it}, \tag{11}$$

where $\delta_{it}$ is a random variable with zero mean. For notational convenience, we will simplify the utility expressions as follows:

$$U_{1t} = U_1 = V_1 + \delta_1 V_1 \tag{12}$$

$$U_{2t} = U_2 = V_2 + \delta_2 V_2.$$

Assume that $\delta_1$ and $\delta_2$ are the only random components of $U_1$ and $U_2$ respectively, and that they are normally distributed. They have means 0, variances $\sigma_1^2$ and $\sigma_2^2$ respectively and correlation ratio .

This model is a special case of the form used by Hausman and Wise (17) and termed by Lerman and Manski (20) the generalised multinomial probit model. The generalised multinomial probit model applies to any number of alternatives, allows for normally distributed taste variation on each coefficient $\beta$ in the function $V_{it} = Z_{it}'\beta$ , and includes an additional set of normally distributed disturbance terms. The random variables ($\delta_1$ and $\delta_2$) represent taste variation in the population being modelled.

Under these assumptions, Clark (8) has shown that:

$$E[Max \ (U_1, U_2)] = V_1 \ \phi(\frac{V_1 - V_2}{a}) + V_2 \phi(\frac{V_2 - V_1}{a}) + a\phi(\frac{V_1 - V_2}{a}) \tag{13}$$

where:

$$a^2 = V_1^2 \sigma_1^2 + V_2^2 \sigma_2^2 - 2\sigma_1 \sigma_2 \, \rho V_1 V_2$$

$\Phi(u)$ denotes the standarised cumulative normal distribution, and

$$\phi(u) = \frac{\partial \Phi(u)}{\partial u}.$$

Noting that $\Phi(-u) = 1 - \Phi(u)$, expression (13) can be simplified as

$$V_2 + (V_1 - V_2) \quad \Phi\left(\frac{V_1 - V_2}{a}\right) + a \, \phi\left(\frac{V_1 - V_2}{a}\right). \tag{14}$$

Now consider

$$\frac{\partial E[\text{Max} \, (U_1, U_2)]}{\partial V_1} = (V_1 - V_2) \phi\left(\frac{V_1 - V_2}{a}\right) \frac{\partial}{\partial V_1}\left(\frac{V_1 - V_2}{a}\right) + \Phi\left(\frac{V_1 - V_2}{a}\right) +$$

$$a\frac{\partial}{\partial V_1} \, \phi\left(\frac{V_1 - V_2}{a}\right) + \frac{\partial a}{\partial V_1} \, \phi\left(\frac{V_1 - V_2}{a}\right). \tag{15}$$

Furthermore

$$a\frac{\partial}{\partial V_1} \, \phi\left(\frac{V_1 - V_2}{a}\right) = a\phi\left(\frac{V_1 - V_2}{a}\right)\left(-\frac{V_1 - V_2}{a}\right)\frac{\partial}{\partial V_1}\left(\frac{V_1 - V_2}{a}\right)$$

$$= -(V_1 - V_2) \, \phi\left(\frac{V_1 - V_2}{a}\right)\frac{\partial}{\partial V_1}\left(\frac{V_1 - V_2}{a}\right). \tag{16}$$

Thus, Equation (15) simplifies to:

$$\frac{\partial E[\text{Max} \, (U_1, U_2)]}{\partial V_1} = \Phi\left(\frac{V_1 - V_2}{a}\right) + \frac{\partial a}{\partial V_1} \, \phi\left(\frac{V_1 - V_2}{a}\right). \tag{17}$$

Since $\Phi\left(\frac{V_1 - V_2}{a}\right)$ and $\phi\left(\frac{V_1 - V_2}{a}\right)$ are by definition positive, Equation (17) can only be negative when:

$$\frac{\partial a}{\partial V_1} < - \frac{\Phi\left(\frac{V_1 - V_2}{a}\right)}{\phi\left(\frac{V_1 - V_2}{a}\right)}. \tag{18}$$

Since,

$$\frac{\partial a}{\partial V_1} = \frac{\partial}{\partial V_1}(V_1^2 \sigma_1^2 + V_2^2 \sigma_2^2 - 2\sigma_1 \sigma_2 \rho V_1 V_2)^{1/2} \tag{19}$$

$$= \frac{1}{a}(V_1 \sigma_1^2 - \sigma_1 \sigma_2 \rho \, V_2)$$

it is quite possible for the expression in Equation (19) to satisfy condition 18. For example, suppose the following hold:

$$\sigma_1 = \sigma_2 = 1$$

$$\rho = 0$$

$$V_1 = -1$$

$$V_2 = 1.$$

In this case

$$\frac{\partial a}{\partial V_1} = \frac{V_1}{a} = \frac{-1}{\sqrt{2}} \cong -.7071 ,$$

while

$$-\frac{\phi(\frac{V_1 - V_2}{a})}{\phi(\frac{V_1 - V_2}{a})} = -\frac{\phi(\frac{-2}{\sqrt{2}})}{\phi(\frac{-2}{\sqrt{2}})} \cong -\frac{.07868}{.1468} \cong -.53598.$$

Thus,

$$\frac{\partial\ E[Max\ (U_1, U_2)]}{\partial V_1} < 0.$$

An interesting question is whether there exists a relatively straightforward set of restrictions on the distribution of the $U_t$ such that the measure of expected maximum utility is monotonic with respect to the mean utilities. One set of useful sufficient conditions was proposed by Williams (34), and Harris and Tanner (16). One way to view conditions on the monotonicity with respect to mean utilities is to consider:

$$\frac{\partial E[k]}{\partial V_{it}} = \frac{\partial}{\partial V_{it}} \int \dots \int k f(U_t)\ dU_t , \tag{20}$$

where k is as defined in Equation (1).

Moving the derivative inside the integral, we can re-write Equation (20) as:

$$\int \dots \int \frac{\partial k}{\partial V_{it}}\ f\ (U_t)\ dU_t + \int \dots \int\ k\ \frac{\partial f(U_t)}{\partial V_{it}}\ dU_t \tag{21}$$

Since

$$\frac{\partial k}{\partial V_{it}} = \begin{cases} \frac{\partial U_{it}}{\partial V_{it}} & \text{when } U_{it} \geq U_{jt}, \text{ for all } j \neq i,\ j \in C_t. \\ 0 & \text{otherwise} \end{cases}$$

this expression can be simplified to

$$\int \ldots \int_{k=U_{it}} \frac{\partial U_{it}}{\partial V_{it}} f(U_t) dU_t + \int \ldots \int k \frac{\partial f(U_t)}{\partial V_{it}} dU_t. \qquad (22)$$

However, if $\frac{\partial U_{it}}{\partial V_{it}} = 1$, the first integral in Equation (22) is simply the probability that $U_{it}$ is greater than all $U_{jt}$, $j \neq i$, $j \varepsilon C_t$, which is the probability that alternative i is chosen. If we write this probability as $P_{it}(C_t)$, then

$$\frac{\partial E[\underset{i \varepsilon C_t}{Max} U_{it}]}{\partial V_{it}} = P_{it}(C_t) + \int \ldots \int k \frac{\partial f(U_t)}{\partial V_{it}} dU_t. \qquad (23)$$

Williams has shown that if the distribution $f(U_t)$ is translationally invariant then:*

$$\frac{\partial f(U_t)}{\partial V_{it}} = 0, \qquad (24)$$

and therefore, Equation (22) reduces to

$$\frac{\partial E[\underset{i \varepsilon C_t}{Max} U_{it}]}{\partial V_{it}} = P_{it}(C_t). \qquad (25)$$

As an obvious consequence of Equation (25), in cases where $\frac{\partial U_{it}}{\partial V_{it}} = 1$ and $\partial f(U_t)/\partial V_{it} = 0$, for all $i \varepsilon C_t$, then

$$\sum_{i \varepsilon C_t} \frac{\partial E[\underset{i \varepsilon C_t}{Max} U_{it}]}{\partial V_{it}} = 1. \qquad (26)$$

---

*The property of translational invariance can be best interpreted by noting that condition 24 implies that

$$f(U_t | V_t) = f((U_t + \partial U_t) | (V_t + \partial V_t)).$$

Thus, increasing the vector of means, $V_t$, by $\partial V_t$ produces a density function which is the same as before the change in $V_t$ except that it is 'shifted over'. The normal distribution with fixed variance-covariance matrix and the Weibull distribution are both translationally invariant, while the exponential and the gamma distributions are not.

This implies that if we increase the utility of every alternative by some amount $\Delta V_t$ then the total accessibility increases by that amount.

For most choice models we can write the utilities as follows:

$$U_{it} = V_{it} + \delta_{it} \, g(V_{it}) + \varepsilon_{it}, \tag{27}$$

where $g(V_{it})$ is any vector of functions, and the vector $_{it}$ and the scalar $\varepsilon_{it}$ are unobserved random variables distributed as $f(\delta_t, \varepsilon_t)$. (The notation $\delta_t$ denotes a vector consisting of all entries in the vectors $\delta_{it}$, $i \varepsilon C_t$.)

Assuming that

$$\frac{\partial f(\delta_t, \varepsilon_t)}{\partial V_{it}} = 0,$$

Equation (22) can now be written as follows:

$$\frac{\partial E[\underset{i \varepsilon C_t}{\text{Max}} \, U_{it}]}{\partial V_{it}} = \int \ldots \int (1 + \frac{\partial}{\partial V_{it}} \, \delta_{it} g(V_{it})) f(\delta_t, \varepsilon_t) \, d\delta_t d\varepsilon_t$$

$$k = U_{it}$$

$$= P_{it}(C_t) + \int \ldots \int \frac{\partial}{\partial V_{it}} \, \delta_{it} \, g(V_{it}) \, f(\delta_t, \varepsilon_t) \, d\delta_t d\varepsilon_t \tag{28}$$

$$k = U_{it} \, .$$

In the simple case where $g(V_{it})$ and $\delta_{it}$ are scalars and

$$g(V_{it}) = V_{it} \tag{29}$$

the restriction that

$$\delta_{it} \geq -1 \tag{30}$$

will result in

$$\frac{\partial E[\underset{i \varepsilon C_t}{\text{Max}} \, U_{it}]}{\partial V_{it}} \geq 0. \tag{31}$$

This case represents a random-utility specification in which $\delta_{it}$ represents unobserved taste variations and $\varepsilon_{it}$ is the unobserved utility which is independent of $V_{it}$. It is reasonable to assume that $f(\delta_t)$ is independent of the Vs and that $\delta_{it}$ has a lower bound of $-1$ such that $\partial U_{it}/\partial V_{it}$ is always positive. This requirement implies that the marginal utility of a positive attribute, including the observed and unobserved utilities components, is always positive. The probit model shown earlier assumes a normal distribution of the $\delta$s which are therefore not bound to be greater than $(-1)$. Thus, if a random-utility model with unobserved taste variation is carefully specified, such that $\partial U_{it}/\partial V_{it}$ is always positive, the desirable property of monotonicity of the expected maximum utility with the mean utilities is always satisfied.

SOME SPECIAL CASES

Fortunately, the accessibility measure can be derived in closed form for the most widely used individual-choice model, multinomial logit. The logit model is characterised by utilities which satisfy

$$U_{it} = V_{it} + \varepsilon_{it} \tag{32}$$

where $\varepsilon_{it}$ independent from all $\varepsilon_{it}$ and every $\varepsilon$ has the following cumulative distribution function

$$P(\varepsilon \leq \omega) = e^{-e^{-\mu(\alpha+\omega)}} \tag{33}$$

Under these assumptions

$$P_{it}(C_t) = \frac{e^{\mu V_{it}}}{\sum_{j \in C_t} e^{\mu V_{jt}}}. \tag{34}$$

The Vs are generally defined such that $\mu = 1$. The function in Equation (33) satisfies the condition

$$\frac{\partial f(\varepsilon_t)}{\partial V_{it}} = 0, \quad \text{and} \quad \frac{\partial U_{it}}{\partial V_{it}} = 1.$$

Hence, by Equation (25)

$$\frac{\partial E[\text{Max } U_t]}{\partial V_{it}} = P_{it}(C_t) = \frac{e^{V_{it}}}{\sum_{j \in C_t} e^{V_{jt}}} \tag{35}$$

This partial differential equation yields the expression:

$$E[\text{Max } U_{it}] = \int \frac{e^{V_{it}}}{\sum_{j \in C_t} e^{V_{jt}}} \, dV_{it}, \tag{36}$$

which when solved implies that

$$E[\text{Max } U_t] = \ln \sum_{j \in C_t} e^{V_{jt}} + \text{Const.} \qquad (37)$$

A more direct approach to derive the expected value of the maximum utility is first to derive the distribution of the maximum utility, defined as k in Equation (1), from the distribution of the vector $U_t$. For the logit model we obtain ((12)p. 63):

$$P(k \leq \omega) = e^{-e^{-(\alpha_t^* + \omega)}}, \qquad (38)$$

where

$$\alpha_t^* = \alpha - \ln \sum_{j \in C_t} e^{V_{jt}}.$$

The expected value is equal to (14):

$$E(k) = -\alpha_t^* + \gamma$$

$$= \ln \sum_{j \in C_t} e^{V_{jt}} - \alpha + \gamma , \qquad (39)$$

where $\gamma$ is Euler's constant ($\approx 0.577$). Since the selection of the parameter $\alpha$ is arbitrary, a logical value is $\alpha = \gamma$, so that the relevant measure for the logit model is:

$$\ln \sum_{j \in C_t} e^{V_{jt}}. \qquad (40)$$

This expression is simply the natural logarithm of the denominator of the multinomial logit model, and can be derived in a number of other ways. See, for example, Williams (34), Domencich and McFadden (12), Ben-Akiva (5), Lerman (20) or Small (31). Since the denominator of the logit function is computed in forecasting travel demand anyway, the measurement of accessibility requires little additional computation.

A second case of interest is the situation in which the utility functions have additive disturbances as in Equation (32), but the $\varepsilon$s are normally distributed with mean 0 and arbitrary (up to a set of restrictions required for the parameters to be identified) variance-covariance matrix $\sum$. As noted in the previous section, this assumption leads to the multinomial probit model.

The multinomial probit model has not been used previously because of the computational problems associated with its estimation and application. However, recent work by Lerman and Manski (20), Hausman and Wise (17), Daganzo, Sheffi and Bouthelier (9), and Andrews and Langdon (3) has made probit for multiple alternatives feasible.

Multinomial probit has the advantage over logit of allowing for the possibility that two alternatives have disturbance terms which are not independent. Thus, apparent anomalies such as the 'red bus-blue bus' paradox do not arise. Furthermore, unlike logit, multinomial probit in its most general form permits measurement of taste variation in the population.

There does not exist a closed-form solution for the expected maximum utility, or the accessibility measure, in the case of the multinomial probit. However, it is possible to use a set of formulas (one of which was used in Equation (13)) for the first two moments of the maximum of two normally distributed random variables and the covariance of the maximum with a third normally distributed variable to approximate the accessibility (8). Clark first proposed that his solution for the moments of the maximum of two normally distributed variables be used as an approximation by assuming that the maximum is itself normally distributed. This approximation could then permit solution of

$$E(Max(U_{1t},\ldots,U_{nt})),$$

by use of a recursive procedure, since

$$Max(U_{1t},\ldots,U_{n-2t},U_{n-1t},U_{nt}) = Max(U_{1t},\ldots,U_{n-2t},$$

$$Max(U_{n-1t},U_{nt})). \tag{41}$$

While Clark recognised that the maximum of two normally distributed variables is obviously not itself normal, some limited empirical evidence indicated that at least for the first moment, it served as a good approximation. Further tests by Daganzo, Sheffi and Bouthelier (9) and subsequently by Manski, Lerman and Albright (23) have provided additional support. The latter group has compared the choice of probabilities for multinomial probit as calculated by simulation techniques with those generated by the Clark approximation, and the results indicate that the absolute error in the Clark algorithm is rarely more than 0.01 or 0.02.

As in the logit model, the value of the accessibility measure for multinomial probit can be computed as part of any forecasting procedure without any major, new computation. In applying the Clark algorithm to finding the choice probability of alternative i, the recursive approximation is used to solve the moments of the maximum utility of all the $j \neq i$ alternatives. The mean and variance of this maximum (as well as its covariance with the utility of alternative i) are then used in a binary comparison, where the choice probability is simply

$$P[U_{it} \geq \max_{\substack{j \in C_t \\ j \neq i}} U_{jt}] .$$

All that is required to compute the expected value of the
maximum utility is one extra application of the Clark approximation for the first moment (Equation (13)) to solve for:

$$E[\text{Max} \ (U_{it}, \ \underset{j \varepsilon C_t}{\text{Max}} \ U_{jt})]$$
$$j \neq i \ .$$

## TRAVEL DEMAND AS A MULTIDIMENSIONAL CHOICE

The previous sections of this chapter dealt with the
properties of random-utility choice models and expected
maximum utilities in terms of a simple choice of an alternative i from a choice set $C_t$. In many applications, the
members of $C_t$ are choices along a simple dimension such as
modes of travel or destinations.  However, a complete travel-
demand model must consider multiple dimensions of choice that
include all household and individual decisions that directly
affect trip-making.  The set of relevant decisions includes
choices with respect to specific trips, such as mode choice
and destination choice, and other choices which are so inter-
related with actual trip-making choices that it is impossible
to separate them from such decisions.  For example, the
choice of residential location is not in itself a trip-making
decision.  However, the combination of worker's employment
location choice and his or her household's residential location decision has as its consequence a trip choice, i.e. a
daily work trip.

For this reason, the general framework from which urban
travel-demand models can be derived begins with a partition
of all possible household and household-member decisions into
two sets, those directly relevant to personal travel and
those which can, for practical purposes, be ignored.  In
formal utility-theory terms, this partition implies that the
total utility function for the decision-maker has an addi-
tively separable form and that the overall income and time
constraints are negligible.  Here, this partition produces
the following vector of travel-related household decisions:

1  employment location (for all workers)
2  residential location
3  housing type
4  car ownership
5  mode to work (for all workers)
6  frequency (for non-work trips of each purpose)
7  destination (for non-work trips of each purpose)
8  time of day (for all trips)
9  mode (for non-work trips of each purpose)
10  route (for all trips)

For most situations, this vector describes the decisions
which a complete travel-demand model must consider.  (There
are possible exceptions.  Frequency and time of day for work
trips may be distinct choices or constrained by the work
situation.  Some decisions may be totally constrained by lack
of a driver's license, though the latter may well be part of
some decision vectors.  Allocation of time to different

activities (or purposes) is also suggested as an element of
this decision vector in a later section of this chapter.)  In
theory, each decision may be dependent on the rest.  For
example, where one chooses to live is obviously linked to the
housing type and the level of car ownership one selects.
Similarly, shopping-trip destination and mode are likely to
be closely linked.

This perspective, if carried through completely, would
produce a model of unmanageable dimensions.  Since the entire
vector of possible choices would result in a virtually limit-
less choice set, useful models would be impossible to develop.
Fortunately, there are some interrelationships among components of this vector which are of a fundamentally different
character than others.  Some of the decisions, such as
residential location choice, have high transaction costs and
are consequently stable over fairly long intervals; other
choices such as social-trip frequency are altered on a daily
basis.  Some decisions are more logically represented as
being made collectively by the household, while others can be
approximated as individual choices.  Thus, it is possible to
formulate explicit behavioural hypotheses, and establish a
structure of the total vector of choices as a logical working
hypothesis.  Such an explicit structure greatly simplifies
the formulation and development of a complete travel-demand
model.  This structure is termed a *hierarchy of choice*.

Figure 30.1 illustrates one possible choice hierarchy.
(See Ben-Akiva (5), Lerman (21) and Adler (1) for more
detailed discussions of alternative choice hierarchies.)  It
defines two distinct sets of choices: the long-run *mobility*
decisions of location, housing, car ownership and mode to
work and the short-run *travel* choices of frequency, mode,
destination, route and time of day for non-work trips.  These
decisions are assumed to be structured hierarchically, i.e.
location and related decisions are made with travel choices
indeterminate, and the travel choices are made, conditional
on the outcome of the mobility-choice process.  *Within each
group of choices, decisions are assumed to be made by a joint
process in which the full range of possible trade-offs is
considered by the household.*

The use of this hierarchical choice structure permits the
development of two different models, one for mobility choices
and one for travel choices.  This set of models can be termed
*block conditional*, where the block of mobility and travel
choices as single units have a conditional structure, while
each block itself has a joint structure.  It is important to
emphasise that block-conditional or other hierarchical
decision structures are based on *explicit behavioural hypotheses* about household or individual behaviour.  Different
choice hierarchies might be proposed for different decision-
makers; since each is essentially an approximation, some
hierarchies may prove useful in some situations but may be
inappropriate in others.  Consider, for example, the
hierarchy depicted in Figure 30.2.  In this case, employment
location has been separated from the remaining mobility
choices, which for convenience are termed the mobility bundle.
This hierarchy might be appropriate for skilled workers,
whose employment location is relatively fixed.  A skilled

employee must generally switch his employer in order to
shift his job location.

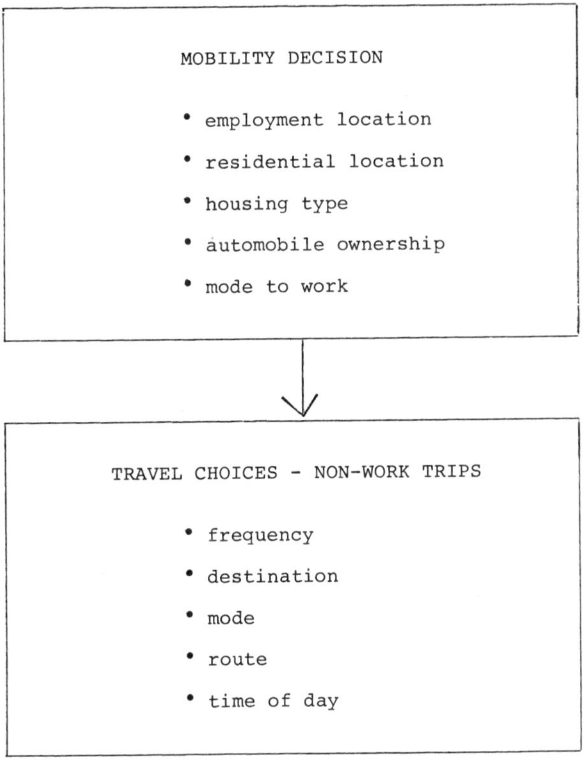

```
+-------------------------------------------------------+
|                                                       |
|            MOBILITY DECISION                          |
|                                                       |
|              • employment location                    |
|                                                       |
|              • residential location                   |
|                                                       |
|              • housing type                           |
|                                                       |
|              • automobile ownership                   |
|                                                       |
|              • mode to work                           |
|                                                       |
+-------------------------------------------------------+

                         │
                         ▼

+-------------------------------------------------------+
|                                                       |
|        TRAVEL CHOICES - NON-WORK TRIPS                |
|                                                       |
|              • frequency                              |
|                                                       |
|              • destination                            |
|                                                       |
|              • mode                                   |
|                                                       |
|              • route                                  |
|                                                       |
|              • time of day                            |
|                                                       |
+-------------------------------------------------------+
```

Figure 30.1  A Simple Choice Hierarchy

MODELLING IMPLICATIONS OF A CHOICE HIERARCHY

To illustrate the implications of the choice hierarchy, let
us consider some choice j from the set J.  Assume that the
choice from set J is part of a hierarchy, and that it is
preceded by a choice of i from set I and followed by a choice
k from set K.  (For notational convenience, the subscript t
denoting the individual decision-maker will be omitted in the
remainder of the chapter.)  Given this situation, the entire
set of variables which influence the choice from J can be
divided into four classes as follows:

1 Those that directly affect the choice of j but do not
directly influence the choice of i or k.
2 Those that directly affect the choice of j as well as

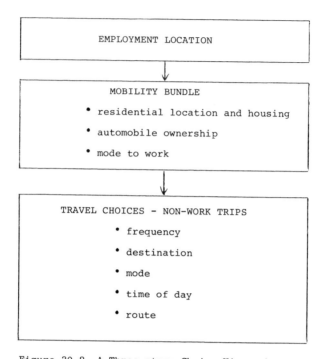

Figure 30.2   A Three-stage Choice Hierarchy

the higher level choice i and the lower level choice k.
3   Those that are determined by the outcome of the higher
level choice i and affect the choice of j, such as an attribute of the chosen i which is treated as a socio-economic
factor affecting the choice among alternatives in J.
4   Those that express expectations from outcome of the lower
level choice from set K; these expectations may vary among
alternatives in J and can be treated as attributes of the
alternatives in J.

The differences among these variables can be clarified by
using the choice hierarchy in Figure 30.2 as an example.  The
factors affecting the choice of mobility bundle (which corresponds to choice of j) are:

1   Direct attributes of the mobility bundle such as rent or
housing price.  This variable, in general, has a different
value for each alternative location and housing-type combination.
2   Joint attributes of the mobility bundle as well as higher
or lower level choices such as level-of-service attributes by
transit for the trip to work.  This variable is an attribute
of an alternative mobility bundle, and is also an attribute

of alternative employment-location alternatives, which
corresponds to I.
3 Variables resulting from the employment-location decisions such as a variable indicating whether or not a worker
in the household is employed in the central business district.
This variable is *predetermined* in the mobility-choice process,
and it is an attribute of the household that does not vary
among alternative mobility bundles.
4 Variables describing travel decisions made conditional on
the mobility choice such as level of service by transit for
shopping trips from a given residential zone. This variable
depends on how the household will choose to travel for shopping, a decision which is made only conditional on the
household's mobility choice. Thus, shopping level of service
is said to be *indeterminate* in the mobility decision. This
does not imply that it does not enter into the household's
evaluation of mobility alternatives; rather, it means that
the household does not know the *specific* shopping trips it
will make. It does have some *composite* picture of the overall shopping level of service it will obtain for various
mobility bundles. For example, if the household were to
choose to own two or more cars in its mobility bundle, it
would generally have a high composite level of service for
shopping associated with suburban locations because it would
typically travel by car. Alternatively, if the household
chose not to own a car, suburban locations would have poor
composite level of service since it would have to shop
exclusively by transit.

For a block-conditional model system to be logically consistent, each of these types of variables must be appropriately
represented. Specific models must have appropriate linkages
to other components of a choice hierarchy in order to reflect
the full sensitivity of the whole range of relevant choices
to variables influencing travel-related decisions.

In formulating such linkings the construction of variables
of the second and fourth types requires further discussion.
In modelling the choice from J, attributes which are specific
to a combination ijk (i.e. a feasible member of I  J  K) as
in the second type of factors or to a jk combination (i.e. a
feasible member of J  K) as in the fourth type of factors,
must be somehow aggregated over alternative choices in K in
order to be included in the model for the choice from J
conditional on the choice of i. This requirement and
alternative formulations which have been used for such composite attributes are discussed in many references, including
Ben-Akiva (5) and Williams (34).

All systems of urban travel-demand models explicitly or
implicitly include such composite variables. Examples would
be the weighted 'inclusive price' used by CRA (7) and the
transit accessibility variable used in a trip-generation
model by the Metropolitan Washington Council of Governments
(32). However, as noted earlier in this chapter, the definition of accessibility or composite attributes is often
arbitrary. The use of expected maximum utilities to define
these composite variables has the advantages cited earlier in
this chapter as well as the advantage of a consistent formulation of a sequence of models based on maximisation of
random utilities.

The assumption of a choice hierarchy for the selection of an alternative from $I \times J \times K$ implies the following decomposition of the utility function:

$$U_{ijk} = U_i + U_{j|i} + U_{k|ij} \qquad (42)$$

where

$U_{ijk}$ = the joint utility of an ijk combination,

$U_i$ = the part of the joint utility which is independent of j and k,

$U_{j|i}$ = the part of the joint utility for a choice of j given that i is selected which is independent of k, and

$U_{k|ij}$ = the part of the joint utility for a choice of k given that i and j are selected.

(Note that if $U_{ijk}$ is not separable, then $U_i$ and $U_{j|i}$ are zero, and by definition $U_{k|ij} = U_{ijk}$.) The model for the lower level of the choice hierarchy, the choice of k, can be expressed as follows:

$$P(k|ij) = \text{Prob}[U_{k|ij} \geq U_{k'|ij}, \; \forall \; k' \; \varepsilon \; K] \qquad (43)$$

where

$P(k|ij)$ = the probability of choosing k given that i and j are selected.

The model for the choice of j is:

$$P(j|i) = \text{Prob}[U_{j|i} + \underset{k\varepsilon K}{\text{Max}} \, U_{k|ij} \geq U_{j'|i} + \underset{k\varepsilon K}{\text{Max}} \, U_{k|ij}, \forall \, j' \varepsilon J], \quad (44)$$

where

$P(j|i)$ = the probability of choosing j given that i is selected.

Thus, the utility for this choice model is:

$$\tilde{U}_{j|i} = U_{j|i} + \underset{k\varepsilon K}{\text{Max}} \, U_{k|ij}. \qquad (45)$$

Similarly, for the choice of i the model can be written as follows:

$$P(i) = \text{Prob}[\tilde{U}_i \geq \tilde{U}_{i'}, \; \forall \; i' \varepsilon I], \qquad (46)$$

where

$P(i)$ = the marginal probability of choosing i, and

$\tilde{U}_i = U_i + \underset{j\varepsilon J}{\text{Max}} \, \tilde{U}_{j|i}.$

Note that

$$\underset{ijk \, \epsilon \, IJK}{\text{Max}} U_{ijk} = \underset{i \epsilon I}{\text{Max}} \tilde{U}_i.$$

Denoting the expected utility components as:

$$E[U_{ijk}] = E[U_i] + E[U_{j|i}] + E[U_{k|ij}]$$

$$= V_i + V_{j|i} + V_{k|ij} \tag{47}$$

we obtain:

$$E[\tilde{U}_{j|i}] = V_{j|i} + E[\underset{k \epsilon K}{\text{Max}} U_{k|ij}] \tag{48}$$

$$E[\tilde{U}_i] = V_i + E[\underset{j \epsilon J}{\text{Max}} \tilde{U}_{j|i}]. \tag{49}$$

This implies that

$$\tilde{V}_{j|i} = V_{j|i} + E[\underset{k \epsilon K}{\text{Max}} U_{k|ij}] \tag{50}$$

and

$$\tilde{V}_i = V_i + E[\underset{j \epsilon J}{\text{Max}} \tilde{U}_{j|i}] \tag{51}$$

are the systematic components of the utilities for the choice models for $P(j|i)$ and $P(i)$, respectively.

We can also derive the disturbances for these choice processes (i.e. the random component of the utility function) as follows:

$$\varepsilon_{k|ij} = U_{k|ij} - V_{k|ij} \tag{52}$$

$$\varepsilon_{j|i} = U_{j|i} - V_{j|i} + \underset{k \epsilon K}{\text{Max}} U_{k|ij} - E[\underset{k \epsilon K}{\text{Max}} U_{k|ij}] \tag{53}$$

$$\varepsilon_i = U_i - V_i + \underset{j \epsilon J}{\text{Max}} \tilde{U}_{j|i} - E[\underset{j \epsilon J}{\text{Max}} \tilde{U}_{j|i}]. \tag{54}$$

Expressing the joint utility function as:

$$U_{ijk} = V_{ijk} + \varepsilon_{ijk}, \tag{55}$$

where

$$V_{ijk} = V_i + V_{j|k} + V_{k|ij}, \tag{56}$$

we obtain:

$$\varepsilon_{ijk} = \varepsilon_i - (\underset{j \epsilon J}{\text{Max}} \tilde{U}_{j|i} - E[\underset{j \epsilon J}{\text{Max}} \tilde{U}_{j|k}])$$

$$+ \varepsilon_{j|i} - (\underset{k \epsilon K}{\text{Max}} U_{k|ij} - E[\underset{k \epsilon K}{\text{Max}} U_{k|ij}]) + \varepsilon_{k|ij}. \tag{57}$$

A LOGIT EXAMPLE OF ACCESSIBILITY IN CHOICE HIERARCHIES

If the $\varepsilon_{ijk}$s are independently and identically distributed (i.i.d) as Weibull, we obtain the following joint logit model:

$$P(ijk) = \frac{e^{V_{ijk}}}{\sum_i \sum_j \sum_k e^{V_{ijk}}} \cdot \qquad (58)$$

Based on the decomposition of the joint-utility function we can calculate the following conditional and marginal probabilities:

$$P(k|ij) = \frac{e^{V_{k|ij}}}{\sum_{k'} e^{V_{k'|ij}}} \qquad (59)$$

$$P(j|i) = \frac{e^{\mu_1 V_{j|i} + \mu_1 \ln \sum_k e^{V_{k|ij}}}}{\sum_{j'} e^{\mu_1 V_{j'|i} + \mu_1 \ln \sum_k e^{V_{k|ij'}}}} \qquad (60)$$

$$P(i) = \frac{e^{\mu_2 V_i + (\mu_2/\mu_1) \ln \sum_j e^{\mu_1 V_{j|i} + \mu_1 \ln \Sigma e^{V_{k|ij}}}}}{\sum_{i'} e^{\mu_2 V_{i'} + (\mu_2/\mu_1) \ln \sum_j e^{\mu_1 V_{j|i'} + \mu_1 \ln \Sigma e^{V_{k|i'j}}}}} \qquad (61)$$

where $\mu_1 = \mu_2 = 1$, and were included for reasons that will be clarified below.

From the derivation of the expected maximum utility for the logit model in an earlier section of this chapter we observe that each higher level model includes the expected maximum utility from the lower level. We note that the joint-logit model implies that the three random components are independent and Weibull distributed with the same parameter as follows:

$$\text{Prob}[\varepsilon \le \omega] = e^{-e^{-\mu(\omega+\alpha)}}$$

where the shift parameter $\alpha$ does not affect the choice probabilities and is not identified, and the scale parameter $\mu$ is assumed to equal one.  A joint logit model was applied by Ben-Akiva (5), Adler and Ben-Akiva (2), Lerman (21) and others.

If we allow $\mu_1$ and $\mu_2$ to be different from unity but require that:

$$0 < \mu_1 \leq 1$$

$$0 < \mu_2 \leq \mu_1$$

we obtain a sequence of logit models which imply that the three random components are independent and Weibull distributed but with three different parameters: 1, $\mu_1$ and $_2$. This condition is required for the following to hold:

$$\frac{\partial P(j\ k|i)}{\partial V_{k'|ij}} \leq 0 \qquad \text{for } k' \neq k$$

$$\frac{\partial P(i\ j)}{\partial \tilde{V}_{j'|i}} \leq 0 \qquad \text{for } j' \neq j$$

where $P(j\ k|i)$ is the joint probability that both j and k are selected conditional on i being chosen and $P(i\ j)$ is the joint probability that both i and j are chosen.

In words, these two conditions imply that:

1  the cross-elasticity of the probability of some choice j and k (given i) with respect to an increase in the utility (given i and j) of alternatives $k' \neq k$ is negative, and
2  the cross-elasticity of the probability of some choice i and j with respect to an increase in the utility of alternatives $j' \neq j$ (given i) is negative.

Except for simple cases with a small number of alternatives it is not feasible in practical applications to estimate this model jointly. A sequential estimation starts with the lowest level choice, a calculation of the estimated expected maximum utility (the natural log of the denominator), estimation of the next higher level choice, and so forth. This model was applied by Ben-Akiva (5), Watanatada and Ben-Akiva (33), Ruiter and Ben-Akiva (28) and Daly and Zachary (11).

McFadden (25) proves that this model can be derived by assuming $\varepsilon_{ijk}$s are jointly distributed with the Oliviera distribution function (18). The marginals of this distribution are Weibull distributions. Daly and Zachary (11) noted that since the variance of the Weibull distribution is inversely proportional to its parameter $\mu$, this model represents a decomposition of the alternatives such that the utilities of the lowest stage of the hierarchy have the lowest variance.

To our knowledge, the system developed for the San Francisco Bay Area by Cambridge Systematics (6) is the only complete operational urban travel—forecasting system which has applied such a hierarchical model system with feedback linkages of expected maximum utilities.

DIRECTIONS FOR FUTURE RESEARCH

The expected maximum utility derived from individual travel-choice models is an operational measure of accessibility which has a direct interpretation and also serves to link models in an hierarchical system and as a measure of benefits. However, several problem areas, which arise directly from limitations of the current state-of-the-art of individual travel and mobility choice models, require further research.

The most obvious problem area, which was alluded to earlier in this chapter, is the inclusion of level of service and opportunities for non-home-based trips in an accessibility measure. The only practical approach to modelling non-home-based trips which has been implemented is a conditional structure in which the choice of non-home-based trip is modelled conditional on a home-based trip (19,28). This structure does not take into account the interdependence among the choices with respect to all trip links in a tour (a round trip that starts and ends at home).

This interdependence was observed in several descriptive studies and was recently modelled by Adler (1) in a logit model which considers an entire daily travel pattern as an alternative. An improvement of the level of service for non-home-based trips is likely to result in fewer tours and as a consequence fewer home-based trips. The number of sojourns may increase or stay the same, implying the consolidation of travel patterns into more multiple-stop tours. This indicates that an accessibility measure that is based solely on home-based trips level of service over-estimates the accessibility in areas with fewer opportunities for trip consolidation and under-estimates the accessibilities in areas with more opportunities for multi-stop tours.

A second important problem area in the measurement of accessibility as the expected maximum utility over a large number of spatial alternatives is a lack of natural definition for the set of alternatives and the constraints that define the choice set. For practical reasons, spatial alternatives are always defined as geographical areas which could be taken as an aggregation of some elemental alternatives. If these areas are not homogeneous with respect to the attributes of the elemental alternatives, the measure of accessibility will in general be biased. The direction of the bias depends on the distribution of the attributes within the areas and the form of the accessibility measure. This implies that a careful definition of elemental alternatives and the enumeration of elemental alternatives in each area is an essential part of spatial-choice modelling.

As a consequence of monotonicity of the expected value of the maximum utility with respect to choice-set size, inclusion of irrelevant alternatives in the choice set will always bias upwards the measure of accessibility. One possible approach to the problem of choice-set definition is to explicitly include time allocation in the decision-making process that is modelled. Individuals' allocation of time to various activities can serve as a constraint on total travel time. Bain (4) has developed a simple model which predicts

the individual's decision of whether or not to participate in
an activity on a given day and the expected level of participation, measured in allocated time.

Finally, the large number of spatial alternatives make the
calculation of any measure of accessibility very costly.  A
useful approach would be to develop procedures for sampling
spatial alternatives which would result in an unbiased
measure of accessibility but require less data and computational effort.

REFERENCES

1. Adler, Thomas J. (1976), 'Modelling non-work travel
patterns', Ph.D. Dissertation, Department of Civil Engineering, MIT, Cambridge, Massachusetts.
2. Adler, Thomas J. and Ben-Akiva, M., 'Joint-choice model
for frequency, destination and travel mode for shopping
trips', *Transportation Research Record 569* (1975).
3. Andrews, R.D. and Langdon, M.G., 'An individual cost
minimising method of determining modal split between three
travel modes', *Transport and Road Research Laboratory Report
698*, Department of the Environment, Crowthorne and Berkshire,
UK (1976).
4. Bain, Joseph J. (1976) 'Activity choice analysis time
allocation and disaggregate travel demand modelling', S.M.
Thesis, Department of Civil Engineering, MIT, Cambridge,
Massachusetts.
5. Ben-Akiva, Moshe (1973) 'Structure of passenger travel
demand models', Ph.D. Dissertation, Department of Civil
Engineering, MIT, Cambridge, Massachusetts.
6. Cambridge Systematics Inc. (1977) Technical report in
progress for the Metropolitan Transportation Commission,
San Francisco, Cambridge, Massachusetts.
7. Charles River Associates, Inc. (1972) *A Disaggregated
Behavioural Model of Urban Travel Demand*, Federal Highway
Administration, US Department of Transportation.
8. Clark, Charles E., 'The greatest of a finite set of
random variables', *Operations Research* (1961).
9. Daganzo, Carlos R., Bouthelier, F. and Sheffi, Y., 'An
efficient approach to estimate and predict with multinomial
probit models', mimeograph, Department of Civil Engineering,
MIT, Cambridge, Massachusetts (1976).
10. Dalvi, M.Q. and Martin, K.M., 'The measurement of
accessibility: some preliminary results', *Transportation*, 5,
1 (1976).
11. Daly, A.J. and Zachary, S., 'Improved multiple choice
models', in *Determinants of Travel Choices*, D.A. Hensher and
M.Q. Dalvi (eds), Teakfield, Farnborough, UK (1978).
12. Domencich, T. and McFadden, D. (1975) *Urban Travel
Demand: A Behavioral Analysis*, North-Holland, Amsterdam.
13. Dunphy, Robert T., 'Transit accessibility as a determinant of automobile ownership', *Highway Research Record 472*
(1973).
14. Gumbel, E.J. (1958) *Statistics of Extremes*, Columbia
University Press, New York.
15. Hansen, W.G., 'How accessibility shapes land use',
*Journal of American Institute of Planners*, 3 (1959).
16. Harris, A.J. and Tanner, J.C., 'Transport demand

models based on personal characteristics', *Transport and Road Research Laboratory Supplementary Report 65 UC*, Department of the Environment, UK (1974).

17. Hausman, Jerry, A. and Wise, David A., 'A conditional probit model for qualitative choice: discrete decisions recognizing interdependence and heterogeneous preferences', *Department of Economics Working Paper No. 173*, Massachusetts Institute of Technology, Cambridge, Massachusetts (1976).

18. Johnson, Norman L. and Kotz, Samuel (1972) *Distributions in Statistics: Continuous Multivariate Distributions*, John Wiley & Sons, Inc., New York.

19. Lerman, Steven R., 'The use of disaggregate choice models in semi-Markov process models of trip chaining behavior', *Center for Transportation Studies Working Paper*, Massachusetts Institute of Technology, Cambridge, Massachusetts (1977).

20. Lerman, Steven R. and Manski, Charles F., 'An estimator for the generalized multinomial probit choice model', *Transportation Research Record* (1978) (in press).

21. Lerman, Steven R., 'A disaggregate behavioral model of urban mobility decisions', *Center for Transportation Studies Report No. 75-5*, Massachusetts Institute of Technology, Cambridge, Massachusetts (1975).

22. Luce, R.D. and Suppes, P., 'Preference, utility and subjective probability', in *Handbook of Mathematical Psychology, Vol. III*, Luce, Bush and Galanter (eds), John Wiley, New York (1965).

23. Manski, Charles F., Lerman, Steven R. and Albright, R., series of technical memos for ongoing project entitled *An Estimator for the Generalized Multinomial Probit Model*, Cambridge Systematics, Inc. (1977).

24. Manski, Charles F. (1973) 'The analysis of qualitative choice', Ph.D. Dissertation, Department of Economics, Massachusetts Institute of Technology, Cambridge, Massachusetts.

25. McFadden, Daniel, 'Quantitative methods for analysing travel behaviour of individuals: some recent developments', (Chapter 13 of the present book).

26. Neuburger, H., 'User benefit in the evaluation of transport and land use plans', *Journal of Transport Economics and Policy*, 5, 1 (1971).

27. Rao, C.R. (1973) *Linear Statistical Inference and Its Applications*, Second Edition, John Wiley and Sons, New York.

28. Ruiter, Earl R. and Ben-Akiva, Moshe, 'A system of disaggregate travel demand models: structure, component models and application procedures', *Cambridge Systematics, Inc. Working Paper*, Cambridge, Massachusetts (1977).

29. Sheffi, Yosef and Daganzo, Carlos, 'Another paradox of traffic flow', *Transportation Research* (1978) (forthcoming).

30. Sherman, L., Barber, B. and Kondo, W., 'Method for evaluating metropolitan accessibility', *Transportation Research Record 499* (1974).

31. Small, Kenneth A., 'Priority lanes on urban radial freeways: an economic simulation', paper presented at the 56th Annual Meeting of the Transportation Research Board, Washington, D.C. (1977).

32. Washington Council of Governments (1974) *Analysis of Zonal Level Trip Generation Relationships*, WCOG Technical Report No. 10.

33. Watanatada, T. and Ben-Akiva, Moshe (1976) *Development*

*of an Aggregate Model of Urbanized Area Travel Demand:
Detailed Specification of Aggregate Travel Demand Model for
Use in TRANS*, draft technical report, Center for Transporta-
tion Studies, Massachusetts Institute of Technology,
Cambridge, Massachusetts.

   34. Williams, H.C.W.L., 'On the formation of travel demand
models and economic evaluation of user benefit', *Institute
for Transport Studies Working Paper 60*, University of Leeds,
UK (1976).

Chapter 31

MOBILITY, ACCESSIBILITY AND TRAVEL IMPACTS OF TRANSPORT
PROGRAMMES FOR THE ELDERLY AND HANDICAPPED

Ronald F. Kirby and Robert G. McGillivray

INTRODUCTION

A variety of programmes at the federal, state and local
levels of government currently provide transport assistance
for elderly and handicapped persons in the US.  These programmes typically earmark funds for particular client groups
and particular transport services according to criteria which
vary greatly from programme to programme.  Each of these
programmes is concerned with bringing about certain kinds of
improvements in the mobility of the client groups served, and
with making certain locations and activities more accessible
to them (8).

One federal programme for which improved mobility for the
elderly and handicapped is an explicit objective is that
administered by the Urban Mass Transportation Administration
(UMTA) and the Federal Highway Administration (FHWA) of the
US Department of Transportation (DOT).  The regulations for
this programme use the term 'elderly and handicapped persons'
to mean:

    those individuals who, by reason of illness, injury, age,
    congenital malfunction, or other permanent or temporary
    incapacity or disability, including those with semi-
    ambulatory capabilities, are unable without special
    facilities or special planning or design to utilize mass
    transportation facilities and services as effectively as
    persons who are not so affected (14).

The *client group* for this programme, then, is composed of
persons who have difficulty using mass transport facilities
because of disabilities.  By comparison, certain other
federal programmes define the elderly as those persons above
a certain age, such as 60 or 65, while still other programmes
limit assistance to those elderly and handicapped below a
certain income level (12).

The DOT programme for the elderly and handicapped provides
assistance only for those transport *services* which qualify as
'mass transport': services which are shared-ride and available to the public on a regular and continuing basis.
Exclusive-ride taxicab services and services restricted to a
particular organisational or institutional clientele
apparently could not receive DOT assistance, for example.
Other federal programmes restrict transport assistance to

680

certain kinds of trips, such as those to and from medical or
educational facilities (12)..

The legislative objectives of these programmes usually
allude to levels of mobility or accessibility to be achieved,
but rarely provide any quantitative measures of those levels:

It is hereby declared to be the national policy that
elderly and handicapped persons shall have the same right
as other persons to utilize mass transportation facilities
and services; that special efforts shall be made in the
planning and design of mass transportation facilities and
services so that the availability to elderly and handicapped persons of mass transportation which they can
effectively utilize will be assured.

... the rates charged elderly and handicapped persons
during non-peak hours for transportation ... financed with
assistance under this section will not exceed one-half of
the rates generally applicable to other persons at peak
hours ... (13).

The Secretary shall require that mobility for elderly
and handicapped persons is available in each urbanized
area requesting a grant or loan under this Act (10).

The above directives from existing and proposed legislation
outlining transport assistance for the elderly and handicapped are all framed in fairly general terms.  Decisions on
just who should qualify for assistance under the programmes
and what levels of mobility and accessibility should be
provided must be made by agencies interpreting and adminis-
tering the programmes at the federal, state and local levels.
Detailed definitions must be developed, for example, of the
kinds of handicaps which inhibit persons from making full use
of transport services, and those wishing to obtain assistance
must demonstrate that they qualify under the definitions
adopted (15).  And once these eligibility criteria have been
established, administrative agencies must decide what kinds
of mobility and accessibility should be provided within the
directives and funding specified in the legislation.

In order to evaluate alternative transport programmes for
the elderly and handicapped, both legislators and administra-
tors need some quantitative measures of the impacts of the
programmes.  In this chapter certain kinds of impacts
associated with these programmes are reviewed, and some
specific approaches to impact assessment are discussed.  Data
from an experimental transport programme in Danville,
Illinois, are used to illustrate some of the issues and
problems which arise in assessing the impacts of these transport programmes.

MOBILITY, ACCESSIBILITY AND TRAVEL BEHAVIOUR

The terms 'mobility' and 'accessibility' are commonly used to
describe objectives and impacts of urban transport programmes.
Mobility is generally associated with particular groups of
urban residents, and describes their ability to travel from

one place to another in an urban area.  Accessibility, on the
other hand, is associated with locations or facilities, and
describes the ease with which they can be reached and enjoyed.
Thus we speak of residents who have limited mobility, for
example, and of certain locations and facilities which are
highly accessible.

It is sometimes suggested that transport planners are
overly concerned with increasing the *mobility* of urban
residents - through highway construction, expanded bus or
rail transit services, or other additions to transport infrastructure and services.  What they should really be concerned
about, it is argued, is increasing the *accessibility* of urban
locations and facilities.  This latter goal could be accomplished in part by changing the distribution of land uses to
reduce the demand for travel, or by changing institutional
arrangements such as uniform working hours which place heavy
peak demands on transport facilities.  To the extent that
these changes could be accomplished, it is claimed, accessibility could be increased without increases in the amount
of travel.

Experience to date suggests, however, that efforts to
encourage or mandate changes in land-use distribution and
institutional arrangements are likely to have only limited
effects on the demand for travel.  Numerous forces continue
to encourage the development of low-density suburban housing.
Employment, shopping, recreation, education and health
facilities tend to be located where they can be reached conveniently by car, the travel mode used by the vast majority
of their clientele.  The result of this continued low-density
development is pressure for more mobility as a means of
maintaining and increasing accessibility.  This pressure is
especially significant from those who rely on public transport services, since the costs of reaching desirable
locations and facilities by public transport increase substantially as land development becomes more oriented to
serving car users.

The task of increasing the accessibility of urban locations and facilities then, has become largely one of
increasing the mobility of urban residents.  Despite the
wishes and efforts of those who would like to see urban areas
structured differently, as a practical matter it seems that
only dramatic new land-use policies or severe restraints on
car travel will effect significant changes in current
patterns of land development and travel demand.  Since such
changes will probably occur very slowly (if at all), the
current focus on mobility as the major means of increasing
accessibility would appear to be quite understandable.  If we
accept this view that improving the mobility of urban residents is the major concern of current transport programmes,
we can turn our attention to the question of how much
mobility should be provided, and for whom.  In order to
address this question adequately, we must attempt to identify
the benefits and costs associated with various levels of
mobility, and the incidence of those benefits and costs
throughout the population of urban residents.  We will
restrict our attention in this chapter, however, to the
*benefits* of improving mobility; we will not attempt to

discuss the costs associated with different ways of achieving mobility improvements.

As noted earlier, mobility refers to the *ability* of urban residents to travel from one place to another in an urban area.  A variety of different measures can be used to characterise the mobility of a particular resident or group of residents: car ownership; possession of driver's licence; highway capacities; speeds, vehicle capacities, headways, route coverage and fares of the bus service; availability and fare structure of taxicab services, and so on.  Transport programmes affect mobility by changing the values of one or more of these mobility measures.  Two kinds of benefits result from these mobility changes: those derived solely from *options* for travel, and those derived from *trips* actually made.

The benefits derived solely from options for travel are difficult to quantify.  How much does an urban resident value a publicly-supported bus service which he plans to use only when his car breaks down, for example?  Since several 'back-up' travel options are nearly always available in an emergency - ride as a car passenger with a friend, borrow a car, call a taxi, or call an ambulance, for example - the benefits of having additional options provided by transport programmes may not be very great.  Experience with fixed-route bus services in small communities suggests that the value of the mobility provided by the services is measured primarily in terms of the number of trips actually served; once that number falls below a certain level, the services are typically discontinued.  Though the option to travel undoubtedly provides some benefits over and above those derived from trips actually made, the magnitude of these 'option' benefits is somewhat questionable.

The most tangible benefits associated with mobility improvements are those derived from trips made by urban residents.  Travellers value these trips at least as much as the time, effort and money they expend in making them. Changes in trip-making patterns which take place as a result of transport programmes therefore provide a firm basis on which to assess the benefits of the programmes [9].

We suggest that the benefits associated with transport programmes for the elderly and handicapped are largely a function of the travel impacts of the programmes.  Changes in mobility and in accessibility occur by definition when these transport programmes are implemented, but the value of these changes lies primarily in the accompanying changes in travel behaviour.  Thus a new bus service may increase both the mobility of certain persons and the accessibility of certain locations and facilities, but the service will not be considered to be of much value unless it is used for a significant number of *trips*.  In the remainder of this chapter, therefore, we will be concerned with the problem of assessing the impacts of transport programmes for the elderly and handicapped on the travel behaviour of this client group.

TRAVEL DEMAND BY THE ELDERLY AND HANDICAPPED

*The Elderly and Handicapped Population in the US*

A number of attempts have been made to estimate the number
and location of elderly and handicapped persons in the US.
Perhaps the best estimates currently available are those
developed by Abt Associates for the US Department of Transportation (15). The Abt estimates are based on population
data from the US census and on 'incidence rates' for
transport handicaps obtained from a 1974 national health
survey conducted by the US Department of Health, Education
and Welfare. As shown in Table 31.1, in 1975 an estimated
8,876,000 persons had handicaps which inhibited them in some
way from using conventional transport modes. An additional
17,851,000 persons were 65 years of age or older in 1975,
giving a total elderly and handicapped population of
26,727,000 - about 12.5 per cent of the total US population.

TABLE 31.1   Abt estimates of 1975 transport-handicapped (TH)
population in the United States

| TH category | Age | | | |
|---|---|---|---|---|
| | Under 18 | 18-64 | 65 & over | Total |
| Chronic | 190,000 | 2,927,000 | 3,791,000 | 6,908,000 |
| Use transit with difficulty | 80,000 | 1,677,000 | 1,719,000 | 3,476,000 |
| Cannot use transit | 110,000 | 1,250,000 | 2,072,000 | 3,432,000 |
| Acute | 104,000 | 419,000 | 71,000 | 594,000 |
| Institutionalised | 81,000 | 370,000 | 923,000 | 1,374,000 |
| TOTAL | 375,000 | 3,716,000 | 4,785,000 | 8,876,000 |

*Source:* US Department of Transportation (15)

An estimate of the elderly and handicapped population of a
particular city or urban area can also be obtained by using
census data and incidence rates for transport handicaps. The
*location* of these persons in the area is much more difficult
to determine, however. Apart from 1970 census tract, city
and county data on elderly persons, virtually no information
is readily available on the location of elderly and handicapped persons in urban areas. This presents a major
obstacle to the estimation of travel demand by the elderly
and handicapped - how are we to identify a representative
sample of these persons for surveys? There are a number of
different techniques which can be used for locating elderly
and handicapped persons in an urban area, though none of them
provides really satisfactory results at low cost. They
include: (1) canvassing on either a random or selective basis;
(2) obtaining client lists from health and social service
agencies, or from private organisations having high

memberships of elderly or handicapped persons; and (3)
so-called 'snowball sampling'.. Canvassing is the most comprehensive way to locate elderly and handicapped persons.
Households are administered a short questionnaire by telephone
or in person to determine whether any elderly or handicapped
persons reside there. Canvassing can be restricted to those
areas (say, census tracts) which are thought to contain high
fractions of the client group. Canvassing is likely to
require several pre-screening interviews, including some
call-backs, per completed interview. Telephone canvassing is
less expensive than door-to-door canvassing, but is also
somewhat less representative since not all households have
telephones.

Lists can be obtained from service organisations or clubs
which represent subsets of the elderly and handicapped
population. There are two difficulties. First, co-operation
and clearance may be difficult to obtain from such organisations due to confidentiality considerations. Second, there
is no assurance that persons represented on the lists
obtained are similar to those not on them, and it is impossible to estimate even the number of those not appearing on
any list. Lists may also be outdated, incorrect and over-
lapping, so that editing would be required before they could
be used for sample selection.

Snowball sampling* could also be used to obtain information on the client group - each person identified would be
asked to provide information on other possible members of the
group. Snowball sampling may be biased in an unknown way,
though it might be a useful adjunct to sampling with lists.

*Travel by the Elderly and Handicapped*

Given the paucity of information on the number and location
of elderly and handicapped persons in the US, it follows that
information on the travel behaviour of this group is even
more limited. A recent study for the US Department of
Transportation concluded that: 'there are no adequate empirical data on the travel behaviour of the transportation
handicapped that would allow for an assessment of their
response to system modifications or the installation of new
systems' (15). The data which do exist have been collected
in several different locations for a variety of special
purposes. These data do provide some insights, however, into
the demand for travel by the elderly and handicapped, and
help to illustrate some of the complexities of this particular travel market.

The first point demonstrated by data collected to date is
that the elderly and handicapped population represents a very
diverse travel market. Bunker, Blanchard and Wachs (1)

---

*Sudman (11) provides a general discussion of snowball
sampling. He does not favour its use in locating rare populations. Its primary value is for obtaining controls for
self-selected members of a population under study (such as
certain kinds of volunteers).

analysed life-styles and travel patterns for elderly residents of Los Angeles County. They used factor analysis to define homogeneous 'life-style groups' among the elderly, and then employed analysis-of-variance to identify differences between travel characteristics of the life-style groups. (Travel data were drawn from the 1967 Los Angeles Regional Transportation Study.) The results of their analyses, shown in Table 31.2, provide evidence of significant differences in daily trip rates between life-style groups: the 'financially secure', for example, appear to average over twice as many vehicular trips per day as the black and Spanish-American communities. Other data collected in Boston by Abt Associates and Wilbur Smith and Associates show daily vehicular trip-making rates for handicapped persons (1.13 trips per day) which differ significantly from those for the general population (2.23 trips per day) (15).

TABLE 31.2   Vehicular travel of different life-style groups among the elderly in Los Angeles County

| Travel variable | Central city dwellers | Financially secure | New suburbanites | Black community | Spanish-American community | Early suburbanites | County | F |
|---|---|---|---|---|---|---|---|---|
| Percent with drivers' licences | 34.55 | 58.64 | 48.14 | 32.89 | 23.43 | 45.61 | 42.49 | 39.24[a] |
| Percent reporting vehicular travel | 40.82 | 55.95 | 47.84 | 33.68 | 31.35 | 47.11 | 44.79 | 19.19[a] |
| Auto driver trips | .66 | 1.38 | 1.04 | .56 | .40 | 1.02 | .91 | 23.86[a] |
| Auto passenger trips | .35 | .52 | .43 | .21 | .23 | .49 | .42 | 8.58[a] |
| Public bus passenger trips | .21 | .09 | .04 | .16 | .19 | .05 | .11 | 26.06[a] |
| Personal business trips | .59 | .92 | .62 | .53 | .44 | .70 | .66 | 12.05[a] |
| Trips for leisure | .23 | .43 | .30 | .13 | .09 | .29 | .27 | 11.99[a] |
| Work trips | .13 | .23 | .23 | .16 | .13 | .19 | .18 | 2.36 |
| Shopping trips | .31 | .46 | .39 | .16 | .21 | .43 | .36 | 13.03[a] |
| Total trips | 1.26 | 2.04 | 1.54 | .98 | .87 | 1.61 | 1.47 | |
| Sample size | 1,528 | 736 | 706 | 387 | 308 | 2,080 | 5,768[b] | |

Source: Bunker, Blanchard and Wachs (1)
a. Statistically significant at the 0.01 level.
b. Includes 23 persons who resided in the Institutionalised Life-style Area.

Overall, then, there is evidence to suggest that certain subgroups of elderly and handicapped persons make significantly fewer trips than other subgroups, and than the general population. This evidence raises the following question: how many more trips would the various subgroups make if their mobility were increased through transport-system improvements? Some attempts have been made to assess this 'latent demand' by attitudinal surveys, in which respondents were asked to estimate how many more trips they would make if certain hypothetical transport improvements were made. Surveys of this kind conducted in Washington, DC and Chicago reported that the transport handicapped would increase their trip-making by 0.53 trips per day (Washington) and 0.34 trips per day (Chicago) if convenient, low-cost transport services were available (15). The validity of these estimates is questionable, however, since the reliability of attitudinal surveys of this type is currently unknown.

The present state of understanding of travel demand by the elderly and handicapped does not permit us to predict with any confidence the usage of various transport improvements which might be made to better serve this client group. This creates difficulties for both the design and evaluation of transport programmes. The design problem is a significant one, though not necessarily crucial to implementing transport improvements. The demand density for these improvements will be quite low in any case, and adjustments can be made to fleet sizes and service levels as experience is gained from the initial operation of a new programme. In the evaluation of alternative transport programmes and policies, however, inadequate understanding of travel demand creates severe problems.

*The Effect of Transport Programmes on Travel by the Elderly and Handicapped*

In order to evaluate alternative transport programmes for the elderly and handicapped, we need to know what effect they will have on travel by the client group. A complete description of the travel behaviour of any particular subgroup of the client group should contain information on the number of trips made by members of the subgroup over a given time period by: (1) service characteristics and price of the travel mode used; (2) trip purpose; (3) time of day, week and month trips are made; and (4) trip origin and destination. In order to determine the effect of a transport programme on the travel behaviour of a particular subgroup, we need to know about significant changes in any of the above descriptors of trip-making. A programme might not increase the total *number* of trips made by a subgroup, for example, but might provide substantial benefits by permitting the trips to be made with reduced travel times, at lower prices or fares or to more desirable destinations.

There are two possible approaches to obtaining a better understanding of the response of subgroups of the elderly and handicapped to different kinds of transport improvements. One approach would collect *attitudinal* information by questioning selected members of the client group about how they would respond to certain hypothetical transport improvements. The other approach would collect *behavioural* information by observing how selected members of the client group actually do respond to the particular transport services and prices available to them.

The first of the two approaches is fraught with uncertainties about the reliability of the information collected: how does what people say they will do under hypothetical circumstances compare with what they actually would do?* The second approach has the advantage that people would be asked to report about actual rather than hypothetical trips, though there is still the question of how accurately they would report their trips. (The only way to check on the accuracy of their reporting would be to assign certain individuals to follow them around without their knowledge - hardly a desirable procedure to adopt.) The major disadvantage of the second procedure is that it is very expensive. For new kinds of transport improvements not currently in place in any urban area, experimental programmes have to be instituted for at least one year and accompanied by detailed measurements of travel responses.

Both of the above approaches have major advantages and disadvantages, and it is not apparent that either is clearly superior to the other. Consequently, it is important that both approaches be pursued. Without continued efforts to collect behavioural travel data, we will be unable to check travel estimates obtained from attitudinal data. And without attitudinal data, it will be a very long time before we obtain any information on likely traveller response to a variety of new kinds of transport improvements which have yet to be placed in operation.

In addition to the general difficulties of collecting and interpreting travel information, there are a number of special problems associated with obtaining a better understanding of the travel behaviour of the elderly and handicapped. The first problem has been discussed earlier - that of locating the elderly and handicapped members of an urban population in order to select a sample for surveys. Even if such a sample can be obtained, however, another problem is presented by the relatively low rates of trip-making observed for certain subgroups of the elderly and handicapped population - often less than one trip per day. Many elderly and handicapped persons do not work, and their travel is limited to less frequent personal business and recreation trips. Though these trips are undoubtedly periodic the periods are likely to be much longer than the period of one day associated with work trips. Behavioural surveys of trip-making by elderly and handicapped persons should be designed, therefore, to obtain a description of each trip made by each surveyed person over an extended period, such as a month. (The first sets of surveys might show that travel by elderly and handicapped persons recurs

---

*Hartgen and Keck (6) discuss discrepancies between attitudinal forecasts and choices actually made for dial-a-bus services in small urban areas.

over shorter periods, though there is little reason to expect
this result from data collected to date.) Some persons
should be surveyed during each month of the year to provide
information on seasonal variations in trip-making. Travel
diaries would be required, and persons participating in the
survey would probably have to be compensated in some way for
completing the diaries. An initial payment when the diary
was first received might be followed by a second, larger
payment when the completed diary was returned. Though this
form of travel survey would be rather expensive and difficult
to carry out, there does not appear to be any alternative way
in which an adequate description of travel behaviour could be
obtained.

A further problem arises because of the diversity of the
elderly and handicapped population. The data given in Table
31.2 suggest that certain transport programmes such as
improved bus services are likely to have a much greater
effect on the travel of some subgroups than on others. The
three life-style subgroups in Table 31.2 with the highest
percentages of driver's licences make significantly more car-
driver trips and significantly less bus trips than the other
three subgroups. Improved bus services might have virtually
no effect on the travel of the former three subgroups but
have a substantial effect on the latter three. In order to
understand the travel effects of this kind of programme,
then, we would like to concentrate our in-depth survey
efforts on those subgroups likely to be the most affected.
In the case of an experimental programme, such subgroups need
to be identified before the programme is introduced, based
upon the best available estimates of the likely impacts of
the programme.

A final difficulty with assessing the impacts of transport
programmes is that of inferring what behaviour might have
been in the absence of the programmes (2,3). Use of transport variables to describe behaviour 'with and without'
transport programmes requires, at the very least, allowances
for exogenous influences which might influence trip-making.
Indicators such as the area's consumer price index, unemployment rate and other measures of major employment or construction changes might suggest temporal shifts in the area's
economic base or land use and hence in transport-demand
patterns. Social programmes, tax laws or transport policy
itself may undergo major revision. Such shifts would require
corrections to eliminate bias. For example, a major increase
in public transport supply might cause shifts from car to
public transport use which are of much greater magnitude than
a specialised programme for the elderly and handicapped.

In summary, the major problems associated with assessing
the travel impacts of transport programmes for the elderly
and handicapped are the following: (1) identifying and locating members of the client group in an urban area; (2)
obtaining individual travel information over a sufficiently
long period of time; (3) identifying in advance those subgroups likely to be the most affected by new programmes; and
(4) controlling for exogenous influences. A transport
programme for the elderly and handicapped recently introduced
in Danville, Illinois, provides some interesting illustrations of these problems.

A TRANSPORT PROGRAMME FOR THE ELDERLY AND HANDICAPPED IN
DANVILLE, ILLINOIS

A demonstration project funded in Danville, Illinois, by the
Service and Methods Demonstration Programme of the Urban Mass
Transportation Administration, US Department of Transportation, provides some interesting insights into the problems of
assessing the effect of transport improvements on the travel
behaviour of the elderly and handicapped. The primary
purpose of this project is actually to test an innovation in
the supply of public transport services: the application of
user-side subsidies to make shared-taxi services available at
low fares to elderly and handicapped persons. The fairly
substantial reduction effected in shared-taxi fares for the
elderly and handicapped provides an interesting example,
however, of traveller response to a major transport improvement. A detailed description of the rationale for this
approach to subsidising public transport has been given by
Kirby and McGillivray [7]. Subsidy techniques are divided
into two categories: 'provider-side' subsidies paid directly
to transport providers for supplying certain transport
services, and 'user-side' subsidies paid directly to transport users in the form of discounted transport vouchers. It
is suggested that though the more common provider-side subsidies may be easier to administer than user-side subsidies,
they have often resulted in dependence of the public on a
relatively small number of providers and services. This
dependence has tended to restrict opportunities for new
providers and to increase costs. User-side subsidies should
encourage greater efficiency in service provision by allowing
users to choose the providers and services which best meet
their needs.

Danville, Illinois, is a relatively small city with a
population of 46,500. At the time the demonstration project
began, 1 December 1975, the city was served by three taxicab
companies: Red Top/Yellow Cab with 19 vehicles, Courtesy Cab
with 10 vehicles, and Brown Cab with one vehicle. The city
has no fixed route bus service. However, eleven specialised
vehicles are operated by social service agencies in providing
transport for their clients [5].

Taxi services in Danville are shared ride: that is, two or
more passengers with differing trip origins or destinations
may share the same taxicab. Fares are based on four concen-
tric zones, with a certain fare associated with each zone.
The fare charged for a trip is that corresponding to the
origin zone or the destination zone, whichever is higher.
For a group of passengers with the same origin and destination, one passenger is charged full fare and each additional
passenger is charged a small flat fee. On 1 December 1975,
the demonstration project introduced a 'charge slip' scheme
by which handicapped persons and persons 65 years of age and
over could purchase up to $20 worth of taxi rides each month
at a discount of approximately 75 per cent on each ride. On
1 January 1977, overall fares were increased, and the payments by elderly and handicapped persons were increased to
approximately 50 per cent of the new fares, with no change in
the $20 monthly limit on the total value of rides taken. The
payments associated with these fare structures are shown in
Table 31.3.

TABLE 31.3  Shared Laxi fares in Danville, Illinois

| Zone | Effective 1 December 1975 | | Effective 1 January 1977 | |
| | Overall fare | Payment by E & H | Actual fare and (% increase) | Payment by E & H and (% increase) |
| --- | --- | --- | --- | --- |
| 1 | $0.75 | $0.25 | $0.85 (13) | $0.45 ( 80) |
| 2 | 1.25 | 0.30 | 1.40 (12) | 0.60 (100) |
| 3 | 1.50 | 0.40 | 1.70 (13) | 0.85 (112) |
| 4 | 1.75 | 0.50 | 2.00 (14) | 1.00 (100) |

Note: For group riding each additional passenger is charged a flat fee of $0.15.

To date in Danville, then, elderly and handicapped persons have experienced two changes in shared-taxi fares: one on 1 December 1975, which amounted to a reduction averaging about 75 per cent, and one on 1 January 1977, which amounted to an increase averaging about 100 per cent.  How could we determine the effect of these quite substantial fare changes on travel by the client group?

As discussed earlier, the first step in measuring the effect of transport improvements on the client group served is to select a representative sample of the group for travel surveys.  The firm monitoring the demonstration project, Crain and Associates, encountered the usual difficulties with identifying and locating the client group, and were able only to develop an estimate of the size of the group.  This estimate was obtained by adding the number of persons reported to be 65 years of age or over by the 1970 census (5,600) to an estimate of the number of handicapped persons under 65 provided by personnel of local rehabilitation agencies (1,900).  The resulting estimate of 7,500 persons constitutes approximately 18 per cent of the population of Danville.

In order to obtain shared-taxi rides at reduced rates in Danville, a member of the client group must register with the City and obtain an identification card containing his or her name, address, signature and identification number.  This card must be shown to the taxi driver each time a trip is made at the reduced rate.  The driver records the passenger's identification number on a charge slip along with the full fare and the reduced fare for the trip.  The passenger then signs the slip and pays the reduced fare (and any tip) in cash.  As a result of this procedure, a record is obtained of each subsidised trip made by each registered member of the client group.

The registration process for members of the client group and the charge slip procedure for taxi trips provide the basis for many of the travel data collected in the Danville

demonstration.  When client group members applied to the City
for identification cards, they were asked to provide certain
information needed to certify their eligibility for the
programme.  Once a person's eligibility was established, and
before an identification card was issued, the person was
interviewed by telephone.  The person was asked a variety of
questions concerning age, sex, race, possession of driver's
licence, car availability, income and nature of handicaps, if
any.  This information, when combined with taxi-trip information from the charge slips, provided a detailed picture of
the use of the reduced-fare programme by various subgroups of
the client population, as shown in Table 31.4.

TABLE 31.4   Danville user-side subsidy demonstration project.
Project person trips per user per month through July 1976 by
subgroups (users are those persons who have used the project
at least once).

| | Fraction of total users | Trips per user/month |
|---|---|---|
| **Age/Handicap** | | |
| 65 and over, handicapped | 0.18 | 3.7 |
| 65 and over, not handicapped | 0.62 | 3.1 |
| Under 65, handicapped | 0.20 | 6.1 |
| **Alternative Transport Available** | | |
| Not driver/receive no rides | 0.18 | 5.9 |
| Not driver/receives rides | 0.60 | 4.1 |
| Driver/auto avail/rides | 0.22 | 1.3 |
| **Ability to Use Taxi vs Bus** | | |
| No difficulty either mode | 0.75 | 3.8 |
| Taxi less difficult than bus | 0.21 | 3.9 |
| Others | 0.04 | 3.7 |
| **Transit Handicapped and Handicapped** | | |
| Problems w/bus and handicapped | 0.18 | 4.2 |
| Problems w/bus and not handicapped | 0.07 | 2.9 |
| No problems w/bus and handicapped | 0.20 | 5.5 |
| No problems w/bus and not handicapped | 0.55 | 3.2 |
| **Type of Primary Handicap** | | |
| Emotionally disturbed | 0.08 | 6.4 |
| Walking problems/aids | 0.07 | 4.3 |
| Arthritis | 0.05 | 4.0 |
| Cardiac ills | 0.03 | 4.6 |
| Mental retardation | 0.03 | 3.4 |
| Blindness | 0.02 | 6.3 |
| **Household Income Per Person** | | |
| Less than $2,500 per person | 0.28 | 4.1 |
| $2,500 to $5,000 per person | 0.62 | 3.8 |
| $5,000 to $10,000 per person | 0.09 | 3.3 |
| Over $10,000 per person | 0.01 | 3.6 |

*Source:* Crain and Associates (5).

An attempt was also made in these registration interviews to obtain information about the travel behaviour of those being registered.  Each person was asked to report all trips made over the three days prior to the interview, by purpose and by mode.  Analysis of these travel data showed that the mean number of trips reported for each day was significantly greater than the mean number reported for the day before.  Crain and Associates concluded that this pattern was due to forgetfulness about trips made more than a day prior to the interview (4).  It was concluded that the travel data could be used only to describe travel for the one day prior to the interview.

Travel data were obtained through these registration interviews for 2,550 of the 2,600 persons who registered.  The data were collected over a registration period of five months, from November 1975 through March 1976.  In August 1976 a random sample of 246 registered persons was selected for a follow-up interview, in which travel data were again collected and analysed.

Had there been any substantial changes in the travel behaviour of project registrants as a result of the project, they might have been detected from a comparison of the two sets of travel data collected for project registrants before and after their identification cards were issued.  The data showed, however, a mean daily trip rate of 1.18 from the registration interview and a rate of 0.99 from the follow-up interview, with a standard deviation of 1.31 in both cases.  These figures, which are based on travel data for the one day prior to the interview, show a great deal of variability in the number of trips reported by those interviewed.  This variability is presumably due in part to variation between individuals, and in part to the fact that one day is too short a period over which to observe travel by elderly and handicapped persons.

Though the travel data collected for Danville do not permit any statements to be made about the overall effects of the project on travel behaviour, information collected by Crain and Associates (5) on participation in the reduced-fare project suggest that there may have been significant impacts on the travel of at least some subgroups of the eligible population.  Of the approximately 7,500 residents of Danville eligible for reduced taxi fares, some 35 per cent have registered.  A random household telephone survey conducted in August 1976 to obtain socio-economic characteristics of the eligible population in Danville showed that users of the project differed significantly from the total eligible population in two major respects: 78 per cent of the users did not drive compared with 51 per cent of all eligibles, and 90 per cent of the users had incomes under $5,000 compared with 52 per cent of all eligibles.

Approximately 80 per cent of those registered (28 per cent of those eligible) have used the project at least once.  Table 31.4 shows that handicapped users under 65 years of age have taken substantially more project trips per month than other users, and that persons who do not drive have taken substantially more trips than those who are drivers.  An

overall increase in taxi ridership of 15 per cent has been
attributed entirely to the project, and fully 30 per cent of
shared-taxi rides in Danville are currently project trips.
These figures suggest that about half of the project trips
are 'new' taxi trips, while the other half are 'old' taxi
trips being made at the lower fares.  Finally, no reduction
has been noticed in the patronage of transport services
provided by social-service agencies in Danville.

In August 1976 approximately 8,000 shared-taxi trips per
month were being made under the reduced-fare project.  This
total represents approximately:

1.1 trips per month per eligible person
3.1 trips per month per registered person
3.8 trips per month per user
4.5 trips per month per non-driver user (derived from
    Table 31.4)
6.1 trips per month per handicapped user under 65 (Table
    31.4)

How great a fraction of total travel do these project trips
represent for the different subgroups listed?  The best
estimate we have of total trip-making is that of about one
trip per day (30 trips per month) obtained from the travel
surveys of registered persons described earlier.  At this
rate project trips would account for about 10 per cent of
total trips for all registered persons, and possibly higher
percentages for all users, for non-driver users, and for
handicapped users under 65.  It seems likely, therefore, that
the Danville reduced-taxifare project has had a substantial
effect on the travel behaviour of certain subgroups of the
eligible population.  Data collected on the travel patterns
of the subgroups are too limited, however, to permit quantification of this effect.  If detailed information about the
impacts of future transport programmes for the elderly and
handicapped is to be obtained, a much more extensive effort
will be needed to measure changes in travel behaviour.

The experience obtained from the Danville project provides
some valuable guidance for monitoring the effects of other
transport programmes for the elderly and handicapped.  Much
greater effort should be devoted to locating eligible persons
using techniques of the type discussed earlier.  The extensive user information obtained in Danville provides a basis
for identifying in advance certain subgroups whose travel
behaviour is likely to be affected to a greater extent than
that of the eligible population as a whole.  The subgroups of
interest should be sampled, and both behavioural and attitudinal surveys conducted.  The behavioural surveys should
collect travel information from each individual over an
extended period, such as a month, while the attitudinal
surveys would ask individuals about how they expected to
respond to the planned transport improvements.  The behavioural surveys would be conducted before and after the implementation of the transport programme, preferably during the same
months of two consecutive years to account for seasonal
effects.  Changes in travel behaviour detected by the
behavioural surveys would then be compared with responses
predicted by the attitudinal surveys.  For transport

programmes like that in Danville involving a substantial
change in transport services or fares for the elderly and
handicapped with virtually no exogenous changes which might
influence travel behaviour significantly, it should be
possible to obtain reliable measures of changes in the travel
behaviour of the subgroups most affected.  Projects of this
type appear likely to have quite substantial impacts on
certain subgroups: impacts which could be measured if sufficient resources were applied to data collection and analysis.


CONCLUSION

Transport programmes for the elderly and handicapped are
designed to increase the mobility of this client group, and
to make a variety of locations and facilities more accessible
to them.  These improvements can be described in terms of
such mobility measures as the travel times and fares associated with travelling between different points in the urban
area in question.  The benefits associated with these
increases in mobility and accessibility are of two kinds:
those derived solely from options for travel, and those
derived from trips actually made.  The benefits of options
for travel, while not insignificant, are difficult to
quantify, and are probably best judged directly by community
decision-makers.  The benefits associated with changes in
travel behaviour are more important and more tangible, though
they cannot be assessed without detailed quantitative information on changes in trip-making by particular subgroups of
the elderly and handicapped population.

This chapter has been concerned with assessing the impact
of transport programmes on the trip-making of the elderly and
handicapped.  We have pointed out that elderly and handicapped persons tend to make fewer trips than the general
population, and that the periodicity in their travel
behaviour is unlikely to be observed if travel information is
obtained only for short periods of no more than a few days.
Careful monitoring of the trip-making of a representative
sample of elderly and handicapped persons over extended
periods will be necessary to increase our understanding of
the travel behaviour of this client group.

Mounting surveys using trip diaries or other means to
obtain behavioural data on trip-making over extended periods
is likely to be a relatively expensive undertaking.  Attitudinal surveys provide a less expensive but also less certain
approach, and in our judgement should be employed in addition
to rather than in place of the behavioural surveys.  Is the
information which would be obtained from these surveys worth
the cost?  The answer to this question depends on how
important it is for decision-makers to know about the benefits of transport programmes to particular groups of
residents.  Relatively inexpensive ridership surveys can and
should be employed to determine who is using a transport
service, how often and for what purpose.  Such surveys cannot
determine, however, what the users would do if the service
were withdrawn or if new services or fare levels were instituted.  Would they take a different number of trips, use
other modes, travel to different destinations, or adopt some
combination of these alternatives?

Obtaining detailed information about the impacts of transport programmes on the trip-making of different resident groups is an essential step in evaluating the programmes. Information of this type will not be available to decision-makers, however, unless research is undertaken to measure travel behaviour of particular resident groups over extended periods of time, and to compare these measurements with travel estimates obtained from attitudinal data. Given the current public concern and involvement with improving the mobility of the elderly and handicapped, such research should be given serious consideration in the near future.

REFERENCES

1. Bunker, James B., Blanchard, Robert D. and Wachs, Martin, *Life-styles and Transportation Patterns of the Elderly: An Empirical Study of Los Angeles*, School of Architecture and Urban Planning, University of California, Los Angeles (1977).

2. Campbell, Donald T. and Stanley, Julian C. (1966) *Experimental and Quasi-Experimental Designs for Research*, Rand McNally and Company, Chicago, Illinois.

3. Charles River Associates (1972) *Measurements of the Effects of Transportation Changes*, Charles River Associates, Cambridge, Massachusetts.

4. Crain and Associates, 'Transit dependent mobility measurement', *Technical Memorandum prepared for the Transportation Systems Center*, US Department of Transportation, Washington, D.C. (1976).

5. Crain and Associates (1977) *User-side Subsidy on a Shared-Ride Taxi Service for Handicapped and Elderly: Danville, Illinois, Demonstration*, Crain and Associates, Menlo Park, California.

6. Hartgen, David T. and Keck, Carol A., 'Forecasting dial-a-bus ridership in small urban areas', *Transportation Research Record 563*, 53 (1976).

7. Kirby, Ronald F. and McGillivray, Robert G., 'Alternative subsidy techniques for urban public transportation', *Transportation Research Record 589*, 25 (1976).

8. Kirby, Ronald F. and Tolson, Francine L., 'Improving the mobility of the elderly and handicapped through user-side subsidies', *Working Paper 5050-4-4*, The Urban Institute, Washington, D.C. (1977).

9. McGillivray, Robert G., 'Some thoughts on the theory of traveller benefit measurement', *Swedish Journal of Economics*, 77, 265 (1975).

10. Senate of the United States, *S.208 - A Bill to Amend the Urban Mass Transportation Act of 1964 to Extend the Authorization for Assistance under such Act, and for other Purposes*, Congress of the United States, Washington, D.C. (1977).

11. Sudman, Seymour (1976) *Applied Sampling*, Academic Press, New York.

12. US Department of Health, Education and Welfare (1976) *Transportation Authorities in Federal Human Services Programs*, Office of the Regional Director, Atlanta, Georgia

13. US Department of Transportation (1975) *Urban Mass Transportation Act of 1964 and Related Laws*, US Department of Transportation, Washington, D.C.

14. US Department of Transportation, 'Transportation for elderly and handicapped persons', *Federal Register*, 41, 85 (1976).

15. US Department of Transportation (1976) *Transportation Problems of the Transportation Handicapped*, 4 volumes edited by Crain and Associates, US Department of Transportation, Washington, D.C.

Chapter 32

BEHAVIOURAL MODELLING, ACCESSIBILITY, MOBILITY
AND TRAVEL NEED

Martin Wachs and Jan Gerard Koenig

SCOPE OF THE WORKSHOP

A basic objective of all transport research and planning is
the provision, within appropriate constraints, of improved
accessibility to the facilities and services which people
require. The workshop investigated measures of mobility and
accessibility in order to arrive at recommendations for
research to improve our understanding of accessibility and
its relationship with travel behaviour and urban form. The
members of the workshop first investigated several current
and potential uses of accessibility and mobility indicators.
Next, the group considered the existing state of knowledge in
relation to these uses. Finally, the workshop arrived at a
list of recommendations for research which is needed to
advance current capabilities with respect to the measurement
and evaluation of accessibility.

SUMMARY OF WORKSHOP DISCUSSIONS

*The Uses of Measures of Accessibility*

Accessibility indicators should constitute an important
component of transport modelling and planning because accessibility is one of the most basic determinants of urban form,
and because the provision of accessibility is usually an
explicit objective of transport planning. Thus, an extremely
important use of accessibility measures would be in the
evaluation of alternative transport networks. If meaningful
and sensitive accessibility indices were available, it is
likely that they would be used alongside cost and volume/
capacity relationships in determining the overall performance
of transport-system designs. In addition, in most metropolitan areas there are differences in the accessibility of
major spatial and demographic groups to a variety of economic
and cultural opportunities. Better accessibility measures
could help to redirect policy and planning towards the equali-
sation of opportunities among such groups. Do citizens of
one neighbourhood have as much access to employment opportunities as do residents of another area? Do the carless have
sufficient access to health-care and recreational opportunities in comparison with car-owning households? Do the
elderly have access to social services which meet some
minimum standard? These are all examples of questions to
which planners are seeking answers through the construction

of summary indices of accessibility.  Because accessibility
can be changed by the location of activities as well as by
the provision of transport services, accessibility measures
can also be useful in the evaluation of locational decisions.
Should a new manufacturing plant or office complex be located
at one site or another?  Should health-care services be
concentrated at a few centralised locations or dispersed
throughout the community?  Many examples were presented in
the workshop of how accessibility measures had actually been
employed in efforts to answer such evaluation questions.
Thus, a primary application of accessibility indicators is in
the evaluation of alternative transport-system configurations,
alternative locations for facilities, and alternative
approaches for meeting the social needs of particular population groups.  In such applications, accessibility measures
are needed to summarise a great deal of information about the
location of one household or one zone in relation to the
entire distribution of urban activities and to the transport
system which connects them.

Because accessibility is a measure of the 'supply' of
opportunities available to a household or to an aggregation
of households it is also likely to be a useful concept in the
development of choice-based models of human behaviour.
Accessibility measures would be useful independent or predictor variables in models where the dependent variables
represent residential choice, car-ownership decisions,
industrial-location decisions, or trip-generation rates.
Indeed, terms representing accessibility have often been used
to represent the properties of transport networks in urban
development models such as the Lowry, Empiric or PLUM models,
and have often been used in travel-demand models to represent
the distribution of activities to which trips were made.  For
this reason, accessibility measures were viewed by members of
the workshop as an important element in the construction of
models representative of the dynamic equilibrium which exists
between transport and urban development.

*Accessibility and Manifest Travel Demand: A Dilemma*

Although the participants in the workshop generally agreed
upon the functions and uses of accessibility measures, there
was considerable discussion and less agreement on the form
which such measures might appropriately take.  Some believed
that accessibility should be specified as a property of a
household or zone in relation to the entire transport network
and to all of the opportunities provided by the distribution
of activities in an urban area.  Thus, the accessibility of
two adjacent units of analysis might be the same regardless
of the fact that one might make several times as many trips
as the other.  Actual trip-making, according to this view, is
not a part of the definition of accessibility.  Rather, the
provision of linkages to *opportunities* is the key concept
defining accessibility.

While this distinction between accessibility and travel
demand is easy to make in concept, it becomes more difficult
to apply in practice.  Some members of the workshop argued
that a citizen is not indifferent between all parks, but

rather he values some parks more than others.  Similarly,
some health-care facilities are more critical to a person
than are others.  Thus, a measure of accessibility must be
sensitive to the value of each travel opportunity to the
traveller as well as to the simple presence of opportunities.
Yet, when consideration is given to the data sets which are
most typically available to planners, there are usually
relatively few ways to measure the value of a travel opportunity to a citizen in relation to the value of another
travel opportunity.  Indeed, often the only way to conclude
that a person values one facility or service above others is
by observing that he or she travelled to that one rather than
to the others.  In this view, the distinction between travel
demand and accessibility becomes less clear.  Given this
uncertainty about the way to incorporate accessibility into
the calculation of indices, some members of the workshop held
to the belief that the value of accessibility and mobility
improvements can be expressed only as a function of changes
in travel behaviour.  Thus, accessibility indicators based
solely upon the distribution of opportunities in travel time
or space cannot be interpreted as measures of benefit of a
public programme or project, and changes in behaviour must be
forecast or monitored if benefits are to be assessed.
Members of the workshop who were principally concerned with
the modelling of choice behaviour and the development of
travel-demand models gave high priority to the development of
techniques and measures which explicitly represent manifest
choices observable in surveys of travel behaviour.  Others,
who were more concerned with system evaluation and social-
investment decisions, placed less weight on the replication
of travel choices in demand models, and tended to give more
emphasis to measures of accessibility to opportunities.  All
of the participants agreed, however, that better understanding of the relationship between accessibility and manifest
travel behaviour was needed as the principal objective of the
research recommendations of the workshop.

*Clarifying the Relationship Between Accessibility and Mobility*

Accessibility, according to the definition proposed in the
resource paper by Dalvi, 'denotes the ease with which any
land-use activity can be reached from a location using a
particular transport system'.  The most commonly used indicators of accessibility, as Dalvi points out, are of two kinds:

*Opportunities weighted by an impedance* ( which is a decreasing function of travel cost or time), for example, the
function in Equation (1).

$$A_i = \sum_{j=1}^{n} O_j \, f(C_{ij}), \qquad\qquad (1)$$

where   $A_i$   = accessibility from zone i to the considered
                type of opportunities

        $O_j$   = opportunities of the considered type in zone
                j (employment places, shops, etc.)

        $C_{ij}$ = generalised (or real) time or cost from i to
                j

$f(C_{ij})$ = impedance function (exponential or power functions are most often used).

*Isochronic definition* (number of opportunities reached within a given travel time).

This indicator can be considered as a particular form of Equation (1) with an impedance function equal to one for travel times less than x or zero beyond x.  It can be criticised, as Ben-Akiva and Lerman explain in their resource paper, because of its obvious deficiencies; but it also has important advantages, as will be seen below.

*Mobility* is described in the resource paper by Kirby and McGillivray as 'the ability to travel associated with a given person or group', measured by such indicators as car ownership, availability and bus or taxi fare.  This concept may seem complementary to that of accessibility.  Mobility measures would reflect the ability of people to use various modes (without considering opportunities that can be reached) and accessibility measures would describe the locations which could be reached by a given mode (without considering the ability of people to really use this mode).  This apparently appealing distinction raises two difficulties.  First, though Kirby and McGillivray suggest that other indicators could be used, Dalvi observes that the number of trips made still remains the indicator of mobility used in most studies.  This practice encourages the planning of transport systems to increase trip rates and overlooks the point that travel is a derived demand - a means to an end.  Second, and more basically, it can be argued that there is no need for two different concepts of mobility and accessibility or, in other words, that both concepts should be merged.  The ability to use transport modes is not a useful measure unless desired destinations can be reached by those modes; and conversely, access to suitable destinations by a given mode is meaningless if this mode is not available to the potential traveller. What is important is the relationship between people and opportunities including consideration, for a given individual, of the set of transport modes available to him.  This leads to the concept that accessibility should be measured *for a given person*, instead of *by a given mode*.

A behavioural basis is needed for such measurement and could be expressed by replacing in Equation (1) the $C_{ij}$ of the considered mode by the $C_{ij}$ of the mode that the traveller would use from i to j; this mode might be different for different destinations among the j destination zones available for travel.  Means to achieve this behavioural calculation of the $C_{ij}$ matrix could be for instance:

selecting for each i-j combination the mode with the lowest generalised travel time among those available for the considered person [18]

calculating for each i-j pair the expected maximum net utility over the different available modes, in the general hierarchical framework proposed by Ben-Akiva and Lerman; this maximum utility could be for example derived from the logit model as proposed by Domencich and McFadden [8].

*Theoretical Basis of Accessibility*

Dalvi states that, though agreement exists on the general form of accessibility indicators shown in Equation (1), there is little or no strong underlying theory to justify their use in policy-making. More exactly, it could be said that most planners and policy-makers do not see exactly what these indicators do reflect, and are not actually convinced that they would be useful in the evaluation of policies. Several theoretical approaches were mentioned in resource papers or during the workshop to justify the form and use of accessibility indicators having slightly different characteristics. They will be reviewed briefly here.

*Common-sense approaches.* As Dalvi reports, accessibility involves two basic factors related to the satisfaction derived from trips: trip ends or opportunities, and the service provided by available transport modes. Thus the formulation of Equation (1) could be derived from two common-sense considerations. The first consideration, as Ben-Akiva and Lerman mention, is the number of opportunities ($O_j$ in zone j) which may be considered as a crude proxy for the satisfaction provided at the destination: the wider the range of choice among opportunities, the higher the probability of finding a satisfying one for fulfilling a given trip purpose. The second consideration is the impedance $f(C_{ij})$, which should reflect the feeling that a 'nearby' destination should have a higher weight than a 'remote' one. An exponential impedance, for example, seems satisfactory as it gives a weight of one to a destination at zero distance and a weight of zero to a destination very far away. Ben-Akiva and Lerman state that this approach may seem too naive and crude to be really considered as relevant. Nevertheless, its interest should not be neglected. More sophisticated approaches, including behavioural approaches, provide a mathematical elaboration upon this first approach but this simple basis for accessibility indicators has the advantage that it can be readily understood by policy planners and decision-makers.

*Axiomatic definition of a social welfare function.* This approach was proposed in 1976 by Weibull (32). He showed that, under some simple axioms (including independence of irrelevant alternatives and insensitivity to elementary partitioning) the social-welfare function should necessarily be a monotonic function of the quantity $A_i$ in Equation (1).

*Consumers' surplus.* In 1971, Neuburger (24) calculated the consumers' surplus in a situation $S_1$ when compared with a reference situation $S_0$, assuming that trip distribution is correctly described by a gravity model with a distribution function of the form:

$$g(C_{ij}) = e^{-\frac{C_{ij}}{x_0}}$$

where $x_0$ is a distribution parameter.

He found that the consumers' surplus is equal to the variation, between situations $S_0$ and $S_1$, if:

$$S = x_o \sum_i T_i \; \text{Log} \; A_i$$

with: $T_i$ = number of trip origins in zone i

and: $A_i = \sum_j O_j \; e^{-\dfrac{C_{ij}}{x_0}}$ .                    (2)

$A_i$ can thus be identified with the similar quantity in Equation (1).

This result is similar to the other two approaches described above, but it is more precise, as it provides a basis for evaluation in economic terms. Members of the workshop wondered why this result seems to have attracted little attention since being published. Ben-Akiva and Lerman suggest that the reason is that it has no direct behavioural interpretation, and thus may have been dismissed as an esoteric mathematical oddity. It might rather be considered as giving a fairly valuable theoretical basis to the use of accessibility indicators, provided the concept of consumers' surplus is accepted, and provided that exponential distribution functions are regarded as the most suitable to reflect trip distribution, which seems to be the case in many cities (5).

*Behavioural utility approach.* Behavioural approaches to accessibility are fully described in the resource paper by Ben-Akiva and Lerman. They usually assume that people associate a cardinal utility with each of the alternatives they are facing (for example: with each destination, travel mode, route, etc.) and make the choice associated with the maximum utility. In addition, since it is not possible for the planner to evaluate all factors affecting the utility associated with each alternative by a given person, this utility can be represented as the sum of a non-random component (for the predictable factors) and a random component (for the non-predictable factors). A behavioural foundation for accessibility can be derived from this general framework when applied to destination choice. In this case, the utility $U^r_{ij}$ associated by an individual r living in zone i, with a randomly selected destination in zone j, can be represented as follows in a simplified way:

$$U^r_{ij} = V^r - C^r_{ij}$$

where $V^r$ = gross utility of considered destination for individual r (random variable)

$C^r_{ij}$ = generalised travel cost or time from i to j for individual r (considered here as predictable and non-random).

Of course, destination choice is only one among the inter-
related choices that people have to make when making transport decisions; Ben-Akiva and Lerman describe in their
resource paper (Chapter 30) how this approach to accessibility
through destination choice could be related to other choices
within a hierarchical model of individual choices.  A problem
exists regarding the determination of a suitable probability
function for the random variable $V^r$, in order to calculate
the expected value of the maximum utility among the utilities
$U^r_{ij}$ associated with each of the potential destinations; this
expected value represents the average benefit derived from a
trip by individual r, as it takes into account both desirability of destinations and trip cost or time.

Two particular methods have been proposed (if we exclude
the probit model, which does not lead to a closed-form
solution).  The first, introduced by Koenig in 1973 (18),
could be called 'HIVEX' (Exponential in area of HIgher Values
of Utility) as it assumes that the probability density
function of $V^r$ is, in the area of the higher values of
utility, an exponential of the following form:

$$p(x) = k\ e^{-\frac{x}{x_0}}$$

The maximum utility, which can be shown to be independent of
the shape of the probability density function in the area of
the lower values of utility, has the following expected value:

$$\overline{U}_i = x_0\ \text{Log}\ A_i + \text{constant} \tag{3}$$

where $A_i$ is the accessibility indicator of Equation (2).

Furthermore, it can be shown (18) that the trip distribution
derived from those assumptions is identical to a gravity
model with an exponential distribution function.  Thus,
provided the concept of individual utility is accepted, and
if exponential gravity models are the best for trip distribution, it might be useful to measure the utility of trip-
making by Equation (3).  This result is consistent with the
three approaches summarised earlier, and it provides a
behavioural significance to the accessibility indicators $A_i$.
Finally, it is possible to take into account the effects
of competition at the destination.  It can be shown (18) that
effects like changes in wages, or changes in the probability
that an applicant for a job is accepted, may be expressed by
using the attraction-correction coefficients calculated in
doubly-constrained gravity models.

In the logit model, the $V^r$ may be assumed as independent
with a Weibull cumulative distribution function (CDF):

$$P(V\leqslant x) = e^{-ke^{-\frac{x}{x_0}}}$$

where $x_0$ and $k$ are constants.

The expected value of maximum net utility was calculated in
1975 by Domencich and McFadden (8); the result was found
identical to that of the HIVEX model.  This could have been
predicted, since the Weibull distribution may be considered a
particular case of the HIVEX distribution.  It is easy to
verify that the probability density function (derivative of
the CDF) of the Weibull distribution is asymptotically
equivalent, in the area of higher values of utility, to an
exponential.

Although these approaches to the derivation of accessibility measures differed in the extent of their behavioural
content, members of the workshop regarded them as sufficient
in combination to establish a theoretical basis for the use
of accessibility indicators.  It was recognised by the
participants, however, that the mathematical complexity of
some of these approaches is a hindrance to their routine use
by planners and transport policy-makers, and that many
problems existed with regard to implementing some of these
indicators given existing data resources.


RECOMMENDATIONS FOR RESEARCH

Although accessibility measures of various forms have already
made a significant contribution to the evaluation of transport alternatives and to the analysis of facility-location
policies, the subject of accessibility and mobility measurement still requires basic as well as applied research.  The
theoretical frameworks for the measurement of accessibility
and mobility which were described above remain incomplete and
imperfect, and there are significant developmental steps
which must be taken before these concepts can be put to
practical use.  The remainder of this report is a summary of
several of the most critical research recommendations arrived
at by the members of the workshop.

*Basic research on the value of accessibility.*  The history of
urbanisation is characterised by constantly increasing
mobility and accessibility.  City forms and patterns of
social and economic interaction have evolved in parallel with
the evolution of transport technology and with systematic
improvements in access.  Yet, little is known about the
ultimate value of accessibility to individuals and to society.
It is not obvious that the quality of life is improved when
access to services and other opportunities is enhanced.  Many
attitudinal surveys have been conducted in which perceptions
of modal attributes and reactions to transport-system
characteristics have been investigated.  Members of the workshop were, however, aware of few studies in which the
perceived value of accessibility to opportunities was
examined.  Many members of the workshop felt that such
studies were warranted given our current level of understanding of accessibility and mobility.  For example, while
current policy gives emphasis to the improvement of accessibility for groups in society which are relatively deprived,
it may well be that access, like many other commodities and
services, is characterised by diminishing marginal utility to
individuals or households.  Similarly, attitude studies might
help to clarify the relationship between accessibility and

mobility, by reflecting how these concepts are related to
one another in the minds of citizens.

*Research on the relationship between accessibility, travel
demand and urban form.*  Although this recommendation is
closely related to the proposal that the value of accessibility be investigated, it is an important subject for
applied and policy-related research in its own right.
Relatively little is known about how trip frequencies, trip-
purpose mixes and destination choices vary as a function of
accessibility.  Some researchers, notably Yehuda Gur(12),
have examined the dependence of trip-generation rates upon
access to a range of travel opportunities, but the dynamic
equilibrium between availability of choice sets and travel
patterns is poorly understood.  There is undoubtedly a
systematic relationship between density of urban development
and accessibility, which in turn yields the often-observed
association between density and travel volumes.  Members of
the workshop felt, however, that little is known about this
association at the aggregate level, and that even less is
known about its implications at the level of individual or
household behaviour.  This is a subject for research which
could yield results of great importance to behavioural travel
modelling.

*Research on the definition and aggregation of choice sets.*
Many applications of accessibility indices have, to date,
simply aggregated all work opportunities or all housing
opportunities in a zone as an indicator of the choice of
opportunities available in that zone.  Yet, as was mentioned
earlier, citizens are not likely to be indifferent among all
work sites or all housing sites.  It is quite probable that
different opportunities are valued differently by each
decision-maker.  While gross accessibility measures can be
constructed without distinguishing among opportunities,
efforts to build choice models require that the value or
attractiveness of all potential destinations be considered.
In models of mode, route, destination, or residential-
location choice, it is critical that an appropriate choice
set be defined.  Such a choice set must be significantly
smaller than the universe of all jobs or all residential
locations.  Many alternative approaches exist for the
stratification of opportunities into those which lie within
and those which are outside an appropriate choice set, but
systematic research is needed to compare and evaluate such
methods.  It is reasonable to include some notion of 'competition at the destination' in the definition of opportunities as well.  That is, it is likely that at some destinations
there are constraints upon the availability of opportunities.
There may be ten jobs at a destination for which ten people
are competing, but there may be ten jobs at a destination
for which 10,000 people are competing.  It seems reasonable
to treat the opportunities differently in these cases.  In
the particular case of employment opportunities, it might be
possible to link some elements of labour-market theory with
travel and locational-choice models in order to account for
this kind of competitive situation.

   Another key issue related to the definition of choice sets
is the problem of aggregation.  In cases where elemental

choice sets are available for each unit of analysis, our
methods are weak in that we also lack effective ways of
aggregating elemental choice sets across many units of
analysis.  Our current inability to define and aggregate
choice sets with precision constitutes one of the most
serious barriers to the development of choice-based behavioural models.  The lack of meaningful choice sets and aggregations of them also makes it very difficult to construct
accessibility measures which are directly useful in disaggregate travel-demand models.  For these reasons, members
of the workshop gave high priority to research on the definition and aggregation of choice sets.

*Accessibility indices which are non-home-based.*  Although
several different forms of accessibility index were presented
earlier, they all determined accessibility to opportunities
or travel impedance from some base point.  In almost every
case known to members of the workshop, indices have been
calculated using accessibility from home, or occasionally,
from the centre of the central business district.  For
example, accessibility indices based upon isochrons of travel
time usually count travel opportunities within x minutes of
travel from home rather than from any other location.
Because many trips are non-home-based, and because many
locational decisions are based upon proximity to non-
residential activities, it is reasonable to conclude that
accessibility measures based solely upon travel impedances
from home are too limited.  It is recommended that research
be conducted to incorporate non-home-based calculations of
accessibility with the home-based measures which are now in
more common use.

*Compound measures of travel impedance.*  Most accessibility
measures constructed to date have employed car travel time as
the sole measure of impedance between an origin and a set of
opportunities.  Members of the workshop felt that research
was needed to develop measures of impedance which incorporate
transit travel times in some realistic combination with car
travel times, in order to present a more realistic picture of
actual travel impedance.  Similarly, it was felt that other
measures of impedance, such as travel cost, should be considered along with the traditional measure of travel time.
Empirical research could be performed to derive the most
appropriate measures of impedance.  Using such criteria as
goodness of fit, choice models could be estimated for which
alternative formulations of an accessibility index could be
tested as independent variables.  The formulation giving rise
to the model of best fit might be a more appropriate
accessibility index in a particular situation than one using
travel time alone.

*Research on accessibility beyond the urban scale.*  Almost all
of the discussion centred upon the study of accessibility
within an urban region, and travel-demand modelling has also
been traditionally carried out most frequently at the
regional scale.  Yet, it is clear that accessibility also
plays an important role in the establishment of economic and
social relationships *among* regions, and can have a critical
impact upon rural areas.  What happens, for example, to small
and moderate-sized towns as rail service is abandoned and

inter-city travel is increasingly handled by air service
which reaches many fewer communities?  What is the real role
of accessibility in regional development?  Members of the
workshop felt that some of the conventional wisdom giving
answers to these questions is based upon assumptions and
theory which have rarely been empirically validated, and that
the theories may be less relevant to developed nations in the
1980s than to underdeveloped ones of past decades.  Research
is needed into the problem of accessibility at the multi-
regional scale in support of centralised development planning
which will characterise many nations of the world during the
coming decades.

*Accessibility indicators as a subject for experimental design.*
Finally, members of the workshop observed that accessibility
indicators have usually been employed in *ad hoc* or applied
studies and have rarely been a subject for systematic
research in and of themselves.  Rarely have researchers compared alternative forms of accessibility indicator to
determine the conditions under which one is superior to
another for some particular purpose.  Similarly, accessibility has rarely been the subject of longitudinal studies.
While major facilities may have been located in response to
explicit considerations of access, and while transport-
network decisions may have been made precisely to enhance
access in a sector or to a particular group, rarely have
accessibility indicators been employed both before and after
the implementation of such an improvement in order to
determine whether the intended change came about, or whether
the accessibility indicators employed were themselves sensitive to the changes in the environment.  Thus, it is
recommended that principles of experimental design be
incorporated in future research on this subject.  Experiments
of a before-and-after nature, comparing alternative forms of
accessibility measure, and measuring the accessibility
impacts of major public investments are needed to improve our
understanding of the role of accessibility in policy-making
and modelling, as well as our understanding of the role of
accessibility and mobility in determining the quality of life.

REFERENCES

1. Baxter, R.S. and Lenzi, G., 'The measurement of
relative accessibility', *Regional Studies*, 9, 15 (1975).
2. Breheny, M.J., 'Towards measures of relative accessibility', *Progress in Planning*, 2 (1974).
3. Burns, L. and Golob, T.F., 'The role of accessibility
in basic transportation choice behaviour', *Transportation*, 5,
175 (1976).
4. Catanese, A.J., 'Home and workplace separation in four
urban regions', *Journal of American Institute of Planners*, 37,
331 (1971).
5. Centre d'Etude des Transport Urbains (1976) *Deplace-
ments des Personnes dans les Villes Françaises*, Paris.
6. Cohen, D.S. and Basner, C., *Accessibility - Its Use as
an Evaluation Criterion in Testing and Evaluating Alternative
Transportation Systems*, Highway Planning Technical Report
No. 28, Network Analysis Series No. 3, Federal Highway
Administration, US Department of Transportation (1972).

7. Dalvi, M.Q. and Martin, K., 'The measurement of accessibility: some preliminary results', *Transportation*, 5 (1976).
8. Domencich, H. and McFadden, D. (1975) *Urban Travel Demand: A Behavioural Analysis*, North Holland, Amsterdam.
9. Dunphy, R.T., 'Transit accessibility as a determinant of automobile ownership', *Highway Research Record No. 472* (1973).
10. Forbes, J., 'Mapping accessibility', *Scottish Geographical Magazine*, 80, 12 (1964).
11. Frye, F.F., 'Some thoughts on evaluating transportation systems by measures of their accessibility', *Report No. PRR-13*, Basic Research Unit, New York State Department of Transportation (1967).
12. Gur, Yehuda (1971) *An Accessibility Sensitive Trip Generation Model*, Chicago Area Transportation Study, Report No. 341, 41.
13. Hansen, W.G., 'How accessibility shapes land use', *Journal of American Institute of Planners*, 3, 73 (1959).
14. Harris, B., 'Notes on accessibility', *Discussion Paper, Institute of Environmental Studies*, University of Pennsylvania (1966).
15. Ingram, D.R., 'The concept of accessibility: a search for an operational form', *Regional Studies*, 5, 101 (1971).
16. Jones, P.M., 'Accessibility, mobility and travel need: some problems of definition and measurement', paper presented at the IBC Transport Geography Study Group Conference (1975).
17. Knudsen, T. and Kanafani, A., 'Definition and measurement of accessibility in urban areas', *Institute of Transportation and Traffic Engineering Research Report No. 54*, University of California, Berkeley (1974).
18. Koenig, J.G., 'A theory of urban accessibility', *PTRC Summer Meeting* (1975). Summary in *Simplified Traffic Models*, OECD (July 1974).
19. Koltnow, P.G. (1970) *Changes in Mobility in American Cities*, Highway Users Federation for Safety and Mobility, Washington, D.C.
20. Lozano, E.D., *et al.*, 'Level of services and degree of accessibility: spatial, urban simulation model', *Regional Studies*, 8, 21 (1974).
21. Mitchell, C.G.B. and Town, S.W., 'Accessibility of various social groups to different activities', paper presented at the 5th British Regional Congress of the Permanent International Association of Road Congresses (1976).
22. Mossman, F. and Farra, A., 'Mobility index based on socio-economic characteristics of households', *Traffic Quarterly*, 29, 347 (1975).
23. Muraco, W.S., 'Inter-urban accessibility', *Economic Geography*, 48, 388 (1972).
24. Neuburger, E., 'User benefits in the evaluation of transport and land use plans', *Journal of Transport Economics and Policy*, V, 1 (1971).
25. O'Sullivan, P.O., 'Accessibility and the spatial structure of the Irish economy', *Regional Studies*, 2, 195 (1968).
26. Savigear, F., 'A quantitative measure of accessibility', *Town Planning Review*, 38, 64 (1967).
27. Sherman, L., Barber, B. and Kondo, W., 'Method for evaluating metropolitan accessibility', *Transportation Research Record No. 499*, 70 (1974).
28. Stegman, M., 'Accessibility models and residential

location', *Journal of American Institute of Planners*, 35, 22 (1969).

29. Vickerman, R.W., 'Accessibility, attraction and potential: a review of some concepts and their use in determining mobility', *Environment and Planning A*, 6, 675 (1974).

30. Volmmer Associates (1975) *Automobile Accessibility and its Relation to Economics of the New York City Central Business District*, New York Chamber of Commerce and Industry.

31. Wachs, M. and Kumagai, T.G., 'Physical accessibility as a social indicator', *Socio-Economic Planning Sciences*, 7 (1973).

32. Weibull, Jörgen W., 'An axiomatic approach to the measurement of accessibility', *Regional Science and Urban Economics*, 6 (1976).

33. Whitbread, M., 'Evaluation in the planning process – the case of accessibility', *Planning Methodology Research Unit, Working Paper No. 10*, School of Environmental Studies, University College, London (1972).

34. Wickstrom, G.W., 'Defining balanced transportation – a question of opportunity', *Traffic Quarterly*, 25, 337 (1971).

35. Wilbanks, T.J., 'Measuring accessibility', *Urban Transportation Institute, Occasional Paper No. 5*, Syracuse University, Syracuse, New York (1970).

36. Zakaria, T., 'Urban transportation accessibility measures: modifications and uses', *Traffic Quarterly*, 28, 467 (1974).

PART TEN

Chapter 33

APPLICATIONS OF PSYCHOLOGICAL MEASUREMENT AND MODELLING
TO BEHAVIOURAL TRAVEL-DEMAND ANALYSIS

Jordan J. Louviere, Eugene M. Wilson and Michael Piccolo

INTRODUCTION

One of the critical issues addressed in transport planning is
that of travel demand or 'choice'. Early origin-destination
studies attacked this issue by treating aggregate travel-
behaviour patterns as a set of sequential choices: generation;
distribution; mode choice; and assignment. Each of these
separate choices was implicitly or explicitly modelled as:
(1) whether to make a trip and for what purpose; (2) given a
purpose, to what destination would it be made; (3) by what
mode; and (4) by what route or routes would the trip be made.
Controversy arose over the aggregate nature of this modelling
process, and amongst the arguments advanced against it was a
so-called behavioural view, typified in the following
statement by Michaels:

... how and why people use transportation alternatives can
be explained only by an analysis of the subjective perceptions of these systems and the psychological measures of
their necessity and sufficiency. This, in turn, suggests
that inadequacy of traditional models based on objective
or extrinsically measurable variables of system performance ... There simply is overwhelming evidence that the
relationship between objectively derived measures of
physical process and behavioral performance is rarely
linear.

Consequently, if one desires to construct a model such
as trip generation, distribution, or mode choice, validity
and reliability are likely to be higher when such a model
is constructed on the basis of behavioral measures rather
than physical measures, especially when the latter are
selected simply because they are convenient or easy to
measure ... (59)

The focus of this chapter is upon the development of
disaggregate models of choices and decisions, employing the
individual as the unit of observation. The work reported in
this chapter differs from recent work in disaggregate modelling in that it is concerned with *direct* rather than *revealed*
estimates of demand. Additionally, rather than each individual supplying a single observation, each individual supplies
multiple observations or estimates of demand. In a sense
then, the models and techniques reported in this chapter are
*totally* disaggregate in that each individual's separate

713

demand function may be estimated from the data.  The data
reported in this chapter, however, are for aggregate or
grouped-demand functions.  Our concern, therefore, is with
developing utility or demand functions for various choice
behaviours of interest to travel behaviour.  The approach
involves the use of theory and methodology from the study of
information processing in judgement and decision-making in
psychology.  None the less, our *intent* and *aims* are very
similar to those working with econometric and stochastic-
choice models: to develop models which describe and predict
choices and decisions of interest to transport authorities.
We discuss theory and methodology appropriate to such modelling and apply the approach in two empirical studies of
choice behaviour: residential choice and shopping-destination
choice.


SOME ISSUES IN 'ATTITUDINAL' MODELLING IN TRANSPORT RESEARCH

At present the burden of proof appears to weigh heavily upon
those who advocate the use of 'attitudinal' measures and
models to demonstrate that they can yield valid and reliable
predictions of human travel behaviour.  Two major reviews,
Fishbein in 1967 (25) and Wicker in 1969 (84), concluded that
there was little evidence that 'attitudinal' predictions
corresponded with overt behaviour, although Fishbein and
Ajzen (26,27) later conclude that attitudinal measures
correlate well with a number of related behaviours, but not
one single action in particular.  Faced with this evidence,
one might well ask why such research in travel behaviour
would be expected to yield results any different from those
in psychology?  Indeed, a search of the travel behaviour
literature fails to turn up much encouragement.  Few studies
have actually investigated attitude-behaviour links and those
which have done so have failed to generate much enthusiasm
over their predictive power (see, e.g., Quarmby (68); Hensher
(37,38); Davidson (17) Westin and Watson (83)).  As Golob
(30) concluded:

> Although these multivariate statistical studies serve to
> validate particular postulates concerning relations
> between cognition, affect, and conation (sic), all reveal
> rather poor connectivities between attribute-level atti-
> tude and actual behavior in the consumer context.

Although Golob goes on to recommend models with assumed
attribute inter-dependence, there is only a limited amount of
evidence available that such models will perform well.  For
example, some success in the use of factor-analytic and
related models such as MDS (multidimensional scaling) is
reported by Burnett (11,13) in her study of consumer spatial-
choice behaviour in Sydney, Australia, and by Cadwallader
(14) and MacKay (54) in separate studies of consumer shopping
behaviour in the US.  More evidence of a replicated nature is
necessary before general conclusions regarding these methods
may be drawn; however, in general, these studies sought to
derive psychological measures for factors hypothesised to
affect the behaviour of interest and to relate these measures
to this behaviour by means of an assumed model.

Thus, a basic question which remains unanswered and which appears to divide researchers who advocate the use of psychological measures in travel-behaviour models is - do attitudes regarding any facet of the transport system affect travel behaviour or is traveller behaviour a response to perceived levels of system variables or both? Although there are some interesting results which bear on this question, it is fair to say that at this time there is insufficient evidence to be conclusive. As suggested earlier, the onus of proof, therefore, is upon researchers to demonstrate that behavioural models which employ psychological measures or methods can *predict* better than traditional models.

This chapter is concerned with providing some empirical evidence that attitudinal models can predict travel choices - in particular the choices of groups other than those from which the models are estimated. The methodology employed is derived from information-integration theory and the method of component ratings in psychology (1-8,45-58,62,63,66,67,74,77) and is based on the conception that the attributes of the system, or at least their perceived levels, are the important influences on behaviour. The objectives of both research efforts, therefore, are several-fold: (1) to identify a set of variables which are hypothesised to be linked to choices; (2) to measure the perceived effects of these variables; (3) to develop a model which predicts choices; and (4) to validate the model with respect to real-world choice behaviour.

PSYCHOLOGICAL MODELLING OF REAL-WORLD CHOICE BEHAVIOUR: TWO VALIDATIONAL CASE STUDIES

Recently, a growing body of literature has applied theory and methodology derived from cognitive psychology to problems in transport and related fields. A representative sample would include the following work: Levin (45,46); Norman and Louviere (62); Norman (63); Louviere *et al.* (47-53,66,67); Donnelly, Howe and DesChamps (24); Golob (30); Golob, Canty, Gustafson and Vitt (29); Margolin, Misch and Dobson (55); and Donnelly (23); Beier (9); Burnett (11-13); Cadwallader (14); Costantino, *et al.* (15,16); Demetsky and Hoel (19); Dobson, *et al.* (21,22); Hartgen and Tanner (32); Hensher, *et al.* (34-38); Nicolaidis, *et al.* (60,61); Spear (75); Sterns (76); Stopher, *et al.* (78-81) and Westin and Watson (83); to name only a few examples. Much of this research was concerned with questions related to mode choice or to differences in judgements of intended patronage related to differences in trip purposes, type of mode, mode attributes and new innovations. These studies were of a preliminary nature and are subject to legitimate questions regarding their validity and applicability to the real world.

This chapter examines one of the above techniques in two detailed case studies to demonstrate the applicability of the theory and methodology to modelling two different types of choice behaviour relevant to transport research:

1 The proportion of people who choose each of several alternative towns as places to shop for clothes, or destination choices from a single origin for a specific shopping purpose.

2 The proportion of non-local employees from various industrial plants who choose each of several alternative towns as places to live, or urban residential choices of non-local employees at each of five different plants in three different states.

## THEORY

The algebraic theory behind the methodological approach adopted in both research projects is detailed in the appendix to this chapter. The assumptions are outlined below. The approach is based on theoretical and empirical research in mathematical psychology and related fields, where there is ample support for the following assumptions:

AX 1. $x_k = f_k (X_k)$                                              (1)

       $x_i = f_i (Q_\ell)$

AX 2. $R = g(x_1, x_2, \ldots, x_k, x_{k+1}, \ldots, x_n)$         (2)

AX 3. $B = h(R)$                                                    (3)

Th 1. $B = h(g(Q_i), f(X_{jk}))$   i = 1, 2, $\ldots, \ell_{ij}$
           = 1, 2, $\ldots$, k.                                        (4)

The $X_k$ are physically measurable or resource values of the system. They represent physical or resource factors which are known to affect travel behaviour or choice. The $x_k$ are the perceived values of $X_k$ that travellers or choosers think the $X_k$ possess. The $Q_i$ are the set of qualitative factors which are perceived and which affect choice. They are 'qualitative' only insofar as we may not understand how to measure the physical referrents of these quantities at this time or else their physical measurement is particularly difficult. The R are the i different responses such as judgements or choices which are observed in some experimental context. Each of the R responses is a consequence of different sets of values of each of the n factors. There are n factors which include the k physically measurable variables and the $\ell$ qualitative factors. The B are the choices or behaviours which are hypothesised to correspond to the response situations or combinations.

AX 1 simply states that the psychological or perceptual values of the k physical system variables are functionally related to the actual measures of the levels of these factors. As Michaels (59) suggested earlier, the functional relationships are generally non-linear. AX 2 simply states that any choice or response behaviour observed in an experiment is functionally related to both quantitative and qualitative measures of the system of interest. AX 3 states that the responses observed in a choice experiment or judgement study are functionally related to real behaviour, B. Such a functional relationship may be non-linear, similar to the relationships postulated in AX 1. The form of the relationship in part will depend upon the level of measurement of R, which is generally interval in nature and not ratio, as well as the nature of the particular behaviour under study. However, we can state that B must be a monotonic function of R to meet our intuitive notion of this relationship.

## SPECIFIC RESEARCH QUESTIONS

The two studies which follow consider the following five questions:

1  What are the important factors which underlie the behaviour of interest?
2  For those factors which are physically measurable, what are the functional relationships between various levels of these factors and corresponding perceptual or psychological estimates of the levels of these factors (Equation (1))?
3  How do the physical, qualitative and psychological values combine to affect choice or response behaviour in an experiment (Equation (2))?
4  How do the response measures from the experiment relate to corresponding real-world behaviour - what is the nature of the functional relationship (Equation (3))?
5  How well can models developed from the logic of Equations (1)-(3) recover real-world choices or behaviour (Equation (4))?

These questions are systematically addressed in the following two studies.

*A Study of Choice of Town as Place of Residence by Non-local Industrial Employees*

There has been considerable research into the journey to work in transport contexts; e.g. reviews are available in Stopher and Meyburg (81) and Dickey *et al.* (20). Considerable empirical success is documented in applying simple gravity-type concepts to the distribution of trips between residence and home.  By extension, residential choice can be conceptualised as a trade-off between the available supply of housing and a distance from work place (in miles).  Thus, we hypothesise that subjects confronted with the task of estimating their preferences for each of a number of town-size and distance alternatives will employ a simple gravity-type model to make their judgements.  Because workers consume more than simply a residence when they choose a town in which to live, a number of one-paragraph descriptions of towns listing the population, the services available and the distance to work place were constructed so as to reflect real-world choices.  This study was designed to reflect the employment/residence situations for isolated plants in Wyoming, North Dakota and Montana.

*Methodology.*  Sets of hypothetical classes of towns were constructed by developing linear regression functions relating population to the number of each of 10 types of facilities such as bars, grocery stores, restaurants, churches, etc. The number of expected functions in each class was predicted from population sizes of 250, 500, 1,000, 1,500, 2,000 and

2,500. This procedure is based on empirical research in central place theory (10). Thus, there are six levels of the composite stimulus (town size + facilities). Six levels of driving distance were chosen by examining actual commuting distances of plant workers (see Old West Commission (64)). The levels chosen were 15, 30, 45, 60, 90 miles. All combinations of towns and distances yield a 6 x 6 factorial design. This design was printed in five different random orders on ditto sheets with four items per sheet; additionally, four filler combinations more extreme than the design combinations were inserted. Thus, the experiment involved 40 combinations on 10 sheets.

The experimental design is similar to that employed by Rushton (70-72) in his studies of revealed preferences. The major difference is that this procedure should permit a *priori* estimates of preferences rather than the *a posteriori* method of Rushton. Filler combinations are used to transfer response bias away from the experimental combination extremes. Subjects respond more extremely to the fillers which they quickly learn are the *best* and *worst* combinations in the design. To test the effects of the 'order' of the combinations in the questionnaires, five different orders were prepared by random draw. Sixty usable questionnaires were obtained from students, faculty and staff at the University of Wyoming who volunteered to participate. These subjects were randomly assigned to the five order conditions, hence, the experiment is a 6 x 6 x 5 x 12 factorial design (towns x distances x orders x subjects).

Subjects were asked to supply a numerical estimate of their degree of preference for each combination by assigning a number between zero (absolutely the *worst* combination imaginable) and 100 (the *best* imaginable). Subjects were shown combinations that pretesting had revealed to be very undesirable and very desirable. They were told that all items to be evaluated were considerably less extreme than these and they were to use these combinations to 'anchor' their numerical judgements of preference. All subjects, therefore, completed all combinations.

Data were first analysed by means of analysis of variance, for a randomised blocks design. The results showed that the effect of order was not significant, while that for distance and town size was, as was their interaction. The traditional gravity model of trip distribution would require a multiplicative relationship between these factors which would yield a significant interaction. The *form* of this interaction, however, is critical.

Graphical evidence of multiplication is necessary to bolster the statistical evidence of a significant interaction. In particular, consider the effect of subtracting the first row of a factorial design from the second row under the assumption of multiplication: $R_{ij} - R_{2j} = $ (Row 1 x column j) - (Row 2 x column j) = Column j (Row 1 - Row 2). If the data are graphed as a function of increasing (or decreasing) column values, it is clear that the difference between each row must increase (or decrease). Thus, the data plot as divergent (convergent) curves. This is approximately true in

Figure 33.1.  Hence, we can tentatively accept the simple
gravity or multiplicative combination rule as a reasonable
approximation to the decision process for this experiment.

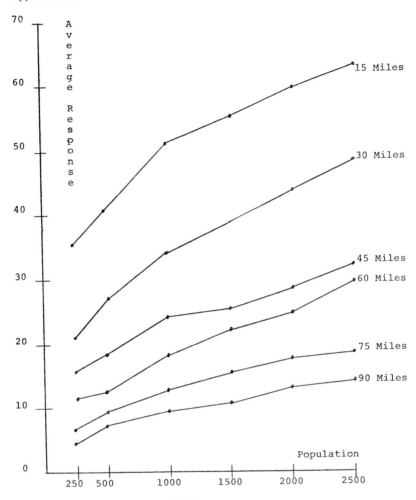

Figure 33.1  Population X Distance

As Anderson has demonstrated (1-8), the marginal row and
column means are interval-scale estimates of the effects of
each of the experimental factors.  Viewed another way, they
are the utility values which correspond to the levels of the
experimental factors, measured in the units of the dependent
variable.  These effects or utilities can only arise from
variation in the experimental values.  Hence,

$$R_{i.} = f_1(\text{Townsize}_i) \tag{5}$$

$$R_{.j} = f_2(\text{Distance}_j). \tag{6}$$

We may assume specific function forms for $f_1$ and $f_2$; from Equation (1), Appendix:

$$R_{i.} = a_1 + b_1 \text{ Townsize}_i{}^{c_1} \tag{7}$$

$$R_{.j} = a_2 10^{-b_2 \text{Distance}_j}, \tag{8}$$

where $R_{i.}$ is the marginal row mean and $R_{.j}$ is the marginal column mean derived from the factorial design. Because these are sums or averages of random variables, it is reasonable to assume them to be normally distributed. The experimental factor values are 'fixed', hence, we have a classical fixed-effects regression case and can estimate the desired parameters via least-squares.

If we assume the multiplicative hypothesis encouraged by Figure 33.1, and the results of the analysis of variance, we can write the well-known relationship:

$$R_{ij} = (R_{i.}) (R_{.j})/R_{..} \tag{9}$$

where $R_{i.}$ and $R_{.j}$ are marginal means and $R_{..}$ is the grand mean. Equation (9) is always true if there is a true multiplicative rule underlying the data. Substituting Equations (7) and (8) into (9) yields:

$$R_{ij} = \frac{1}{R_{..}} (a_1 + b_1 \text{ Townsize}_i{}^{c_1}) (a_2 10^{-b_2 \text{Distance}_j}). \tag{10}$$

By expanding (10) and combining constants, we have the expectation of the following equation:

$$\tag{11}$$

$$R_{ij} = k_0 + k_1 \text{ Distance}_j{}^{c_2} + k_2 (\text{Townsize}_i{}^{c_1} \; 10^{-b_2 \text{Distance}_j}).$$

Thus, if $c_1$ and $c_2$ are known (and we can estimate them by observing for townsize (from (8)):

$$\log (R_{i.} - a_1) = \log b_1 + c_1 \log \text{Townsize}_i. \tag{12}$$

$c_1$ and $c_2$ are therefore best least-squares estimates and can be used to transform the physical measures in the following manner:

let $X_{1i} = \text{Townsize}_i{}^{c_1}$

let $X_{2i} = 10^{-b_2 \text{Distance}_j}$

let $X_{3i}$ = Townsize$^{c_1}$ $\times$ 10$^{-b_2 Distance_j}$.

Then we have:

$$R_i = k_0 + k_1 X_{1i} + k_2 X_{2i} + k_3 X_{3i}. \tag{13}$$

The log of the marginal means (minus a constant = 7) was then plotted against the log of townsize. (It seemed obvious that the graph intercepted somewhere in the neighbourhood of an $R_i$ value of 7). An acceptable fit was obtained with the following empirical equation:

$$\log (R_i - 7) = \log -0.504 + 0.562 \log Townsize_i. \tag{14}$$

The R-square for Equation (14) is 0.997, with an F-value of 1263.6 (1,4 df), confirming a highly significant linear relationship.

The $R_{.j}$ function decreased at a decreasing rate and it seemed appropriate to approximate it with a negative exponential because it must have an upper bound. A plot of log $R_{.j}$ versus distance$_j$ yielded a highly linear relationship. Least-squares regression yielded the empirical equation:

$$\log R_{.j} = \log 1.83 - 0.0093 Distance_j. \tag{15}$$

The R-square is 0.994 and the F-value is 698.3 (1,4 df), again far beyond significance.

Substitution of (14) and (15) into Equation (10) yields:

$$R_{ij} = \frac{1}{R_{..}} (a_1 + b_1 Townsize_i^{0.562}) (a_2 . 10^{-0.0093 Distance_j}) \tag{16}$$

collecting terms and constants, we have (logs to the base ten were used):

$$R_{ij} = k_1 (10^{-0.0093 Distance_j}) + k_2 (Townsize_i^{0.562} \times 10^{-0.0093 Distance_j}). \tag{17}$$

Applying least-squares regression yields:

$$R_{ij} = 24.76 (10^{-0.0093 Distance_j}) + 1.313 (Townsize_i^{0.562} \times 10^{-0.0093 Distance_j}). \tag{18}$$

The intercept was -0.565, which seems close enough to zero to be acceptable. This equation accounts for 98.6 per cent of the variance in the experimental cell means. The two independent factors in Equation (17) were both highly significant and the overall F (2,33) for the equation was 1154.8, which is highly significant. Hence, we tentatively retain Equation (15) as an approximation to the decision process on trade-off function employed in the experiment.

*Real-world Validation.* A set of data suitable for testing the model was available from the Old West Commission (64). Employees at one soda ash and four power plants in Wyoming, North Dakota and Montana were sampled regarding their place of residence. The proportions living in each of 42 different towns near each of the five plant locations and the approximate mileages to each plant from each town were calculated. Population estimates for each town were obtained from census and state estimates. Equation (18) was applied directly to the population and distance data to yield a vector of 42 values of $\hat{R}_i$, where $\hat{R}_i$ is an estimate of the response or utility value that subjects would have assigned the real-world population-distance combination, had it been present in the experiment. The $\hat{R}_i$s were then plotted against the observed proportions (Figure 33.2): the graph exhibits considerable monotonic correspondence with a curvilinear trend, and the rank-order correlation is 0.93. We regressed log $\hat{R}_i$ against the log of the proportion data. The derived regression equation accounted for 84 per cent of the variance in the residential choice proportions and, of course, is highly statistically significant. The empirical equation estimated from the data is:

$$\text{Log } P_i = \log -1.86 + 1.595 \ (\log \hat{R}_i). \tag{19}$$

Equation (19) was used to transform the cell averages from the 6 × 6 experiment into expected choice proportions. The 6 × 6 design was then reanalysed in a manner similar to the original analysis of the response data: functions were estimated on the two sets of marginal means and then these functions were algebraically combined into a multiplicative model. The marginal relations have the following empirical form:

$$\hat{P}_{i.} = 1.18 + 0.0013175 \ \text{Townsize}_i \tag{20}$$

$$\hat{P}_{.j} = 11.852e^{-0.0337 \ \text{Distance}_j} \tag{21}$$

where $\hat{P}_{i.}$ and $\hat{P}_{.j}$ are respectively the row and column marginal means for the expected choice proportions. The correlation-squared for Equation (20) is 0.9887 and that for Equation (21) is 0.9934. Both equations are statistically significant. Combining Equations (20) and (21) multiplicatively yields (after dividing through by the grand mean):

$$\hat{P}_{ij} = 4.85e^{-0.0337 \ \text{Distance}_j} + 0.0054156 \ \text{Townsize}_i$$

$$e^{-0.00337 \ \text{Distance}_j} \tag{22}$$

where $\hat{P}_{ij}$ is the expected choice proportion observed in the ij cells.

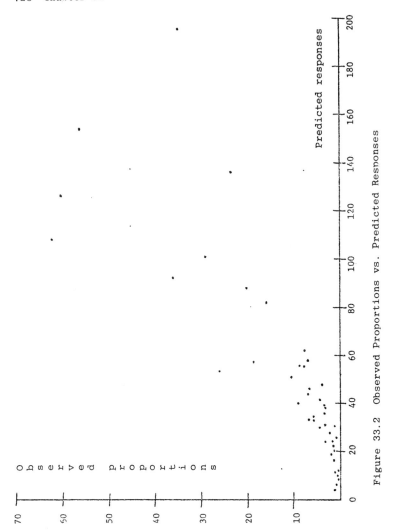

Figure 33.2   Observed Proportions vs. Predicted Responses

Equation (22) permits us to estimate the proportion of
workers who will choose a town of any size and distance
*anywhere*. That is, this equation is derived from a controlled experiment in which town size and distance combinations
were manipulated such that the combinations could be anywhere.
Similar results could be obtained for mode choice, destination choice or route choice situations, given a comparable
design. To compare this fit, two models of the following
form were fit to the choice data ($P_i$ = proportions):

$$P_i = \alpha_1 (\text{Population}_i^{b_1} \cdot \text{Distance}_i^{b_2}) \qquad (23)$$

$$P_i = \alpha_2 (\text{Population}_i^{b_1} \cdot e^{-b_2\text{Distance}}). \qquad (24)$$

Equation (21) accounts for 74.6 per cent of the variance in the choice proportions while Equation (24) accounts for 83.1 per cent. Neither equation recovers the data as well as the laboratory model. The estimates from Equation (24) are (for comparison):

$$\ln P_i = -0.22 + 0.47 \ln \text{population}_i - 0.015 \text{ Distance}_i. \qquad (25)$$

These results confirm a finding by Louviere that students often constitute as good a sample as the general public for studying decision processes. More significantly, the results show that a model developed by studying the behaviour of one group of individuals can be applied to the behaviour of a second group. Moreover, the model is a cause-and-effect model, rather than one based on statistical associations.

## A Study of Out-of-town Shopping Behaviour

The business community of Laramie, Wyoming, has a long-standing interest in out-of-town shopping behaviour. Considerable trade loss is evident to the chamber of commerce and its causes are the object of this study. There is a long history of research into consumer shopping behaviour over space in geography and related fields (see Rushton [70-72]; Golledge and Rushton [28]), and even the simplest models provide a fair account of the data. As in the previous study, these models are mainly versions or extensions of gravity models which have been frequently used to predict trip distribution in transport plans. In the initial phase of this study, we designed an experiment to examine trade-offs in selection (number of stores) and travel time because previous studies in Laramie had shown these two variables to be key influences on behaviour [56]. Five levels of selection (30, 50, 100, 200, 400 stores) and travel time ($\frac{1}{2}$, 1, 2, 3, 4 hours) were combined into a 5 × 5 factorial design. These combinations were arranged in several different orders on ditto sheets and assembled into packets along with five filler items more extreme than the experimental items.

One hundred-and-six student volunteers served as subjects. Subjects were told that each selection and travel-time combination represented a potential out-of-Laramie shopping opportunity. Respondents judged each combination on a 1-20 scale, anchored by a *best* (20) and a *worst* (1) shopping opportunity which were supplied as standards. The response data were treated as a 5 × 5 × 106 between-subjects experimental design and analysed by means of procedures similar to those detailed above in the residential-choice study.

Results showed significant main and interaction effects. Yet, graphical plots of the data could not support the multiplicative combination rule hypothesis (Figure 33.3). Because

there was no strong curve divergence, we could not accept
multiplication; none the less, following the procedures outlined in the first study, we fit a multiplying function by
estimating the marginal mean relationships. They are as
follows:

$$\text{Log } R_{i.} = \log b_1 + 0.274 \text{ log Stores}_i \tag{26}$$

$$\text{Log } R_{.j} = \log b_2 - 0.135 \text{ Distance}_j. \tag{27}$$

As before, a negative exponential distance function was
fitted because the data plotted linearly on semi-log paper.
The intercept for the 'stores' function was approximately
zero; hence, a simple power function was estimated. The
empirical equation for the laboratory data becomes:

$$R_{ij} = \alpha(\text{Stores}_i^{0.274} \cdot 10^{-0.135 \text{ Distance}_j}). \tag{28}$$

Therefore, the expectation is that:

$$R_{ij} = k_1 X_{ij} \tag{29}$$

where $X_{ij} = \text{Stores}_i^{.274} \cdot 10^{-0.135 \text{ Distance}}.$

The empirical model is:

$$R_{ij} = 1.633 \ X_i. \tag{30}$$

The intercept was equal to -0.165, again close to zero.
Equation (30) accounts for 96.6 per cent of the variance in
the 25 cell means, and the F values, of course, are highly
significant.

*Real-world validation - first analysis.* A validational check
with real-world data was sought. Two hundred and five
students present at Greek Day activities in April 1976 at the
University of Wyoming were asked how many shopping trips they
had made to buy clothes to each of 13 cities in Wyoming,
Nebraska, Colorado and Utah in the previous two months.
Being suspicious of such an unrepresentative sample, a random
sample of 300 non-students was drawn from the Laramie telephone directory by selecting 600 names in order to replace
disconnections, non-contactables and refusals. Individuals
were called until a total of 300 was obtained. They were
asked whether they had made a trip to one of the 13 cities in
the last two weeks primarily to shop for clothes. Data were
converted to choice proportions by dividing the number of
trips made to each town by the total number of trips.

The choice proportions for both samples were graphed
against one another and correlated. The graph was linear,
with origin about zero and slope about unity: the correlation
was 0.97. Clearly, both samples provided reasonable estimates of the trip proportions. Data on the number of
clothing stores in each city was obtained from census sources
and travel times were calculated by reference to maps.

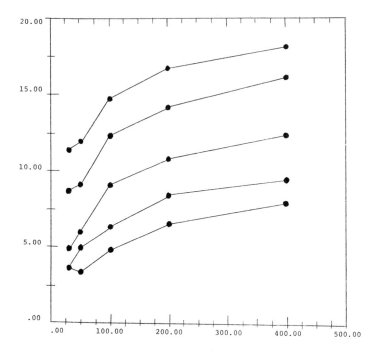

Figure 33.3  Number of Stores

Equation (30) was used to generate expected estimates, $\hat{R}_i$, as before. These were plotted against the pooled proportions from the two samples (Figure 33.4): there was a monotonic relationship and the correlation was 0.73. This relationship is not as good as that in the previous study; hence, we were left with the following possibilities: (1) the laboratory model is wrong, or biased, or both; (2) there are missing factors; or (3) both the model and the factors have been mis-specified.

*Real-world validation - second analysis.*  In the previous study, the non-student respondents were asked why they chose the cities they did for shopping: price and selection were almost always given as reasons.  Another experiment including price as a factor was then designed.

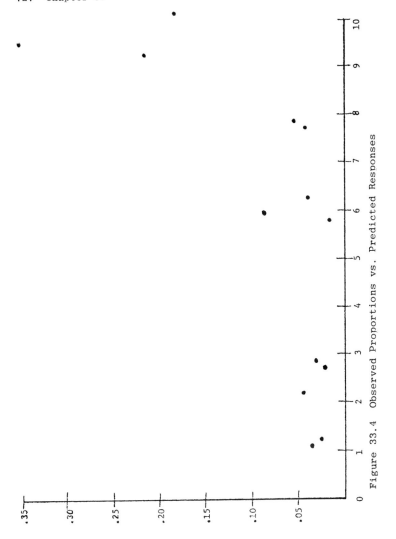

Figure 33.4   Observed Proportions vs. Predicted Responses

Sixteen subjects volunteered to participate.  The experiment
had two conditions: a 'big ticket' (a stereo set) and a
'small ticket' (a blouse or a shirt) item.  Five levels each
of travel time, selection and price difference were combined
so as to produce an orthogonal design with 25 combinations.
Because of the design, some interactions are confounded with
others and cannot be reliably estimated.  Hence, comparisons
are based on correlations of a linear and a log-linear model
with data, a far less exacting test than might be desired.

In both conditions the log-linear model form correlated more highly with the 25 cell means than its linear competitor ($R^2 = 0.88$ vs. $0.94$). Hence, we again tentatively propose that the multiplicative hypothesis is worthy of further consideration.

*Real-world validation - third analysis.*  Validation of the previous results requires data on price differences.  Subjective estimates of price differences relative to Laramie were obtained from a sample of 72 students in several introductory classes in geography.  Subjects made judgements as to whether clothing prices in Laramie were 'much lower', 'lower', 'the same as', 'higher' or 'much higher' than other places.  Mean ratings for each city were compiled and used as $x_k$ estimates (values of -2, -1, 0, +1, +2 were used).

Two analyses were initially performed: (1) a regression function employing the logarithm of the physical measures of the number of stores and the travel times was fitted to the log of the choice proportions.  This model accounted for 61 per cent of the variance and was statistically significant; and (2) a similar model using the physical measures of stores and travel time and adding the price estimates was fitted to the data.  This model did little better than the first.  In a step-wise sense, price estimates enter the equation first and are highly significant, but when selection enters, price estimates become non-significant.  This suggests that selection values and price estimates must be related.  Price data were plotted against the store data and a high correlation (0.89) was obtained: the graph was not linear, but increased at a decreasing rate.  Hence, we may tentatively hypothesise that price estimates are keyed to size of city - larger places imply lower prices.  One might speculate that similar results would obtain for intra-urban shopping, as well.

Residuals from the regression analysis were mapped and it was evident that cities with intervening opportunities were being highly over-predicted.  Price estimates were significantly related to behaviour in the second experiments, but the collinear nature of the real world was preventing us from inferring this in empirical data because price estimates were correlated with size of city.

This result suggested two additional analyses that could be performed.  Because the empirical model from the first experiment (Stores x Distance) has selection values already included with distance, it was possible that price would be statistically significant if included as a variable along with a vector of predicted values from the first model (Equation (30)).  In addition, this should be a better test of the model in contrast to the previous test which yielded a correlation with behaviour of 0.73.  That is, perhaps expected responses generated by the laboratory model applied to the real-world data would approximate a *ceteris paribus* condition.  If this is true, we would not expect the expected responses or utilities to correlate highly with the choice proportions because price and possibly intervening opportunities have not been accounted for.  To test this possibility, both price judgements and a physical measure of the cumulative number of stores up to but not including those in a

particular city (IO) were tested by means of a log-linear
regression analysis. Letting $\hat{P}_i$ represent the predicted
response to city i as generated by the laboratory model, and
$B_i$ represent the actual pooled choice proportions:

$$\text{Log } B_i = -1.054 + 0.535 \log \hat{P}_i - 2.452 \log \text{Prices}_i$$

$$- 0.136 \log \text{IO}_i. \tag{31}$$

Equation (31) accounts for over 80 per cent of the variance
in the dependent variable and is significant at the 0.01
level. Also, the cumulative number of intervening opportunities and the price judgements were significant at the 0.01
level. The expected responses ($\hat{P}_i$) were significant at the
0.05 level.

*Real-world validation - fourth analysis.* These results
suggest that a log-linear regression model employing physical
values of travel time, number of stores and cumulative intervening opportunities is the desired physical variable model.
Applying least-squares regression yields the empricial form
of the model:

$$P_i = 10^{-0.92334} (\text{Stores}_i^{1.05} \cdot \text{Time}_i^{-0.99}$$

$$\cdot \text{Opportunities}_i^{-0.97}) \tag{32}$$

The overall F (3,9) is 34.76 which is significant beyond the
0.01 level and the correlation-squared is 0.92. The individual F values are 78.1 (stores), 22.0 (time) and 35.3
(opportunities). These F-values, therefore, are also highly
significant. Note further that all the exponents are
approximately unity. Indeed, stores and opportunities are in
the same units, so the units of the constant of proportion-
ality are simply in minutes, or alternatively, miles.
Moreover, none of the inter-correlations are high; hence,
estimates should be relatively stable.

Equation (32) is significant in several respects. It
demonstrates that laboratory experiments can be very useful
in developing a new theoretical model. Moreover, it suggests
that the results of laboratory studies can provide important
insight into functional form and parameters. Most importantly, the combination of laboratory and real-world research
assisted in the formulation and testing of a new model which
combined *both* the gravity and intervening opportunities
hypotheses in a single framework. It differs from the
traditional intervening opportunities model by discarding the
notion of distance bands and the cumulation of opportunities
within a band (81). It makes explicit the relative locations
of places with respect to one another by cumulating opportunities only along direct or shortest travel routes between
places. Hence, direction and connection are explicitly
incorporated by cumulating the number of opportunities only
along the routes that connect places. Hence, a new, richer
hypothesis emerges from a blend of theory and empirical work
in the laboratory and in the real-world.

DISCUSSIONS AND CONCLUSIONS

This chapter has postulated and tested several simple models
of spatial behaviour.  As in previous research (46), correspondence of the laboratory models with real-world
behaviour is more encouraging.  Burnett (12) challenged
researchers in the judgement area to model complex real-world
behaviour.  We believe that challenge has been amply answered
and that the methodology described, combined with the
paradigm outlined in Equations (1)-(13) in the Appendix,
appears to offer a powerful approach to hypothesis testing
and model building.  Additionally, it seems to provide a
rigorous tool for 'boot-strapping' to new, more comprehensive
theory.

The research reported in this chapter demonstrates the
powerful leverage one gains on empirical reality through
controlled experimentation.  It is growing increasingly clear
that social scientists can have laboratories and can experiment as in the physical sciences.  This opens up exciting
possibilities for the use of tools not well known in social
science, such as analysis of variance and functional and
conjoint measurement  (1-8,14,18,23,24,28,31,40-54,58,62,63,
66,67,69-72,74,77,85).  We hope that the results reported in
this chapter will encourage others to consider carefully the
potential of the approach discussed here.

For researchers in integration theory, the results of the
research suggest that integration theory has empirical
validity.  Correspondence with real behaviour has been rare
in social-psychological enquiry and it is significant that
researchers in integration theory have compiled an impressive
record of empirical successes in both laboratory and field
tests.  For these researchers, it is hoped that the results
will encourage more field tests and applications so that
limitations as well as empirical regularities can be documented.  The potential appears obvious; but more work is
needed.  For researchers in travel behaviour, the results
suggest that both psychological and physical or resource
measures can be used to model choice behaviour.  The physical
values, however, must be transformed in an appropriate manner
given by the empirical relationships which one finds when
estimating Equation (1).  This chapter and the empirical
results provide no evidence as to whether 'attitudes' have
any role to play in developing models of travel behaviour.
However the results suggest that psychological modelling of
the type reported deserves serious consideration by travel-
demand modellers as a means of direct utility estimation and
modelling.

In particular, the results of the town-choice study
suggest that causal models which predict well can be
developed without recourse to revealed behaviour data from
the system itself.  That is, it may be that the procedures
outlined in this chapter can assist in the development of
models which will have the potential to predict anywhere;
i.e. models that are independent of the spatial structure
from which they are estimated and have transferable parameters.  Unlike the combinations in the experiments, which
represent choice alternatives that can be anywhere, when one

estimates a transport model to a particular city, one only
observes what behaviour has occurred over the range of
alternatives presently available.  There is no information
about what behavioural changes would occur with different
sets of alternatives - hence, one must re-estimate.  The
models in this chapter describe choices over a wide range of
alternatives available in a variety of places.  Considerably
more research, however, is necessary to establish this as a
regularity.

If the results are replicable, these models have a number
of important advantages which are deserving of attention;
(1) they permit one to control the effects of the real world
and create a *ceteris paribus* situation in which the decision
process can be analytically examined; (2) the experiments can
be designed such that the collinearity of the real world can
be removed and a greater variance and range of values of the
independent variables can be examined; (3) the form of the
utility function can be directly identified and specified.
One can uncover linear and non-linear combination rules and
effects and specify them; (4) one can transform experimental
data into real-world estimates and reanalyse the data in the
context of the experimental design.  It is then possible to
'see' the real-world process operating just as one might in a
controlled experiment.  It is hoped that the results of the
studies will arouse sufficient interest in their potential to
encourage more applications in urban transport contexts.

APPENDIX A: MATHEMATICAL CONSIDERATIONS IN THE ANALYSIS AND
MODELLING OF TRAVEL-DECISION PROCESSES

AX 1.  For any observed travel behaviour there exists a set
of independent factors which are lawfully connected to its
occurrence or the magnitude of its occurrence.  Each factor
may be either quantitative or qualitative in nature.  We
shall denote the set of k quantitative factors by
$\{S_1, S_2, \ldots, S_k\}$ and the set of $\ell$ qualitative factors by
$\{Q_1, Q_2, \ldots, Q_\ell ;\} k + \} = n.$

AX 2.  Associated with each quantitative and qualitative
factor is a corresponding 'estimate' of its magnitude which
may be obtained by one of several psychological measurement
procedures.  We shall let the 'estimate' of this quantity
provided by one or a group of subjects be $\{\Psi_1, \Psi_2, \ldots, \Psi_n\}$.
Because there may be i different values of corresponding
'estimates' for each of the n factors, we may represent the
estimates as $\Psi_{in}$.  Formally, we postulate that

$$\Psi_k = f_k (S_k) \tag{1}$$

$$\Psi_\ell = f_\ell (Q_\ell).$$

AX 3.  In an experimental context we observe a response to a
combination of $\{S_1, S_2, \ldots, S , Q_1, \ldots, Q_\ell\}$ on a psychological
measurement scale.  We assume that this response measure is
lawfully connected to the 'estimate' of the manipulated

experimental factors according to some algebraic combination rule.  If we agree to let R represent the response to the i sets of manipulations, each containing n different values of S and Q, we can write:

$$R = g\ (\Psi_1,\ \Psi_2,\ldots,\Psi_k,\ \Psi_{k+1},\ldots,\Psi_n).\tag{2}$$

Ax 4.  The response (R) is lawfully connected to the observed travel behaviour by means of some algebraic function.  Hence, if we agree to call the observed behaviour B and assert that it is related to the observed response, then we can write:

$$B = h\ (R).\tag{3}$$

Then by substitution:

Th.1.   $\begin{aligned} B &= h(R) \\ &= h(g(\Psi)) \\ &= h(g(f(S,Q))). \end{aligned}$   (4)

This is too general a formulation for modelling purposes. Hence, we proceed to make explicit assumptions about f, g and h and deduce their consequences.  The results lead to a general paradigm for the analysis of travel behaviour which has growing empirical support (45,46).

AX 1.A.   $\Psi_i = a_i + b_i\ S_1^{c_i}$   i = 1, 2,...,k.   (5)

This axiom expresses the 'estimates' as a power function of the physical, experimental factors.  An intercept term is included to allow for an arbitrary zero in the 'estimate' scale, should one be required.  (See Anderson, (4) for empirical support of this assumption.)

AX 1.B.   $\Psi_j = a_j + b_j\ Q_j$   j = 1, 2,...,$\ell$.   (6)

Here we assume that qualitative variable estimates are linearly related to their counterparts.  Because we can only measure such factors in units of $\Psi_j$ to begin with, it seems reasonable to assume that they are linearly related.

AX 2.A.   $R = a + b\ (\Psi_1 + \Psi_2 + \ldots + \Psi_n).$   (7)

This simply postulates an additive or linear combination rule for R as a function of $\Psi_i$.  (i = 1, 2,...,n.)  There is ample support for this formulation in Anderson's work (1-7).

    Alternative functional forms are possible; however, we restrict our attention to the class of functions which permit intercepts or non-zero values for y when the value of x or the independent variable is zero.  Hence, exponential functions of the form $y = ae^{bx}$, or log functions of the form $y = a + b \log X$ are permitted.  This chapter provides empirical examples of their application, although many other forms are possible.

AX 2.B.   $R = a' + b' \ (\Psi_1 \cdot \Psi_2, \ldots, \Psi_n)$. (8)

This is simply the multiplicative version of AX 2.A.   Support for it may be found in Anderson (1-7).

AX 3.A.   $B = a'' + b'' \ R^{c'}$. (9)

This postulates a relationship between behaviour and response of the same form as AX 1.A.   There is support for this in the work of Louviere, Piccolo, Meyer and Duston (46), although other monotonic transformations are possible.

Then by substituting the right side of Equations (5), (6) and (7) into (9):

Th.1.A.   $B = a'' + b'' \{a' + b' \ (\Sigma_i (a_i + b_i \ S_i^{c_i})$

$$+ \sum_j (a_j + b_j \ Q_j))\}^{c'}. \tag{10}$$

And by substituting the right side of Equations (5), (6) and (8) into (9):

Th.1.B.   $B = a'' + b'' \{a' + b' \ (\Pi_i (a_i + b_i \ S_i^{c_i})$

$$\cdot \Pi_j (a_j + b_j \ Q_j))\}^{c'}. \tag{11}$$

Hence, Theorems 1.A. and 1.B. express expected relationships between quantitative/qualitative variables and behaviour.   If there are no qualitative variables, as in gravity-model studies, the system reduces to that portion including only physically measured variables.   If the estimated relationships go through a real zero, all of the intercept terms drop out and the equations simplify to:

Th.2.A.   $B = k_1 \ \Sigma_i b_i \ S_i^{c_i} + d_2 \ \Sigma_j \ b_j \ Q_j$ (12)

Th.2.B.   $B_{in} = k' \ (\Pi_i \ S_i^{c_i} \cdot \Pi_j \ Q_j)$. (13)

Thus, the estimation problem is made simple by the establishment of ratio-scaled relationships which intercept at the origin.   Although such ratio scales are desirable, in practice our measures of 'estimates' are usually of interval quality and Ths 1.A. and 1.B. result.   These theorems constitute direct—demand or utility models for travel behaviour. They express relationships between dependent and independent variables in a fixed way.   There is no concept called 'weight' or 'importance'.   There is merely the empirical law.   The analyses reported in the body of the chapter are based upon these algebraic derivations.   Similar derivations may be found in Keeney and Raiffa (42).

REFERENCES

1. Anderson, N.H., 'Functional measurement and psycho-physical judgment', *Psychological Review*, 77, 153-70 (1970).
2. Anderson, N.H., 'Cross-task validation of functional measurement', *Perception and Psychophysics* 12, pp. 389-95 (1972).
3. Anderson, N.H., 'Partial analysis of high-way factorial designs', *Behaviour Research Methods and Instrumentation*, 1, pp. 2-7 (1968).
4. Anderson, N.H., 'Information integration theory: a brief survey', in *Contemporary Developments in Mathematical Psychology*, D.H.Krantz, R.C. Atkinson, R.D. Luce and P.Suppes (eds), Vol. 2, W.H. Freeman, San Francisco (1974).
5. Anderson, N.H., 'Algebraic models in perception', in *Handbook of Perception*, E.C. Carterette and M.P. Friedman (eds), Vol. 2, Academic Press, New York (1974).
6. Anderson, N.H., 'Cognitive algebra', in *Advances in Experimental Social Psychology*, L. Berkowitz (ed.), Vol. 7, Academic Press, New York (1974).
7. Anderson, N.H., 'Social perception and cognition', *Center for Human Information Processing Technical Report CHIP 62*, Center for Human Information Processing, University of California at San Diego, La Jotta (1976).
8. Anderson, N.H., 'How functional measurement can yield validated interval scales of mental qualities', *Journal of Applied Psychology* (1976).
9. Beier, J.F., 'The attitudes of drivers towards mass-transit', US Department of Commerce, National Technical Information Service (1971).
10. Berry, B.J.L. (1967) *Geography of Market Centers and Retail Distribution*, Prentice Hall, Englewood Cliffs.
11. Burnett, K.P., 'The dimensions of alternatives in spatial choice processes', *Geographical Analysis*, 5, pp. 181-204 (1973).
12. Burnett, K.P., 'Disaggregate behavioural models of travel decisions other than mode choice: a review and con-tribution to spatial choice theory', *Transportation Research Board, Special Report 149* (1974).
13. Burnett, K.P., 'The dimensions of alternatives in spatial choice processes: a reply', *Geographical Analysis*, 7, pp. 327-34 (1975).
14. Cadwallader, M., 'A behavioural model of consumer spatial decision-making', *Economic Geography*, 51, pp. 339-49 (1975).
15. Costantino, D.P., Dobson, R. and Canty, E.T., 'An investigation of model choice for dual mode transit, people mover, and personal rapid transit systems', paper presented to the International Conference on Dual Mode Transportation, Washington, D.C. (1974).
16. Costantino, D.P., Golob, T.F. and Stopher, P.R., 'Consumer preferences for automated public transportation systems', *Transportation Research Record 527* (1974).
17. Davidson, J.D., 'Forecasting traffic on STOL', *Operational Research Quarterly*, 24, 4, pp. 561-9 (1975).
18. Dawes, R., 'Graduate admissions variables and future success', *Science*, pp. 721-3 (1975).
19. Demetsky, M.J. and Hoel, L.A., 'Modal demand: a user perception model', *Transportation Research 6*, pp. 293-308 (1972).

20. Dickey, J.W. *et al.* (1975) *Metropolitan Transportation Planning*, McGraw-Hill, New York.

21. Dobson, R. and Kehoe, J., 'Disaggregated behavioural views of transportation attributes', *Transportation Research Record 627* (1974).

22. Dobson, R. and Tischer, M.L., 'Beliefs about buses, carpools, and single occupant autos: a market segmentation approach', *Proceedings: Transportation Research Forum*, XVII, 1 (1976).

23. Donnelly, E.P., 'Formulation and evaluation of alternative state transit operating assistance programs: a quantitative preference technique', *Preliminary Research Report No. 90*, New York State DOT (1975).

24. Donnelly, E.P., Howe, S.M. and DesChamps, G.A., 'Trade-off analysis: theory and applications to transportation policy planning', *Preliminary Research Report No. 103*, New York State DOT (1976).

25. Fishbein, M., 'Attitudes and the prediction of behaviour', in *Readings in Attitude Theory and Measurement*, M. Fishbein (ed.), Wiley, New York (1967).

26. Fishbein, M. and Ajzen, I., 'Attitudes and opinions', *Annual Review of Psychology*, 23, pp. 487-544 (1972).

27. Fishbein, M. (1975) *Belief, Attitude, Intention and Behaviour*, Addison-Wesley, Reading, Massachusetts.

28. Golledge, R.G., Revizzigno, V.L. and Spector, A., 'Learning about a city: analysis by multidimensional scaling', in *Spatial Choice and Spatial Behaviour*, R.G. Golledge and G. Rushton (eds), Ohio State University Press, Columbus (1976).

29. Golob, T.F., Canty, E.T., Gustafson, R.L. and Vitt, J.E., 'An analysis of consumer preferences for a public transportation system', *Transportation Research*, 6, pp. 81-102 (1972).

30. Golob, T.F., 'Attitudinal models', Urban Travel Demand Forecasting, *Transportation Research Board Special Report 143* (1973).

31. Harman, E.J. and Betak, J.F., 'Behavioural geography, multidimensional scaling, and the mind', in *Spatial Choice and Spatial Behaviour*, R.G. Golledge and G. Rushton (eds), Ohio State University Press, Columbus (1976).

32. Hartgen, D.T. and Tanner, G.H., 'Individual attitudes and family activities: a behavioural model of traveller mode choice', *High Speed Ground Transportation Journal*, 4, pp. 439-67 (1970).

33. Helson, H. (1964) *Adaptation Level Theory*, Harper and Row, New York.

34. Hensher, D.A., 'Market segmentation as a basis of allowing for variability in traveller behaviour', *Transportation*, 5 (1976).

35. Hensher, D.A., 'Use and application of market segmentation', in *Behavioural Travel-Demand Models*, P.R. Stopher and A.H. Meyburg (eds), Lexington Books, Lexington (1976).

36. Hensher, D.A., 'Perception and commuter modal choice - an hypothesis', *Urban Studies*, 12, pp. 101-4 (1975).

37. Hensher, D.A., McLeod, P.B. and Stanley, J.K., 'Usefulness of attitudinal measures in determining the influences on mode choice', *International Journal of Transport Economics*, 2 (1975).

38. Hensher, D.A. and McLeod, P.B., 'Towards an integrated approach to the identification and evaluation of the transport determinants of travel choices', *Transportation Research*, 10, 2 (1977).

39. Horton, F., Reynolds, D. and Louviere, J., 'Mass transit utilization: individual response data inputs', *Economic Geography*, 49, pp. 122-33 (1973).

40. Huber, G.P., 'Multi-attribute utility models: a review of field and field-like studies', *Management Science*, 20, pp. 1393-402 (1974).

41. Kelley, E.L. and Fiske, D.W. (1951) *The Prediction of Performance in Clinical Psychology*, University of Michigan Press, Ann Arbor.

42. Keeney, R.L. and Raiffa, H. (1976) *Decisions with Multiple Objectives: Preferences and Value Tradeoffs*, John Wiley, New York.

43. Krantz, D.H., Luce, R.D., Suppes, P. and Tversky, A. (1971) *Foundations of Measurement*, Academic Press, New York.

44. Krantz, D.H. and Tversky, A., 'Conjoint measurement analysis of composition rules in psychology', *Psychological Review*, 78, pp. 151-69 (1971).

45. Levin, I.P. and Gray, M.J., 'Analysis of human judgment in transportation', paper presented to the Special Sessions on Human Judgement and Spatial Behaviour. Great Plains/Rocky Mountains AAG Meetings, Manhattan, Kansas (1976).

46. Levin, I.P., 'The development of attitudinal modelling approaches in transportation research', (Chapter 35 of present volume).

47. Louviere, J.J., Beavers, L.L., Norman, K.L. and Stetzer, F., 'Theory, methodology and findings in mode choice behaviour', *Working Paper No. 11*, The Institute of Urban and Regional Research of the University of Iowa (1973).

48. Louviere, J.J., Meyer, R., Stetzer, F. and Beavers, L.L., 'An experiment to derive predictive models of public response to policy manipulations in public bus transportation', *Technical Report No. 35*, The Institute of Urban and Regional Research of the University of Iowa (1974).

49. Louviere, J.J., 'The dimensions of alternatives in spatial choice processes: a comment', *Geographical Analysis*, 7, pp. 315-27 (1975).

50. Louviere, J.J., Ostresh, L.M., Henley, D. and Meyer, R., 'Travel demand segmentation: some theoretical considerations related to behavioural modelling', in *Behavioural Travel Demand Models*, P.R. Stopher and A.H. Meyburg (eds), Lexington Books, Lexington (1976).

51. Louviere, J.J. and Norman, K.L., 'Applications of information processing theory to the analysis of urban travel demand', *Environment and Behaviour* (1977).

52. Louviere, J.J., 'Psychological measurement of travel attributes', in *Determinants of Travel Choice*, D.A. Hensher and M.Q. Dalvi (eds), Teakfield, Farnborough, England (1978).

53. Louviere, J.J., Piccolo, J.M., Meyer, R.J. and Duston, W., 'Theory and empirical evidence in real-world studies of human judgment: three shopping behaviour examples', *Center for Behavioural Studies, Research Paper No. 1*, Institute for Policy Research, University of Wyoming (1977).

54. MacKay, D.B., Olshavsky, R.W. and Sentell, G., 'Cognitive maps and spatial behaviour of consumers', *Geographical Analysis*, 7, pp. 19-34 (1975).

55. Margolin, J.B., Misch, M.R. and Dobson, R., 'Incentives and disincentives to ridesharing behaviour: a progress report', paper presented to the Transportation Research Board Meetings, Washington, D.C. (1976).

56. McDermott, P., 'Retail leakages from Laramie, Wyoming',

paper presented to the Laramie Area Chamber of Commerce (1976).
57. McLynn, J.M., 'The simulation of travel choice behaviour without the independence of irrelevant alternatives hypothesis', paper presented at the Summer Simulation Conference, Washington, D.C. (1976).
58. Meyer, R., 'An experimental analysis of student apartment selection decisions under uncertainty', paper presented to the Special Session on Human Judgment and Spatial Behaviour. Great Plains/Rocky Mountains AAG Meetings, Manhattan, Kansas (1976).
59. Michaels, R.M., 'Behavioural measurement: an approach to predicting travel demand', *Transportation Research Board, Special Report No. 149*, pp. 51-7 (1974).
60. Nicolaidis, G.C., 'Quantification of the comfort variable', *Transportation Research*, 9, pp. 55-66 (1975).
61. Nicolaidis, G.C., Wachs, M. and Golob, T.F., 'Evaluation of market segmentations for transportation planning', paper presented to the Transportation Research Board Meeting, Washington, D.C. (1977).
62. Norman, K.L. and Louviere, J.J., 'Integration of attributes in public bus transportation: two modelling approaches', *The Journal of Applied Psychology*, 59, pp. 753-8 (1974).
63. Norman, K.L., 'Attributes in bus transportation: importance depends on trip purpose', *Journal of Applied Psychology*, 62, 2, pp. 164-70 (1977).
64. Old West Commission (1975) *Construction Worker Profile*, Billings, Montana.
65. Olsen, W.T. and Smith, S., 'Variations in psychological responses to characteristics of bus service', *Transportation Research Record*, 513 (1974).
66. Ostresh, L.M. and Louviere, J.J., 'A simple model for multiple stimulus integration and some support in geography and psychology', paper presented to the Special Session on Human Judgment and Spatial Behaviour. Great Plains/Rocky Mountains AAG Meetings, Manhattan, Kansas (1976).
67. Piccolo, J.M. and Louviere, J.J., 'Information integration theory applied to real-world choice behaviour: validational experiments involving shopping and residential choice', paper presented to the Special Session on Human Judgment and Spatial Behaviour. Great Plains/Rocky Mountains AAG Meetings, Manhattan, Kansas (1976).
68. Quarmby, D.A., 'Choice of travel mode for the journey to work: some findings', *Journal of Transport Economics and Policy*, 1, pp. 1-42 (1967).
69. Rohrbaugh, J., 'Cognitive maps: describing the policy ecology of a community', paper presented to the Special Sessions on Human Judgment and Spatial Behaviour. Great Plains/Rocky Mountains AAG Meetings, Manhattan, Kansas (1976).
70. Rushton, G., 'Analysis of spatial behaviour by revealed space preferences', *Annals of the Association of American Geographers*, 59, pp. 31-40 (1969).
71. Rushton, G., 'Preference and choice in different environments', *Proceedings of the Association of American Geographers*, 3, pp. 146-9 (1971).
72. Rushton, G., 'Decomposition of space preference functions', in *Spatial Choice and Spatial Behaviour*, R.G. Golledge and G. Rushton (eds), Ohio State University Press (1976).
73. Sherret, A. and Wallace, J.P., 'Product Attributes',

*Highway Research Board, Special Report 143* (1972).
74. Slovic, P., Fischoff, B. and Lichtenstein, S., 'Behavioural decision theory', *Annual Review of Psychology* (1977).
75. Spear, B.D., 'Attitudinal modelling: its role in travel demand forecasting', *Behavioral Travel-Demand Models*, P.R. Stopher and A.H. Meyburg (eds), Lexington Books, Lexington (1976).
76. Stearns, M.D. (1975) *Proceedings: TSC Workshop on Attitudinal Surveys for Transportation Planning and Evaluation*, US Department of Transportation, UMTA, Office of Transit Planning.
77. Stewart, T.R. and Gelberd, L., 'Capturing judgment policies: a new approach for citizen participation in planning', *Proceedings*, URISA Conference, San Francisco (1972).
78. Stopher, P.R., 'On the application of psychological measurement techniques to travel demand estimation', paper presented to the American Psychological Association Meeting, New Orleans, Louisiana (1974).
79. Stopher, P.R., Spear, B.D. and Sucher, P.D., 'Toward the development of measures of convenience for travel modes', *Transportation Research Record*, 527 (1974).
80. Stopher, P.R. and Watson, P.L., 'Destination choice modelling: an application of psychometric techniques', paper presented to the Annual Meeting of the American Psychological Association, Chicago, Illinois (1975).
81. Stopher, P.R. and Meyburg, A.H. (1976) *Urban Transportation Modelling and Planning*, Lexington Books, D.C. Heath and Co., Lexington.
82. Tversky, A., 'Elimination by aspects: a theory of choices', *Psychological Review*, 79, pp. 281-99 (1972).
83. Westin, R.B. and Watson, P.L., 'Reported and revealed preferences as determinants of mode choice behaviour', *Journal of Marketing Research*, 12, pp. 282-9 (1975).
84. Wicker, A.W., 'Attitudes versus actions: the relationship of verbal and overt behavioural responses to attitude objects', *Journal of Social Issues*, 25, pp. 41-78 (1969).
85. Wilkie, W.L. and Pessemier, E.A., 'Issues in marketing's use of multi-attribute attitude models', *Journal of Marketing Research*, 10, pp. 428-91 (1973).

Chapter 34

ATTITUDE-BEHAVIOUR RELATIONSHIPS IN
TRAVEL-DEMAND MODELLING

Thomas F.Golob, Abraham D.Horowitz and
Martin Wachs

SUMMARY

A conceptual dynamic model of the relation between attitudes
and travel choice (behaviour) can be developed which distinguishes between antecedent variables (characteristics of
choice alternatives and of the consumer) and process
variables (attitudes and perceived availability constraints).
While antecedent variables determine the process variables,
the latter are in a complex mutual-adjustment mechanism with
behaviour. That is, attitudes determine and are determined
by behaviour as a function of availability constraints.

The effect of choice upon attitudes can be studied within
the framework of the psychological theory of cognitive
dissonance. The theory asserts that after an individual
makes a choice between alternatives, he will align his stated
attitudes to his choice, upgrading the satisfaction with both
the positive and negative attributes of the chosen alternative and downgrading those of the rejected alternatives.
The theory further asserts that the magnitude of these
distortions (cognitive dissonance reductions) are a function
of the importance of the attributes in the decision-making
process.

Specific hypotheses covering the relation between travel-
mode choice and attitudes towards the modes can be derived
from cognitive dissonance theory. The first two hypotheses
refer to individuals for whom there are no constraints on
availabilities of the modal alternatives; the hypotheses
predict certain relations between attitudes, as a function of
choices actually made. A third hypothesis relates attitudes
to choice through the effect of perceived availability constraints. Employing data from Ottawa, Canada, the hypotheses
have been confirmed.

The bi-directional attitude-choice causality should be
used in providing transport—planning information. Attitudes,
though affected by past behaviour, appear to be indicators
of potential behavioural change when the availability and/or
characteristics of choice alternatives change. Attitudes
are therefore critical to development of innovative transport
concepts and effective communications to promote advantages
of certain transport alternatives.

739

INTRODUCTION

Urban transport planning models generally predict trip interchanges, modal shares and network flows on the basis of
relationships between supply and demand which are estimated
using information from origin-destination surveys.  It is
typically assumed in developing these models that traveller
choices can be predicted on the basis of 'manifest preferences'.  Manifest preferences are those which travellers
reveal through choices which they actually make when
confronted with real alternatives.  Cross-sectional information is used to establish relationships between socio-economic
status and choices of particular options.

Beginning about a decade ago, a number of authors have
argued that the study of travellers' attitudes, or 'conceived
preferences', might provide planners with better information
about the values or priorities of travellers.  Attitudes are
defined by psychologists as 'predispositions to respond in a
particular way toward a specified class of objects' (36), or
'the sum total of man's inclinations and feelings, prejudice
or bias, preconceived notions, ideas, fears, threats, and
concerns about any specific topic' (41).

Analyses of early attitude surveys, performed using
multivariate statistical techniques new to transport planning,
demonstrated that attitudes toward travel modes (29,8,32)
and route choices (42) revealed interpretable structures and
could be systematically related to the socio-economic characteristics of travellers and to their travel experiences.
Since then, researchers have applied a variety of techniques
to the analysis of attitudinal data dealing with travel
demand; reviews of these works are provided by Hartgen and
Tanner (18), Golob (15), Golob and Dobson (16), Johnson (25),
McFadgen (30), and Stearns (38).  Much of this research has
been conducted without application to the specific problems
of developing travel-demand models.  Demand modellers have
basically adhered to the belief that only manifest preference
- observations of actual behaviour - are useful in the
estimation of predictive models.  Recently, a few researchers
have begun to incorporate a variety of attitudinal variables
into models of modal choice, and have found encouraging
goodness-of-fit (1,19,21,23,35,43,44).  However, almost all
of the travel-demand models described in the literature,
both manifest-preference and conceived-preference models,
have been based upon correlations between variables at a
single point in time for each decision-maker in a sample;
they are not concerned with the decision processes which led
to the observed state.  It is the contention of the present
authors that studies of choice processes are required before
the potential benefits of attitudinal variables to travel-
demand analysis can be realised.

In the next section of this chapter a conceptual model of
travel decision-making behaviour is described.  The description begins with an overview, then traces linkages leading
from antecedent variables to process variables and from
process variables to choice.  Feedback from choice to
process variables is investigated.

One theory incorporated in the conceptual model - cognitive dissonance - is relatively new to travel-demand modelling. Specific travel-demand hypotheses are developed from this theory and results from empirical tests of these hypotheses are presented. Implications for travel-demand modelling are discussed in the final section of this chapter. Presented together with these implications are recommendations for future research directions.

A CONCEPTUAL MODEL OF TRAVEL DECISION-MAKING

*Overview*

The conceptual model is outlined in Figure 34.1. Variables explaining decision-making behaviour are divided in this model into two types: *antecedent variables* and *process variables*. Antecedent variables are those which are not under the daily control of the individual decision-maker. These variables are only affected indirectly by the cumulative results of choices made by many individual decision-makers or are the long-run consequences of such major decisions as choice of residence. Process variables are influenced by antecedent variables and both influence and are influenced by choice. Antecedent variables are usefully divided into two types: *characteristics of the individual decision-maker* and *characteristics of the choice alternatives*. Decision-maker characteristics typically include sex, ethnicity, stage in the family life cycle, social class and social role functions. Life cycle is an interactive combination of marital status, age, number of children, age distribution of children and whether or not children are living at home. Social class is frequently measured in terms of occupation and income. Social role functions can be classified for transport planning purposes (6)into four complexes: work-career, interpersonal-social, household-family and leisure recreation.

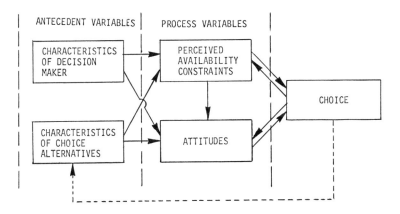

Figure 34.1  A Conceptual Model of Travel Decision-making

Traveller responses to the characteristics of the choice
alternatives are measured along interrelated dimensions called
attributes.  While a number of travel-demand studies have
focused on reductions in the inter-correlations among
attribute measures included in choice models (e.g.,(17))
there has been little work concerning how attribute lists are
generated in the first place.  Wilkie and Pessemier (45)
cite a number of market research attitude studies which
emply group interviews for such attribute-generation purposes.

Process variables also are usefully divided into two
types: *perceived availability constraints* and *attitudes*.
Availability constraints cover all those variables which a
decision-maker perceives to be blocking his or her realisation of choice.  In a mode-choice situation such constraints
might be time to access transit (over and above some acceptable threshold) and availability of a car.  In a car-
ownership choice situation the constraints might include a
threshold level of income.  The perceived availability
constraints of the present conceptual model are referred to
as 'psychological situational variables' or 'psychological
environmental variables' in the general field of consumer
research.  Together with so-called objective situational
variables classified here within the two types of antecedent
variables, these variables are receiving recent attention by
consumer researchers in their attempts to improve understanding of consumer behaviour (2,37).

It is useful for transport planning purposes to divide
attitudes into two main classes: *attitudes toward an object*
and *attitudes toward an act*.  Attitudes toward an object can
be further divided into two subclasses: *beliefs* (extents to
which an object such as a travel mode is perceived to possess
a particular attribute, often called 'cognitions' or
'perceived instrumentalities') and *evaluations* (extents to
which an object satisfies an individual on a particular
attribute, or the overall like or dislike of an object,
often called 'affect').  Attitudes toward an act can also be
divided into two subclasses: *importances* (saliences of
particular attributes in making a choice) and *behavioural
intention* (an individual's intention to make a particular
choice prior to any overt act of choosing or not choosing the
alternative).  The relative effectiveness of each of these
classes and subclasses of attitudes as precursors of
behaviour is discussed briefly in the last section of this
chapter.

*Relations of Antecedent to Process Variables*

In general, both types of antecedent variables affect both
types of process variables (Figure 34.1).  The links between
antecedent variables and attitudes are the strongest.  Social
psychologists have argued convincingly that attitude
structures are a function of an individual's ethnicity, sex,
social status and stage in the life cycle.  Moreover, there
is a strong link between attitudes and social-role functions.
While there is causality in this latter link in both
directions, stability of decision-makers' role commitments
justifies stratifying the cause and effect relationships:
the link from social-role function to attitudes is direct
(a solid arrow in Figure 34.1), while the linkage from
attitudes to social-role function is assumed to be a long term
or indirect effect.

Next, there is the link from characteristics of choice
alternatives to attitudes. Most transport policies, except
those limited only to promotional and advertising strategies,
affect characteristics of choice alternatives; they do not
affect attitudes directly. For example, transit planners
might vary bus headways or fare levels, which would be
expected to affect potential bus users' beliefs and satisfactions with the bus alternative on these attributes. Thus,
the link from characteristics of choice alternatives to
attitudes is key to the usefulness of attitudes in transport
planning. Unfortunately, there are no well developed
theories covering these relationships, but some supporting
empirical evidence is available (22,27,43).

Perceived-availability constraints, like attitudes, are in
general affected by characteristics of both choice alternatives and the decision-maker. The effect attributed to
choice alternatives is the greater of the two. Determination
of an unambiguous mapping from system characteristics to
choice constraints is slightly less critical to the usefulness of attitudes in transport planning than the previously
discussed mapping of system characteristics to attitudes. It
is critical because of the importance attributed to choice
constraints in the conceptual model; they mediate between
attitudes and behaviour, and in some cases effectively block
any realisation of preferences into actual choice. On
balance, however, they are less critical than the link to
attitudes because some transport policies or exogenous
changes in travel conditions will not substantially affect
choice constraints but will have some effect on attitudes.

Finally, the link from characteristics of decision-makers
to perceived availability constraints accounts for idiosyn-
crasies in perceptions. Individuals of different ethnicity,
sex, social status, stage in life cycle, or with different
social-rank structures, can be expected to weigh the various
components of choice constraints differently. There is only
scattered empirical evidence for this because most previous
studies have not isolated choice constraints explicitly.
However, many of the social and behavioural science theories
relating decision-maker characteristics to attitudes are
also relevant here. Indeed, the similar subjective nature
of attitudes and perceived-availability constraints is the
definitional determinant of their common 'process variable'
labelling in the conceptual model.

In summarising the considerations of relationships between
antecedent and process variables in developing attitudinal
travel-demand models, the most important relationship is the
one between characteristics of the choice alternatives and
attitudes. The two next most important relationships are
those between characteristics of the decision-maker and
attitudes and characteristics of the choice alternatives and
choice constraints. Finally, but still important in most
choice situations, is the relationship between characteristics
of the decision-maker and choice constraints.

*Interrelations Among Process Variables and Choice*

The conceptual model of Figure 34.1 specifies a complex mutual-adjustment mechanism between attitudes, choice constraints and choice. Attitudes determine behaviour, but only in the presence of choice constraints. Behaviour, in turn, conditions attitudes and choice constraints. Feedback, particularly from choice to attitudes, can be attributed to the effects of experience on perception and to the presence of the psychological phenomenon of cognitive dissonance.

With regard to the link between perceived-availability constraints and choice, behaviour is so dominated in many situations by perceived availability that observed choices reflect these constraints almost to the exclusion of true preferences of the travellers. For example, correlational relationships based upon the finding in an origin-destination study that commuters living in a particular residential community do not travel to work by carpool could be interpreted by planners to mean that these individuals prefer to travel by other modes. However it is possible that there are few trips by carpool because the work locations of the residents are scattered throughout the metropolitan area.

The explicit consideration of availability constraints is important in all choice-process models, regardless of whether or not attitudinal variables are included. However, as will be argued in remaining sections of this chapter, they are particularly important when attitudinal variables are included. Even in correlational models expressing choice directly in terms of antecedent variables, the exclusion of availability constraints usually has an insidious effect. Explanatory power more properly attributed to availability constraints has often been transferred to variables measuring the service levels of alternatives. Since most service variables appear in linear-model structures, availability constraints are then compensatory to other attributes of competing alternatives (that is, decision-makers are supposedly able to trade off availability constraints for levels of other attributes such as comfort or cost). This is judged to be unrealistic from a choice process viewpoint. For example, if a bus stop is located so far from a decision-maker's residence so as to make transit service essentially unavailable in that individual's eyes, no level of bus seating comfort or express priority travel time can compensate for the low availability. Linear-model structures are inconsistent with hierachical or multiplicative effects judged to better represent the roles of choice constraints in decision processes.

The next strongest link is the bi-directional one between attitudes and choice. This bi-directional link complicates analyses aimed at identifying the relative strength of the two causal sequences: attitudes to behaviour and behaviour to attitudes. As noted in this context by Hartgen (19), Lovelock (28), Johnson (24,25) and Tardiff (39), causal inferences cannot be made from correlations alone. It is necessary instead to apply one or more *theories* which will lead to testable hypotheses, and statistical evidence supporting the hypotheses can then be used to reach

conclusions regarding causality. The remainder of this
section explores such a theory and test results are provided
as part of the next section.

*Cognitive-dissonance theory*, as formulated by Festinger (14),
is aimed at explaining dynamic post-behaviour cognitive
processes within individuals. It states that decision-makers
rationalise their behaviour by bringing into consonance relevant cognitions which become dissonant as a result of the
behaviour. For example, a traveller is faced with choice
between two modes, say car and bus. Prior to choosing, he or
she might believe that a number of attributes, say waiting
time and riding time, would favour choice of car. Other
attributes, say out-of-pocket cost and opportunity to relax,
would favour choice of bus. Assume that the traveller
chooses bus. His or her cognitions concerning out-of-pocket
cost and opportunity to relax are then *consonant* with the
decision, while cognitions concerning waiting time and riding
time are *dissonant* with the decision. Awareness of this dissonant state is amplified when the traveller is asked to
complete a questionnaire eliciting his or her attitudes
toward the modes.

Cognitive dissonance has received substantial attention in
the general field of consumer research. Cummings and
Venkatesan (9) provide a comprehensive review and conclude
that with regard to post-purchase attitude change and
tendency to repurchase: 'There is no other single explanation
- other than cognitive-dissonance theory - that can account
for the results of these studies' (p. 306). While the
studies reviewed by Cummings and Venkatesan date back to the
late 1960s, there have been apparently no applications of
cognitive dissonance in travel-demand modelling. The first
recognition of such an application can be attributed to
McLeod ((31) p. 148): 'Theories of cognitive dissonance and
studies of reaction tendency ... will possibly be important
in the study of mode choice.'

Cognitive dissonance is very similar to self-perception
theory (3,4). According to Bem (4), 'Individuals come to
know their own attitudes, emotions, and other internal states
partially by inferring them from observations of their own
overt behaviour and/or the circumstances in which the
behaviour occurs'. Discussion of the distinction between
cognitive dissonance and self-perception is not within the
scope of the present chapter and does not affect recommendations. Moreover, the present focus is consistent with the
distinction proposed by Fazio, *et al.* (13): self-perception
theory applies to early stages of attitude development where
there is a low level of familiarity with choice alternatives,
while cognitive-dissonance theory applies when an individual
is more certain of his or her feelings. The latter situation
is deemed more relevant to the majority of travel choices.

The following notation is helpful in expressing hypotheses
derived through application of cognitive-dissonance principles to travel-demand modelling. In order to simplify the
presentation it is assumed that the travel choice under consideration is one of mode choice, and there are two modes
available, A and B (car and bus, say). Extension to other

travel choice situations and to multiple alternatives is straightforward. Let $A^k$ and $B^k$ denote an individual's evaluation of modes A and B respectively on attribute k. Assume that more favourable evaluations correspond to higher values of $A^k$ and $B^k$. Furthermore, if an individual chooses mode A, denote his or her evaluations of modes A and B as $A_a^k$ and $B_a^k$, respectively. Likewise, if mode B is chosen denote these evaluations as $A_b^k$ and $B_b^k$, respectively.

Hypothesis I:

$A_a^k > A_b^k$ and $B_b^k > B_a^k$ for all k.

That is, an individual will evaluate an alternative more favourably if he or she has chosen that alternative, implicitly rejecting the non-chosen alternative. Figure 34.2 graphically depicts a manifestation of Hypothesis I on a hypothetical data set. Attributes in Figure 34.2 are separated into those for which mode A is commonly evaluated as being superior (say riding time for an autobus example) and those for which mode B is commonly evaluated as being superior (say out-of-pocket cost).

Hypothesis II:

Case I: $A_a^k - A_b^k < B_b^k - B_a^k$ for all k for which A is judged superior to B, and

Case II: $A_a^k - A_b^k > B_b^k - B_a^k$ for all k for which B is judged superior to A.

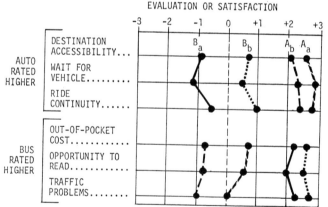

EVALUATION OR SATISFACTION

Figure 34.2  Hypothetical Attitudinal Profile

That is, for a given attribute and mode, dissonance is more pronounced if the chosen mode is *inferior* to the alternative with respect to the attribute than if it is superior. Possible reasons for this are two-fold. First, attributes on which a chosen alternative is inferior to a non-chosen alternative are purported to be more stress-producing than attributes on which the chosen alternative is superior; more dissonance reduction on the former attributes can thus be expected. Second, realities associated with commonly accepted beliefs of the dominance of one mode on a particular attribute (e.g., for Case I, high destination accessibility of the car)set limits on dissonance reduction. However, dissonance reduction is not so bounded by realism when an individual evaluates a mode which is inferior on the same attribute. An example of a Case II attribute is out-of-pocket cost; the bus mode is accepted as being relatively satisfactory but cost of travel by car can be variously interpreted, allowing opportunity for rationalisation leading to dissonance reduction.

The final important links among process variables and choice are those from perceived availability to attitudes (Figure 34.1). These links are both direct and through choice. The direct link simply means that individuals faced with different perceived constraints on availabilities of choice alternatives are likely to possess different attitude structures; availability influences attitudes. Consequently, when analyses aimed at determining simple structure (e.g. factor analyses, multi-dimensional scaling) are applied to attitudinal data, the appropriate samples of individuals are segments which are relatively homogeneous with respect to perceived-availability constraints. In general the attitude structures of segments formed on availability constraints will be substantially different from one another, and thus each can be expected to be different from the structure found for the total sample. In addition, goodness-of-fit of the structure solutions should be better for the segments as opposed to the total sample.

The link from perceived availability to attitudes through choice provides a further opportunity for the application of cognitive-dissonance theory. A mode-choice example is again helpful. Consider two types of travellers: one for whom both car and bus are readily available and a traveller for whom bus alone is available (i.e. a 'captive' bus rider in travel-demand forecasting parlance). According to cognitive-dissonance theory, the first type of traveller would encounter more psychological pressure to realign his or her attitudes after a choice was made. The second type of traveller would find little reason to do the same. This argument leads to the last testable hypothesis.

Let $A_{b,y}^{k}$ denote an individual's evaluation of car on attribute k when that individual chooses bus and both modes are available (y = 'yes' for availability). Let $A_{b,n}^{k}$ denote evaluation of car when bus is chosen and car is not available. Similarly, $B_{b,y}^{k}$ and $B_{b,n}^{k}$ represent evaluations of bus when bus is chosen and car is and is not available, respectively.

Hypothesis III:

$$A^k_{b,y} < A^k_{b,n} \text{ and } B^k_{b,y} > B^k_{b,n} \text{ for all } k.$$

That is, an individual with more than one alternative available will upgrade evaluations of his or her chosen alternative and downgrade evaluations of the non-chosen alternatives more than will an individual who has only one alternative readily available.

In the following section results from empirical testing of the cognitive-dissonance hypotheses are presented and discussed. Implications of various tenets of the conceptual model are presented in the last section.

## TESTS OF COGNITIVE-DISSONANCE HYPOTHESES

Hypotheses I and II are supported by results from reanalyses of attitudinal data collected in San Francisco (28), Los Angeles (11), Chicago (23), and Ottawa, Canada (35). The Los Angeles study elicited evaluations of car, bus and carpool; the Chicago study elicited evaluations of car and carpool only; and the San Francisco and Ottawa studies focused on car and bus. Results are similar for all four data sets, even though different attributes and scaling semantics were used. Consequently, only the results for Ottawa are presented for reasons of brevity. Documentation of the complete results and comparisons on a detailed attribute-by-attribute basis is underway.

As shown in Figure 34.3, Hypothesis I holds for every one of the 25 attributes. Hypothesis II holds for 21 of the 25 attributes. Moreover, the differences in cognitive dissonance levels for the two modes implied by the inequalities of Hypothesis II are significantly different from zero at the 0.001 chance probability confidence level for 8 of these 21 attributes; the differences are significantly different from zero at the 0.05 chance probability confidence level for 12 of the 21 attributes.

Attributes in Figure 34.3 are listed in order of the magnitude of the cognitive dissonance on the mode rated inferior (i.e. in order of $B_b-B_a$ in the top group of attributes and $A_a-A_b$ in the bottom group). The $B_b-B_a$ differences for the first eleven attributes in the first group and the $A_a-A_b$ difference for the first attribute in the second group were significant according to t-tests at the 0.999 confidence level; these attributes are marked with asterisks to the right side of the figure. The first eight of these attributes in the first group and the single attribute with such a significant difference in the second group are precisely those attributes found by Recker and Golob (17) to have significant coefficients in logit models of choice between the car and bus modes for the work trip. Only one attribute not marked by an asterisk was found to have such a significant coefficient. Moreover, the first eight of these attributes make up the 'bus service' factor found by Recker and Golob

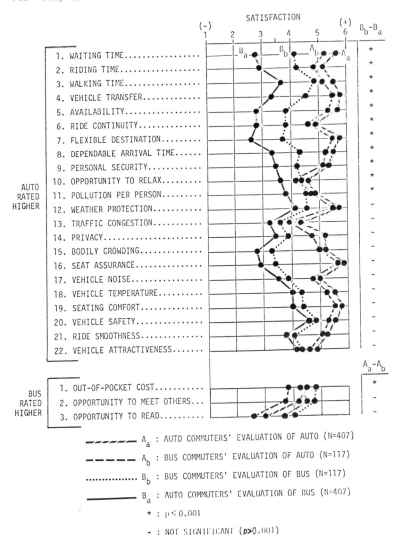

Figure 34.3  Ottawa-Carleton Attitudinal Profile
*Source:* Recker and Golob (35)

(34).  The authors' interpretation is that the magnitude of
the cognitive dissonance determined for a particular attribute is related to the power which that attribute has in
explaining choice.  This is consistent with principles
advanced by Brehm and Cohen (7) and Festinger (14).

Figures 34.4 and 34.5 are relevant to Hypothesis III.  The
data again are from Ottawa, and the development of segments
of travellers with high perceived availabilities of both car
and bus (called the 'mobile segment') and with low perceived
availability of car (called the 'carless segment') is documented in Recker and Golob (35).

Hypothesis III holds for 22 of the 25 relevant attributes
for evaluations of the bus mode (Figure 34.4).  Note that the
largest differences $B_{b,y}-B_{b,n}$ occur mainly for attributes
that yielded large differences of the $B_b-B_a$ type (Figure 34.3).
Hypothesis III holds also for 20 of the 25 relevant attributes for evaluations of the car mode (Figure 34.5).  The
differences for each attribute separately are not significant,
according to t-tests at the 0.001 level.  The confirmation
of Hypothesis III is therefore weaker than that for the first
two hypotheses.  Additional testing of Hypothesis III awaits
analyses of additional data which contains choice-constraint
information.

The inversion of the expected direction of the difference
$A_{b,n}-A_{b,y}$ for 'Availability' (Figure 34.5) is explained in
terms of the realism of the situation: bus users for whom car
is available are more satisfied with respect to 'Car Availability' than those bus users for whom car is not available.

Thus the empirical validation of each of the three
cognitive-dissonance hypotheses has been demonstrated on one
data set.  Additional testing of the hypothesis relating
perceived-availability constraints and attitudinal cognitive
dissonance awaits procurement of other data sets with such
choice-constraint information.

## IMPLICATIONS AND RECOMMENDATIONS

Attitudes are not simply precursors of behaviour.  Rather
attitudes, behaviour and availability constraints are elements in a dynamic adjustment process, mutually influencing
one another and being influenced by characteristics of the
choice alternatives.  They are also conditioned by the
relatively stable characteristics of the decision-maker.  The
implications of this dynamic mutual adjustment process for
travel-demand modelling are of course profound.  This final
section of the chapter outlines a few of these implications
and recommends directions for research aimed at developing
improved modelling methodologies.

The conceptual model states that decision-makers'
attitudes are influenced by characteristics of the choice
alternatives, characteristics of the decision-makers themselves, perceived constraints on availabilities of choice
alternatives, and actual choices.  It is proposed that in

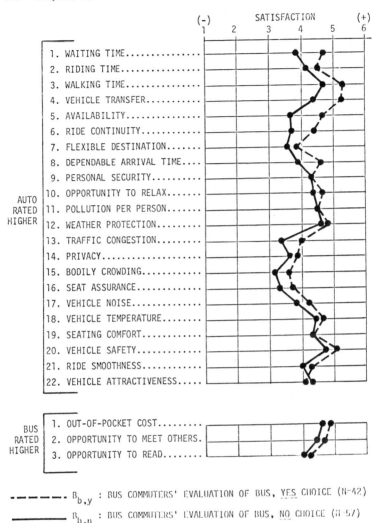

SATISFACTION

(-)  1  2  3  4  5  6  (+)

AUTO RATED HIGHER
1. WAITING TIME...............
2. RIDING TIME...............
3. WALKING TIME..............
4. VEHICLE TRANSFER..........
5. AVAILABILITY..............
6. RIDE CONTINUITY...........
7. FLEXIBLE DESTINATION......
8. DEPENDABLE ARRIVAL TIME....
9. PERSONAL SECURITY.........
10. OPPORTUNITY TO RELAX......
11. POLLUTION PER PERSON......
12. WEATHER PROTECTION........
13. TRAFFIC CONGESTION........
14. PRIVACY...................
15. BODILY CROWDING...........
16. SEAT ASSURANCE............
17. VEHICLE NOISE.............
18. VEHICLE TEMPERATURE.......
19. SEATING COMFORT...........
20. VEHICLE SAFETY............
21. RIDE SMOOTHNESS...........
22. VEHICLE ATTRACTIVENESS.....

BUS RATED HIGHER
1. OUT-OF-POCKET COST.........
2. OPPORTUNITY TO MEET OTHERS.
3. OPPORTUNITY TO READ........

− − − − − − $B_{b,y}$ : BUS COMMUTERS' EVALUATION OF BUS, YES CHOICE (N=42)

─────── $B_{b,n}$ : BUS COMMUTERS' EVALUATION OF BUS, NO CHOICE (N=57)

Figure 34.4  Ottawa-Carleton Bus Commuters' Attitudinal Profile, Mobile vs. Carless Segments
*Source:* Recker and Golob (35)

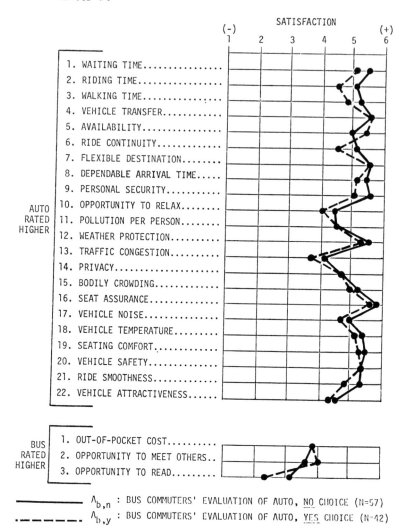

Figure 34.5  Ottawa-Carleton Bus Commuters' Attitudinal Profile, Mobile vs. Carless Segments

*Source:* Recker and Golob (35)

planning studies and forecasting models the influences of
decision-maker characteristics and perceived-availability
constraints be accounted for through the use of market segmentation (10) or similar techniques capable of handling
simultaneous effects of large numbers of interrelated
variables. Equation systems treating such variables as
purely additive effects are not appropriate.

A two-stage market-segmentation analysis is suggested by
results obtained by Nicolaidis, et al. (33) in a comparative
study of market-segmentation data bases for use in transport
planning. Such a two-stage segmentation would first group
individuals into segments which are homogeneous with respect
to sex, ethnicity, stage in the family life cycle, social
class and social-role function. The resulting segments would
then be further segmented on the basis of perceived constraints on availabilities of the choice alternatives being
studied. The final segments would represent samples on which
to formulate and estimate attitude-choice models.

Two problems are immediately evident with the two-stage
approach. First, there is the problem of sufficient sample
sizes. If a particular minimum sample size (e.g. 75 individuals) is needed for attitude-choice model estimation, and if
there were four equal-sized choice-constraint segments and
four equal-sized decision-maker characteristic segments, a
large initial sample size would be required (e.g. 1200
individuals). Unequal distributions among segments would
necessitate even larger initial samples. Another problem is
that hypothetical changes made in characteristics of the
choice alternatives for sensitivity analysis may change
perceived-availability constraints so that some individuals
are reclassified from one choice-constraint segment to
another.

Both of these problems can be overcome if attitude-choice
is studied in a micro-environment in which all subjects have
the same constraints on choice of alternatives and are
similar with respect to some of their more complex characteristics such as social-role functions. This implies that data
collection may appropriately be limited spatially and in
other ways dictated by knowledge gained in a broad study of
temporal factors (e.g. hours of employment) or socio-economic
factors (e.g. income) which are influential in determining
availability constraints. Advantages of micro-environment
studies have been suggested by Johnson (24) and Everett (12).

Assuming that influences of decision-maker characteristics
and perceived-availability constraints can be eliminated in
this manner, attitude-choice can then be studied. The study
design should involve change over time. A dynamic study
design allows the following questions to be addressed: How
do changes in the characteristics of choice alternatives
influence choice through attitudes? How do changes in choice
consequently influence adjustments in attitudes? Through
what process is a dynamic equilibrium approached by continued
readjustment based upon stochastic cues?

Cognitive-dissonance theory, and empirical information of
the sort presented in this chapter, demonstrate that

single-equation static-choice models are inappropriate for such a study. Models developed to describe learning theory and brand switching in marketing could be useful. Also, simultaneous equations could be used to test some specific hypothesis regarding the equilibrium which exists between choices, availability constraints and attitudes. One such structure was proposed by Tardiff (40), but data problems and problems associated with attributing additive influences to choice constraints prevented development of convincing results. It is proposed that influences of availability constraints be accounted for by multiplicative terms in choice models, such as the 'conditional probability' concept suggested by Recker and Golob (34).

Decisions must be made concerning which types of attitudinal variables (beliefs, evaluations, affect, importances or behavioural intentions) should be measured in such a dynamic study. It is proposed that measurement of behavioural intention and affect should be included in every study. It would be extremely useful to establish 'discount factors' which relate behavioural intention to actual choice under various choice situations. This will allow the use of readily obtained behavioural-intention information as a guide to more detailed planning (26). Measurement of affect is useful in studies aimed at determining attributes which are salient in the development of promotional strategies (23) and new products and services.

Comparative analyses need to be performed regarding beliefs, evaluations and importances. It appears that beliefs and evaluations are highly related in individuals' minds, so that solicitation of belief about a particular choice alternative with respect to a particular attribute will elicit a substantial amount of evaluative information, and vice versa. However, a specific relationship needs to be established. The same need for direct comparison concerns the measurement of importances. Wilkie and Pessemier (45) reviewed 21 market-research studies which addressed the question of including both importances and beliefs or evaluations in attitude models and found conflicting evidence. Travel-demand studies using importances as explicit scale-transformation weights on satisfaction levels (21,44) generally have not reported any comparison of results with those obtained using beliefs or evaluations alone.

An important implication of cognitive-dissonance theory for disaggregate travel-demand models using perceived time and cost data is that such subjective measures could be distorted by dissonance. Measures of waiting time, walking time, transfer time, riding time and various cost components are susceptible to such a systematic bias. It is recommended that dissonance hypotheses be tested in studies relating subjective and objective time and cost measures. Indeed, a recent study found systematic biases in reported bus walking time and transfer time and suggested that 'one possible explanation for these discrepancies, which show the same pattern of bias throughout, is that of *ex post rationalisation* ... empirical studies of modal choice based on stated information (on time and cost) ... lead to a significantly biased result' ((20) p. 23).

Finally, it is recommended that the bi-directional
attitude-choice causality which complicates development of
attitudinal demand models be put to use in providing indirect
transport-planning information.  The recommendation is to use
aggregate data (mean values of beliefs, evaluations or
importances) to identify key attributes of choice alternatives where policy considerations should be focused.  This
approach is cost-effective in terms of analysis and takes
advantage of the high reliability of aggregate attitude
measures in comparison to the modest reliability of
disaggregate attitudinal measures (5).

The first step is to organise attributes into separate
profiles, one for each choice alternative.  Each profile
would represent those attributes for which a given alternative is rated superior in terms of mean values.  Attributes
in each profile are then ordered with respect to the level of
cognitive dissonance observed.  The attributes at the top of
each profile are those which are candidate planning subjects.

The cognitive-dissonance interpretation of attitudinal
data may lead erroneously to the conclusion that because
attitudes are a product of behaviour, attitudes are not
predictors of future behaviour.  While attitudes, as opera-
tionally measured in surveys, reflect past and present
behaviour, they indirectly are indicators of potential
behaviour change when the availability and/or characteristics
of choice alternatives change.  Such a dynamic approach to
the question of travel-behaviour modelling requires further
research.  There is no doubt in the authors' minds that
attitude analyses are critical to development of (1) innovative transport concepts and (2) effective communications to
promote advantages of certain transport alternatives.

REFERENCES

1. Allen, W.B. and Isserman, A., 'Behavioural modal split',
High Speed Ground Transportation Journal, 6, pp. 179-99 (1972).
2. Belk, R.W., 'Situational variables and consumer
behavior', Journal of Consumer Research, 2, pp. 157-64 (1975).
3. Bem, D.J., 'Self-perception: an alternative interpretation of cognitive dissonance phenomena', Psychological
Review, 74, pp. 183-200 (1967).
4. Bem, D.J., 'Self-perception theory', in Advances in
Experimental Social Psychology 6, L. Berkowitz (ed.),
Academic Press, New York (1972).
5. Best, R.J., Hawkins, D.I. and Albaum, G., 'Reliability
of measurement beliefs in consumer research', Advances in
Consumer Research, 4, pp. 19-23 (1976).
6. Boston College (1977) New Approaches to Understanding
Travel Behavior: Phase I Summary Report, National Co-operative
Highway Research Program Report, Transportation Research
Board, National Research Council, Washington.
7. Brehm, J.W. and Cohen, A.R. (1962) Explorations in
Cognitive Dissonance, Wiley, New York.
8. Brunner, G.A., et al. (1966) User Determined Attributes
of Ideal Transportation Systems, Department of Business
Administration, University of Maryland.
9. Cummings, W.H. and Venkatesan, M., 'Cognitive dissonance

and consumer behavior: a review of the literature', *Journal of Marketing Research*, 13, pp. 303-8 (1976).

10. Dobson, R., 'Market segmentation: a tool for transport decision making', (Chapter 10 of this book).

11. Dobson, R. and Tischer, M.L., 'Beliefs about buses, carpools, and single occupant autos: a market segmentation approach', *Transportation Research Forum Proceedings*, XVII,1 (1976).

12. Everett, P.B., 'Accommodating or planning travel behaviors', paper presented at 83rd Annual Convention of The American Psychological Association, Chicago (1975).

13. Fazio, R.H., Zanna, M.P. and Cooper, J., 'Dissonance versus self-perception: an integrative view of each theory's proper domain of application', *Journal of Experimental Social Psychology* (1977).

14. Festinger, L. (1957) *A Theory of Cognitive Dissonance*, Stanford University Press, Stanford.

15. Golob, T.F., 'Attitudinal models', *Highway Research Board Special Report No. 143*, Urban Travel Demand Forecasting, (1973).

16. Golob, T.F. and Dobson, R., 'Assessment of preferences and perceptions toward attributes of transportation alternatives', Behavioral Demand Modeling and Valuation of Travel Time, *Transportation Research Board, Special Report 149* (1974).

17. Golob, T.F. and Recker, W.W., 'Mode choice prediction using attitudinal data: a procedure and some results', *Transportation*, 6 (1977).

18. Hartgen, D.T. and Tanner, G., 'Individual attitudes and family activities', *High Speed Ground Transportation Journal*, 4, pp. 439-67 (1970).

19. Hartgen, D.T., 'Attitudinal and situational variables influencing urban mode choice: some empirical findings', *Transportation*, 4, pp. 377-92 (1974).

20. Heggie, I.G., 'A diagnostic survey of urban journey-to-work behaviour', in *Modal Choice and the Value of Travel Time*, I.G. Heggie (ed.), Oxford University Press, Oxford (1976).

21. Hensher, D.A., McLeod, P.B. and Stanley, J.K., 'Usefulness of attitudinal measures in investigating the choice of travel mode', *International Journal of Transport Economics*, 2, 51-75 (1975).

22. Hensher, D.A. and McLeod, P.B., 'Towards an integrated approach to the identification and evaluation of the transport determinants of travel choices', *Transportation Research*, X, 2 (1977).

23. Horowitz, A.D. and Sheth, J.N., 'Ridesharing to work: a psychosocial analysis', paper presented at 56th Annual Meeting of Transportation Research Board, National Research Council, Washington (1977).

24. Johnson, M.A. (1974) 'Travel attitudes and the choice between automobiles and public transportation', unpublished Doctoral Dissertation, University of California, Berkeley.

25. Johnson, M.A., 'Psychological variables and choices between auto and transit travel: a critical research review', *Urban Travel Demand Forecasting Project Working Paper No. 7509*, Institute of Transportation and Traffic Engineering, University of California, Berkeley (1975).

26. Louviere, J.J. and Norman, K.L., 'Applications of information processing theory to the analysis of urban travel demand', *Environment and Behavior* (1977).

27. Louviere, J.J., Piccolo, J.M., Meyer, R. and Duston,W.,

'Theory and empirical evidence in real-world studies of human judgement: three shopping behavior examples', *Center for Behavioral Studies Research Paper No. 1*, University of Wyoming, Laramie (1977).

28. Lovelock, C.H. (1973) 'Consumer oriented approaches to marketing urban transit', unpublished Doctoral Dissertation, Stanford University.

29. Mahoney, Joseph F. (1964) *A Survey to Determine Factors Which Influence the Public's Choice of Mode of Transportation*, Joseph Napolitan Associates, Boston.

30. McFadgen, D.G., 'Transportation mode-choice research: recent contributions from the social sciences', *Urban Travel Demand Forecasting Project Working Paper No. 7502*, Institute of Transportation and Traffic Engineering, University of California, Berkeley (1975).

31. McLeod, P.B., 'The role of subjective data in modelling mode choice', in *Urban Travel Choice and Demand Modelling*, D.A. Hensher (ed.), Special Report No. 12, Australian Road Research Board, Melbourne (1974).

32. McMillan, R.K. and Assael, H. (1968) *National Survey of Transportation Attitudes and Behavior: Phase 1, Summary Report*, Report No. 49, National Co-operative Highway Research Project, Transportation Research Board, National Research Council, Washington.

33. Nicolaidis, G.C., Golob, T.F. and Wachs, M., 'Evaluation of alternative market segments for transportation planning', paper presented at 56th Annual Meeting of Transportation Research Board, Washington.

34. Recker, W.W. and Golob, T.F., 'A behavioral travel demand model incorporating choice constraints', *Advances in Consumer Research*, 3, pp. 416-24 (1975).

35. Recker, W.W. and Golob, T.F., 'An attitudinal modal choice model', *Transportation Research*, 10, pp. 299-310 (1976).

36. Rosenberg, J., 'Cognitive structure and attitudinal affect', *Journal of Abnormal and Social Psychology*, 53, pp. 367-72 (1956).

37. Russell, J.A. and Mehrabian, A., 'Environmental variables in consumer research', *Journal of Consumer Research*, 3, pp. 62-3 (1976).

38. Stearns, M.D. (1975) *Proceedings: TSC Workshop on Attitudinal Surveys for Transportation Planning and Evaluation*, Report No. UMTA-MA-06-0049-75-1, US Department of Transportation, Washington.

39. Tardiff, T.J. (1974) 'The effects of socioeconomic status on transportation attitudes and behavior', Doctoral Dissertation, University of California, Irvine.

40. Tardiff, T.J., 'Causal inferences involving transportation attitudes and behavior', *Department of Civil Engineering Working Paper*, University of California, Davis (1976).

41. Thurstone, L.L., 'Attitudes can be measured', *American Journal of Sociology*, 33, 529-54 (1928).

42. Wachs, M., 'Relationships between drivers' attitudes toward alternate routes and driver and route characteristics', *Highway Research Record No. 197*, pp. 70-87 (1967).

43. Wallace, J.P. and Sherret, A., 'Estimation of product attributes and their importances', *Lecture Notes in Economics and Mathematical Systems No.89*, Springer-Verlag, New York (1973).

44. Westin, R.B. and Watson, P.L., 'Reported and revealed preferences as determinants of mode choice behavior', *Journal of Marketing Research*, 12, pp. 282-9 (1975).

45. Wilkie, W.L. and Pessemier, E.A., 'Issues in marketing's use of multi-attribute attitude models', *Journal of Marketing Research*, 10, pp. 428-41 (1973).

Chapter 35

THE DEVELOPMENT OF ATTITUDINAL MODELLING
APPROACHES IN TRANSPORT RESEARCH

Irwin P. Levin

INTRODUCTION

In recent years a number of transport researchers have come
to view traveller behaviour as a function of subjective
perceptions of system characteristics, satisfaction levels
of attributes for alternative modes, personal value systems
and individualised decision rules (7,8,12,19,20,21,25,28,30,
36,39,42). The search for such decision rules based on
subjective perceptions and evaluations is called *attitudinal
modelling*. Specific research goals generated by the
attitudinal-modelling approach include scaling of subjective
system characteristics such as comfort, safety and convenience (15,28,34,42), identifying dynamic processes in
traveller behaviour (16,32), defining individual-difference
factors in evaluating alternative transport systems (7,25),
and determining the role of interpersonal factors in
traveller behaviour (28). Such research can be contrasted
with earlier transport research in terms of the breadth of
variables studied and the variety of research designs.

Attitudinal modelling of traveller behaviour has evolved
from earlier attitude research by social scientists. The
reason for this is clear if we consider Thomas's (45)
statement that attitude measurement is intended to replace
knowledge of an individual's past history and represents an
attempt to assess probable future behaviour. Transport
researchers are most certainly interested in predicting
future behaviour on the basis of simple and inexpensive
measuring devices which bypass detailed analyses of past
history.

This chapter describes the problems faced by psychologists
in defining and measuring attitudes, and relating attitudes
to behaviour. It is shown that attitudinal modelling of
traveller behaviour has a greater chance of achieving its
goals than would be expected from the clouded history of
research on attitudes and behaviour. A schema is presented
for describing the goals of attitudinal modelling. Several
approaches to the attitudinal modelling of traveller behaviour are discussed, and an experimental study of mode choice
is described which illustrates a relatively new approach to
attitudinal modelling of traveller behaviour. The validity
of this approach is analysed by predicting actual mode split
from a combination of situational constraints and attitudinal
measures derived from experimentally controlled trade-offs.

*Attitudes: Definition, Measurement and*
*the Relationship Between Attitudes and Behaviour*

There are almost as many definitions of *attitude* as there
are researchers working in the area. Here are a few examples.
Thurstone (46) defined attitude as 'the affect for or
against a psychological object'. Allport (1) defined an
attitude as 'a mental and neural state of readiness, organised through experience, exerting a directive or dynamic
influence upon the individual's response to all objects and
situations with which it is related'. Doob (11) defined
attitude as 'an implicit, drive-producing response considered socially significant in the individual's society'.
Rhine (40) likened attitude to 'a concept where one of the
mediators is an evaluative reaction'.

Conceptualisations of attitudes can generally be classified into two categories: those, like Thurstone's, that are
uni-dimensional and stress the affective nature of attitudes,
and those, like Allport's, that are multi-dimensional and
include affective, cognitive and behavioural components. The
attitude measurement itself is usually uni-dimensional - e.g.
a scale of 'degree of favourableness'. Regardless of their
formal conceptualisation of attitude, most attitude researchers imply that attitudes have some directive influence on
behaviour and that, therefore, there should be a direct
relationship between attitudes and behaviour. However,
Fishbein (13) summarised 75 years of attitude research with
the conclusion that 'there is still little, if any, consistent evidence supporting the hypothesis that knowledge of an
individual's attitude toward some object will allow one to
predict the way he will behave with respect to the object'.
More specifically, Wicker (51) examined scores of previous
studies dealing with the attitude-behaviour relationship and
concluded that attitudinal data rarely account for as much as
10 per cent of the variance in overt behavioural measures.

If the attitude-behaviour relationship is so tenuous,
then it would appear that researchers who are interested in
predicting behaviour should avoid wasting their time and
energies in measuring attitudes. However, as we examine the
causes of previous failures to establish an attitude-
behaviour relationship, we may discover that the prognosis
is more favourable for attitudinal modelling of traveller
behaviour.

There appear to be two main causes of the failure of
social scientists to relate attitude towards an object with
behaviour directed at the object. The first is that the
stimulus objects under study have been insufficiently
specific. Attitudes are often measured to a class of objects (e.g., a particular racial group), whereas behaviour
is directed at a specific example of this class (47). The
second is that an attitude may be but one of a number of
determinants of behaviour, and if other situational factors
interact with attitudes to affect behaviour, a given attitude
may not be correlated with a given measure of behaviour (13).

Transport researchers are in a good position to overcome
these problems by obtaining reactions to specific combinations of system attributes (e.g., bus fare = 25¢, transit
time = 30 min., etc.) rather than to broad classifications
(e.g.,'buses'). Furthermore, transport researchers generally
deal with situations in which the major determinants of
behaviour can be specified before predictions are made. Thus,
transport researchers may be in a better position than most
behavioural scientists to delineate and measure the situational factors affecting the behaviour of an individual, and use
these in conjunction with attitude measures to predict
traveller behaviour. These factors might include the availability of transport options and the characteristics of
travellers which define homogeneous population segments.

In addition, the goals of attitude research in transport
are often different from goals in the social sciences.
Planners are interested in the *predictive validity* of
attitudinal models. If predictions with a certain absolute
level of accuracy can be achieved, planners can utilise this
information to achieve more efficient and effective transport
systems. Psychologists, on the other hand, are interested in
the *processes* by which attitudes are formed and utilised.
Thus, a given attitudinal model may be more readily acceptable to a transport researcher - because of its pragmatic
features - than to a psychologist who is more deeply concerned with deciding which of several competing models best
represents the underlying processes of attitude formation and
utilisation. For an example see Anderson's (2) critical
analysis of the algebraic formulation of Fishbein's (13)
attitude-belief model. One point to be made in this paper is
that in the long run transport researchers would do well to
apply a broader set of criteria in evaluating attitudinal
modelling efforts. Attitude measurement should be tied to
specific formulations of how system attributes are evaluated
and combined, and the usefulness of the attitudinal-modelling
approach should be related to the extent to which a given
class of attitudinal measures can explain and predict
behaviour in a variety of contexts. This is illustrated
later.

*Attitudinal Modelling of Traveller Behaviour*

Transport researchers have generally adopted a multi-
dimensional conceptualisation of attitude. For example,
Hartgen (19) describes three distinct components: cognition -
beliefs concerning an object; affect - feelings of like or
dislike toward the object; conation - tendencies to act with
respect to the object. Hartgen and other transport researchers have also developed a multi-dimensional attitude-
measurement system. In some mode-choice studies, respondents
have been asked to rate the importance of an attribute like
cleanliness and rate their level of satisfaction with that
attribute for each mode. The importance rating is said to
represent the cognitive component of attitude and the
satisfaction ratings are said to represent the affective
component.

Attitudinal modelling of traveller behaviour is seen as a way of linking individuals' preferences and perceptions to the planning and evaluation of transport-system alternatives. More broadly conceived, attitudinal modelling can have long-range benefits if it can further the understanding of basic decision processes which underlie traveller behaviour (8,25, 28).

Dobson (8) has suggested a three-phase strategy for developing attitudinal models. The phases can be summarised as follows: attitudinal measurement for predictor variables; derivation of psycho-physical functions for attitudinal predictors; and estimation of a model. While these three phases are usually considered sequentially, a more efficient system would deal with two or more phases concurrently. Figure 35.1 presents a schema for describing attitudinal modelling in which the measurement of attitudes is a component part of, rather than a precursor to, model development and estimation.

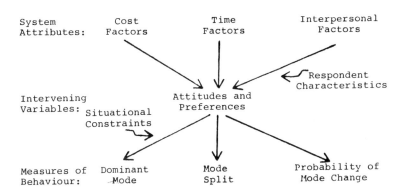

Figure 35.1 Attitude Measurement and the Conceptualisation of Attitudes as Mediators of Behaviour

It can be seen that this schema for describing attitudinal modelling has two sets of linkages: the links between system attributes and implied attitudinal determinants of behaviour, and the links between implied (intervening) attitudinal variables and measures of behaviour. Since the ultimate goal of attitudinal modelling is to study traveller-decision processes, the term *decision-process model* can be used to describe any particular application of the schema presented in Figure 35.1. The example listed is from a mode-choice study described in more detail later.

Both sets of linkages in Figure 35.1 are important in evaluating the usefulness of attitudinal modelling of traveller behaviour. The top set represents attitude measurement. Without proper operational definitions linking

attitudes and preferences to combinations of manipulable
system attributes, these terms have no reliable explanatory
basis.  That is, one cannot 'explain' behaviour with concepts
that themselves need to be explained.  However, the manner in
which system attributes are weighted and combined to
determine attitudes and preferences may depend on respondent
characteristics.  As is illustrated later, empirically
obtained preference functions may be used to define homogeneous subgroups or population segments.

   The bottom set of linkages addresses the validity of the
attitudinal—modelling approach.  It highlights one aspect of
attitudinal modelling that has perhaps not received
sufficient attention.  The usefulness of assuming intervening
processes linking measures of behaviour to their antecedent
conditions is directly related to the extent that a small
number of such processes can be tied to a variety of
behaviours.  For example, a high preference for situations
that are perceived to maintain privacy would lead a person
to avoid various forms of mass transit and ride sharing.
Within this general framework, situational constraints can be
identified and shown to interact with attitudes to affect
specific target behaviours.


*Modelling Approaches*

A number of modelling approaches may be adequate for
traveller-behaviour forecasting based on extrapolations from
the model with minimal changes in system characteristics.
However, the long-range benefits of traveller-behaviour
modelling depend on the ability to predict the effects of
possible substantial changes in system characteristics or
even the establisment of new systems.  Models of greater
flexibility and generality are needed.  The discussion of
attitudinal modelling in the preceding section was meant to
show that the attitudinal-modelling approach has great
potential for meeting these needs.  Affective and cognitive
processes underlie traveller behaviour in a variety of
situations ranging from mode choice to highway driving speed.
The measurement of attitudes reflecting these affective and
cognitive processes can be an important component of
traveller-behaviour predictions and can serve to tie
together a variety of traveller-behaviour patterns.  Traditional modelling procedures included attitude measurement
as a surrogate for a particular system attribute.  New
approaches allow for a more comprehensive analysis of how
attributes combine to determine attitudes towards complete
transport systems.

   As seen in Hartgen's comprehensive literature review (18),
early attempts to include attitudinal factors in predicting
traveller behaviour used survey instruments to measure
attitudes towards the quality of alternative transport modes.
Typically, 7-point scales were used to obtain importance
ratings of specific attributes and ratings of satisfaction
levels for different modes.  These measures were then
incorporated into a regression of utility model (14).  Since
satisfaction ratings were based on existing attributes of
specific transport systems, the models reflect circumstantial

rather than caus... relationships between predictor variables and behavioural measures. Researchers have reported success in predicting mode split with such models (37). However, the mathematical structure of the model was often assumed on an *a priori* basis - e.g. additive relationships between predictor variables and logistic transformations for generating probabilities. While the level of prediction achieved in a single application of such an approach may be satisfactory, additional work is necessary to define a set of causal factors and describe how they combine to affect traveller behaviour under changing circumstances.

Another problem arises in the way attitudes towards transport attributes are measured. The contribution of a given factor in a multi-factor judgement or decision depends not only on its level of intrinsic importance but also on the range of values over which it varies (27). Time may be a potentially important factor in mode choice, but if the total time for a trip to work is nearly the same for car and bus, workers will base their choice on other factors. Furthermore, ratings of satisfaction of a respondent's current mode may be inflated due to 'cognitive dissonance' effects (16). Thus, importance ratings and satisfaction ratings have questionable validity as measures of the actual degree to which different attributes affect behaviour.

Another approach uses a paired-comparisons method where respondents are presented pairs of attributes and are asked to choose the preferred attribute. Attributes are segmented into groups such as vehicle-design characteristics, level-of-service variables and convenience factors. Uni-dimensional and/or multi-dimensional models are then applied to separate respondents on the basis of preference patterns. Dobson (7) reports success in linking socio-economic-respondent characteristics to preference dimensions uncovered by a multi-dimensional-scaling model. While such information can be useful to planners who can then attempt to meet the needs of specific population segments, it still leaves unanswered the basic question of how specific attributes are evaluated and combined to determine traveller behaviour.

Recent approaches to attitudinal modelling of traveller behaviour address some of these problems. Approaches based on the conjoint-measurement and information-integration theories of mathematical psychology incorporate systematic manipulation of system attributes to arrive at the 'trade-off' of competing factors in traveller behaviour. Since both approaches attempt to study individual judgemental and decision-making processes, they fit in quite nicely with the 'disaggregate' approach of traveller-behaviour modelling (39).

*Conjoint measurement*. Conjoint measurement is a technique by which the various levels of each of several variables can be scaled simultaneously or conjointly (23,31). Data are obtained from experiments where respondents are required to rank order a series of stimulus combinations. Consumer and marketing researchers are often interested in the 'trade-off' among levels of a specified pair of attributes of a product and thus have adopted a conjoint-measurement technique to accomplish a 'trade-off' of consumer preferences (17,22).

The interests and goals of transport researchers often parallel those of marketing researchers, and it is small wonder that tools used in marketing research have been adopted by transport researchers. The conjoint-measurement approach is appealing to transport planners because respondents can be asked to evaluate new as well as existing attribute combinations. The data can thus be used in policy planning as well as in policy evaluation. Several recent studies illustrate the application of trade-off analysis and conjoint measurement to transport-research problems (9,10,41).

Donnelly *et al.* (9) describe the use of conjoint measurement with trade-off data in transport applications. The input data are in the form of matrices, where each matrix represents a trade-off among the levels of a pair of policy attributes such as ways of encouraging transit (lowering fares, improving service, making cars more costly) and ways of pricing transit (same cost for everyone, less for special groups). Each cell of the matrix represents a combination of one level from each of the two attributes and the respondent is asked to rank-order these combinations. A preference utility is then computed for each level of each attribute for a given respondent. These utilities are computed on the basis of the 'best fit' of the trade-off data for each respondent. The utilities for a given respondent tell which attribute is most important and what levels of a given attribute are preferred by that respondent. An aggregation of individual preferences can then be used to determine the appeal of alternative policies to the overall population or to different market segments.

The strength of the conjoint-measurement approach is that individual and group preferences are based on *combinations* of system attributes rather than relying on reactions to single attributes in unknown contexts. The information-integration approach described below shares this property and has the additional advantage of testing alternative combination rules in describing decision processes.

*Information-integration theory.* The primary goal of information-integration theory is developing simple models to describe how information is combined or integrated in human judgement and decision-making situations. The theory was developed by Anderson (3,4,5,6) and has had a history of success in describing complex cognitive processes in a variety of sub-areas in psychology (24). It has recently been applied to multi-factor judgements related to traveller behaviour, including mode choice (25,36), trip purpose (35), estimates of travel expense (25), and inter-personal factors in carpooling (12).

The theory assumes that each attribute within a given system or set of attributes represents a piece of information to be integrated into the overall evaluation of that system. When a number of different factors or attributes have to be taken into account when making a judgement or decision, each piece of information (level of a given attribute) can be characterised by two parameters: a *scale value* corresponding to the subjective evaluation of the information along the dimension of judgement (e.g. the favourability of 30-minute transit time) and a *weight* representing the importance of the information for the judgement or decision to be made. It will be recognised that this is analogous to the attitude-measurement procedures described earlier where respondents are asked to make *separate ratings* of the importance of an attribute (the cognitive component of attitude) and their level of satisfaction (the affective component of attitude) of that attribute for a given response alternative. However, studies of information integration require only a *single rating* for a given set of system attributes representing the respondent's overall evaluation of that set of attributes. The net effect of a given attribute within a system is the product of its weight and scale value; the integrated judgement or decision is assumed to be represented by an algebraic function of the weights and scale values of the various attributes.

When factorial designs are employed where each respondent receives all possible combinations of attribute levels, goodness-of-fit tests for alternative algebraic models such as adding, averaging, and multiplying models are available through analysis-of-variance techniques (6). The validation of a given model serves to confirm the validity of the measurement scale used. For example, an additive model will fit the rating data only if an additive process describes how the factors combine *and* the ratings form an interval scale of measurement.

When support for a particular form of algebraic model is obtained, simple procedures, known collectively as 'functional measurement', are available for estimating the weights of the different factors (3). The relative weights of each of several factors estimated in a multi-factor information-integration task are the weights that are *functional* in the decision-making process. In a similar vein, the psycho-physical function relating subjective values to actual values of each factor is also obtained directly from the multi-factor judgements. The algebraic model provides the framework for scaling and measurement, which is another reason why the term 'functional measurement' has been used to describe the scaling and measurement aspects of the information-integration approach. For example, if an additive or a multiplicative model fits the data, the marginal means of the factorial design can be taken as the scale values for the various factors, and a plot of these values for a given factor will describe its psycho-physical function. Thus, the phases of attitude measurement, psycho-physical scaling, and model tests are considered concurrently in the application of the information-integration approach to attitudinal modelling.

There are two major differences between the conjoint-measurement approach and the information-integration or functional-measurement approach. First, only rank-order data are admissable with conjoint measurement, whereas both rank-order data and rating data can be handled with functional measurement. Functional measurement places primary emphasis on using the metric information contained in numerical-response measures such as those obtained with simple rating scales. With suitable precautions, rating methods have been shown to provide valid interval scales in a variety of applications (5,6,24). Thus, information-integration studies can offer information about the absolute as well as relative rating of alternative systems. Second, the usual application of conjoint measurement in trade-off analysis lacks an adequate test of goodness-of-fit and often assumes additivity of factors. The problem arises from the lack of a satisfactory error theory for handling the response variability in aggregated data. Information-integration theory, on the other hand, is more empirically oriented and uses formal goodness-of-fit tests to consider a variety of model forms. The importance of distinguishing between additive and non-additive processes is that a single extremely unfavourable factor can be balanced by other favourable factors if the factors are additive, but a single extremely unfavourable factor can have an exaggerated influence if factors are non-additive, as was illustrated in the Norman and Louviere (36) and Levin and Corry (26) mode-choice studies.

These distinctive features of the information-integration approach were important in Dueker and Levin's study of inter-personal factors in carpooling (12). First, inter-action (non-additivity) effects were found that showed the inhibiting influence of forming carpools with strangers to depend on the sex of the strangers. Second, examination of absolute response levels revealed that the addition of a single acquaintance to a carpool of non-acquaintances raised the respondents' ratings from a highly unfavourable level to a slightly favourable level. The policy implication of these findings is that carpool 'matching' procedures will be improved if they are designed to include at least one acquaintance for each rider.

Another brief illustration is Levin et al.'s study of perceived highway driving safety (28). Respondents rated perceived safety for a variety of hypothetical situations described by varying weather conditions, driving speeds, time of day and hours of consecutive driving. Safety ratings were shown to vary systematically over combinations of these factors and a simple algebraic model was shown to describe these variations. The specification of a model to describe this combinatorial relationship is an example of attitude measurement - relating qualitative attitudinal variables to system variables - as represented in the top portion of Figure 35.1. The study went further by requiring respondents to indicate a variety of behavioural intentions. These were shown to be highly correlated with the safety ratings and illustrate the bottom portion of Figure 35.1.

Since the information-integration approach is probably unfamiliar to most transport researchers, it is illustrated more fully by describing a recent mode-choice study conducted in the author's laboratory (33). This study also serves to illustrate the schema for attitudinal modelling of traveller behaviour given in Figure 35.1.

ATTITUDINAL MODELLING IN A MODE-CHOICE STUDY

This study was designed to describe decision processes underlying mode-choice trade-offs and to relate these to actual
mode choices.  A questionnaire was mailed to a sample of 250
employees at the University of Iowa.  Of this original set,
130 were returned; 99 of these were fully completed in a
usable form.  The initial and main part of the questionnaire
consisted of a series of mode choices to hypothetical trade-
off situations.  The design of this series is described
below.  The questionnaire also included sections designed to
assess the worker's personal background, work schedule and
distance from work, present mode split and availability,
satisfaction level, and ratings of importance of a variety
of factors related to mode choice.  Finally, in an effort to
gain information about constraints which may have influenced
actual mode choices, the following open-ended question was
inserted at the end of the questionnaire: 'What are the most
compelling reasons why you personally choose the method of
travel you use to get to and from work?'

*Design of Mode Choice Task*

Respondents were presented with descriptions of 27 hypothetical situations described by the factors Time Difference
(0,15 or 45 minutes per day longer for bus than for car),
Cost Difference (0,25 or 75¢ per day more for car than for
bus), and Number of Riders in the car with the driver (0,1 or
3).  Each given situation was described by one level of each
of two factors - Time Difference and Cost Difference, Time
Difference and Number of Riders, or Cost Difference and
Number of Riders.  One complete factorial design was formed
for each pair of factors; three 3 x 3 factorial designs
were formed overall, and a separate 'treatments-by-subjects'
analysis of variance was performed to analyse the results
for each design.

Respondents were told to assume that both a car and a bus
were available to them in each hypothetical situation and
that they were to respond on the basis of the information
presented assuming these availabilities.  The purpose of
this instruction was to elicit a car-bus propensity abstract
from an availability constraint.  For each hypothetical
situation, the respondent was asked to rate the relative
likelihood of taking the bus or car.  A 20-point rating scale
was used, where '0' represented 'certain to take car' and
'20' represented 'certain to take bus'.  They were to use
numbers between 0 and 20 to represent varying degrees of
preference for car or bus.  This car-bus preference scale was
used earlier by Levin and Corry [26] and provides information
about *degree of preference* as well as binary mode choice.

*Results and Discussion*

Description of the results will be divided into three phases.
In Phase 1, analyses of responses to the hypothetical mode-
choice situations will be presented.  Phase 2 will explore
the relationships between decision processes identified in

Phase 1 and group differences in socio-economic, behavioural
and situational characteristics.   In Phase 3, responses in
the experimental task will be related to actual mode choices.

*Phase 1.*   Because each respondent completed only one replication of the experiment, decision models could not be
tested at the level of the individual respondent.   However,
by grouping respondents who exhibited similar arrays of
responses into 'segments', inferences about *individual*
decision-making processes could be made at a minimum risk of
ecological fallacy.   In order to derive homogeneous decision-
making segments, the raw responses of each of the 99 respondents in the experimental task were subjected to the minimum-
euclidean distance-grouping algorithm of Ward (48).   Three
distinct groups were identified.   Group 1 (with 30 members)
had a grand mean of 6.3; Group 2 (with 51 members) had a
grand mean of 10.5; Group 3 (with 18 members) had a grand
mean of 15.3.   Recalling that responses were recorded on a
20-point rating scale with '0' representing 'certain to take
car' and '20' representing 'certain to take bus', it was
clear that Group 1 is a *car-biased* group and Group 3 is a
*bus-biased* group.   Group 2 is in the middle and, for reasons
that will be made clear later, is defined as an *unbiased*
group.

   Plots of mean values for each cell of the three sub-
designs for each group are shown in Figures 35.2 to 35.4.
Several things should be kept in mind when examining these
figures.   Parallel lines show that the two factors being
plotted combine in an additive fashion to determine car-bus
preference ratings for that group.   The slopes and separa-
tions of the lines reflect the relative weights of the two
factors.   The comparative spacing of the lines in a given
panel and the shape of each line provide information about
the psycho-physical functions, i.e. the relationships between
objective attribute values and their subjective counterparts.
If the lines in a given panel converge at a particular level
of the variable plotted on the abscissa, this shows differential weighting (non-additivity) with that particular level
having greater weight than other levels of that variable.

   Figure 35.2 displays the results of the car-biased group.
The top panel shows this group's mean response to each
combination of Time Difference and Cost Difference; the
middle panel shows the mean response to each combination of
Cost Difference and Number of Riders; and the bottom panel
shows the mean response to each combination of Time Difference and Number of Riders.   The top panel shows that
preference for the car increased as time savings for car
over bus increased and decreased as cost savings for bus over
car increased.   Cost Differences were more important
(contributed a greater percentage of the variance) than Time
Differences for this group.   The non-parallelism suggests a
non-additive combination rule for Time Differences and Cost
Differences in determining car-bus preferences for the car-
biased group.   This is confirmed by a significant interaction
in the analysis of variance.   Convergence of the lines at a
Time Difference of 45 minutes indicates that Cost Difference
had a lesser effect at high Time Differences than at low
Time Differences.   This finding replicates and extends the

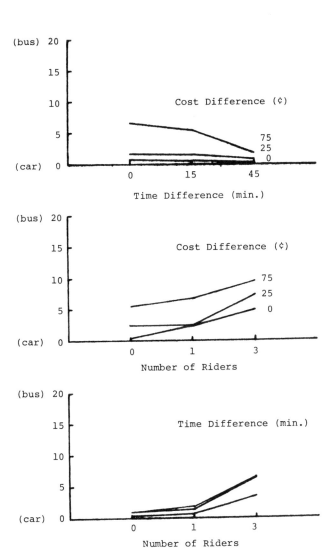

Figure 35.2   Mean Rating Responses of Car-biased Group

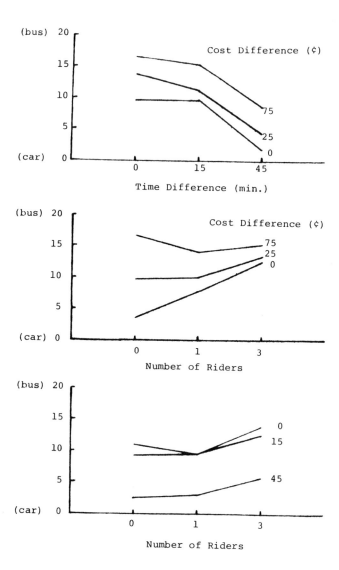

Figure 35.3   Mean Rating Responses of Unbiased Group

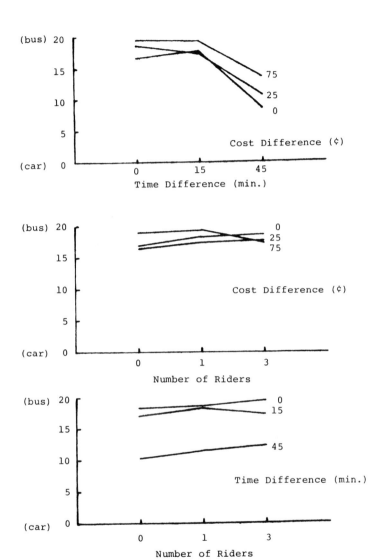

Figure 35.4   Mean Rating Responses of Bus-biased Group

generality of the Levin and Corry study (26) and has been interpreted by Levin (25) to indicate that respondents place greatest weight on those pieces of information that support their initial biases.  In this case, car-biased respondents placed great weight on large Time Differences favouring car over bus even when there were large Cost Differences favouring bus over car.

The middle panel of Figure 35.2 shows lines that are approximately parallel, and the lack of a significant interaction supports an additive relationship between Cost Difference and Number of Riders for the car-biased group. Car preference decreased as Number of Riders increased.  The bottom panel again shows that car preference decreased as Number of Riders increased and that car preference increased as Time Differences increased.  These two factors interacted, with the lines converging at 0 Riders.  This suggests that car-biased respondents were particularly apt to prefer the car when they could drive alone, even in the absence of time savings over the bus.  Across sub-designs, it appears that Cost Differences and Number of Riders were of about equal importance in affecting car-bus preferences for the car-biased group, with each of these factors being more important than Time Differences.  However, the interactions show that Time Differences increased in importance when time savings for car over bus were substantial.

Figure 35.3 displays the results for the unbiased group. These curves show that the manipulated factors had large and systematic effects on the car-bus preferences of the unbiased group.  The approximately neutral mean response of this group reflects a balancing or trade-off of factors rather than a lack of responsiveness to the variations.  The label 'unbiased' thus seems appropriate for this group.  Time Differences were more important than Cost Differences, with these two factors combining in an additive fashion.  Overall, Number of Riders was less important, but there was an interaction between Cost Difference and Number of Riders (middle panel).  Cost Difference had its largest effect when Number of Riders was 0, with preference for the car occurring at 0 Cost Difference and preference for the bus occurring at 75¢ Cost Difference.  When Number of Riders was 3, respondents in this group tended to prefer the bus at all Cost Differences.

Results for the bus-biased group are shown in Figure 35.4. Time Differences were more important than Cost Differences for this group.  Number of Riders was not a significant source of variance, nor did this factor interact with other factors.  In other words, preferences for bus over car were not affected by the Number of Riders who would be in the car. This can be seen in the flatness of the lines in the middle and bottom panels of Figure 35.4.  The top panel shows a small (but statistically significant) interaction between Cost Differences and Time Differences, with Cost Differences tending to have a lesser effect at low Time Differences than at high Time Differences.  This result is consistent with the Levin and Corry study (26).  The bottom panel shows a non-linear relationship between designated Time Differences and car-bus preferences.  There is little difference in response between 0 and 15 minutes Time Difference, but 45 minutes Time Difference produced preferences much closer to the 'car' end of the scale. A similar pattern of results can be detected for the other groups (see bottom panels of Figures 35.2 and 35.3).

Across groups and sub-designs, the nine different analyses of variance produced four significant interactions and five non-significant interactions. Some of these interactions might be interpreted as ceiling or floor effects, but since realistic levels were chosen for the factors and the ends of the scale refer to certainty of mode choice, it would be premature to dismiss the interactions as scaling 'artefacts'. The most parsimonious interpretation of this pattern of results is that the manipulated factors combine by an averaging process to determine car-bus preferences. For additive factors, the averaging process can be described as one where the weight of a given factor is constant across levels of that factor. For non-additive factors, the averaging process can be described as one where extreme values of the time or cost factor receive increased weight when they are consistent with the respondent's initial bias towards car or bus.

The three groups identified by the minimum-euclidean distance-grouping algorithm differed in terms of the relative weighting of factors manipulated in the experiment, as well as in overall preference for car or bus. Number of Riders was an important factor for the car-biased group, with preference for the car decreasing as Number of Riders increased, but this factor was not important for the bus-biased group. Cost Differences were weighted more than Time Differences in the car-biased group, but Time Differences were weighted more than Cost Differences in the bus-biased group. At first thought, it might seem counter-intuitive that respondents who preferred the money-saving mode (i.e. bus) would be more influenced by time factors and respondents who preferred the time-saving mode (i.e.car) would be more influenced by cost factors. However, what this means is that *degree of preference for the bus* by respondents in the bus-biased group was heavily influenced by the amount of extra time involved in taking the bus, and *degree of preference for the car* by respondents in the car-biased group was heavily influenced by the amount of added cost involved in driving a car. Viewed in terms of these trade-off processes, the results are interesting and informative. The unbiased group showed some of the same effects as each of the other two groups, e'.g. a large Time Difference effect like the bus-biased group and a significant Riders effect like the car-biased group.

The next phase of data analysis examines the relationship between group differences found in Phase 1 and differences in other factors measured in the experimental questionnaire.

*Phase 2.* Table 35.1 summarises results for various parts of the questionnaire when respondents were divided into the groups identified in Phase 1. Two analyses were conducted on these data: a qualitative examination of group differences on various questionnaire items; and a regression analysis to specify the extent to which groupings could be related to variations in socio-economic, demographic and situational characteristics.

TABLE 35.1   Group means on questionnaire items

| Item | Car-biased group | Unbiased group | Bus-biased group |
|---|---|---|---|
| Age (yrs) | 37.4 | 36.3 | 39.1 |
| Sex (0 = female; 1 = male) | 0.47 | 0.49 | 0.61 |
| Income (7 categories) | 5.3 | 4.6 | 4.8 |
| Yrs employed at | | | |
| University of Iowa | 8.5 | 5.5 | 8.9 |
| Home-work distance (mi.) | 3.8 | 7.8 | 2.5 |
| Home-bus stop distance (mi.) | 0.96 | 4.40 | 0.93 |
| Variability of work shift | | | |
| (0 = no; 1 = yes) | 0.57 | 0.45 | 0.28 |
| Importance ratings | | | |
| (20-point scale): | | | |
| Travel time | 15.3 | 13.0 | 11.5 |
| Cost | 7.9 | 11.8 | 12.0 |
| Amenities | 7.9 | 4.6 | 1.1 |
| Convenience | 13.6 | 9.7 | 6.7 |
| Privacy | 12.6 | 12.6 | 11.1 |
| Conserving energy | 9.1 | 11.2 | 12.6 |
| Satisfaction with | | | |
| current mode | | | |
| (20-point scale) | 17.5 | 14.9 | 17.4 |

It can be seen that car-biased individuals tended to be slightly younger and of higher income than bus-biased individuals. A higher proportion of males tended to be in the bus-biased group. The unbiased group had, on the average, been employed at the university for a shorter period of time than the other groups and lived farther from work and from the nearest bus stop than the other groups. The bus-biased group had the shortest average home-work distance and the least variable work schedule.

Ratings of importance of various factors differed between groups. Ratings for the unbiased group were generally inter-mediate to the other two groups but this group had the lowest level of satisfaction with their current mode. As would be expected, amenities, convenience and privacy were rated more important by car-biased respondents than by bus-biased respondents, while conserving energy was rated more important by bus-biased respondents. The high rating of privacy by the car-biased group is consistent with the large effect of number of riders observed for that group in Phase 1. This group obviously prefers a single-occupant vehicle. Travel time was rated more important by car-biased respondents and cost was rated more important by bus-biased respondents.

However, these ratings were made in the 'abstract' and did not actually involve trade-offs of specified levels of competing factors. Car-biased individuals may presently choose car over bus because of perceived time savings; this does not necessarily mean that they would be unresponsive to changes in cost factors. The functional-measurement procedure used in Phase 1 revealed trade-off relationships that operate in car-bus preferences and would seem to provide more information about decision processes underlying mode choice than would simple importance ratings. In particular, the non-linear functions obtained in the trade-off analyses show that cost and time factors increase in importance when they reach extreme values.

A regression analysis was conducted to relate quantitatively the grouping assignments to the following socio-economic measures, demographic characteristics and transport constraints obtained in the questionnaire: home-to-work distance, home-to-bus-stop distance, work time (day vs. night), type of work shift (fixed vs. variable), variability in work schedule, need of car for business and personal reasons (ratings), convenience of parking at work place (rating), work place (code), age, sex and income. The ability of a linear combination of these variables to predict group membership was measured by two statistics: the overall proportion of variance ($R^2$) of grouping explained by the linear combination, and the proportion of cases correctly assigned to each bias group (analogous to discriminant analysis). The resulting overall regression was significant beyond the 0.01 level, but it explained only 27 per cent (corresponding to R = 0.52) of the variance in grouping. This corresponded to 67 per cent of cases being correctly assigned to bias groups. While this result is disappointing from the point of view of identifying bias groups on an *a priori* basis, it is consistent with a number of other studies finding weak relationships between attitudes and socio-demographic characteristics measured at one point in time (43,44). However, when responses to the experimental task are included in Phase 3 to predict actual mode split, accuracy is greatly improved.

*Phase 3.* It is of interest in this phase of the analysis to relate responses to the experimental task to actual mode-choice behaviour. Ideally, this would be done by taking a model of mode choice derived under hypothetical trade-offs for individuals and substituting in measures of the individuals' real-world transport environment. Each of these 1-point predictions would then be compared with frequency of patronage of car and bus. In the present case, neither individual models nor accurate estimates of car-bus cost differences and time differences for individual respondents were available. Hence, the following simplified regression model was used to predict mode patronage:

$$\text{Freq}_B = (a(\text{MEAN}) + b(\text{HWD}) + c(\text{MIN})) \ (\text{AVAIL}) \qquad (1)$$

where $\text{Freq}_B$ = number of work trips by bus during the month prior to receiving the questionnaire,
   MEAN = mean response on experimental task,

$$HWD = \text{home-work distance,}$$
$$MIN = \begin{cases} 1 & \text{if car rated cheaper than bus} \\ 0 & \text{if otherwise,} \end{cases}$$
$$AVAIL = \begin{cases} 0 & \text{if no bus available} \\ 1 & \text{otherwise.} \end{cases}$$

The 'availability' factor (obtained from responses to the questionnaire) was entered as a multiplier in the regression equation because if it were at a '0' level for a given individual, then frequency of bus patronage would be zero. There were 27 respondents in this category; 20 were in the unbiased group, and 7 were in the car-biased group.

The prediction was good, explaining over 78 per cent (based on R = 0.885) of the variance in the number of bus trips for different respondents. An examination of residuals revealed some tendency for over-prediction of low values and under-prediction of high values, but overall, the model appears to provide a reasonable description of the data, especially in light of the crudity of measurement of some of the predictor variables.

The single factor, mean response on the experimental task, accounted for over 70 per cent of the variance in the number of bus trips. It is quite likely that had actual travel-time and cost-difference measures been available, the overall predictive ability of the model would have been even higher than at present.

In order for the predictive ability of the present model to be compared to more traditional models such as the logit, respondents were divided into two groups - car riders and bus riders, depending on the most frequently used mode. The present model was then applied in a discriminant analysis to determine its classificatory ability. The result was that 94 per cent of the cases were correctly classified, a result comparable to those obtained in more successful applications of logit-type analyses (29,38,49). For a further comparison of traditional methods, a linear combination of 'importance' ratings of various factors measured in the questionnaire was used in place of the bias measure in Equation (1) to predict bus patronage. The following importance ratings were used: travel time, cost, amenities, convenience, privacy and energy conservation. The resulting regression model was highly significant, but the predictive ability of 59 per cent was clearly less than that obtained with the bias measure from the experimental task.

SUMMARY AND CONCLUSIONS

Since a major goal of attitudinal modelling in transport is to use attitudinal measures to predict behaviour, it is disturbing to note the lack of success of social scientists in relating attitudes and behaviour. However, if transport researchers can be sufficiently specific in describing systems and estimating models, accurate predictions of traveller behaviour can be expected, and useful descriptions of traveller-decision processes can be provided.

In evaluating the usefulness of attitudinal models in transport it is important to distinguish between short-term and long-term benefits. Most applications have had short-term goals - evaluating current transport systems or comparing specific alternative systems and predicting responses to them. It is clear that the inclusion of measures reflecting subjective perceptions and evaluations of system attributes can enhance the ability of planners to provide effective and efficient systems. Perceptual and affective analogues of system characteristics are more directly relevant to individual judgements and decisions than are objective values (50).

The study reported here used an experimental task based on hypothetical mode-choice situations generated by the information-integration paradigm to examine subjective impressions and evaluations of system attributes. Attitude measurement was achieved by relating mode-choice preferences to combinations of system attributes. Since these combinations were not confined to current attribute levels, they demonstrate the potential for generating predictions of travel behaviour in novel situations. The validity of the measurement procedure was established by demonstrating that the attitudinal measure (mean car-bus preference response on the experimental task), when combined with information about individual constraints on mode choice, accurately predicted actual mode choice. This shows the usefulness of attitudinal modelling based on an experimental methodology that examines responses to combinations of system attributes. Additional support for this methodology is provided by Louviere et al. (30). The observed interactive relationship between attitudinal and situational factors is consistent with Fishbein's (13) conceptualisation of the attitude-behaviour relationship and Golob et al.'s (16) conceptualisation of travel decision-making. Through a combination of experimental design and demographic analysis, a variety of qualitative and quantitative variables affecting traveller behaviour can be studied directly, and other factors can be shown to exert their influence through individual differences in response bias and the weighting of information.

The present study shows how respondents can be divided into homogeneous subgroups of decision-makers on the basis of attitudinal measures. The differing models for the different subgroups are consistent with the principle that individuals place greatest weight on those factors that support their current bias towards the car or bus. This might be labelled as an 'inertia' effect in that it results in a maintenance of current attitudes and behaviour. The implication of this is that individuals will be unlikely to change modes unless substantial changes are made in the attribute levels and the perceived advantages of their current mode are diminished. This is related to Golob et al.'s (16) notion that choices affect attitudes as well as vice versa. For either conceptualisation, it would be of interest to collect longitudinal data for studying the attitude *change*-behaviour *change* relationship and the effect of *habit* on traveller decisions.

Market segmentation based on attitudinal modelling of mode-choice decision processes represents a departure from usual methods based on socio-economic and demographic

measures.  However, the segments identified in the present
study differed systematically on such measures.  This
suggests that predictive ability can be greatly enhanced if
the following sequence is followed: (1) identify homogeneous
subgroups of respondents on the basis of attitudinal modelling of mode-choice decision processes in a controlled
experimental task; (2) identify differences in socio-economic
and demographic characteristics for different subgroups of
decision-makers; (3) when future predictions are needed for a
given target population, give the experimental task to a
sample from the population, identify homogeneous subgroups of
decision-makers, and determine the relative proportion of
individuals in each subgroup.  When time constraints make
this inconvenient, socio-economic and demographic measures
can be used in conjunction with the differences obtained in
'2' to assign individuals to subgroups (note, however, that
such assignment is less than optimal, and the researcher will
have to trade-off accuracy of prediction for economy in data
collection); (4) use the models derived in '1' in conjunction
with knowledge of availability constraints to make predictions for individual consumers and aggregate across
individuals in the population.

The stage appears to be set for the most useful long-term
application of the attitudinal-modelling approach.  Decision-
process models derived through carefully-designed experimental tasks and attitude measurements can be used to predict
responses to new transport systems and environments.  The
models must be continually updated and refined to reflect
rapidly changing systems and values, but this can be done
within the modelling framework developed here.  For example,
current societal needs make it mandatory that future studies
of traveller behaviour examine the issue of energy conservation.  This would appear to be a problem especially well
suited to attitudinal modelling.  New variables can be
incorporated into the attitudinal-modelling procedures
described in this chapter to reflect the importance of subjective evaluations of energy expenditures for alternative
transport modes.  By modelling the decision processes of
suppliers as well as consumers, the impact of new programmes
designed to reduce energy consumption in transport can be
evaluated.  It is hoped that the present description and
discussion of attitudinal modelling procedures - particularly
those based on a carefully-controlled experimental methodology such as provided by information-integration theory -
will serve to encourage their application to a variety of
such problems in transport.

REFERENCES

1. Allport, G.W., 'Attitudes', in A Handbook of Social
Psychology, C. Murchison (ed.), Clark University Press,
Worcester, Massachusetts (1935).
2. Anderson, N.H., 'Averaging versus adding as a stimulus-
combination rule in impression formation', Journal of
Experimental Psychology, 70, 394 (1965).
3. Anderson, N.H., 'Functional measurement and psycho-
physical judgment', Psychological Review, 77, 153 (1970).
4. Anderson, N.H., 'Integration theory and attitude change',
Psychological Review, 78, 171 (1971).

5. Anderson, N.H., 'Algebraic models in perception', in *Handbook of Perception* (vol. 2), E.C. Carterette and M.P. Friedman (eds), Academic Press, New York (1974).

6. Anderson, N.H., 'Information integration theory: a brief survey', *Contemporary Developments in Mathematical Psychology* (vol. 2), D.H. Krantz, R.C. Atkinson, R.D. Luce and P. Suppes (eds), Freeman, San Francisco (1974).

7. Dobson, R., 'Challenges for psychological research and theory in urban transportation planning', *Meeting of American Psychological Association*, New Orleans (1974).

8. Dobson, R., 'Uses and limitations of attitudinal modelling', in *Behavioral Travel-Demand Models*, P.R. Stopher and A.H. Meyburg (eds), Lexington Books, Lexington (1976).

9. Donnelly, E.P., Howe, S.M. and DesChamps, G.A., 'Trade-off analysis: theory and applications to transportation policy planning', *Preliminary Research Report No. 103*, New York State Department of Transportation, Albany, New York (1976).

10. Donnelly, E.P., Weiss, D.L., Cohen, G.S., Liou, P.S. and Holthoff, W.C., 'Statewide public opinion survey on public transportation', *New York State Department of Transportation, Technical Report No. 80*, Planning and Research Bureau, Albany, New York (1975).

11. Doob, L.W., 'The behavior of attitudes', *Psychological Review*, 54, 135 (1947).

12. Dueker, K.J. and Levin, I.P., 'Carpooling: Attitudes and participation', *Institute of Urban and Regional Research, Technical Report No. 81*, University of Iowa, Iowa City (1976).

13. Fishbein, M., 'Attitude and the prediction of behavior', in *Readings in Attitude Theory and Measurement*, M. Fishbein (ed.), Wiley, New York (1967).

14. Golob, T.F., 'The survey of user choice of alternative transportation modes', *High Speed Ground Transportation Journal*, 4, 103 (1970).

15. Golob, T.F., Canty, E.T., Gustafson, R.L. and Vitt, J.E., 'An analysis of consumer preferences for a public transportation system', *Transportation Research*, 6, 13 (1972).

16. Golob, T.F., Horowitz, A.D. and Wachs, M., 'Attitude-behaviour relationships in travel demand modelling' (Chapter 34 of the present book).

17. Green, P.E. and Rao, V., 'Conjoint measurement for quantifying judgmental data', *Journal of Marketing Research*, 8, 355 (1971).

18. Hartgen, D.T., 'Mode choice and attitudes: a literature review', *Preliminary Research Report No. 21*, New York State Department of Transportation, Albany, New York (1970).

19. Hartgen, D.T., 'The influence of attitudinal and situational variables on urban mode choice', *Preliminary Research Report No. 41*, New York State Department of Transportation, Albany, New York (1973).

20. Horowitz, A.D. and Sheth, J.N., 'Ridesharing to work: a psychosocial analysis', *Research Publication GMR-2216*, General Motors Research Laboratories, Warren, Michigan (1975).

21. Johnson, M.A., 'Psychological variables and choices between auto and transit travel: a critical research review', *Urban Travel Demand Forecasting Project Working Paper No. 7509*, Institute of Transportation and Traffic Engineering, University of California, Berkeley (1975).

22. Johnson, R.M., 'Trade-off analysis of consumer values', *Journal of Marketing Research*, 11, 121 (1974).

23. Krantz, D.H. and Tversky, A., 'Conjoint-measurement analysis of composition rules in psychology', *Psychological Review*, 78, 151 (1971).

24. Levin, I.P., 'Information integration in numerical judgments and decision processes', *Journal of Experimental Psychology: General*, 104, 39 (1975).

25. Levin, I.P., 'Information integration in transportation decisions', in *Human Judgment and Decision Processes: Applications in Problem Settings*, M.F. Kaplan and S. Schwartz (eds), Academic Press, New York (1977).

26. Levin, I.P. and Corry, F.A., 'Information integration models of transportation decisions', *Transportation Research Record* (1977).

27. Levin, I.P., Kim, K.J. and Corry, F.A., 'Invariance of the weight parameter in information integration', *Memory & Cognition*, 4, 43 (1976).

28. Levin, I.P., Mosell, M.K., Lamka, C.M., Savage, B.E. and Gray, M.J., 'Measurement of psychological factors and their role in travel behavior', *Transportation Research Record* (1977).

29. Liou, P.S. and Talvitie, A.P., 'Disaggregate access mode and station choice models for rail trips', *Transportation Research Record*, 526, 42 (1974).

30. Louviere, J.J., Wilson, E.M. and Piccolo, J.M., 'Psychological modelling and measurement in travel demand: a state-of-the-art review with applications' (Chapter 33 of the present book).

31. Luce, R.D. and Tukey, J.W., 'Simultaneous conjoint measurement: a new type of fundamental measurement', *Journal of Mathematical Psychology*, 1, 1 (1964).

32. Margolin, J.B., Misch, M.R. and Dobson, R., 'Incentives and disincentives to ridesharing behavior: a progress report', *Meeting of Transportation Research Board*, Washington, D.C. (1976).

33. Meyer, R.J. and Levin, I.P., 'Functional determinants of mode choice behavior', *Institute of Urban and Regional Research Technical Report No. 88*, University of Iowa, Iowa City (1977).

34. Nicolaidis, G.C., 'Quantification of the comfort variable', *Transportation Research*, 9, 55 (1975).

35. Norman, K.L., 'Attributes in bus transportation: importance depends on trip purpose', *Journal of Applied Psychology*, 62, 164 (1977).

36. Norman, K.L. and Louviere, J.J., 'Integration of attributes in bus transportation: two modelling approaches', *Journal of Applied Psychology*, 59, 753 (1974).

37. Quarmby, D.A., 'Choice of travel mode for the journey to work: some findings', *Journal of Transport Economics and Policy*, 1, 1 (1967).

38. Recker, W.W. and Golob, T.F., 'An attitudinal modal choice model', *Transportation Research*, 10, 299 (1976).

39. Reichman, S. and Stopher, P.R., 'Disaggregate stochastic models of travel-mode choice', *Highway Research Record*, 369, 91 (1971).

40. Rhine, R.J., 'A concept-formation approach to attitude acquisition', *Psychological Review*, 65, 362 (1958).

41. Ross, R.B., 'Measuring the effects of soft variables on travel behavior', *Traffic Quarterly*, 29, 333 (1975).

42. Stopher, P.R., 'On the application of psychological measurement techniques to travel demand estimation', *Meeting*

of American Psychological Association, New Orleans (1974).

43. Stopher, P.R., 'The development of market segments of destination choice', Transportation Research Record (1977).

44. Tardiff, J.J., 'Comparison of effectiveness of various measures of socio-economic status in models of transportation behavior', Transportation Research Record, 534, 1 (1975).

45. Thomas, K. (ed.) (1971) Attitudes and Behaviour, Penguin Books, Middlesex, England.

46. Thurstone, L.L., 'The measurement of social attitudes', Journal of Abnormal and Social Psychology, 61, 110 (1960).

47. Triandis, H.C. and Triandis, L.M., 'Race, social class, religion and nationality as determinants of social distance', Journal of Abnormal and Social Psychology, 61, 110 (1960).

48. Ward, J.H., 'Hierarchical grouping to optimize an objective function', 'Journal of the American Statistical Association, 58, 236 (1963).

49. Watson, P.L., 'Comparison of the model structure and predictive power of aggregate and disaggregate models of intercity modal choice', Transportation Research Record, 527, 59 (1974).

50. Westin, R.B. and Watson, P.L., 'Reported and revealed preferences as determinants of mode choice behavior', Journal of Marketing Research, 12, 282 (1975).

51. Wicker, A.W., 'Attitudes versus actions: the relationship of verbal and overt behavioral responses to attitude objects', Journal of Social Issues, 25, 41 (1969).

Chapter 36

ATTITUDES, ATTITUDINAL MEASUREMENT AND THE RELATIONSHIP
BETWEEN ATTITUDES AND BEHAVIOUR

Jordan J. Louviere

SCOPE OF WORKSHOP

The major theme of the workshop was the relationship between
attitudes towards issues (as measured in terms of characteristics) and behaviour. Is there any stable relationship
between the two? How can attitudinal research assist in
understanding behaviour and in predicting the likely response
to a policy proposal? The work in information-integration
theory and utility measurement needs further appraisal.

SUMMARY OF WORKSHOP RECOMMENDATIONS

The workshop members argued that the so-called attitudinal
area appears to be drifting amidst a bewildering array of
techniques, pseudo-theory and misunderstanding. Thus, a
major portion of the discussion focused upon the theoretical
role of psychological theory and measurement procedures as it
relates to existing theory and practice in travel-demand
modelling. The main issues highlighted were the relationships between physical or engineering measures and self-
reports or psychological values, and the relationship of
these psychological values to choice. Thus, of major concern
were the following functional relationships:

$$p_i = f_i(EV_i) \qquad (1)$$

$$U_i = g(p_i) \qquad (2)$$

where $p_i$ is the perceived value of the $i^{th}$ physically
measurable variable, $EV_i$; and $U_i$ is the utility attached to
the $i^{th}$ variable. Hence,

$$U_i = g(f_i(EV_i)) \qquad (3)$$

$$P(a/A) = F_i(U_i) \qquad (4)$$

where $P(a/A)$ is the probability of selecting a from A and F
is some function defined over the utilities of the i independent variables. The workshop focused attention on theory and
methodology for deriving expectations for Equations (1)-(4).
The major recommendations and conclusions were as follows:

1 Understanding and quantification of choice and decision processes should lead to improved prediction and increased sensitivity to policy questions. In this regard, policy questions can be built into choice experiments very easily so that answers can be obtained very quickly and cheaply. These areas of research are ready for immediate implementation.

2 Defining and measuring qualitative or 'soft' variables can have immediate impact on the precision of existing models. Such results would also interface well with alternative modelling procedures suggested in this chapter. This research area is ready for immediate implementation.

3 The use of controlled experiments in laboratory and real-world contexts must lead to improved understanding of the effects of alternative policies and system configurations. We are ready to implement such experiments now, and they should greatly improve understanding compared to existing procedures.

4 Exogenous and/or future change factors can be included directly in choice experiments or studies and their effects anticipated *a priori*, rather than relying upon 'revealed' behaviour to estimate effects *a priori*. This is ready for immediate implementation.

5 Plan Evaluation and Impact Assessment can be approached through decision studies in which the public or subgroups thereof evaluate the alternatives. This is ready for immediate implementation.

6 Qualitative and interpersonal factors can be assessed in decision studies and built into existing models or included in new choice models. The effects of these factors on decisions can be examined. This is ready for immediate implementation.

7 Although of academic interest, the effects of multi-collinearity, and functional and variable mis-specification can be studied in the laboratory using procedures outlined in this chapter. This is ready for immediate implementation.

8 Market segmentation can be accomplished *a priori* or *a posteriori* on the basis of similarity of decision parameters. In particular, the decision parameters can be used as attributes upon which to classify individuals or the observed choices can be used as *a priori* groupings. One may then try to make inferences regarding the relationship between decision processes and socio-economic characteristics. Likewise, one may try to infer whether a relationship exists between the *a priori* choice groupings and the decision parameters. Little work has been evident in this area and the workshop recommended that at least some pilot work be begun in this direction.

9 Although currently of academic interest, dynamic and interactive effects must be assessed. Assessment in the laboratory and in the field can begin immediately.


DISCUSSION OF RECOMMENDATIONS AND CONCLUSIONS

*Introduction*

The attitudinal area has been the subject of considerable research in the recent past. The major areas of research interest include the following:

1  What are the factors or variables that affect the travel behaviour or demand of interest?
2  How do these factors interrelate or combine to affect the behaviour of interest?
3  Which of these factors are actually under the control of policy-makers, either directly or indirectly?
4  What is the effect of manipulating the policy-controllable variables on the behaviour of interest?
5  How do the variables and/or the theoretical structure relating them to behaviour change with respect to time and changes in the system and the choice behaviour?
6  What is the effect of new technology on the preceding questions?
7  How can technology be modelled?
8  And/or what factors now exogenous to the system or considered 'unimportant' will become influential in the future during the planning horizon?  And how are their effects to be modelled?
9  Finally, do different consumer groups respond differently to each of the preceding questions?  That is, can the population be segmented into relatively homogenous groups *vis-à-vis* one or more of the above issues?

These questions are particularly important in transport planning because reliable future estimates of demand and/or choice require that they be adequately answered.  Although considerable research effort has been expended to answer these questions, few generalisations have been forthcoming and there have been no major breakthroughs with the possible exception of work directed towards identification of decision factors and development and testing of direct demand models for individuals (see the Louviere *et al.* (29) and Levin (23) chapters in this volume).  Some of the relevant literature in each of these areas was reviewed by the workshop and areas in need of research were identified (see References).  The major areas of discussion are now briefly outlined.

*Topics of Immediate Concern to the Workshop*

*How does one model the choice process?*  The workshop ident-ified a number of decision or choice processes, some of which are well-known in the psychological literature but perhaps less well-known in transport research.  These processes include:

1  Algebraic utility-type processes such as adding, averaging, multiplying and combinations of these strategies (multi-attribute utility models).
2  Elimination-by-aspects choice processes (EBA).
3  Conjunctive and disjunctive decision processes.
4  Hierarchical models.
5  Satisficing models.
6  Stochastic choice models.

It was recommended that more attention be directed towards the empirical testing of these models and their potential for incorporation into individual choice models.

Given the attention that has been directed towards discrete choice models, current data-collection procedures preclude the development and testing of alternative models. This is a particularly unfortunate situation in that if we do not look for any other models, we obviously conclude that discrete choice models are in some sense 'right'. Moreover, because such thinking has biased our data-collection procedures to the exclusion of any and all non-multiple-choice models requiring zero/one data, it ensures that discrete choice models will not be challenged empirically. This is one of the more blatant cases of a self-fulfilling academic prophecy and is clearly anti-theoretical and stifling of further research or the development of alternative hypotheses. The workshop strongly urged that additional models be considered and that data-collection procedures be structured so as to permit their estimation and testing.

*A New Approach to Data Collection and Model Estimation Suggested by the Workshop*

The above conclusions, of course, beg the question that the data are suitable for testing the models in the first place. We have already argued that they are not and proceed now to outline a several-stage approach to the development and testing of choice models discussed and recommended by the workshop:

*Stage I: the identification of independent variables.* A number of procedures were suggested and discussed in the workshop: (1) focus-group interviews; (2) open-ended questionnaires or interviews; (3) structured questionnaires; (4) combinations of all the preceding. Focus-group interviews involve bringing samples of the target population or segments thereof into group-interview sessions in which various alternatives and the factors which underlie them are discussed. In particular, a discussion leader attempts to get the group members to concentrate on their reasons for their current and projected behaviour patterns in a given context of interest and to discuss factors that might cause them to alter their behaviour patterns. These sessions are tape-recorded and transcribed and then subjected to a content analysis in which the factors which seem to underlie the behaviour are classified and their absolute and relative frequencies coded according to how often they arise in the discussions.

Open-ended questionnaires or interviews are similar to the focus-group interviews, except that they are normally administered to single individuals or single families. The analytical procedures (content analysis) are the same as those reported above (19). This procedure may be administered in the form of a questionnaire in which individuals are asked to supply open-ended responses to questions of interest. These questions focus upon the reasons underlying various behaviour patterns of interest. Likewise, a personal interview format would elicit the same end-product as the questionnaire, except that the interviewer has control over the interview and can follow up questions and responses so as to clarify points or seek deeper understanding.

Structured questionnaires require respondents to estimate
the influence that various factors have upon the behaviour of
interest. Usually such responses are obtained through some
series of 'importance' judgements or a semantic differential
technique. 'Importance' is indicated by the average score on
each of the factors of interest or from a statistical analysis which attempts to infer 'importance' from the amount of
variance accounted for either in some behaviour, some overall
response, or within a reduction of a matrix of scores.
Factors which the researchers hypothesise to affect the
behaviour of interest are listed in sets of structured
questions and respondents are required to supply some numerical criterion regarding the influence of each on some
behaviour pattern of concern. Normally, standard parametric
statistical procedures are employed in analysis.

The workshop concluded and recommended that a combination
of these procedures be employed in any attempt to infer the
so-called 'relevant' factors. It was recommended that open-
ended or focus-group interviews be employed first to obtain a
set of factors which could be investigated along with others
of interest to the researchers in a more structured format.
It was also recommended that other procedures such as Kelley's
(21) 'repertory-grid' technique be investigated because it
combined the major features of all the suggested approaches
in a single technique.

The workshop cautioned against a number of methodological
pitfalls. These include the difference between cognitive
and statistical independence of factors and real-world
independence of factors. It is clear that individuals may
use a number of dimensions upon which to base decisions or
choices; these dimensions are likely to be highly correlated
in any real-world system. Attempts to infer importance or
use of factors from real-world, collinear data clearly lead
to biased results and incorrect assessments of relevance of
factors. A good example of this problem and a potential
solution is given by Shanteau (35). Cognitive independence
means that the individuals act 'as if' they treat the
factors independently regardless of whether the factors are
independent in the real world. The pitfall is that most
models which attempt to infer 'relevant' factors from real-
world data run the high risk of mistakenly inferring that
factors are treated as clumps or combinations of factors or
that one or the other of two factors (but not both) are
significant influences on behaviour. While it is true that
such assumptions might permit reasonably accurate short-run
predictions of behaviour, it is clear that they involve no
real understanding of process, and hence must fail to
predict well as soon as the current covariance structure
changes. Because covariance structures change over time and
space, it is clear that one source of the apparent lack of
'transferability' of choice models is that they are based on
covariance and not understanding. Because covariance
structures as well as the distribution and ranges of the
values of the independent variables differ from place to
place and time to time, it is clear that the parameters must
also vary even if the same underlying process is inherent in
all. Moreover, models which are based on simplistic
assumptions about independent factors, such as those which

employ only time and cost measures must certainly be subject
to errors of specification which cannot be accounted for by
assuming that the remaining effects are absorbed into the
'constant'. The workshop concluded that research must be
directed at the joint problem of system closure and inference
in the face of dependencies in real-world data. A potential
solution to this problem is offered in the next section.

In another vein, it is absolutely essential that the
system be closed to the extent possible and that all factors
now exogenous but potentially endogenous in the near future
be included. A pertinent current example would include
building in various ramifications of the proposed fossil—
energy conservation programme in the US (1977). Hence,
response to new levels of petrol prices, concurrent with
service and pricing changes in public transit would need to
be examined. To the extent that such factors are excluded,
the models will be biased and it is difficult to imagine any
model based on revealed behaviour that would not fail. In
line with this observation is the problem to be pursued
next: all revealed behaviour models are estimated on data
that tell us what choices were made, given what was available. Such data *always* exclude information on choices that
were not made as well as choices that would have been made
under different values of the alternatives. An excellent
example of the problems with such a covariance structure is
given by Dawes' (9) careful study of the problems of
modelling real-world choices among applicants to graduate
programmes. An analagous parallel holds in transport choice
modelling.

*Stage II: design of experiments to model the choice process.*
The factors uncovered in Stage I are employed to design a
data collection procedure which avoids many of the problems
discussed above. In particular, methods are available
(1,25) to construct experimental designs that specify the
exact combinations of values (levels) of the factors and
what effects can be estimated. To elaborate briefly, each
factor (say, ten) is assigned a set of values or levels
which encompass the current and projected range of the
variables. The levels are then combined in accordance with
some experimental plan specified by the nature of the
information one wishes to obtain. Each combination of
values constitutes one cell or observation in the experimental design. Thus, there are n observations and m factors.
Because the factors are 'fixed' in this type of experiment
they are measured without error - a concern previously
neglected in the so-called 'random-effects' utility models
currently popular. For example, it is clearly doubtful that
times and costs are measured without error and there has
been little attention paid to the parameter bias problems
associated.

Because the entire range of factor values is included in
the design, one can observe choices that are made given the
entire range of possible alternatives. Moreover, such a
design permits the testing of a wide range of alternative
models. The basic procedure is as follows:

1  Depending upon the number of combinations, each individual interviewed responds to all combinations or a subset of all combinations designed to encompass the entire range of alternatives.  Thus, rather than a single random observation to random values of the independent factors, each individual supplies information about all or some specific subset of the alternatives.

2  Each individual estimates his behaviour with respect to each alternative on a continuous numerical scale.  As repeatedly demonstrated by Anderson (1,2,3), Louviere, et al. (24-29), and others (10,20,32,34,37), interval-scale utility or response measures can be obtained directly from people provided reasonable care is exercised in experimental methods. Moreover, Louviere, et al. (26,27) have demonstrated in a series of studies that these continuous response measures are monotonically related to real-choice behaviour.  Of course, once a continuous numerical measure or observation is available, it can be transformed to all measurement levels of lower metric content: ordinal (ranks), binary (yes/no), or nominal (categorical choice).  This permits a wide variety of choice and decision models to be developed and tested.

3  Diagnosis and testing of choice models proceeds directly from this type of design.  Two procedures are available to diagnose and test algebraic models, given an experimental design of the type suggested above: Functional Measurement and Conjoint Measurement (1-3,22,25,27).  It is possible to determine whether individual—choice processes are additive, multiplicative or some other.  A number of statistical tests are available to permit such diagnosis or to test an a priori hypothesis of one of these processes.  Thus, the algebraic form of  the process can be directly determined from the data.  It is important to note that this approach may be referred to as 'Totally Disaggregate' in that a separate decision or choice model may be built for each separate individual.  Thus, considerably more information is made available by recourse to a different method of design and collection of data.

4  The choice models may be estimated or used directly to predict aggregate choice proportions.  To estimate, the individual is simply asked to estimate the number of trips for various purposes that were made in some recent time-period (say two weeks) by each of several modes.  Then, the individual is required to estimate the values of the independent factors used in the experiment for each mode for each trip purpose.  Obviously, not all individuals need to provide all of this information.  It can be randomly assigned to individuals across all respondents.  Hence, the amount of information requested of any person can be kept to manageable proportions.

To estimate, the values of the independent variables as given or estimated by the individuals are plugged into the individual—choice equation to generate the 'expected' response of that individual.  The observed choice behaviour or proportion of trips is then plotted against the 'expected' response and a transformation is estimated.  In this manner, each cell of the original data matrix of responses can be transformed to 'expected' choice proportions and the model can be re-estimated so as to predict choice proportions. An example of this procedure is provided in the Louviere, Wilson and Piccolo chapter in this volume (Chapter 33). This may be accomplished at either the aggregate or individual level.

5 Direct use of totally individual-choice models. The individual-choice functions may be directly employed to predict aggregate choice proportions. This is made possible because there is a monotonic relationship between model predictions and real behaviour (see the Louviere, Wilson and Piccolo chapter in this volume for evidence). Hence, one need only plug in the values of the system change of interest and generate an 'expected' response. This is compared to corresponding responses generated by plugging in the values of the alternatives, given the system change. The individual is simply assigned to the alternative with the highest 'expected' response. In this manner, the proportion of the total sample choosing each alternative may be determined by totalling the number assigned to each mode.

*Other Areas of Immediate Application Considered by the Workshop*

*Identification and measurement relationships between physical or engineering values and perceived values (the psycho-physical link).* The workshop concluded that we are in a position to identify the relationship(s) between physical-system measures and perceived measures. The methodology is similar to that suggested earlier for choice models: (1) individuals are interviewed regarding the factors that constitute 'convenience', 'comfort', and the like. (2) these factors are assigned values and combined in an experimental design. The response observed in this type of experiment is 'how convenient' or 'how comfortable' is each alternative. The definition and measurement of these perceived variables is given by the experimentally-derived function of the experimental variables. In this way, a value on each of these dimensions can be assigned to each individual or aggregates thereof by means of the functional definition.

These definitions can be tested with recourse to existing data or by small samples of additional individuals who can supply suitable data. Thus, just as force is defined as $F = ma$, so can convenience be defined as some function of a set of experimental factors. Good examples of the application of this procedure may be found in Levin, *et al.* (24). Towards this end, the workshop recommended that considerable attention be given to the potential for controlled experimentation. The potential exists for the establishment of market panels, new-product panels and product clinics which could be co-sponsored by government and private industry. Private industry, in particular, should sponsor such research because their products are used in and by the public sector and they should be concerned with their impact. Of course, information derived from such experiments would be useful for design of and marketing of transport products.

*Inclusion of non-traditional and interpersonal factors in experimental designs to assess their effects on choice behaviour.* A good example of the role of interpersonal factors is given by Levin in his study of the propensity of individuals to try a carpool match for the first time. His results clearly show that factors not usually considered in transport research have considerable influence on the initial decision to *try* a carpooling experience. Hence, the choice of carpool as an alternative mode is clearly conditional upon a number of factors not previously considered. Such decisions can be easily studied in experimental situations; but would be difficult, if not impossible, to be studied via traditional analytical methods used in transport planning. This provides a good example of misunderstanding why carpools are rejected as alternatives by many, and, hence, leads to erroneous conclusions about the effects of policy decisions on carpool choice.

*Plan evaluation and environmental impact assessment.* These appear to be other areas in which decision research can provide insight. In particular, as previously argued, all plans involve inputs and outputs. If these can be specified and assigned values or levels, alternative plans can then be evaluated by the public and/or teams of experts *vis-à-vis* the desirability of their implementation or how closely they come to achieving some previously outlined objective(s). The trade-offs that the public and/or particular interest groups would be willing to make can be determined and made explicit. Thus, any particular plan can be given an overall numerical value which represents its values *vis-à-vis* all inputs and outputs.

*Design of demonstration projects which can be guided by* a *priori decision research.* Because any demonstration project involves changes in the alternatives open to people through a change in the values of certain system variables, decision or choice experiments designed to uncover the potential choices of the public as the factors change can provide insight to guide design of the demonstration. As discussed above, the procedures for assessing potential public reaction to the project are similar and will not be detailed at this point.


AREAS TARGETED FOR PRIORITY RESEARCH

*Dynamic and Interactive Effects*

The workshop strongly urged that research be directed towards longitudinal and interactive effects in choice processes. Clear evidence of such effects is provided in the Golob *et al.* chapter in this volume (13) and the study of learning effects by Burnett (5). Much more research needs to be directed towards such effects: a set of factors affects behaviour which, in turn, affects the variable set. The choice set is also assumed to be in a state of dynamic flux: the number of alternatives available or perceived to be available increases over time. One possible assumption is that the number of alternatives perceived to be available follows a logistic distribution with respect to time spent

in an area.  In turn, there is a distribution of time spent
in the area.  The average time spent in an area could tell us
a great deal about the number of perceived alternatives
available if such basic data were available.  If we assume
that both distributions are a function of socio-demographic
factors, these relationships are also a target for research.
Such research potentially can tell us much more about choice
sets and constraints than is currently known.

Additionally, it is possible that both the factors underlying behaviour, their causal relationship, and their
psycho-physical relations change over time or as a result of
choice changes.  If this is the case then we need to identify
such changes and begin to incorporate them into models.
The Golob *et al.* chapter (13) suggests such changes occur
and are important in measurement and modelling.  More
research is needed in this area.

Considerable previous research (23-28,31,32) has demonstrated interactive or dependency effects in mode-choice
behaviour.  Similar effects have been shown in destination
choice (26,27).  Such dependency effects preclude the use of
linear models to describe such relationships.  These findings
involve relationships in which the dependent variable is
measured at the interval level and/or the expected independent arguments are not of a form such that if the independent
variable equals zero, the dependent variable also equals
zero or both.  Given that utility is an interval level
measure (arbitrary zero) and that most expected relationships
are not of the 'Y = 0, when X = 0' form, the model forms are
intrinsically non-linear.  That is, they are not transform-
able by logarithms or any other method to linear form.  Much
more research is needed in the area of dependency effects and
model specification under these conditions.  For example,
when travel time or cost differences equal zero, probability
is not zero.  If the combination rule is not linear in the
parameters, other methods must be found to handle it.  The
Louviere, Wilson and Piccolo chapter (in this volume) provides one way of handling the model for a continuous
dependent variable in destination choice.  There needs to be
more work in similar problems in mode and route choice.

Thus, the workshop recommended that research priority be
given to dynamic and interactive effects and that the
utility of simultaneous-equation methods (including differential equations) and learning models be investigated.  Other
areas for research concern include modelling 'cognitive-
dissonance' effects (see the Golob, *et al.* chapter in this
volume) and adaptation effects (15).  The latter provides a
framework for understanding the shifts in value of factors
as we change contexts: e.g. the same weight seems heavier in
the presence of a light weight and lighter in the presence
of a heavy weight.  The same cost may be perceived higher or
lower in a car context than a bus context.  Helson's (15)
theory provides a quantitative framework for analysing such
effects.

*Change and Newness as Factors*

It is unclear whether the effect of change itself in the
system or the newness of an innovation are independent
factors in their own right.  If they are, they are not
currently being handled adequately in our models.  The workshop recommended that research be undertaken to investigate
the effect, if any, of these factors on behaviour.  The
workshop suggested that this could be done experimentally in
the context of new-product clinics or in association with
decision experiments in laboratory contexts.

*Inter-regional and Intra-urban Comparisons*

The workshop recommended that research priority be given to
comparisons of results between and within cities and regions.
The lack of transferability of models may be due to failure
to control for the effects of variables that should be
endogenous, or the failure to understand linkages to measures
of urban and regional system differences.  There is a
priority need to standardise data-collection procedures such
that these analytic comparisons can be undertaken.

Related to this issue is that of real-world experimentation.  The workshop recommended that priority attention be
given to the possibilities for conducting experiments between
routes, zones, cities or other disaggregates thereof.  In the
case of bus systems, the workshop noted that different system
variables could be set at different levels on different
routes or in different areas and the effects systematically
analysed.  In this way, a systematic set of findings could be
amassed over time as a product of controlled experimentation.

REFERENCES

1. Anderson, N.H., 'Functional measurement and psycho-
physical judgment', *Psychological Review*, 77 pp. 153-70 (1970).
2. Anderson, N.H., 'Information integration theory: a
brief survey', in *Contemporary Developments in Mathematical
Psychology Vol. 2*, D.H. Krantz, R.C. Atkinson, R.D. Luce and
P. Suppes (eds), W.H. Freeman, San Francisco (1974).
3. Anderson, N.H., 'How functional measurement can yield
validated interval scales of mental qualities', *Journal of
Applied Psychology*, 61, 6, pp. 677-92 (1976).
4. Burnett, K.P., 'The dimensions of alternatives in
spatial choice processes', *Geographical Analysis*, 5, pp. 181-
204 (1973).
5. Burnett, K.P., 'Tests of a linear learning model on
destination choice: applications to shopping travel by a
heterogeneous population', *Geografiska Annaler* (1976).
6. Burnett, K.P., 'Disaggregate behavioural models of
travel decisions other than mode choice: a review and contribution to spatial choice theory', *Transportation Research
Board, Special Report 149* (1974).
7. Costantino, D.P., Golob, T.F. and Stopher, P.R.,
'Consumer preferences for automated public transportation
systems', *Transportation Research Record 527* (1974).
8. Davidson, J.D., 'Forecasting traffic on STOL', *Operational Research Quarterly*, 24, 4, pp. 561-9 (1975).

9. Dawes, R., 'Graduate admissions variables and future success', *Science*, 28 February, pp. 721-3 (1975).
10. Demetsky, J.J. and Hoel, L.A., 'Modal demand: a user perception model', *Transportation Research*, 6, pp. 293-308 (1972).
11. Dobson, R. and Kehoe, J., 'Disaggregated behavioural views of transportation attributes', *Transportation Research Record 627* (1974).
12. Golob, T.F., 'Attitudinal models', Urban Travel Demand Forecasting, *Transportation Research Board Special Report No. 143*, pp. 130-45 (1973).
13. Golob, T.F., Horowitz, A.D. and Wachs, M., 'Attitude-behaviour relationships in travel demand modelling' (Chapter 34 of present book).
14. Hartgen, D.T. and Tanner, G.H., 'Individual attitudes and family activities: a behavioural model of traveller mode choice', *High Speed Ground Transportation Journal*, 4, pp. 439-67 (1970).
15. Helson, H. (1964) *Adaptation Level Theory*, Harper and Row, New York.
16. Hensher, D.A., 'Use and application of market segmentation', in *Behavioural Travel-Demand Models*, P.R. Stopher and A.H. Meyburg (eds), Lexington Books, Lexington (1976).
17. Hensher, D.A., 'Perception and commuter modal choice - an hypothesis', *Urban Studies*, 12, pp. 101-4 (1975).
18. Hensher, D.A., McLeod, P.B. and Stanley, J.K., 'Usefulness of attitudinal measures in determining influence on mode choice', *International Journal of Transport Economics*, 2 (1975).
19. Hills, P., 'Summary of downtown Leeds transport changes', *University of Leeds Transport Institute, Technical Report* (1976).
20. Keeney, R.L. and Raiffa, H. (1976) *Decisions with Multiple Objectives: Preferences and Value Trade-offs*, John Wiley, New York.
21. Kelley, E.L. and Fiske, D.W. (1951) *The Prediction of Performance in Clinical Psychology*, University of Michigan Press, Ann Arbor.
22. Krantz, D.H. and Tversky, A., 'Conjoint measurement analysis of composition rules in psychology', *Psychological Review*, 78, pp. 151-69 (1971).
23. Levin, I.P. and Gray, M.J., 'Analysis of human judgment in transportation', paper presented to the Special Sessions on Human Judgment and Spatial Behaviour. Great Plains/Rocky Mountains American Association of Geographers Meetings, Manhattan, Kansas (1976).
24. Levin, I.P., 'The development of attitudinal modelling approaches in transportation research' (Chapter 35 in the present book).
25. Louviere, J.J., 'Psychological measurement of travel attributes', in *Determinants of Travel Choice*, D.A. Hensher and M.Q. Dalvi (eds), Teakfield, Farnborough, England (1978).
26. Louviere, J.J., Ostresh, L.M., Henley, D. and Meyer, R., 'Travel demand segmentation: some theoretical considerations related to behavioural modelling', in *Behavioural Travel Demand Models*, P.R. Stopher and A.H. Meyburg (eds), Lexington Books, Lexington (1976).
27. Louviere, J.J., Piccolo, J.M., Meyer, R.J. and Duston, W., 'Theory and empirical evidence in real-world studies of human judgment: three shopping behaviour examples', *Center for Behavioral Studies Research Paper No. 1*, Institute for Policy Research, University of Wyoming (1977).

28. Louviere, J.J. and Norman, K.L., 'Applications of information processing theory to the analysis of urban travel demand', *Environment and Behaviour* (1977).

29. Michaels, R.M., 'Behavioural measurement: an approach to predicting travel demand', *Transportation Research Board, Special Report No. 149*, pp. 51-7 (1974).

30. Nicolaidis, G.C., 'Quantification of the comfort variable', *Transportation Research*, 9, pp. 55-66 (1975).

31. Norman, K.L., 'Information integration theory and the attributes of bus service', *Journal of Applied Psychology* (1977).

32. Norman, K.L. and Louviere, J.J., 'Integration of attributes in public bus transportation: two modelling approaches', *The Journal of Applied Psychology*, 59, pp. 753-8 (1974).

33. Rohrbaugh, J., 'Cognitive maps: describing the policy ecology of a community', paper presented to the Special Sessions on Human Judgment and Spatial Behaviour. Great Plains/Rocky Mountains American Association of Geographers Meetings, Manhattan, Kansas (1976).

34. Rushton, G., 'Analysis of spatial behaviour by revealed space preferences', *Annals of the Association of American Geographers*, 59, pp. 31-40 (1969).

35. Shanteau, J. and Phelps, R.H., 'Livestock judges: how much information can the experts use?', *Human Information Processing (KSU-HIPI) Report No. 76-15*, Department of Psychology, Kansas State University (1976).

36. Stewart, T.R. and Gelberd, L., 'Capturing judgment policies: a new approach for citizen participation in planning', *Proceedings, URISA*, Conference, San Francisco (1972).

37. Stopher, P.R., Spear, B.D. and Sucher, P.D., 'Toward the development of measures of convenience for travel modes', *Transportation Research Record*, 527 (1974).

38. Westin, R.B. and Watson, P.L., 'Reported and revealed preferences as determinants of mode choice behaviour', *Journal of Marketing Research*, 12, pp. 282-9 (1975).

PART ELEVEN

Chapter 37

EVALUATING THE SOCIAL AND ENVIRONMENTAL IMPACTS
OF TRANSPORT INVESTMENT

Donald Appleyard

INTRODUCTION

This chapter deals with the conceptual issues involved in
evaluating the impacts of transport systems. Like most other
planning fields, the present state of impact assessment is
fragmented. Techniques developed to measure the impacts of
freeways concentrated on analysis of cheaply obtainable
secondary data from the census, assessed values, accident
statistics, crime statistics, and anything else available.
In the real political world, freeways were frequently halted
on the basis of  community opposition defending socially or
environmentally valued areas, with only a minimum of impact
measurement (e.g. houses or jobs taken). As impact analysis
evolved, a few direct environmental measures such as those of
noise, sedimentation in the water, air pollution, visual
intrusion or likelihood of flooding have been added to the
checklists. In many cases these measures have been selected
on the basis of professional judgement, availability of data
or ease of measurement.

More recently research has moved towards measuring the
psychological and behavioural responses of surrounding
residents to transport systems or changes. Among these have
been studies of the impacts of the BART rapid-transit system,
of freeways, of traffic and of airports.

There are now therefore three parallel measurement trends
in impact analysis, the first searching for economic indices
such as land value, rents or other indicators of migration
behaviour, the second focusing on environmental measures, and
the third on attitudinal responses. My concern is to develop
a conceptual model of the chain of social-environmental and
functional interactions that take place around transport
facilities, in order to relate these current measurement
points in the impact process, and to show how inadequate
present measures are in gaining a true picture of the impact
process. No doubt this model is also incomplete and will
benefit from extensive discussion. Others may see it as
unnecessarily detailed, and beyond the budget of any current
impact study. If, however, we can lay out all the impact
relationships we may see more clearly at which points
measurement is more effective. Strategies of measurement
selection are discussed at the end of the chapter.

A CONCEPTUAL MODEL OF ENVIRONMENTAL IMPACTS

Consider the impacts of a hypothetical transport system first
under relatively stable conditions, then under conditions of
change.

*The Stable Impact Model*

The model diagrammed in Figure 37.1 describes transport-
environment relationships at two points in time.  The left-
hand column describes relationships under relatively stable
conditions; the second, later in time, describes them under
conditions of change which may have been initiated in several
ways.  Each column graphs the *seven* main components of
environmental impact; the *traveller needs and preferences*
that create the transport system in the first place, the
*transport system* under study, the *transport and land use
generated* by the studies system and that derived from other
*external causes*, the *emissions* created directly and indirectly
by the system, the environmental factors which *mitigate or
magnify* the impacts, the *environmental impacts* themselves,
and the *impacted populations*.  These entities interact back
and forth in a feedback system in which the results of
traveller needs and preferences impact surrounding populations,
and those populations endeavour to minimise the negative
impacts.  The model is, of course, complicated by the fact that
travellers and impacted populations are sometimes the same
people, torn by internal conflict.

*Traveller needs and preferences*.  The desire for mobility is
what creates transport facilities, whether it be a neighbour
walking past one's front door, or an airport which impacts a
sizeable area of landscape.  Impacts can therefore be affected as much by changes in traveller demand as by a facility
itself, and the impact of most systems is directly related to
the number of travellers they have to carry.

*The transport route*.  The transport system itself contains a
number of environment-affecting elements.  *People and vehicles*
(e.g. cars, buses, trains, airplanes, bicycles, pedestrians)
are the primary and mobile elements of the system; *channels
and change-points* (e.g. streets, freeways, track tunnels,
airports, off-ramps, bus stops, stations, ports) make up the
support and enclosure facilities.  These combine into *systems*,
whose impact we treat only briefly in this chapter.

It should be noted that while the people and vehicles on
the transport system can change gradually over time, growing,
diminishing or changing composition, facilities are lumpy
investments, have lumpy impacts when constructed, and have
therefore been the focus of most controversies.  Traffic
creeps up on communities slowly and imperceptibly.

*Generated and extraneous systems*.  Transport often generates
transport, especially at change points.  Bus routes generate
pedestrians at bus stops, transit stations can generate cars,
buses, bicycles and pedestrians.  As with transit systems,
the secondary transport impacts may be more negative than the
direct impacts of  the facility.  An underground BART line
may have no impact on the above residential neighbourhood,
but the station may have powerful effects.

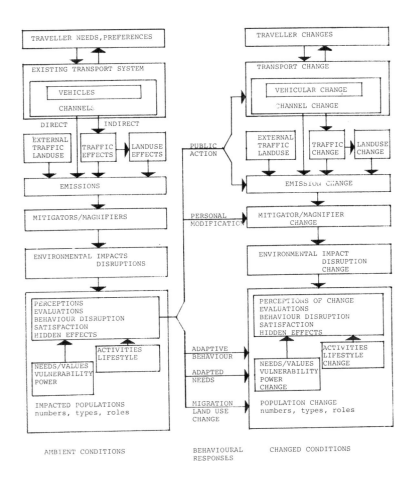

Figure 37.1   The Socio-environmental Impacts of Transport
Change

Transport also provides a market. It generates land uses, again usually at change points. Shopping centres have clustered at freeway intersections, commercial uses have gathered around transit stations, industry around airports. However, it is extremely difficult to separate out the causal effects of land-use generation, since land uses locate for combinations of reasons, and since most urban areas are served by mixed transport systems. Transport systems extraneous to the one studied may generate their own transport and land uses. All of these may impact on a social and physical environment. In complex urban situations such as downtown San Francisco, the effects of BART on such land uses as office construction remains a matter of opinion (20). In simpler suburban situations these impacts may be possible to sort out.

*Emissions.* While the more commonly identified emissions are those of noise, vibration and air pollution, emissions such as danger, visual images, glare, darkness and the presence of travellers with prying eyes are of similar character. Emissions, at the source point of the transport system, may or may not impact directly the surrounding population, and therefore may not necessarily require measurement unless identification is required for their direct control, as in the case of controls over vehicle emissions of various kinds.

*Mitigators and magnifiers.* Between the transport route and the impacted populations the emissions can be dampened or accentuated. Channels and change points can in many ways mitigate the impact of transport vehicles. Streets by their width, channel design, parking, barriers or landscaping can modify the impact of traffic. Distance from the facility itself modifies impact. On the other hand, glare-creating sound-reflecting channel walls can accentuate discomforts. Facilities that are designed to be compatible with the neighbourhood are likely to be more acceptable than those which affront its character. Identification of these elements can be crucial in situations where emissions cannot be controlled and compromise is necessary.

*Environmental impacts and disruptions.* The point of impact or disruption is the point where it is likely to hurt. There is an environmental interpretation of this point and a social interpretation. Both are necessary to understand complete impacts. The environmental interpretation tends to measure the spatial distribution of impact over an impact zone. A typical measure is that of noise level. Noise contours are now commonly measured around transport facilities. Similarly, it is possible to measure air-pollution levels, erosion, water pollution and a chain of ecological effects. The disruption of existing environmental connections and systems, such as the blockage of watercourses, the disruption and siphoning of wind patterns, the interruption of views, and neighbourhood character can also be recorded. These impacts can be measured at any point in the environment surrounding the transport facility. Impact analyses of transport facilities travelling through natural landscapes have commonly employed a series of contour maps, sometimes computer-generated, which have recorded existing environmental qualities and in some cases predicted environmental impact (19).

A social interpretation of environmental impact concentrates more on where people are or are likely to be in the surrounding environment, and takes measures at these points. Instead of treating the environment as the equi-important space, the places where people live, meet or carry on other activities are the primary points of environmental measurement. Hence, in the BART impact study we planned to carry out measures at individual houses, on street blocks, at community facilities, and on local access paths (1), and in Lassiere's work on evaluating transport plans, impacts are measured at each home site (13). Similarly, in a study of the impacts of high-rise buildings in San Francisco, Dornbusch calculated the number of pedestrian exposures to high noise levels from the extra traffic generated by predicted high-rise buildings. Such measures assume that the environmental impacts on facilities or activities will be a cause of disturbance, and in many cases assumed perceptions are extrapolated from other research studies.

*Impacted populations*. The most critical impacts are on people, the environments they value, the resources that they command and their sense of power or powerlessness. In the final calculus the question is who has gained and who has lost, in what ways, by how much, for how long?

*Population characteristics*. The social and psychological impacts of transport systems ultimately depend on which populations are being impacted, how, and in what degree. Neither are people all alike. Some are more susceptible to impact than others.

The most important strategy is to search for those most *intensely impacted*, that is, those who experience the most intense negative environmental impacts and social disruption for the longer duration, and to search for those most *vulnerable* to impacts. Vulnerability, a term coined in the Buchanan Report, refers to those most sensitive and least able to cope with negative environments, for instance, families with small children who are especially vulnerable to traffic accidents, to the old and infirm who may be most sensitive to local access, to low-income groups least able to move away when conditions worsen. Who belongs to the most vulnerable groups is still an empirical question. It is therefore necessary to look at a number of personal characteristics of the impacted population, particularly income, age and family composition but also length of residence, time spent in the home, travel modes and patterns, home ownership, health problems and infirmities, ethnic and sex differences.

Any one person will be playing a number of *roles*. Assessments must therefore be careful to avoid double-counting. In a politicised context, populations may form interest or neighbourhood or block groups, with coherent viewpoints about the impacts on their constituency. Perceptions of the impacts on activists will have to be taken into account, though there is no substitute for assessing the impacts on the silent majority. The actual and perceived *power* of impacted populations may also affect their response to a transport facility. Around San Francisco airport more complaints are received from affluent residents who live in the

hills at a distance from the airport, than from the poorer
groups who live under the flight patterns.  Power may depend
on income and education, or on group cohesion.  Those who
have a sense of power or real power are more likely to
articulate any negative impacts.

An accounting of the numbers of different groups impacted
by a transport facility will begin to clarify its distributional effects.  But the question remains how much and in what
ways are people impacted.  Moderate income groups, for
instance, may feel more vulnerable to impacts that affect rent,
local access and neighbourhood facilities, while affluent
groups may be more concerned about impacts on the natural
landscape.  There is no single index of impact.

*Activities and travel.*  The activities in which people
engage or desire to engage may affect their vulnerability to
negative environmental impact.  People are interrupted by
traffic more when trying to sleep, when talking or watching
television, than when eating (2).  Ball-playing in the street
is more vulnerable to traffic impact than gardening or
chatting.  Residential land use in general is more susceptible
to negative impacts than are commercial or industrial uses.
Similarly, walking trips are more likely to be disrupted than
car trips.

Indoor activites are defended from impacts by protecting
facilities such as homes, gardens, or any building which
provides shelter.  The numbers of people engaged in different
activities and the degree of activity disturbance or
disruption can thus become another measure of social impact.
The British, for instance, have concentrated on 'severance'
as an index of the disruption of neighbourhood access and
interaction.  It has been measured by changes in length of
pedestrian trips (3) across a freeway alignment, based on
origin-destination surveys.  More recently they have
concentrated on changes in perceived distance (14).  The
activities people engage in can be assessed   through behavioural observations and land-use surveys.

*Perceptions, evaluations and behavioural disruption.*
Perceptions and evaluations arise from the interaction of
environmental impacts and the population's characteristics,
usual activities and travel patterns.  Important also is
each individual's background experience, especially recent
experience, his past and present expectations and value
dispositions.  For instance, more-educated groups have higher
expectations and are thus more likely to evaluate environments critically.  Educated groups are also likely to be more
aware of negative impacts, hidden effects and better
alternatives than are the less educated.  Even so, some
environmental impacts will probably go unperceived, due to
behavioural adaptations or their invisible nature.

The range of perceived environmental impacts and behavioural disruptions is impressive.  Appendix A identifies a number
of those that have been mentioned in resident surveys.
These impacts are described in *negative* terms as increasing
with the presence of a transport system.  This is only one
side of impact, however.  Impacts can also be positive.

They can decrease hazards, improve micro-climate, orientation and reduce noise levels. The critical measure is the *change* from previous levels.

The list could be more finely divided or more compactly grouped. These are matters of judgement. Under stable conditions the losses, destruction and disruptions will probably be forgotten and therefore be seen as unimportant, whereas in conditions of change such impacts will come to the fore. These impacts and disruptions are usually listed as separate, yet there are several causal relationships between them. Environmental conditions, for instance, affect social linkages, privacy and sense of turf. It is beyond the capacity of this chapter, however, to outline such relationships.

The most important information concerning these impacts is how *significant* the impacted populations consider the impacts to be, which impacts are of most concern, i.e. to which ones are they most *vulnerable*, and which impacts do they consider to be the most *intensive*.

In appearance, responses and behaviour are a much softer area of impact assessment than environmental measurement. Subjective judgements are more likely to vary between groups, there will probably be no consensus, and yet they may well change over time. It seems like quicksand. Yet the chance of keeping in close contact with what the impacted population feels is politically less hazardous than the pursuit of professional measurements that may have no popular relevance. Besides, there are scaling techniques, sampling and survey-research methods that provide a level of reliability in assessing responses even to some of the more intangible kinds of impacts.

*Satisfaction, disturbance and acceptance.* In impact measurement there is frequently a quest for the single index of evaluation. Are people generally satisfied, how much are they disturbed, how much are they willing to accept? People can respond to such questions but their responses vary, for instance, depending on how subjective or objective the question is. If they are asked their personal feelings about an environment, responses are more likely to be affected by *personal dispositions*. If they are asked to evaluate the environment, they respond more to *environmental conditions* (6). If asked to compare their situations with others, their responses will differ from their direct response to one situation.

Even if we can obtain some reliable estimate of satisfaction, for instance, by creating an index from a number of questions, there is the still more fundamental question of how serious do we consider dissatisfaction to be, and how many have to be dissatisfied before we consider the impact to be unacceptable? In impact studies by agencies, who wish to construct facilities, it is tempting to minimise dissatisfaction by trying to determine levels of acceptability. Opponents, on the other hand, are after higher levels of environmental quality than mere satisficing. The establishment of standards of acceptability will necessarily be a

matter of debate.  Proposals for standards of traffic capacity
for residential streets presently range from 2,000 vehicles
per day to 14,000 per day (2)!

*Power.*  Perhaps the most significant effects of transport
investments are their differential impacts on the economic
and political power of  various groups.  The effects of
environmental impacts on the economic resources of different
groups, on their employment, markets and land values, and the
effects on their political power, community or group solidar-
ity, must be recognised.  These are seen as beyond the scope
of this chapter, but must not be ignored for those reasons.

*Hidden effects.*  Some impacts may only be discerned by
scientists or by professionals who measure impact.  Health
hazards are the most obvious of these.  The effects of air
pollution, prolonged exposure to high noise levels, perceptual
overload, or lack of exercise may all have serious effects on
health, effects which have yet to be discovered.  No list of
perceived and measured impacts can therefore be taken as final
and definitive.

*Behavioural responses.*  While we have now described the
possible impacts of a transport facility, we have so far not
dealt with people's responses to these conditions.  These
modes of response must be taken into account, for they may be
instrumental in dampening or accentuating levels of dissatisfaction.

*Adaptation.*  People may simply adapt their perceptions and
behaviour to an intrusion.  They may psychologically screen
out noises, either by not mentioning them, or by habituating
themselves so well to them that they pass unnoticed.  They
may engage in adaptive behaviour, by moving to the back of
the house, discontinuing use of the garden, prohibiting their
children from playing on the street, accompanying them to
school.  Finally, they may adapt their value systems to the
situations, modifying their demands for a better environment,
resigning themselves to poor conditions.

Adaptation-level theory may help explain this latter
phenomenon.  It argues that people adapt to certain levels of
environmental stimulus, depending on their previous 'pooled
experience'.  They tend to become indifferent to these
conditions, reacting more strongly as situations deviate from
the adaptation level.  Under low exposure to changed stimuli,
adaptation level will change (11).  This theory suggests that
an assessment of adaptation levels, and adaptability, be a
matter of attention in environmental impact research.

One kind of adaptation is to be found frequently in
interviewing, when residents act defensively in front of a
strange interviewer.  Admission that chosen residential
conditions are less than acceptable would be an admission of
poor judgement, or lack of resources.  Some of this behaviour
can be examined with the theory of cognitive dissonance
((10) and Chapter 34 of present book).  When people have made
a decision, they tend to reduce or screen out any dissonant
or negative factors in that choice in order to maintain a
consistent view  of their condition.  This is somewhat

similar to the 'halo' effect, wherein the beneficial factors in a situation so colour overall attitudes that negative factors are discounted. These attitudes are probably much more common among less-educated groups who engage in what anthropologist Rosalie Cohen calls 'relational thinking' as contrasted to 'analytical thinking' (5).

Adaptation is not always painless. Some of the more severe environmental impacts, especially those which cause loss of the home, or valued neighbourhood characteristics, can cause grief. It will be important to look behind the facades and into histories of adaptation if the true impacts are to be measured (15). For as Rene Dubos (9) reminded us, man adapts too easily to adverse environmental conditions, at unknown cost. Karl Marx used the term 'false consciousness' to describe working-class acceptance of such conditions. It takes a long time to cast off such adaptations. On one London street reported in Britain's *Community Action Journal*, four children were killed by passing traffic before the residents finally protested.

*Modification*. Personal modifications to the environment to minimise or mitigate impacts are a common occurrence. Closing windows (35 per cent of our San Francisco respondents kept their windows closed because of traffic), drawing blinds, locking doors, erecting fences, planting vegetation or installing outdoor lighting are just some of the modifications people make. Some of these are to their own houses, some to the surrounding property.

*Public action*. In the case of severe impacts, people may become more active. Some engage in environmental behaviour such as erecting barricades, milling-in on the street, sitting in front of the bulldozers, or chaining themselves to trees. Others protest through social channels, signing petitions, attending meetings, voting for politicians, organising neighbourhood groups. The intent of these actions is to bring about more powerful modifications to the environmental impact of the transport facility, such as reducing vehicle volumes, diverting trucks, the construction or elimination of a new facility or some other change such as emission controls.

*Migration and land-use change*. If conditions deteriorate beyond a certain point and protests are ineffective, residents and others may move out, although environmental conditions are not the only reasons for moving. In some cases they are forced to do so by the transport agency. But when people move out, others move in. Such changes in population and land use are a common consequence of a transport facility. Studies of BART have shown that incomes rise with distance from the tracks, although this may be due more to the influence of the existing railroad rights of way and arterial streets than to BART itself. Similarly lower-income groups tend to live around airports. On the other hand, in Britain no correlation was found between traffic volumes and the income of residents on those streets (10). There is, however, a clear tendency for commercial land uses to locate and be zoned for more heavily-trafficked streets.

These four kinds of adaptive behaviour seek to work out
satisfactory solutions to an environmental conflict.  While
three of them are personal adjustments to the system, public
action attempts to modify the system by prevention or
amelioration of its negative effects.  The crucial point for
impact analysis is that this kind of adaptive behaviour
reduces dissatisfaction and makes the impact of the transport
facility appear to be less negative than it actually is.
Those who remain may be screening out the impacts at some
psychological or financial costs, others who would not
tolerate it have been forced to leave.  It will be tempting
for transport agencies to ignore these hidden impacts, but it
is essential that such behaviour be monitored if true measures
of impact are to be obtained.

A stable situation is one in which these adaptive adjustments have been made as far as possible within the constraints
in which people operate.  Since people are not entirely free,
we will always find some who have been unable to adapt.  The
'deprived' who lack the wish to or cannot move, frequently
are among those forced to remain living by transport
facilities.  We have found old people on heavily-trafficked
streets unable to sleep at night for the traffic, yet unable
to move.

*Impacts of a Transport Change*

We have discussed a relatively stable situation that evolves
around transport facilities when they themselves stabilise
their volumes of traffic.  When a new transport facility, or
a dramatic change in vehicle characteristics or volume is
planned, the impacts must be seen against such ambient
conditions.

The conditions of a transport facility will change and to
some extent accumulate over time, as the facility itself
changes.  There appear to be four main phases in the impact
pattern of a new facility: the *planning phase*, when planning
is under way before physical changes have yet occured; the
*construction phase*, which, in the case of large investments,
may take many years; the *opening phase*, when construction is
complete and vehicles begin to run or fly; and finally the
new *stabilisation phase*, when major changes cease to take
place.

*The planning phase.*  During this phase when plans are underway, no primary environmental impacts occur beyond those that
are in the minds of the impacted populations, although these
may lead to various kinds of satisfaction, and adaptive and
anticipatory behaviour resulting in secondary impacts.  During
this phase, the important factors are the simulations made
of the proposed investments, the degree of diffusion throughout the impacted population, and the population's previous
experience with similar investments.

The *simulations* include verbal reports, such as environmental impact reports as well as visual presentations, graphs,
diagrams, plans, perspectives or models.  Many of these are
subject to distortion.  The proponents of transport

investments seek to minimise any suggestion of negative
impacts.  Robert Caro in his book on Robert Moses (4) for
instance, described in detail how the presentation perspective of his Lower Manhattan bridge was shown from a bird's eye
view which portrayed the bridge as a thin matchstick-like
structure touching down on the Battery Park, whereas in fact,
the piers were to be six-storey massive concrete piers,
blocking the view of all the office property along the park
front.  An opponent's presentation, on the other hand, took a
viewpoint from below ground level using equivalent exaggeration to depict the worst aspects of the project.

The institutionalisation of environmental—impact reports
has increased substantially the amount of information on
projected impacts of transport projects, even though they have
been criticised for irrelevancy, inaccuracy and deliberate
distortion (8).  Such impact reports unfortunately rely on the
shaky base of the few empirical studies of impacts that have
been carried out.  Their predictions are inevitably open to
question since each situation is unique.  Whereas physical
impacts such as shadowing visual intrusion and description of
access are predictable, behaviour, its emissions and response,
may depend on unpredictable events such as accidents or crimes,
political elections, economic recession or other changes in
the historical or local climate of values.

The *diffusion* of information about the project is another
issue.  For many years, if projects were quietly developed
behind the scenes, they could often be announced at the last
moment, with little participation from any impacted populations.  This might lead to shock and last-minute protests.
More commonly today, information and participation diffuse
information about investments early in the planning stages.
Even so, it is not uncommon for opponents or losers to be
left out of the participatory process.  In either case the
efforts are to stress the positive aspects of preferred
schemes.  In the anticipatory stage, then, impacts are subject
to distortion, rumour and exaggeration.  Controversy may
bring out different opinions on impacts and increase the
quality of impact prediction.  However, this will depend on
the degree of expertise available on each side of the
controversy.

Analogies from previous transport planning experience and
the credibility of the proponent agency will do much to create
or allay suspicion among the impacted populations.  Prior to
the opening of BART, Bay Area attitudes towards the system
were very positive, yet few in our interviews had ever travelled on a transit system except for Disney Land's monorail.
Such lack of experience meant that BART publicity was
virtually the sole source of information about the system, a
highly unpredictable situation.

An example of anticipatory impacts was the spate of office
building that took place in San Francisco before and during
the construction of BART.  Anticipatory attitudes were to be
found around other BART stations, where substantial numbers
feared that future development would impact their neighbourhoods in a negative manner.  Expectation and fears are
therefore the primary psychological impacts in this phase and
can cause migration, change of land use, protests or other
forms of response.

*The clearance and construction phase.* This is frequently the most negative phase of a transport project. In larger schemes, sites have to be cleared, people may be moved from their homes, work places eliminated, whole neighbourhoods with shopping centres and other valued facilities may be demolished and replaced. This first part of the phase is predominantly destructive with few compensating features, unless the area being destroyed is universally disliked or viewed with neutral feelings.

Clearance and destruction also generate trucks and other heavy vehicles which create dirt, dust, noise, vibration and other unpleasant environmental impacts that lower the social image of an area. In the construction phase the new facility emerges from the noise, mud and dust. Although this phase is usually minimised in transport planning and forgotten once the new system is open, it is frequently so painful that impacted populations migrate or simply go out of business. The construction of a major subway or freeway system can change the patterns of traffic, and therefore, business for numbers of years - lengths of time during which some businesses are unable to survive. During this time the appearance of the surrounding areas can be blighted, crime may increase and permanent damage can be done.

During this phase opposition may peak as opponents try to prevent construction taking place. While some have sat in front of bulldozers to stop freeways, others have attacked workmen installing street barriers to protect neighbourhoods from through traffic. Attitudinal surveys taken during this phase may find little praise for the facility except that which looks forward to its ultimate completion. Towards the end of construction, on larger projects like fixed-rail transit systems, there is sometimes a hiatus while the finishings are complete and before the system is opened.

*The opening phase.* When vehicles begin to run, the purpose of the investment is fulfilled, the feared or hoped-for impacts are actually felt. This too can be a highly charged period. It could be a 'honeymoon', or a disaster. There may be surprise or disappointment. If the investment is large it will capture the attention of the media, and if controversial, may stay in the public eye for several months or years. This will be a time when people are most aware of and articulate about environmental impacts, since change has just occurred. Traffic, noise, nuisance levels, strangers, parking problems and visual description can suddenly materialise. Later on people may adapt to these changes, habituate themselves to new routines and screen out the negative impacts, but meanwhile it can be an important time to interview people for they are especially articulate about the impacts at this time. Transport agencies, however, are usually unhappy about this prospect, frequently arguing that attitudes in the opening phase are 'unfair' to the project since the bugs will not have been worked out. Yet, the environmental impacts of this opening phase may balance negative with positive. While the negative intrusions may be seen at their most acute, the newness and cleanliness of the facility usually cast it in a positive light.

*The stabilisation phase.* This phase may be the most difficult to identify, since stabilisation may never quite take place. It may be characterised by a stabilisation of attitudes, of behavioural adaptations, of physical modifications or populations and use change. All these phenomena, including patronage of the facility and surrounding vegetation may continue to change without ever stabilising. The only way to assess this will be repeated surveys over time.

## THE COMPARATIVE NATURE OF IMPACT ASSESSMENT

Clearly transport impacts are relative in nature. They change over time and are obfuscated by extraneous influences. All this makes accurate assessment of environmental impacts an almost impossible and very costly task.

### The Impact Profile

There is rarely any clearly definable *before* situation to compare with what may never be a stabilised *after* situation. Since impacts may peak at certain points in the process, particularly after construction begins, and at the time of the opening, continuous or at least intermittent impact measures may be the only way to catch the longitudinal profile of impact.

### Secular Effects

Unfortunately the transport-impact process is not isolated from the rest of the world. Several other transportation, land-use, population and environmental events may be taking place in the impact zone of the facility during the period of measurement. For example, in the case of BART, if automobile traffic increases in the Bay Area generally, with its attendant congestion, noise, air pollution, BART-area residents may benefit - relative to others - if use of BART maintains automobile traffic at its present level in their neighbourhoods, or even if the increase in their local traffic is less extreme than it is in areas distant from BART. In order to gain an accurate assessment of transport impacts, it is necessary therefore to evaluate environmental and social changes in areas identical to the impacted zone, in places where no equivalent transport facility has been placed. In the BART study, residential neighbourhoods equivalent to two of the impacted neighbourhoods were selected as control sites to assess these secular changes.

### What Would Have Been?

Finally, comparisons should be made between what actually happens and 'what would have been' had there been no transport facility - the No-build Alternative - (7) or had alternative investments been made. In the BART impact programme there was extensive debate over the characteristics of what was called the NO-BART alternative. A decision was finally made, for instance, to assume that a fourth Bay

Bridge crossing would not have been built.  As can be
imagined, this decision will have profound effects on assessments of BART's relative impact.


## STRATEGIES FOR IMPACT ASSESSMENT

This rather complex picture of the environmental and social
impact process creates formidable measurement problems.  If
all cannot be measured, how do we select samples?  Selection
will vary depending on whether the assessment is to be an
empirical study of an existing system, or a prediction of
future impacts.


### Criteria for Selection of Impact Measures

In order to arrive at priority measures some criteria are
necessary.  The following appear to be the most critical:

*Relevance to impacted populations.*  The most important
criterion from a social and environmental viewpoint is relevance to the population impacted, whether they are aware or
not.  The question then is relevance to whom?  This criterion
is more often forgotten than one would imagine.  In too many
impact studies, lists of professionally-selected environmental measures are taken without any reference to populations.
In several known cases these measures have been taken only to
find out later that the impacted populations have different
concerns.

*Comprehensiveness.*  Measures should not omit impacts that are
or may be significant to some group.

*Reliability.*  Measures should have some reliability, and
their gathering should be the object of careful objective and
explicit sampling and measurement.

*Public comprehension.*  In most impact studies public comprehension of the results is extremely important, especially
if policy is to be affected.  Measures which are meaningful
to the public are therefore more powerful than those which
are obscure, technical or otherwise incomprehensible.

*Cost.*  The criterion that ultimately and, in most cases,
primarily determines the selected impact measures and the
sampling procedures is cost.  Unfortunately, this often
results in taking the easiest measures rather than the most
relevant.


### Priority Measures

Intuitively weighing the above criteria, especially relevance,
the following might be the order of measurement priority.
Note that these are points of measurement on the impact model
presented in Figure 37.1.  The methods of measurement are
shown in the parentheses.

*Population groups.*  Identify the principal interest groups

among the population who are likely to be most intensely
impacted and most vulnerable to impacts and their relative
resources.

*Population values.* Following the criterion of relevance, the
first measures should estimate what the impacted population
holds to be of most environmental and social value, which
environmental qualities they judge to be most important
(interviews, meetings with cross-sectional sample).

*Derived impact measures.* From the above desired qualities,
environmental quality measures could be derived. This was
the strategy of the BART Residential Impact Study (1). The
principal impact measures can be taken, preferably at points
of population vulnerability, such as homes, activity centres,
access routes (field surveys, aerial photo, map analysis).

*Distributive effects.* The distribution of the population in
the zones of various impacts can provide some crude measures
of the numbers being impacted. The numbers of residents
subject to different traffic volumes is, for instance, a
useful measure, even though the correlation of traffic volume
with environmental impacts is complicated (2) (map analysis,
census, traffic flows).

*Perceived impacts.* Directly asking the population to
evaluate impacts will provide the most relevant and publicly
comprehensible information, even though it is more costly.
It will also act as a check on the environmental measures
(interviews, meetings).

*Behavioural impacts.* Changes in population composition and
land use will be useful but subject to time lag and census
collection. Micro-behavioural impacts, like accidents, also
take time before patterns emerge. Interference with activities such as street play are more difficult to obtain (census,
observation, interview).

*Impact sources, mitigators and magnifiers.* The identification of impact sources and the separation of impacts from
different facilities begins to be very complicated. Carefully sampling different situations becomes necessary
probably using quasi-experimental research designs. Environmental impact depends here on land-use impact methods.

*Hidden effects.* Health effects require extremely careful
epidemiological or laboratory studies.


CONCLUSION

Where budgets are constrained it may only be possible to
obtain measures of the first four categories. Even so this
would entail meetings and interviews with selected members of
the impacted population. This is still an unusual procedure
in most environmental-impact assessments, which rely primarily on Derived-impact Measures. The goals of basic research
in the field should be to endeavour to validate and correlate
the various measures in order to identify the most relevant
and reliable. This would involve taking samples of the whole

array of measurements and correlating them with each other.
An attempt was made on the BART Residential Impact Study to
carry out such a wide array of measures.

APPENDIX A: THE NEGATIVE IMPACTS OF TRANSPORT SYSTEMS; AN
INCOMPLETE LIST

| | |
|---|---|
| Hazards | impacts which *increase* the chance of earthquake, flood, fire, unfenced heights, deep water, etc. |
| Traffic hazards | danger due to vehicles, live rails, etc. |
| Crime | danger due to criminal activity, street assault, burglary, car theft, etc. |
| Health | threats to health through air pollution, lack of sunlight, water pollution, garbage and trash. |
| Exercise | constraints on physical exercise through lack of athletic facilities, walking difficulties, etc. |
| Overcrowding | lack of space to carry out desired activities, loss of parking facilities,etc. |
| Loss | loss of homes, work places, community or other facilities. |
| Noise, vibration | noise and vibration levels which interfere with desired activities and are unpleasant. |
| Air pollution | fumes, smells, smoke, dust and smog which are unpleasant to breathe, a nuisance to clean and which obscure desirable views. |
| Darkness, glare | blockage of light into buildings, shadowing of public and private spaces, glare of headlights into windows, bright reflective surfaces. |
| Micro-climate | unacceptable increase in wind levels, exposure to rain, snow. |
| Disruption of access | local vehicular access to various parts of the community, cycle and pedestrian access, public access to valued environments and facilities such as shorelines, beaches, lakes, rivers, viewpoints and other amenities. |
| Disorientation | invisibility and poor signing of significant and desirable destinations. |
| Destruction of valued places | destruction of, intrusion upon, erosion and deterioration of valued natural, historic and community environments. |

| Territorial invasion | invasion of territory which individuals, streets, neighbourhoods and communities feel 'belong' to them, for which they care and feel responsible even if they are not owned. |
|---|---|
| Privacy intrusion | from prying eyes, the presence of strangers, distracting events. |
| Disruption of neighbourhood | reduction of contacts on the street, through noise, presence of strangers, unpleasant conditions. |
| Oppressiveness | environments which through their vast scale and authoritarian nature diminish human identity. |
| Reduction of choice | loss of activities, uses and places that lent diversity and life to a community. |
| Ugliness | environments which are unpleasant to the senses of sight, sound, smell or touch, or are littered with trash, dilapidated. |
| Visual disruption, intrusion | environments that disrupt the character of an existing landscape, neighbourhood or city, that block desired views. |
| Regional image | reduction in the quality of the regional image. |
| Natural loss | loss of wildlife habitats, ecological riches and other valued natural features. |
| Artificiality | environments which lack natural materials, outdoor views, open windows, vegetation. |

## REFERENCES

1. Appleyard, D. and Carp, F., 'The BART residential impact study: a longitudinal study of environmental impacts', in *Improving Environmental Impact Assessment: Guidelines and Commentary*, T. Dickens and K. Domeny (eds), University of California Press, pp. 121-6 (1974).
2. Appleyard, D. with Gerson, M.S. and Lintell, M. (1976) *Liveable Urban Streets: Managing Auto Traffic in Residential Neighborhoods*, US Department of Transportation, Government Printing Office, Washington, D.C.
3. Bor, W. and Roberts, J., 'Urban motorway impact', *Town Planning Review*, LXIII, 4, pp. 299-321 (1972).
4. Caro, R.A. (1974) *The Power Broker: Robert Moses and the Fall of New York*, Vintage Books, New York.
5. Cohen, R.A., 'Conceptual styles, culture conflict and non-verbal tests of intelligence', *American Anthropologists*, 71, 5 (1969).
6. Craik, K.H. and Zube, H.E. (eds) (1976) *Perceiving Environmental Quality: Research and Applications*, Plenum Press, New York.
7. Crane, D.A. and Partners (1975) *The No-Build*

*Alternative: Social Economic and Environmental Consequences of Not Constructing Transportation Facilities*, National Cooperative Highway Research Program, Washington, D.C.

8. Dickert, T., 'Approaches to environmental impact assessment', *DMG-DRS Journal Design Research and Methods*, G.1, pp. 10-15 (January-May 1975).

9. Dubos, R. (1965) *Man Adapting*, Yale University Press, New Haven.

10. Festinger, L. (1957) *A Theory of Cognitive Dissonance*, Row, Peterson, Evanston, Illinois.

11. Hedges, B. (1973) *Road Traffic and Environment: Preliminary Report*, Social and Community Planning Research, London.

12. Helson, H. (1964) *Adaptation Level Theory*, Holt, Rinehart and Winston, New York.

13. Lassiere, A. (1974) *The Evaluation of Transport Plans at the Strategy Level*, UK Department of the Environment, London.

14. Lee, T. and Tags, S., 'The social severance effects of major urban roads', in *Transportation Planning for a Better Environment*, P. Stringer and H. Wenzel (eds), Plenum Press, New York (1976).

15. Marris, P. (1974) *Loss and Change*, Institute of Community Studies, Routledge and Kegan Paul, London.

16. McHarg, I.L. (1971) *Design with Nature*, Doubleday, New York.

17. Perloff, H. (ed.) (1969) *The Quality of the Urban Environment*, Johns Hopkins Press, Baltimore.

18. San Francisco Planning and Urban Renewal Association (1975) *History of Intensive High Rise Development in San Francisco*, Final Report.

19. Steinitz, C., Murray, T., Sinton, D. and Way, D. (1969) *A Comparative Study of Resource Analysis Methods*, Department of Landscape Architectural Research Office, Graduate School of Design, Harvard University.

20. Webber, M.M., 'The BART experience – what have we learned?', *Institute of Urban and Regional Development, Monograph No. 26*, University of California, Berkeley (1976).

Chapter 38

BEHAVIOURAL MODELLING AND THE EVALUATION OF SOCIAL AND
ENVIRONMENTAL IMPACTS OF TRANSPORT INVESTMENT

Peter Hills and Ross King

SCOPE OF THE WORKSHOP

The aim of the workshop was to examine suitable conceptual
and model frameworks for identifying, measuring and
evaluating the effects on the community, both positive and
negative, of transport investments. It was concerned with
the changing nature of these effects (or 'impacts') over
time: e.g. at the stages of policy generation, preliminary
planning, construction or implementation and eventual
commissioning; and at the various stages in the operating
life of a policy or investment. It was concerned also with
the wider questions of conflict and competition, alienation,
disamenity and severance which the implementation of major
transport plans can bring about within an otherwise stable
community.

MAIN CONCLUSIONS REACHED

*The Relevance of Behavioural Modelling to Evaluation*

To say that every individual exists within an environment
implies that each holds some kind of perspective on his or
her place in that environment, and that each is capable of
construing objectives for adapting or improving the characteristics and use of that environment. Those objectives can
virtually always be analysed in terms of utility (i.e. preferred activities or uses of time) or ideas of equity and
the 'ethics of means' (i.e. the fairness with which material
satisfactions are distributed throughout the community and
the justice inherent in the decision rules which govern that
distribution). The fact that these objectives are sought by
individuals creates eternal (and sometimes insoluble) problems for those responsible for providing public goods and
services which materially alter the environment - for only
very rarely do individuals' objectives coincide. For the
most part, they conflict and occasionally contradict to the
extent that no community consensus (of any meaning) can be
found.

Transport systems must be regarded as typical of such
provisions within the broader environment. Certainly they
make considerable impacts upon the environment, both
positive and negative; and these impacts are judged in terms
of benefits and costs by each affected person on the basis

815

of his own valuations, which in turn are a function of his
particular perspective and the objectives he seeks.  Transport investment, therefore, can only be analysed in a context
of the overall environment and of often-conflicting social
objectives conceived at the broadest level by the community
whose needs it is supposed to serve.  Just how one can tackle
the problem of evaluation in the presence of conflicting
objectives has exercised the minds of welfare economists and
political scientists for many years.  Even the comparatively
trivial problem of determining appropriate values of time
preference, in order to discount future costs and benefits,
defies objective solution and remains a matter of academic
debate.

By contrast, behavioural modelling presupposes the
existence of a consensus (at least across significant subsets
of the population) and assumes that, once identified, that
consensus can yield the basis for predicting future social
values and preferences.  Inevitably, the scope for
behavioural modelling has to be confined to those areas where
its assumptions are valid; and above all, this assumption of
consensus, together with the implicit acceptance of consumers'
sovereignty, must limit the use of behavioural modelling to
only a rather narrow range of technically-specified problems
within the sphere of transport planning and operation.  Thus
if evaluation of transport plans is correspondingly confined
(by definition) to the comparison of valuations based upon
individuals' revealed preferences, then behavioural modelling
has an important role to play.  If, however, the evaluation
task is construed more widely, to encompass all the significant impacts on the environment, assessed and valued from an
appropriate community viewpoint, then behavioural modelling
has little to offer - or worse, may actually bias the outcome.  This led us whimsically to suggest that the next
conference be almost entirely devoted to evaluation with only
a minor role assigned to the problems of behavioural modelling.

*Identifying Transport Problems and Selecting Alternative
Solutions*

In this Trojan spirit, one can argue that the scope of
evaluation is even wider yet!  Assessment of impact is, after
all, merely a reactive process of testing the performance of
the various alternatives proposed against the objectives
previously set for them.  The initiative lies with the
definition of the problem in the first place and in the
subsequent narrowing down of the range of alternative
'solutions' whose performance will be subject to test.

To say that a problem (of any kind) exists is to judge
that the situation fails (in certain respects) to meet a
preconceived set of values, expectations or norms.  In the
case of transport problems, that judgement rests upon social
values and is therefore inescapably political in nature:
nevertheless, judgement depends upon evaluation.  Likewise,
in reducing the near-infinity of feasible solutions down to
a manageable number of alternatives - on entirely reasonable
grounds such as the cost of data collection, the time

required for computing and so on - judgements abound.  In
selecting a short-list of alternatives, even decisions as to
what is 'feasible', requiring so-called technological
evaluation, are based upon judgements about resource allocation.  Within the design space regarded as feasible, the
reduction process is aided  further by the use of 'engineering' standards and often steered towards solutions known to
be acceptable.  These and the standards on which they are
based are usually very strongly loaded politically, dealing
as they do with such matters as safety, convenience and
comfort.  Thus to rule out an otherwise competent solution on
the grounds that its safety in operation falls below some
predetermined level, implies both evaluation and political
judgement.  For, when a 'formal' evaluation is performed at
the end of a long-winded process of specifying and quantifying what is already a highly select short-list of alternatives, the chances of finding a solution anywhere near a
'social optimum' are arguably quite slim.  And yet, ironic-
ally, the techniques of evaluation are more frequently
criticised than the circumstances in which they are employed.

Perhaps one of the main reasons for this trap into which
planners and decision-makers have fallen lies in their
respective roles as they have evolved in the framework of a
mixed economy.  In the UK and Australia, for example, these
roles are typified as 'client' and 'consultant'.  The client,
usually a political body representing some constituency
within the community, seeks the advice of the consultant, who
is professionally qualified and almost invariably skilled in
the techniques of problem-solving.  Since this advice is by
no means free, it is only sought when conditions are (or
threaten to be) 'inadequate', 'substandard', 'dangerous', or
in some other way 'unacceptable'.  These value-loaded terms
reveal not only the political nature of problem identification but the lack of rigour and the dearth of relevant data
on which it is often based.  Similarly, the choice of
alternatives suffers through the same division of responsibility.  Either the alternative solutions are given by the
client - who may refer to established policies, apply
principles of least objection or most electoral appeal, even
compound previous errors of political judgement - or they are
advised by the consultant, who may seek to limit his costs
of data collection and analysis, favour solutions narrowly
based upon his own discipline, apply rigid and arbitrary
standards in what he regards as the sphere of technical
design.  None of these distortions need be intentional for
their combined effect to relegate the formal evaluation
procedure and its ensuing decision to a remote corner of the
field of possibility.

*Measuring Social and Environmental Impacts*

Even within the confined space conventionally set aside for
evaluation, unwitting reductionism is commonplace, mainly
due to the tendency of professionals to construe evaluation
as a mere extension of the technical process of analysis.
The objection to standard-setting is not just that it
precludes otherwise feasible options or that it leads to a
misallocation of resources, but that it is not accountable

politically.  A similar objection can be raised to the way in
which the 'system boundaries' are defined for the purposes of
evaluation; significant effects can often be excluded from
the comparison of equiibrium states by being ascribed to non-
users or attributed to an external source of agency.  The
widespread use of the term 'externalities' betrays this bias
in viewpoint.  Where evaluation is intended to inform
decisions on resource allocation, the externalities most
usually referred to are environmental and social (except
where these have become 'internal' costs through specific
adherence to standards and minimum legal requirements).  It
is of interest here that, by effecting the reduction of
options on the grounds of economic or technical feasibility,
'efficiency' objectives are implicitly given preference over
environmental and social ones.  This may not matter so long
as the environment is seen simply as the context of a problem
and the community as capable of absorbing the changes imposed
by the various solutions.  Increasingly however in the developed economies, communities are concerned with environmental
and social objectives *per se*.  In that case, it is not
enough merely to measure 'impacts', however carefully they
may be weighed, in pursuit of economic efficiency as a
maximand.  Indeed, when the benefits from an investment are
almost wholly intangible (or tangible but difficult to
measure) - many improvements in amenity, health and other
aspects of welfare fall into this category - the task is that
of assessing the impact on the economy of achieving the
desired social objectives.  This suggests a shift in emphasis
away from cost-benefit analysis towards cost-effectiveness
criteria within a wider framework of environmental evaluation.
The Leitch Committee in the UK is currently advising the
Department of Transport on precisely this subject.

To return to the question of impact, where this is still
appropriate, the three main problems of identification,
measurement and valuation remain.  Conventionally, a handful
of environmental/social aspects are included in evaluation
(almost always noise; usually air pollution, severance; more
rarely land-take, buildings affected, visual intrusion, etc.)
whilst the check-list of *possible* effects is in scores if not
hundreds.  Appleyard in his resource paper identified 23
aspects under fairly broad headings, such as effects on micro-
climate, alterations to access, privacy, risk of criminal
assault, etc.  Although only a few may be relevant to any
particular scheme, identifying those relevant becomes a task
which is crucial to evaluation.  For, to include an effect
which is irrelevant (even though it may be measurable) could
be as serious as leaving out a relevant effect on the grounds
that it cannot be measured.

Just as with problem identification, far too little effort
seems to go into identifying what are the social and
environmental effects that should be measured.  The tendency
is to choose indicators of change only for those effects
which are easily measurable in physical terms; this assumes
that the physical measure of a change is a suitable and
valid proxy for its perceived impact.  As Appleyard pointed
out, to reduce the incidence of common assault in a subway
is one thing, but to reduce the fear of it happening may be
another.  Indicators which accurately reflect perceived

changes are needed, similar to those devised for measuring
subjective responses to noise as a nuisance. These can only
come from carefully controlled psycho-physical investigations
of environmental change (e.g. Clyde (7) and Hills (16).

*Dealing with the Problems of Valuation and Equity*

The drawbacks of relying upon market prices as the basis for
valuation of costs and benefits arising from any transport
investment are well known (see (3,25,31)), and yet the market
mechanism could be described as the quintessence of a
behavioural model. The problem lies not just in the doubt
that the values revealed in the market do not reflect the
appropriate weightings of social preference for various goods.
The manifest imperfections of the market mechanism are
sufficient for this. It concerns also the fact that a system
boundary defined for evaluation purposes rarely, if ever,
coincides with the boundary of the market catchment (see
(18)). In addition, there are the eternal problems of fixing
exchange rates for imported goods, eliminating transfers
between subsections of the community, and so on.

   Above all, there is the acceptance of the validity of
interpersonal comparison - the assumption that, in evaluation, we are entitled to 'trade' one individual's gain for
another's loss. This lies at the heart of the equity issue.
The theoretical problems of equity are well known. Welfare
economists have for a long time dealt with them - Pigou (30)
and Little (23) and Winch (38), for example, have all written
on the subject. The relevance of equity to cost-benefit
analysis is brought out in Prest and Turvey (31) and the
philosophical background history is summarised in Phelps (29).
The handling of equity in the wider realm of public policy is
even more problematic however, and no less the subject of
value judgements (McGuire and Garne (24); Mishan (25)).
Three general approaches can be distinguished.

1  Questions of equity can be kept wholly external to the
evaluation of a policy of investment in the hope that
decisions can be made on 'efficiency' grounds alone. Then
either financial compensation or specific welfare policies
can be implemented to remedy any resulting inequity arising
from the maldistribution of income or other impacts. As a
corrollary to this approach, transport projects should not be
seen as a means of achieving welfare objectives (and vice
versa).
2  Income distributional, environmental and other effects of
policies and projects can be acknowledged and outputs of an
evaluation can be modified by a whole series of politically-
determined (so-called) equity weightings to assist the
decision-taker in the comparison of alternatives.
3  Specific income-distributional effects can be sought
(i.e. where the investment is intended to assist one group
relative to another) and these effects can be 'internalised'
by extensive use of shadow-prices within the analysis (37).

The choice between these approaches is a philosophical one,
at least in the realm of political economy. Thus, for
instance, Harvey (14) argues that a 'liberal' critique would

demand approaches (1) or (2), while a more 'socialist' one
would lead to (3) being adopted.  It is also worth observing
that economic power is only one aspect of the overall command
of resources.  The result of investment and its attendant
confrontations can also affect the distribution of political
power - the movement of wealthier commuters to suburbs beyond
the city limits is an example of this kind of effect.

RECOMMENDATIONS FOR AN IMPROVED EVALUATION PROCEDURE

In an endeavour to achieve some constructive outcome from the
deliberations of the workshop, some time was devoted to the
drafting of a set of guidelines for the improvement of the
evaluation procedure.  Seven major points were distinguished:

1   Define explicitly the system boundary appropriate to the
scale of the problem identified and relevant to the investment or project.  That definition should be open to public
scrutiny and capable of being challenged at an early stage.
2   Strive to achieve comprehensiveness, in the range of
effects or 'impacts' investigated.  Alternatively, make
explicit any arbitrary cut-off adopted to narrow the scope
of  the evaluation.
3   Stipulate the viewpoint from which the evaluation is
being undertaken, the nature of the value judgements inherent
in each valuation (i.e. in each weighting of cost and benefit
items within the analysis), and the nature of the evaluation
criteria or decision rules involved in each decision or
reduction of alternatives.
4   Adopt a method of evaluation appropriate to the stage at
which the reduction of alternatives is taking place.
Appleyard refers to 'quick and dirty methods' which are
frequently available and adequate; they should be recognised
and used.
5   Open all aspects of policy development, planning and
evaluation to public scrutiny and, as far as possible,
incorporate public participation interactively with the
analysis and evaluation processes.
6   Demonstrate clearly the redistributive consequences on
real income of each feasible alternative, the responsiveness
of each to declared interests of minority groups, and the
effects of any equity weightings built into the evaluation.
7   Recognise explicitly the limitations of reliability of
all values subject to prediction, given the uncertainty
which attends any forecast of the future.  Demonstrate the
effects that errors of forecasting would have on the outcome
by means of sensitivity analysis.

One point, not included here, but which kept recurring in our
discussions concerned the *scale* of projects.  Where the
investment proposed is relatively modest, wholly within a
particular sector and likely to have only marginal influence
on consumption, then the well-tried techniques of evaluation
still have much to commend them.  Nowadays, the supposed
economies of scale coupled with a growing technological
capability lead to increasingly vast projects which can
transform patterns of activity in whole regions.  For these,
conventional evaluation techniques are woefully inadequate -
especially when  the *need* for the project in the first place

is seldom open to question. The limit of application of
cost-benefit analysis, for example, must surely have been
reached by the Roskill Commission in their search for a Third
London Airport site (19,26).

## FURTHER DISCUSSION OF THE SOCIO-POLITICAL NATURE OF EVALUATION

### An Idealogical Perspective

At its simplest level, the history of public policy-making
can be used to explore and understand trends and changes in
the nature of policy; but, for real understanding of the
social context, it is necessary to consider some models of
the social process underlying public policy. Harvey (14),
for example, suggests two such formulations, which he distinguished as 'liberal' and 'socialist'. The liberal one, he
argues, is where the social system is recognised not to be in
a state of equilibrium: indeed, its various members and
groups are in 'differential disequilibrium' with their
environment and their adjustment to changes in that environment occur at different rates depending on their relative
command of personal resources of wealth, education and
political power. In terms of transport, the more advantaged
groups exercise their advantage by advocating broad average-
cost pricing (for example of road space and of fuel) and
behaviourally-determined values, for example of time-savings
or improved comfort.

The effects of applying these values and prices is often
to support investment in projects such as commuter-rail
improvements and freeways to serve the activities and the
suburbs of those already privileged and to bisect suburbs
where less-privileged groups live, demolish 'poor' housing
for car parks, and so on. In short, many schemes thought
previously to be socially or economically beneficial are now
regarded as 'regressive' in terms of their impact on the
redistribution of income within the community. Of particular
relevance, in this context, is the argument as to whether
behaviourally-determined values of travel-time savings (often
strongly income-dependent) should be used instead of a
standard 'equity' value.

Harvey's 'socialist formulation' rejects the dualism
between production and distribution implicit in his 'liberal'
model (and argues against behavioural determinism in these
matters), insisting instead that 'efficiency *is* equity in
distribution'. A model of this kind searches for explanations
in the nature of production, the historic division of labour
and its related social alienation, and in the modes of
economic integration or co-ordination. Conflict and
competition in such a formulation still underlie all transport
investments and pricing, but the roots of that conflict and
competition are regarded as fundamental to the processes of
social evolution. Other similarly comprehensive perspectives
are to be found in Rawls (32), Runciman (34) and Passmore
(27).

Public policy and investment can only really be evaluated in the light of these (or similar) sorts of understanding. Stretton (35), for example, argues that some vision of *future* social evolution, and some value judgements concerning future social objectives, are implicit in every evaluation. The analyst in this view, far from being a technocrat, must be both a social historian and an ideologist. But, whatever his skills, he must above all be accountable to those on whose behalf the analysis is performed.

*Variety of Perspectives and Variety of Objectives*

In view of these different ideas of the division of labour and consequent differential command of resources, the so-called 'differential disequilibrium', and the conflict and competition for environmental resources, it is necessary to understand the conflicting perspectives and objectives of the various antagonists in the battles over public policy and investment. However to understand it is not enough, since evaluation to be meaningful requires a viewpoint to be specified. For that viewpoint to be a 'community' one implies that conflicting views are reconcilable and that social values are distributed continuously over a range of perspectives. As soon as perspectives are polarised, consensus values have no meaning and a community viewpoint can no longer be established. Nowadays we can find various groups advocating 'firm government', and 'law and order' groups opposing the opposers and arguing against the freedom to dissent. On the other hand, there are groups with a genuine concern about threats to the environment (though often seeing them in radically different ways) - for example, in Australia the 'greenies' are sometimes concerned more about effects on future options than about present effects of policies and projects. Beyond them are to be found the doom predictors, for whom any policy or project which is consumption-oriented is seen to bear out their beliefs in impending doom and environmental catastrophe.

In addition to these perspectives of individuals or groups within the community, there are also institutional perspectives. These reflect the inherent vested interest, of firms and institutions, in maintaining some aspect of the *status quo*. Institutions have been known to apply pressure on public decision-takers in the past, to beguile or to coerce, or otherwise threaten with dire warnings of unemployment, future scarcity and high prices. Where industries by their operations have posed direct threats to public health and safety or to the environment generally, many have responded reluctantly and minimally to concerted public pressure. Often the response is only to legislative restrictions. By contrast, other firms and institutions may strive to promote and inform the social good, to enforce planning, even to run ahead of 'public opinion' in these matters.

Perspectives, therefore, come from numerous viewpoints: from those to whom 'the environment' is a nuisance, even an obstacle, standing in the way of development and (as they see it) progress; to those for whom it may represent no more than an electoral convenience by which the kudos and votes in

'the environment' can be tapped; to those for whom the
environment is of paramount importance. Depending on the
perspective, so are the individual's, or group's or institution's objectives determined and value judgements made.
Recognition of this is the crux of evaluation in the social
context.

## Consensus or Conflict

Social objectives are legitimately concerned with (1)
individual's preferences for activities, consumption of goods
and uses of time; (2) concerns about equity or 'distributive
justice', and (3) ideas about the ethics of means of altering
the distribution. Theory about the first two has probably
been best developed by Winch (38), where it is argued that
there is no scope in public policy for value judgements
concerning commodities or environmental conditions directly:
these are merely the variables determining individuals'
utility levels, but the utility or satisfaction actually
derived from consumption of a particular commodity entirely
depends upon the individuals' preferences (utility function).
Public policy from the community viewpoint is concerned only
with the 'welfare significance' or weighting of that
individual's utility.

Arrow's (1,2) classic argument, however, is that consensus
is impossible where wide variety in preferences is to be
found. If we exclude, on his argument, the possibility of
interpersonal comparisons of utility, then the only methods
of passing from individual preferences to social preferences
for alternative social policies, that will be logically
consistent, will be either imposed ('dictated') or presumed.
Hence the problem of conflict about the ethics of means. The
usual solution, of course, is one of benign bureaucratic
dictatorship, often assisted by restriction of information
both about the effects and costs of the favoured scheme and
about other feasible options. In more recent times, restriction is maintained by confining 'public involvement' to
choices between narrowly-defined alternatives within a
previously determined strategy, and by giving opponents
appropriate roles in a formal context. But increasingly,
with the explosion of information, the growing community
awareness about environmental effects and social costs and
the mounting scepticism about 'expert' opinion, the diversity
of social objectives is beginning to show through. For this
reason, the development of a more open and responsive
procedure for evaluating and decision-taking in the transport
sector is  urgently needed.

There is a further, practical aspect of this dilemma. The
problem with remote or bureaucratic decision-making is not
just that demonstrably 'poor' decisions can be taken (i.e.
not conforming with individuals' preferences and heedless of
the equity or conflict questions), but that even 'good' or
conforming decisions are not understood because the process
of consultation or involvement has not been experienced. The
result is frequently resentment and increased alienation,
ultimately conflict and rejection. That people must always
be part of the process of identifying their own problems and
defining the solutions, is the conclusion reached by Rittel

and Webber (33). When we then add the equity dimension and
the problems of conflicting objectives - acknowledging that
every policy or investment benefits some individuals or groups
relative to others, and that some will have their problems
made worse rather than eased by the decision - then the
exercise of social choice and public policy becomes even more
intractable.

*Impacts on Life-styles*

In addition to the more direct impacts on relative command of
resources, transport policies and investments also affect
individual's allocations of time and money, thereby influencing their social ties, use of institutions and of services
and indeed all those aspects of social behaviour sometimes
characterised as 'life-style'. Transport changes and severance effects can markedly alter the constraints on preferred
activities and uses of time. This suggests that these
preferred activites (and their associated perceptions, preferences and values) are best studied within the context of
life-style differences and their resilience in the face of
change.

Certainly within social classes, populations appear to
exhibit wide variations in life-style. Gans (10) for example,
in his work on lower-class Bostonians, identified four quite
distinct patterns of life-style. He named them as *routine-
seekers*, *action-seekers*, *the maladapted* and *the middle-class
mobiles*. The patterns he identified seem to related to
dimensions that are quite widely encountered in most Western
societies and might be expected to moderate differentially
the impact of social changes.

Forced relocation and other forms of severance, for
instance, have frequently been shown to have particularly
severe impacts on working-class families: there is a rupture
of social networks, which are generally more locality-based
than are those of more middle-class households. Of the four
groups, the *routine-seekers* and the *maladapted* seem to suffer
most. Nevertheless, working-class communities are changing
anyway (as are all communities), so how 'important' are these
disruptions in the broader context of social change? Buttimer
(6) suggests that this contextual change can be described
under three main headings:

1  *Economic*: rising income and living standards which appear
to give workers an entree to middle-class patterns and levels
of consumption;
2  *Technological*: changes in industrial technology which
alter the nature of work and the monetary rewards for labour
and are associated with changing attitudes towards work and
the consequent restructuring of social relationships at
shop-floor level and;
3  *Ecological*: the movement of workers from rural to urban
and from central urban to suburban residential environments,
with the consequent dilution of old life-styles and place-
oriented ties.

Various traditions of community studies (related to the sorts

of historical and ideological perspectives of social process
discussed earlier) stress various interpretations of these
changes.  So, the Marxist emphasises enduring social
alienation of workers despite superficial changes in life-
style.  The embourgeoisement theorist suggests that these
sorts of changes have led large sections of the working class
to adopt a middle-class outlook and way of life; attitudes
and life-styles are seen by him to be economically determined.
However, various studies have found that the values and
relational aspects of class do persist, regardless of superficial changes in life-style (12).  Instead, any change in
values and normative orientations affects only marginal groups
of both the working class and the middle class, the change
being in the same direction for both groups, i.e. towards
similarity of outlooks and aspirations (6).  Otherwise, most
studies (5,39,40) tend to indicate that normative orientations and preferences persist despite relocation of working-
class families.

These considerations of life-style orientations and of
social class raise more generally the problem of relative
vulnerability.  That is to say, different groups within the
community affected by a transport investment will be vulnerable to different degrees of impact, and these differences
cannot be related merely to physical conditions.  They
concern inherited values, class affiliation and life-style,
command of resources and above all the ability to respond
and adapt to change.

The point of this is that the benefits enjoyed by any
individual and/or the costs borne by him, resulting from
environmental change, may vary substantially with his particular characteristics.  These differences are entirely
independent of the differential weightings applied to
benefits derived and costs endured consequent on some overall
value judgement of equity.  Evaluation techniques must be
extended to cope with both (see for example (7)).

Clearly this whole field of life-style, social differentiation, relative vulnerability to the deleterious effects
of environment and of environmental change, methods of
assessing that impact and methods of ameliorating it, still
lies ahead of social research.


THE PROBLEM OF CHOICE, ACTUAL AND PERCEIVED

Finally, we must return to the problem of social choices.
In a notable attack on Arrow's general impossibility theorem
discussed previously, Little (22) rejected two necessary
conditions of Arrow's theory: (1) the *independence of
irrelevant alternatives* and (2) *non-dictatorship*.  Little
finds the idea of non-dictatorship 'nonsense'.  He argues
that dictatorship in public decision-making is essential,
the only problem being to decide upon the dictator.  Phelps
(29), following Rawls (32), supports this: non-dictatorship
must be abandoned because the least advantaged must be free
to determine the ship's course in relation to the dimensions
of their deprivation.  Given the widely differing command of
resources within most communities, even where participation
is encouraged, some element of dictatorship is probably
inevitable.

The condition of independence of irrelevant alternatives is also questioned by Little (22) on logical grounds. Information on irrelevant alternatives - i.e. non-feasible alternatives, lying outside the design space - should be used in decision-making because it can help to clarify the underlying nature of the differences in preference between conflicting antagonists. One could question it on more practical grounds: the consideration and public debate of currently non-feasible alternatives may indeed clarify options falling 'within' the design space, but it is also the way to extend that design space ... to render the non-feasible, feasible. For example, substitutes for the use of cars for many purposes may now be regarded as non-feasible, but that could be because the implications have not been adequately thought through.

In a way, these arguments are sterile as far as the dedicated behavioural modeller is concerned. For, to make use of relationships derived from observed behaviour requires him to presume: (1) that individuals are fully aware of the options available; (2) that they **are** unconstrained in their choice between them; and (3) that the surplus gained by any individual is wholly reflected as an increase in social welfare. The last two of these are subjects of unending debate; but the first, in many ways the most important of the three, is rarely acknowledged. The trouble is that choice is a *perceived* situation - that is, although options for an individual may exist, they are only valid as choices when the individual is aware of them. Moreover, merely to provide information about available options may not be enough; they must be perceived as feasible substitutes for the purpose at hand, if they are to influence the choices that people make. Even in the relatively straightforward matter of modal split, bus use may not be regarded as a feasible substitute by many car owners: thus, to construe the balance of mode in terms of a choice-mechanism is to build a fatal flaw into the model, with possibly serious consequences for transport planning and policy.

We conclude that the real world is exceedingly complex and that something of that complexity must be reflected in the processes of evaluation that are applied to investment plans and policies in the public sector. Every time a reduction is made by imposing boundaries or applying standards or otherwise limiting the range of feasible alternatives, formal evaluation and the political participation exercises that accompany it are rendered less and less meaningful, to the point where the whole process risks being challenged (and, in notable instances, overthrown) by an increasingly aware, concerned and sceptical community. As professionals, we must realise the fundamentally political nature of the value judgements inherent in the analysis of transport problems, and strive to raise the level of public involvement and accountability in their solution. This will necessitate more than behavioural modelling.

REFERENCES

1. Arrow, K.J. (1951) *Social Choice and Individual Values*, (2nd edn. 1963), Wiley, New York.
2. Arrow, K.J., 'Values and collective decision-making', in *Philosophy Politics and Society*, P. Laslett and W.G. Runciman (eds), vol. 3, Blackwell, London (1967).
3. Barrell, D. and Hills, P.J., 'The application of cost-benefit analysis to transport investment projects in Britain', *Transportation*, 1, 1 (1972).
4. Beesley, M.E. and Foster, C.D., 'The Victoria Line: social benefit and finances', *Journal of the Royal Statistical Society, (Series A)*, 128, 1, pp. 67-88 (1965).
5. Berger, B.M. (1960) *Working Class Suburb*, University of California Press, San Francisco.
6. Buttimer, A., 'Sociology and planning', *Town Planning Review*, 42, 2, pp. 154-68; reprinted in *The City*, M. Stewart (ed.), Penguin Interdisciplinary Readings (1972).
7. Clyde, C.A., 'Determination of perceived attributes of the pedestrians' environment', in *Urban Transportation Planning: Current Themes and Future Prospects*, P.W. Bonsall, M.Q. Dalvi and P.J. Hills (eds), Abacus Press Ltd, Tunbridge Wells (1977).
8. Dupuit, J., 'On the measurement of the utility of public works', *Annales des Ponts et Chausees*, 8, 2 (1884); reprinted in English, *Penguin Modern Economics: Transport*, London.
9. Gans, H.J., 'The human implications of current redevelopment and relocation planning', *Journal of American Institute of Planners*, 25, pp. 15-25 (1959).
10. Gans, H.J. (1962) *The Urban Villagers*, The Free Press, New York.
11. Gans, H.J. (1972) *People and Plans*, Penguin, Harmondsworth.
12. Goldthorpe, J.H. *et al.* (eds) (1969) *The Affluent Worker in the Class Struggle*, Cambridge University Press, Cambridge.
13. Harrison, A.J. (1974) *The Economics of Transport Appraisal*, Croom Helm, London.
14. Harvey, D. (1973) *Social Justice and the City*, Edward Arnold, London.
15. Hill, M., 'A method for the evaluation of transportation plans', *Highway Research Record No. 180*, 21-34 (1967).
16. Hills, P.J., 'People's responses to pedestrianisation schemes', in *Transportation Planning for a Better Environment*, P. Stringer and H. Wenzel (eds) Plenum Publishing Corporation, New York (1976).
17. King, R.J., 'Project development and the social environment', *Search*, 7, 6, pp. 246-9 (1976).
18. Layard, R. (ed.) (1973) *Cost-Benefit Analysis*, Penguin Readings, London.
19. Lichfield, N., 'Cost-benefit analysis in planning - a critique of the Roskill Commission', *Regional Studies*, 5, 157-83 (1971).
20. Lichfield, N., Kettle, P. and Whitbread, M. (1975) *Evaluation in the Planning Process*, Pergammon Press, Oxford.
21. Lichfield, N., Quarmby, D. and others, 'Special issue on evaluation - methods in urban and regional planning', *Regional Studies*, 4, 2 (1970).
22. Little, I.M.D., 'Social choice and individual values', *Journal of Political Economy*, 60, pp. 422-32 (1952).

23. Little, I.M.D. (1958) *A Critique of Welfare Economics*, 2nd edn., Oxford University Press, Oxford.

24. McGuire, M.C. and Garne, M.A., 'The integration of equity and efficiency in public project selection', *Economic Journal*, LXXXIX, December (1969).

25. Mishan, E.J. (1972) *Cost-Benefit Analysis*, Allen and Unwin, London.

26. Nwaneri, V.C., 'Equity in cost-benefit analysis. A case study: third London Airport', *Journal of Transport Economies and Policy*, V, 3 (1970).

27. Passmore, J. (1974) *Man's Responsibility for Nature: Ecological Problems and Western Traditions*, Duckworth, London.

28. Peters, G.H., 'Cost-benefit analysis and public expenditure', *Eaton Paper 8* (2nd edn.), Institute of Economic Affairs, London (1968).

29. Phelps, E.S. (1973) 'Editor's introduction', *Economic Justice*, Penguin Modern Economics Readings, London.

30. Pigou, A.C. (1932) *The Economics of Welfare*, Macmillan, London.

31. Prest, A.R. and Turvey, R., 'Cost benefit analysis: a survey', *Economic Journal*, LXXXV, 300, pp. 169-72 (1965).

32. Rawls, J. (1971) *A Theory of Justice*, Harvard University Press, Harvard.

33. Rittle, H.W.J. and Webber, M.M., 'Dilemmas in a general theory of planning', *Institute of Urban and Regional Development Working Paper 194*, Berkeley, California (1972).

34. Runciman, W.G. (1966) *Relative Deprivation and Social Justice*, Routledge and Kegan Paul for Institute of Community Studies, London.

35. Stretton, H. (1976) *Capitalism, Socialism and the Environment*, Cambridge University Press, Cambridge.

36. Thomas, E.N. and Schofer, J.L., 'Strategies for the evaluation of alternative transportation plans', *National Cooperative Highway Research Programme, Report 96*, Highway Research Board, Washington (1970).

37. Weisbrod, B.A., 'Income redistribution effects and benefit-cost analysis', in *Problems in Public Expenditure Analysis*, S.B. Chase, Jr., (ed.), 177-209, The Brookings Institute, New York (1968).

38. Winch, D.M. (1971) *Analytical Welfare Economics*, Penguin Modern Economics, London.

39. Young, M. and Willmott, P. (1963) *The Evolution of a Community: Dagenham after Forty Years*, Routledge and Kegan Paul, London.

40. Young, M. and Willmott, P. (1957) *Family and Kinship in East London*, Routledge and Kegan Paul, London.

PART TWELVE

## Chapter 39

SUMMARY OF MAJOR FINDINGS AND RECOMMENDATIONS

Peter R.Stopher and David A.Hensher

## PURPOSE

In this chapter, we intend to summarise the principal findings and recommendations of the ten workshops that were set
up for the conference. These findings and recommendations
are the result of deliberations that took place during seven
sessions, each of which lasted about three hours. In this
chapter, we have attempted to put into summary form the
various reports by each of the workshop chairman, so that a
coherent overview can be obtained of the deliberations of the
conference, and a fairly clear picture can be drawn of the
recommended courses of action for the next few years in this
important area of behavioural travel modelling. We have
also attempted to draw together a few common themes in a
final section of the chapter, in order to re-emphasise some
of those points that seem to recur from various different
workshops.

## MAJOR FINDINGS AND RECOMMENDATIONS

*Workshop A: New Approaches to Understanding Traveller
Behaviour*

This workshop was an exploratory one, designed to permit a
relatively free discussion of a range of possible new
approaches to understanding travel behaviour and developing
forecasting models. The principal emphasis of the workshop
was on activity theory as it has been developed by human
geographers. The principal notion within this theory is that
travel behaviour is the result of certain decisions made by
members of a household concerning the activities which they
wish to undertake. Thus, the approach conforms with notions
of travel as a derived demand, but pays considerably more
than the lip-service that has been accorded to this notion
in travel-forecasting work up to the present time. The
workshop indentified four areas of relevant policy application. The first of these is the analysis of the impact of
transport systems and policies on users and non-users
through various levels of activity substitution within the
household, e.g. re-scheduling of activities creating more
leisure. The second area of policy application was seen to
be that of determining latent activity needs, with a view to
finding out which ones are being met and which are not, thus
leading to formulation of policies that would permit more
equitable transport provision. The third area of policy

application was seen to be that of understanding the interactions between travel and land use, an area which has
generally not been handled successfully by other approaches.
The final policy application identified is that of the
possibilities of changes in the activity framework as a
means for achieving various social and transport goals, or as
a means for determining how to set social goals, e.g. the
need for activity rearrangement to achieve a desired
redistribution of income in accordance with the normative
position.

The workshop drew six principal conclusions and recommendations.  First, they concluded that the activity approach is
presently a verbalisation of a desirable framework for
travel-behaviour modelling.  Second, they recommend that
models should be developed to represent the formation of
activity sets and to demonstrate how activities are carried
out within the constraints of time and space.  Third, they
recommend that choice-based models need appraisal in an
activity framework to determine how appropriate the theory of
choice processes is to activity-set formation.  In a similar
vein, the fourth conclusion was that a number of theoretical
and/or modelling procedures currently used elsewhere need
investigation in the light of activity theories.  These
include, in particular, queueing theory, conjoint models,
learning theory, learning models, graph theory and gaming
simulation.  Fifth, the workshop considered that multi-trip
and multi-purpose journeys are relatively important but have
generally been neglected in the development of travel-
forecasting procedures in the past.  While Markov chains
have frequently been considered as potential methods for the
development of chained trips, the workshop concluded that
their deficiencies outweighed their usefulness in this particular application.  They recommended that graph theory be
considered as a better way of identifying activities in trip
chains.  The final recommendation of the workshop was that
appropriate activity data require alternative data-gathering
methods of questionnaire design from the traditional transport tools, and that research should be undertaken to
develop appropriate methods for the acquisition of the
required data.

*Workshop B: Equilibrium Modelling*

This workshop concentrated on the interdependencies between
the land-use system, the economic system and the transport
system.  The workshop concluded that the primary issue to be
dealt with was the need to develop travel-forecasting
procedures that can be applied to a situation that is
generally in disequilibrium, but in which interdependencies
exist between demand and supply.  Thus, the workshop discard-
ed the notion of attempting to develop models that place
demand and supply in equilibrium with each other.  Towards
the development and implementation of such procedures, the
workshop made six recommendations.  First, they recommended
that there is an urgent need to extend solutions for equilibrium regardless of the mathematical problems that may be
encountered.  They considered that this would probably be
achievable only by applying various *ad hoc* approximation

methods.  They concluded also that the inclusion of individual travel choice models in *ad hoc* network equilibrium represents no major problems.  However, disaggregation of the network itself, for equilibrium solution, is too expensive given current knowledge and capabilities in this field. Third, the workshop considered that route choice, as the principal connector between demand and supply, requires further investigation.  To this end, data are required on the decision process of the individual in making route choices. Fourth, land-use modelling systems need to be standardised. The workshop noted that there is no standard planning package at present, for land-use modelling.  It was felt that the development of such a standardised package would help considerably in the development of improved techniques for investigating interdependencies between land use and the transport system.  Fifth, the workshop noted that travel-demand models are relatively sophisticated, particularly compared with models currently used for land-use and economic systems.  Therefore, the workshop recommended that a better understanding be developed of the mathematical models that might be appropriate for land-use and economic systems. Finally, the workshop recommended that relatively more re-sources should be channelled into the investigation of the interdependency between individual-choice models and the transport network.

*Workshop C: Consumer Segmentation*

The workshop began by defining segmentation as being rather loosely any relevant classification for transport modelling or planning that aids in the understanding and prediction of travel behaviour.  The workshop also identified the major categories for application as being demand forecasting, marketing and the evaluation of impacts.  Seven conclusions and recommendations were made by the workshop.  First, they recommended that a document be developed that contains the range of segmentation techniques that are available to the transport planner and that the document also show examples of their application.  The workshop felt strongly that insufficient information was available to the transport planner at the present time on how he should go about segmentation and apply it.  Second, they concluded that no particular hierarchy for a base for segmentation seems more appropriate than any other.  Rather, the selection of a segmentation base should be considered upon the needs of the models, and the interface between model specification and segmentation. This area requires much further study.  Third, the workshop noted that much work in transport on segmentation has concentrated on the use of socio-economic variables as the basis for segmentation.  The workshop recommended that these variables not be considered as the primary basis for segmentation, since the interdependencies that exist between the level of transport service, socio-economic characteristics of households and individuals, and individual tastes generally restrict independent segmentation in terms of socio-economic variables.  It was recommended that segmentation be based on subjective judgements rather than socio-economic variables. Fourth, the workshop noted (and warned) that the range of techniques for clustering (e.g. multi-dimensional scaling,

etc.), may be over-sophisticated for transport-planning
requirements. They recommended, rather, that simple and
somewhat less sophisticated techniques be applied. Fifth,
the workshop recommended that alternative data-collection
procedures should be investigated when segmentation is to be
carried out. They noted that standard data-collection
procedures used in transport planning are generally not
appropriate for identifying segmentation bases, and recommended that transport planners investigate the use of such
procedures as focus groups, in-depth interviews, media exposure and political outlooks. They recommended that
investigation into these methods be carried out by the use
of small pilot studies. Commenting on some recent work in
individual-choice models, the workshop considered that
simplified segmentation processes are not generally a good
substitute for improving model specification. They also
considered that segmentation would generally have a low pay-
off as a procedure for reducing aggregation bias. Finally,
the workshop recommended that considerable research be
undertaken in the area of segmentation in the immediate
future, since such work has lagged far behind the research
into travel-demand model specification.

*Workshop D: Theoretical and Conceptual Developments in
Demand Modelling*

The workshop noted that considerable research had been undertaken since the second International Conference at Asheville
into demand-model specification, potentials of transferability of the models, and procedures for data collection. The
primary role of this workshop was seen to be to consolidate
and extend this considerable research activity. The workshop
made three recommendations for immediate implementation, and
made seven recommendations for future research. In terms of
current implementation, the workshop recommended that choice-
based sampling procedures be used to a greater extent in
the development of data bases for transport planning. This
recommendation was made on the basis that choice-based
sampling procedures are as valid as traditional random-
sampling procedures, but produce significant cost savings.
It was noted that a distinction must be drawn between
sampling procedures and survey procedures, since choice-
based sampling does not dictate whether or not data are
collected by interview, self-administered questionnaire or
other methods. Thus, the recommendation is for a sampling
procedure not for a survey procedure. Second, the workshop
recommended that structured-logit analysis should be
implemented if a relaxation of the independence of irrelevant
alternatives (IIA) assumption between sets of alternatives
is desired. To permit this recommendation to be carried out,
the workshop recommended that a non-technical summary of the
procedure be sponsored and developed. Third, the workshop
noted that, currently reweighting of coefficients is seen as
the appropriate method for transferring models between
locations. They considered that further research and
empirical testing is necessary for this type of activity,
with a view to establishing a general set of criteria for
transferability. The workshop also noted that the greatest
area of location sensitivity in the models is in the constant
term, and that further work is required in the transference
of this term.

The workshop also recommended that further research be
undertaken in a number of areas.  First, the relationship
between consumer-surplus estimation and individual-choice
models requires both theoretical and sensitivity justification.  The consistency between individual-choice models and
utility maximisation should also be investigated, together
with the conditions required for aggregating consumer-surplus
measures across individuals in order to achieve meaningful
measures of social welfare.  Second, research is needed into
problems of robust estimation and sensitivity of the conventional methods of estimation of individual-choice models,
particularly with respect to data outliers and coding errors.
Third, in line with the third recommendation for immediate
implementation, further investigation is required into
transferability.  In this matter, in particular, the workshop
noted that there is a need for common reporting procedures,
and a common basis of experimentation.  The workshop also
recommended that further research be done into mathematical
structures that do not incorporate the IIA property.  Fifth,
the workshop recommended that retrospective surveys be
undertaken in order to assess the temporal variation of
variables.  Six, there is a need to tie together demand and
supply, in particular to take account of the constraints that
exist on individual actions, thus influencing the choices
being made.  Finally, the workshop recommended that research
be undertaken into the definition of choice sets.

*Workshop E: The Role of Disaggregate Modelling Tools in the
Policy Arena*

This workshop, together with the following one, was concerned
with aspects of the policy role of individual-choice models.
It sought to identify those types of policy issues which may,
potentially, be addressed and investigated with the aid of
individual-choice models.  The workshop attempted to address
the issue of the interface between planning tools and policy.
The workshop developed three major findings.  First, they
recommended that senior government advisers should explain
to modellers what they perceive to be the type of advice
which they are expected to provide.  Second, the workshop
concluded that disaggregate models have a definite role as
short-run policy models, but the workshop was unclear as to
their role in long-run policy formulation.  Third, the workshop noted that individual-choice models are theoretically
and conceptually superior to aggregate models.  However,
their future depends in large measure on a number of aspects.
These include their potential for transferability, the
horizon for short-run implementation, the possibilities of
fast turn-around and quick responses at a course level.

The workshop made one recommendation for further research,
namely the application of disaggregate models to long-run
policy concerns, particularly emphasising interactions with
land use.

*Workshop F: The Relationship between Behavioural Models, Evaluation, Forecasting and Policy*

Whereas workshop E considered the general issue of the role of disaggregate models in policy, this workshop focused on several specific issues. These issues included the policy framework within which the models would operate, the conditions for transferability, situations in which different types of models are applicable, and the treatment of the variability of input data over time. The workshop recommended that considerably greater communication should take place between the people involved in evaluation, policy advice, policy-making and modelling research. Such communication is essential in order that research approaches be responsive to institutional settings, constraints and priorities.

The workshop developed seven conclusions and recommendations. First, they concluded that individual-choice models have an important role in the assessment of short-run policies, but noted that their role in long-range planning and strategic planning is unclear. Second, they recommended that where individuals in a study area have not experienced a particular type of project or policy, it may be preferable to transfer data from another location, where such a project or policy has been experienced rather than to use direct attitudinal data. Third, considering this notion of transferability, the workshop recommended that empirical investigations be carried out to determine and control conditions for such transferability. The workshop recommended that, in the case of mode choice, the control conditions be the time of day, income level, trip frequency and the chosen mode. The appropriateness of these control conditions beyond mode choice also needs investigation. Fourth, the workshop recommended that resources should be redirected towards studies of goods movements and the interaction between goods and passenger movements. Fifth, the workshop also recommended that resources be redirected towards the study of supply, in particular supply constraints which influence an individual's choice. The workshop considered that individual-choice models can offer an alternative basis for identifying projects and policies, given the inherent relationship in the models between user benefits in evaluation and the underlying attributes of travel choice. Finally, the workshop noted that the value of leisure time over time is a variable input, the time trend in values of time needs more research, since to apply a constant annual adjustment appears questionable. This recommendation was made with a note that the value of time appeared to be a more pertinent issue of concern in Britain and Australia than in the United States.

*Workshop G: Goods Movement*

Goods movement is a relatively neglected topic in conferences on behavioural approaches to understanding travel behaviour. The modelling of goods movement requires some fundamental differences in approach to those used for passenger movements. This is particularly so because the unit moving, i.e. the consignment, is distinct from the decision-maker, i.e. the

shipper or owner.  This workshop recognised the limited knowledge of the behavioural process and hence concentrated on
the determination and choice context of the goods-movement
process.  In particular, the workshop concentrated on the
decision process and deemed discussion of the physical process inappropriate to a conference on behaviour.  Particular
models and policy evaluation are areas for future research.
The workshop made three principal recommendations.  First,
the workshop recommended that resources be directed towards
understanding the decision process, given the current state
of knowledge.  The workshop also recommended that the development of models is not appropriate until the roles of the
shipper, the operator, the receiver, the government (planner)
and the public are understood.  Second, the workshop
recommended that work be concentrated on the short-range
issues first, before proceeding to consider long-range
aspects.  Finally, the workshop concluded that individual-
choice models do have applications to goods movement.  Once
the rudiments of the process are established, the workshop
recommended that a number of simplified subprocess models
should be investigated, suggesting particularly a shipper
level-of-service choice model, an operator-choice model and
a receiver-choice model.

*Workshop H: Behavioural Modelling, Accessibility, Mobility*
*and Travel Need*

The workshop decided that the central issue was that of
accessibility, which was defined as a property of people,
opportunities and space, and as being independent of trips
actually made.  The workshop noted that mobility is measured
by actual trip-making, while accessibility measures the
potential for trip-making.  Despite the difficulties of
attaining an early consensus on the meaning attached to such
notions as opportunity, need, accessibility and mobility, six
research recommendations were arrived at.

    First, the workshop recommended that basic research be
carried out into the value and perception of accessibility to
the individual and to society.  The workshop noted that
increased accessiblity may not always be either an obvious or
a desirable goal of society.  Second, the workshop recommended that a hierarchy of accessibility measurements appears to
promise greatest returns in research.  A useful hierarchy is:
level one, supply of access in physical terms (isochrons);
level two, supply of accessibility and perceived terms (the
difficulty of getting through the network as the individual
perceives it); level three, perceived value of accessibility
(utility of combinations of opportunities in time and space);
and level four, demand for travel as a manifestation of
utility attainable through the transport network.  Third,
the workshop recommended that policy research should be
undertaken to assess how trip frequencies, choices,etc. varied by accessibility, how the relationship between
accessibility and density of development occurs at the
disaggregate level, and what is the relationship between
home- and non-home-based trips in the measurement of accessibility.  Fourth, the workshop recommended that research
should be undertaken into the competition between

destinations, and that such a measure should be included in a generalised accessibility index. The workshop recommended that for certain types of trips, labour-mobility theory should be consulted.

The workshop's fifth recommendation was that accessibility measures should be studied with more attention to experimental design. Finally, the workshop recommended that research be undertaken on the role of accessibility in activities beyond the urban context.

*Workshop I: Attitudes, Attitudinal Measurement and the Relationship between Behaviour and Attitudes*

This workshop was concerned with the identification of individual decision processes, and the translation of these processes into algebraic terms. The workshop made six recommendations, most of which are somewhat divergent from current research. First, the workshop recommended that a broad set of models be developed to enable future planning to select an appropriate model given a policy issue, rather than restrict the policy outcomes according to the limitations of a smaller set of models. A number of different types of models were suggested, details of which may be found in the workshop report. Second, the workshop felt that too much attention has been given to stochastic-choice models. As a result, analysis and data collection have been conditioned to this particular type of model. The workshop recommended that data collection should be reoriented to test a much wider range of models. Third, the workshop recommended that laboratory experimentation is a desirable first step given the need to understand the choice process. Fourth, the workshop recommended that research be undertaken into dynamic shifts. Within this, newness, change and situational factors should all be considered as variables themselves that could easily be incorporated in experimental designs and used to anticipate behaviour under conditions not specified at present. Fifth, the workshop recommended that longitudinal data collection should be given priority. Finally, the workshop recommended that factors such as habit, disruption, thresholds and interpersonal variables, should be included in decision models.

*Workshop J: Behavioural Modelling and Evaluation of Social and Environmental Impacts of Transport Investment*

This workshop was concerned with assessing the role of non-user attributes in a study of individual behaviour, in contrast to the primary concern of past work of assessing user attributes in travel demand. The workshop found, however, that any discussion along the lines of the role of non-user attributes led to a reappraisal of the evaluation framework in which individual choice models have been studied. The workshop suggested that non-user social and environmental attributes could not simply be added to the evaluation framework of the past, since these dimensions would still be swamped by the traditional user benefits. Along these lines, the workshop made four principal

recommendations with respect to the appropriateness of the
behavioural models and the role of social and environmental
attributes in evaluation. First, the workshop concluded that
it was necessary to develop measures of perceived impact,
since perceptions of impact may frequently differ rather
extensively from actual measurable impacts. Second, the
workshop recommended that, in order to understand and predict
the extent of impacts, there is a need to expand behavioural
travel modelling into activity analysis. Third, the workshop
recognised that existing behavioural models may be very
restricted in application because they concentrate on travel
users and do not study non-users. However, in principal
they point the way to improved prediction and they have
greater potential when expanded into activity analysis.
Fourth, the workshop noted that choice is a perceived situation. It is, therefore, necessary to distinguish between the
presence of options and the perception that options exist.
Until better understanding of this is incorporated in
behavioural models, the workshop concluded that such models
will have limited application with respect to impact analysis.


SUMMARY

In the foregoing summaries of each individual workshop, a
number of recurrent recommendations appear to be made. First
and foremost, a number of the workshops have called for a
greater degree of communication between the modeller, the
researcher and the policy-maker. This is certainly an
appropriate recommendation to come from a conference that was
required to consider more carefully the policy relevance of
recent research in travel forecasting. However, it does not
seem possible to over-emphasise this point. It is quite
clear that continued research into travel-forecasting
procedures without interaction with the decision-makers must
condemn that research to a relatively sterile product. It
is only by a concentrated effort to develop strong lines of
communication between the users of the product and the
developers of the product, that useful products can be
produced. It should be taken, therefore, that a major
recommendation of this conference is that a better dialogue
be developed between the various actors in the development
and use of travel-forecasting procedures.

A second common theme that appears to evolve from a
number of the workshops is an allied one. This is the
concern for improved channels of communication. Not only is
communication lacking between the policy-maker and the
planner, but the planner frequently does not know or understand what the researcher is doing. Several workshops have
recommended that a primary activity be the development of
manuals or documents explaining current developments in
research work. From the Asheville Conference, such a
document was produced in the form of a publication of the
Federal Highways Administration on the use of individual-
choice models in the policy area. Clearly, similar efforts
to this are required in areas such as market segmentation
and sampling procedures, to name a few.

For the research community, several common recommendations
also arise.  First, several of the recommendations from the
workshops voiced a need to study the formulation of choice
sets.  This is seen as a very important step in the
continued development of choice models, and in their application to various types of policy areas.  Second, a number of
workshops have noted the need to investigate further activity
analysis.  Again, the recommendation is couched in terms
which suggest that failure to undertake such research will
doom the broader applications and extended development of
individual-choice models.  Third, there is a strong call made
for some degree of standardisation in approaches to research.
In particular, several workshops noted the previous lack of
general formulations for reporting the results of experiments,
for data collection, etc.  It is clear that there is a need
for a better scientific approach to experimentation and
empirical testing than has been shown up till now in much of
the research in this area.  Several workshops also noted the
need for increased activity in the area of land-use and
transport interactions.  Suggestions were made that a more
appropriate model structure be developed for this type of
work, and that the interface with individual-choice models
should be of prime concern.  Finally, several workshops
concluded that the role of individual-choice models in travel-
forecasting work was fairly clear in short-run policy issues.
However, these same workshops expressed some doubts concerning the application of the techniques to long-run policy
issues.  This is clearly an area that requires future study.
It is certainly not reasonable to assert that individual-
choice procedures should be applicable for long-run planning
issues.  However, it is necessary to consider this topic
rather more carefully, in order to determine whether or not
long-run applications are feasible from individual-choice
models, and if not to propose improved techniques for handling such long-run applications.

Bearing in mind the primary recommendations of the
conference, concerning better channels of communication
between the various communities involved in transport
planning, we recommend two separate directions that should be
pursued.  First, on a national basis, conferences and semin-
ars should be set up to foster a two-way communication
process between the various communities involved in the
development and use of travel-forecasting procedures.  On the
one side, researchers need to communicate the available
methods to transport planners and policy advisers and explore
the ways in which these methods may assist policy formulation
and the generation of alternatives.  On the other hand,
policy advisers and transport planners need to communicate
their problems and constraints to researchers, so that
research can, where necessary, be redirected into more policy-
responsive channels.  The second direction is that at least
one more International Research Conference is warranted, to
continue the process of the last three in defining research
priority areas and in providing a major communications
channel among the growing community of researchers in
individual travel-choice models throughout the world.  In
the light of  this latter recommendation, a fourth
International Conference on Travel Behaviour is in the
planning stages for mid-1979.  This recommendation, however,

should not be taken as suggesting that a series of such conferences should be held to perpetuity.  Rather, as has been
the case so far, a decision on whether or not a subsequent
conference is warranted should be made on each occasion that
a conference is held.  It should not be taken as a foregone
conclusion that future conferences will be arranged.  This
should continue only so long as there is a demonstrable need,
and that benefits could be gained from the holding of a
further conference.

Finally, it is hoped that the material contained within
this book, embodying the proceedings and resource documents
used for the third International Conference, will provide a
stimulus and a reference document for research and
development over the next several years.  As was stated in
the Asheville Proceedings, the success of this conference
can only be judged in retrospect after there has been an
opportunity for its recommendations to be carried through
into practice.  However, considering the record of the
preceding conferences, the likelihood of success of this
conference seems high.  That success, however, rests at this
stage in the hands of the research community, the transport-
planning community and the policy-makers who need to act
upon the various recommendations made by the conference.

Appendix I

WORKSHOP MEMBERS

WORKSHOP A   NEW APPROACHES TO UNDERSTANDING TRAVELLER
             BEHAVIOUR (INCLUDING NATURE OF TRAVEL AND THE
             STRUCTURE OF JOURNEYS)

*Chairman*

Mr N. Thrift,
Department of Geography,
Leeds,
Yorkshire LS2 9JT
England

*Resource Papers*

Professor S. Hanson,
Department of Geography,
State University of New York,
Buffalo,
New York State 14226
United States of America

Mr P. Jones,
Transport Studies Unit,
University of Oxford,
11 Bevington Road,
Oxford   OX2 6NB
England

Mr K. Kobayashi,
Central Research Laboratory,
Mitsubishi Electric Corporation,
80 Nakano,
Minami Shimizu,
Amagasaki,
Hyogo 661,
Japan

*Members*

Mr W. Brög,
Sozialforschung Brög,
Schongauer Strasse 8,
8 München 70,
Munich
Germany

Professor P. Burnett,
Department of Geography,
University of Texas,
Austin,
Texas 78712
United States of America

Professor E. Kutter,
Lehrstuhl Fur Verkenhrsplanung,
Technical University of Berlin,
1 Berlin 10,
Salzuser 17-19,
Gerbaude SG9 Erdgesch,
West Germany

Dr V. Vidakovic,
Urban Research Division,
Department of Public Works,
Wilbauthuis,
Room 8026,
Amsterdam,
Holland

Professor Dipl. Math. Manfred Wermuth,
Institut Fuer Baulingenieur Weser VI,
Technische Universitaet Muenchen,
Arcisstr 21,
8000 Muenchen 2,
West Germany

Dr D.W. Bennett,
Department of Civil Engineering,
University of Melbourne,
Parkville,
Victoria 3052
Australia

Mr R.E. Brindle,
Loder and Bayley,
79 Power Street,
Hawthorn,
Victoria 3122
Australia

Mr B.R. Leavens,
Urban Transport Study Group,
P.O. Box 609,
North Sydney,
New South Wales 2060
Australia

Dr D. Parkes,
Department of Geography,
University of Newcastle,
Newcastle,
New South Wales 2308
Australia

Dr R.J. Stimson,
Department of Geography,
Flinders University,
Adelaide,
South Australia 5000
Australia

Dr M.A.P. Taylor,
Division of Building Research,
C.S.I.R.O.,
P.O. Box 56,
Highett,
Victoria 3190
Australia

Professor G. Mills,
Department of Economics,
University of Sydney,
Sydney 2006
Australia

WORKSHOP B   EQUILIBRIUM MODELLING

*Chairman*

Mr E. Ruiter,
Cambridge Systematics,
Room 310,
Kendall Square Building,
238 Main Street,
Cambridge,
Massachusetts 02142
United States of America

*Resource Papers*

Professor A.E. Anderson,
Department of Economics,
Goteborgs Universitet,
Fack 400,
10 Goteborg 6
Sweden

Dr M. Gaudry,
Centre de Reserche des Transportes,
Universite de Montreal,
Case Postale 6128,
Succursale 'A',
Montreal
Canada

Professor A.G. Wilson,
Department of Geography,
University of Leeds,
Leeds  LS2 9JT
England

*Members*

Dr R. Dial,
Software Systems Division,
Office of Research and Development,
Urban Mass Transportation Administration,
US Department of Transportation,
400 7th Street, S.W.,
Washington D.C. 20590
United States of America

Mr D.P. Bowyer,
Australian Road Research Board,
P.O. Box 156 (Bag 4),
Nunawading,
Victoria 3131
Australia

Dr M. Fitzpatrick,
Bureau of Transport Economics,
Civic Permanent Building,
Allara Street,
Canberra,
Australian Capital Territory 3168
Australia

Mr R.M.V. Morris,
P.G. Pak-Poy & Associates,
170 Greenhill Road,
Parkside,
Adelaide,
South Australia 5063
Australia

Mr R.J. Nairn,
R.J. Nairn & Partners,
13 Tanumbirini Street,
Hawker,
Australian Capital Territory 2614
Australia

Dr J. Paterson,
John Paterson Urban Systems Limited,
150 Albert Road,
South Melbourne,
Victoria 3205
Australia

Dr J. Roy,
Building Design and Urban Research Unit,
C.S.I.R.O.,
Graham Street,
Highett,
Victoria 3190
Australia

WORKSHOP C   CONSUMER SEGMENTATION

*Chairman*

Mr W. Tye,
Charles River Associates,
1050 Massachusetts Avenue,
Cambridge,
Massachusetts 02138
United States of America

*Resource Papers*

Dr R. Dobson,
Charles River Associates,
1050 Massachusetts Avenue,
Cambridge,
Massachusetts 02138
United States of America

Mr R. Ian Kingham,
Transportation Research Board,
2101 Constitution Avenue N.W.,
Washington D.C. 20418
United States of America

*Members*

Mr R. Bassett,
Transport Section,
Department of Civil Engineering,
University of Melbourne,
Parkville,
Victoria 3052
Australia

Mr W. Godfrey,
P.A. Consulting Services Pty. Limited,
38 Currie Street,
Adelaide,
South Australia 5000
Australia

Mr W.H. Saggers,
Country Roads Board,
60 Denmark Street,
Kew,
Victoria 3101
Australia

847   Appendix I

WORKSHOP D   THEORETICAL AND CONCEPTUAL DEVELOPMENTS IN
             DEMAND MODELLING

*Chairman*

Professor R. Westin,
Centre for Urban and Community Studies,
150 St George Street,
University of Toronto,
Toronto,
Ontario   M5S 1A1
Canada


*Resource Papers*

Mr N. Bruzelius,
School of Economics,
University of Stockholm,
Stockholm,
Sweden

Mr A.J. Daly,
Local Government Operational Research Unit,
201 King's Road,
Reading   RG1 4LH
England

Dr J.F. Brotchie,
C.S.I.R.O.,
Graham Street,
Highett,
Victoria 3190
Australia


*Members*

Professor C. Manski,
School of Urban and Public Affairs,
Carnegie-Mellon University,
Pittsburgh,
Pennsylvania 15213
United States of America

Professor S. Lerman,
Department of Civil Engineering,
Massachusetts Institute of Technology,
Cambridge,
Massachusetts 02139
United States of America

Mr K. Train,
Institute of Transportation Studies,
University of California,
Berkeley,
California,
United States of America

Mr R.G. Bullock,
Bureau of Roads,
P.O. Box 372F,
Melbourne,
Victoria 3001
Australia

Mr J. Fields,
R.J. Nairn & Partners Pty. Limited,
13 Tanumbirini Street,
Hawker,
Australian Capital Territory 2614
Australia

Mr K.Y. Loong,
Department of Civil Engineering,
Queensland Institute of Technology,
P.O. Box 246,
North Quay,
Queensland 4000
Australia

Mr J. Neil,
Department of Applied Mathematics,
University of Adelaide,
G.P.O. Box 498,
Adelaide,
South Australia 5001
Australia

Mr A.J. Richardson,
Department of Civil Engineering,
Monash University,
Clayton,
Victoria 3168
Australia

Mr A.B. Smith,
Bureau of Transport Economics,
Civic Permanent Building,
Allara Street,
Canberra,
Australian Capital Territory 2600
Australia

Dr R. Wilson,
John Paterson Urban Systems Limited,
P.O. Box 1,
Clifton Hill,
Victoria 3068
Australia

WORKSHOP E   THE ROLE OF MODELLING TOOLS IN THE POLICY ARENA

*Chairman*

Dr A. Sherrett,
Peat, Marwick, Mitchell and Company,
San Francisco International Airport,
P.O. Box 8007,
San Francisco,
California 94128
United States of America

*Resource Papers*

Mr F. Reid,
Travel Demand Forecasting Project,
Institute of Transportation Studies,
109 McLaughlin Hall,
University of California,
Berkeley,
California 94720
United States of America

Mr K.G. Rogers,
U.N.D.P.,
P.O. Box 1555,
Tehran,
Iran

*Members*

Dr P.S. Liou,
Maryland Department of Transport,
Maryland,
Virginia
United States of America

Professor C. Gannon,
Department of Economics,
Monash University,
Clayton,
Victoria 3168
Australia

Mr L. Henning,
Department of Economics,
University of Queensland,
St Lucia,
Queensland 4067
Australia

Mr M.J. Hutchinson,
P.G. Pak Poy & Associates,
170 Greenhill Road,
Parkside,
Adelaide,
South Australia 5063
Australia

Mr J. Huggett,
Commonwealth Bureau of Roads,
G.P.O. Box 372F,
Melbourne,
Victoria 3001
Australia

Mr J. Lathrop,
GHD Parsons-Brinkernoff,
216 Northbourne Avenue,
Canberra,
Australia Capital Territory 2601
Australia

Dr D. Scrafton,
Director-General of Transport,
Administration Centre,
Victoria Square,
Adelaide,
South Australia 5000
Australia

Ms M. Starrs,
Office of Director-General of Transport,
Administration Centre,
Victoria Square,
Adelaide,
South Australia 5000
Australia

Mr J.O.C. White,
Department of Transport,
P.O. Box 367,
Canberra City,
Australian Capital Territory 2601
Australia

Mr A. Williams,
Department of Business,
Queensland Institute of Technology,
George Street,
Brisbane,
Queensland 4000
Australia

WORKSHOP F   RELATIONSHIP BETWEEN BEHAVIOURAL MODELS,
             EVALUATION, FORECASTING AND POLICY

*Chairman*

Mr R. Carruthers,
R. Travers Morgan Planning Partnership,
Wellington House,
125 The Strand,
London   WC2R 0AR
England

*Resource Papers*

Professor M.E. Beesley,
Department of Economics,
London Graduate School of Business,
Sussex Place,
Regent's Park,
London   NW1 4SA
England

Mr J. Stanley,
Commonwealth Bureau of Roads,
G.P.O. Box 372F,
Melbourne,
Victoria 3001
Australia

*Members*

Dr J. Guttman,
Department of Economics,
University of California,
405 Hilgard Avenue,
Los Angeles,
California 90024
United States of America

Dr D.J. Buckley,
Planning Division,
N.S.W. Public Transport Commission,
Margaret Street,
Sydney,
New South Wales 2000
Australia

Mr M. Chaffin,
Operations Research Department,
Shell (Australia),
155 William Street,
Melbourne,
Victoria 3000
Australia

Dr J. Grant,
Town and Country Planning Board,
235 Queen Street,
Melbourne,
Victoria 3000
Australia

Mr I Richards,
Office of Director-General of Transport,
G.P.O. Box 3599,
Adelaide,
South Australia 5000
Australia

Mr W. Ryan,
Department of Economics,
University of Queensland,
St Lucia,
Queensland 4067
Australia

Mr R. Simpson,
Robert Simpson & Associates Pty. Limited,
Suite 10,
9th Floor,
570 Bourke Street,
Melbourne,
Victoria 3000
Australia

Dr J.H.E. Taplin,
Department of Transportation,
Civic Permanent Building,
Allara Street,
Canberra,
Australian Capital Territory 2600
Australia

Dr R. Toakley,
C.S.I.R.O.,
Graham Street,
Highett,
Victoria 3190
Australia

Mr H. Souter,
Bureau of Transport Economics,
Civic Permanent Building,
Allara Street,
Canberra,
Australian Capital Territory 2600
Australia

Ms A. Walker,
Department of Transport,
P.O. Box 367,
Canberra City,
Australian Capital Territory 2601
Australia

WORKSHOP G   GOODS MOVEMENT

*Chairman*

Professor A.H. Meyburg,
Department of Environmental Engineering,
Cornell University,
Ithaca,
New York 14850
United States of America

*Resource Papers*

Dr P. Rimmer,
Department of Human Geography,
Research School of Pacific Studies,
Australian National University,
P.O. Box 4,
Canberra,
Australian Capital Territory 2600
Australia

Dr M.R. Wigan,
Australian Road Research Board,
P.O. Box 156 (Bag 4),
Nunawading,
Victoria 3131
Australia

*Members*

Mr K.J.G. Smith,
Transport Operations Research Group,
University of Newcastle-upon-Tyne,
Newcastle upon Tyne,
England

Mr D. Zavattero,
Chicago Area Transportation Study,
300 West Adams Street,
Chicago,
Illinois 60606
United States of America

Dr J. Black,
School of Transportation and Traffic,
University of New South Wales,
P.O. Box 1,
Kensington,
New South Wales 2033
Australia

Mr J.W. Hutchinson,
Director-General of Transport Office,
Administration Building,
Victoria Square,
Adelaide,
South Australia 5000
Australia

Mr H.T. Loxton,
Bureau of Roads,
60 Collins Street,
Melbourne,
Victoria 3000
Australia

Dr K.W. Ogden,
Department of Civil Engineering,
Monash University,
Clayton,
Victoria 3168
Australia

Mr P.G. Hooper,
Commonwealth Bureau of Roads,
G.P.O. Box 372F,
Melbourne,
Victoria 3001
Australia

WORKSHOP H   BEHAVIOURAL MODELLING, ACCESSIBILITY, MOBILITY
AND TRAVEL NEED

*Chairman*

Professor M. Wachs,
School of Architecture and Urban Planning,
University of California,
405 Hilgard Avenue,
Los Angeles,
California 90024
United States of America

*Resource Papers*

Professor M. Ben-Akiva,
Department of Civil Engineering,
Massachusetts Institute of Technology,
Cambridge,
Massachusetts 02139
United States of America

Professor M.Q. Dalvi,
Senior Transport Adviser,
United Nations Planning Commission,
Parliament Street,
New Delhi 1
India

Dr R. Kirby,
Urban Institute,
2100 M Street, N.W.,
Washington D.C. 20037
United States of America

*Members*

Professor A. Karlquist,
Department of Mathematics,
Royal Institute of Technology,
10044 Stockholm 70,
Sweden

Dr G. Koenig,
Department de la Savoie,
Direction Departementale de l'Equipment,
Chambery Le,
297 Route du Bourget,
73011 Chambery,
France

Mr G.K.R. Reid,
Bureau of Transport Economics,
Civic Permanent Building,
Allara Street,
Canberra,
Australian Capital Territory 2600
Australia

Mr K. Davidson,
Transport Division,
N.C.D.C.,
P.O. Box 373,
Canberra City,
Australian Capital Territory 2601
Australia

Mr D.G. Keech,
Urban Transport Study Group,
P.O. Box 609,
North Sydney,
New South Wales 2060
Australia

Dr R. Sharpe,
C.S.I.R.O.,
Graham Street,
Highett,
Victoria 3190
Australia

Mr J.B. Locke,
Main Roads Department,
State Offices,
10 Murray Street,
Hobart,
Tasmania 7000
Australia

Mr R.A. Smith,
Commonwealth Bureau of Roads,
G.P.O. Box 372F,
Melbourne,
Victoria 3001
Australia

WORKSHOP I   ATTITUDES, ATTITUDINAL MEASUREMENT AND THE
             RELATIONSHIP BETWEEN BEHAVIOUR AND ATTITUDES

*Chairman*

Professor J. Louviere,
Department of Geography,
University of Wyoming,
Laramie,
Wyoming,
United States of America


*Resource Papers*

Mr T. Golob,
Transportation Research Department,
General Motors Research Laboratories,
General Motors Technical Centre,
Warren,
Michigan 48090
United States of America

Professor I. Levin,
Department of Psychology,
University of Iowa,
Iowa City,
Iowa 52242
United States of America


*Members*

Mr T. Atherton,
Cambridge Systematics Inc.,
Kendall Square Building,
238 Main Street,
Cambridge,
Massachusetts 02142
United States of America

Mr P. Barnard,
Department of Economics,
University of Adelaide,
Adelaide,
South Australia 5000
Australia

Ms S. Young,
M.S.J. Keys Young and Associates,
35 Richards Avenue,
Surrey Hills,
New South Wales 2010
Australia

Mr G.L. Baxter,
Division of Building Research,
C.S.I.R.O.,
P.O. Box 56,
Highett,
Victoria 3190
Australia

Mr H.P. Brown,
Department of Civil Engineering,
University of Melbourne,
Parkville,
Victoria 3052
Australia

Dr P.B. McLeod,
Department of Economics,
University of Western Australia,
Perth,
Western Australia 6000
Australia

Dr D. Nix-James,
Traffic Accident Research Unit,
Department of Motor Transport,
P.O. Box 28,
Sydney,
New South Wales 2001
Australia

WORKSHOP J  BEHAVIOURAL MODELLING AND EVALUATION OF SOCIAL
            AND ENVIRONMENTAL IMPACTS OF TRANSPORT INVESTMENT

*Chairman*

Professor P. Hills,
Transport Operations Research Group,
University of Newcastle-upon-Tyne,
Newcastle   NE1 7RU
England

*Resource Paper*

Professor D. Appleyard,
College of Environmental Design,
Department of Landscape Architecture,
202 Worster Hall,
University of California,
Berkeley,
California 94720
United States of America

*Members*

Mr N.F. Clark,
N. Clark and Associates,
P.O. Box 524,
South Melbourne
Victoria 3205,
Australia

Mr W. Counsell,
Commonwealth Bureau of Roads,
G.P.O. Box 372F,
Melbourne,
Victoria 3001
Australia

Mr R. King,
Centre for Environmental Studies,
University of Melbourne,
Parkville,
Victoria 3052
Australia

Mr A. Rattray,
Commonwealth Bureau of Roads,
G.P.O. Box 372F,
Melbourne,
Victoria 3001
Australia

Mr B. Wight,
Department of Environment, Housing and Community Development,
McArthur House,
Northbourne Avenue,
Canberra,
Australian Capital Territory 2600
Australia

Appendix II

THE THIRD INTERNATIONAL CONFERENCE ON BEHAVIOURAL TRAVEL
MODELLING

John Paterson

David Hensher and Peter Stopher have asked me to explain
briefly to the visitors amongst us this evening, just what
this Conference is all about. The Conference is organised
into 10 workshops, and each participant is allocated one
workshop for the duration, which may not be unilaterally
varied on pain of death. Workshop A is grappling with the
question of new approaches to understanding travel behaviour
(nature of travel and structure of journeys). This general
approach seeks to set travel in the context of daily
activities, and is producing rich new insights by treating
travel explicitly as a derived demand. Spectacular advances
are hoped for; some workers have reported success in forecasting choice of mode for the journey-to-work in terms of
what the traveller had for breakfast.

Workshop B is dealing with questions of equilibrium
modelling, or in other words, the problem of 'when is enough,
enough'. It appears at this moment that the answer is either
in 'you know you're at equilibrium because it feels so good',
or alternatively, 'equilibrium is established in road traffic
at zero pricing when nothing on the road moves'.

Workshop C is dealing with consumer segmentation. This
approach seeks to pre-classify consumers in terms of salient
attributes. Which kind of attributes is a matter of debate,
but for sake of simplicity, one approach is to classify
individuals in terms of personal circumstances, for example
'little old lady', 'conscientious housewife', 'drunk', etc.
One then sets about building separate choice models for each
class. It will immediately be seen that this simplifies the
modelling task, for example it can immediately be predicted
that the little old lady model will have a distinctive
pattern of choice behaviour in regard to riding high-powered
motorbikes. In the case of drunks, it is possible to use
prior information to devise model structure; in this case the
random walk is suggested.

Workshop D is dealing with theoretical and conceptual
developments; this subject is no joking matter.

Workshop E is dealing with the role of disaggregate models
in the policy arena. Here I want to report a local success.
Application of disaggregate models to work-mode choice has
indicated the effectiveness of parking policy as a determinant of mode split. On this basis, one Australian state with
a particularly zealous police force has introduced a pilot

859

test on use of parking policy. The first trick is to
introduce random error into parking meters, and when the
parker returns to the expired meter, the cop beats the shit
out of him.

Workshop F deals with the relationship between behavioural
modelling, forecasting and policy. Here the triumph of the
logit model is complete. It is so sensitive to specification
error that having decided on your policy, you can get any
result you want to support it. The problem is that in these
days of profound technological agnosticism, we no longer know
what we want, and our political masters aren't about to take
the rap by telling us.

Workshop G - goods movement. A considerable success has
been reported in development of models to predict joint
origin, shipment size, frequency and mode choice behaviour.
The pioneering work in the United States was funded by the
Mafia and in Australia, local highjackers have expressed
interest. The price of coloured television sets, bulk
cigarettes and booze in the unofficial market has fallen; the
sponsors are considering funding a new project on second
round effects.

Workshop H is dealing with behavioural modelling,
accessibility, mobility and travel need: this workshop has
been devoted entirely to semantic issues. The results are to
be published in the Journal of Philosophical Semantics.

Workshop I. This workshop is dealing with attitudes,
attitudinal measurement and the relationship between
behaviour and attitudes. This new field derived major
insights from a powerful theoretical development which
suggests that one way to predict what people will do is to
ask them. Such an approach may not seem radical to the
psychologist, but those of us reared in the economics tradition take the view that you can't rely on the average
consumer to verbalise his choice behaviour in a way which is
consistent with the axiomatic basis of utility maximisation.
When it comes to the traveller, the economist doesn't believe
him, and the engineer doesn't care. The cross-disciplinary
fertilisation which has occurred in the travel field has led
to a number of approaches being used for 'asking' the
consumers. The approach may be complex or naive. One
approach has relied on the semantic differential; respondents
are given a list of 25 paired alternatives, such as cars are
fast-slow, hot-cold, fat-thin, etc. Factors or scales are
isolated which are found to explain as much variance as
models based on level of service and observed choice
behaviour: this leads to the question of just what have we
achieved in the use of revealed preferences? Others have
used techniques which owe much to the professional skills of
the psychologist, such as psychoanalysis of a sample of
travellers. Interviews are conducted under clinical conditions and deep probing is used to elucidate earliest
memories of trains, buses and cars. Yet another approach has
been called naive interlocution, 'what do you think of ...'.
The response is frequently rich in information and colour
which can be brought into prediction of mode split by the
skilled analyst. For example, the respondent says, 'I

wouldn't ride your (expletive deleted) trains if you paid me'. This indicates to the qualified researcher, first, mode aversion, second, free fares will cause a 23 per cent rise in patronage, third, not all traders face up to their responsibilities, and fourth, in line with national policy in promoting public transport and reducing rail deficits, the respondent is recommended for behaviour-modifying psychosurgery.

Workshop J has been dealing with evaluation of social and environmental impacts of transport investment. This was a remedial workshop for people lacking basic numeracy.

Obviously the published proceedings of the Conference will be in wide demand, and the organisers are relying on their selection as Book of the Month to meet the Conference deficit. Film rights are still to be negotiated.